BIOLOGICAL SCIENCE

ABOUT THE PHOTO:

Great tiger moth, *Arctia caja,* as photographed by
Joseph Scheer. Scheer, of New York's Alfred University,
is the author of *Night Visions: The Secret Designs of Moths*
(Prestel, 2003).

BIOLOGICAL SCIENCE
SECOND EDITION

SCOTT FREEMAN
University of Washington

CONTRIBUTORS

Healy Hamilton
*California Academy of Science,
Center for Biodiversity*

Sara Hoot
University of Wisconsin, Milwaukee

Greg Podgorski
Utah State University

James M. Ryan
Hobart and William Smith Colleges

Sally Sommers Smith
Boston University

Carol Trent
Western Washington University

Charles Walcott
Cornell University

D. Scott Weigle
University of Washington

PEARSON
Prentice Hall

Upper Saddle River, New Jersey 07458

Library of Congress Cataloging-in-Publication Data

Freeman, Scott
  Biological science / Scott Freeman.— 2nd ed.
    p. cm.
  ISBN 0-13-140941-7 (hardcover)
  1. Biology. I. Title.

  QH308.2.F73 2005
  570—dc22
              2004027312

Publisher: Sheri L. Snavely
Senior Development Editor: Karen Karlin
Production Editor: Donna Young
Project Manager: Karen Horton
Senior Media Editor: Patrick Shriner
Manager of Electronic Composition: Allyson Graesser
Electronic Production Specialists/Electronic Page Makeup:
  Karen Stephens, Vicki Croghan, Julita Nazario,
  Richard Foster, Jim Sullivan
Project Art Director: Kenny Beck
Editor in Chief: John Challice
Editor in Chief of Development: Carol Trueheart
Senior Marketing Manager: Shari Meffert
Marketing Manager: Andrew Gilfillan
Director of Science Marketing: Linda Taft MacKinnon
Executive Managing Editor: Kathleen Schiaparelli
Director of Creative Services: Paul Belfanti
Managing Editor, Audio/Video Assets: Patricia Burns
Manufacturing Buyer: Alan Fischer
Editorial Assistants: Nancy Bauer, Lisa Tarabokjia

Marketing Assistant: Laura Rath
Assistant Managing Editor, Science Media: Nicole Jackson
Assistant Managing Editor, Science Supplements: Becca Richter
Media Production Editor: Aaron Reid
Copy Editor: Chris Thillen
Proofreader: Brian I. Baker
Cover and Interior Designer: Joseph Sengotta
AV Editor: Connie Long
Illustrators: Pearson Artworks, Quade Paul, Imagineering
National Sales Director for Key Markets: David Theisen
Photo Researcher: Yvonne Gerin
Director, Image Resource Center: Melinda Reo
Manager, Rights and Permissions: Zina Arabia
Manager, Visual Research: Beth Boyd-Brenzel
Manager, Cover Visual Research and Permissions: Karen
  Sanatar
Image Permission Coordinator: LaShonda Morris
Cover photo: © Joseph Scheer
Other image credits appear in the backmatter.

© 2005, 2002 Pearson Education, Inc.
Pearson Prentice Hall
Pearson Education, Inc.
Upper Saddle River, NJ 07458

Printed in the United States of America
10 9 8 7 6 5 4 3 2

ISBN 0-13-140941-7 (Student Edition)
ISBN 0-13-141050-4 (Instructor's Edition)
ISBN 0-13-150293-X (Volume 1)
ISBN 0-13-150295-6 (Volume 2)
ISBN 0-13-150296-4 (Volume 3)

Pearson Education LTD., *London*
Pearson Education Australia PTY, Limited, *Sydney*
Pearson Education Singapore, Pte. Ltd
Pearson Education North Asia Ltd, *Hong Kong*
Pearson Education Canada, Ltd., *Toronto*
Pearson Educación de Mexico, S.A. de C.V.
Pearson Education—Japan, *Tokyo*
Pearson Education Malaysia, Pte. Ltd

Brief Contents

About the Author

Scott Freeman received his Ph.D. in Zoology from the University of Washington and was subsequently awarded an Albert Sloan Postdoctoral Fellowship in Molecular Evolution at Princeton University. His research publications explore a range of topics, including the behavioral ecology of nest parasitism and the molecular systematics of the blackbird family. Scott teaches the majors' general biology course as a Lecturer at the University of Washington. He assisted in the groundbreaking and influential redesign of the course, which emphasizes an inquiry-based approach and the logic of experimental design. With Jon Herron, Scott is co-author of the standard-setting *Evolutionary Analysis*, which over 50,000 students have used to explore evolution with the same spirit of inquiry. He is currently conducting research on how active learning and peer teaching techniques affect student learning.

CONTRIBUTORS

The author and publisher are grateful to the large number of teachers and experts in all fields of biology who shared their experiences and offered advice at every stage of planning and writing the second edition. In particular, we'd like to thank the following contributors for lending their expertise and providing new ideas and material for this revision.

Healy Hamilton
California Academy of Science, Center for Biodiversity

Sara Hoot
University of Wisconsin, Milwaukee

Greg Podgorski
Utah State University

James M. Ryan
Hobart and William Smith Colleges

Sally Sommers Smith
Boston University

Carol Trent
Western Washington University

Charles Walcott
Cornell University

D. Scott Weigle
University of Washington

ILLUSTRATOR

Kim Quillin combines training and experience in biology and art to create effective and scientifically accurate visual representations of biological principles. She received her B.A. in Biology at Oberlin College and her Ph.D. in Integrative Biology from the University of California, Berkeley, and

has taught undergraduate biology at both schools. Students and instructors alike have praised Kim's illustration programs for *Biology: A Guide to the Natural World*, by David Krogh, and *Biology: Science for Life*, by Colleen Belk and Virginia Borden, for their success at clearly conveying complex biological ideas in a visually appealing manner.

Preface

Cultural evolution can be defined as a change in the frequency of ideas and practices over time. Introductory biology courses and *Biological Science* are textbook examples.

The courses we design for our majors are changing in response to selection pressure from two sources: the knowledge explosion in biology and dramatic advances in research on how introductory students learn. The knowledge explosion has made it less and less viable to teach an introductory biology course that emphasizes the memorization of facts. At the same time, research on student learning has shown that introductory students struggle to differentiate key unifying concepts from supporting details and that the greatest gains in understanding occur when students have to apply the facts and concepts they are learning to new situations. In a recent article, Handelsman et al.[1] noted that "There is mounting evidence that supplementing or replacing lectures with active learning strategies and engaging students in discovery and scientific process improves learning and knowledge retention."

Instead of being satisfied with memorization, instructors are training students to use facts. The goal is to have students demonstrate a mastery of content and concepts by applying them in new contexts.

The second edition of *Biological Science* is designed to make the transition to active, higher-level learning easier for both professors and students. Every sentence and figure in the text has been revised with that goal in mind.

Ease of Use

To make the transition to inquiry-based active learning easier, I made two major changes to the second edition: I increased the amount of content coverage to give you more flexibility in the topics you emphasize, and I added study aids to help students with the task of stepping up to a college-level biology course.

Increased Content Coverage

Compared with the first edition of *Biological Science*, this book contains much more content. Recommendations from well over 500 instructors guided decisions on which topics and terms to add. The goal was to provide students with more core coverage and vocabulary, and thereby provide instructors with more flexibility in designing a syllabus and better support for organizing lectures and labs. Experiments still play a central role in this edition, but I trimmed the overall number so that the re-

maining experiments could be developed more thoroughly and with a clearer focus on the concept they illustrate. Throughout, the text retains its commitment to presenting topics in the context of questions, hypotheses, tests, and conclusions. Facts are tools for understanding—not ends in themselves.

New Study Tools

As introductory biology instructors, one of our most important jobs is to help our students become better students. As my colleague Mary Pat Wenderoth says, "We need to help them learn how to learn." The students in our courses are novices in biology. Like novices in any field, they have a difficult time distinguishing important points from unimportant points. They also struggle with self-diagnosis—to recognize that they do not understand something well. To help students get better at studying biology, and to take some of the burden for doing so off you, this edition offers several new features:

- **Key Concepts** are listed at the start of the chapter and then revisited in the Summary of Key Concepts. Each chapter's "big ideas" are laid out at the start, developed in detail, and then summarized.

- **Check Your Understanding** boxes appear at the ends of key sections within each chapter. These features briefly summarize one or two fundamental points and then present two to three tasks that students should be able to complete in order to demonstrate a mastery of the material. These boxes are checkpoints—a way for students to make sure that they understand what is going on before they move ahead.

- **Diversity Boxes** serve as the capstone for each of the chapters on biodiversity (Chapters 27–34). Their goal is to present a focused summary of features in key lineages. The detailed information about each group is tied to (1) where it occurs on the tree of life and (2) how and why the featured lineage diversified. Instead of swamping students with details during a traditional "march through lineages," the diversity boxes present selected information in a well-developed conceptual context.

The Forest and the Trees: Helping Students Synthesize and Unify

In addition to coping with an enormous amount of content in this course, instructors have to manage its diversity. In *Biological Science*, the emphasis on inquiry and experimentation provides a unifying theme from biochemistry through ecosystem ecology. In addition, the text highlights the fundamental how and why

[1]Handelsman, J., D. Ebert-May, R. Beichner, P. Bruns, A. Chang, R. De-Haan, J. Gentile, S. Lauffer, J. Stewart, S. M. Tilghman, and W. B. Wood. April 23, 2004. Scientific teaching. *Science* 304(5670): 521–522.

questions of biology. How does this event or process occur at the molecular level? In an evolutionary context, why does it exist?

Most chapters include at least one case history of an analysis done at the molecular level. Natural selection is introduced by exploring the evolution of antibiotic resistance via point mutations in the RNA polymerase gene of *Mycobacterium tuberculosis*. A box in the chapter on behavior features research on alleles that influence fruit-fly foraging behavior.

Similarly, evolutionary analyses do not begin or end with the unit on evolution. Concepts such as adaptation, homology, natural selection, and phylogenetic thinking are found in virtually every chapter. Unit 1, for example, presents traditional content in biochemistry—ranging from covalent bonding to the structure and function of macromolecules—in the context of chemical evolution and the origin of life. Meiosis is analyzed in terms of its consequences for generating genetic variation and hypotheses to explain the evolution of sex.

The overriding idea is that molecular and evolutionary analyses can help unify introductory biology courses, just as molecular tools and evolutionary questions are helping to unify many formerly disparate research fields within biology.

Supporting Visual Learners

Clear, attractive, and extensive graphics are critical to our success in the classroom. The second edition offers a major improvement in the visual presentation of the material. Kim Quillin has revised virtually every figure in the book to increase clarity, accuracy, and visual appeal and to tighten the focus on the central teaching point. Compared with the first edition, this book has 350 additional diagrams and 325 additional photographs.

To support active learning and conceptual understanding, the figures contain several important features:

- **Caption Questions and Exercises** challenge students to critically examine the information in the figure—not just absorb it.

- **Experiment Boxes** offer a standardized design to help students see how biologists answer questions by posing hypotheses and testing predictions, and to give students practice with interpreting data. In some experiment boxes, space is left blank for the null hypothesis, predicted outcomes, or conclusion. Students are challenged to fill them in.

- **Figure Pointers** act like your hand at the whiteboard so that students can easily find a figure's central teaching point.

Throughout the revision, the goal was to build an art program that supports the book's focus on thinking like a biologist. Color is used judiciously to distinguish important points from supporting details and general context. Layouts flow from top to bottom or left to right, and extensive labeling lets students work through each figure in a step-by-step manner. The overall look and feel of the art is clean, clear, accessible, and inviting.

Serving a Community of Teachers

As instructors, we have at least four major texts available that are essentially well organized, well written, and beautifully illustrated encyclopedias of the life sciences. *Biological Science* is different. By de-emphasizing the encyclopedic approach to learning biology and focusing more on the questions and experimental tools that make the science come alive, my aim is to offer a book that is more readable, attractive, and contemporary than traditional texts. Learning concepts well enough to apply them to new examples and data sets may be more challenging for some students than simply memorizing facts, but also it is more compelling. By motivating the presentation with questions and then using facts as tools to find answers, students of biology may come to think and feel more like the people who actually do biology.

Thank you for your devotion to biology, for your commitment to teaching, and for considering *Biological Science*.

Scott Freeman
University of Washington

Acknowledgments

Contributors

First and foremost, I'd like to acknowledge the second edition contributors for lending their expertise and for providing new ideas and material for this revision. Their commitment to scholarship and their passion for teaching resonated throughout the contributed chapters and had an enormous impact on the published version. In writing an introductory text, it is challenging to appeal to biologists of all specialties yet focus on what a student needs. The contributors were invaluable. They made the material more accurate and teachable and gave me much-needed partners in this endeavor.

Healy Hamilton, *California Academy of Science,*
   Center for Biodiversity
Sara Hoot, *University of Wisconsin, Milwaukee*
Greg Podgorski, *Utah State University*
James M. Ryan, *Hobart and William Smith Colleges*
Sally Sommers Smith, *Boston University*
Carol Trent, *Western Washington University*
Charles Walcott, *Cornell University*
D. Scott Weigle, *University of Washington*

Focus Group Participants

I have been fortunate to be the beneficiary of advice and inspiration from biology instructors attending a series of workshops at Sundance, Utah, since the inception of this book. The first and second editions are influenced by the experiences and wisdom of these visionaries. The focus group attendees read through chapters and helped me make countless critical decisions about content that should be added or deleted or handled differently.

Michel Bellini, *University of Illinois, Urbana-Champaign*
Peter Berget, *Carnegie Mellon University*
Jack Burk, *University of California, Fullerton*
Ruth Buskirk, *University of Texas, Austin*
Patrick Carter, *Washington State University*
Thomas Christianson, *University of Chicago*
Jim Colbert, *Iowa State University*
William Collins, *Stony Brook University*
Mark Decker, *University of Minnesota*
Kathryn L. Edwards, *Kenyon College*
Jeffrey Feder, *University of Notre Dame*
Ross Feldberg, *Tufts University*
Michael Gaines, *University of Miami*
Miriam Golbert, *College of the Canyons*
Harry Greene, *Cornell University*
Judith Heady, *University of Michigan*
Jean Heitz, *University of Wisconsin*
Carole Kelley, *Cabrillo College*
Kevin Kelley, *California State University, Long Beach*
Stephen Kelso, *University of Illinois, Chicago*
Judith Kjelstrom, *University of California, Davis*
Jeff Klahn, *University of Iowa*
Loren Knapp, *University of South Carolina*
Karen Koster, *University of South Dakota*
Dan Krane, *Wright State University*
Mary Rose Lamb, *University of Puget Sound*
Andrea Lloyd, *Middlebury College*
Jim Manser, *Harvey Mudd College*
Mike Meighan, *University of California, Berkeley*
Robert Newman, *University of North Dakota*
Harry Nickla, *Creighton University*
Shawn Nordell, *Saint Louis University*
Julie Palmer, *University of Texas, Austin*
Marc Perkins, *Orange Coast College*
Randall Phillis, *University of Massachusetts, Amherst*
Carol Reiss, *Brown University*
Amanda Schivell, *University of Washington*
Tom Sharkey, *University of Wisconsin*
Fred Singer, *Radford University*
Sally Sommers Smith, *Boston University*
Lori Stevens, *University of Vermont*
Briana Timmerman, *University of South Carolina*
Carol Trent, *Western Washington University*
Barbara Wakimoto, *University of Washington*
Charles Walcott, *Cornell University*
D. Scott Weigle, *University of Washington*
John Whitmarsh, *University of Illinois*
Susan Whittemore, *Keene College*
David Wilson, *Parkland College*
Dan Wivagg, *Baylor University*

Media and Supplements Contributors

The media and support materials that accompany the second edition were created by a team of talented and dedicated introductory biology instructors who brought an extraordinarily high level of creativity, experience, and ability to their re-spective projects. Our goal was to provide an innovative and tightly focused support package that addresses the unique challenges facing instructors and students in introductory biology today. I thank the instructors who attended workshops in which the critical roles of assessment and media in this course were carefully considered and discussed and that lead to guidelines inspiring the creation of content throughout the textbook's support package.

Media Contributors

Jennifer Butler, *Willamette University*
Carol Chihara, *University of San Francisco*
Cheryl Ingram-Smith, *Clemson University*
James M. Ryan, *Hobart and William Smith Colleges*
Eric Stavney, *DeVry University*
Mark D. Decker, *University of Minnesota*

Supplements Contributors

Marc Albrecht, *University of Nebraska, Kearney*
Brian Bagatto, *University of Akron*
Jay Brewster, *Pepperdine University*
Warren Burggren, *University of North Texas*
Sharon Eversman, *Montana State University*
Michelle Fay
Michael Gaines, *University of Miami*
Vanessa Handley, *University of California, Berkeley*
Harry Nickla, *Creighton University*
Laurel Hester, *University of South Carolina*
Christopher Keller, *Minot State University*
Marc Perkins, *Orange Coast College*
Randall Phillis, *University of Massachusetts, Amherst*
Greg Podgorski, *Utah State University*
Susan Rouse, *Brenau University*
Elena Shpak, *University of Washington*
Sally Sommers Smith, *Boston University*
Ellen Smith
Briana Timmerman, *University of South Carolina*
David Wilson, *Parkland College*
Cindy Wedig, *University of Texas, Pan American*

The Book Team

The production team for this edition brought a high level of experience and expertise to bear and was characterized by a single-minded focus on quality. The tenacity, work rate, and attention to detail of Senior Development Editor Karen Karlin were both instrumental and exemplary. Production Editor Donna Young held the reins with an expert's touch as a large team moved forward at high speed. Illustrator Kim Quillin acted as lead on the figure program and is a talent that comes along once in a generation. In addition to coordinating what is probably the largest review program in the history of textbook publishing, Project Manager Karen Horton managed focus groups, contributors, and media and supplements authors. Senior Media Editor Patrick Shriner worked tirelessly to revise and improve the quality of the media program—never losing sight of the fact that media must solve problems for professors and students. Photo Researcher Yvonne Gerin again provided superb photo research and was particularly effective at contacting scientists for images of research results that are not available from stock agencies. Formatters Karen Stephens, Vicki Croghan, and Julita Nazario worked patiently under relentless deadlines to create pages for the book and ensure that the layout works for students. Research Scientist Kathleen Hunt, of the University of Washington, is responsible for the dramatically improved Glossary. The entire team was assembled and inspired by Publisher Sheri Snavely, who has been the driving force behind this project since its inception. Her commitment to innovative biology publishing and devotion to meeting the needs of instructors are the reasons that this book exists.

The art program was executed by the talented crews at Pearson Artworks, Quade Paul, and Imagineering; a special note of thanks goes to Managing Editor Patricia Burns and AV Editor Connie Long. I'm particularly grateful to Lee Wilcox, of the University of Wisconsin, who worked with Kim Quillin to upgrade the photo and art program dramatically for all of the plant chapters, and to Robin Manasse, of RMBlueStudios, who worked with Kim to improve the illustrations in the anatomy and physiology unit. Project Art Director Kenny Beck managed multiple rounds of revision on the design and cover and added key creative input. Designer Joseph Sengotta is responsible for creating a clear and accessible text design and a striking cover.

A textbook can help students and professors only if it ends up on their desks. The marketing and sales efforts for this edition are directed by Director of Science Marketing Linda Taft MacKinnon, who was instrumental in making the first edition the most successful launch in the history of majors' biology. Sincere thanks to Senior Marketing Manager Shari Meffert and to Marketing Manager Andrew Gilfillan for their work on the thoughtful preview booklet and their continued tireless efforts in the field on behalf of *Biological Science*. I extend a special thank you to Director of ESM Sales Programs Meghan O'Donnell and to the Sales Directors—especially Don O'Neal, Rebecca Bersagel, Kate Brousseau, Brian Buckley, Megan Donnelly, Meghan Duffy, Christine Henry, Tom Johnson, and Michelle Renda—for their input, travel, and commitment to this effort. I'm also deeply grateful to Dave Theisen, National Sales Director for Key Markets, for his tactical skill and devotion to this book.

Finally, I thank my students at the University of Washington for inspiration, Barb Radin for invaluable help with cataloging and organizing reviewer comments, and Ben and Peter Freeman for love and support.

This book has two dedications. The first is to the memory of Bill Keeton and Neil Campbell, whose books inspired two generations of introductory biology students. As teachers and authors, they are the giants whose shoulders I try to stand on. The second is to my wife, Susan. After 24 years together, I have one thing to say: I am the luckiest man alive.

Reviewers

The review program for the second edition was even more rigorous than that for the first edition. The chapters were reviewed three times as they moved through the revision process. Reviewers included star instructors who addressed issues such as level, pacing, accuracy, and student comprehension. Other reviewers were experts in particular fields who focused primarily on making sure that the details are correct and that chapters are authoritative and current. In addition, all 55 chapters underwent a fourth and final review for accuracy just prior to publication. To a person, our reviewers supplied exemplary attention to detail, expertise, and empathy for students. I am deeply indebted to all of the colleagues who reviewed chapters of *Biological Science*; it is not possible to overstate how crucial these individuals are to the success of this book. Their effort reflects a deep commitment to excellence in teaching and a profound belief in the importance of introductory courses for training the next generation of professionals.

Julie Aires, *Florida Community College at Jacksonville*
Marc Albrecht, *University of Nebraska, Kearney*
Terry C. Allison, *University of Texas, Pan American*
Jorge E. Arriagada, *St. Cloud State University*
David Asai, *Harvey Mudd College*
David Asch, *Youngstown State University*
Karl Aufderheide, *Texas A & M University*
Christopher Austin, *Louisiana State University*
Ellen Baker, *Santa Monica College*
Christopher Beck, *Emory University*
Robert Beckman, *North Carolina State University*
Patricia Bedinger, *Colorado State University*
Peter Bednekoff, *Eastern Michigan University*
Michel Bellini, *University of Illinois*
Carl Bergstrom, *University of Washington*
John Bishop, *Washington State University, Vancouver*
Meredith Blackwell, *Louisiana State University*
Andrew Blaustein, *Oregon State University*
Dona F. Boggs, *Eastern Michigan University*
Barry Bowman, *University of California, Santa Cruz*
Jerry Brand, *University of Texas, Austin*
Angela Brown, *University of Idaho*
Albert Burchsted, *College of Staten Island*
Warren Burggren, *University of North Texas*
John Burr, *University of Texas, Dallas*
Scott Burt, *Truman University*
David Byres, *Florida Community College at Jacksonville*
Jeff Carmichael, *University of North Dakota*
Patrick Carter, *Washington State University*
David Champlin, *University of Southern Maine*
Jung Choi, *Georgia Institute of Technology*
Thomas Christianson, *University of Chicago*
Cynthia Church, *Metropolitan State College*
Vitaly Citovsky, *Stony Brook University*
Michael Clancy, *Boston University*

Anne B. Clark, *Binghamton University*
Jim Colbert, *Iowa State University*
Jerry L. Cook, *Sam Houston State University*
Scott Cooper, *University of Wisconsin, LaCrosse*
Erica Corbett, *Southeastern Oklahoma State University*
David Craig, *Williamette University*
Sarah Cunningham, *University of California, Berkeley*
Elizabeth Dahlhoff, *Santa Clara University*
David Dalton, *Reed College*
Sandra Davis, *University of Louisiana, Monroe*
Neta Dean, *Stony Brook University*
Lynda Delph, *Indiana University*
Charles F. Delwiche, *University of Maryland*
Jean DeSaix, *University of North Carolina, Chapel Hill*
Kathryn Dodd, *University of Texas, Pan American*
John Downing, *Iowa State University*
Marvin Druger, *Syracuse University*
Ernest F. Dubrul, *University of Toledo*
John Dudley, *University of Illllinois Champaign-Urbana*
Charles Duggins, Jr., *University of South Carolina*
William Eckberg, *Howard University*
Jean Everett, *College of Charleston*
Sharon Eversman, *Montana State University*
Stephanie Fabritius, *Southwestern University*
Scott Fay, *University of California, Berkeley*
Ross Feldberg, *Tufts University*
Siobhan Fennessy, *Kenyon College*
Anne Findley, *University of Louisiana, Monroe*
Jon Fischer, *Saint Louis University*
Teresa G. Fischer, *Indian River Community College*
Robert Fogel, *University of Michigan, Ann Arbor*
Don Fontes, *Community College of Rhode Island*
Larry J. Forney, *University of Idaho*
Irwin Forseth, *University of Maryland*
Marty Fox, *Edinboro University*
Krista Frankenberry, *Marshall University*
Michael Gaines, *University of Miami*
Gary Galbreath, *Northwestern University*
George Gilchrist, *College of William and Mary*
John Godwin, *North Carolina State University*
Miriam Golbert, *College of the Canyons*
Walter M. Goldberg, *Florida International University*
Sara V. Good-Avila, *Acadia University*
John Graham, *Bowling Green State University*
Eileen Gregory, *Rollins College*
Patricia A. Grove, *College of Mount Saint Vincent*
Alan Gubanich, *University of Nevada*
Cary Guffey, *Our Lady of the Lake University*
Bill Hamilton, *Washington and Lee University*
Samuel Hammer, *Boston University*
Vanessa Handley, *University of California, Berkeley*
Alice Harmon, *University of Florida*
Carla Hass, *Pennsylvania State University*
Stephen Hauschka, *University of Washington*
Judith Heady, *University of Michigan, Dearborn*
Albert Herrera, *University of Southern California*
Karen Hicks, *Kenyon College*

Jay Hirsch, *University of Virginia*
William Hoese, *California State University, Fullerton*
Mark Holbrook, *University of Iowa*
John Hoogland, *University of Maryland*
Sara Hoot, *University of Wisconsin, Milwaukee*
Margaret Horton, *University of North Carolina, Greensboro*
Kelly Howe, *University of New Mexico*
Lawrence E. Hurd, *Washington and Lee University*
Erin Irish, *University of Iowa*
Donald Jackson, *Brown University*
Rebecca Jann, *Queens University of Charlotte*
Lee Johnson, *Ohio State University*
Walter S. Judd, *University of Florida*
Thomas C. Kane, *University of Cincinnati*
Elizabeth A. Kellogg, *University of Missouri, St. Louis*
Stephen Kelso, *University of Illinois, Chicago*
Chris Kennedy, *Simon Fraser University*
Gwendolyn Kinnebrew, *John Carroll University*
John Kiss, *Miami University of Ohio*
Jeff Klahn, *University of Iowa*
Helen Koepfer, *Queens College*
John La Claire, *University of Texas at Austin*
Mary Rose Lamb, *University of Puget Sound*
Thomas Lehman, *Morgan Community College*
Paula Lemons, *Duke University*
Lynn Lewis, *Mary Washington College*
Andi Lloyd, *Middlebury College*
Paula Lovett, *University of Maryland*
John Lugthart, *Dalton State College*
Robert D. Lynch, *University of Massachusetts, Lowell*
Richard Malkin, *University of California, Berkeley*
Charles Mallery, *University of Miami*
Paul Manos, *Duke University*
James Manser, *Harvey Mudd College*
Diane Marshall, *University of New Mexico*
Steven L. Matzner, *Augustana College*
Michael Mazurkiewicz, *University of Southern Maine*
Richard E. McCarty, *John Hopkins University*
Kelly McLaughlin, *Tufts University*
Mona Mehdy, *University of Texas*
Eli Meier, *Simbiotic, Inc.*
Michael Meighan, *University of California, Berkeley*
Madeline Mignone, *Dominican College*
Molly Morris, *Ohio University*
Dale Mueller, *Texas A & M University*
Leann Naughton, *University of Wyoming*
Jacalyn S. Newman, *University of Pittsburgh*
Robert Newman, *University of North Dakota*
Karen Bushaw Newton, *American University*
Harry Nickla, *Creighton University*
Shawn Nordell, *Saint Louis University*
Amanda Norvell, *The College of New Jersey*
Deborah O'Dell, *Mary Washington College*
John Olsen, *Rhodes College*
John Osterman, *University of Nebraska*
Norman Pace, *University of Colorado, Boulder*
Julie Palmer, *University of Texas, Austin*

Matthew B. Parks, *University of Idaho*
C. O. Patterson, *Texas A & M University*
Andrew Pease, *Villa Julie College*
Craig L. Peebles, *University of Pittsburgh*
Curtis Pehl, *University of California, Berkeley*
Marc Perkins, *Orange Coast College*
Gary Peterson, *South Dakota State University*
Patricia Phelps, *Austin Community College*
Randall Phyllis, *University of Massachusetts*
Greg Podgorski, *Utah State University*
Frank Polanowski, *Elizabethtown College*
Donald Potts, *University of California, Santa Cruz*
F. Harvey Pough, *Rochester Institute of Technology*
Jerry Purcell, *San Antonio College*
Jonathan Reed, *Community College of Southern Nevada*
Stuart Reichler, *University of Texas, Austin*
Robin Richardson, *Winona State University*
Jared Rifkin, *Queens College*
Bruce Riley, *Texas A & M University*
John Romeo, *University of South Florida*
Peter Russell, *Reed College*
James M. Ryan, *Hobart and William Smith Colleges*
Brody Sandel, *University of California, Berkeley*
Amanda Schivell, *University of Washington*
Brian G. Scholtens, *College of Charleston*
Susan Schreier, *Villa Julie College*
Robert Seagull, *Hofstra University*
Eli Seigel, *Tufts University*
Marty Shankland, *University of Texas, Austin*
Thomas B. Shea, *University of Massachusetts, Lowell*
Allen Shearn, *Johns Hopkins University*
Tim Sherman, *University of South Alabama*
Richard M. Showman, *University of South Carolina*
Michele Shuster, *New Mexico State University*
Amanda Simcox, *Ohio State University*
Anne Simon, *University of Maryland*
Sally Sommers Smith, *Boston University*
William Smith, *Wake Forest University*
James Sniezek, *Montgomery College*
Phillip Sokolove, *University of Maryland, Baltimore County*
James Staples, *University of Western Ontario*
Eric Stavney, *University of Washington*

David Steen, *Andrews University*
William Stein, *State University of New York, Binghamton*
Lori Stevens, *University of Vermont*
Charles Stinemetz, *Rhodes College*
Sarah H. Swain, *Middle Tennessee State University*
Kevin Teather, *University of Prince Edward Island*
Ethan J. Temeles, *Amherst College*
Joshua Tewksbury, *University of Washington*
Briana Timmerman, *University of South Carolina*
Albert Torzill, *George Mason University*
Carol Trent, *Western Washington University*
John True, *Stony Brook University*
Nancy Trun, *Duquesne University*
Stephen Turnbull, *University of New Brunswick*
Elizabeth Van Volkenburgh, *University of Washington*
Neal J. Voelz, *St. Cloud State University*
Susan Waaland, *University of Washington*
Charles Walcott, *Cornell University*
Margaret Wallace, *John Jay College of Criminal Justice*
Jennifer Warner, *University of North Carolina, Charlotte*
Cindy Wedig, *University of Texas, Pan American*
D. Scott Weigle, *University of Washington*
Elizabeth A. Weiss, *University of Texas, Austin*
Barry Welch, *San Antonio College*
Sue Simon Westendorf, *Ohio University*
Susan Whittemore, *Keene College*
David Wilkes, *Harvey Mudd College*
Judy Williams, *Southeastern Oklahoma State University*
David Wilson, *Parkland College*
Eric Winkler, *Blinn College*
Bob Winning, *Eastern Michigan University*
Dan Wivagg, *Baylor University*
C. B. Wolfe, *University of North Carolina, Charlotte*
Lorne Wolfe, *Georgia Southern University*
Denise Woodward, *Pennsylvania State University*
Shawn Wright, *Albuquerque TVI*
Richard P. Wunderlin, *University of South Florida*
Peter H. Wyckoff, *University of Minnesota, Morris Campus*
Todd Christian Yetter, *Cumberland College*
Stephen Yezerinac, *Reed College*
Donna Young, *University of Winnipeg*
Gregory Zimmerman, *Lake Superior State University*

Print and Media Resources for Instructors and Students

For the Instructor

Lecture Presentation Tools

Instructor Resource Center on CD/DVD

As the demands of teaching introductory biology continue to grow, it becomes increasingly important for instructors that the right textbook be accompanied by the right tools to aid them during every stage—in lecture preparation and presentation, in assessment, and in the lab, as well as in the overall management of the course. The Freeman *Biological Science* Instructor Resource Center, provided in both CD and DVD formats, offers a rich suite of electronic tools designed to help instructors make the most of their limited time. Features include:

- As JPEG files, the entire textbook illustration program: all line drawings with labels individually enhanced for optimal projection results (as well as unlabeled versions), all tables, and all photos, as well as additional photos not found in the textbook

- The entire textbook illustration program pre-loaded into comprehensive PowerPoint presentations for each chapter

- A second set of PowerPoint presentations consisting of a thorough lecture outline for each chapter, augmented by key text illustrations

- An impressive series of concise instructor animations, Web Tutorials, and video clips adding depth and visual clarity to the most important topics and dynamic processes described in the text

- These same illustrative animations, Web Tutorials, and video clips pre-loaded into PowerPoint presentations for each chapter

- PowerPoint presentations containing a comprehensive set of in-class CRS questions for each chapter

- In Word files, a complete set of the assessment materials and study questions and answers from the Test Bank for Assessment, the textbook's in-chapter questions, and the student media practice questions, as well as files containing the entire Instructor's Guide

- Finally, to help instructors keep track of all that is available in this media package, a printable Media Integration Guide consisting of a PDF file that lists the media offerings for each chapter

Transparency Package and Instructor Resource Kit

Transparencies are an effective way to reinforce your lecture presentation visually. Every illustration from the text is available on four-color transparency acetates. The transparency set is three-hole-punched and organized in manila folders, which are stored in an Instructor Resource Kit file box along with the printed lecture tools from the Instructor's Resource Guide. Labels and images have been enlarged and modified to ensure optimal readability in a large lecture hall.

Instructor's Guide

Susan Rouse (Brenau University) with Brian Bagatto (University of Akron)

The Instructor's Guide includes traditional instructor support tools—lecture outlines, answers to end-of-chapter questions, vocabulary lists—as well as innovative activities to help motivate students and reinforce their understanding of the material. All of the content in the Instructor's Guide is available electronically on the Instructor Resource Center on CD/DVD.

Assessment Tools

Test Bank for Assessment and TestGen EQ

Edited by Harry Nickla (Creighton University) and Marc Perkins (Orange Coast College)

Contributors: Marc Albrecht (University of Nebraska, Kearney), Brian Bagatto, (University of Akron), Warren Burggren (University of North Texas), Sharon Eversman (Montana State University), Michelle Fay, Michael Gaines (University of Miami), Vanessa Handley (University of California, Berkeley) Laurel Hester (University of South Carolina), Christopher Keller (Minot State University), Randall Phillis (University of Massachusetts, Amherst), Greg Podgorski (Utah State University), Elena Shpak (University of Washington), Sally Sommers Smith (Boston University), Briana Timmerman (University of South Carolina), David Wilson (Parkland College), and Cindy Wedig (University of Texas, Pan American)

Our assessment materials were guided by a team of faculty who participated in an assessment workshop. The group developed guidelines for writing effective test questions for students and set our standards for quality and accuracy. The Test Bank features questions that:

- are based on the major biological themes and processes covered in each chapter rather than the minutiae

WEB TUTORIAL

- require critical thinking or application/integration of pre-existing knowledge, not rote memorization and recall

- are biologically accurate, with only one clearly correct answer choice

- are relatively resistant to test-taking strategy

- have been peer reviewed and student tested

To make this test bank more useful, we have made a few significant changes compared with the typical test bank, including:

- *In-class questions*: For each chapter, we feature three to five questions that could be used in class to help motivate discussion and formative assessment during class sessions.

- *Literature questions*: Questions based on data or ideas contained in a relevant peer-reviewed journal article. These questions extrapolate from material covered in the chapter.

Laboratory Support

Symbiosis

We have assembled a set of inquiry-based experiments to complement the approach of *Biological Science*. Symbiosis, the state-of-the-art Pearson Custom Publishing service, allows you to choose from hundreds of other experiments and/or include your own. The result is a well-integrated lab manual customized to your needs.

Course Management Tools

Prentice Hall's **OneKey** online course management content offers instructors using *Biological Science* everything they need to plan and administer their course. It includes the best teaching and learning resources all in one place. Organized by textbook chapter, the compiled resources help you save time and help your students reinforce and apply what they have learned in class. All resources from the Instructor Resource Center on CD/DVD and the Companion Website are included. OneKey is available for institutions using **WebCT** or **Blackboard** and is also available in Prentice Hall's nationally hosted **Course Compass** course management system. If desired, WebCT and Blackboard cartridges containing only the Test Bank for Assessment are available for download. Visit http://cms.prenhall.com for details.

For The Student

Companion Website (www.prenhall.com/freeman/biology)

Student study time is an increasingly valuable resource in introductory biology courses, and the greatest challenge many students face is making sure they use it wisely. The Companion Website has been designed to allow students using *Biological Science* to zero in quickly on the chapter sections and topics where they need review or further explanation. Its **Online Study Guide** offers concise lesson summaries punctuated with key illustrations, as well as probing review questions that offer hints and feedback on correct and incorrect responses. **Web Tutorials** offer students the opportunity to visualize complex topics and dynamic processes. The media's strict adherence to both the principles and specific lessons of the textbook means that study time is not being wasted. Tabs like the one in the margin of this page alert students that a Web Tutorial exists on a topic related to the coverage on that page.

Additional features of the Companion Website include an **Art Notebook**, hundreds of section-specific **Weblinks** to aid in online research; and **Textbook Answers** for all end-of-chapter and figure-caption questions.

Accelerator CD-ROM

The Accelerator CD-ROM that accompanies all student editions of *Biological Science,* Second Edition, offers the same **Web Tutorials** that are found on the Companion Website but in a stand-alone CD format.

Student Study Guide

Edited by Warren Burggren (University of North Texas) Contributors: Jay Brewster (Pepperdine University), Laurel Hester (University of South Carolina), and Brian Bagatto (University of Akron)

The Student Study Guide helps students focus on the fundamentals chapter by chapter. Each chapter presents a breakdown of key biological concepts, difficult topics, and quizzes. In addition, the Study Guide features four introductory, stand-alone chapters: Introduction to Experimentation and Research in the Biological Sciences, Presenting Biological Data, Understanding Patterns in Biology and Improving Study Techniques, and Reading and Writing to Understand Biology.

Contents

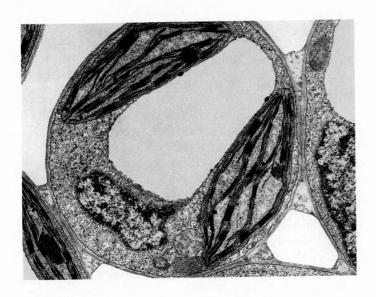

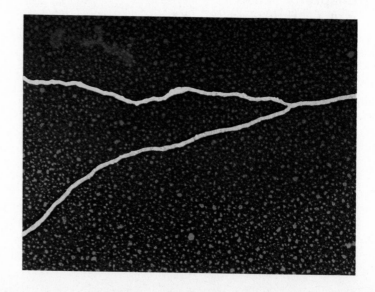

UNIT 7
How Plants Work 803

Chapter 35
Plant Form and Function 804

Chapter 36
Water and Sugar Transport in Plants 828

Chapter 37
Plant Nutrition 852

Biology and the Tree of Life

A biologist measuring an Atlantic puffin chick on an island off the coast of Norway. The data he's collecting will help him test hypotheses about factors that influence the growth and survival of seabirds.

KEY CONCEPTS

▪ Biological science was founded with the development of (1) the cell theory, which proposes that all organisms are made of cells and that all cells come from preexisting cells, and (2) the theory of evolution by natural selection, which maintains that the characteristics of species change through time—primarily because individuals with certain heritable traits produce more offspring than do individuals without those traits.

▪ A phylogenetic tree is a graphical representation of the evolutionary relationships among species. Phylogenies can be estimated by analyzing similarities and differences in traits. Species that share many traits are closely related and are placed close to each other on the tree of life.

▪ Biologists ask questions, generate hypotheses to answer them, and design experiments that test the predictions made by competing hypotheses.

In essence, biology is a search for ideas and observations that unify the incredible diversity of life. Chapter 1 is an introduction to this search. Its goals are to introduce the amazing variety of life-forms alive today, consider some fundamental traits shared by all organisms, and explore how biologists go about answering questions about life. Appreciating the diversity of life, understanding its underlying unity, and learning how to think like a biologist are themes that will resonate throughout this book.

We begin by examining two of the greatest unifying ideas in all of science: the cell theory and the theory of evolution by natural selection. When these concepts emerged in the mid-1800s, they revolutionized the way that biologists understand the world. The cell theory proposed that all organisms are made of cells and that all cells come from preexisting cells. The theory of evolution by natural selection maintained that species have changed through time and that all species are related to one another through common ancestry. The theory of evolution by natural selection established that bacteria, mushrooms, roses, and robins are all part of a family tree, similar to the genealogies that connect individual people.

A **theory** is an explanation for a very general class of phenomena or observations. The cell theory and the theory of evolution provided a foundation for the development of modern biology because they focused on two of the most general questions possible: How are organisms structured? Where did they come from? Let's begin by tackling the first of these two questions.

1.1 The Cell Theory

The initial conceptual breakthrough in biology—the cell theory—emerged after some 200 years of work. In 1665 Robert Hooke used a crude **microscope** to examine the structure of cork (a bark tissue) from an oak tree. The instrument magnified objects to just 30 times (30×) their normal size, but it allowed Hooke to see something extraordinary. In the cork he observed small, pore-like compartments that were invisible to the naked eye (**Figure 1.1a**). These structures came to be called cells.

Soon after Hooke published his results, Anton van Leeuwenhoek succeeded in developing much more powerful microscopes, some capable of magnifications up to 300×. With these instruments, Leeuwenhoek inspected samples of pond water and made the first observations of single-celled organisms like the *Paramecium* in **Figure 1.1b**. He also observed and described the structure of human blood cells and sperm cells.

In the 1670s a researcher who was studying the leaves and stems of plants with a microscope concluded that these large, complex structures are composed of many individual cells. By the early 1800s enough data had been accumulated for a biologist to claim that *all* organisms consist of cells. This conclusion was a classic example of inductive reasoning. Scientists made a broad generalization only after making thousands of supporting observations.

(a) The first view of cells: Robert Hooke's drawing from 1665

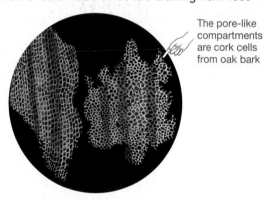

The pore-like compartments are cork cells from oak bark

(b) Anton van Leeuwenhoek was the first to view single-celled "animalcules" in pond water.

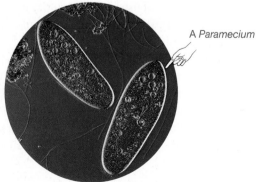

A *Paramecium*

FIGURE 1.1
The Discovery of Cells

Are *All* Organisms Made of Cells?

The smallest organisms known today are bacteria that are barely 200 nanometers wide, or 200 *billionths* of a meter. (See the Appendix at the back of this book to review the metric system and its prefixes.) Lining up bacteria end to end would take 5000 of these organisms to span a millimeter—the distance between the smallest hash marks on a metric ruler. In contrast, sequoia trees can be over 100 meters tall—the equivalent of a 20-story building. Bacteria and sequoias are composed of the same fundamental building block, however—the cell. Bacteria are unicellular (one-celled) organisms; sequoias are multicellular (many-celled) organisms.

Biologists have become increasingly dazzled by the diversity and complexity of cells as advances in microscopy have made it possible to examine cells at higher magnifications. The basic conclusion made in the 1800s is intact, however: As far as is known, all organisms are made of cells. Today, a **cell** is defined as a highly organized compartment that is bounded by a thin, flexible structure called a plasma membrane and that contains concentrated chemicals in an aqueous (watery) solution. The chemical reactions that sustain life take place inside cells. Most cells are also capable of reproducing by dividing—in effect, by making a copy of themselves.

The realization that all organisms are made of cells was fundamentally important, but it formed only the first part of the cell theory. In addition to understanding what organisms are made of, scientists wanted to understand how cells come to be.

Where Do Cells Come From?

Most scientific theories have two components: The first describes a pattern in the natural world, and the second identifies a mechanism or process that is responsible for creating that pattern. The early workers had articulated the pattern component of the cell theory. In 1858 Rudolph Virchow added a process component by stating that all cells arise from preexisting cells. The complete **cell theory**, then, can be stated as follows: All organisms are made of cells, and all cells come from preexisting cells.

This claim was a direct challenge to the prevailing explanation, called **spontaneous generation**. At the time, most biologists believed that organisms arise spontaneously under certain conditions. For example, the bacteria and fungi that spoil foods such as milk and wine were thought to appear in these nutrient-rich media of their own accord—meaning they spring to life from nonliving materials. Spontaneous generation was a **hypothesis**: a proposed explanation. The all-cells-from-cells hypothesis, in contrast, maintained that cells do not spring to life spontaneously but are produced only when preexisting cells grow and divide.

Soon after the all-cells-from-cells hypothesis appeared in print, Louis Pasteur set out to test its predictions experimentally. Pasteur wanted to determine whether microorganisms could arise spontaneously in a nutrient broth or whether they appear only when a broth is exposed to a source of preexisting cells.

To address the question, he created two treatment groups: a broth that was not exposed to a source of preexisting cells and a broth that was. The spontaneous generation hypothesis predicted that cells would appear in both treatments. The all-cells-from-cells hypothesis predicted that cells would appear only in the treatment exposed to a source of preexisting cells.

Figure 1.2 shows Pasteur's experimental setup. Note that the two treatments are identical in every respect but one. Both used glass flasks filled with the same amount of the same nutrient broth. Both were boiled for the same amount of time to kill any existing organisms such as bacteria or fungi. But because the flask pictured in Figure 1.2a had a straight neck, it was exposed to preexisting cells after sterilization by the heat treatment. These preexisting cells are the bacteria and fungi that cling to dust particles in the air. They could drop into the nutrient broth because the neck of the flask was straight. In contrast, the flask drawn in

Figure 1.2b had a long swan neck. Pasteur knew that water would condense in the crook of the swan neck after the boiling treatment and that this pool of water would trap any bacteria or fungi that entered on dust particles. Thus, the swan-necked flask was isolated from any source of preexisting cells even though it was still open to the air. The experimental setup was effective because there was only one difference between the two treatments and because that difference was the factor being tested—in this case, a broth's exposure to preexisting cells.

The result? As Figure 1.2 shows, the treatment exposed to preexisting cells quickly filled with bacteria and fungi. This treatment was important because it showed that the heat sterilization step had not altered the nutrient broth's capacity to support growth. But the treatment in the swan-necked flask remained sterile. Even when the broth was left standing for months, no organisms appeared in it.

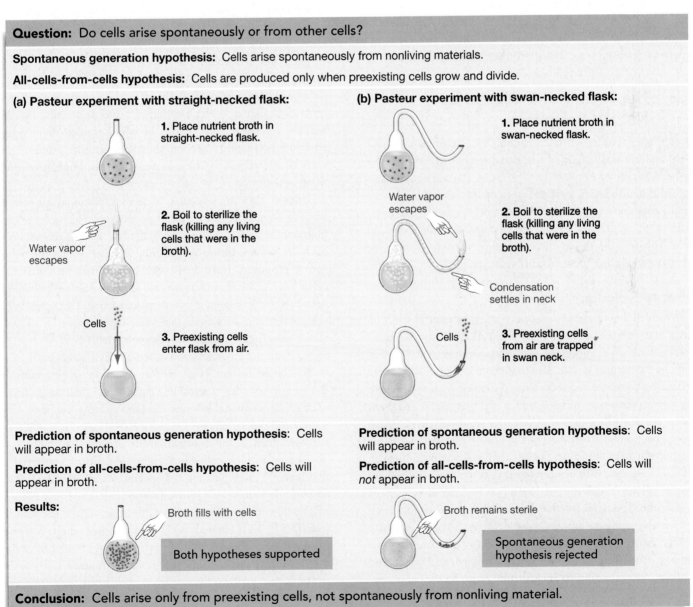

FIGURE 1.2 The Spontaneous Generation Hypothesis Was Tested Experimentally

Because Pasteur's data were in direct opposition to the predictions made by the spontaneous generation hypothesis, the results persuaded most biologists that the all-cells-from-cells hypothesis was correct.

The success of the cell theory's process component had an important implication: If all cells come from preexisting cells, then all individuals in a population of single-celled organisms must be related by common ancestry. Similarly, in a multicellular individual such as you, all of the cells present are descended from preexisting cells, tracing back to a fertilized egg. A fertilized egg is a cell created by the fusion of sperm and egg—cells that formed in individuals of the previous generation. In this way, all of the cells in unicellular and multicellular organisms are connected by common ancestry.

The second great founding idea in biology is similar, in spirit, to the cell theory. It also happened to be published the same year as the all-cells-from-cells hypothesis. This was the realization, made independently by Charles Darwin and Alfred Russel Wallace, that all *species*—meaning all distinct, identifiable types of organisms—are connected by common ancestry.

1.2 The Theory of Evolution by Natural Selection

In 1858 short papers written separately by Darwin and Wallace were read to a small group of scientists attending a meeting of the Linnean Society of London. In their essays Darwin and Wallace argued that all species, past and present, are related by descent from a common ancestor. Their hypothesis was that species come from other, preexisting species and that species change through time. This was the theory of evolution by natural selection, or what Darwin called "descent with modification."

What Is Evolution?

Like the cell theory, the theory of evolution by natural selection has a pattern and a process component. Darwin and Wallace's theory made two important claims concerning patterns that exist in the natural world. The first claim was that species are related by common ancestry. This contrasted with the prevailing view in science at the time, which was that species represent independent entities that were created separately by a divine being. The second claim was equally novel. Instead of accepting the popular hypothesis that species remain unchanged through time, Darwin and Wallace proposed that the characteristics of species can be modified from generation to generation.

Evolution, then, means that species are not independent and unchanging entities, but are related to one another and can change through time. This part of the theory of evolution—the pattern component—was actually not original to Darwin and Wallace. Several scientists had already come to the same conclusions about the relationships among species. The great insight by Darwin and Wallace was in proposing a process, called **natural selection**, that explained *how* evolution occurs.

What Is Natural Selection?

Natural selection occurs whenever two conditions are met. The first condition is that individuals within a population vary in characteristics that are heritable. A **population** is defined as a group of individuals of the same species living in the same area at the same time. Darwin and Wallace had studied natural populations long enough to realize that variation among individuals is almost universal. In wheat, for example, some individuals are taller than others. **Heritable** traits are characteristics that can be passed on to offspring. As a result of work by wheat breeders, Darwin and Wallace knew that short parents tend to have short offspring. Subsequent research has shown that heritable variation exists in most traits and populations. The second condition of natural selection is that in a particular environment, certain versions of these heritable traits help individuals survive better or reproduce more than do other versions. For example, if tall wheat plants are easily blown down by wind, then in windy environments shorter plants will tend to survive better and leave more offspring than tall plants will.

If certain heritable traits lead to increased success in producing offspring, then those traits become more common in the population over time. In this way, the population's characteristics change as a result of natural selection acting on individuals. In wheat, populations of wheat that grow in windy environments tend to become shorter from generation to generation. A change in the characteristics of a population, over time, is evolution.

Darwin also introduced some new terminology to identify what is happening during natural selection. He used the term **fitness** to mean the ability of an individual to survive and reproduce. In biology, fitness is measured in units of number of offspring produced. Individuals with high fitness produce many offspring. A trait that increases the fitness of an individual in a particular environment is called an **adaptation**. Once again consider wheat: In windswept habitats, wheat plants with short stalks have higher fitness than do individuals with long stalks. Short stalks are an adaptation to windy environments.

To clarify further how natural selection works, consider the origin of the vegetables called the "cabbage family plants." Broccoli, cauliflower, Brussels sprouts, cabbage, kale, savoy, and collard greens descended from the same species—the wild plant in the mustard family pictured in **Figure 1.3a**. To create the plant called broccoli, horticulturists selected individuals of the wild mustard species with particularly large and compact flowering stalks. In mustards, the size and shape of the flowering stalk is a heritable trait. When the selected individuals were mated with one another, their offspring turned out to have larger and more compact flowering stalks, on average, than the original population (**Figure 1.3b**). By repeating this process over many generations, horticulturists produced a population with extraordinarily large and compact flowering stalks. The derived population has been artificially selected for the size and shape of the flowering stalk;

WEB TUTORIAL 1.1
Artificial Selection

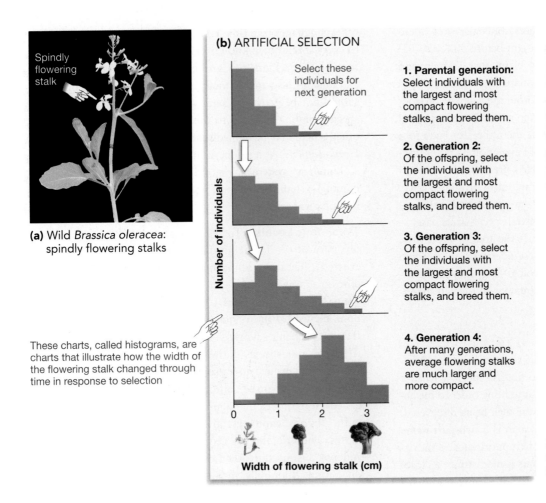

(a) Wild *Brassica oleracea*: spindly flowering stalks

Spindly flowering stalk

(b) ARTIFICIAL SELECTION

Select these individuals for next generation

Number of individuals

These charts, called histograms, are charts that illustrate how the width of the flowering stalk changed through time in response to selection

Width of flowering stalk (cm)

1. **Parental generation:** Select individuals with the largest and most compact flowering stalks, and breed them.

2. **Generation 2:** Of the offspring, select the individuals with the largest and most compact flowering stalks, and breed them.

3. **Generation 3:** Of the offspring, select the individuals with the largest and most compact flowering stalks, and breed them.

4. **Generation 4:** After many generations, average flowering stalks are much larger and more compact.

Large, compact flowering stalks

(c) Broccoli: extremely large, compact flowering stalks

FIGURE 1.3 Artificial Selection Can Produce Dramatic Changes in Organisms

as **Figure 1.3c** shows, it barely resembles the ancestral form. Note that during this process, the size and shape of the flowering stalk in each individual plant did not change within its lifetime—the change occurred in the characteristics of the population over time.

Darwin pointed out that natural selection changes the characteristics of a wild population over time, just as the deliberate manipulation of "artificial selection" changes the characteristics of a domesticated population over time. But no horticulturist is involved in the case of natural selection. Natural selection occurs naturally, simply because certain individuals in wild populations have heritable traits that allow them to leave more offspring than do individuals without those traits. Evolution, or change in the population over time, is the outcome of this process.

Since Darwin and Wallace published, biologists have succeeded in measuring hundreds of examples of natural selection in wild populations and have accumulated a massive body of evidence documenting that species have changed through time.

Together, the cell theory and the theory of evolution provided the young science of biology with two central, unifying ideas:

1. The cell is the fundamental structural unit in all organisms.

2. All species are related by common ancestry and have changed over time in response to natural selection.

✓ CHECK YOUR UNDERSTANDING

Natural selection occurs when heritable variation in certain traits leads to improved success in reproduction. Because individuals with these traits produce many offspring with the same traits, these traits increase in frequency and evolution occurs. Evolution is simply a change in the characteristics of a population over time. Although these ideas appear simple, they are often misunderstood. Research has shown that some biology students think that evolution is progressive, meaning that species always get larger, more complex, or "better" in some sense. In addition, it is common for students to think that *individuals* as well as populations change when natural selection occurs, or that individuals with high levels of fitness are stronger or bigger or "more dominant." None of these ideas are correct. Using the example of selection on height of wheat stalks, you should be able to explain why each of these three common misconceptions is wrong.

1.3 The Tree of Life

In Section 1.2 we focused on how individual populations change through time in response to natural selection. But over the past several decades, biologists have also documented dozens of cases in which natural selection has caused populations of one species to diverge and form new species. This divergence process is called **speciation**. In several instances, biologists are documenting the formation of new species right before our eyes.

Research on speciation supports a claim that Darwin and Wallace made over a century ago—that natural selection can lead to change *between* species as well as within species. The broader conclusions are that all species come from preexisting species and that all species, past and present, trace their ancestry back to a single common ancestor. If the theory of evolution by natural selection is valid, biologists should be able to reconstruct a **tree of life**—a family tree of organisms. If life on Earth arose just once, then such a diagram would describe the genealogical relationships among species with a single, ancestral species at its base.

Has this task been accomplished? If the tree of life exists, what does it look like? To answer these questions, we need to step back in time and review how biologists organized the diversity of organisms *before* the development of the cell theory and the theory of evolution.

Linnaean Taxonomy

In science, the effort to name and classify organisms is called **taxonomy**. This branch of biology began to flourish in 1735 when a botanist named Carolus Linnaeus set out to bring order to the bewildering diversity of organisms that were then being discovered.

The building block of Linnaeus' system is a two-part name unique to each type of organism. The first part indicates the organism's **genus** (plural: **genera**). A genus is made up of a closely related group of species. For example, Linnaeus put humans in the genus *Homo*. Although humans are the only living species in this genus, several extinct organisms, all of which walked upright and made extensive use of tools, were later also assigned to *Homo*. The second term in the two-part name identifies the organism's species. In Section 1.1 we defined a species as a distinct, identifiable type of organism. More formally, a **species** is made up of individuals that regularly breed together or have characteristics that are distinct from those of other species. Linnaeus gave humans the specific name *sapiens*.

An organism's genus and species designation is called its scientific name or Latin name. Scientific names are always italicized. Genus names are always capitalized, but species names are not: for instance, *Homo sapiens*. Scientific names are based on Latin or Greek word roots or on "Latinized" words from other languages (see **Box 1.1**). Linnaeus gave a scientific name to every species then known. (He also latinized his own name—from Karl von Linné to Carolus Linnaeus.)

Linnaeus also maintained that different types of organisms should not be given the same genus and species names. Other species may be assigned to the genus *Homo*, and members of other genera may be named *sapiens*, but only humans are named *Homo sapiens*. As a result, each scientific name is unique.

Linnaeus' system has stood the test of time. His two-part naming system, or **binomial nomenclature**, is still the standard in biological science.

Taxonomic Levels   To organize and classify the tremendous diversity of species being discovered in the 1700s, Linnaeus created a hierarchy of taxonomic groups: From the most specific grouping to the least specific, the levels are **species, genus, family, order, class, phylum** (plural: **phyla**), and **kingdom. Figure 1.4** shows how this nested, or hierarchical, classification scheme works, using humans as an example. Although our species is the sole living member of the genus *Homo*, humans are now grouped with the orangutan, gorilla, common chimpanzee, and pygmy chimpanzee in a family called Hominidae. Linnaeus grouped members of this family with gibbons, monkeys, and lemurs in an order called Primates. The Primates are grouped in the class Mammalia with rodents, bison, and other organisms that have fur and produce milk. Mammals, in turn, join other animals with structures called notochords in the phylum Chordata, and all other animals in the kingdom Animalia. Each of these named groups—primates, mammals, or *Homo sapiens*—can be referred to as a **taxon** (plural: **taxa**). The essence of Linnaeus' system is that lower-level taxa are nested within higher-level taxa.

Aspects of this hierarchical scheme are still in use. As biological science matured, however, several problems with Linnaeus' original proposal emerged.

How Many Kingdoms Are There?   Linnaeus proposed that species could be organized into two kingdoms: plants and animals. According to Linnaeus, organisms that do not move and that produce their own food are plants; organisms that move and acquire food by eating other organisms are animals.

BOX 1.1   Scientific Names and Terms

Scientific names and terms are often based on Latin or Greek word roots that are descriptive. For example, *Homo sapiens* is derived from the Latin *homo* for "man" and *sapiens* for "wise" or "knowing." The yeast that bakers use to produce bread and that brewers use to brew beer is called *Saccharomyces cerevisiae*. The Greek root *saccharo* means "sugar," and *myces* refers to a fungus. *Saccharomyces* is aptly named "sugar fungus" because yeast is a fungus and because the domesticated strains of yeast used in commercial baking and brewing are often fed sugar. The specific name of this organism, *cerevisiae*, is Latin for *beer*. Loosely translated, then, the scientific name of brewer's yeast means "sugar fungus for beer."

Most biologists find it extremely helpful to memorize some of the common Latin and Greek roots. To aid you in this process, new terms in this text are often accompanied by a reference to their Latin or Greek word roots in parentheses.

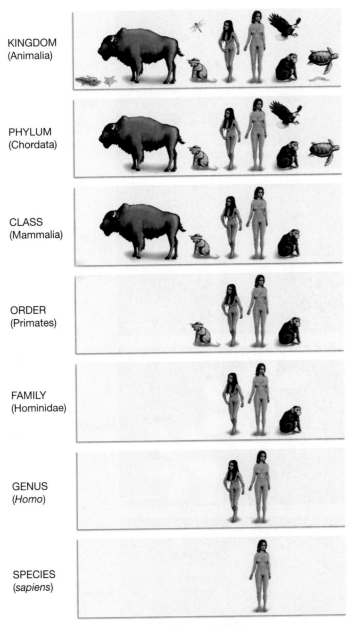

KINGDOM
(Animalia)

PHYLUM
(Chordata)

CLASS
(Mammalia)

ORDER
(Primates)

FAMILY
(Hominidae)

GENUS
(*Homo*)

SPECIES
(*sapiens*)

FIGURE 1.4 Linnaeus Defined Taxonomic Levels
In the Linnaean system, each animal species is placed in a taxonomic hierarchy with seven levels. Lower levels are nested within higher levels. Linnaeus proposed that these levels reflected the natural order of organisms.

Not all organisms fall neatly into these categories, however. Molds, mushrooms, and other fungi survive by absorbing nutrients from dead or living plants and animals. Even though they do not make their own food, they were placed in the kingdom Plantae because they do not move. The tiny, single-celled organisms called bacteria also presented problems. Some bacteria can move, and many can make their own food. Initially they, too, were thought to be plants. Eventually, though, it became clear that the two-kingdom system was simply inadequate.

Further, the development of the theory of evolution suggested a new goal for taxonomy. As evidence for evolution mounted, biologists concentrated on understanding how classification sys-tems such as the one invented by Linnaeus could be modified to reflect the genealogical relationships among organisms—not just how similar they are in appearance or other broad characteristics. The goal of taxonomy became an attempt to reflect **phylogeny** (meaning "tribe-source")—the true historical relationships among types of organisms.

Linnaeus proposed the two-kingdom system because he thought it reflected a fundamental pattern in the natural world. His two-kingdom system was a hypothesis—a proposed explanation for an observed phenomenon. In this case, the pheno-menon that Linnaeus and other biologists were trying to explain was the relationship among the diverse forms of life on Earth.

For example, Linnaeus proposed that the fundamental division in organisms was between plants and animals. But when advances in microscopy allowed biologists to study the contents of individual cells in detail, a different fundamental division emerged and his hypothesis was rejected. In plants, animals, and the organisms that taxonomists call protists, cells contain a prominent component called a nucleus (**Figure 1.5a**). But in bacteria, cells lack this kernel-like structure (**Figure 1.5b**). Organisms with a nucleus are called **eukaryotes** ("true-kernel"); organisms without a nucleus are called **prokaryotes** ("before-kernel"). The vast majority of prokaryotes are unicellular ("one-celled"), many eukaryotes; are multicellular ("many-celled").

(a) Eukaryotic cells have a membrane-bound nucleus.

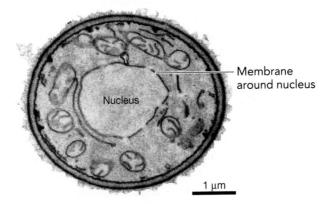

Membrane
around nucleus

Nucleus

1 μm

(b) Prokaryotic cells do *not* have a membrane-bound nucleus.

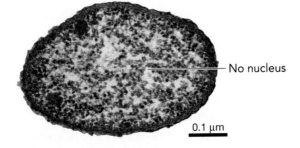

No nucleus

0.1 μm

FIGURE 1.5 Eukaryotes and Prokaryotes
Cross sections of **(a)** a eukaryotic and **(b)** a prokaryotic cell.
EXERCISE Study the scale bars; then draw two ovals that accurately represent the relative sizes of a eukaryotic cell and a prokaryotic cell.

When data began to conflict with Linnaeus' original scheme, biologists proposed alternative hypotheses. In the late 1960s one researcher suggested that a system of five kingdoms best reflects the patterns observed in nature. This five-kingdom system is depicted in **Figure 1.6**. Although the scheme has been widely used, it represents just one proposal out of many. Other biologists proposed that organisms are organized into three, four, six, or eight kingdoms. But it was still not clear which of these schemes, if any, accurately described the phylogeny of organisms.

About the time that the five-kingdom proposal was published, however, Carl Woese (pronounced "woes") and colleagues began working on the problem from a radically different angle. Instead of assigning organisms to kingdoms on the basis of characteristics such as the presence of a nucleus or the ability to move or to manufacture food, these researchers attempted to understand the relationships among organisms by analyzing their chemical components.

Using Molecules to Understand the Tree of Life

Woese and his co-workers had an explicit goal: to estimate where major branches occurred on the tree of life by analyzing the molecular components of cells. To accomplish this goal they needed to study a molecule that is found in all organisms. The molecule they selected is called small subunit rRNA. It is an essential part of the machinery that all cells use to grow and reproduce.

Although rRNA is a large and complex molecule, its underlying structure is simple. The rRNA molecule is made up of sequences of four smaller chemical components called ribonucleotides. Ribonucleotides are symbolized by the letters A, U, C, and G. In rRNA, ribonucleotides are connected to one another linearly, like boxcars of a freight train (**Figure 1.7**).

Why might rRNA be useful for understanding the relationships among organisms? The answer is that the ribonucleotide sequence in rRNA is a trait, similar to the height of wheat stalks or the size of flowering stalks of broccoli, that can change during the course of evolution. Although rRNA performs the same function in all organisms, the sequence of ribonucleotide building blocks in this molecule is not identical among species. In land plants, for example, the molecule might start with the sequence A-U-A-U-C-G-A-G. In green algae, which are closely related to land plants, the same section of the molecule might contain A-U-A-U-G-G-A-G. But in brown algae, which are not closely related to green algae or to land plants, the same part of the molecule might consist of A-A-A-U-G-G-A-G.

The research program that Woese and co-workers pursued was based on a simple premise: If the theory of evolution is correct, then rRNA sequences should be very similar in closely related organisms but less similar in organisms that are less closely related. The rRNA sequences of two species that diverged from each other long ago should be quite different, while rRNA sequences from two species that diverged from each other more recently should be much more alike.

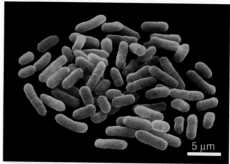

KINGDOM MONERA
(includes all prokaryotes)

KINGDOM PROTISTA
(includes several groups of unicellular eukaryotes)

KINGDOM PLANTAE

KINGDOM FUNGI

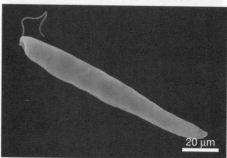

KINGDOM ANIMALIA

FIGURE 1.6 The Five-Kingdom Scheme
For decades, most biologists accepted the hypothesis that organisms naturally fall into the five kingdoms illustrated here. **QUESTION** How many times bigger is a fruit fly than one of the prokaryotic cells pictured here?

In rRNA four types of ribonucleotides (A, U, C, and G) are arranged in a linear sequence. The complete molecule contains about 2000 ribonucleotides; just eight are drawn here.

The sequence of ribonucleotides may vary among species. If the above sequence is observed in land plants, the sequence below might be found at the same location in the rRNA molecule of green algae.

FIGURE 1.7 RNA Molecules Are Made Up of Smaller Molecules
The four smaller molecules that make up an rRNA molecule are symbolized A, U, C, and G. The sequence of A, U, C, and G subunits in rRNA varies among species. QUESTION Suppose that in the same portion of rRNA, molds and other fungi have the sequence A-U-A-U-G-G-A-C. According to these data, are fungi more closely related to green algae or to land plants? Explain your logic.

To put this insight to work, the researchers determined the sequence of ribonucleotides in the rRNA of a wide array of species. Then they considered what the similarities and differences in the sequences implied about relationships among the species. The goal was to produce a diagram that described the phylogeny of the organisms in the study. A diagram that depicts evolutionary history in this way is called a **phylogenetic tree.** Just as a family tree shows relationships among individuals, a phylogenetic tree shows relationships among species. On a phylogenetic tree, branches that are close to one another represent species that are closely related; branches that are farther apart represent species that are more distantly related.

The rRNA Tree To construct a phylogenetic tree, researchers use a computer to find the arrangement of branches that is most consistent with the similarities and differences observed in the data—in this case, in the sequences of ribonucleotides observed in rRNA. The tree produced by comparing these sequences is shown in **Figure 1.8**. Because this tree includes species from many different kingdoms and phyla, it is often called the universal tree, or the tree of life.

The tree of life implied by rRNA data astonished biologists. According to data from this molecule:

- The fundamental division in organisms is not between plants and animals or even between prokaryotes and eukaryotes. Rather, *three* major groups occur: (1) the Bacteria; (2) another group of prokaryotic, single-celled organisms called

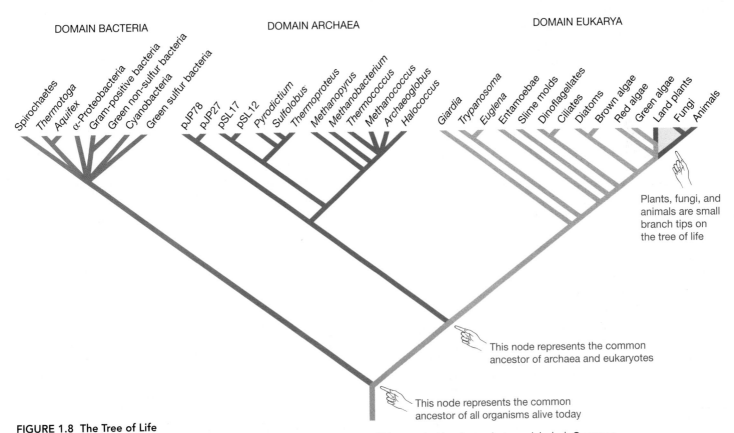

FIGURE 1.8 The Tree of Life
"Universal tree" estimated from rRNA sequence data. The three domains of life revealed by the analysis are labeled. Common names are given for most lineages in the domains Bacteria and Eukarya. Genus names are given for members of the domain Archaea, because most of these organisms have no common names. Archaean species with labels such as "pSL17" have not yet been given scientific names. EXERCISE Circle the branches and tips that represent prokaryotes.

the Archaea; and (3) the eukaryotes. To accommodate this new perspective on the diversity of organisms, Woese created a new taxonomic level called the **domain**. As Figure 1.8 indicates, the three domains of life are now called the Bacteria, Archaea, and Eukarya.

- Some of the kingdoms that had been defined earlier do not reflect how evolution actually occurred. For example, recall that Linnaeus grouped the multicellular eukaryotes known as fungi with plants. But the rRNA data indicate that fungi are much more closely related to animals than they are to plants.

- Bacteria and Archaea are much more diverse than anyone had imagined. If the differences among animals, fungi, and plants warrant placing them in separate kingdoms, then dozens of kingdoms exist among the prokaryotes.

The Tree of Life Is a Work in Progress   The rRNA tree was a hypothesis that inspired a flurry of research. Biologists in laboratories all over the world tested the conclusions by determining the sequences of other molecules found in cells and by reanalyzing older data in light of the new findings. In general, these studies have confirmed the major features of the tree in Figure 1.8. The discovery of the Archaea and the placement of lineages such as the fungi qualify as exciting breakthroughs in our understanding of life's diversity.

Work on the tree of life continues at a furious pace, however, and the location of certain branches on the tree is hotly debated. As databases expand and as techniques for analyzing data improve, the shape of the tree of life presented in Figure 1.8 will undoubtedly change.

✓CHECK YOUR UNDERSTANDING

A phylogenetic tree is analogous to a family tree. A phylogenetic tree shows the evolutionary relationships among species, just as a family tree shows the genetic relationships among individuals.

To infer where species belong on a phylogenetic tree, biologists examine the characteristics of the species involved. Closely related species should have similar characteristics, while less closely related species should be less similar. A characteristic that has been used extensively in phylogenetic analyses is the sequence of subunits in a molecule called rRNA, which is found in all species. You should be able to examine the sequences given below and draw a phylogenetic tree showing the relationships among Species A, B, and C that these data imply.

Species A:	A	A	C	T	A	G	C	G	C	G	A	T
Species B:	A	A	C	T	A	G	C	G	C	C	A	T
Species C:	T	T	C	T	A	G	C	G	G	T	A	T

1.4   Doing Biology

This chapter has introduced some of the great ideas in biology. The development of the cell theory and the theory of evolution by natural selection provided cornerstones when the science was young; the tree of life is a relatively recent insight that has revolutionized the way researchers understand the diversity of life on Earth.

These theories are considered great because they explain fundamental aspects of nature, and because they have consistently been shown to be correct. They are considered correct because they have withstood extensive testing. How do biologists test ideas about the way the natural world works? The answer is that they test the predictions made by alternative hypotheses, often by setting up carefully designed experiments. To illustrate how this approach works, let's consider two questions currently being addressed by researchers.

Why Do Giraffes Have Long Necks? An Introduction to Hypothesis Testing

If you were asked why giraffes have long necks, you might say that long necks enable giraffes to reach food that is unavailable to other mammals. This hypothesis is expressed in African folktales and has traditionally been accepted by many biologists. The food competition hypothesis is so plausible, in fact, that for decades no one thought to test it. Recently, however, Robert Simmons and Lue Scheepers assembled data suggesting that the food competition hypothesis is only part of the story. Their analysis supports an alternative hypothesis—that long necks allow giraffes to use their heads as effective weapons for battering their opponents.

How did biologists test the food competition hypothesis? What data support their alternative explanation? Before we answer these questions, it's important to recognize that hypothesis testing is a two-step process. The first step is to state the hypothesis as precisely as possible and list the predictions it makes. The second step is to design an observational or experimental study that is capable of testing those predictions. If the predictions are accurate, then the hypothesis is supported. If the predictions are not met, then researchers do further tests, modify the original hypothesis, or search for alternative explanations.

The Food Competition Hypothesis: Predictions and Tests

Stated precisely, the food competition hypothesis claims that giraffes compete for food with other species of mammals. When food is scarce, as it is during the dry season, giraffes with longer necks can reach food that is unavailable to other species and to giraffes with shorter necks. As a result, the longest-necked individuals in a giraffe population survive better and produce more young than do shorter-necked individuals, and average neck length of the population increases with each generation. To use the terms introduced earlier in this chapter, long necks are adaptations that increase the fitness of individual giraffes during competition for food. This type of natural selection has gone on so long that the population has become extremely long necked.

(a) Most feeding is done below neck height.

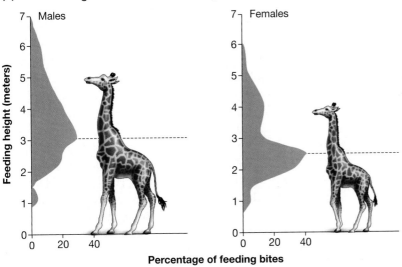

(b) Typical feeding posture in giraffes

FIGURE 1.9 Giraffes Do Not Usually Extend Their Necks to Feed
(a) The proportion of bites that male and female giraffes of average size take at different heights. **(b)** Although it is common to see photos of giraffes straining to reach leaves high in trees, giraffes usually feed at the heights shown here.

The food competition hypothesis makes several explicit *predictions*. A prediction is something that can be measured and that must be correct if a hypothesis is valid. For example, the food competition hypothesis predicts that (1) neck length is variable among giraffes; (2) neck length in giraffes is heritable; and (3) giraffes feed high in trees, especially during the dry season, when food is scarce and the threat of starvation is high.

The first prediction is clearly correct. Studies in zoos and natural populations confirm that neck length is variable among individuals.

The researchers were unable to test the second prediction, however, because they studied giraffes in a natural population and were unable to do breeding experiments. As a result, they simply had to accept this prediction as an assumption. In general, though, biologists prefer to test every assumption behind a hypothesis.

What about the prediction regarding feeding high in trees? According to Simmons and Scheepers, this is where the food competition hypothesis breaks down. Consider, for example, data collected by a different research team about the amount of time that giraffes spend feeding in vegetation of different heights. **Figure 1.9a** shows that in a population from Kenya, both male and female giraffes spend most of their feeding time eating vegetation that averages just 60 percent of their full height. Studies on other populations of giraffes, during both the wet and dry seasons, are consistent with these data. Giraffes usually feed with their necks bent (**Figure 1.9b**).

These data cast doubt on the food competition hypothesis, because one of its predictions does not appear to hold. Biologists have not abandoned this hypothesis completely, though, because feeding high in trees may be particularly valuable during extreme droughts, when a giraffe's ability to reach leaves far above the ground could mean the difference between life and death. Still,

Simmons and Scheepers have offered an alternative explanation for why giraffes have long necks. The new hypothesis is based on the mating system and social behavior of giraffes.

The Sexual Competition Hypothesis: Predictions and Tests
Giraffes have an unusual social system. Breeding occurs year round rather than seasonally. To determine when females are coming into estrus (or "heat") and are thus receptive to mating, the males nuzzle the rumps of females. In response, the females urinate into the males' mouths. The males then tip their heads back and pull their lips to and fro, as if tasting the liquid. Biologists who have witnessed this behavior have proposed that the males taste the females' urine to detect whether estrus has begun.

Once a female giraffe enters estrus, males fight among themselves for the opportunity to mate. Combat is spectacular. The bulls stand next to one another, swing their necks, and strike thunderous blows with their heads. Researchers have seen males knocked unconscious for 20 minutes after being hit and have cataloged numerous instances in which the loser died. Giraffes are the only animals that fight in this way.

These observations inspired a new explanation for why giraffes have long necks. The sexual competition hypothesis is based on the idea that longer-necked giraffes are able to strike harder blows during combat than can shorter-necked giraffes. In engineering terms, longer necks provide a longer moment arm. A long moment arm increases the force of the impact. (Think about the type of sledge hammer you'd use to bash down a concrete wall—one with a short handle or one with a long handle?) Thus, longer-necked males should win more fights and, as a result, father more offspring than do shorter-necked males. If neck length in giraffes is inherited, then the average neck length in the population should increase over time. Under the sexual

WEB TUTORIAL 1.2 Introduction to Experimental Design

competition hypothesis, long necks are adaptations that increase the fitness of males during competition for females.

Although several studies have shown that long-necked males are more successful in fighting and that the winners of fights gain access to estrous females, the question of why giraffes have long necks is not closed. With the data collected to date, most biologists would probably concede that the food competition hypothesis needs further testing and refinement and that the sexual selection hypothesis appears promising. It could also be true that both hypotheses are correct. But clearly, more work needs to be done.

In many cases in biological science, "more work" involves experimentation. Experimenting on giraffes is difficult. But in the case study considered next, biologists were able to test an interesting hypothesis experimentally.

Why Are Chili Peppers Hot? An Introduction to Experimental Design

Experiments are a powerful scientific tool because they allow researchers to test the effect of a single, well-defined factor on a particular phenomenon. Experiments that test the effect of neck length on food and sexual competition in giraffes have yet to be done. Instead, as an example of how experiments are designed, let's consider a different phenomenon: Why do chili peppers taste so spicy?

The jalapeño, anaheim, and cayenne peppers used in cooking descended from a wild shrub that is native to the deserts of the American Southwest. As **Figure 1.10a** shows, wild chilies produce fleshy fruits with seeds inside, just like their domesticated descendants. In both wild chilies and the cultivated varieties, the "heat" or pungent flavor of the fruit and seeds is due to a molecule called capsaicin. In humans and other mammals, capsaicin binds to heat-sensitive cells in the tongue and mouth.

In response to this binding, signals are sent to the brain that produce the sensation of burning. Similar signals would be transmitted if you drank boiling water. Asking why chilies are hot, then, is the same as asking why chilies contain capsaicin.

Josh Tewksbury and Gary Nabhan proposed that the presence of capsaicin is an adaptation that protects chili fruits from being eaten by animals that destroy the seeds inside. To understand this hypothesis, it's important to realize that the seeds inside a fruit have one of two fates when the fruit is eaten. If the seeds are destroyed in the animal's mouth or digestive system, then they never germinate (sprout). In this case, "seed predation" has occurred. But if seeds can travel undamaged through the animal, then they are eventually "planted" in a new location along with a valuable supply of fertilizer. In this case, seeds are dispersed. Here's the key idea: Natural selection should favor fruits that taste bad to animal species that act as seed predators. But these same fruits should not deter species that act as seed dispersers. This proposal is called the *directed dispersal hypothesis*.

Does capsaicin deter seed predators, as the directed dispersal hypothesis predicts? To answer this question, the researchers captured some cactus mice (**Figure 1.10b**) and birds called curve-billed thrashers (**Figure 1.10c**). These species are among the most important fruit- and seed-eating animals in habitats where chilies grow. Based on earlier observations, the biologists predicted that cactus mice destroy chili seeds but that curve-billed thrashers disperse them effectively.

To test the directed dispersal hypothesis, the biologists offered both cactus mice and curve-billed thrashers three kinds of fruit: hackberries, chilies from a strain of plant that can't synthesize capsaicin, and pungent chilies that have lots of cap-

(a) Chilies produce fruits that contain seeds.

(b) Cactus mice are seed eaters.

(c) Curve-billed thrashers eat chili fruits.

FIGURE 1.10 Chilies … and Chili Eaters?
(a) Chili fruits are hot because they contain a molecule called capsaicin. **(b)** Cactus mice and **(c)** curve-billed thrashers are common in habitats where chilies grow.

saicin. The non-pungent chilies are about the same size and color as normal chilies and have similar nutritional value. The hackberries don't look anything like chilies, however, and contain no capsaicin. The three fruits were present in equal amounts. For each animal tested, the researchers recorded the percentage of hackberry, non-pungent chili, and pungent chili that was eaten during a specific time interval. Then they calculated the average amount of each fruit that was eaten by five test individuals from each species.

The directed dispersal hypothesis predicts that seed dispersers will eat the pungent chilies readily but that seed predators won't. Recall that a prediction specifies what we should observe if a hypothesis is correct. Good scientific hypotheses make testable predictions—predictions that can be supported or rejected by collecting and analyzing data. If the directed dispersal hypothesis is wrong, however, then there shouldn't be any difference in what various animals eat. This latter possibility is called a **null hypothesis**. A null hypothesis specifies what we should observe when the hypothesis being tested doesn't hold. These predictions are listed in **Figure 1.11**.

Do the predictions of the directed dispersal hypothesis hold? To answer this question, look at the results plotted in Figure 1.11. View the data for each type of fruit to see if different amounts were eaten by different types of predators. Then fill in the following table:

Type of fruit	Did the two predators eat about the same amount or very different amounts of this fruit?	If the amount eaten was very different, comment on the nature of the difference.
Hackberry		
Non-pungent chilies (no capsaicin)		
Pungent chilies (lots of capsaicin)		

Based on your analysis of the data, decide whether the results support the directed dispersal hypothesis or the null hypothesis. Use the conclusion stated in Figure 1.11 to check your answer.

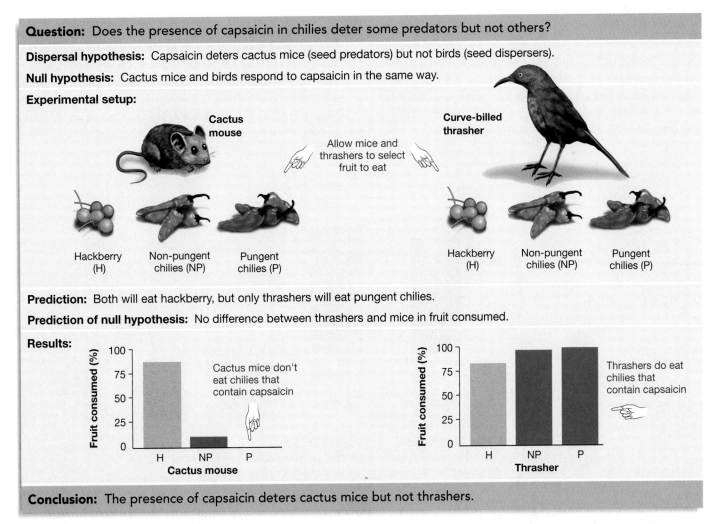

FIGURE 1.11 An Experimental Test: Does Capsaicin Deter Some Fruit Eaters?

In relation to designing effective experiments, this study illustrates several important points:

- *It is critical to include control groups.* A **control** checks for factors, other than the one being tested, that might influence the experiment's outcome. For example, if hackberries had not been included as a control, it would have been possible to claim that the cactus mice in the experiment didn't eat pungent chilies simply because they weren't hungry. But the not-hungry hypothesis can be rejected because all of the animals ate hackberries.

- *The experimental conditions must be carefully controlled.* The investigators used the same feeding choice setup, the same time interval, and the same definitions of predator response in each test. Controlling all of the variables except one—the types of fruits presented—is crucial because it eliminates alternative explanations for the results. For example, what types of problems could arise if the cactus mice were given less time to eat than the thrashers, or if the test animals were always presented with hackberries first and pungent chilies last?

- *Repeating the test is essential.* It is almost universally true that larger sample sizes in experiments are better. For example, suppose that the experimenters had used just one cactus mouse instead of five, and that this mouse was unlike other cactus mice because it ate almost anything. If so, the resulting data would be badly distorted. By testing many individuals, the amount of distortion or "noise" in the data caused by unusual individuals or circumstances is reduced.

To test the assumption that cactus mice are seed predators and curve-billed thrashers are seed dispersers, the researchers did a follow-up experiment. They fed fruits of the non-pungent chili to each type of predator. When the seeds had passed through the animals' digestive systems and were excreted, the researchers collected and planted the seeds—along with 14 uneaten seeds. Planting uneaten seeds served as a control treatment, because it tested the hypothesis that the seeds were viable and would germinate if they were not eaten. **Figure 1.12** shows the percentage of seeds that germinated. The data indicate that seeds pass through curve-billed thrashers unharmed but are destroyed when eaten by cactus mice.

Based on the outcomes of these two experiments, the researchers concluded that curve-billed thrashers are efficient seed dispersers and are not deterred by capsaicin. The cactus mice, in contrast, refuse to eat chilies. If they ate chilies, the mice would kill the seeds. These are exactly the results predicted by the directed dispersal hypothesis. The biologists concluded that the presence of capsaicin in chilies is an adaptation that keeps their seeds from being destroyed by mice. In habitats that contain cactus mice, the production of capsaicin increases the fitness of individual chili plants.

These experiments are a taste of things to come. In this text you will encounter hypotheses and experiments on questions ranging from how water gets to the top of 100-meter-tall sequoia trees to why the bacterium that causes tuberculosis has become resistant to antibiotics. A commitment to tough-minded hypothesis testing and sound experimental design is a hallmark of biological science. Understanding their value is an important first step in becoming a biologist.

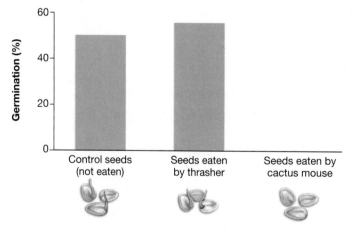

FIGURE 1.12 Mice Destroy Seeds, but Thrashers Do Not
Histograms showing the percentage of seeds that germinated after being planted directly into the ground versus being planted after passing through the digestive system of a curve-billed thrasher or cactus mouse. **QUESTION** Seven thrashers and five cactus mice were used in this experiment. Why wan't just one individual of each species used?

ESSAY Where Do Humans Fit on the Tree of Life?

Given the vast diversity of organisms that make up the tree of life, where do humans fit in? Most of the major branches on the tree diagrammed in Figure 1.8 represent unicellular organisms. Familiar multicellular species such as grasses and mushrooms are part of the twigs labeled "Plants" and "Fungi," respectively. At the tree's tip, the label "Animals" identifies the kingdom that appeared most recently among the eukaryotes, about 800 million years ago. Humans are found on this branch, along with millions of other species ranging from sponges and corals to insects and fish.

Our species is a tiny new twig on an enormous and ancient tree of life.

Just how recently did our species appear? In Chapter 2 we examine a technique called radiometric dating that allows geologists to estimate the age of rocks containing bones, teeth, shells, and other traces of organisms that lived in the past. According to this technique, the first traces left by our species appear about 100,000 years ago. To clarify just how recent this is in the sweep of Earth history, consider the calendar in **Figure 1.13**. The 12 months shown are scaled to represent the 4.6 billion years of Earth history. At this scale, 7 seconds make up a millennium; each hour denotes a span of 525,000 years; and each day represents an interval of 12.6 million years. Note that the first unicellular organisms appear in late March and the first multicellular life in early October. Hominids (members of the family Hominidae) walk upright for the first time on mid-afternoon of New Year's Eve. *Homo sapiens* appears about 15 minutes before the stroke of midnight.

The message of these analyses is clear: Our species is a tiny new twig on an enormous and ancient tree of life.

Given that our species evolved so recently, some other interesting questions arise. For example, a typical species of mammal lasts about 2.5 million years in the fossil record. In your opinion, how likely is it that our species will continue to leave fossil evidence for another 2.4 million years? What factors might contribute to the long-term survival of *Homo sapiens*? What factors might contribute to our demise?

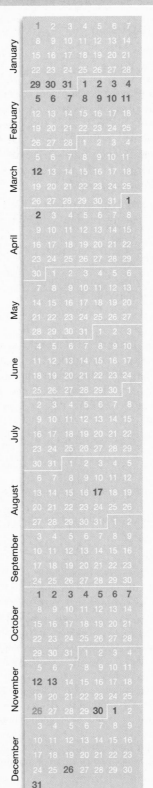

1: Earth forms

29–11: Oldest known rocks

12: Oldest chemical evidence of life

1–2: Oldest fossil cells

17: First eukaryotes

1–7: First multicellular organisms (algae)

12–13: First animals with shells and limbs

26: First animals with vertebrae

30: First land plants

26: Extinction of dinosaurs

1: First land animals

31: *Homo sapiens* appears one hour before midnight. Humans set foot on Moon ¼ second before midnight

FIGURE 1.13 Earth's History as a Calendar

QUESTION   If the length of Earth's history (4.6 billion years) were to correspond to the length of a football field (100 yards), where on the football field does the first fossil evidence of humans (100,000 years ago) appear?

1 day = 12.6 million years
1 second = 143 years

CHAPTER REVIEW

Summary of Key Concepts

Biologists have been discovering traits that unify the spectacular diversity of living organisms for over two hundred years.

■ **Biological science was founded with the development of (1) the cell theory, which proposes that all organisms are made of cells and that all cells come from preexisting cells, and (2) the theory of evolution by natural selection, which maintains that the characteristics of species change through time—primarily because individuals with certain heritable traits produce more offspring than do individuals without those traits.**

The cell theory is an important unifying principle in biology, because it identified the fundamental structural unit common to all life. The theory of evolution by natural selection is another key unifying principle, because it states that all organisms are related by common ancestry. It also offered a robust explanation for why species change through time and why they are so well adapted to their habitats.

Web Tutorial 1.1 Artificial Selection

■ **A phylogenetic tree is a graphical representation of the evolutionary relationships among species. Phylogenies can be estimated by analyzing similarities and differences in traits. Species that share many traits are closely related and are placed close to each other on the tree of life.**

The cell theory and the theory of evolution predict that all organisms are part of a genealogy of species, and that all species trace their ancestry back to a single common ancestor. To reconstruct this phylogeny, biologists have analyzed the sequence of components in a molecule called rRNA, which is found in all cells. A tree of life, based on similarities and differences in these sequences, has recently been constructed. According to the information contained in rRNA, the tree of life has three major lineages: the Bacteria, Archaea, and Eukarya. Analyses of other molecules and traits have largely supported the conclusions made from rRNA data.

■ **Biologists ask questions, generate hypotheses to answer them, and design experiments that test the predictions made by competing hypotheses.**

Another unifying theme in biology is a commitment to hypothesis testing and to sound experimental design. Analyses of neck length in giraffes and the capsaicin found in chilies are case studies in the value of testing alternative hypotheses and conducting experiments. Biology is a hypothesis-driven, experimental science.

Web Tutorial 1.2 Introduction to Experimental Design

Questions

Content Review

1. Anton van Leeuwenhoek made an important contribution to the development of the cell theory. How?
   a. He articulated the pattern component of the theory—that all organisms are made of cells.
   b. He articulated the process component of the theory—that all cells come from preexisting cells.
   c. He invented the first microscope and saw the first cell.
   d. He invented more powerful microscopes and was the first to describe the diversity of cells.

2. Suppose that a proponent of the spontaneous generation hypothesis claimed that cells would appear in Pasteur's swan-necked flask eventually. According to this view, Pasteur did not allow enough time to pass before concluding that life does not originate spontaneously. Which of the following is the best response?
   a. The spontaneous generation proponent is correct: Spontaneous generation would probably happen eventually.
   b. Both the all-cells-from-cells hypothesis and the spontaneous generation hypothesis could be correct.
   c. If spontaneous generation happens only rarely, it is not important.
   d. If spontaneous generation did not occur after weeks or months, it is not reasonable to claim that it would occur later.

3. What does the term *evolution* mean?
   a. The strongest individuals produce the most offspring.
   b. The characteristics of an individual change through the course of its life, in response to natural selection.
   c. The characteristics of populations change through time.
   d. The characteristics of species become more complex over time.

4. What does it mean to say that a characteristic of an organism is heritable?
   a. The characteristic evolves.
   b. The characteristic can be passed on to offspring.
   c. The characteristic is advantageous to the organism.
   d. The characteristic does not vary in the population.

5. In biology, what does the term *fitness* mean?
   a. how well trained and muscular an individual is, relative to others in the same population
   b. how slim an individual is, relative to others in the same population
   c. how long a particular individual lives
   d. the ability to survive and reproduce

6. Could *both* the food competition hypothesis and the sexual selection hypothesis explain why giraffes have long necks? Why or why not?
   a. No. In science, only one hypothesis can be correct.
   b. No. Observations have shown that the food competition hypothesis cannot be correct.
   c. Yes. Long necks could be advantageous for more than one reason.
   d. Yes. All giraffes have been shown to feed at the highest possible height and fight for mates.

Conceptual Review

1. The Greek roots of the term *taxonomy* can be translated as "arranging rules." Explain why these roots were an appropriate choice for this term.

2. It was once thought that the deepest split among life-forms was between two groups: prokaryotes and eukaryotes. Draw and label a phylogenetic tree that represents this hypothesis. Then draw and label a phylogenetic tree that shows the actual relationships among the three domains of organisms.

3. Why was it important for Linnaeus to establish the rule that only one type of organism can have a particular genus and species name?

4. What does it mean to say that an organism is adapted to a particular habitat?

5. Compare and contrast natural selection with the process that led to the divergence of a wild mustard plant into cabbage, broccoli, and Brussels sprouts.

6. The following two statements explain the logic behind the use of molecular sequence data to estimate evolutionary relationships:

   "If the theory of evolution is true, then rRNA sequences should be very similar in closely related organisms but less similar in organisms that are less closely related."

   "On a phylogenetic tree, branches that are close to one another represent species that are closely related; branches that are farther apart represent species that are more distantly related."

   Is the logic of these statements sound? Why or why not?

Group Discussion Problems

1. A scientific theory is a set of propositions that defines and explains some aspect of the world. This definition contrasts sharply with the everyday usage of the word *theory*, which often carries meanings such as "speculation" or "guess." Explain the difference between the two definitions, using the cell theory and the theory of evolution by natural selection as examples.

2. Turn back to the tree of life shown in Figure 1.8. Note that Bacteria and Archaea are prokaryotes, while Eukarya are eukaryotes. On the simplified tree below, draw an arrow that points to the branch where the structure called the nucleus originated. Explain your reasoning.

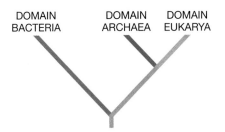

3. The proponents of the cell theory could not "prove" that it was correct in the sense of providing incontrovertible evidence that all organisms are made up of cells. They could state only that all organisms examined to date were made of cells. Why was it reasonable for them to conclude that the theory was valid?

4. How do the tree of life and the taxonomic categories created by Linnaeus (kingdom, phylum, class, order, family, genus, and species) relate to one another?

Answers to Multiple-Choice Questions   1. d; **2.** d; **3.** c; **4.** b; **5.** d; **6.** c

Evolutionary Processes and Patterns

A tiger beetle. Beetles are the most species-rich lineage on the tree of life, with well over 300,000 species named to date.

23

Evolution by Natural Selection

KEY CONCEPTS

- Populations and species evolve, meaning that their characteristics change through time. More precisely, evolution is defined as changes in allele frequencies over time.

- Natural selection occurs when individuals with certain alleles and traits produce the most offspring in a population. An adaptation is a genetically based trait that increases an individual's ability to produce offspring in a particular environment.

- Evolution by natural selection is not progressive, and it does not change the characteristics of the individuals that are selected—it changes only the characteristics of the population. Not all traits are adaptive. All adaptations are constrained by genetic and historical factors.

Alpine skypilots grow at high elevations in the Rocky Mountains of North America. Populations of skypilots that are pollinated by bumblebees have large, sweet-smelling flowers, but populations that are pollinated by flies have small, skunky-smelling flowers. Recent experiments have shown that the differences in these populations are due to evolution by natural selection.

This chapter is about one of the great ideas in science. The theory of evolution by natural selection, formulated independently by Charles Darwin and Alfred Russel Wallace, explains how organisms have come to be adapted to environments ranging from arctic tundra to tropical wet forest. As an example of a revolutionary breakthrough in our understanding of the world, the theory of evolution by natural selection ranks alongside Copernicus's theory of the Sun as the center of our solar system, Newton's laws of motion and theory of gravitation, and Einstein's general theory of relativity.

Like most scientific breakthroughs, this one did not come easily. When Darwin published his theory in 1859 in a book called *On the Origin of Species by Means of Natural Selection*, it unleashed a firestorm of protest throughout Europe. At that

time, the leading explanation for the diversity of organisms was a theory called special creation. This theory held that all species were created independently, by God, perhaps as recently as 6000 years ago. The theory of special creation also maintained that species were immutable, or incapable of change, and thus had been unchanged since the moment of their creation. Darwin's ideas were radically different. He proposed that life on Earth was ancient and that species change through time.

To understand the contrast between the theory of special creation and the theory of evolution by natural selection more thoroughly, recall from Chapter 1 that scientific theories usually have two components: a pattern and a process. The first component is either a claim about a pattern that exists in nature or a statement that summarizes a series of observations about the natural world. In short, the pattern component is about facts—about how things *are* in nature. The second component of a scientific theory is a process that produces that pattern or set of observations. For example, the pattern component in the theory of special creation was that species were seen to be created independently of one another and that they do not change through time. The process that explained this pattern was the instantaneous and independent creation of living organisms by a supernatural being.

This chapter begins by examining the pattern component of Darwin's theory. (**Box 23.1** discusses why the theory of evolu-

tion became associated primarily with Darwin's name.) Specifically, Section 23.1 considers the evidence behind the claim that species are not independent but are instead related and that they are not immutable but instead have changed through time. Darwin proposed that natural selection explains this pattern; in Section 23.2 we analyze the process. The concluding sections review two recent studies of evolution by natural selection. These case studies illustrate how biologists test Darwin's theory by studying evolution in action. The case studies also help clarify some common misunderstandings about how evolution by natural selection works.

BOX 23.1 Why Darwin Gets Most of the Credit

Although Charles Darwin and Alfred Russel Wallace formulated the same explanation for how species change over time, Darwin's name is much more prominently associated with the theory of evolution because he developed the idea more thoroughly and provided massive evidence for it in *On the Origin of Species*. But historians of science speculate about whether Darwin would have published his theory at all had Wallace not threatened to scoop him (**Figure 23.1**).

Darwin wrote a paper explaining evolution by natural selection in 1842—a full 17 years before the first edition of *On the Origin of Species* came out. He never submitted the work for publication, however. Why? Darwin claimed that he needed time to document all of the arguments for and against the theory and to examine its many implications. There is probably an element of truth in this—Darwin was a remarkably thorough thinker and writer. But many historians of science argue that he held off largely out of fear. Because his theory was inconsistent with the creation story in the Bible's Book of Genesis, Darwin knew that he would be exposed to scathing criticism from religious and scientific leaders. He was also an extremely private person, had a strong religious upbringing, and was frequently in poor health. He responded to stress or personal attacks by suffering long bouts of debilitating illness. The prospect of

fighting for his ideas against the most powerful men in Europe was daunting. But Wallace forced his hand.

Wallace was also a native of England, but he had been making a living by collecting butterflies and other natural history specimens in Malaysia and selling them to private collectors. While recuperating from a bout of malaria there in 1858, he wrote a brief article outlining the logic of evolution by natural selection. He sent a copy to Darwin, who immediately recognized that they had formulated the same explanation for

how populations change through time. The two had their papers read together before the Linnean Society of London, and Darwin then rushed *On the Origin of Species* into publication a year later. The first edition sold out in a day.

Fortunately for Darwin's health, a fellow biologist and friend named Thomas Huxley publicly defended the theory against criticism, which came from both scientific and religious quarters. Darwin continued to live quietly on his estate in Down, England, and actively continued a brilliant research career.

(a) Charles Darwin

(b) Alfred Russel Wallace

FIGURE 23.1 The Codiscoverers of Evolution by Natural Selection
(a) Charles Darwin in 1840, four years after he returned from the voyage of the Beagle and two years before he drafted his first paper explaining evolution by natural selection. (b) Alfred Russel Wallace, who in 1858 independently formulated the theory of natural selection.
[(b) Alfred Russel Wallace by unknown artist, after a photograph by Thomas Sims, fl. 1860s. Reg. No.: 1765. National Portrait Gallery, London.]

23.1 The Pattern of Evolution: Have Species Changed through Time?

In *On the Origin of Species*, Darwin repeatedly used the phrase **"descent with modification"** to describe evolution. By this he meant that the species existing today have descended from other, preexisting species and that species are modified or change through time. This view was clearly a radical departure from the pattern of independently created and immutable species proposed by the theory of special creation. But what evidence do we have that Darwin was correct? How can biologists infer that species have changed through time instead of being static entities and that they are indeed related—not independent? Let's consider each question in turn.

Evidence for Change through Time

When Darwin began his work, biologists and geologists had just begun to assemble and interpret the fossil record. A **fossil** is any trace of an organism that lived in the past. These traces range from bones and branches to shells, tracks or impressions, and dung (**Figure 23.2**). The **fossil record** consists of all the fossils that have been found and described in the scientific literature.

Initially, fossils were organized according to their relative ages. This was possible because most fossils are found in **sedimentary rocks**, which form from sand or mud or other materials deposited at locations such as beaches or river mouths, and because sedimentary rocks are known to form in layers. Fossils from rocks underneath other rocks were judged to be older than the fossils found above them. In this way, researchers began putting fossils in an older-to-younger sequence. They also began naming different periods of geologic time, creating the sequence of eons, epochs, and periods called the **geologic time scale**. After the discovery of radioactivity in the early 1900s, researchers used radiometric dating techniques (introduced in Chapter 2) to assign absolute ages to the relative ages in the geologic time scale. According to these data, Earth is about 4.6 billion years old, and the earliest signs of life appear in rocks that formed about 3.85 billion years ago. Instead of being 6000 years old, life on Earth is ancient.

The fossil record continues to expand in size and quality. Three additional observations about this data set—regarding extinction, transitional forms, and environmental changes—support the hypothesis that species have changed through time.

Extinction In the early nineteenth century, researchers began discovering fossil bones, leaves, and shells that were unlike structures from any known animal or plant. At first, many scientists insisted that living examples of these species would be found in unexplored regions of the globe. But as research continued and the number and diversity of fossil collections grew, the argument became less and less plausible. After Baron Georges Cuvier published a detailed analysis of an **extinct** species—that is, a species that no longer exists—called the Irish "elk" in 1812, most scientists accepted extinction as a reality. This gigantic deer was judged to be too large to have escaped discovery and too distinctive to be classified as a large-bodied population of an existing species.

Advocates of the theory of special creation argued that the fossil species were victims of the flood at the time of Noah. Darwin, in contrast, interpreted them as evidence that species are not static, immutable entities, unchanged since the moment of special creation. His reasoning was that if species have gone extinct, then they have changed through time. Recent analyses of the fossil record support the claim that more species have gone extinct than exist today. The data also indicate that species have gone extinct throughout Earth's history—not just in one or even a few catastrophic events.

Transitional Forms Long before Darwin published his theory, researchers reported striking resemblances between the fossils found in the rocks underlying certain regions and the living species found in the same geographic areas. The pattern was so widespread that it became known as the "law of succession." The general observation was that extinct species in the fossil record were succeeded, in the same region, by similar species

(a) 110-Million-year-old ammonite shell

(b) 50-Million-year-old bird tracks

(c) 20,000-Year-old sloth dung

FIGURE 23.2 A Fossil Is Any Trace of an Organism That Lived in the Past
In addition to **(a)** body parts such as shells or bones or branches, fossils may consist of **(b)** tracks or impressions, or even **(c)** pieces of dung.

(**Figure 23.3a**). Early in the nineteenth century, the pattern was simply reported and not interpreted. But later, Darwin pointed out that it provided strong evidence in favor of the hypothesis that species had changed through time. His idea was that the extinct forms and living forms were related—that they represented ancestors and descendants.

As the fossil record improved, researchers discovered transitional forms that broadened the scope of the law of succession. A **transitional form** is a fossil species with traits that are intermediate between those of older and younger species. For example, intensive work over the past several decades has yielded fossils that document a gradual change over time from land-dwelling mammals that had limbs to ocean-dwelling mammals that had reduced limbs or no limbs (**Figure 23.3b**). All of the species in this sequence have distinctive types of ear bones that researchers use to identify whales. The oldest whale fossils found to date are from fox-sized animals that had eyes located at the tops of their heads. Based on this observation and the fact that the fossils were found in rocks that form only in ocean deposits, biologists suggest that the earliest whales were semiaquatic animals not unlike hippopotamuses. Over the subsequent 12 million years, the limbs of whale species became more reduced. These observations support the hypothesis that whales gradually became more strictly aquatic and more like today's whales in appearance and lifestyle. Whale species have clearly changed through time.

Similar sequences of transitional forms document changes that led to the evolution of feathers and flight in birds, stomata and vascular tissue in plants, upright posture and large brains in humans, jaws in vertebrates, limbs in amphibians and other vertebrates, and other traits. The discovery of the transitional forms provides strong evidence for change through time.

Environmental Change   Several of the whale fossils pictured in Figure 23.3b were found high in the Himalayan Mountains of southern Asia. The rocks surrounding the fossils originated as beach and mud deposits made by oceans, yet are now found thousands of meters above sea level. Similarly, fossilized shells from extinct clams and other ocean-dwelling species are found high in Europe's Alps and in the rock layers that line the Grand Canyon in North America's southwestern desert. In addition, geologists have compiled massive evidence in support of the *theory of plate tectonics*, which holds that Earth's crust is made up of plates and that the positions of the continents change almost continuously (see Chapter 24).

Observations such as these support the hypothesis that Earth's topography and environment have changed dramatically over time. If the Earth itself has changed, it is logical to infer that organisms have changed as well.

To summarize, data from the fossil record indicate that life is ancient and that species have changed through the course of Earth's history. Our planet and its species are dynamic—not static, (unchanging) as claimed by the theory of special

(a) Living species "succeed" fossil species in the same region.

Fossil sloth   Present-day sloth

(b) Transitional forms during the evolution of whales

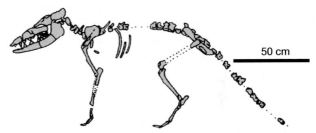

50 cm

Pakicetus, about 50 million years old

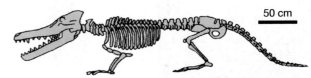

50 cm

Ambulocetus, about 49 million years old

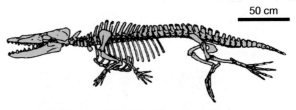

50 cm

Rhodocetus, about 47 million years old

50 cm

Basilosaurus, about 38 million years old

FIGURE 23.3 Evidence That Species Have Changed through Time **(a)** Fossil and living sloths are found only in Central America and South America. Darwin interpreted similarities among fossil and living species in the same geographic area as evidence that species had changed through time. **(b)** Transitional forms document the changes that occurred as whales evolved from terrestrial mammals to the aquatic mammals of today.

creation. Two observations about living species support the same conclusion:

1. Darwin was the first to describe and interpret **vestigial traits**—reduced or incompletely developed structures that have no function or reduced function but are clearly similar to functioning organs or structures in other species. Bowhead whales and rubber boas have tiny hip and leg bones that do not help them swim or slither; ostriches have reduced wings and cannot fly. Humans have a variety of vestigial traits. Our appendix is a reduced version of the cecum—an organ found in other vertebrates that functions in digestion. The human coccyx, illustrated in **Figure 23.4a**, is a reduced form of the tailbone found in monkeys and other primates. Even goose bumps, shown in **Figure 23.4b**, are a vestigial trait. Many mammals, including primates, are able to erect their hair when they are cold or excited. But our sparse fur does little to keep us warm, and goose bumps are largely ineffective in signaling our emotional state. In addition, our genome includes dozens of pseudogenes—the functionless DNA sequences introduced in Chapter 20. To make sense of these observations, biologists argue that the structure and function of these traits have changed over time.

2. Researchers have now documented hundreds of instances in which contemporary populations have changed in response to changes in their environment. Bacteria have become resistant to drugs; insects have become resistant to pesticides; and weedy plants have become resistant to herbicides. In several cases, biologists are even documenting the formation of new species, right before our eyes. Change through time continues and can be measured directly.

Evidence That Species Are Related

Data from the fossil record and contemporary species refute the hypothesis that species are immutable. What about the claim that species were created independently—meaning that they are unrelated to each other?

Charles Darwin began to formulate an answer during a five-year voyage he took aboard the English naval ship HMS *Beagle*. While fulfilling its mission to explore and map the coast of South America, the *Beagle* spent considerable time in the Galápagos Islands off the coast of present-day Ecuador. Darwin had taken over the role of ship's naturalist and gathered extensive collections of the plants and animals found in these islands. Among the birds he collected were what came to be known as the Galápagos mockingbirds, pictured in **Figure 23.5a**.

Several years after Darwin had returned to England, a naturalist friend in London pointed out that the mockingbirds Darwin had collected on different islands were distinct species, based on differences in coloration and beak size and shape. This struck Darwin as remarkable. Why would species that inhabit neighboring islands be so similar, yet clearly distinct? This turns out to be a very general pattern: In island groups across the globe, it is routine to find very similar, but different, species on neighboring islands.

Darwin realized that this pattern—puzzling when examined as a product of special creation—made perfect sense when interpreted in the context of evolution, or descent with modification. He proposed that the mockingbirds were similar because they had descended from the same common ancestor. If so, then the mockingbird species are part of a **phylogeny**—a family tree of populations or species. Further, the mockingbirds can be placed on a **phylogenetic tree**, a branching diagram that describes the relationships among species, much as a genealogy describes the relationships among individual humans. **Figure 23.5b** illustrates this idea. Darwin's hypothesis was that, instead of being created independently, mockingbird populations that colonized different islands had changed through time and formed new species.

But island forms are not the only species that show strong similarities. Resemblances among species are widespread. Today, these resemblances can be identified and analyzed at three levels, each of which relates to the others. Resemblances among species are structural, developmental, and genetic in nature.

(a) The human tailbone is a vestigial trait.

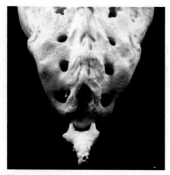

Human coccyx

Vervet monkey tail
(used for balance)

(b) Goose bumps are a vestigial trait.

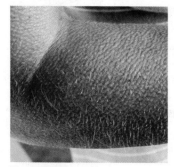

Human goose bumps

Erect hair on chimp
(insulation, emotional display)

FIGURE 23.4 Vestigial Traits Are Reduced Versions of Traits in Other Species

(a) The tailbone and **(b)** goose bumps are human traits that have reduced function. They are similar to larger, fully functional structures in other species.

(a) Similar mockingbird species on different Galápagos islands

Nesomimus parvulus ■   *Nesomimus melanotis* ■

Nesomimus trifasciatus ■   *Nesomimus macdonaldi* ■

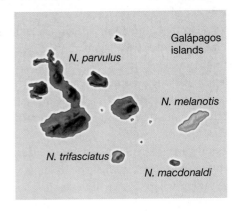

(b) Darwin reasoned that they share a common ancestor.

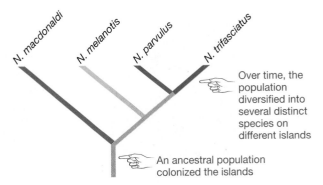

Over time, the population diversified into several distinct species on different islands

An ancestral population colonized the islands

FIGURE 23.5 Close Relationships among Island Forms Argue for Shared Ancestry
(a) Darwin collected mockingbirds from several islands of the Galápagos. **(b)** Phylogeny illustrating Darwin's explanation for why mockingbirds from different islands are similar yet distinct.

Structural Homologies Biologists routinely document similarities in the limbs, shells, or flowers of different species. Morphological traits that are similar are termed **structural homologies**. (Translated literally, *homology* means "the study of likeness.") A classic example is the common structural plan observed in the limbs of vertebrates (**Figure 23.6**). In Darwin's own words, "What could be more curious than that the hand of a man, formed for grasping, that of a mole for digging, the leg of the horse, the paddle of the porpoise, and the wing of the bat, should all be constructed on the same pattern, and should include the same bones, in the same relative positions?"

Darwin's point was that an engineer would never use the same underlying pattern to design the structure of a grasping tool, a digging implement, a walking device, a propeller, and a wing. But if all mammals descended from a common ancestor,

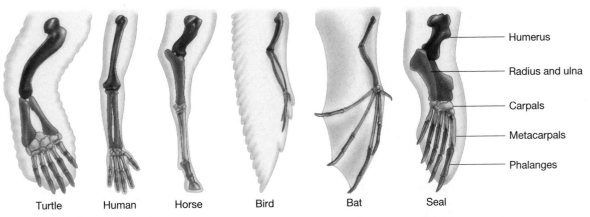

Turtle   Human   Horse   Bird   Bat   Seal

Humerus
Radius and ulna
Carpals
Metacarpals
Phalanges

FIGURE 23.6 Structural Homology: Limbs with Different Functions Have the Same Underlying Structure
Even though their function varies, all vertebrate limbs are modifications of the same number and arrangement of bones. Darwin interpreted structural homologies like these as a product of descent with modification. (These limbs are not drawn to scale.)

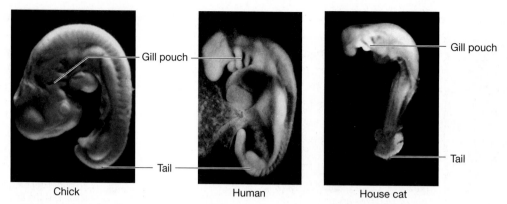

FIGURE 23.7 Developmental Homology: Structures That Appear Early in Development Are Similar
The early embryonic stages of a chick, a human, and a cat, showing a strong resemblance.

and if that ancestor had a limb with the same ground plan shown in Figure 23.6, then it would logical to observe that its descendants had modified forms of the same design.

Developmental Homologies When newly developing embryos of different species have a similar trait, they are said to exhibit a **developmental homology**. Developmental homologies are routinely observed at two levels: in the overall **morphology** (that is, form) of embryos and in the fate of particular embryonic tissues.

Figure 23.7 illustrates the strong general resemblance among the embryos of **vertebrates**—animals with backbones. Early in development, gill pouches and tails form in chicks, humans, and cats. Later in development, gill pouches are lost in all three species and tails are lost in humans. To explain this observation, biologists hypothesize that gill pouches and tails exist in chicks, humans, and cats because they existed in the fish-like species that was the common ancestor of birds and mammals.

Developmental homologies are also observed at the level of specific tissues. Even though the structure of the adult jaw is different in fish and mammals, the same group of embryonic cells develops into the jaw structure in both groups. This observation is logical if fish and mammals descended from the same common ancestor, and if this ancestor also had a jaw that developed from the same population of embryonic cells.

The general point is that, in many cases, traits are similar in different species because the species in question are related to each other by common descent. If species were created independently of one another, these types of similarities would not occur.

Genetic Homologies Today, biologists recognize that structural homologies are observed because developmental homologies exist. The causal connection is logical, because adult structures develop from embryonic structures and processes. In addition, biologists now understand that both structural and developmental homologies occur because genetic homology exists. A **genetic homology** is a similarity in the DNA sequences of genes from different species. Developmental sequences and adult structures cannot be similar in different species unless their genes are similar.

As an example of genetic homology, consider the *eyeless* gene in fruit flies and the *Aniridia* gene in humans. Both genes act in determining where eyes will develop. The genes are so similar in DNA sequence that they code for proteins that are nearly identical in amino acid sequence (**Figure 23.8**). This observation is interesting because eye structure is so different in the two species—fruit flies have a compound eye with many lenses, and humans have a camera eye with a single lens. In most cases, species that have similar alleles also have

Gene: Amino acid sequence (single-letter abbreviations):
Aniridia (Human)   LQRNRTSFTQEQIEALEKEFERTHYPDVFARERLAAKIDLPEARIQVWFSNRRAKWRREE
eyeless (Fruit fly)   LQRNRTSFTNDQIDSLEKEFERTHYPDVFARERLAGKIGLPEARIQVWFSNRRAKWRREE

Only six of the 60 amino acids in these sequences are different.
The two sequences are 90% identical.

FIGURE 23.8 Genetic Homology: Genes from Different Species May Be Similar in DNA Sequence or Other Attributes
Amino acid sequences from a portion of the *Aniridia* gene product found in humans and the *eyeless* gene product found in *Drosophila*. QUESTION What is the relationship among genetic homologies, developmental homologies, and structural homologies?

similar developmental sequences and similar embryonic or adult structures. In this instance, however, homology exists at the genetic level but *not* at the structural level. To explain this observation, biologists propose that fruit flies and humans descended from a common ancestor that had a gene similar to *eyeless* and *Aniridia*, and that this gene was involved in the formation of a simple light-gathering organ. (See Chapter 18 for more detail on homology in genes.)

Perhaps the ultimate similarity among organisms is the genetic code. With a few minor exceptions, the same 64 mRNA codons specify the same amino acids in all organisms that have been studied. To explain the existence of the universal genetic code, biologists hypothesize that today's code also existed in the common ancestor of all organisms alive today. Similarly, all organisms living today have a plasma membrane consisting of a phospholipid bilayer with interspersed proteins, transcribe the information coded in DNA to RNA via RNA polymerase, use ribosomes to synthesize proteins, employ ATP as an energy currency, and make copies of their genome via DNA polymerase. Like the genetic code, these traits undoubtedly existed in the cell that gave rise to all the species alive today.

Although genetic homologies were discovered fairly recently and were unknown to Darwin, structural and developmental homologies were recognized and studied even in the 1700s. Before the theory of evolution by natural selection was developed, no one was sure why striking similarities existed among certain organisms but not others. Darwin, in contrast, saw them as a consequence of descent with modification. In fact, biologists now reserve the term **homology** for similarities that exist *because* of common descent. This distinction is important because traits found in different species can be similar even though they were *not* inherited from a common ancestor. Characteristics that are similar but that did not exist in a common ancestor are said to be **analogous traits**, or **convergent traits**.

How Do Biologists Distinguish Homology from Analogy?

Homology is an exceptionally powerful concept in biology. For example, biomedical researchers study cancer in rats and vision in cats because the genes involved in rat cancers and cat vision are homologous with genes involved in human cancers and human vision. Research findings that improve cancer therapy in rats or that correct vision problems in cats can be successfully transferred to humans; the same underlying mechanisms are involved. The underlying mechanisms are shared because they existed in a common ancestor. Clearly, it is often vital to determine whether a homologous relationship exists among genes and other traits found in different species.

Determining whether homology exists is not always a simple matter, however, because not all similarities among organisms result from shared ancestry. The aquatic reptiles called ichthyosaurs, for example, were strikingly similar to modern dolphins (**Figure 23.9a**, page 502). Both are large marine animals with streamlined bodies and large dorsal fins. Both chase down fish and capture them between elongated jaws filled with dagger-like teeth. But no one would argue that ichthyosaurs and dolphins are similar because the traits they share existed in a common ancestor. As the phylogeny in Figure 23.9a shows, analyses of other traits show that ichthyosaurs are reptiles whereas dolphins are mammals.

Based on these data, it is logical to argue that the similarities between ichthyosaurs and dolphins result from convergent evolution. **Convergent evolution** occurs when natural selection favors similar solutions to the problems posed by a similar way of making a living. But convergent traits do not occur in the common ancestor of the similar species. Streamlined bodies and elongated jaws filled with sharp teeth help *any* species—whether it is a reptile or a mammal—chase down fish in open water. Convergent evolution produces analogous traits. Analogous traits are similar in structure and function, but they are not homologous. They are shared by species even though they were not present in their common ancestor.

In many cases, homology and convergence are much more difficult to distinguish than in the ichthyosaur and dolphin example. How do biologists recognize homology in such cases? As an example, consider the *HOM* genes of insects and the *Hox* genes of vertebrates introduced in Chapter 22. Even though insects and vertebrates last shared a common ancestor some 600–700 million years ago, biologists argue that *HOM* and *Hox* genes are derived from the same ancestral sequences. There are several lines of evidence to support this hypothesis:

- The genes are organized in a similar way. **Figure 23.9b** (page 502) shows that these genes in both insects and vertebrates are found in gene complexes, with similar genes found adjacent to one another on the chromosome. Recall from Chapter 20 that genes with these characteristics are called gene families. The organization of the gene families is nearly identical among insect and vertebrate species.

- All of the *HOM* and *Hox* genes share a 180-base-pair sequence called the *homeobox*, introduced in Chapter 22. The polypeptide encoded by the homeobox is almost identical in insects and vertebrates and has a similar function: It binds to DNA and regulates the expression of other genes.

- The products of the *HOM* and *Hox* genes have similar functions: They specify the locations of cells in embryos. They are also expressed in similar patterns in time and space.

In addition, many other animals, on lineages that branched off between insects and mammals, have similar genes. This is a crucial observation: If genes found in distantly related lineages are indeed related by common ancestry, then similar genes should be found in many intervening lineages on the tree of life. (Examine the phylogeny in Figure 23.9a and ask yourself

(a) Analogous traits: Similarities result from convergent evolution.

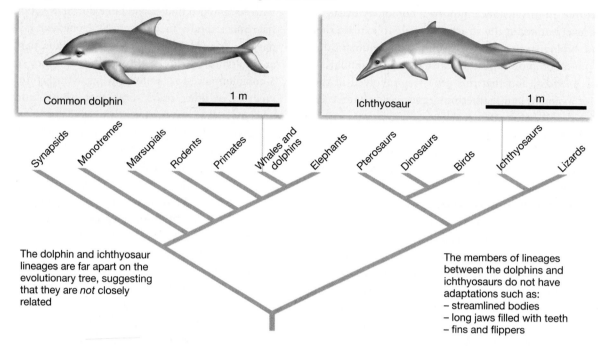

The dolphin and ichthyosaur lineages are far apart on the evolutionary tree, suggesting that they are *not* closely related

The members of lineages between the dolphins and ichthyosaurs do not have adaptations such as:
– streamlined bodies
– long jaws filled with teeth
– fins and flippers

(b) Homologous traits: Similarities are inherited from a common ancestor.

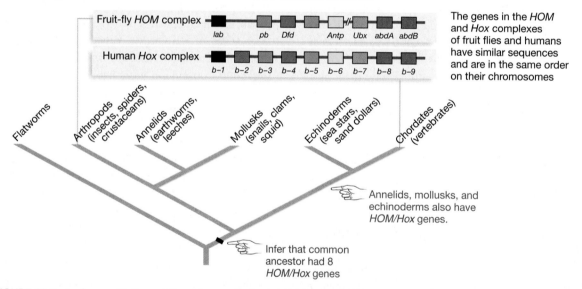

The genes in the *HOM* and *Hox* complexes of fruit flies and humans have similar sequences and are in the same order on their chromosomes

Annelids, mollusks, and echinoderms also have *HOM/Hox* genes.

Infer that common ancestor had 8 *HOM/Hox* genes

FIGURE 23.9 Analogous Traits and Homologous Traits Are Similar for Different Reasons
(a) Dolphins and ichthyosaurs look similar but are not closely related—dolphins are mammals; ichthyosaurs are reptiles. Because their similar traits did not exist in the common ancestor of mammals and reptiles, biologists infer that traits such as streamlined bodies, sharp teeth, and flippers evolved independently in these two groups. **(b)** All of the animal groups illustrated on this phylogeny have *Hox/HOM* complexes that are similar to those illustrated for fruit flies and humans.

whether the analogous traits found in ichthyosaurs and dolphins fulfill this criterion.)

To summarize, *HOM* and *Hox* genes are considered homologous because they are similar in organization, function, and composition. It is extremely improbable that convergent evolution could produce such a high degree of similarity.

Darwinism and the Pattern Component of Evolution

Biologists draw upon data from several sources to challenge the hypothesis that species are immutable and were created independently. Instead, the data support the idea that species have descended, with modification, from a common ancestor. **Table 23.1** summarizes this evidence.

TABLE 23.1 Evidence for Evolution

Prediction 1: Species Are Not Static, but Change through Time	Prediction 2: Species Are Related, Not Independent
• Many species have gone extinct.	• Closely related species often live in the same geographic area.
• Fossil species frequently resemble living species found in the same area.	• Homologous traits are common and exist at three levels:
• Transitional forms document change in traits through time.	1. structural (morphological traits)
• Earth is ancient; environments and landscapes have changed through time.	2. developmental (embryonic structures and processes)
• Vestigial traits are common.	3. genetic (gene structure and the genetic code)
• Populations and species can be observed changing today.	

Even before genetic homologies had been described and many of the most important transitional forms discovered, the vast majority of biologists were convinced that the pattern component of the theory of evolution was valid and that the theory of special creation was incorrect. Among biologists, controversy over the fact of evolution ended more than 120 years ago. As you analyze the evidence, though, recognize that no single observation or experiment instantly "proved" the fact of evolution and swept aside belief in special creation. Rather, Darwin and others argued that the pattern called evolution was much more consistent with the data. Stated another way, descent with modification was a more successful and powerful scientific theory because it explained observations—such as vestigial traits and the close relationships among species on neighboring islands—that special creation could not.

What about the process component of the theory of evolution by natural selection? If the limbs of bats and humans were not created independently and recently, how did they come to be?

23.2 The Process of Evolution: How Does Natural Selection Work?

Darwin's greatest contribution did not lie in recognizing the fact of evolution. Several other researchers had already proposed evolution as a pattern in nature long before Darwin began his work. (**Box 23.2** introduces one of these thinkers.) Instead, Darwin's crucial insight lay in recognizing a process, called natural selection, that could explain the pattern of descent with modification.

BOX 23.2 Evolutionary Theory before Darwin

In the introduction to *On the Origin of Species*, Darwin outlined some of the evidence for the fact of evolution, including many of the points summarized in Section 23.1. Then he wrote that a biologist "might come to the conclusion that each species had not been independently created, but had descended ... from other species. Nevertheless, such a conclusion even if well founded, would be unsatisfactory, until it could be shown *how* the innumerable species inhabiting this world have been modified" (emphasis added). Most scientists would agree with Darwin—recognizing a pattern in nature without understanding the process that caused it means that more work needs to be done.

In 1809, decades before Darwin began his work, Jean-Baptiste de Lamarck proposed that species are not static and independently created entities, but have changed through time and are related by common ancestry. Lamarck was one of several researchers who advocated the reality of evolution before Darwin published *On the Origin of Species*.

More important, Lamarck offered a mechanism to explain how species evolved. He proposed that simple forms of life are continuously being created by spontaneous generation. Lamarck reasoned that these simple cells then evolved to progressively "higher" forms through a process known as the **inheritance of acquired characters**. Specifically, he proposed that evolution was progressive in the sense of always producing larger or more complex or "better" species and that it occurs because individuals change as they develop, in order to meet challenges posed by the environment. Lamarck also proposed that these phenotypic changes are passed on to offspring. A classic scenario is that giraffes that stretched their necks to reach leaves high in treetops would develop longer necks and that they would then produce offspring with elongated necks.

Lamarck's ideas were roundly rejected by the leading scientists of the day, partly because he was not as thorough as Darwin in amassing evidence for the fact of evolution. Today biologists reject Lamarck's ideas because they know that changes in phenotypes that develop as a result of an individual's actions do not produce changes in genotypes and cannot be transmitted to offspring. For example, if you gained increased upper body strength through weight lifting, your offspring would not tend to be stronger as a result. Still, Lamarck deserves credit for bringing evolution to the forefront of scientific debate. In this way, he helped to pave the way for Darwin.

The process of **natural selection** can be broken down into four simple postulates, or steps, in a logical sequence:

1. The individual organisms that make up a population vary in the traits they possess, such as their size and shape.

2. Some of the trait differences are passed on to offspring. For example, tall parents may tend to have tall offspring. This means that the trait has some genetic basis and is heritable.

3. Only a subset of the offspring that are produced in each generation survives long enough to reproduce, and of the individuals that reproduce, not all produce the same number of offspring. Thus, some individuals in the population produce more offspring than others do.

4. The subset of individuals that produce the most offspring is not a random sample of the population. Instead, individuals with certain traits are more likely to produce the greatest number of offspring in a given environment. The individuals with these traits are "naturally selected."

Because the frequency of the selected traits increases from one generation to the next, the outcome of these four steps is **evolution**—a genetically based change in the characteristics of a population over time.

Biologists often combine postulates 1 and 2 by observing that heritable variation exists for most traits in most populations. Postulates 3 and 4 can also be combined by noting that individuals experience differential reproductive success. Evolution, or change in the characteristics of a population over time, results when differential reproductive success is based on heritable variation. The traits that contribute to high reproductive success increase in frequency in the population over time.

To explain the process of natural selection, Darwin referred to successful individuals as "more fit" than other individuals. He gave the word *fitness* a definition quite different from its everyday English usage. **Darwinian fitness** is the ability of an individual to produce offspring, relative to that ability in other individuals in the population. This is a measurable quantity. Researchers study populations in the lab or in the field and estimate the relative fitness of each individual by counting how many offspring it produces relative to other individuals.

The concept of fitness, in turn, provides a compact way of formally defining adaptation. The biological meaning of adaptation, like the biological meaning of fitness, is quite different from its normal English usage. In biology, an **adaptation** is a heritable trait that increases the fitness of an individual in a particular environment relative to individuals lacking the trait. Adaptations increase fitness—the ability to produce offspring.

To help you understand each step in the process of natural selection, Section 23.3 is devoted to data—specifically, to recent studies of how natural selection works in actual populations. Biologists accept the theory of evolution by natural selection not only because of its explanatory power but also because evolution by natural selection has been observed directly.

23.3 Evolution in Action: Recent Research on Natural Selection

Darwin's theory of evolution by natural selection is testable. If the theory is correct, biologists should be able to test the validity of each postulate and actually observe evolution in natural populations.

This section summarizes three examples in which evolution by natural selection has been, or is being, observed in nature. Literally hundreds of other case studies are available, involving a wide variety of traits and organisms. The examples here were chosen because they demonstrate *how* biologists go about studying evolution by natural selection. To begin, let's explore the evolution of drug resistance—one of the great challenges facing today's biomedical researchers and physicians.

How Did *Mycobacterium tuberculosis* Become Resistant to Antibiotics?

Mycobacterium tuberculosis, the bacterium that causes **tuberculosis**, or TB, has long been a great scourge of humankind (**Figure 23.10**). TB was responsible for almost 25 percent of all deaths in New York City in 1804; in nineteenth-century Paris, the figure was closer to 33 percent. To put these num-

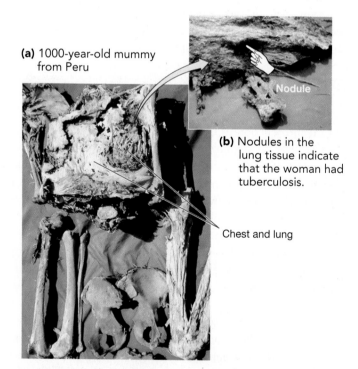

(a) 1000-year-old mummy from Peru

Nodule

(b) Nodules in the lung tissue indicate that the woman had tuberculosis.

Chest and lung

FIGURE 23.10 Tuberculosis Is an Ancient Disease of Humans (a) A mummy of a middle-aged woman, from a tomb in the coastal deserts of Peru. (The mummy's chest and upper arms are at the top.) **(b)** A close-up of lung tissue from the mummy. The nodules in the tissue are strikingly similar to those found in contemporary TB victims. DNA extracted from nodules in the mummy contained sequences diagnostic of *M. tuberculosis*. This is strong evidence that TB was present in the New World long before the arrival of Europeans.

bers in perspective, consider that all types of cancer, combined, currently account for about 30 percent of the deaths that occur in the United States. TB was once as great a public health issue as cancer is now.

Although tuberculosis still kills more adults than any other viral or bacterial disease in the world, TB attracted relatively little attention in the industrialized nations between about 1950 and 1990. During that time, TB was primarily a disease of developing nations.

The decline of tuberculosis in Western Europe, North America, Japan, Korea, and Australia is one of the great triumphs of modern medicine. In these countries, sanitation, nutrition, and general living conditions began to improve dramatically in the early twentieth century. When people are healthy and well nourished, their immune systems work well enough to stop most *M. tuberculosis* infections quickly—before the infection can harm the individual and before the bacteria can be transmitted to a new host. Equally important, antibiotics such as rifampin started to become available in the industrialized countries in the early 1950s. These drugs allowed physicians to stop even advanced infections, and they saved millions of lives.

In the late 1980s, however, rates of *M. tuberculosis* infection surged in many countries, and in 1993 the World Health Organization (WHO) declared TB a global health emergency (**Figure 23.11**). Physicians were particularly alarmed, because the strains of *M. tuberculosis* responsible for the increase were largely or completely resistant to rifampin and other antibiotics that were once extremely effective. How and why did the evolution of drug resistance occur? The case of a single patient—a young man with AIDS who lived in Baltimore—will illustrate what is happening all over the world.

The story begins when the HIV-positive individual was admitted to the hospital with fever and coughing. Chest X-rays, followed by bacterial cultures of fluid ejected from the lungs, showed that he had an active TB infection. He was given a battery of antibiotics for six weeks, followed by twice-weekly doses of rifampin and isoniazid for an additional 33 weeks. Ten months after therapy started, bacterial cultures from his chest fluid indicated no *M. tuberculosis* cells. His chest X-rays were also normal. The antibiotics seemed to have cleared the infection.

Just two months after the TB tests proved normal, however, the young man was readmitted to the hospital with a fever, severe cough, and labored breathing. Despite being treated with a variety of antibiotics, including rifampin, he died of respiratory failure 10 days later. Samples of material from his lungs showed that *M. tuberculosis* was again growing actively there. But this time the bacterial cells were completely resistant to rifampin.

Drug-resistant bacteria had killed this patient. Where did they come from? Is it possible that a strain that was resistant to antibiotic treatment evolved *within* him? To answer this question, a research team analyzed DNA from the drug-resistant strain and compared it with stored DNA from *M. tuberculosis* cells that had been isolated a year earlier from the same patient. After examining extensive stretches from each genome, the biologists were able to find only one difference: a point mutation in a gene called *rpoB*. This gene codes for a component of the enzyme RNA polymerase. Recall from Chapter 16 that RNA polymerase transcribes DNA to mRNA and that a point mutation is a single base change in DNA. In this case, the mutation changed a cytosine to a thymine, altering the normal codon TCG to a mutant one, TTG. As a result, the RNA polymerase produced by the drug-resistant strain had leucine instead of serine at the 153rd amino acid in the polypeptide chain.

This result is meaningful. The drug that was being used to treat the patient works by binding to the RNA polymerase of *M. tuberculosis*. When the drug enters an *M. tuberculosis* cell

WEB TUTORIAL 23.1
Natural Selection for Antibiotic Resistance

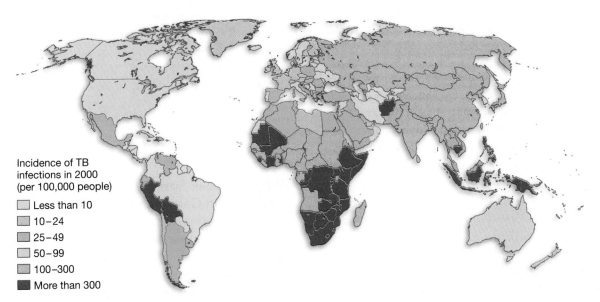

FIGURE 23.11 Rates of TB Infection Are Highest in Developing Nations
In developing nations, TB is responsible for 7 percent of all deaths (children and adults).

Incidence of TB
infections in 2000
(per 100,000 people)

- Less than 10
- 10–24
- 25–49
- 50–99
- 100–300
- More than 300

and binds to its RNA polymerase, it interferes with transcription. If sufficient quantities of rifampin are present for long enough and if the drug binds tightly, bacterial cells will not be able to make proteins efficiently and will die. But apparently the substitution of a leucine for a serine prevents rifampin from binding efficiently. Consequently, cells with the C → T mutation continue to grow even in the presence of the drug.

These results suggest that a chain of events led to this patient's death. The researchers hypothesized that, by chance, one or a few of the cells present early in the course of the infection happened to have an *rpoB* gene with the C → T mutation. Under normal conditions, mutant forms of RNA polymerase do not work as well as the more common form, so cells with the C → T mutation would not produce many offspring and would stay at low frequency—even while the overall population grew to the point of inducing symptoms that sent the young man to the hospital.

At that point, therapy with rifampin began. In response, cells in the population with normal RNA polymerase began to grow much more slowly or to die outright. As a result, the overall bacterial population declined in size so drastically that the patient appeared to be cured—his symptoms began to disappear. But cells with the C → T mutation had an advantage in the new environment. They began to grow more rapidly than the normal cells and continued to increase in number after therapy ended. Eventually the *M. tuberculosis* population regained its former abundance, and the patient's symptoms reappeared. However, drug-resistant cells now dominated the population. This is why the second round of rifampin therapy was futile. If health-care workers or the patient's family had contracted TB from him, rifampin therapy would have been useless on them, too, and the disease would have continued to spread.

Does this sequence of events mean that evolution by natural selection occurred? One way of answering this question is to review the four postulates listed in Section 23.2 and determine whether each was tested and verified:

1. *Did variation exist in the population?* The answer is yes. Due to mutation, both resistant and nonresistant strains of TB were present prior to administration of the drug. Most *M. tuberculosis* populations, in fact, exhibit variation for the trait; studies on cultured *M. tuberculosis* show that a mutation conferring resistance to rifampin is present in one out of every 10^7 to 10^8 cells.

2. *Was this variation heritable?* The researchers showed that the variation in the phenotypes of the two strains—from drug susceptibility to drug resistance—was due to variation in their genotypes. Because the mutant *rpoB* gene is copied before a *Mycobacterium* replicates, the allele and the phenotype it produces—drug resistance—are passed on to offspring.

3. *Was there variation in reproductive success?* That is, did some *M. tuberculosis* individuals leave more offspring than other *M. tuberculosis* individuals? Clearly, only a tiny fraction of *M. tuberculosis* cells in the patient survived the first round of antibiotics long enough to reproduce—so few that, after the initial therapy, his chest X-ray was normal and his fluid sample contained no *M. tuberculosis* cells. In contrast, drug-resistant bacterial cells were able to survive and keep reproducing after the onset of drug treatment.

4. *Did selection occur?* That is, did a nonrandom subset of the population produce the most offspring? The answer is yes. The *M. tuberculosis* population present early in the infection was different from the *M. tuberculosis* population present at the end. This could have occurred only if cells with the drug-resistant allele had higher reproductive success when rifampin was present than did cells with the normal allele. *Mycobacterium tuberculosis* individuals with the mutant *rpoB* gene had higher fitness in an environment where rifampin was present. The mutant allele produces a protein that is an adaptation when the cell's environment contains the antibiotic.

This study verified all four postulates and confirmed that evolution by natural selection had occurred. The example also neatly illustrates the contemporary definition of **evolution**: a change in allele frequencies in a population and thus its genetic makeup over time. The characteristics of the *M. tuberculosis* population changed over time because the mutant *rpoB* gene increased in frequency. But note that the individual cells themselves did not evolve. When natural selection occurred, the individual cells did not change through time; they simply survived or died, or produced more or fewer offspring. This is a fundamentally important point: Natural selection acts on *individuals*, but only *populations* evolve.

The events just reviewed have occurred many times in other patients. Recent surveys indicate that drug-resistant strains now account for about 10 percent of the *M. tuberculosis*-causing infections throughout the world. And the emergence of drug resistance in TB is far from unusual. Resistance to a wide variety of insecticides, fungicides, antibiotics, antiviral drugs, and herbicides has evolved in hundreds of insects, fungi, bacteria, viruses, and plants. In many cases, the specific mutations that lead to a fitness advantage and the spread of the resistance alleles are known.

Why Are Beak Size, Beak Shape, and Body Size Changing in Galápagos Finches?

The TB example is particularly satisfying, because the molecular basis of both heritable variation and differential reproductive success is understood. But can biologists study natural selection when the alleles responsible for variation are not known and when the molecular mechanism of adaptation has not been identified? The answer is yes. As an example of how this work is done, let's review research led by Peter and Rosemary Grant. These biologists have been investigating changes in beak size, beak shape, and body size that have occurred in finches native to the Galápagos Islands.

The medium ground finch, pictured in **Figure 23.12a**, makes its living by eating seeds. Finches crack seeds with their beaks. For over three decades, the population of medium ground finches on Isle Daphne Major of the Galápagos has been

(a) Medium ground finch

(b) Natural selection during a drought

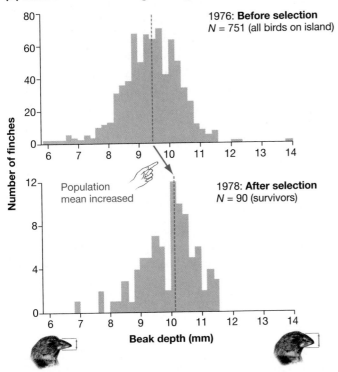

FIGURE 23.12 During a Drought, Finches with Deep Beaks Survived Better
(a) Medium ground finches live only on the Galápagos Islands.
(b) Histograms showing the distribution of beak depth in medium ground finches before and after natural selection occurred, due to changes in food availability, during the drought of 1977. *N* is the sample size. **QUESTION** Why was the sample size so much smaller in 1978?

studied intensively by the Grants' team. Because Daphne Major is small—about the size of 80 football fields—the researchers have been able to catch, weigh, and measure all individuals and mark each with a unique combination of colored leg bands.

Early studies of the finch population established that beak size and shape and body size vary among individuals, and that beak morphology and body size are heritable. Stated another way, parents with particularly deep beaks tend to have offspring with deep beaks. Large parents also tend to have large offspring.

Not long after the team had established these results, a dramatic selection event occurred. In the annual wet season of 1977, Daphne Major received just 24 mm of rain instead of the 130 mm that normally falls. During the drought, few plants were able to produce seeds, and 84 percent (about 660 individuals) of the medium ground finch population disappeared.

Two observations support the hypothesis that most or all of these individuals died of starvation. The researchers found a total of 38 dead birds, and all were emaciated. Further, none of the missing individuals were spotted on nearby islands, and none reappeared once the drought had ended and food supplies returned to normal.

The research team realized that the die-off presented an opportunity to study natural selection. Were the survivors different from nonsurvivors? When the biologists analyzed the characteristics of each group, they found that survivors tended to have much deeper beaks than did the birds that died. This was an important finding, because the type of seeds that were available to the finches had changed dramatically as the drought continued. At the drought's peak, the tough fruits of a plant called *Tribulus cistoides* served as the finches' primary food source. These fruits are so difficult to crack that they are ignored in years when food supplies are normal. Grant's group hypothesized that individuals with particularly large and deep beaks were more likely to crack these fruits efficiently enough to survive.

As the data in **Figure 23.12b** show, natural selection led to an increase in average beak depth in the population. When breeding resumed in 1978, the offspring that were produced had beaks that were half a millimeter deeper, on average, than the population that existed before the drought. In only one generation, natural selection had led to measurable evolution—a change in the characteristics of the population over time. Alleles that led to the development of deep beaks must have increased in frequency. Large, deep beaks were an adaptation for cracking large fruits and seeds.

In 1983, however, the environment changed again. Over a 7-month period, a total of 1359 mm of rain fell. Plant growth was luxuriant, and finches fed primarily on small, soft seeds that were being produced in abundance. During this interval, small individuals with small, pointed beaks had exceptionally high reproductive success. As a result, the characteristics of the population changed again. In fact, the Grants have documented continued evolution in

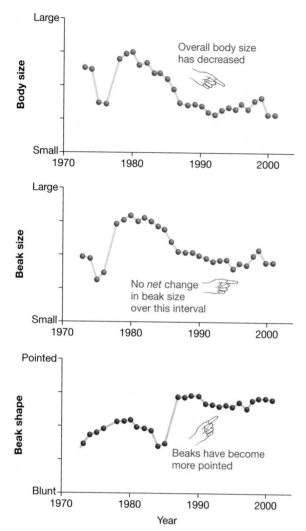

FIGURE 23.13 Body Size, Beak Size, and Beak Shape in Finches Have Changed over 30 Years
Graphs summarizing 30 years of data on changes in the traits of medium ground finches. **EXERCISE** Label the drought in 1978 and the wet year in 1983. Circle years when (1) average body size increased, (2) average beak size declined, and (3) beaks became pointier.

response to continued changes in the environment. **Figure 23.13** documents changes that have occurred in average body size, beak size, and beak shape over the past 30 years. Although beak size has shown no net change when we compare the start and end of this interval, beak shape has changed dramatically. On average, finch beaks have gotten much pointier over the course of the study. In addition, overall body size has gotten smaller.

Long-term studies such as this have been powerful, because they have succeeded in documenting natural selection in response to changes in the environment. But researchers are also using experimental approaches to study natural selection in action.

Can Natural Selection Be Studied Experimentally?

Candace Galen began to do research on variation in the flower size of a Rocky Mountain wildflower after she made a simple but astute observation: Individual alpine skypilot plants found

in the treeless, tundra habitats above timberline have flowers that are substantially larger than flowers on individuals growing among the stunted trees at timberline. The tundra-dwellers also have longer stalks, and most have sweet-smelling flowers. Flowers on plants growing at timberline have shorter stalks, and most have an aroma that biologists describe as "skunky."

Galen also noticed that individuals growing above the timberline are pollinated primarily by bumblebees, while individuals below timberline are pollinated primarily by flies. Because bumblebees are much larger than flies and are attracted to sweet smells, Galen thought that the differences in pollinators might be responsible for the difference in skypilot flowers. Specifically, she proposed that natural selection, caused by the bumblebees' preference to land on larger, sweet-smelling flowers and the flies' preference to land on smaller, skunky-smelling flowers, had acted on variation in flower size and produced the differences she observed in the two populations.

To test the hypothesis that large flower size has evolved in the tundra population in response to natural selection by bumblebees, Galen performed the experiment outlined in **Figure 23.14**. Examine the figure and note that she established two treatment groups, called controls and experimentals. The goal of the control treatment was to observe what happens when bumblebees are *not* agents of natural selection; the goal of the experimental treatment was to observe what happens when bumblebees do cause natural selection.

To create the control, Galen pollinated a large group of randomly selected timberline skypilots by hand, using pollen from randomly selected individuals. She collected the resulting seeds, germinated them in the greenhouse, and planted the seedlings out into randomly assigned locations in the timberline habitat. When these individuals matured, she compared their flower sizes with the sizes in offspring of individuals that had been pollinated by bumblebees.

The result? Examine the two histograms in Figure 23.14. They show that the flowers from offspring of bumblebee-pollinated plants were significantly larger than the flowers from offspring of hand-pollinated plants. Because previous experiments had confirmed that flower size is heritable, Galen could conclude that the experimental population, pollinated by bumblebees, was genetically different from the control population, which was randomly pollinated. Natural selection can be documented experimentally.

✓ CHECK YOUR UNDERSTANDING

If individuals with certain genetically based traits produce the most offspring in a population, then the alleles responsible for those traits increase in frequency over time, resulting in evolution. You should be able to explain each of Darwin's four postulates by using the examples of selection by drugs on the TB bacterium, changes in seed availability in Galápagos finches, and pollinator preferences in alpine skypilots.

Question: Do pollinators cause natural selection on flower characteristics?

Hypothesis: Bumblebees prefer large flowers. Therefore, natural selection leads to the evolution of large flowers.

Null hypothesis: Bumblebees do not prefer large flowers. Therefore, bumblebees do not cause natural selection on flower size.

Experimental setup:

Experimental group

1. Allow bees to pollinate a large sample of skypilots.

2. Collect seeds; germinate in greenhouse.

3. Plant seedlings in randomly assigned locations in the field.

4. Measure flower sizes and plot data.

Control group

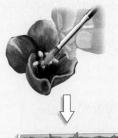

1. Hand pollinate a random sample of skypilots.

2. Collect seeds; germinate in greenhouse.

3. Plant seedlings in randomly assigned locations in the field.

4. Measure flower sizes and plot data.

Prediction: Offspring of bumblebee-pollinated flowers will be larger than offspring of randomly pollinated flowers.

Prediction of null hypothesis: Offspring of bumblebee-pollinated and randomly pollinated flowers will have the same size flowers.

Results:

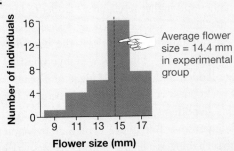

Average flower size = 14.4 mm in experimental group

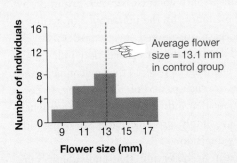

Average flower size = 13.1 mm in control group

Conclusion: On average, flowers are larger in offspring of bumblebee-pollinated plants than in offspring of hand-pollinated plants. Natural selection by bumblebees favors increased flower size.

FIGURE 23.14 Experimental Evidence for Natural Selection in Action
The histograms show the distribution of flower size in the offspring of alpine skypilots randomly pollinated by hand and in the offspring of those pollinated by bumblebees. The average size of flowers in the experimental treatment is over 1 mm larger than the average size of flowers in the control treatment.
QUESTION Why was it important that hand pollination was carried out randomly with respect to flower size?

23.4 The Nature of Natural Selection and Adaptation

Natural selection appears to be a simple process, but appearances can be deceiving. Research has shown that it is often misunderstood. To help clarify how natural selection works, let's consider some of the more common misconceptions about natural selection in light of data on drug resistance in the TB bacterium, changes in finch populations, and selection on flower size in alpine skypilots.

Selection Acts on Individuals, but Evolutionary Change Occurs in Populations

Perhaps the most important point to clarify about natural selection is that individuals do not change during the process—only the population does. During the drought, the beaks of individual finches did not become deeper. Rather, the average beak depth in the population increased over time, because deep-beaked individuals produced more offspring than shallow-beaked individuals did. Natural selection acted on individuals, but the evolutionary change occurred in the characteristics of the population. In the same way, individual bacterial cells did not change when rifampin was introduced to their environment—only the characteristics of the bacterial population changed. Further, the mutation in the *rpoB* gene that conferred high fitness in the new environment was not induced by the presence of rifampin; the gene occurred randomly and just happened to be advantageous. Alpine skypilots did not get bigger and sweeter smelling because bumblebees landed on them. Rather, bumblebees preferred to land on the larger and sweeter-smelling individuals. As a result, individuals with large, sweet-smelling flowers were pollinated frequently and produced the most offspring in habitats where bumblebees are common.

This point highlights a sharp contrast between evolution by natural selection and evolution by the inheritance of acquired characters—the hypothesis promoted by Jean-Baptiste de Lamarck (see Box 23.2). Lamarck proposed that individuals change in response to challenges posed by the environment and that the changed traits are then passed on to offspring. In contrast, Darwin pointed out that individuals do not change when they are selected—they simply produce more surviving offspring than other individuals do.

Lamarck was correct in observing that individuals may change over the course of their lifetime in response to changes in environmental conditions, however. To capture this point, biologists use the terms *adaptation* and *acclimatization* to distinguish between changes that occur in populations and changes that occur in individuals. **Adaptation** occurs when a population changes in response to natural selection. **Acclimatization** occurs when an individual changes in response to a change in environmental conditions. For example, wood frogs native to northern North America are exposed to extremely cold temperatures as they overwinter. When ice begins to form in their skin, their bodies begin producing a sort of natural antifreeze—molecules that protect their tissues from being damaged by the ice crystals. These individuals are acclimatizing to cold temperatures.

In frogs, the ability to produce antifreeze molecules is an adaptation,[1] but actually producing the molecules is acclimatization. You may have observed changes in your own body as you acclimatized to high elevation or to particularly hot or cold environments. You did not adapt to these environments in the evolutionary sense, however. Acclimatization is a short-term change, and this change is not passed on to offspring.

Evolution Is Not Progressive

It is tempting to think that evolution by natural selection is progressive—meaning organisms have gotten "better" over time. (In this context, *better* usually means more complex.) It is true that the groups appearing later in the fossil record are often more morphologically complex or "advanced" than closely related groups that appeared earlier. Flowering plants are considered more complex than mosses, and mammals are more complex than turtles. But there is nothing predetermined or absolute about this tendency. In fact, complex traits are routinely lost or simplified over time as a result of evolution by natural selection. Populations that became parasitic are particularly prone to this trend. Tapeworms, for example, lack any sort of gut. As parasites that live in the intestines of humans and other mammals, they absorb nutrients directly from their environment. But tapeworms evolved from species with a sophisticated digestive tract. Tapeworms lost their digestive tract as a result of evolution by natural selection.

The nonprogressive nature of evolution by natural selection contrasts sharply with the hypothesis of evolution by inheritance of acquired characters (see Box 23.2). According to Lamarck's conception of the evolutionary process, organisms progress over time to higher and higher levels on a "ladder of life." The lowest rung of the ladder was occupied by species that had just appeared via spontaneous generation—the creation of life from nonlife (see Chapter 1). Humans were hypothesized to be at the top of the ladder, and fungi and algae and plants and animals that lack backbones occupied the rungs in between (**Figure 23.15a**). Under this hypothesis, it is sensible to refer to "higher" and "lower" organisms. But under evolution by natural selection, there is no such thing as a higher or lower organism (**Figure 23.15b**). Green algae may be a more ancient group than flowering plants, but neither group is higher or lower than the other. Green algae simply have a different suite of adaptations than do flowering plants, so they thrive in different types of environments.

---

[1] In some species of frogs, so much extracellular fluid freezes during cold snaps that individuals appear to be frozen solid. Their hearts also stop beating. When temperatures warm in the spring, their hearts start beating again, their tissues thaw, and they resume normal activities.

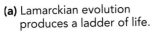

(a) Lamarckian evolution produces a ladder of life.

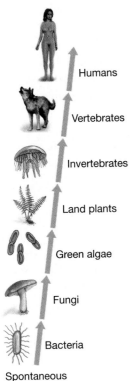

Humans

Vertebrates

Invertebrates

Land plants

Green algae

Fungi

Bacteria

Spontaneous generation

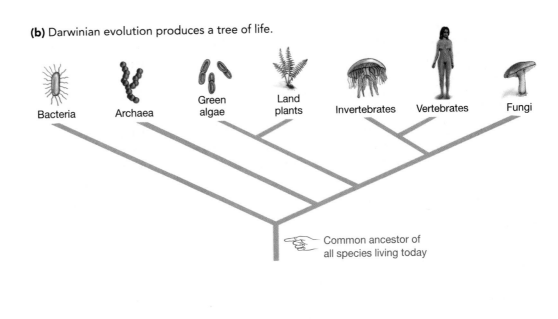

(b) Darwinian evolution produces a tree of life.

Bacteria   Archaea   Green algae   Land plants   Invertebrates   Vertebrates   Fungi

Common ancestor of all species living today

FIGURE 23.15 Evolution Produces a Tree of Life, Not a Progressive Ladder of Life
(a) Under evolution by the inheritance of acquired characters, species are related according to their position on a ladder of life. As evolution continues, simpler species evolve to become more complex. **(b)** Under evolution by natural selection, species are related according to their position on a tree of life. As evolution continues, species may become simpler or more complex.

The general message here is that all populations have undergone natural selection based on their ability to gather resources and produce offspring. All organisms are adapted to their environment. Evolution by natural selection is not progressive, so no organism is any "higher" than any other organism.

To drive this point home, recall what happened when torrential rains fell on Daphne Major. Instead of continuing to increase in size, the average beak size of the finch population *declined*. Natural selection did not produce a progression of ever-larger beaks. Instead, the population simply responded to whatever change happened to occur in the environment. In many bacterial and viral populations, the frequency of drug-resistant individuals declined dramatically when the drug in question was discontinued. If bumblebees disappeared from alpine habitats and flies began to pollinate tundra alpine skypilots, then tundra individuals with smaller, skunkier-smelling flowers would increase in frequency.

Natural selection is not progressive. It simply favors individuals that happen to be better adapted to their environment.

Not All Traits Are Adaptive

Although organisms are often exquisitely adapted to their environment, adaptation is far from perfect. Vestigial traits such as the human coccyx (tailbone), goose bumps, and appendix do not increase the fitness of individuals with those traits. Many cave-dwelling animals still have structures that correspond to eyes even though the structures have no function and the organisms live in complete darkness. The structures are not adaptive in these populations. They exist simply because they were present in the ancestral population.

Vestigial traits are not the only types of structures with no function. Some adult traits exist as holdovers from structures that appear early in development. For example, human males have rudimentary mammary glands. The structures are not adaptive. They exist only because nipples form in the human embryo before sex hormones begin directing the development of male organs instead of female organs.

Perhaps the best example of nonadaptive traits involves evolutionary changes in DNA sequences. Recall from Chapter 16 that mutation may change a base in the third position of a codon without changing the amino acid sequence of the protein encoded by that gene. Changes such as these are said to be *neutral*, or silent. They occur because of the redundancy of the genetic code (see Chapter 13). Neutral changes in DNA sequences are extremely common, yet not adaptive.

The general point here is that not all traits are adaptive. In addition, structures that do have functions are constrained in a variety of important ways. Let's take a closer look.

Genetic Constraints   The Grants' team analyzed data on the characteristics of finches that survived the 1977 drought, and the team made an interesting observation: Although individuals with deep beaks survived better than individuals with shallow beaks, birds with particularly narrow beaks survived better than individuals with wider beaks. This observation made sense because finches crack *Tribulus* fruits by twisting them and because narrow beaks should concentrate the twisting force more efficiently. But narrower beaks did not evolve in the population. To explain why, the biologists noted that parents with deep beaks tend to have offspring with beaks that are both deep and wide. This is a common pattern. Many alleles that affect body size have an effect on all aspects of size—not just one structure or dimension. As a result, selection for increased beak depth overrode selection for narrow beaks, even though a deep and narrow beak would have been more advantageous.

The general point here is that selection was not able to optimize all aspects of a trait. In the case of the finches, wider beaks were not the best possible beak shape for individuals living in an arid habitat. Wider beaks evolved anyway, due to a type of constraint called a **genetic correlation**. In this case, selection on alleles for one trait (increased beak depth) caused a correlated, though suboptimal, increase in another trait (beak width).

Genetic correlations are not the only genetic constraint on adaptation. Lack of genetic variation is also important. Consider that salamanders have the ability to regrow severed limbs and that some eels can sense electric fields. Even though these traits might confer increased reproductive success in humans, they do not exist—because the requisite genes are lacking.

Historical Constraints In addition to being constrained by genetic correlations and lack of genetic variation, adaptations are constrained by history. The reason is simple: All traits have evolved from previously existing traits. For example, the tiny hammer, anvil, and stirrup bones found in your middle ear evolved from bones that were part of the jaw and braincase in the ancestors of mammals. These bones now function in the transmission and amplification of sound from your outer ear to your inner ear. Biologists routinely interpret these bones as adaptations that improve your ability to hear airborne sounds. But are the bones a "perfect" solution to the problem of transmitting sound from one part of the ear to the other? The answer is no. They are the best solution possible, given an important historical constraint. Natural selection was acting on structures that originally had a very different function. Other vertebrates have different structures involved in transmitting sound to the ear.

To summarize, not all traits are adaptive, and even adaptive traits are constrained by genetic and historical factors. In addition, natural selection is not the only process that causes evolutionary change. Chapter 24 introduces three other processes that change allele frequencies over time. Compared with natural selection, these processes have very different consequences.

✓CHECK YOUR UNDERSTANDING

Using the examples of selection by drugs on the TB bacterium, changes in seed availability in Galápagos finches, and pollinator preferences in alpine skypilots, you should be able to explain (1) why individuals do not change when natural selection occurs, (2) why evolution is not progressive, and (3) why genetic and historical constraints prevent adaptations from being "perfect."

ESSAY The Debate over "Scientific Creationism"

The theory of evolution by natural selection is the cornerstone of modern biology. Controversy over the pattern component of the theory (descent with modification) ended in the late nineteenth century, when mounting evidence for the fact of evolution silenced Darwin's critics. Controversy over the process component (natural selection) ended when biologists successfully integrated Mendelian genetics into the theory. The discovery of mutation and genetic recombination explained why heritable variation exists in populations, and Mendel's laws clarified how new alleles and new combinations of alleles are passed on to offspring. This extension of Darwin's theory, called the "Modern Synthesis," occurred in the 1930s. Before the Modern Synthesis, biologists did not know how heritable variation could be maintained in populations so that evolution by natural selection would continue over time.

Although scientific controversy over the theory of evolution by natural selection largely ended over 70 years ago, political and social controversy are alive and well. In the United States, advocates of "scientific creationism" and "intelligent design theory" lobby for a ban on teaching evolution in public schools, for equal time devoted to teaching the theory of special creation, or for disclaimers in textbooks declaring that evolution is "just a theory." Intelligent design theory is based on the claim that, because organisms seem to be well designed or well adapted, they must have been created by an intelligent designer.

Science and religion are compatible because they address different types of questions.

This controversy is puzzling, given that there is no inherent conflict between accepting the validity of the theory of evolution by natural selection and believing in God. Pope John Paul II has stated that evolution by natural selection is compatible with traditional Christian understandings of God, and mainstream Protestant denominations have issued resolutions agreeing with this view.

Science and religion are compatible because they address different types of questions. Biologists ask questions that can be answered by measuring things—by collecting data. Science is about formulating hypotheses and finding evidence that supports or conflicts with the predictions of those hypotheses. Religious faith, in contrast, addresses questions that cannot be answered by data. The questions addressed by the world's great religions focus on why we exist and how we should live.

Because social and political controversy continue in the United States and elsewhere, it will be helpful to review the nature of the arguments promoted by creationists. In objecting to the theory of evolution by natural selection, creationists may stress one or more of the following claims:

1. *Earth is only about 6000 years old.* The assertion here is that life has not existed long enough for natural selection to produce the diversity of organisms seen today. Many creationists do not accept the physical and chemical principles behind radiometric dating, reviewed in Chapter 2, or the principle known as uniformitarianism. **Uniformitarianism** states that the laws of the universe are constant over time and throughout space. In geology, uniformitarianism holds that rock-forming processes occurring today, such as the slow deposition of shells to form limestone, were responsible for rock formations produced in the past, such as the white cliffs of Dover, England, made of massive limestone deposits. Geologists working in the late 1700s realized that under uniformitarianism, Earth must be extremely old.

2. *Organisms are "irreducibly complex."* According to this argument, adaptations such as the vertebrate eye and metabolic pathways are so complex and well integrated that they could not have arisen from a gradual, incremental process such as natural selection. Proponents of this view downplay the importance of the many fossils with characteristics that show transitions between simpler and more complex traits, and ignore analyses that show how complex metabolic pathways have evolved from simpler ancestral pathways.

3. *The theory is unproven.* Creationists argue that the evidence for evolution is inherently "soft" or unsatisfactory because no one was present when life began and because no one actually witnessed major events such as the origin of the flower in plants or the evolution of limbs in vertebrates. In doing so, creationists deny that inferring historical events from contemporary evidence is a valid research program in science. They insist that scientific evidence must be firsthand or direct in nature. To evaluate this viewpoint, consider that for some 150 years after the atomic theory was proposed, no one had actually seen an atom. Yet the theory was widely accepted as correct during this time.

Where does the controversy stand today? The U.S. Supreme Court has repeatedly ruled that legislation banning the teaching of evolution in public schools is unconstitutional on the basis of the First Amendment—the separation of church and state. The Court's opinion is that scientific creationism and intelligent design theory promote a specific religious belief, because they are founded on religious tenets codified in the Bible.

Despite the Court's rulings, however, and even though many religious leaders as well as many scientists see no conflict between evolution and religious faith, the controversy continues.

CHAPTER REVIEW

Summary of Key Concepts

■ **Populations and species evolve, meaning that their characteristics change through time. More precisely, evolution is defined as changes in allele frequencies over time.**

Evidence for the fact of evolution—that species are related through shared ancestry and have changed over time—has accumulated since Lamarck, Darwin, Wallace, and others began research on the topic over 200 years ago. An array of observations is inconsistent with the alternative theory that species were formed instantaneously, recently, and independently, and have remained unchanged through time. These observations include the geographic proximity of closely related species such as the Galápagos mockingbirds; the existence of structural, developmental, and genetic homologies; the near universality of the genetic code; resemblances of modern to fossil forms; transitional fossils; the fact of extinction; and the presence of vestigial traits. These data support the hypothesis that populations and species change through time.

■ **Natural selection occurs when individuals with certain alleles and traits produce the most offspring in a population. An adaptation is a genetically based trait that increases an individual's ability to produce offspring in a particular environment.**

In addition to articulating evidence for the pattern called evolution, Darwin and Wallace independently conceived of a process responsible for it. Natural selection occurs whenever genetically based differences among individuals lead to differences in their ability to reproduce. Alleles or traits that increase the reproductive success of an individual are said to increase the individual's fitness. A trait that leads to higher fitness, relative to individuals without the trait, is an adaptation. If a particular allele increases fitness and leads to adaptation, the allele will increase in frequency in the population.

Evolution is an outcome of natural selection. It is defined as changes in allele frequencies that occur in populations from one generation to the next. Evolution by natural selection has been confirmed by a wide variety of studies and has long been considered to be the central organizing principle of biology.

Web Tutorial 23.1   Natural Selection for Antibiotic Resistance

Web Tutorial 23.2   Natural Selection in Alpine Skypilots

■ **Evolution by natural selection is not progressive, and it does not change the characteristics of the individuals that are selected—it changes only the characteristics of the population. Not all traits are adaptive. All adaptations are constrained by genetic and historical factors.**

Individuals that are naturally selected are not changed by the process—they simply produce more offspring than other individuals do. Traits that increase in frequency under natural selection do so because they improve fitness, not because they are necessarily larger or more complex. Because natural selection acts only on existing traits, it does not lead to perfection.

Questions

Content Review

1. How can Darwinian fitness be estimated?
   a. Document how long different individuals in a population survive.
   b. Count the number of offspring produced by different individuals in a population.
   c. Determine which individuals are strongest.
   d. Determine which phenotype is the most common one in a given population.

2. Why are some traits considered vestigial?
   a. They improve the fitness of an individual who bears them, compared with the fitness of individuals without those traits.
   b. They change in response to environmental influences.
   c. They existed a long time in the past.
   d. They are reduced in size, complexity, and function compared with traits in related species.

3. What is an adaptation?
   a. a trait that improves the fitness of its bearer, compared with individuals without the trait
   b. a trait that changes in response to environmental influences within the individual's lifetime
   c. an ancestral trait—one that was modified to form the trait observed today
   d. a trait that is reduced in size or complexity but increases the fitness of its bearer

4. Why does the presence of extinct and transitional forms in the fossil record support the pattern component of the theory of evolution by natural selection?
   a. It supports the hypothesis that individuals change over time.
   b. It supports the hypothesis that weaker species are eliminated by natural selection.
   c. It supports the hypothesis that species evolve to become more complex and better adapted over time.
   d. It supports the hypothesis that species have changed through time.

5. Why are homologous traits similar?
   a. They are derived from a common ancestor.
   b. They are derived from different ancestors.
   c. They result from convergent evolution.
   d. They result from genetic and developmental similarities.

6. Which of the following statements is correct?
   a. When individuals change in response to challenges from the environment, their altered traits are passed on to offspring.
   b. Species are created independently of each other and do not change over time.
   c. Populations—not individuals—change when natural selection occurs.
   d. The Earth is young, and most of today's landforms were created during the floods at the time of Noah.

Conceptual Review

1. Some biologists encapsulate evolution by natural selection with the phrase "mutation proposes, selection disposes." Explain what they mean, using the formal terms introduced in this chapter.

2. Review the section on the evolution of drug resistance in *Mycobacterium tuberculosis*.

   • In *M. tuberculosis*, how does heritable variation arise for the trait of drug resistance?

   • What evidence do researchers have that a drug-resistant strain evolved in the patient analyzed in their study, instead of having been transmitted from another infected individual?

   • If the antibiotic rifampin were banned, would the mutant *rpoB* gene have lower or higher fitness in the new environment? Would strains carrying the mutation continue to increase in frequency in *M. tuberculosis* populations?

3. Consider the experiment that tested whether the offspring of hand-pollinated plants and bumblebee-pollinated plants differed in flower size. Why did the randomly pollinated population serve as an experimental control?

4. How has the conflict between the everyday and scientific usage of the word *theory* contributed to the controversy over the theory of evolution by natural selection?

5. The evidence supporting the pattern component of the theory of evolution can be criticized on the grounds that it is indirect. For example, no one has directly observed the formation of a vestigial trait over time. Due to the indirect nature of the evidence, it could be argued that structural and genetic homologies are coincidental and do not result from common ancestry. Is indirect evidence for a scientific theory legitimate? Are you persuaded that modification with descent is the best explanation available for the data reviewed in Section 23.1? Why or why not?

6. Why isn't evolution by natural selection progressive? Why don't the strongest individuals in a population always produce the most offspring?

Group Discussion Problems

1. The geneticist James Crow wrote that successful scientific theories have the following characteristics: (1) They explain otherwise puzzling observations; (2) they provide connections between otherwise disparate observations; (3) they make predictions that can be tested; and (4) they are heuristic, meaning that they open up new avenues of theory and experimentation. Crow added two other elements that he considered important on a personal, emotional level: (5) They should be elegant, in the sense of being simple and powerful; and (6) they should have an element of surprise.

   How well does the theory of evolution by natural selection fulfill these six criteria? Think of a theory you've been introduced to in another science course—for example, the atomic theory or the germ theory of disease—and evaluate it by using this list.

2. The average height of humans has increased steadily for the past 100 years in industrialized nations. This trait has clearly changed over time. Most physicians and human geneticists believe that the change is due to better nutrition and a reduced incidence of disease. Has human height evolved?

3. Genome sequencing projects may dramatically affect how biologists analyze evolutionary changes in quantitative traits. For example, suppose that the genomes of many living humans are sequenced and that genomes could be sequenced from many people who lived 100 years ago. (That might be possible with preserved tissue.) If 20 genes have been shown to influence height, how could you use the sequence data from these genes to test the hypothesis that human height has evolved in response to natural selection?

4. Examine Figure 23.11. How would you expect the data on this map to change over the next several decades as drug-resistant strains of *M. tuberculosis* increase in frequency in North America and Western Europe?

Answers to Multiple-Choice Questions   1. b; 2. d; 3. a; 4. d; 5. a; 6. c

www.prenhall.com/freeman is your resource for the following: Web Tutorials; Online Quizzes and other Online Study Guide materials; Answers to Conceptual Review Questions; Solutions to Group Discussion Problems; Answers to Figure Caption Questions and Exercises; and Additional Readings and Research.

24 Evolutionary Processes

KEY CONCEPTS

▪ Each of the four evolutionary mechanisms has different consequences. Only natural selection produces adaptation. Genetic drift causes random fluctuations in allele frequencies. Gene flow equalizes allele frequencies between populations. Mutation introduces new alleles.

▪ The Hardy-Weinberg principle acts as a null hypothesis when researchers want to test whether evolution or nonrandom mating is occurring at a particular gene.

▪ Inbreeding changes genotype frequencies but does not change allele frequencies.

This chapter focuses on how natural selection and other processes are causing cliff swallows and other species to evolve.

Chapter 22 defined evolution as a change in allele frequencies over time. One of the key concepts from that chapter was that even though natural selection acts on individuals, evolutionary change occurs in **populations**. A population is a group of individuals from the same species that live in the same area and regularly interbreed.

Natural selection is not the only process that causes evolutionary change, however. There are actually four mechanisms that shift allele frequencies in populations:

1. *Natural selection* increases the frequency of certain alleles—the ones that contribute to improved reproductive success.

2. *Genetic drift* causes allele frequencies to change randomly. In some cases, drift may even cause alleles that decrease fitness to increase in frequency.

3. *Gene flow* occurs when individuals immigrate into (enter) or emigrate from (leave) a population. Allele frequencies may change when gene flow occurs, because arriving individuals introduce alleles to their new population and emigrating individuals remove alleles from their old population.

4. *Mutation* modifies allele frequencies by continually introducing new alleles. The alleles created by mutation may be beneficial, detrimental, or have no effect on fitness.

This chapter has two fundamental messages: Natural selection is not the only agent responsible for evolution, and each of the four evolutionary processes has different consequences. Natural selection is the only mechanism that consistently results in adaptation. Mutation, gene flow, and genetic drift do not favor certain alleles and rarely produce adaptation. Mutation and drift introduce a nonadaptive component into evolution.

Let's take a closer look at the four evolutionary processes by examining a null hypothesis—what happens to allele frequencies when the evolutionary mechanisms are *not* operating.

24.1 Analyzing Change in Allele Frequencies: The Hardy-Weinberg Principle

To study how the four evolutionary processes affect populations, biologists take a three-pronged approach. First they create mathematical models that track the fate of alleles over time. Then they collect data to test predictions made by the models' equations. Finally, they apply the results to solve problems in conservation biology or human genetics.

This research strategy began with work published independently in 1908 by G. H. Hardy and Wilhelm Weinberg. At the time, it was commonly believed that changes in allele frequency occur simply as a result of sexual reproduction—meiosis followed by the random fusion of gametes (eggs and sperm) to form offspring. Some biologists claimed that dominant alleles would inevitably increase in frequency, while others predicted that two alleles of the same gene would eventually reach a frequency of 0.5.

To test these hypotheses, Hardy and Weinberg analyzed what happens to the frequencies of alleles when many individuals in a population mate and produce offspring. Instead of thinking about the consequences of a mating between two parents with a specific pair of genotypes, as we did with Punnett squares in Chapter 13, Hardy and Weinberg wanted to know what happened in an entire population, when *all* of the individuals—and thus all possible genotypes—bred.

To analyze the consequences of matings among all of the individuals in a population, Hardy and Weinberg invented a novel approach: They imagined that all of the gametes produced in each generation go into a single bin called the **gene pool**. To determine which genotypes would be present in the next generation and in what frequency, they simply had to calculate what happened when two gametes were plucked at random out of the gene pool, many times, and each of these gamete pairs was then combined to form offspring. These calculations would predict the genotypes of the offspring that would be produced, as well as the frequency of each genotype.

The researchers began by analyzing the simplest situation possible—that just two alleles of a particular gene exist in a population. Let's call these alleles A_1 and A_2. We'll use p to symbolize the frequency of A_1 alleles in the gene pool and q to symbolize the frequency of A_2 alleles in the same gene pool. Because there are only two alleles, the two frequencies must add up to 1; that is, $p + q = 1$. Although p and q can have any value between 0 and 1, let's suppose that the initial frequency of A_1 is 0.7 and that of A_2 is 0.3 (**Figure 24.1**, step 1). Thus, 70 percent of the gametes in the gene pool carry A_1, and 30 percent carry A_2 (Figure 24.1, step 2).

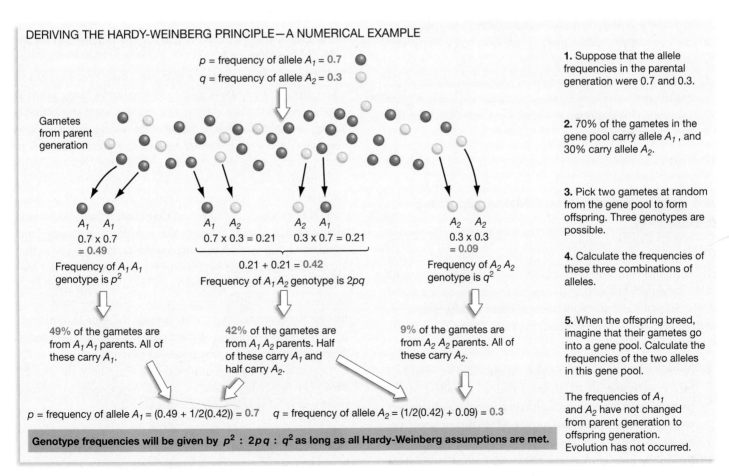

DERIVING THE HARDY-WEINBERG PRINCIPLE—A NUMERICAL EXAMPLE

p = frequency of allele A_1 = 0.7
q = frequency of allele A_2 = 0.3

1. Suppose that the allele frequencies in the parental generation were 0.7 and 0.3.

Gametes from parent generation

2. 70% of the gametes in the gene pool carry allele A_1, and 30% carry allele A_2.

A_1 A_1
0.7 x 0.7
= 0.49

A_1 A_2 A_2 A_1
0.7 x 0.3 = 0.21   0.3 x 0.7 = 0.21

A_2 A_2
0.3 x 0.3
= 0.09

3. Pick two gametes at random from the gene pool to form offspring. Three genotypes are possible.

Frequency of $A_1 A_1$ genotype is p^2

0.21 + 0.21 = 0.42
Frequency of $A_1 A_2$ genotype is $2pq$

Frequency of $A_2 A_2$ genotype is q^2

4. Calculate the frequencies of these three combinations of alleles.

49% of the gametes are from $A_1 A_1$ parents. All of these carry A_1.

42% of the gametes are from $A_1 A_2$ parents. Half of these carry A_1 and half carry A_2.

9% of the gametes are from $A_2 A_2$ parents. All of these carry A_2.

5. When the offspring breed, imagine that their gametes go into a gene pool. Calculate the frequencies of the two alleles in this gene pool.

p = frequency of allele A_1 = (0.49 + 1/2(0.42)) = 0.7 q = frequency of allele A_2 = (1/2(0.42) + 0.09) = 0.3

Genotype frequencies will be given by $p^2 : 2pq : q^2$ as long as all Hardy-Weinberg assumptions are met.

The frequencies of A_1 and A_2 have not changed from parent generation to offspring generation. Evolution has not occurred.

FIGURE 24.1 A Numerical Example of the Hardy-Weinberg Principle

In this population, then, three genotypes are possible: A_1A_1, A_1A_2, and A_2A_2 (Figure 24.1, step 3). What will the frequency of these three genotypes be in the next generation? Figure 24.1, step 4, explains the logic of Hardy's and Weinberg's result:

- The frequency of the A_1A_1 genotype is p^2.
- The frequency of the A_1A_2 genotype is $2pq$.
- The frequency of the A_2A_2 genotype is q^2.

The genotype frequencies in the offspring generation must add up to 1, which means that $p^2 + 2pq + q^2 = 1$. In our numerical example, $0.49 + 0.42 + 0.09 = 1$. Figure 24.1, step 5, shows how the frequencies of alleles A_1 and A_2 are calculated from these genotype frequencies. Hardy and Weinberg showed that, in the next generation, the frequency of allele A_1 is still p and the frequency of allele A_2 is still q. In our example, the frequency of allele A_1 is still 0.7 and the frequency of allele A_2 is still 0.3. Thus, no allele frequency change occurred. Even if A_1 is dominant to A_2, it does not increase in frequency. And there is no trend toward both alleles reaching a frequency of 0.5.

This result is called the **Hardy-Weinberg principle**. It makes two fundamental claims:

1. If the frequencies of alleles A_1 and A_2 in a population are given by p and q, then the frequencies of genotypes A_1A_1, A_1A_2, and A_2A_2 will be given by p^2, $2pq$, and q^2 for generation after generation.

2. When alleles are transmitted according to the rules of Mendelian inheritance, their frequencies do not change over time. For evolution to occur, some other factor or factors must come into play.

What are these other factors?

The Hardy-Weinberg Model Makes Important Assumptions

The Hardy-Weinberg model is based on important assumptions about how populations and alleles behave. Specifically, for a population to conform to the Hardy-Weinberg principle, none of the four mechanisms of evolution can be acting on the population. In addition, the model assumes that mating is random with respect to the gene in question. Thus, the five conditions that must be met are:

1. *No natural selection* at the gene in question. In step 2 of Figure 24.1, the model assumed that all members of the parental generation survived and contributed equal numbers of gametes to the gene pool, no matter what their genotype.

2. *No genetic drift, or random allele frequency changes*, affecting the gene in question. We avoided this type of allele frequency change in step 4 of Figure 24.1 by assuming that we drew alleles in their exact frequencies p and q, and not

at some different values caused by chance. For example, allele A_1 did not "get lucky" and get drawn more than 70 percent of the time. Stated another way, there is no random change due to a sampling effect.

3. *No gene flow.* No new alleles were added by immigration or lost through emigration anywhere in Figure 24.1. As a result, all of the alleles in the offspring population came from the original population's gene pool.

4. *No mutation.* We didn't consider that new A_1s or A_2s or other, new alleles might be introduced into the gene pool in step 2 or step 5 of Figure 24.1.

5. *Random mating* with respect to the gene in question. We enforced this condition by picking gametes from the gene pool at random in step 3 of Figure 24.1. We did not allow individuals to choose a mate based on their genotype at this locus.

The Hardy-Weinberg principle tells us what to expect if none of the four evolutionary mechanisms are acting and if mating is random. If there is no selection, no genetic drift, no gene flow, and no mutation affecting a gene, *and* if mating is random with respect to that gene, then the genotypes A_1A_1, A_1A_2, and A_2A_2 should be in the Hardy-Weinberg proportions p^2, $2pq$, and q^2, and no evolution will occur.

How Does the Hardy-Weinberg Principle Serve as a Null Hypothesis?

Recall from Chapter 1 that a null hypothesis specifies what a researcher should observe when the hypothesis being tested is wrong. Biologists often want to test whether natural selection is acting on a particular gene or whether nonrandom mating is occurring or one of the other evolutionary mechanisms is at work. In addressing such questions, the Hardy-Weinberg principle functions as a null hypothesis. Given a set of allele frequencies, it tells us what genotype frequencies will be when natural selection is not affecting that gene; whether mutation, genetic drift, and gene flow are also not affecting the gene; and whether mating is random with respect to the genotype at that locus. If biologists observe genotype frequencies that do *not* conform to Hardy-Weinberg proportions based on the original allele frequencies, something interesting is going on—nonrandom mating is occurring, or at least one of the four evolutionary mechanisms is operating. Further research would be needed to determine which of the four processes is acting.

Let's consider two examples to illustrate how the Hardy-Weinberg principle is used as a null hypothesis: MN blood types and *HLA* genes, both in humans.

MN Blood Types in Humans One of the first genes that geneticists could analyze in natural populations was the MN blood group of humans. Most human populations have two alleles, designated M and N, at this locus. Because the gene

codes for a protein found on the surface of red blood cells, researchers could determine whether individuals are *MM*, *MN*, or *NN* by treating blood samples with antibodies to each protein (this technique was first introduced in Chapter 8). To estimate the frequency of each genotype in a population, geneticists obtain data from a large number of individuals, then divide the number of individuals with each genotype by the total number of individuals in the sample.

Table 24.1 shows the genotype frequencies for populations from throughout the world and illustrates how observed genotype frequencies are compared with the genotype frequencies expected if the Hardy-Weinberg principle holds. The analysis is based on the following steps:

1. Estimate genotype frequencies by observation—in this case, by testing many blood samples for the *M* and *N* alleles. These frequencies are given on the lines labeled "observed" in Table 24.1.

2. Calculate observed allele frequencies from the observed genotype frequencies. In this case, the frequency of the *M* allele is the frequency of *MM* homozygotes plus half the frequency of *MN* heterozygotes; the frequency of the *N* allele is the frequency of *NN* homozygotes plus half the frequency of *MN* heterozygotes. (You can review the logic behind this calculation in step 5 of Figure 24.1.)

3. Use the observed allele frequencies to calculate the genotypes expected according to the Hardy-Weinberg principle. Under the null hypothesis of no evolution and random mating, the expected genotype frequencies are $p^2 : 2pq : q^2$.

4. Compare the observed and expected values. Although using statistical testing is beyond the scope of this text, researchers must use statistical tests to determine whether the differences between the observed and expected geno-

type frequencies are small enough to be due to chance or large enough to reject the null hypothesis of no evolution and random mating.

Note that in every case the observed and expected genotype frequencies are almost identical. (A statistical test shows that the small differences observed are probably due to chance.) For every population surveyed, genotypes at the *MN* locus are in Hardy-Weinberg proportions. As a result, biologists conclude that the assumptions of the Hardy-Weinberg model are valid for this locus. The data imply that the *M* and *N* alleles are not affected by the four evolutionary mechanisms and that mating is random with respect to this locus—meaning that humans do not choose mates on the basis of their *MN* genotype.

Before moving on, however, it is important to note that a study such as this does not mean that the *MN* gene has never been under selection or subject to nonrandom mating or genetic drift. Even if selection has been very strong for many generations, one generation of no evolutionary forces and of random mating will result in genotype frequencies that conform to Hardy-Weinberg expectations. The Hardy-Weinberg principle is used to test the hypothesis that currently no evolution is occurring at a particular gene and that, in the previous generation, mating was random with respect to the gene in question.

***HLA* Genes in Humans** A research team recently collected data on the genotypes of 124 individuals from the Havasupai tribe native to Arizona. These biologists were studying two genes that are important in the functioning of the human immune system. More specifically, the genes that they analyzed code for proteins that help immune system cells recognize and destroy invading bacteria and viruses. Previous work had shown that different alleles exist at both the *HLA-A* and *HLA-B* genes,

TABLE 24.1 Do Genotype and Allele Frequencies in the MN Blood Group of Humans Conform to the Hardy-Weinberg Model?

EXERCISE Fill in the values for allele frequencies and expected genotype frequencies for the Ainu people of Japan.

Population and Location		Genotype Frequencies			Allele Frequencies	
		MM	MN	NN	M	N
Inuit (Greenland)	Observed	0.835	0.156	0.009	0.913	0.087
	Expected	0.834	0.159	0.008		
Native Americans (U.S.)	Observed	0.600	0.351	0.049	0.776	0.244
	Expected	0.602	0.348	0.050		
Caucasians (U.S.)	Observed	0.292	0.494	0.213	0.540	0.460
	Expected	0.290	0.497	0.212		
Aborigines (Australia)	Observed	0.024	0.304	0.672	0.178	0.824
	Expected	0.031	0.290	0.679		
Ainu (Japan)	Observed	0.179	0.502	0.319		
	Expected					

and that the alleles at each gene code for proteins that recognize slightly different disease-causing organisms. Like the *M* and *N* alleles, *HLA* alleles are *codominant*—meaning that both are expressed and create the phenotype (Chapter 13). As a result, the research group hypothesized that individuals who are heterozygous at one or both of these genes may have a strong fitness advantage. The logic is that heterozygous people have a wider variety of HLA proteins, so their immune systems can recognize and destroy more types of bacteria and viruses. They should be healthier and have more offspring than homozygous people do.

To test this hypothesis, the researchers used their data on observed genotype frequencies to determine the frequency of each allele present. When they used these allele frequencies to calculate the expected number of each genotype according to the Hardy-Weinberg principle, they found that the observed and expected values did not match (**Table 24.2**). Specifically, there were too many heterozygotes and not enough homozygotes compared with the predicted values. Statistical tests show that it is extremely unlikely that the difference between the observed and expected numbers could occur purely by chance.

These results supported the team's prediction and indicated that one of the assumptions behind the Hardy-Weinberg principle was being violated. But which assumption? The researchers argued that mutation, migration, and drift are negligible in this case and offered two competing explanations for their data:

1. *Mating may not be random with respect to the* HLA *genotype.* Specifically, people may subconsciously prefer mates with *HLA* genotypes unlike their own and thus produce an excess of heterozygous offspring. This hypothesis is plausible. For example, experiments have shown that college students can distinguish each others' genotypes at loci related to *HLA* on the basis of body odor. Individuals in this study were more attracted to the smell of genotypes unlike their own. If this is true among the Havasupai, then nonrandom mating would lead to an excess of heterozygotes compared with the proportion expected under Hardy-Weinberg.

TABLE 24.2 Do *HLA* Genotype Frequencies of Humans Conform to the Hardy-Weinberg Model?

	Observed Number	Expected Number
HLA-A		
Homozygotes	38	48
Heterozygotes	84	74
HLA-B		
Homozygotes	21	30
Heterozygotes	101	92

Source: T. Markow et al., *HLA* polymorphism in the Havasupai: Evidence for balancing selection. *American Journal of Human Genetics* 53 (1993): 943–952.

2. *Heterozygous individuals may have higher fitness.* This hypothesis is supported by data collected by a different research team, studying Hutterite people living in South Dakota. In this population, married women who have the same *HLA*-related alleles as their husbands have more trouble getting pregnant and experience higher rates of spontaneous abortion than do women with *HLA*-related alleles different from those of their husbands. The data suggest that homozygous fetuses have lower fitness than do fetuses heterozygous at these loci. If this were true among the Havasupai, selection would lead to an excess of heterozygotes relative to Hardy-Weinberg expectations.

Which explanation is correct? It is possible that both are. But the fact is, no one knows. Using the Hardy-Weinberg principle as a null hypothesis allowed biologists to detect an interesting pattern in a natural population. Research continues on the question of why the pattern exists.

✓ CHECK YOUR UNDERSTANDING

The Hardy-Weinberg principle functions as a null hypothesis when researchers test whether nonrandom mating or evolution is occurring at a particular gene. Given a set of genotype frequencies or the numbers of various genotypes observed in a population, you should be able to (1) calculate observed allele frequencies, (2) calculate the genotype frequencies expected according to the Hardy-Weinberg principle, and (3) compare the observed and expected genotype frequencies and suggest whether the data support the null hypothesis of random mating and no selection, genetic drift, gene flow, or mutation.

24.2 Natural Selection and Sexual Selection

Natural selection occurs when individuals with certain phenotypes produce more offspring than individuals with other phenotypes do. If certain alleles are associated with the favored phenotypes, they increase in frequency while other alleles decrease in frequency. The result is evolution.

These concepts were introduced in Chapter 23. It is important to recognize that natural selection occurs in a wide variety of patterns, however, and that each type of natural selection has different causes and consequences. The data in Table 24.2, for example, document a pattern of natural selection called **heterozygote advantage**: When heterozygous individuals have higher fitness than homozygous individuals do, natural selection maintains genetic variation in populations. **Genetic variation** refers to the number and relative frequency of alleles that are present in a particular population.

Lack of genetic variation in a population is usually a bad thing. To understand why, recall from Chapter 23 that selection can occur only if heritable variation exists in a population. If genetic variation is low and the environment changes—perhaps due to the emergence of a new disease-causing virus, a rapid change in climate, or a reduction in the availability of a particular food source—it is unlikely that any alleles will be present that have high fitness under the new conditions. As a result, the average fitness of the population will decline. If the environmental change is severe enough, the population may even be faced with extinction.

Let's examine some of the different types of natural selection with this question in mind: How do they affect the level of genetic variation in the population?

Directional Selection

According to the data introduced in Chapter 23, natural selection has increased the frequency of drug-resistant strains of the tuberculosis bacterium and caused changes in beak shape and body size in medium ground finches. This type of natural selection is called **directional selection**, because allele frequencies changed in one direction.

Figure 24.2a illustrates how directional selection works when the trait in question has a bell-shaped, normal distribution in a population. Recall from Chapter 13 that when many different genes influence a trait, the distribution of phenotypes in the population tends to form a bell-shaped curve. In such cases, directional selection is acting on many different genes at once. In the case of selection on drug resistance in the TB bacterium, however, selection was acting on a single gene.

Most often, directional selection tends to reduce the genetic diversity of populations. If directional selection continues over time, the favored alleles will eventually reach a frequency of 1.0 while disadvantageous alleles will reach a frequency of 0.0. Alleles that reach a frequency of 1.0 are said to be *fixed*; those that reach a frequency of 0.0 are said to be *lost*.

Fixation and loss may not always occur under directional selection, however. To appreciate why, consider recent data on the body size of cliff swallows native to the Great Plains of North America. In 1996 a population of cliff swallows endured a six-day period of exceptionally cold, rainy weather. Cliff swallows are migratory and feed by catching mosquitoes and other insects in flight. Insects disappeared during this cold snap, however, and the biologists recovered the bodies of 1853 swallows that died of starvation. As soon as the weather improved, the researchers caught and measured the body size of 1027 survivors from the same population. As the histograms in **Figure 24.2b** show, survivors were much larger on average than the birds that died. Directional selection, favoring large body size, had occurred. To explain this observation, the investigators suggest that larger birds survived because they had larger fat stores and did not get as cold as the smaller birds. As a result, the larger birds were less likely to die of exposure to cold and more likely to avoid starvation until the weather warmed up and insects were again available.

(a) Directional selection changes the average value of a trait.

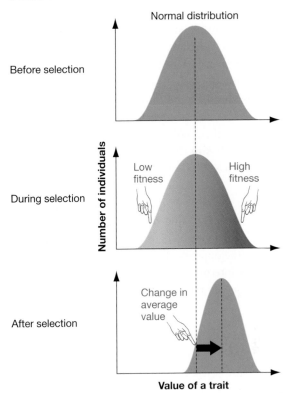

(b) For example, directional selection caused average body size to increase in a cliff swallow population.

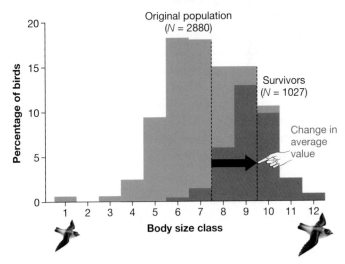

FIGURE 24.2 Directional Selection
(a) When directional selection acts on traits that have a normal distribution, individuals with one set of extreme values experience poor reproductive success. **(b)** The light-green histogram shows the distribution of overall body size in cliff swallows prior to an extended cold snap that killed many individuals. (The various body-size classes were calculated from measurements of wing length, tail length, leg length, and beak size.) The dark green histogram shows the size distribution of individuals from the same population that survived the cold spell. Here N indicates the sample size.

If this type of directional selection continued, alleles that contribute to small body size would quickly be eliminated from the cliff swallow population. It is not clear that this will be the case, however, because directional selection is rarely constant throughout a species' range and through time. By examining weather records, the researchers established that cold spells as severe as the one that occurred in 1996 are rare. Further, research on other swallow species suggests that smaller birds are more maneuverable in flight and thus more efficient when they feed. If so, then selection for feeding efficiency could counteract selection by cold weather. When this is the case, individuals with intermediate body size should be favored. Opposing patterns of directional selection will help maintain genetic variation for this trait.

When studies on a wide array of populations and species are considered, it is common to find that one cause of directional selection on a trait is counterbalanced by a different factor that causes selection in the opposite direction. In such cases, the optimal phenotype is intermediate. The same pattern can result from an entirely different type of natural selection, called stabilizing selection.

Stabilizing Selection

When cliff swallows were exposed to cold weather, selection greatly reduced one extreme in the range of phenotypes and resulted in a directional change in the average characteristics of the population. But selection can also reduce both extremes in a population, as illustrated in **Figure 24.3a**. This pattern of selection is called **stabilizing selection**. It has two important consequences: There is no change in the average value of a trait over time, and genetic variation in the population is reduced.

Figure 24.3b shows a classical data set in humans illustrating stabilizing selection. Biologists who analyzed birth weights and mortality in 13,730 babies born in British hospitals found that babies of average size (slightly over 7 pounds) survived best. Mortality was high for very small babies and very large babies. This is persuasive evidence that birth weight is under strong stabilizing selection in humans.

Disruptive Selection

One of the most important patterns of selection observed in nature has the opposite effect of stabilizing selection. Instead of favoring phenotypes near the average value and eliminating extreme phenotypes, **disruptive selection** eliminates phenotypes near the average value and favors extreme phenotypes (**Figure 24.4a**). As a result, disruptive selection tends to maintain the amount of genetic variation in a population.

Biologists have recently shown that disruptive selection maintains the striking bills of black-bellied seedcrackers, pictured in **Figure 24.4b**. The data plotted in the figure show that

(a) Stabilizing selection reduces the amount of variation in a trait.

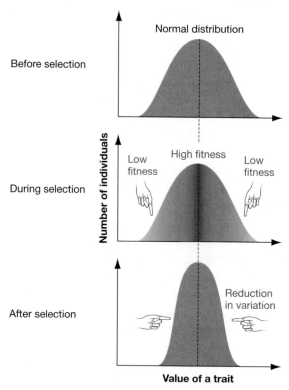

(b) For example, very small and very large babies are the most likely to die, leaving a narrower distribution of birth weights.

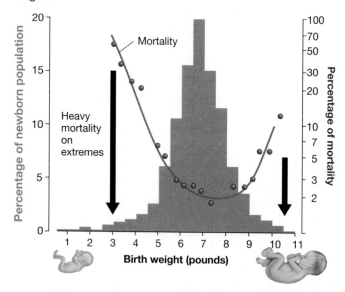

FIGURE 24.3 Stabilizing Selection
(a) When stabilizing selection acts on normally distributed traits, individuals with extreme phenotypes experience poor reproductive success. **(b)** Histogram showing the percentages of newborns with various birth weights on the left-hand axis. The purple dots indicate the percentage of newborns in each weight class that died, plotted on the logarithmic scale shown on the right. The purple line is a function that fits the data points.

(a) Disruptive selection increases the amount of variation in a trait.

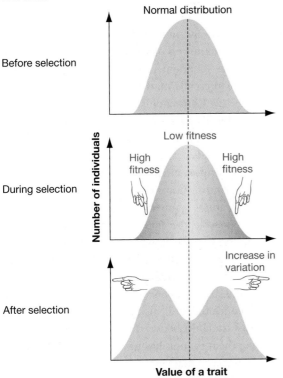

(b) For example, only juvenile black-bellied seedcrackers that had very long or very short beaks survived long enough to breed.

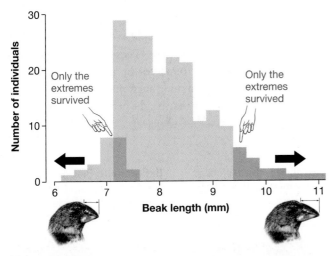

FIGURE 24.4 Disruptive Selection
(a) When disruptive selection occurs on traits with a normal distribution, individuals with extreme phenotypes experience high reproductive success. **(b)** Histogram showing the distribution of beak length in a population of black-bellied seedcrackers from Cameroon, West Africa. The light orange bars represent all juveniles; the dark orange bars, juveniles that survived to adulthood. Only juveniles with the most extreme phenotypes survived long enough to breed.

individuals with either very short or very long beaks survive best and that birds with intermediate phenotypes are at a disadvantage. In this case, the agent that causes natural selection is food. At a study site in south-central Cameroon, West Africa, a researcher found that only two sizes of seed are available to the seedcrackers: large and small. Birds with small beaks crack and eat small seeds efficiently. Birds with large beaks handle large seeds efficiently. But birds with intermediate beaks have trouble with both. Disruptive selection maintains high overall variation in this population.

Disruptive selection is important because it can lead to **speciation**, or the formation of new species. If small-beaked seedcrackers began mating with other small-beaked individuals, all of their offspring would be small beaked and would feed on small seeds. Similarly, if large-beaked individuals chose only other large-beaked individuals as mates, they would produce large-beaked offspring that would feed on large seeds. In this way, selection would result in two distinct populations and eventually two new species. The process of species formation, based on disruptive selection and other mechanisms, is explored in detail in Chapter 25.

In many cases, the process of speciation depends on how individuals choose mates. The process of choosing mates, in turn, causes a distinctive type of natural selection called sexual selection. Let's take a closer look at what sexual selection is and how it works.

Sexual Selection

If female peacocks choose males with the longest and most iridescent tails as mates, then the frequency of alleles that contribute to long, iridescent tails will increase in the population. Charles Darwin was the first biologist to recognize that this form of natural selection, called **sexual selection**, is a mechanism of evolutionary change. Sexual selection occurs when individuals within a population differ in their ability to attract mates. It is a special case of natural selection—selection for enhanced ability to obtain mates.

In 1948 A. J. Bateman contributed a fundamental insight about how sexual selection works. His idea was elaborated by Robert Trivers in 1972. The Bateman-Trivers theory contains two elements: a claim about a pattern in the natural world and a mechanism that causes the pattern.

The pattern component of their theory is that sexual selection usually acts on males much more strongly than on females. As a result, traits that attract members of the opposite sex are much more highly elaborated in males. The mechanism that Bateman and Trivers proposed to explain this pattern can be summarized with a quip: "Eggs are expensive, but sperm are cheap." That is, the energetic cost of creating a large egg is enormous, whereas a sperm contains few energetic resources. Thus, in most species, females invest much more in their offspring than do males. This phenomenon is called the **fundamental asymmetry of sex**. It is characteristic of almost all sexual species and has two important consequences:

1. Because eggs are large and energetically expensive, females produce relatively few young over the course of a lifetime. A female's fitness is limited primarily by her ability to gain the resources needed to produce eggs and rear young—not by the ability to find a mate.

2. Sperm are so simple to produce that a male can father an almost limitless number of offspring. For males, fitness is limited not by the ability to acquire the resources needed to produce sperm, but by the number of females they can mate with.

The theory of sexual selection makes strong predictions. If females invest a great deal in each offspring, then they should be choosy about whom they agree to mate with. Conversely, if males invest little in each offspring, then they should compete with each other for mates. Finally, sexual selection should be much stronger on males than on females. As a result, traits that result from sexual selection—meaning traits that are useful only in courtship or in competition for mates—should be found primarily in males. Do data from experimental or observational studies agree with these predictions? Let's consider each of them in turn.

Sexual Selection Via Female Choice   Female birds invest a great deal of time and energy in each offspring. Brown kiwis, which are flightless, ground-dwelling birds that live in New Zealand, offer an extreme example. Female brown kiwis weigh about 2.7 kg (6 lb) and lay an egg that weighs 0.45 kg (1 lb). As **Figure 24.5** shows, each egg represents a considerable investment of resources.

Females that invest this much in each offspring should be choosy about the males they mate with. If so, what criteria do females use to make their choice? Recent experiments have shown that, in several bird species, females prefer to mate with males that are well fed and in good health. These experiments were motivated by three key observations: (1) In many bird species, the existence of brightly colored feathers or a bright beak is due to the presence of the red and yellow pigments called carotenoids. (2) Carotenoids, which were introduced in Chapter 10, function as accessory pigments in plant chloroplasts, where they help to harvest light energy used in photosynthesis. But in animals, carotenoids stimulate the immune system to fight disease more effectively. Their presence also prevents tissue damage from free radicals that form naturally in cells. (3) Animals cannot synthesize carotenoids—these molecules must be obtained in the diet. Taken together, these three observations suggest that birds with the brightest beaks and feathers are those that are well fed. Further, beaks and feathers can be bright only if an individual is healthy enough to allocate carotenoids to these structures instead of using the molecules to fight disease or reduce free radicals.

Is there really a trade-off between using carotenoids to fight disease and using them to look bright and colorful? To answer this question, a research team did an experiment with European blackbirds (**Figure 24.6**). Males of this species have a bright-orange bill due to the presence of carotenoids. When the biologists injected a group of male blackbirds with red blood cells from sheep, it provoked a strong response by the immune system. In addition, beak color in these males dimmed. There was no change in beak color, however, in a control group of males that was injected with a simple salt solution, which did not provoke an immune response. These results suggest that there is indeed a trade-off between the use of carotenoids in beak coloration and in immune function. The results are consistent with the hypothesis that bright beaks are an honest signal of male health.

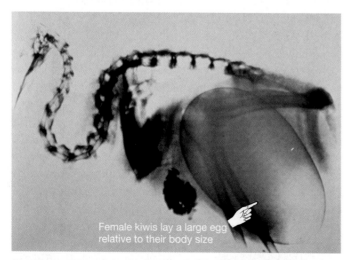

FIGURE 24.5 In Many Species, Females Make a Large Investment in Each Offspring
X-ray of a female kiwi, ready to lay an egg.

Control: Blackbird injected with harmless salt solution   **Experiment:** Blackbird injected with cells that stimulate immune system

Bill is brighter   Bill is more yellow

FIGURE 24.6 When the Immune Systems of Male Blackbirds Are Challenged, Their Beaks Dull in Color
These two male blackbirds are brothers. The individual on the right was injected with sheep red blood cells, which activate the immune system in much the same way as a disease-causing bacterium or parasite does. The male on the right is not actually sick, however.

To test the hypothesis that females prefer to mate with bright-beaked birds, a different team of researchers experimented with zebra finches (**Figure 24.7a**). As **Figure 24.7b** shows, they fed one group of male zebra finches a diet that was heavily supplemented with carotenoids. A second group of male zebra finches (the control group), consisting of brothers of males in the first group, was fed a diet that was similar in every way except for the additional carotenoids. As predicted, the males eating the carotenoid-supplemented diet developed brighter beaks than the males fed the carotenoid-poor diet. When females were given a choice of spending time with the two brothers, 9 out of 10 females tested preferred the brighter-beaked male. These results are strong evidence that females of this species are choosy about their mates and that they prefer to mate with healthy, well-fed males.

Enough experiments have been done on other bird species to support a general conclusion: Bright male coloration, along with songs and dances and other types of courtship displays, carry the message "I'm healthy and well fed because I have good alleles. Mate with me." By choosing a colorful male as the father of her offspring, a female is likely to have offspring with alleles that will help the offspring fight disease and feed efficiently.

(a) Zebra finches have bright beaks.

Male

Female

(b) Males fed carotenoids get brighter beaks.

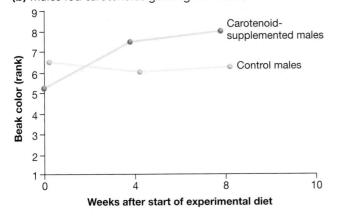

Carotenoid-supplemented males

Control males

Beak color (rank)

Weeks after start of experimental diet

FIGURE 24.7 If Male Zebra Finches Are Fed Carotenoids, Their Beaks Get Brighter

(a) Male and female zebra finches look markedly different—males have brighter feathers. **(b)** In males, beaks brighten in response to an increased presence of carotenoids in the diet.

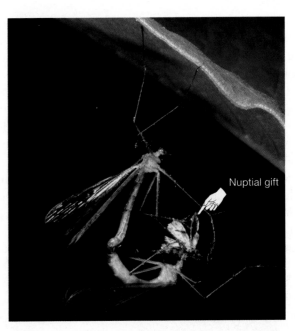

Nuptial gift

FIGURE 24.8 In Some Insects, Males Present Food to Females as a "Nuptial Gift"

Hanging scorpion flies copulating while the female (below) eats a prey presented to her by the male (above). The larger the prey, the longer she will eat and the longer she will copulate with the male.

QUESTION Most female insects copulate with more than one male before laying eggs. Why would it be advantageous for a male to copulate with a female for a long time?

Extensive experimentation has also shown that in many species, females prefer to mate with males who provide some of the resources required to produce eggs or care for young. In insects, for example, it is common to observe that females will not mate with a male unless he brings a "nuptial gift"—a piece of food that the female devours as they copulate (**Figure 24.8**). In some fish species, females prefer to mate with males that protect a nest site and care for the eggs until they hatch. In humans and many species of birds, males provide food, protection, and other resources required for rearing young.

To summarize, females may choose mates on the basis of physical characteristics that signal male genetic quality, resources provided by males, or both. In some species, however, females do not have the luxury of choosing a male. Instead, competition among males is the primary cause of sexual selection.

Sexual Selection via Male-Male Competition   As an example of research on how males compete for mates, consider data from a long-term study of a northern elephant seal population breeding on Año Nuevo Island, off the coast of California. Elephant seals feed on marine fish and spend most of the year in the water. But when females are ready to mate and give birth, they haul themselves out of the water onto land. Females prefer to give birth on islands, where newborn pups are protected from terrestrial and marine predators. Because elephant seals have flippers that are ill suited for walking, females can haul themselves out of the water only on the few beaches that have gentle slopes. As a result, large numbers of females congregate in tiny areas to breed.

Male elephant seals establish territories on breeding beaches by fighting (**Figure 24.9a**). A **territory** is an area that is actively defended and that provides exclusive use by the owner. Males that win battles with other males monopolize matings with the females residing in their territories. Males that lose battles are relegated to territories with few females or are excluded from the beach. Fights are essentially slugging contests and are usually won by the larger male. The males stand face to face, bite each other, and land blows with their heads.

Based on these observations, it is not surprising that male northern elephant seals frequently weigh three tons (2700 kg) and are over four times more massive, on average, than females. The logic here runs as follows: Males that own beaches with large congregations of females will father large numbers of offspring. Males that lose fights will father few or no offspring. As a result, the alleles of territory-owning males will rapidly increase in frequency in the population. If the ability to win fights and produce offspring is determined primarily by body size, then alleles for large body size will have a significant fitness advantage, leading to the evolution of large male size. The fitness advantage is due to sexual selection.

Figure 24.9b provides evidence for intense sexual selection in males. Biologists have marked many of the individuals in the seal population on Año Nuevo to track the lifetime reproductive success of a large number of individuals. As the data show, in this population a few males father a large number of offspring, while most males father few or none. Among females, variation in reproductive success is much lower (**Figure 24.9c**). In this species, most sexual selection is driven by male-male competition rather than female choice.

What Are the Consequences of Sexual Selection?

The process component of the theory of sexual selection contends that sexual selection is most intense in the sex that makes the *least* investment in offspring. The biologists who studied the northern elephant seal population have data that is relevant to this prediction. By capturing and weighing females and seal pups throughout the breeding season, the investigators found that adult females weigh about 650 kg on average and give birth to one pup, weighing 50 kg, each year. In their first five weeks of life, the pups routinely gain over 100 kg—entirely from mother's milk. The energy expenditure required to accomplish this growth takes a tremendous toll on the mother. The researchers found that mothers typically lost up to 200 kg, or almost a third of their body mass, during nursing. To breed the following year, females had to regain their lost body weight. The father's parental energy investment, in contrast, was limited to a few minutes spent in copulation and the almost negligible energetic cost of producing sperm.

(a) Males compete for the opportunity to mate with females.

(b) Variation in reproductive success is high in males.

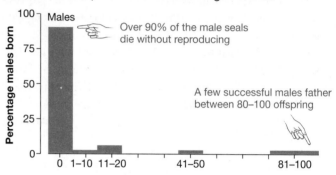

(c) Variation in reproductive success is relatively low in females.

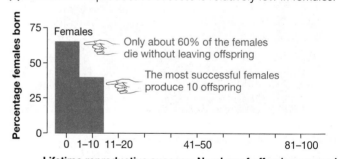

FIGURE 24.9 Sexual Selection on Male Elephant Seals Is Strong (a) Two bull northern elephant seals compete for the opportunity to mate with females that will breed on this beach. The histograms show that variation in lifetime reproductive success is even higher **(b)** in male northern elephant seals than it is **(c)** in females. Alleles that lead to high reproductive success in males will rapidly increase in frequency.

These data support the hypothesis that the lifetime reproductive ability of elephant seal females is limited by their ability to acquire enough food to support pregnancy and nursing, not by their ability to persuade a male to copulate. In fact, in most animal species, every female that survives to adulthood gets a mate. In contrast, many males do not get any mate at all.

(a) Beetle   **(b)** Scarlet tanager   **(c)** Lion

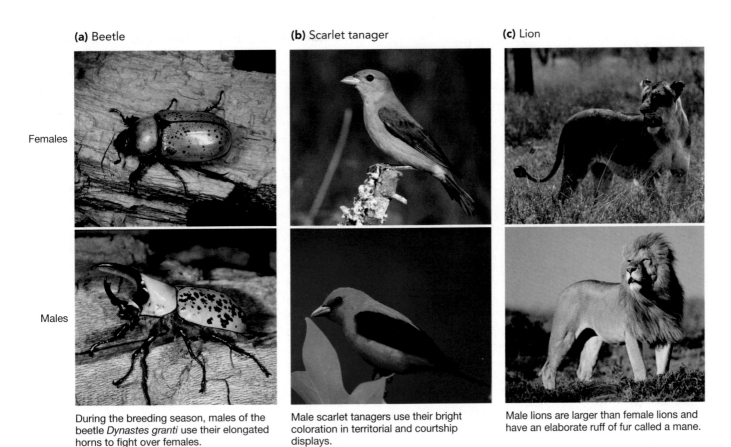

Females

Males

During the breeding season, males of the beetle *Dynastes granti* use their elongated horns to fight over females.

Male scarlet tanagers use their bright coloration in territorial and courtship displays.

Male lions are larger than female lions and have an elaborate ruff of fur called a mane.

FIGURE 24.10 Sexually Selected Traits Are Used to Compete for Mates
Males often have exaggerated traits that they use in fighting or courtship. In many species, females lack these traits.   [(b) top ©B. Schorre/VIREO; (b) bottom ©R. & A. Simpson/VIREO] QUESTION How do traits that differ in male and female humans help individuals compete for mates?

The overall message of this analysis is that sexual selection is usually much more intense in males than in females. As a result, it is common to observe that sexually selected traits differ sharply between the sexes. **Sexual dimorphism** ("two-forms") refers to any trait that differs between males and females. **Figure 24.10** illustrates sexually dimorphic traits. They range from weapons that males use to fight over females, such as antlers and horns, to the elaborate ornamentation and behavior used in courtship displays. Humans are sexually dimorphic in size, distribution of body hair, and many other traits. Sexual selection often results in sexual dimorphism. It is a form of natural selection based on the ability to attract mates.

✔ CHECK YOUR UNDERSTANDING

Selection can be directional, stabilizing, or disruptive. You should be able to explain how each pattern of natural selection affects genetic variation, and why. Sexual selection occurs on traits that are used in courtship and is based on differential success in obtaining mates. You should be able to define the fundamental asymmetry of sex and explain (1) why males are usually the sex with exaggerated traits used in courtship and (2) how male-male competition for mates differs from female choice of mates.

24.3   Genetic Drift

Both natural selection and sexual selection are directed processes. Natural selection is directed by the environment and results in adaptation. Sexual selection is directed by female choice or male-male competition and results in traits that lead to mating success. **Genetic drift**, in contrast, is undirected. It is any change in allele frequencies that is due to chance in a population. The process is aptly named, because it causes allele frequencies to drift up and down randomly over time. When drift occurs, allele frequencies change due to blind luck—what is formally known as **sampling error**.

To understand why genetic drift occurs, consider the group of people who founded the present population of Pitcairn Island in the South Pacific. The founding event occurred in 1789, when a small band of mutineers led by Fletcher Christian took over the British warship HMS *Bounty* and fled to Pitcairn. The six sailors were joined by two Tahitian men and six Tahitian women. Suppose that six couples formed and raised two children each. How will allele frequencies compare in the founding population that bred versus the next generation?

To answer this question, let's focus on a hypothetical gene A with alleles A_1, A_2, and A_3. Suppose that the six males and six females in the original couples have the genotypes given in

TABLE 24.3 A Thought Experiment on Genetic Drift

The genotypes of the children in this table were generated by flipping a coin to simulate which alleles combined during fertilization. The frequencies of alleles A_1, A_2, and A_3 differ in the two generations, due to genetic drift.

	Father	Mother	Child 1	Child 2
Couple 1	$A_3 A_3$	$A_2 A_3$	$A_3 A_3$	$A_3 A_3$
Couple 2	$A_2 A_3$	$A_2 A_3$	$A_2 A_3$	$A_3 A_2$
Couple 3	$A_1 A_2$	$A_2 A_3$	$A_2 A_3$	$A_1 A_2$
Couple 4	$A_1 A_1$	$A_1 A_2$	$A_1 A_2$	$A_1 A_1$
Couple 5	$A_1 A_2$	$A_3 A_3$	$A_1 A_3$	$A_1 A_3$
Couple 6	$A_1 A_2$	$A_1 A_3$	$A_2 A_3$	$A_2 A_1$

Table 24.3. The gametes formed by each of these parents have an equal chance of carrying either allele. Further, eggs and sperm combine at random when fertilization occurs—meaning that each type of egg and sperm has an equal probability of combining, irrespective of its genotype. To simulate the fertilization process, you can flip a coin to decide which gametes combine. Let heads stand for the first allele listed in the parent's genotype, while tails represents the second allele listed. Doing these coin flips resulted in the offspring genotypes given in the "Child 1" and "Child 2" columns of Table 24.3.

Now let's count up alleles and calculate allele frequencies. Because there are 12 breeding individuals and 12 offspring, the total number of alleles present in each generation (ignoring the two founding individuals who did not breed) is 24. From the data in Table 24.3, the allele frequencies are as follows:

	A_1	A_2	A_3
Allele frequencies in the parents	7/24 = 29.2%	8/24 = 33.3%	9/24 = 37.5%
Allele frequencies in the offspring	7/24 = 29.2%	7/24 = 29.2%	10/24 = 41.6%

Although the frequency of allele A_1 did not change, the frequency of A_2 declined by over 4 percent and the frequency of A_3 increased by over 4 percent, purely by chance. Allele A_3 did not confer higher fitness. Instead, it just got lucky. Random chance caused evolution—a change in allele frequencies in a population. Instead of each allele being sampled in exactly its original frequency when offspring formed, a chance sampling error occurred.

This exercise helps to illustrate several important points:

• *Genetic drift is random with respect to fitness.* The allele frequency changes it produces are not adaptive.

• *Genetic drift is most pronounced in small populations.* If the original population had included only couples 1 and 2, then the frequency of allele A_2 would have declined by 12.5 percent and the frequency of A_3 would have increased by 12.5 percent. The smaller the sample, the larger the sampling error. Conversely, if the population had consisted of 500 couples who produced 1000 offspring (necessitating 2000 coin flips), it is extremely unlikely that drift would produce a change in allele frequency as large as 4 percent. Sampling error is usually insignificant when the sample is large.

• *Over time, genetic drift can lead to the random loss or fixation of alleles.* If allele A_1 continued to be unlucky generation after generation, it would eventually drift to a frequency of 0.0 and be lost. If allele A_3 continued to be lucky for several more generations, it might drift to a frequency of 1.0. When random loss or fixation occurs, genetic variation in the population declines.

Experimental Studies of Genetic Drift

To understand the long-term consequences of genetic drift in small populations, consider an experiment conducted by Warwick Kerr and Sewall Wright in the mid-1950s. These biologists started with a large laboratory population of fruit flies that contained a **genetic marker**—a specific allele that causes a distinctive phenotype. In this case, the marker was the morphology of leg bristles. Fruit flies have bristles on their legs that can be either straight or bent. This difference in leg bristle phenotype depends on a single gene, and Kerr and Wright's lab population contained just two alleles—normal (straight) and "forked" (bent).

To begin the experiment, the researchers set up 96 cages in their lab. Then they placed four adult females and four adult males of the fruit fly *Drosophila melanogaster* in each. They chose individual flies to begin these experimental populations so that the frequency of the normal and forked alleles in each of the 96 starting populations was 0.5. The two alleles do not affect the fitness of flies in the lab environment, so Kerr and Wright could be confident that if changes in the frequency of normal and forked phenotypes occurred, they would not be due to natural selection.

After these first-generation adults bred, Kerr and Wright reared their offspring. Then they randomly chose four males and four females—meaning that they simply grabbed individuals without regard to whether their leg bristles were normal or forked—from each of the 96 offspring populations and allowed them to breed and produce the next generation. The researchers repeated this procedure until all 96 populations had undergone a total of 16 generations. During the entire course of the experiment, no migration from one population to another occurred, and previous studies had shown that mutations from normal to forked are rare. The only evolutionary process operating during the experiment was genetic drift.

Their result? After 16 generations, the 96 populations fell into three groups. Forked leg bristles were found on all of the individuals in 29 of the experimental populations. Due to drift, the forked allele had been fixed in these 29 populations and the normal allele had been lost. In 41 other populations, however, the opposite was true: All individuals had normal bristles. In these populations, the forked allele had been lost due to chance. Both alleles were still present in 26 of the populations. The message of the study is startling: In 73 percent of the experimental populations (70 out of the 96), genetic drift had reduced allelic diversity to zero. Genetic drift decreased genetic variation within populations and increased genetic differences between populations. Is drift important in natural populations as well?

Genetic Drift in Natural Populations

The sampling process that occurs during fertilization and that caused changes in allele frequencies in Pitcairn islanders and fruit flies occurs in every population in every generation. It is particularly important in small populations, where sampling error tends to be high. This is of enormous concern to conservation biologists, because populations all over the planet are being drastically reduced in size by habitat destruction and other human activities. Small populations that occupy nature reserves or zoos are particularly susceptible to genetic drift. If drift leads to a loss of genetic diversity, it could darken the already bleak outlook for some endangered species.

Given enough time, however, drift can be an important factor even in moderately large populations. To drive this point home, consider two types of alleles that were introduced in earlier chapters and that have no effect on fitness. Recall from Chapter 16 that alleles containing silent mutations, usually in the third position of a codon, do not change the gene product. As a result, they have no effect on the phenotype. Yet these alleles routinely drift to high frequency or even fixation over time. Similarly, recall from Chapter 20 that pseudogenes do not code for a product. Although their presence does not affect an individual's phenotype, dozens of pseudogenes in the human genome have reached fixation, due to drift.

It is important to realize that because drift is caused by sampling error, it can occur by *any* process or event that involves sampling—not just the sampling of gametes that occurs during fertilization.

How Do Founder Effects Cause Drift?

When a group of individuals emigrates to a new geographic area and founds (establishes) a new population, a **founder event** is said to occur. If the group is small enough, the allele frequencies in the new population are almost guaranteed to be different from those in the source population, due to sampling error. A change in allele frequencies that occurs when a new population is established is called a **founder effect**. As soon as the mutineers and Tahitians set foot on Pitcairn Island and took up residence, evolution had occurred due to drift, via a founder effect. The allele frequencies in the new population were different from those in England or in Tahiti. (Once breeding took place, drift continued to act via sampling errors that occurred during fertilization.)

Fishermen on the island of Anguilla in the Caribbean recently witnessed a founder event involving green iguanas. A few weeks after Hurricane Luis and Hurricane Marilyn swept through the region in September 1995, a large raft composed of downed logs tangled with other debris floated onto a beach on Anguilla. The fishermen noticed green iguanas on the raft and several on shore. Because green iguanas had not previously been found on Anguilla, the fishermen notified biologists. The researchers were able to document that at least 15 individuals had arrived; two years later they were able to confirm that at least some of the individuals were breeding. It is extremely unlikely that allele frequencies in the Anguilla population of green iguanas match those of the source population, thought to be on the islands of Guadeloupe.

Founder events like these have been the major source of populations that occupy islands all over the world, as well as island-like habitats such as mountain meadows, caves, and ponds. Each time a founder event occurs, a founder effect is likely to accompany it, changing allele frequencies via genetic drift.

How Do Population Bottlenecks Cause Drift?

If a large population experiences a sudden reduction in size, a **population bottleneck** is said to occur. The term comes from the metaphor of a few individuals passing through the neck of a bottle, by chance. Disease outbreaks, natural catastrophes such as floods or fires or storms, or other events can cause population bottlenecks.

Genetic bottlenecks follow population bottlenecks, just as founder effects follow founder events. A **genetic bottleneck** is a sudden reduction in the number of alleles in a population. Drift occurs during genetic bottlenecks and causes a change in allele frequencies. As an example of a genetic bottleneck, consider the human population of Pingelap Atoll in the South Pacific. On this island, only about 20 people out of a population of several thousand managed to survive the effects of a typhoon and a subsequent famine that occurred around 1775.

The survivors apparently included a person who carried a loss-of-function allele at a gene called *CNGB3*, which codes for a protein involved in color vision. When this allele is homozygous, it causes a serious vision deficit called *achromatopsia*. The condition is extremely rare in most populations, with the homozygous genotype and the affliction occurring in about 0.005 percent of the population. If genotypes at this locus are in Hardy-Weinberg proportions, then $q^2 = 0.00005$ and the frequency of the loss-of-function allele in most populations is about $\sqrt{0.00005}$, or 0.7 percent. In the population that survived the Pingelap Atoll disaster, however, the loss-of-function allele was at a frequency of about 1/40, or 2.5 percent. If the allele was at the typical frequency of 0.7 prior to the population bottleneck, then a huge frequency change occurred during the bottleneck, due to drift.

In today's population on Pingelap Atoll, about 1 in 20 people is afflicted with achromatopsia and the allele is at a frequency of over 20 percent. Because it is extremely unlikely that the loss-of-function allele is favored by directional selection or heterozygote advantage, researchers hypothesize that its frequency in this small population has continued to increase over the past 230 years due to drift.

✓ CHECK YOUR UNDERSTANDING

Genetic drift occurs any time allele frequencies change due to chance. Drift occurs during many different types of events, including random fusion of gametes at fertilization, founder events, and population bottlenecks. You should be able to explain why drift leads to a random loss or fixation of alleles and why drift is particularly important as an evolutionary force in small populations.

24.4 Gene Flow

Gene flow is the movement of alleles from one population to another. It occurs when individuals leave one population, join another, and breed. As an evolutionary mechanism, gene flow has one outcome: It equalizes allele frequencies between the source population and the recipient population. When alleles move from one population to another, the populations become more alike.

To see the consequences of gene flow in action, consider data from prairie lupine populations on Mount St. Helens in the state of Washington. Prairie lupine is a perennial, meaning that individuals grow from the same root system year after year. They are fairly short lived, however, with each individual rarely living for more than 5 years. Prairie lupines are most commonly found living at high altitude on volcanoes, in habitats like that shown in **Figure 24.11a**.

The explosion of Mount St. Helens in 1980 created thousands of hectares of mudflows and ash deposits that are ideal habitat for prairie lupines. In 1981 a single individual founded a population on an extensive ash plain in the area most highly affected by the blast. This new population was located over 4 km from the nearest surviving population. In the blast zone near this new population, dozens of other populations eventually appeared.

To study genetic diversity in these populations, biologists collected tissue samples from 532 individuals in 32 populations. The age of each of the study populations was known. The researchers made sure they analyzed tissues from the oldest individuals in populations that had been established just 1 to 3 years prior to the start of the study. These individuals were of special interest, because they were undoubtedly the founders of the new populations. Once tissues from all 532 individuals had

(a) Lupines colonize sites and form populations.

(b) Gene flow reduces genetic differences among populations.

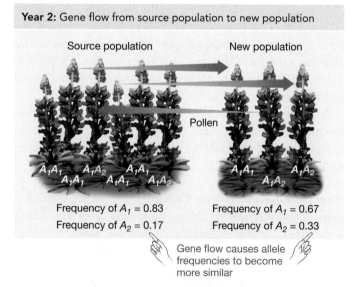

FIGURE 24.11 Gene Flow Equalizes Allele Frequencies between Populations

(a) Several species of lupine grow on the slopes of volcanoes. **(b)** The movement of alleles between populations tends to equalize allele frequencies over time.

been collected, the investigators used the polymerase chain reaction (see Chapter 19) to amplify the alleles of two genes in each individual. These data allowed them to estimate the number of alleles present in each population, as well as the frequency of each allele.

Their results? The data showed that the oldest individuals in newly established populations had allele frequencies that were very different from allele frequencies in the source populations. To interpret this observation, the biologists hypothesize that strong founder effects occurred each time a new lupine population became established. But their data also showed that as populations got older, allele frequencies became progressively more like those in the oldest populations—specifically, like the populations that had survived the eruption. Because selection due to differences in soil and moisture conditions, diseases, and predators present in each population would tend to make populations less similar over time, the researchers doubted that natural selection played a large role in equalizing allele frequencies. Instead, they propose that a continuous flow of alleles between populations, primarily via pollen carried by bees, was responsible for making the populations genetically more alike over time (**Figure 24.11b**).

What impact did gene flow have on the average fitness of individuals in these populations? The answer is not known, but theory suggests that it might vary. If a population had lost alleles due to genetic drift, then the arrival of new alleles via gene flow should increase genetic diversity. In this way, gene flow might increase the average fitness of individuals. But if natural selection had resulted in a population that was highly adapted to a specific habitat, then gene flow from other populations might introduce alleles that have low fitness in that particular environment. In this case, gene flow would have a negative effect on average fitness in the population.

To summarize, the fitness effects of gene flow depend on the situation. But a movement of alleles between populations always tends to reduce genetic differences between them. This latter generalization is particularly important in our own species right now. Large numbers of people from Africa, the Middle East, Mexico, Central America, and Asia are emigrating to the countries of the European Union and the United States. Because individuals from different cultural and ethnic groups are intermarrying frequently and having offspring, allele frequencies in human populations are becoming more similar.

24.5 Mutation

To appreciate the role of mutation as an evolutionary force, let's return to one of the central questions that biologists ask about an evolutionary mechanism: How does it affect genetic variation in a population? Gene flow can increase genetic diversity in a recipient population if new alleles arrive with immigrating individuals. But gene flow can instead decrease genetic variation in the source population if alleles leave with

emigrating individuals. The impact of genetic drift on genetic variation is much more clear cut than the impact of gene flow. Genetic drift tends to decrease genetic diversity over time, as alleles are randomly lost or fixed. Similarly, most forms of selection favor certain alleles and lead to a decrease in overall genetic variation.

If most of the evolutionary mechanisms lead to a loss of genetic diversity over time, what restores it? In particular, where do entirely new alleles come from? The answer to both of these questions is **mutation**.

As discussed in Chapter 16, errors by DNA polymerase that occur as a DNA molecule is copied result in changes in the sequence of deoxyribonucleotides. If a mutation occurs in a stretch of DNA that codes for a protein, the changed codon may result in a polypeptide with a novel amino acid sequence. Because errors are inevitable, mutation constantly introduces new alleles into all populations at all genes. Mutation is an evolutionary mechanism that increases genetic diversity in populations.

The second important point to recognize about mutation is that, with respect to the fitness of the affected allele, mutation is a random process. Mutation just happens. Changes in DNA do not occur in a way that tends to increase fitness or decrease fitness. Because most organisms are well adapted to their current habitat, random changes in genes usually result in products that do not work as well as the alleles that currently exist. Stated another way, most mutations result in the production of **deleterious alleles**—alleles that lower fitness. On rare occasions, however, mutation produces a beneficial allele—an allele that allows individuals to produce more offspring. Beneficial alleles should increase in frequency in the population due to natural selection.

Because mutation produces new alleles, it can in principle change the frequencies of alleles through time. That is, the existence of mutation by itself should be able to change allele frequencies and cause evolution. But does mutation occur often enough to make it an important factor in changing allele frequencies of a particular gene? The short answer is no.

Mutation as an Evolutionary Mechanism

To understand why mutation is not a significant mechanism of evolutionary change by itself, consider that the highest mutation rates that have been recorded at individual genes in humans are on the order of 1 mutation in every 10,000 gametes produced by an individual. This rate means that, for every 10,000 alleles produced, on average one will have a mutation. When two gametes combine to form an offspring, then, at most about 1 in every 5000 offspring will carry a mutation at a particular gene. Will mutation affect allele frequencies in a species such as ours? To answer this question, suppose that 195,000 humans live in a population; that 5000 offspring are born one year; and that, at the end of that year, the population numbers 200,000. In a population this size, there is a total of 400,000 copies of each gene. Only one of them is a new allele created by mutation, however.

Over the course of a year, the allele frequency change introduced by mutation is 1/400,000, or 0.0000025 (2.5×10^{-6}). At this rate, it would take 4000 years for mutation to produce a change in allele frequency of 1 percent.

These calculations support the conclusion that mutation does little to change allele frequencies on its own. As an evolutionary mechanism, mutation is extremely slow compared with selection, genetic drift, and gene flow.

Does this conclusion mean that mutation plays *no* role in evolution? The answer is no. For example, because humans have over 30,000 genes, each individual carries 60,000 alleles. Although the rate of mutation per allele may be very low, the total number of alleles is high. Multiplying the estimated number of genes in a human by the average mutation rate per gene suggests that an average person contains 1.6 new alleles created by mutation. In addition, recall from Chapter 12 that these new alleles and existing alleles are constantly being shuffled into new combinations during meiosis and crossing over. Thus, each individual also represents a unique combination of alleles.

The message here is that mutation introduces new alleles into every individual in every population in every generation. And in species that undergo sexual reproduction, meiosis and genetic recombination create variation in terms of the allele combinations present in each individual. Even if selection and drift are eliminating genetic diversity, mutation will act to restore it.

What Role Does Mutation Play in Evolutionary Change?

To get a better appreciation for how mutation affects evolution, consider an experiment conducted by Richard Lenski and colleagues. The experiment was designed to evaluate the role that mutation plays in many genes over many generations. The researchers focused on *Escherichia coli*, a bacterium that, as we have seen, is a common resident of the human intestine. **Figure 24.12** diagrams how Lenski's group set up a large series of populations, each founded with a single cell. Half of the cells contained an allele that allowed them to metabolize the sugar arabinose. The populations founded by these cells were designated *Ara+*. The other cells carried an allele that did not allow them to metabolize arabinose. The populations founded by these cells were designated *Ara−*. When *Ara+* cells are grown on special plates, they produce colonies that look red. When *Ara−* colonies are grown on the same plates, they produce colonies that look white. Under the temperature and nutrient conditions used in the experiment, however, neither *Ara−* nor *Ara+* cells had a selective advantage. The presence of the *Ara* allele simply gave the researchers a way to identify particular populations by color. What happened to the descendants of these founding cells over time?

To answer this question, the researchers transferred a small number of cells from each of the populations into a new batch of the same growth medium, under the same light and tempera-

Question: How does average fitness in a population change over time?

Hypothesis: Average fitness increases over time.

Null hypothesis: Average fitness does not increase over time.

Experimental setup:

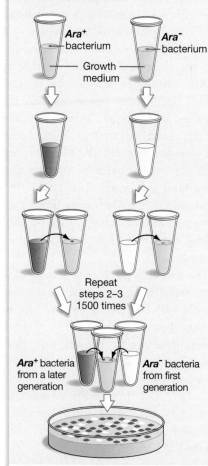

1. Place 10 mL of identical growth medium into many replicate tubes with one bacterium in each, some *Ara+* and some *Ara−* so that they can be distinguished by color when grown on special plates.

2. Incubate overnight. Average population in each tube is now 5×10^8 cells. (Cells are red or white when grown on special plates.)

3. Remove 0.1 mL from each tube and move to 10 mL of fresh medium. Freeze remaining cells for later analysis.

4. Put an equal number of cells from generation 1 and a later generation in fresh growth medium (competition experiment).

5. Incubate overnight and count the cells. Which are more numerous?

Prediction: Descendant populations have higher average fitness. (There will be more individuals from descendant than ancestral populations on the plates.)

Prediction of null hypothesis: There will be no difference in fitness between descendant and ancestral populations.

Results:

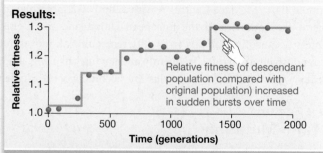

Relative fitness (of descendant population compared with original population) increased in sudden bursts over time

Conclusion: Descendant populations have higher fitness than do ancestral populations.

FIGURE 24.12 An Experiment to Test Changes in Fitness Over Time

TABLE 24.4 A Summary of Evolutionary Mechanisms

Process	Definition and Notes	Effect on Genetic Variation	Effect on Average Fitness
Selection	Certain alleles are favored	Can lead to maintenance, increase, or reduction	Can produce adaptation
Genetic drift	Random changes in allele frequencies: most important in small populations	Tends to reduce, via loss or fixation of alleles	Usually reduces
Gene flow	Movement of alleles between populations; reduces differences between populations	May increase by introducing new alleles; may decrease by removing alleles	May increase by introducing high-fitness alleles or decrease by introducing low-fitness alleles
Mutation	Production of new alleles	Increases by introducing new alleles	Random with respect to fitness; most mutations lower fitness

ture conditions, every day for over four years. In this way, each population grew continuously. Over the course of the experiment, the researchers estimated that each population underwent a total of 10,000 generations. This is the equivalent of over 200,000 years of human evolution. In addition, the biologists saved a sample of cells from each population at regular intervals and stored them in a freezer. The frozen cells served as a fossil record of cells that existed over the 10,000-generation time interval. But because frozen *E. coli* cells resume growth when they are thawed, the "fossil" individuals could be brought back to life.

The strain of *E. coli* used in the experiment is completely asexual and reproduces by cell division. Thus, mutation was the only source of genetic variation in these populations. Although no gene flow occurred, both selection and genetic drift were operating in each population.

Were cells from the older and newer generations of each population different? This question was addressed via competition experiments. *Ara*$^+$ cells from one generation and *Ara*$^-$ cells from a different generation were put together in the same flask. After allowing the cells to grow and compete, the researchers grew them on special plates and counted the number of red versus white colonies. The more numerous population had grown faster, meaning that it was better adapted to the experimental environment. In this way the researchers could measure the fitness of descendant populations relative to ancestral populations. If relative fitness was greater than 1, it meant that recent-generation cells outnumbered older-generation cells when the competition was over.

The data from a series of competition experiments are graphed at the bottom of Figure 24.12. Note that relative fitness increased dramatically—almost 30 percent—over time. But note also that fitness increased in fits and starts. This pattern is emphasized by the solid line on the graph, which represents a mathematical function fitted to the data points.

What caused this stair-step pattern? Lenski's group hypothesizes that genetic drift was relatively unimportant in this experiment because population sizes were so large. Instead, they propose that each jump was caused by a novel mutation that conferred a fitness benefit. Their interpretation is that cells that

happened to have the beneficial mutation grew rapidly and came to dominate the population. After a beneficial mutation occurred, the fitness of the population stabilized—sometimes for hundreds of generations—until another random but beneficial mutation occurred and produced another jump in fitness.

The experiment makes an important point: Mutation is the ultimate source of genetic variation. If mutation did not occur, evolution would eventually stop. Without mutation, there is no variation for natural selection to act on.

To summarize, mutation alone is usually inconsequential in changing allele frequencies at a particular gene. When considered across the genome and when combined with natural selection, however, it becomes an important evolutionary mechanism. Mutation is the ultimate source of the genetic variation that makes evolution possible.

Table 24.4 summarizes the four evolutionary forces and their consequences. If one or more of these processes affects a gene, then genotypes will not be in Hardy-Weinberg proportions. But we have yet to consider the effects of another assumption in the Hardy-Weinberg model—that mating takes place at random with respect to the gene in question. What happens when this assumption is violated?

24.6 Inbreeding

In the Hardy-Weinberg model, gametes were picked from the gene pool at random and paired to create offspring genotypes. In nature, however, matings between individuals may not be random with respect to the gene in question. For example, in small populations that are isolated from gene flow, matings between relatives become common. Mating between relatives is called **inbreeding**. Because relatives are likely to share alleles that they inherited from a common ancestor, matings between relatives are nonrandom with respect to all of the genes in the genome. Inbreeding always violates the assumptions of the Hardy-Weinberg principle.

To understand how inbreeding affects populations, let's follow the fate of alleles and genotypes when it occurs. We'll again focus on a single locus with two alleles, A_1 and A_2. Suppose that these

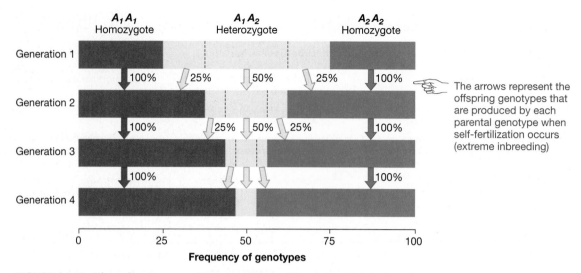

FIGURE 24.13 **Inbreeding Increases Homozygosity and Decreases Heterozygosity**
The width of the boxes corresponds to the frequency of each genotype. Note that homozygosity has almost doubled in just four generations.

alleles initially have equal frequencies of 0.5. In **Figure 24.13**, the width of the boxes represents the frequency of the three genotypes, which start out at the Hardy-Weinberg ratio of $p^2 : 2pq : q^2$. But now let's imagine that these individuals don't produce gametes that go into a gene pool. Instead, they self-fertilize. Many flowering plants, for example, contain both male and female organs and routinely self-pollinate. As the arrows in Figure 24.13 show, homozygous parents produce all homozygous offspring. Heterozygous parents, in contrast, produce homozygous and heterozygous offspring in a 1:2:1 ratio. As a result, the homozygous proportion of the population increases each generation, while the heterozygous proportion is halved. At the end of the four generations illustrated in the figure, heterozygotes are rare. No evolution has occurred, however, because allele frequencies have not changed in the population as a whole.

Self-fertilization, or selfing, is the most extreme form of inbreeding; but the same outcome occurs, more slowly, with less extreme forms of inbreeding. Although inbreeding does not change allele frequencies, it does change genotype frequencies. Inbreeding increases the frequency of homozygotes and decreases the proportion of heterozygotes. In essence, inbreeding takes alleles from heterozygotes and puts them into homozygotes.

This outcome is important because of a phenomenon called inbreeding depression. **Inbreeding depression** is a loss of fitness that takes place when homozygosity increases and heterozygosity decreases. Inbreeding depression results from two processes:

1. Many recessive alleles represent loss-of-function mutations. Because these alleles are usually rare, there are normally very few homozygous recessive individuals in a population. Instead, most loss-of-function alleles exist in heterozygous individuals. The alleles have little or no effect when they occur in heterozygotes, because one normal allele usually produces enough functional protein to support a normal phenotype.

But inbreeding increases the frequency of homozygous recessive individuals. Loss-of-function mutations are usually deleterious or even lethal when they are homozygous.

2. Many genes—especially those involved in fighting disease—are under intense selection for heterozygote advantage. If an individual is homozygous at these genes, then fitness declines.

As a result, the offspring of inbred matings are expected to have lower fitness than the progeny of outcrossed matings. This prediction has been verified in a wide variety of species, often through laboratory or greenhouse studies that compare the fitnesses of offspring from controlled matings (**Figure 24.14**). As **Table 24.5** shows, inbreeding depression is also pronounced in humans.

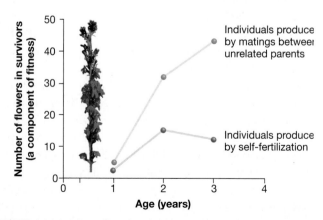

FIGURE 24.14 **Inbreeding Depression Occurs in *Lobelia cardinalis***
Inbreeding depression is the fitness difference between non-inbred and inbred individuals. **EXERCISE** Label the parts on this graph that indicate inbreeding depression. Does inbreeding depression increase with age in this species or remain constant throughout life?

Although inbreeding does not change allele frequencies, it is an indirect cause of evolutionary change. Specifically, it increases the rate at which natural selection eliminates deleterious recessive alleles. When natural selection against homozygous individuals and in favor of heterozygous individuals is strong, allele frequencies in an inbred population change rapidly.

Sexual selection also violates the assumption of random mating that underlies the Hardy-Weinberg principle, but its evolutionary consequences are completely different from inbreeding. Inbreeding affects all genes, while sexual selection affects only genes involved in competition for mates. Inbreeding produces changes in genotype frequencies only, but sexual selection produces changes in allele and genotype frequencies. In terms of its evolutionary consequences, then, sexual selection is a form of natural selection rather than a form of nonrandom mating.

Because inbreeding has such deleterious consequences in humans, it is not surprising that many contemporary human societies have laws that forbid marriages between individuals who are more closely related than first cousins. As the end-of-chapter essay shows, analyzing evolutionary processes affects other aspects of human health and welfare as well.

TABLE 24.5 Inbreeding Reduces Fitness in Humans

The percentages reported here give the mortality rate of children produced by first-cousin marriages versus marriages between nonrelatives. In every study, children of first-cousin marriages have a higher mortality rate.

Age	Period	Deaths Children of First Cousins (%)	Deaths Children of Non-relatives (%)
Children under 20 (U.S.)	18th–19th century	17.0	12.0
Children under 10 (U.S.)	1920–1956	8.1	2.4
At/before birth (France)	1919–1950	9.3	3.9
Children (France)	1919–1950	14.0	10.0
Children under 1 (Japan)	1948–1954	5.8	3.5
Children 1–8 (Japan)	1948–1954	4.6	1.5

Source: C. Stern, *Principles of Human Genetics* (San Francisco: Freeman, 1973).

ESSAY Evolutionary Theory and Human Health

Even though this chapter is sprinkled with examples of how evolutionary theory can address public health issues, it is still legitimate to ask whether evolutionary theory has practical implications. For example, can efforts to understand the four evolutionary mechanisms help researchers, physicians, and conservationists grapple with real-life problems more successfully?

Later chapters will explore how evolutionary theory is assisting efforts to save endangered species and preserve biodiversity. Here let's examine a case in which an understanding of evolutionary processes helped biomedical researchers better understand an important disease. The illness in question is the genetic disease *cystic fibrosis* (CF), which is caused by a defect in a membrane protein that is important in the functioning of lungs and other organs. The essay at the end of Chapter 6 points out that, in human populations of northern European descent, CF is the most common genetic disease. In these populations, one in 2500 infants is born with CF, and alleles that cause CF are at a frequency of 0.02. Although until recently people with CF rarely lived past their teens and thus were unlikely to reproduce, the incidence of the disease is very high.

Models developed in the mid-1990s suggested that the incidence of 1 in 2500 could not be caused by new mutations

Does evolutionary theory have practical implications?

alone or by gene flow from other populations. Then why are "defective" alleles at such a relatively high frequency? After all, they are strongly selected against when homozygous. Based on models of allele frequency change, researchers hypothesized that the mutant alleles must have some selective advantage in northern European populations when found in a heterozygous state.

Shortly after the mathematical analyses of CF's incidence were published, Gerald Pier and coworkers provided dramatic support for the prediction of heterozygote advantage. These biologists showed that the bacterium that causes **typhoid fever**, called *Salmonella typhi*, is not able to enter cells in the gut and cause disease if the cells express mutant forms of the membrane protein involved in CF. Typhoid fever is a potentially fatal illness that is acquired by eating food or drinking water contaminated with feces from an infected person. In the Pier team's experiments, cells in the gut that are homozygous for CF alleles did not allow any *S. typhi* cells to enter, while heterozygous cells admitted only a fraction of

(Continued on next page)

(Essay continued)

the *S. typhi* cells admitted by cells with two copies of the normal allele. The team's experiments suggest that CF alleles are under two distinct types of selection. They are advantageous in the heterozygous state because they protect people against typhoid fever. But they are disadvantageous in the homozygous state, because they cause cystic fibrosis. The observed frequency of 2 percent represents a balance between opposing forms of selection.

The story of CF and typhoid fever is just one of many examples for which an understanding of evolutionary processes has led to a deeper understanding of important diseases. Other chapters in this text highlight evolutionary aspects of the battles against malaria, HIV and AIDS, and antibiotic resistance. What people are calling "Darwinian medicine" is a new but growing topic in biological science.

CHAPTER REVIEW

Summary of Key Concepts

■ **Each of the four evolutionary mechanisms has different consequences. Only natural selection produces adaptation. Genetic drift causes random fluctuations in allele frequencies. Gene flow equalizes allele frequencies between populations. Mutation introduces new alleles.**

The four mechanisms of evolution cause allele frequencies to change in populations, but only natural selection results in adaptation. Genetic drift, gene flow, and mutation may introduce a nonadaptive component into evolutionary change.

■ **The Hardy-Weinberg principle acts as a null hypothesis when researchers want to test whether evolution or nonrandom mating is occurring at a particular gene.**

Biologists study the consequences of the different evolutionary mechanisms through a combination of mathematical modeling and experimental or observational research. The Hardy-Weinberg principle serves as a null hypothesis in the study of evolution, because it specifies what genotype and allele frequencies are expected to be if mating is random with respect to the gene in question and none of the four evolutionary processes is operating on that gene.

Natural selection occurs in both a wide variety of patterns and a wide variety of intensities. Directional selection may lead to certain alleles becoming fixed—and thus reduces allelic diversity in populations. Stabilizing selection eliminates phenotypes with extreme characteristics and decreases allelic diversity in populations. The rate of evolution under natural selection depends on both the intensity of selection and the amount of genetic variation available.

Sexual selection is a type of selection that occurs when certain traits help males succeed in contests over mates or when certain traits are attractive to females. It is responsible for the evolution of phenotypic differences between males and females.

Genetic drift results from random sampling and is an important evolutionary force in small populations. Drift leads to the random fixation of alleles and tends to reduce overall allelic diversity.

Gene flow is the movement of alleles between populations. Gene flow tends to homogenize allele frequencies and decrease differentiation among populations, but it can also serve as an important source of new variation in populations.

Mutation is too infrequent to be a major cause of allele frequency change. But because mutation continually introduces new alleles at all genes, it is essential to evolution. Without mutation, natural selection and genetic drift would eventually eliminate genetic variation, and evolution would cease.

Web Tutorial 24.1   The Hardy-Weinberg Principle

Web Tutorial 24.2   Three Modes of Natural Selection

■ **Inbreeding changes genotype frequencies but does not change allele frequencies.**

Inbreeding, or mating among relatives, is a form of nonrandom mating. Inbreeding does not change allele frequencies, so it is not an evolutionary mechanism. It does, however, change genotype frequencies by leading to an increase in homozygosity and a decrease in heterozygosity. These patterns can accelerate natural selection and can cause inbreeding depression.

Questions

Content Review

1. Why isn't inbreeding considered an evolutionary mechanism?
   a. It does not change genotype frequencies.
   b. It does not change allele frequencies.
   c. It does not occur often enough to be important in evolution.
   d. It does not violate the assumptions of the Hardy-Weinberg principle.

2. Why is random genetic drift aptly named?
   a. It causes allele frequencies to drift up or down randomly.
   b. It is the ultimate source of genetic variability.
   c. It is an especially important mechanism in small populations.
   d. It occurs when populations drift into new habitats.

3. How do sexual selection and inbreeding differ?
   a. Unlike inbreeding, sexual selection changes allele frequencies and affects only genes involved in attracting mates.
   b. Unlike sexual selection, inbreeding changes allele frequencies and involves any mating between relatives—not just self-fertilization.
   c. Unlike inbreeding, sexual selection results from the random fusion of gametes during fertilization. It is particularly important in small populations, where few mates are available.
   d. Inbreeding occurs only in small populations, while sexual selection can occur in any size population.

4. What does it mean when an allele reaches "fixation"?
   a. It is eliminated from the population.
   b. It has a frequency of 1.0.
   c. It is dominant to all other alleles.
   d. It is adaptively advantageous.

5. In what sense is the Hardy-Weinberg principle a null hypothesis, similar to the control treatment in an experiment?
   a. It defines what genotype frequencies should be if nonrandom mating is occurring.
   b. Expected genotype frequencies can be calculated from observed allele frequencies and then compared with the observed genotype frequencies.
   c. It defines what genotype frequencies should be if natural selection, genetic drift, gene flow, or mutation is occurring *and* if mating is random.
   d. It defines what genotype frequencies should be if evolutionary mechanisms are *not* occurring.

6. Mutation is the ultimate source of genetic variability. Why is this statement correct?
   a. DNA polymerase (the enzyme that copies DNA) is remarkably accurate.
   b. "Mutation proposes and selection disposes."
   c. Mutation is the only source of new alleles.
   d. Mutation occurs in response to natural selection. It generates the alleles that are required for a population to adapt to a particular habitat.

Conceptual Review

1. Create a table with three columns and six rows, modeled after Table 24.4. Title the columns "Event," "Effect on Populations," and "Example." In the rows of the "Event" column, write "Mutation," "Gene Flow," "Genetic Drift," "Inbreeding," "Selection," and "None (Hardy-Weinberg Conditions)." Fill in the rest of the table.

2. Directional selection can lead to the fixation of favored alleles. When this occurs, genetic variation is zero and evolution stops. Explain why this rarely occurs.

3. Why does sexual selection often lead to sexual dimorphism? Why are males usually the sex that exhibits exaggerated characteristics?

4. Genotype frequencies occur in Hardy-Weinberg proportions when mutation, selection, drift, and gene flow are not operating on a particular gene, and when mating is random with respect to that gene. Is it possible for one gene in a population *not* to have genotype frequencies that are in Hardy-Weinberg proportions, even though all other genes in the population are in Hardy-Weinberg proportions? Explain why or why not.

5. Explain why small populations become inbred.

6. Explain why genetic drift is much more important in small populations than large populations.

Group Discussion Problems

1. In humans, albinism is caused by loss-of-function mutations in genes involved in the synthesis of melanin, the dark pigment in skin. Only people homozygous for a loss-of-function allele have the relevant phenotype. In Americans of northern European ancestry, albino individuals are present at a frequency of about 1 in 10,000 (or 0.0001). Knowing this genotype frequency, we can calculate the frequency of the loss-of-function alleles. If we let p_2 stand for this frequency, we know that $p_2^2 = 0.0001$; therefore $p_2 = \sqrt{0.0001} = 0.01$. By subtraction, the frequency of normal alleles is 0.99. If these loci conform to the conditions required by the Hardy-Weinberg principle, what is the frequency of "carriers"—or people who are heterozygous for this condition? Your answer indicates the percentage of Caucasians in the United States who carry an allele for albinism.

2. Conservation managers frequently use gene flow, in the form of transporting individuals or releasing captive-bred young, to counteract the effects of drift on small, endangered populations. Explain how gene flow can also mitigate the effects of inbreeding.

3. In 1789 a small band of mutineers led by Fletcher Christian took over the British warship HMS *Bounty*. They fled first to Tahiti and then to tiny Pitcairn Island in the South Pacific, along with a small population of native Tahitians. Some of the sailors married Polynesians there and raised families. All contemporary Pitcairn islanders can trace at least part of their family tree back to the colonization event. What evolutionary forces were at work in this small, isolated population during and after the arrival of the mutineers and Polynesians?

4. You are a conservation biologist charged with creating a recovery plan for an endangered species of turtle. The turtle's habitat has been fragmented into small, isolated but protected areas by suburbanization and highway construction. Some evidence indicates that certain turtle populations are adapted to marshes that are normal, whereas others are adapted to acidic wetlands or salty habitats. Further, some turtle populations number less than 25 breeding adults, making genetic drift and inbreeding a major concern. In creating a recovery plan, the tools at your disposal are captive breeding, the capture and transfer of adults to create gene flow, or the creation of habitat corridors between wetlands to make migration possible. Write a two-paragraph essay outlining the major features of your proposal.

Answers to Multiple-Choice Questions   **1.** b; **2.** a; **3.** a; **4.** b; **5.** d; **6.** c

25 Speciation

KEY CONCEPTS

- Speciation occurs when populations of the same species become genetically isolated by lack of gene flow and then diverge from each other due to selection, genetic drift, or mutation.

- Populations can become genetically isolated from each other if they occupy different geographic areas, if they use different habitats within the same area, or if one population is polyploid and cannot breed with the other.

- Populations can be recognized as distinct species if they are reproductively isolated from each other, if they have distinct morphological characteristics, or if they form independent branches on a phylogenetic tree.

- When populations that have diverged come back into contact, several outcomes are possible.

The Bullock's oriole was once considered a member of the same species as Baltimore orioles, because the two populations interbreed in certain regions. Recent analyses, based on the phylogenetic species concept introduced in this chapter, have shown that Bullock's orioles are an independent species.

Although Darwin called his masterwork *On the Origin of Species by Means of Natural Selection*, he actually had little to say about how new species arise. Instead, his data and analyses focused on the process of natural selection and the changes that occur within populations over time. He spent much less time considering changes that occur *between* populations.

Since Darwin, however, biologists have realized that populations of the same species may diverge from each other when they are isolated in terms of gene flow. Recall from Chapter 24 that gene flow equalizes allele frequencies between populations. When gene flow ends, allele frequencies in the isolated populations are free to diverge—meaning that the populations begin to evolve independently of each other. If mutation, selection,

and genetic drift cause isolated populations to diverge sufficiently, distinct types, or species, form—that is, the process of **speciation** takes place. Speciation is usually a splitting event that creates two or more distinct species from a single ancestral group (**Figure 25.1**). When speciation is complete, a new branch has been added to the tree of life.

In essence, then, speciation results from genetic isolation and genetic divergence. Isolation results from lack of gene flow, and divergence occurs because selection, genetic drift, and mutation proceed independently in the isolated populations. How does genetic isolation come about? And how do selection, drift, and mutation cause divergence?

This chapter is devoted to exploring these questions. Our first task is to examine how species are defined and identified.

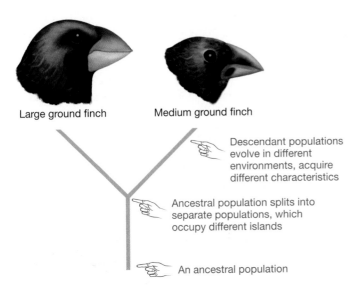

Large ground finch   Medium ground finch

Descendant populations evolve in different environments, acquire different characteristics

Ancestral population splits into separate populations, which occupy different islands

An ancestral population

FIGURE 25.1 Speciation Creates Distinct Populations
The large ground finch and medium ground finch are derived from the same ancestral population. This ancestral population split into two populations that occupied different islands in the Galápagos and were isolated by lack of gene flow. Because the populations began evolving independently, they acquired the distinctive characteristics observed today.

Subsequent sections focus on how speciation occurs in two contrasting situations: when populations occupy the same geographic area and when they are separated into distinct regions. The chapter concludes with a look at a classical question in speciation research: What happens when populations that have been isolated from one another come back into contact? Do they interbreed and merge back into the same species, or do they remain independent and form new species?

25.1 Defining and Identifying Species

Species are distinct types of organisms and represent evolutionarily independent groups. Like the Galápagos finches in Figure 25.1, species are distinct from one another in appearance, behavior, habitat use, or other traits. These characteristics differ among species because their genetic characteristics differ. Genetic distinctions occur because mutation, selection, and drift act on each species independently of what is happening in other populations.

What makes one species "evolutionarily independent" of other species? The answer begins with *lack* of gene flow. As explained in Chapter 24, gene flow eliminates genetic differences among populations. Allele frequencies in populations, and thus the populations' characteristics, become more alike when gene flow occurs between them. If gene flow between populations is extensive and continues over time, it eventually causes even highly distinct populations to coalesce into the unit known as a species. Conversely, if gene flow between populations stops, then mutation, selection, and drift begin to act on the populations independently. If a new mutation creates an allele that changes the phenotype of individuals in one population, there

is no longer any way for that allele to appear in the other population. As a result, allele frequencies and other characteristics in the populations diverge. When allele frequencies change sufficiently over time, populations become distinct species.

Formally, then, a **species** is defined as an evolutionarily independent population or group of populations. Even though this definition sounds straightforward, it can be exceedingly difficult to put into practice. How can evolutionarily independent populations be identified in the field and in the fossil record? There is no single, universal answer. Even though biologists agree on the definition of a species, they frequently have to use different sets of criteria to identify them. Three criteria for identifying species are in common use: (1) the biological species concept, (2) the morphospecies concept, and (3) the phylogenetic species concept.

The Biological Species Concept

According to the **biological species concept**, the critical criterion for identifying species is reproductive isolation. This is a logical yardstick because no gene flow occurs between populations that are reproductively isolated from each other. Specifically, if two different populations do not interbreed in nature, or if they fail to produce viable and fertile offspring when matings take place, then they are considered distinct species. Groups that naturally or potentially interbreed, and that are reproductively isolated from other groups, belong to the same species. Biologists can be confident that reproductively isolated populations are evolutionarily independent.

Reproductive isolation can result from a wide variety of events and processes. To organize the various mechanisms that stop gene flow between populations, biologists distinguish (1) **prezygotic** ("before-zygote") **isolation**, which prevents individuals of different species from mating, and (2) **postzygotic** ("after-zygote") **isolation**, in which the offspring of matings between members of different species do not survive or reproduce. In prezygotic isolation, reproductive isolation occurs before mating can occur. In postzygotic isolation, interspecies mating does occur, but any offspring produced have low fitness. **Table 25.1** (page 540) summarizes some of the more important mechanisms of prezygotic and postzygotic isolation.

Although the biological species concept has a strong theoretical foundation, it has disadvantages. The criterion of reproductive isolation cannot be evaluated in fossils or in species that reproduce asexually. In addition, it is difficult to apply when closely related populations do not happen to overlap with each other geographically. In this case, biologists are left to guess whether interbreeding and gene flow would occur if the populations happened to come into contact.

The Morphospecies Concept

How do biologists identify species when the criterion of reproductive isolation cannot be applied? Under the **morphospecies** ("form-species") **concept**, researchers identify evolutionarily

TABLE 25.1 Mechanisms of Reproductive Isolation

Type	Description	Example
Prezygotic Isolation		
Temporal	Populations are isolated because they breed at different times	Bishop pines and Monterey pines release their pollen at different times of the year.
Habitat	Populations are isolated because they breed in different habitats.	Parasites that begin to exploit new host species are isolated from their original population.
Behavioral	Populations do not interbreed because their courtship displays differ.	To attract male fireflies, female fireflies give a species-specific sequence of flashes.
Gametic barrier	Matings fail because eggs and sperm are incompatible.	In sea urchins, a protein called bindin allows sperm to penetrate eggs. Differences in the amino acid sequence of bindin cause matings to fail between closely related populations.
Mechanical	Matings fail because male and female genitalia are incompatible.	In many insects, the male copulatory organ and female reproductive canal fit like a "lock and key." Changes in in either organ initiate reproductive isolation.
Postzygotic Isolation		
Hybrid viability	Hybrid offspring do not develop normally and die as embryos.	When ring-necked doves mate with rock doves, less than 6% of eggs hatch.
Hybrid sterility	Hybrid offspring mature but are sterile as adults.	Eastern meadowlarks and western meadowlarks are almost identical morphologically, but hybrid offspring are largely infertile.

independent lineages by differences in size, shape, or other morphological features. The logic behind the morphospecies concept is that distinguishing features are most likely to arise if populations are independent and isolated from gene flow.

The morphospecies concept is compelling simply because it is so widely applicable. It is a useful criterion when biologists have no data on the extent of gene flow, and it is equally applicable to sexual, asexual, or fossil species. Its disadvantage is that the features used to distinguish species are subjective. In the worst case, different researchers working on the same populations disagree on the characters that distinguish species. For example, some researchers who work on the fossil record of humans argue that the specimens currently named *Homo habilis* and *Homo rudolfensis* (**Figure 25.2**) actually belong to the same species. Disagreements like these often end in a stalemate, because no independent criteria exist for resolving the conflict.

The Phylogenetic Species Concept

The **phylogenetic species concept** is a recent addition to the tools available for identifying evolutionarily independent lineages. It is based on reconstructing the evolutionary history of populations and is increasingly popular among biologists. Proponents of this approach argue that it is widely applicable and precise.

To understand the reasoning behind the concept, recall Darwin's claim, explored in Chapter 23, that all species are related by common ancestry. That chapter also introduced the phylogenetic trees used to represent the genealogical relationships

(a) *Homo habilis*

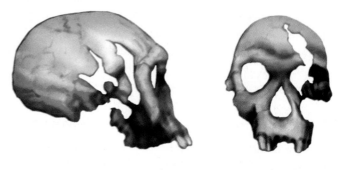

(b) *Homo rudolfensis*

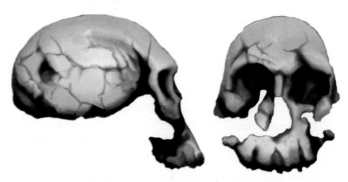

FIGURE 25.2 Morphospecies May Be Difficult to Distinguish from Each Other

Fossils from **(a)** *Homo habilis* and **(b)** *Homo rudolfensis* have been recovered from the same region in Africa, in rocks of the same age. Biologists argue over whether the two populations were distinct enough morphologically to be considered separate species or whether they should be considered the same species.

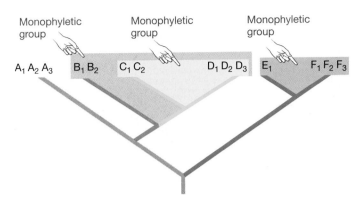

FIGURE 25.3 Monophyletic Groups
The color-coded lineages on this phylogenetic tree are monophyletic
because they contain a common ancestor and all its descendants.
EXERCISE *This tree has several monophyletic groups other than
those indicated. Identify one by picking any branch point (node) on
the tree. Then draw a circle around it and all of the branches
descending from that point.*

among populations. **Figure 25.3** shows such a tree. Here the
tips of the branches represent phylogenetic species (A, B, C, D,
E, and F), and the clusters at some of the tips (A_1, A_2, A_3, etc.)
are populations within those species. On any given evolutionary tree, there are many monophyletic ("one-tribe") groups. A
monophyletic group consists of an ancestral population, all of
its descendants, and *only* those descendants. Three monophyletic groups are color coded on the tree in Figure 25.3, including one (containing species B, C, and D) that overlaps
another monophyletic group (the one containing just species C
and D). Under the phylogenetic species concept, a species is defined as the smallest monophyletic group in a phylogenetic tree.
The populations within each species may be separated geographically, but they are so similar that they do not form independent twigs on the tree.

The phylogenetic species concept has two distinct advantages:
(1) It can be applied to any population (fossil, asexual, or sexual), and (2) it is logical because populations are monophyletic
only if they are independent of one another and isolated from
gene flow. The approach has a distinct disadvantage, however:
Carefully estimated phylogenies are available only for a tiny

(though growing) subset of populations on the tree of life. Critics of this approach also point out that it would probably lead to
recognition of many more species than either the morphospecies
or biological species concept. Proponents counter that, far from
being a disadvantage, the recognition of increased numbers of
species might better reflect the extent of life's diversity.

In actual practice, researchers use all three species concepts
summarized here (**Table 25.2**). Conflicts have occurred, however, when different species concepts are applied to the real
world. To appreciate this point, let's consider the case of the
dusky seaside sparrow.

Species Definitions in Action: The Case of the Dusky Seaside Sparrow

Seaside sparrows live in salt marshes along the Atlantic and Gulf
coasts of the United States. Recall from Chapter 1 that scientific
names consist of a genus name followed by a species name; the
scientific name of this species is *Ammodramus maritimus*. Researchers had traditionally named a variety of seaside sparrow
"subspecies" under the morphospecies concept. **Subspecies** are
populations that have distinguishing features, such as coloration
or calls, but are not considered distinct enough to be called
separate species.

Because salt marshes are often destroyed for agriculture or
oceanfront housing, by the late 1960s biologists began to be concerned about the future of some seaside sparrow populations. A
subspecies called the dusky seaside sparrow (*Ammodramus
maritimus nigrescens*) was in particular trouble; by 1980 only six
individuals from this population remained. All were males.

At this point government and private conservation agencies
sprang into action under the auspices of the Endangered Species
Act, a law whose goal is to prevent the extinction of species. The
law uses the biological species concept to identify species and
calls for the rescue of endangered species through active management. Because current populations of seaside sparrows are
physically isolated from one another, and because young seaside
sparrows tend to breed near where they hatched, researchers believed that little to no gene flow occurred among populations.
Under the biological species concept and morphospecies

TABLE 25.2 A Summary of Species Concepts

Species Concept	Criterion for Recognizing Species	Advantages	Disadvantages
Biological	Reproductive isolation between populations (they don't breed and don't produce viable offspring)	Reproductive isolation = evolutionary independence	Not applicable to asexual or fossil species; difficult to assess if populations do not overlap geographically
Morphospecies	Populations are morphologically distinct	Widely applicable	Subjective (researchers often disagree about how much or what kinds of morphological distinction = speciation)
Phylogenetic	Smallest monophyletic group on phylogenetic tree	Widely applicable; based on testable criteria	Relatively few well-estimated phylogenies are currently available

(a) Each subspecies of seaside sparrow has a restricted range.

(b) The six subspecies form two monophyletic groups when DNA sequences are compared.

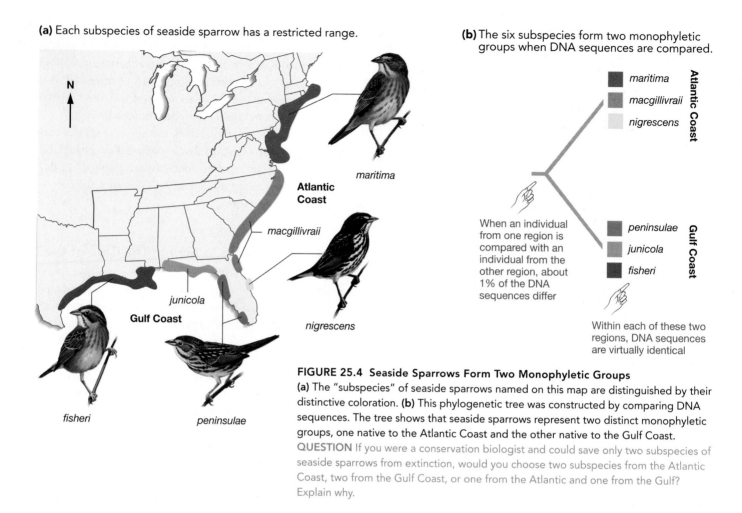

When an individual from one region is compared with an individual from the other region, about 1% of the DNA sequences differ

Within each of these two regions, DNA sequences are virtually identical

FIGURE 25.4 Seaside Sparrows Form Two Monophyletic Groups
(a) The "subspecies" of seaside sparrows named on this map are distinguished by their distinctive coloration. **(b)** This phylogenetic tree was constructed by comparing DNA sequences. The tree shows that seaside sparrows represent two distinct monophyletic groups, one native to the Atlantic Coast and the other native to the Gulf Coast.
QUESTION If you were a conservation biologist and could save only two subspecies of seaside sparrows from extinction, would you choose two subspecies from the Atlantic Coast, two from the Gulf Coast, or one from the Atlantic and one from the Gulf? Explain why.

concept, there may be as many as six species of seaside sparrow (**Figure 25.4a**). The dusky seaside sparrow subspecies became a priority for conservation efforts because it was reproductively isolated.

To launch the rescue program, the remaining male dusky seaside sparrows were taken into captivity and bred with females from a nearby subspecies: *A. maritimus peninsulae*. Officials planned to use these hybrid offspring as breeding stock for a reintroduction program. The goal was to preserve as much genetic diversity as possible by reestablishing a healthy population of dusky-like birds. The plan was thrown into turmoil, however, when a different group of biologists estimated the phylogeny of the seaside sparrows by comparing gene sequences. This tree, shown in **Figure 25.4b**, shows that seaside sparrows represent just two distinct monophyletic groups: one native to the Atlantic Coast and the other native to the Gulf Coast. Under the phylogenetic species concept, only two species of seaside sparrow exist. Far from being an important, reproductively isolated population, the phylogeny showed that the dusky sparrow is genetically indistinguishable from other Atlantic Coast sparrows. Further, officials had unwittingly crossed the dusky males with females from the Gulf Coast lineage. Because the goal of the conservation effort was to preserve existing genetic diversity, this was the wrong population to use.

The researchers who did the phylogenetic analysis maintained that the biological and morphospecies concepts had misled a well-intentioned conservation program. Under the phylogenetic species concept, officials should have allowed the dusky sparrow to go extinct and then concentrated their efforts on simply preserving one or more populations from each coast. In this way, the two monophyletic groups of sparrows—and the most genetic diversity—would be preserved. Under the morphospecies concept, however, officials did the right thing by preserving distinct types.

When conservation funding is scarce, life-and-death decisions like these are crucial. Now our task is to consider an even more fundamental question: How do isolation and divergence produce the event called speciation?

✓ CHECK YOUR UNDERSTANDING

Species are evolutionarily independent because no gene flow occurs between them and other species. You should be able to describe the advantages and disadvantages of the biological, morphological, and phylogenetic species concepts and explain how each of these approaches to identifying species helps biologists identify populations as evolutionarily independent.

25.2 Isolation and Divergence in Allopatry

Speciation begins when gene flow between populations is reduced or eliminated. Genetic isolation happens routinely when populations become physically separated. Physical isolation, in turn, occurs in one of two ways: dispersal or vicariance. As **Figure 25.5a** illustrates, a population can disperse to a new habitat, colonize it, and found a new population. Alternatively, a new physical barrier can split a widespread population into two or more subgroups that are physically isolated from each other (**Figure 25.5b**). A physical splitting of habitat is called **vicariance**. Speciation that begins with physical isolation via either dispersal or vicariance is known as **allopatric** ("different-homeland") **speciation**. Populations that live in different areas are said to be in **allopatry**.

The case studies that follow address two questions: How do colonization and range-splitting events occur? Answering this question takes us into the field of **biogeography**—the study of how species and populations are distributed geographically. Once populations are physically isolated, how do genetic drift and selection produce divergence?

Dispersal and Colonization Isolate Populations

Peter Grant and Rosemary Grant witnessed a colonization event while working in the Galápagos Islands off the coast of South America. Recall from Chapter 23 that they had been studying medium ground finches on the island of Daphne Major since 1971. In 1982 five members of another species,

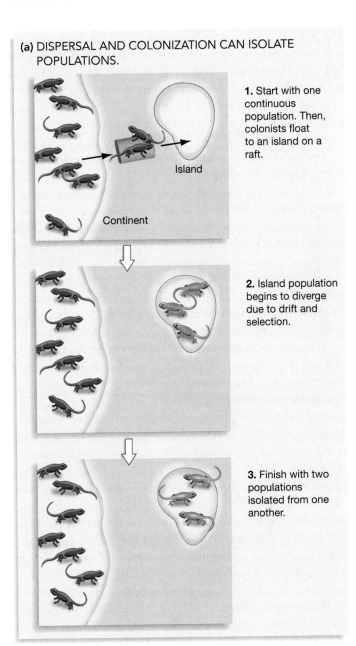

(a) DISPERSAL AND COLONIZATION CAN ISOLATE POPULATIONS.

1. Start with one continuous population. Then, colonists float to an island on a raft.

Island

Continent

2. Island population begins to diverge due to drift and selection.

3. Finish with two populations isolated from one another.

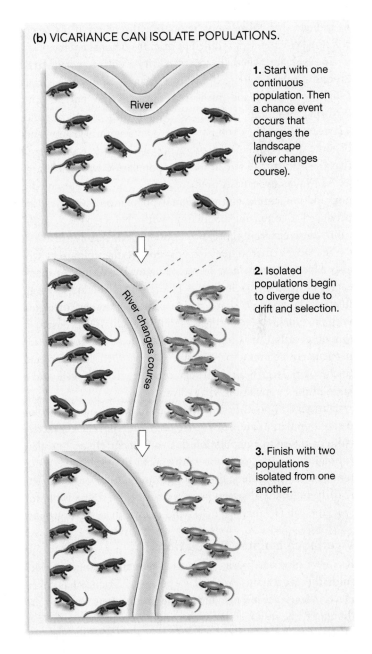

(b) VICARIANCE CAN ISOLATE POPULATIONS.

River

1. Start with one continuous population. Then a chance event occurs that changes the landscape (river changes course).

River changes course

2. Isolated populations begin to diverge due to drift and selection.

3. Finish with two populations isolated from one another.

FIGURE 25.5 Allopatric Speciation Begins via Dispersal or Vicariance
(a) When dispersal occurs, colonists establish a new population in a novel location. **(b)** In vicariance, a widespread population becomes fragmented into isolated subgroups.

called the large ground finch, arrived and began nesting. These colonists had apparently dispersed from a population that lived on a nearby island in the Galápagos. Because finches normally stay on the same island year-round, the colonists represented a new population, allopatric with their source population.

The large ground finches' arrival gave the researchers a chance to test a long-standing hypothesis about dispersal and colonization. Decades ago Ernst Mayr suggested that colonization events are likely to trigger speciation, for two reasons: (1) The physical separation between populations reduces or eliminates gene flow, and (2) genetic drift will cause the old and new populations to diverge rapidly. Drift occurs during the colonization event itself via the founder effect, introduced in Chapter 24. And if the number of individuals in the new population remains small for several generations, genetic drift will continue to alter allele frequencies. In addition, natural selection may cause divergence if the newly colonized environment is different from the original habitat.

To evaluate whether genetic drift occurred when large ground finches colonized Daphne Major, Grant and Grant caught, weighed, and measured most of the parents and offspring produced on Daphne Major over the succeeding 12 years. When they compared these data with measurements of large ground finches in other populations, they discovered that the average bill size in the new population was much larger. As Mayr's hypothesis predicted, the colonists represented a nonrandom sample of the original population. Genetic drift produced a colonizing population with characteristics significantly different from those of the source population.

The novel environment experienced by the colonizers will also expose them to new forms of natural selection. In many cases, the agent of selection is changes in available food. In finches, the size and shape of the beak are closely correlated with the types of seeds or insects that individuals eat. Based on this observation, it is logical to predict that if large seeds are particularly common on Daphne Major, then large ground finches with large beaks will survive and reproduce well, and a large-beaked population will evolve.

The general point here is that the characteristics of a colonizing population are likely to be different from the characteristics of the source population due to founder effects and that subsequent natural selection may extend the rapid divergence that begins with genetic drift. Colonization, followed by genetic drift and natural selection, is thought to be responsible for speciation in Galápagos finches and many other island groups.

Vicariance Isolates Populations

If a new physical barrier such as a mountain range or river splits the geographic range of a species, vicariance has taken place. Vicariance events during the most recent ice age are thought to be responsible for the origin of many of the species observed today. Over the past several million years, glaciers that covered large regions of the northern continents advanced and retreated repeatedly. During most of the glacial advances,

the growing ice fields fragmented existing forest and grassland habitats into smaller regions that were isolated from each other by expanses of ice. If populations of the same species had occupied these isolated regions, then an inability of individuals to migrate over the ice fields would have left the populations genetically isolated. The advancing ice would have split turtle, flowering plant, insect, and fish species into geographically separated populations that then might have undergone speciation.

Another example of speciation by vicariance involves the group of large, flightless birds called the ratites. You probably are most familiar with the ostrich; kiwis, emus, rheas, and cassowaries are also ratites. Today, ratites are found in South America, Africa, Australia, and New Zealand. Unfortunately, habitat destruction by humans and hunting recently extinguished ratites called elephant birds that lived on the island of Madagascar, off the southeast coast of Africa, as well as 11 species of moas that were native to New Zealand. The elephant bird is the largest bird species ever recorded, with a maximum height of 3.5 m (11 ft), a maximum mass of 454 kg (1000 lbs), and eggs with a volume of up to 7 L (2 gallons). Moas may have been the tallest birds that ever lived, with some species possibly reaching 4 m (13 ft) when they held their necks erect.

The earliest ratites in the fossil record lived about 150 million years ago, on a landmass called Gondwana. As **Figure 25.6a** shows, Gondwana was made up of a number of physically distinct landmasses. The **theory of plate tectonics** holds that Earth's entire crust—the layer at its surface—is made up of distinct blocks of rock called *plates*. Landmasses, including Gondwana, make up continental plates; crustal blocks that are overlain by the oceans are oceanic plates. All of these plates are in constant motion—probably driven by heat that rises to the crust from Earth's interior. The term **continental drift** is used to describe the motion of continental plates through time, and this process is still going on.

Although the "supercontinent" Gondwana existed for tens of millions of years, the continents in it began to drift apart about 140 million years ago (**Figure 25.6b**). Geologic data indicate that, first, a landmass consisting of today's South America and Africa split off from a slightly smaller landmass composed of Antarctica, Madagascar, India, and Australia. Later, each of these large landmasses split up to form today's configuration of islands and continents. Each splitting event isolated populations of ratites. As the continents moved, climates and the composition of plant and animal species changed. The initial splitting event could have triggered changes in ratite populations via genetic drift. These differences would have been extended by natural selection as environments changed, leading to the evolution of the species observed today and their current distributions (**Figure 25.6c**). In this way, a series of vicariance events triggered a series of speciation events.

To summarize, physical isolation of populations via dispersal or vicariance produces genetic isolation—the first requirement of speciation. When genetic isolation is accompanied by genetic divergence due to mutation, selection, and genetic drift, speciation results.

(a) Gondwana was the original home of ratites.

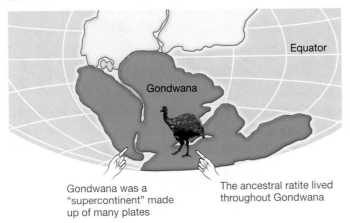

Gondwana was a "supercontinent" made up of many plates

The ancestral ratite lived throughout Gondwana

(b) Gondwana began to break up into separate continents.

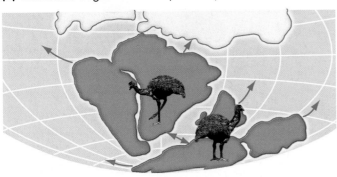

(c) Ratites speciated as the continents moved apart.

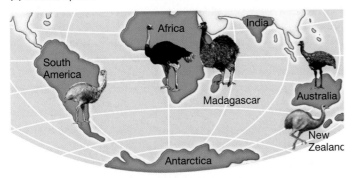

FIGURE 25.6 Continental Drift Caused Vicariance in Ratite Birds
(a) The ancestors of today's ratites lived on the supercontinent of Gondwana. **(b)** Continental drift led to the breakup of Gondwana. This was a vicariance event that isolated ratite populations. **(c)** Continued continental drift brought the continents and islands into their present positions. Ratites in each area diverged in response to mutation, selection, and drift.

25.3 Isolation and Divergence in Sympatry

When populations or species live in the same geographic area, or at least close enough to one another to make interbreeding possible, biologists say that they live in **sympatry** ("together-homeland"). Traditionally, researchers have predicted that speciation could *not* occur among sympatric populations, because

gene flow is possible. The prediction was that gene flow would easily overwhelm any differences among populations created by genetic drift and natural selection.

To illustrate this point, consider research on the water snakes native to the Lake Erie region of North America. Unbanded snakes predominate on islands in the lake (**Figure 25.7a**), but on mainland habitats near the lake, the vast majority of water snakes have a banded coloration (**Figure 25.7b**). Why? In the late 1950s biologists noticed that island snakes bask on limestone rocks. They hypothesized that unbanded snakes are more difficult for predators to distinguish against this surface. A

(a) Unbanded water snake

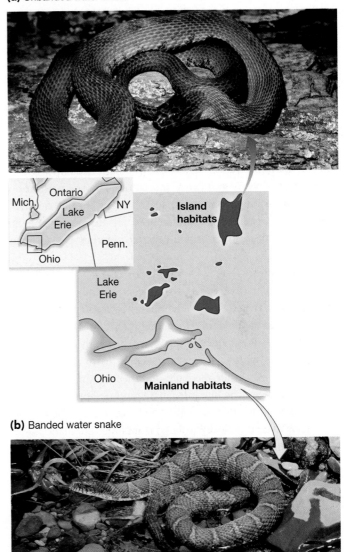

(b) Banded water snake

FIGURE 25.7 Gene Flow Prevents Speciation in Island and Mainland Populations of Water Snakes
Water snakes occupy most of the larger islands in Lake Erie, along with mainland habitats around the lake. **(a)** On island habitats, unbanded individuals are common. **(b)** All or most of the water snakes found in mainland habitats are strongly banded. **QUESTION** Banded snakes are most common on the island closest to shore. Why?

researcher recently provided data supporting this hypothesis by showing that, on islands, young unbanded snakes survive much better than do young banded snakes. In mainland habitats, however, limestone rocks are rare. On the shrubby, muddy shores of the mainland, banded snakes may be better camouflaged than unbanded snakes and thus harder for predators to detect.

If natural selection favors different coloration patterns in the two populations, why haven't they diverged to become separate species? The answer is gene flow. Individuals from the mainland swim out and join the island populations on a regular basis. The alleles they introduce when they breed keep the banded coloration pattern at a reasonably high frequency on islands and maintain the two populations as a single species. In this way, gene flow overwhelms the diversifying force of natural selection and prevents speciation. Is this always the case?

Can Natural Selection Cause Speciation Even When Gene Flow Is Possible?

Recently, several well-documented examples have upset the traditional view that **sympatric speciation**—speciation that occurs without physical isolation—is rare or nonexistent. These studies are fueling a growing awareness that, under certain circumstances, natural selection that causes populations to diverge can overcome gene flow and cause speciation. The key realization is that even though populations are not physically isolated, they may be isolated by preferences for different habitats. As an example, let's consider research on the speciation of soapberry bugs.

The soapberry bug is a species of insect, illustrated in **Figure 25.8a**, native to the south-central and southeastern United States. The bugs make their living by feeding on plants in a family called Sapindaceae, including the soapberry tree, serjania vine, and balloon vine. As the figure shows, the bugs feed by piercing fruits with their beaks, reaching in to penetrate the coats of the seeds inside the fruit, and then sucking up the contents of the seeds. The bugs also mate on their host plants. As a result, populations that specialize on a particular type of host plant tend to be genetically isolated from other populations.

The soapberry bug's story began to get interesting when horticulturists brought three new species of sapindaceous plants to North America from Asia in the twentieth century. Two of these *exotic*, or nonnative, species are now cultivated as ornamentals, and one grows as a weed. Soon after these plants were introduced to the New World, soapberry bugs began using them as food. As Figure 25.8a shows, the fruits of the nonnative species are much different from the fruits of native species. Did the arrival of new host plants lead to genetic isolation? If so, have soapberry bugs begun to diverge?

In soapberry bug populations that feed on native host plants, beak length corresponds closely to the size of the host fruit. For example, bugs that feed on species with big fruit tend

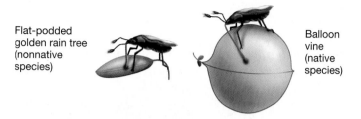

(a) Beak length correlates with fruit size.

Flat-podded golden rain tree (nonnative species)

Balloon vine (native species)

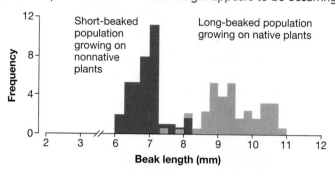

(b) Disruptive selection on beak length appears to be occurring.

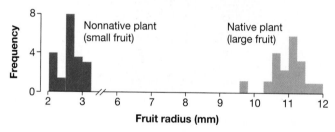

Short-beaked population growing on nonnative plants

Long-beaked population growing on native plants

Nonnative plant (small fruit)

Native plant (large fruit)

FIGURE 25.8 Beak Length in Soapberry Bugs Correlates with Host Fruit Size

(a) A soapberry bug feeding on a nonnative plant called the flat-podded golden rain tree (left) and one feeding on a native plant called the balloon vine (right). **(b)** Soapberry bug populations that feed on these two host plants have very different beak lengths, which correspond to differences in the size of the host fruit.

to have long beaks. The correlation between fruit size and beak length is logical, because it should allow individuals to reach the seeds inside the fruit efficiently. It also prompted a biologist to ask a simple question: In populations of soapberry bugs that exploit the introduced plant species, have beak lengths evolved to match the sizes of the new fruits? If so, it would imply that populations have become genetically isolated on different host plants and that natural selection is currently causing soapberry bug populations to diverge.

To answer this question, researchers measured large samples of bugs found on both native and nonnative hosts. Some of the resulting data are shown in **Figure 25.8b**. The top histogram shows that bugs collected on native plants growing in south Florida have much longer beaks than those collected on nonnative plants growing in central Florida. The bottom histogram confirms that the fruits of the native species are much larger than the fruits of the introduced species. The data support the argument that soapberry bug populations exploiting exotic species have indeed changed—presumably in response to natural selection for efficient use of host fruits. More specifically, the data support the hypothesis that disruptive selection has occurred on beak length ever since some soapberry bug populations switched to new host plant species.

Will the populations that exploit native and exotic host plants continue to diverge and eventually form new species? Only time, and further research, will tell. But the researchers who are following the story expect that the answer is yes. Because soapberry bugs mate on or near their host plants, switching to a new host species should have reduced gene flow among populations at the same time that it set up disruptive selection. As a result, natural selection may be able to overwhelm gene flow and cause speciation, even when populations are sympatric.

Although the soapberry bug's story might seem localized and specific, the events may be common. Biologists currently estimate that over 3 million insect species exist. Most of these species are associated with specific host plants. Based on these observations, it is reasonable to hypothesize that switching host plants, as soapberry bugs have done, has been a major trigger for speciation throughout the course of insect evolution.

How Can Polyploidy Lead to Speciation?

Based on the theory and data reviewed thus far, it is clear that gene flow and natural selection play important roles in speciation. Can a third evolutionary process—mutation—influence speciation as well? The answer might appear to be no. Chapter 24 emphasized that even though mutation is the ultimate source of genetic variation in populations, it is an inefficient mechanism of evolutionary change. If populations become isolated, it is unlikely that mutation, on its own, could cause them to diverge appreciably.

This view turns out to be naive, however. There is a particular type of mutation, relatively common in plants, that can trigger speciation among sympatric populations. The key is that the mutation reduces gene flow between mutant and normal, or wild-type, individuals. It does so because mutant individuals have more than two sets of chromosomes. This condition is known as **polyploidy**.

To understand why polyploid individuals are genetically isolated from wild-type individuals, consider the mating between a normal individual and a polyploid individual as diagrammed in **Figure 25.9**. In the example given here, the normal individual is diploid, or *2n*, and the polyploid mutant is **tetraploid**, or *4n*. By meiosis, the normal individuals produce haploid gametes, while the mutant individuals produce diploid gametes. These gametes unite to form a **triploid** (*3n*) zygote. Even if this offspring develops normally and reaches sexual maturity, it is rare that it will be able to form functional gametes. The sketch at the bottom of Figure 25.9 illustrates why: When meiosis occurs

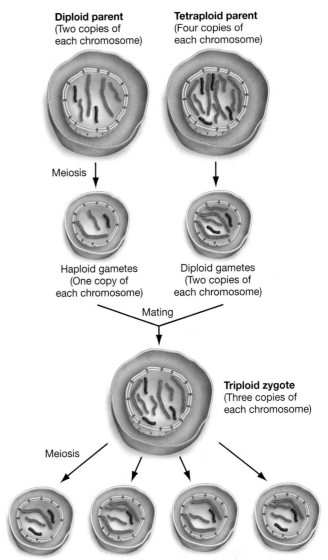

Diploid parent
(Two copies of each chromosome)

Tetraploid parent
(Four copies of each chromosome)

Meiosis

Haploid gametes
(One copy of each chromosome)

Diploid gametes
(Two copies of each chromosome)

Mating

Triploid zygote
(Three copies of each chromosome)

Meiosis

When these gametes combine, most offspring do not have 2, 3, or 4 of each type of chromosome.

FIGURE 25.9 Polyploidy Can Lead to Reproductive Isolation
The mating diagrammed here illustrates why tetraploid individuals are reproductively isolated from diploid individuals. **EXERCISE** Identify gamete combinations that *would* result in viable offspring. Would the likelihood of triploid individuals producing viable gametes increase or decrease if the number of chromosomes was 10 instead of 4?

in a triploid individual, homologous chromosomes cannot synapse and separate correctly. Thus, they are not distributed to daughter cells evenly, and virtually all of the gametes produced by the triploid individual end up with an uneven number of chromosomes. Because its gametes contain a dysfunctional set of chromosomes, the triploid individual is virtually sterile. As a result, the tetraploid and diploid individuals rarely mate and produce fertile offspring. The tetraploid and diploid populations are reproductively isolated.

How do polyploid individuals form? There are two general mechanisms:

1. **Autopolyploid** ("same-many-form") individuals are produced when a mutation results in a doubling of chromosome number.

2. **Allopolyploid** ("different-many-form") individuals are created when parents that belong to different species mate and produce an offspring with a polyploid chromosome number.

Let's consider specific examples to illustrate how speciation by polyploidy occurs.

Autopolyploidy Although autopolyploidy is thought to be much less common than allopolyploidy, biologists recently documented autopolyploidy in the maidenhair fern. This plant inhabits woodlands across North America. **Figure 25.10** shows that, during the normal life cycle of a fern, individuals alternate between a haploid (n) stage and a diploid ($2n$) stage. Biologists initially set out to do a routine survey of allelic diversity in a population of these ferns. They happened to be examining individuals in the haploid stage and found several individuals that had *two* versions of each gene instead of just one. These individuals were diploid even though they had the "haploid" growth form shown in Figure 25.10. The biologists followed these individuals through their life cycle and confirmed that when the ferns mated, they produced offspring that were tetraploid ($4n$). The researchers had stumbled upon a polyploid mutant within a normal population.

To follow up on the observation, they located the parent of the mutant individuals. This plant turned out to have a defect in meiosis. Instead of producing normal, haploid cells as a result of meiosis, the mutant individual produced diploid cells. These diploid cells eventually led to the production of diploid gametes. Because maidenhair ferns can self-fertilize, they can produce tetraploid offspring. The tetraploid offspring could then self-fertilize or mate with their tetraploid parent. If the process continued, a polyploid population of maidenhair ferns would be established. The polyploid individuals would be genetically isolated from the parental population and thus evolutionarily independent, because tetraploid individuals can breed with other tetraploids but not with diploids. If genetic drift and selection then caused the two populations to diverge, speciation would be under way.

It is no exaggeration to say that this autopolyploidy study documented the critical first step in speciation—the establishment of genetic isolation. In this population of maidenhair ferns, as in soapberry bugs, speciation is under way right before our eyes.

Allopolyploidy New tetraploid species may be created when two diploid species hybridize. **Figure 25.11** illustrates the sequence of events involved. If a diploid offspring that forms from a mating between two different species has chromosomes that do not pair normally during meiosis, the offspring is sterile. But if a mutation occurs that doubles the chromosome number, then homologs synapse, meiosis can proceed, and

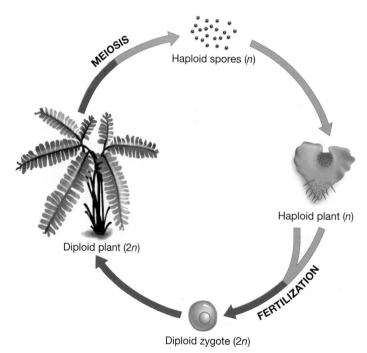

FIGURE 25.10 The Life Cycle of Maidenhair Ferns
Like all ferns, maidenhair ferns have a haploid stage and a diploid stage during their life cycle.

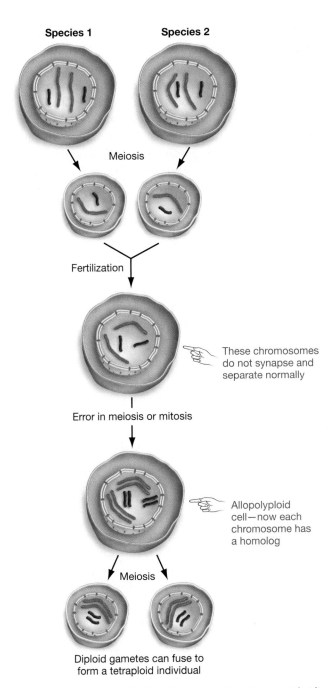

Species 1   Species 2

Meiosis

Fertilization

These chromosomes do not synapse and separate normally

Error in meiosis or mitosis

Allopolyploid cell—now each chromosome has a homolog

Meiosis

Diploid gametes can fuse to form a tetraploid individual

FIGURE 25.11 Allopolyploidy May Occur after Two Species Hybridize Allopolyploid individuals contain chromosomes from two different species.

diploid gametes are produced. When diploid gametes fuse during self-fertilization, a tetraploid individual results.

Exactly this train of events occurred repeatedly after three European species of weedy plants in the genus *Tragopogon* were introduced to western North America in the early 1900s. In 1950 a biologist described the first of two tetraploid species that have been discovered. Both were clearly the descendants of the introduced diploids. Follow-up work has shown that at least one of the new tetraploid species is expanding its geographic range. Further, the nature and amount of genetic variation in the tetraploid species suggests that the allopolyploid

sequence occurred repeatedly—meaning each tetraploid species of *Tragopogon* originated independently multiple times.

Because many diploid species of plants have close relatives that are polyploid, speciation by polyploidization is thought to be particularly important in plant evolution. Two properties of plants have been noteworthy in making this mode of speciation possible: (1) In plants, reproductive cells and somatic cells are not separated early in development, as they are in animals. Instead, plant somatic cells that have undergone many rounds of mitosis can undergo meiosis and produce gametes. If sister chromatids fail to separate during anaphase of one of these mitotic divisions, it can result in a tetraploid cell line that later undergoes meiosis to form diploid gametes. (2) The ability of some plant species to self-fertilize makes it possible for diploid gametes to fuse and create genetically isolated tetraploid populations.

To summarize, mutation can create instant species in the form of polyploids. Speciation by polyploidization breaks two long-held "rules" about speciation: It occurs in sympatry, and it is not slow. Speciation by polyploidy is virtually instantaneous.

✓ CHECK YOUR UNDERSTANDING

Speciation occurs when populations become isolated genetically and then diverge due to selection, genetic drift, or mutation. You should be able to (1) give an example of at least three events that lead to the genetic isolation of populations and (2) explain why selection and drift cause the populations in each example to diverge.

25.4 What Happens When Isolated Populations Come into Contact?

Suppose two populations that have been isolated come into contact again. If divergence has taken place and if divergence has affected when, where, or how individuals in the populations mate, then it is unlikely that interbreeding will take place. In cases such as this, prezygotic isolation exists. When it does, mating between the populations is rare, gene flow is minimal, and the populations continue to diverge.

But what if prezygotic isolation does not exist, and the populations begin interbreeding? The simplest outcome is that the populations fuse over time, as gene flow erases any distinctions between them. Several other possibilities exist, however. Let's explore three of them: reinforcement, hybrid zones, and speciation by hybridization.

Reinforcement

If two populations have diverged extensively and are distinct genetically, it is reasonable to expect that their hybrid offspring will have lower fitness than their parents. The logic here is that if organisms have evolved distinctive developmental sequences or reproductive systems, then a hybrid offspring will not be able to develop or reproduce normally. Recall from Table 25.1 that hybrid offspring may die early in development or survive to sexual maturity but be infertile. In such cases, postzygotic isolation exists. When it occurs, there should be strong natural selection against interbreeding. The hypothesis is that hybrid offspring represent a wasted effort on the part of parents. Individuals that do not interbreed, due to a different courtship ritual or pollination system or other form of prezygotic isolation, should be favored because they produce more viable offspring.

Natural selection for traits that isolate populations in this way is called **reinforcement**. The name is descriptive because the selected traits reinforce differences that developed while the populations were isolated from one another.

Some of the best data on reinforcement come from laboratory studies of closely related fruit fly species in the genus *Drosophila*. Researchers recently analyzed a large series of experiments that tested whether members of closely related fly species are willing to mate with one another. The biologists found an interesting pattern. If closely related species are sympatric—meaning that they live in the same area—individuals from the two species are seldom willing to mate with one another. But if the species are allopatric—meaning that they live in different areas—then individuals are often willing to mate with one another. This is exactly the pattern that we would expect to observe if reinforcement is occurring. The pattern is logical because natural selection can act to reduce mating between species only if their ranges overlap. Thus, it is reasonable to find that sympatric species exhibit prezygotic isolation but that allopatric species do not. There is a long-standing debate, however, over just how important reinforcement is in groups other than the genus *Drosophila*.

Hybrid Zones

Hybrid offspring are not always dysfunctional. Frequently they are perfectly capable of mating and producing offspring and have features that are intermediate between those of the two parental populations. When this is the case, hybrid zones can form. A **hybrid zone** is a geographic area where interbreeding occurs and hybrid offspring are common. Depending on the fitness of hybrid offspring and the extent of breeding between parental species, hybrid zones can be narrow or wide, and long or short lived. As an example of how researchers analyze the dynamics of hybrid zones, let's consider recent work on two bird species.

Townsend's warblers and hermit warblers live in the coniferous forests of North America's Pacific Northwest. In southern Washington state, where their ranges overlap, the two species hybridize extensively. As **Figure 25.12a** shows, hybrid offspring have characteristics that are intermediate relative to the two parental species. To explore the dynamics of this hybrid zone, a team of biologists examined gene sequences in the mitochondrial DNA (mtDNA) of a large number of Townsend's, hermit, and hybrid warblers collected from forests throughout the region. The team found that each of the parental species has certain species-specific mtDNA sequences. This result allowed the researchers to infer how hybridization was occurring. To grasp the reasoning here, it is critical to realize that mtDNA is maternally inherited in most animals and plants. If a hybrid individual has Townsend's mtDNA, its mother had to be a Townsend's warbler while its father had to be a hermit warbler. In this way, identifying mtDNA types allowed the research team to infer whether Townsend's females were mating with hermit males, or vice versa, or both.

Their data presented a clear pattern: Most hybrids form when Townsend's males mate with hermit warbler females. One of the investigators followed up on this result with experiments showing that Townsend's males are extremely aggressive in establishing territories and that they readily attack hermit warbler males. Hermit males, in contrast, do not challenge Townsend's males. The hypothesis, then, is that Townsend's males invade hermit territories, drive off the hermit males, and mate with hermit females.

The team also found something completely unexpected. When they analyzed the distribution of mtDNA types along the Pacific Coast and in the northern Rocky Mountains, they found that many Townsend's warblers actually had hermit mtDNA. **Figure 25.12b** shows that in some regions—such as the larger islands off the coast of British Columbia—*all* of the warblers had hermit mtDNA, even though they looked like full-blooded Townsend's warblers. To explain this result, the team hypothesized that hermit warblers were once found as far north as Alaska and that Townsend's warblers have gradually taken over their range. Their logic is that repeated mating with Townsend's warblers over time made the hybrid offspring look more and more like Townsend's, even while maternally inherited mtDNA kept the genetic record of the original hybridization event intact.

If this hypothesis is correct, then the hybrid zone should continue moving south. If it does so, hermit warblers may eventually become extinct. In many cases, however, hybridization does not lead to extinction but rather leads to the opposite—the creation of new species.

New Species through Hybridization

A team of researchers recently examined the relationships of three sunflower species native to the American West: *Helianthus annuus*, *H. petiolaris*, and *H. anomalus*. The first two of these species are known to hybridize in regions where their ranges overlap. The third species, *H. anomalus*, resembles these hybrids. In fact, because some gene sequences in *H. anomalus* are remarkably similar to those found in *H. annuus*, while other gene sequences are almost identical to those found in *H. petiolaris*, biologists have suggested that *H. anomalus* originated in hybridization between *H. annuus* and *H. petiolaris* (**Figure 25.13**).

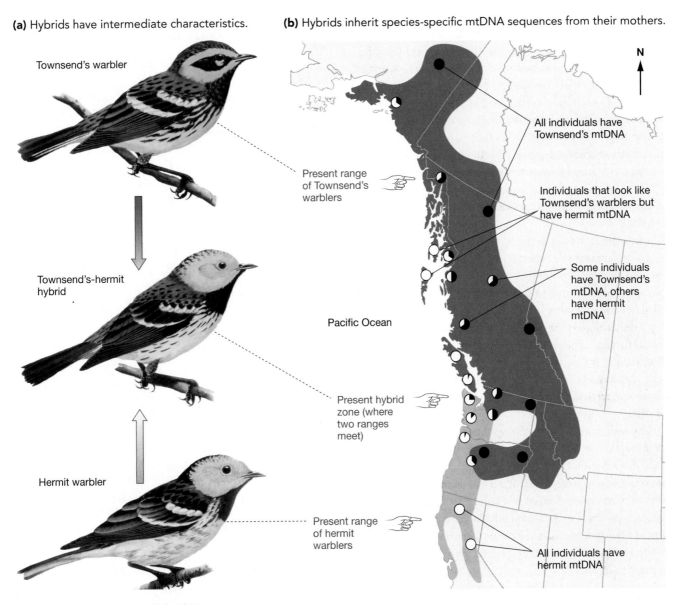

(a) Hybrids have intermediate characteristics.

Townsend's warbler

Townsend's-hermit hybrid

Hermit warbler

Present range of Townsend's warblers

Present hybrid zone (where two ranges meet)

Present range of hermit warblers

Pacific Ocean

(b) Hybrids inherit species-specific mtDNA sequences from their mothers.

N

All individuals have Townsend's mtDNA

Individuals that look like Townsend's warblers but have hermit mtDNA

Some individuals have Townsend's mtDNA, others have hermit mtDNA

All individuals have hermit mtDNA

FIGURE 25.12 Analyzing a Hybrid Zone
(a) When Townsend's warblers and hermit warblers hybridize, the offspring have intermediate characteristics.
(b) Map showing the current range of Townsend's and hermit warblers. The small pie charts show the percentage of individuals with Townsend's warbler mtDNA (in black) and hermit warber mtDNA (in white).

Hybridization hypothesis for the origin of a new species:

Helianthus annuus   x   *H. petiolaris*   ?   *H. anomalus*

FIGURE 25.13 Hybridization Occurs between Sunflower Species
The sunflower *Helianthus anomalus* may have originated in hybridization events between *H. annuus* and *H. petiolarus*.

All three species have the same number of chromosomes, so neither allopolyploidy nor autopolyploidy was involved.

If this interpretation is correct, then hybridization must be added to the list of ways that new species can form. The specific hypothesis here is that *H. annuus* and *H. petiolaris* were isolated and diverged as separate species, and later began interbreeding. The hybrid offspring created a third, new species that had unique combinations of alleles from each parental species and therefore different characteristics. This hypothesis is supported by the observation that *H. anomalus* grows in much drier habitats than either of the parental species and is distinct in appearance. A unique combination of alleles allowed *H. anomalus* to thrive in certain habitats.

Biologists set out to test the hybridization hypothesis by trying to re-create the speciation event experimentally (**Figure 25.14**). Specifically, they mated individuals from the two parental species and raised the offspring in the greenhouse. When these hybrid individuals were mature, the researchers either mated the plants to other hybrid individuals or "backcrossed" them to *H. annuus* individuals. This breeding program continued for four more generations before the experiment ended. Ultimately, the experimental lines were backcrossed twice, and they were mated to other hybrid offspring three times.

The experimental hybrids looked like the natural hybrid species, but did they resemble them genetically? To answer this question, the research team constructed genetic maps of each population, using a large series of genetic markers, similar to the types of markers introduced in Chapter 19 and Chapter 20. Because each parental population had a large number of unique markers in their genomes, the research team hoped to identify which genes found in the experimental hybrids came from which parental species.

Their results are diagrammed in the "Results" section of Figure 25.14. The bottom bar in the illustration represents a region called S in the genome of the naturally occurring species *Helianthus anomalus*. As the legend indicates, this region contains three sections of sequences that are also found in *H. petiolaris* and two that are also found in *H. annuus*. The top bar shows the composition of this same region in the genome of the experimental hybrid lines. The key observation here is that the genetic composition of the synthesized hybrids matches that of the natural species.

In effect, the researchers had succeeded in re-creating a speciation event. Their results provide strong support for the hybridization hypothesis for the origin of *H. anomalus*. The data also highlight the dynamic range of possible outcomes as a result of secondary contact of two populations: fusion of the populations, reinforcement of divergence, founding of stable hybrid zones, extinction of one population, or the creation of new species.

Question: Can new species arise by hybridization between existing species?

Hypothesis: *Helianthus anomalus* originated by hybridization between *H. annuus* and *H. petiolaris*.

Null hypothesis: *Helianthus anomalus* did not originate by hybridization between *H. annuus* and *H. petiolaris*.

Experimental setup:

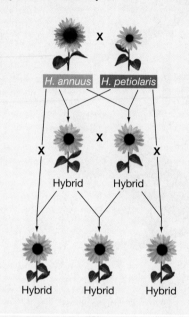

1. Mate *H. annuus* and *H. petiolaris* and raise offspring.

2. Mate F$_1$ hybrids or backcross F$_1$s to parental species; raise offspring.

3. Repeat for four more generations.

Prediction: Experimental hybrids will have the same mix of *H. annuus* and *H. petiolaris* genes as natural *H. anomalus*.

Prediction of null hypothesis: Experimental hybrids will not have the same mix of *H. annuus* and *H. petiolaris* genes as natural *H. anomalus*.

Results:

DNA comparison of a chromosomal region called S:

(Only colored portions of S region were analyzed)

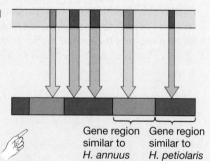

Experimental hybrid (cross of *H. annuus* and *H. petiolaris*)

H. anomalus (naturally occurring species)

Gene region similar to *H. annuus*

Gene region similar to *H. petiolaris*

Genetic composition of experimental hybrids matches that of *H. anomalus*

Conclusion: New species may arise via hybridization between existing species.

FIGURE 25.14 Experimental Evidence That New Species Can Originate in Hybridization Events

ESSAY Human Races

This chapter is built around one fundamental theme: When populations are isolated by lack of gene flow, they can diverge as a result of mutation, genetic drift, and natural selection. Human populations are sometimes called **races** if they are differentiated by physical characteristics such as facial features, the color and texture of hair, skin color, and in some cases height or body build. Presumably, these distinctions arose due to a lack of gene flow in the past. How does the divergence observed among human populations compare with the amount of divergence measured among populations of other organisms? In a biological sense, how profound is the divergence that has occurred among human populations?

To quantify the amount of genetic variation that exists among populations, researchers routinely collect DNA samples from individuals that represent populations from throughout the range of a species. In humans, for example, biologists usually survey genetic diversity by obtaining DNA from aboriginal African, Asian, Australian, European, New Guinean, and Native American people (**Figure 25.15**). Then they analyze the base sequence at one or more loci in the nuclear or mitochondrial genomes of these individuals. To summarize the genetic differences that exist, the researchers calculate the sequence divergence. This number represents the average proportion of bases that are different in two randomly chosen individuals from the populations.

Because of the intense interest in genetic diversity in our own species, numerous studies of humans have been done by using an array of techniques and focusing on an array of loci. And because measuring the amount of genetic divergence among populations is fundamental to understanding a wide variety of questions about speciation and evolution, data sets from dozens, if not hundreds, of plant and animal species are available to compare with data sets of human genes.

Two important conclusions have emerged from these studies:

1. Genetic divergence among human populations is extremely low compared with that found in other species of animals.

    … how profound is the divergence that has occurred among human populations?

    For example, when Rebecca Cann and colleagues examined mitochondrial DNA in humans from all over the world, they observed an average sequence divergence of just 0.32 percent. (This means that, on average, only one-third of 1 percent of bases are different in different populations.) When John Avise and co-workers did a similar study of deer mice from North America, they observed an average sequence divergence of 3.1 percent. Deer mouse populations have experienced 10 times as much genetic divergence as human populations.

2. Populations within sub-Saharan Africa exhibit more genetic diversity than do non-African populations. For example, when Sarah Tishkoff and associates examined the distribution of alleles at the *CD4* gene in nuclear DNA, they found that all of the alleles present in humans can occur in African individuals. Populations in other parts of the world have subsets of the array of alleles found in Africa, but no unique alleles. Based on data such as these, Svante Pääbo comments that "in a genetic sense, everyone on this planet looks like an African."

What is the overall message of these studies? Compared with genetic differences observed within other species, genetic differences among human populations are tiny.

(a) African   **(b)** Asian   **(c)** Australian   **(d)** European   **(e)** New Guinean   **(f)** North American

FIGURE 25.15 At the Genetic Level, How Different Are Aboriginal Populations of Humans?
These individuals represent aboriginal populations from around the world.

CHAPTER REVIEW

Summary of Key Concepts

■ **Speciation occurs when populations of the same species become genetically isolated by lack of gene flow and then diverge from each other due to selection, genetic drift, or mutation.**

When gene flow does not occur between populations, natural selection, genetic drift, and mutation act on the populations independently. As a result, the genetic and physical characteristics of the populations change over time. Eventually, the populations become so different that they are recognized as distinct species—evolutionarily independent populations. Speciation is a splitting event in which one lineage gives rise to two or more independent descendant lineages.

■ **Populations can become genetically isolated from each other if they occupy different geographic areas, if they use different habitats within the same area, or if one population is polyploid and cannot breed with the other.**

Speciation often begins when small groups of individuals colonize a new habitat or when a large, continuous population becomes fragmented into isolated habitats. Colonization is thought to be a major mode of speciation on islands, while range splitting is thought to be a major mode of speciation in glaciated areas and in species that occupied the supercontinent Gondwana during its breakup.

Contrary to traditional expectations, evidence is mounting that speciation can occur due to disruptive selection even when populations are sympatric. Similarly, mutations that produce polyploidy can trigger rapid speciation in sympatry

because they lead to reproductive isolation between diploid and tetraploid populations.

Web Tutorial 25.1   Allopatric Speciation

Web Tutorial 25.2   Speciation by Changes in Ploidy

■ **Populations can be recognized as distinct species if they are reproductively isolated from each other, if they have distinct morphological characteristics, or if they form independent branches on a phylogenetic tree.**

Researchers use several criteria to test whether populations represent distinct species. The biological species concept focuses on the degree of hybridization between species to determine whether gene flow is occurring. The morphospecies concept infers that speciation has occurred if populations have distinctive morphological traits. The phylogenetic species concept defines species as the smallest monophyletic groups on evolutionary trees.

■ **When populations that have diverged come back into contact, several outcomes are possible.**

If prezygotic isolation exists, populations that come back into contact will probably continue to diverge. Alternatively, interbreeding can cause diverged populations to fuse into the same species. Researchers have also documented that secondary contact can lead to reinforcement and complete reproductive isolation, the formation of hybrid zones, or the creation of a new hybrid species.

Questions

Content Review

1. What distinguishes a morphospecies?
   a. It has distinctive characteristics, such as size, shape, or coloration.
   b. It represents a distinct twig in a phylogeny of populations.
   c. It is reproductively isolated from other species.

2. When does vicariance occur?
   a. when small populations coalesce into a large, continuous population
   b. when a large population is fragmented into isolated subpopulations
   c. when individuals colonize a novel habitat
   d. when individuals disperse and found a new population

3. Why is "reinforcement" an appropriate name for the concept that natural selection should favor divergence and genetic isolation if populations experience postzygotic isolation?
   a. Selection should reinforce high fitness for hybrid offspring.
   b. Selection should reinforce the fact that they are "good species" under the morphological species concept.
   c. Selection acts because hybrid offspring do not develop at all or are sterile when mature.
   d. It reinforces selection for divergence that began when the species were geographically isolated.

4. The biological species concept can be applied only to which of the following groups?
   a. bird species living today   b. dinosaurs
   c. bacteria   d. archaea

5. Why are genetic isolation and genetic divergence occurring in soapberry bugs, even though populations occupy the same geographic area?
   a. Members of the different populations feed and mate on different types of fruit.
   b. One population recently became tetraploid, and hybrid offspring cannot undergo meiosis correctly.
   c. A vicariance event occurred when nonnative host plants were introduced.
   d. Beak length has changed due to disruptive selection.

6. When the ranges of different species meet, a stable "hybrid zone" occupied by hybrid individuals may form. How is this possible?
   a. Hybrid individuals may have intermediate characteristics that are advantageous in a given region, relative to traits in the different species.
   b. Hybrid individuals are always allopolyploid and are thus unable to mate with either of the original species. Allopolyploidy forms new species.
   c. Hybrid individuals may have reduced fitness and thus be strongly selected against.
   d. One species has a selective advantage, so as hybridization continues, the other species will go extinct.

Conceptual Review

1. Because speciation is a historical event, it can be difficult to study. Make an outline listing the sections and subsections in this chapter, and then fill it in by describing the experimental and analytical approaches used in the case studies provided for each topic. In your opinion, which approaches to studying speciation described in this chapter are most powerful? Why?

2. In the case of the seaside sparrow, how did the species identified by the biological species concept, the morphospecies concept, and the phylogenetic species concept conflict?

3. Explain how isolation and divergence are occurring in soapberry bugs, even though populations occupy the same geographic area. Of the four evolutionary processes (mutation, gene flow, drift, and selection), which two are most important in causing this event?

4. Why would genetic drift be an especially important evolutionary process causing divergence during colonization events? If colonists occupy a different habitat from that of the source population, why would natural selection also be an important process causing divergence? Explain your answers.

5. Summarize the possible outcomes when populations that have been separated for some time come back into contact and hybridize. In each case, explain why the fitness of hybrid offspring determines the outcome.

Group Discussion Problems

1. A large amount of gene flow is now occurring among human populations due to intermarriage among people from different ethnic groups and regions of the world. Is this phenomenon increasing or decreasing racial differences in our species? Explain.

2. In plants, reproductive cells do not differentiate until late in life. Stated another way, plant cells do not undergo meiosis and produce gametes until they have gone through mitosis hundreds or thousands of times, as the plant grows into an adult. In contrast, animal reproductive cells go through mitosis only a few times before they undergo meiosis and produce gametes. Explain how these observations relate to the following fact: Plants are much more likely to produce diploid gametes and produce polyploid offspring than are animals.

3. Ellen Censky and co-workers recently documented a colonization event. After two hurricanes passed through the Caribbean Sea in the fall of 1995, a large raft made up of fallen trees and other debris, along with green iguanas, drifted to the shore of the island of Anguilla. At least 15 individuals of the green iguana left the raft and went onto the island. Iguanas were not found on Anguilla prior to this event. Based on the wind patterns during the storms, the researchers propose that the 15 iguanas originally lived on the island of Guadalupe.

   • Outline a short-term study designed to test the hypothesis that genetic drift produced allele frequency differences in the two populations (the "old" green iguana population of Guadalupe and the new population on Anguilla).

   • Outline a long-term study designed to test the hypothesis that natural selection will produce changes in the characteristics of the two populations over time.

4. All over the world, natural habitats are being fragmented into tiny islands as suburbs, ranches, and farms expand. In effect, vicariance is occurring at a scale never before seen in the history of life. Based on the data presented in this chapter, do you predict that speciation will take place in the newly isolated populations? Or do you predict that they will be wiped out before speciation can occur?

26 Phylogenies and the History of Life

Skull from a species called *Homo erectus* ("upright man"). This individual is from a population that is closely related to the ancestors of modern humans.

KEY CONCEPTS

▨ Phylogenies and the fossil record are the major tools that biologists use to study the history of life.

▨ The Cambrian explosion was the rapid morphological and ecological diversification of animals that occurred during the Cambrian period.

▨ The new field of "evo-devo" is providing insights into how major events in the history of life occurred, by revealing the genetic mechanisms involved.

▨ Adaptive radiations are a major pattern in the history of life. They are instances of rapid diversification associated with new ecological opportunities and new morphological innovations.

▨ Mass extinctions have occurred repeatedly throughout the history of life. They rapidly eliminate most of the species alive in a more or less random manner.

This chapter is about the tree of life—the wonderful diversity of organisms that have been produced since life began. More specifically, this chapter is about how the tree of life has grown and changed over time. Life has existed on Earth for some 3.8 billion years. What patterns can be discerned as species appeared and disappeared over the course of history? Why do patterns exist?

Chapters 1 and 2 introduced the size and shape of the tree of life by examining the geologic time scale, radiometric dating, and life's three domains: the Bacteria, Archaea, and Eukarya. Chapter 25 considered speciation, or how individual twigs on the tree of life form. The focus of this chapter falls between the broad scope of the early chapters and the fine level of Chapter 25. The task is to find common themes and patterns in the history of life.

The goal of Sections 26.1 and 26.2 is to introduce the data and analytical tools that researchers use to reconstruct *what* happened in the history of life. Section 26.3 focuses on *how* key

changes documented in the fossil record occurred—specifically, how gene duplication events, changes in gene expression, and natural selection cause major evolutionary innovations. A prominent pattern in the tree of life, called adaptive radiation, is the subject of Section 26.4. The chapter closes with a look at what may be the most influential of all events in the history of life: the ecological disasters known as mass extinctions.

26.1 Tools for Studying History: Phylogenies and the Fossil Record

Biologists have two major tools to use when studying the past: phylogenies and the fossil record. A **phylogeny** is the evolutionary history of a group of organisms, and it is usually summarized and depicted in the form of a phylogenetic tree. A **phylogenetic tree** shows the ancestor-descendant relationships among populations or species, just as a genealogy or family tree shows the ancestor-descendant relationships among a group of individuals. A **fossil**, in contrast, is a physical trace of an organism that lived in the past. The **fossil record** is the total collection of fossils that have been found throughout the world. The fossil record is housed in thousands of private and public collections.

Fossils document what organisms looked like in the past and when various groups existed. Phylogenies clarify who is related to whom. Let's first consider how phylogenies are read and how they are estimated and then analyze how the fossil record is collected and interpreted. The remainder of this chapter will explore what these two tools have revealed about the story of life on Earth.

Using Phylogenies

Evolutionary trees have been introduced at various points in earlier chapters—often with just enough information to help you understand a new concept. Now let's focus on the trees themselves. How do biologists read a finished tree, and how are trees put together in the first place? We'll consider each question in turn.

A Field Guide to Reading Phylogenetic Trees Phylogenetic trees are an extremely effective way of summarizing data on the evolutionary history of a group of organisms. They are unusual diagrams, however, and it can take practice to interpret them correctly. To understand how evolutionary trees work, consider **Figure 26.1**. Note that a phylogenetic tree consists of branches, nodes, and tips. **Branches** represent populations through time. **Nodes** (or *forks*) occur where an ancestral group splits into two or more descendant groups (see point B in Figure 26.1). If more than two descendant groups emerge from a node, the node is called a **polytomy** (see node C). **Tips** (or *terminal nodes*) are the tree's endpoints, which represent groups living today or a dead end—a branch ending in extinction. Groups that occupy adjacent branches on the tree are called **sister taxa**. Recall from Chapter 1 that a **taxon** (plural: *taxa*) is any named group of organisms. A taxon could be a single species, such as *Homo sapiens*, or a large group of species, such as Primates.

The phylogenetic trees used in this text are all **rooted**—meaning the bottom, or most basal, node on the tree is the most ancient. To determine where the root on a tree occurs, biologists include an outgroup species when they are collecting data to estimate a particular phylogeny. An **outgroup** is a taxonomic group that is known to have diverged prior to the rest of the taxa in the study. In Figure 26.1, "Taxon 1" is an outgroup of the monophyletic group consisting of taxa 2–6. (Recall from Chapter 25 that a **monophyletic group** consists of an ancestral species and all of its descendants. Monophyletic groups may also be called **clades** or **lineages**.) If you were estimating the phylogeny of human species that lived over the past 5–6 million years, you might use chimpanzees or gorillas as an outgroup in the analysis. Chimps and gorillas would then be the first groups to have split off in the resulting tree, with the various species of humans forming a monophyletic group descended from the common ancestor at the tree's bottom node (**Figure 26.2**, page 558).

To practice how to read a tree, put your finger at the root of the diagram in Figure 26.1 and begin to work your way up. At node A, the ancestral population split into two groups. One of these lineages eventually evolved into Taxon 1. The other lineage gave rise to taxa 2–6. Now continue moving your finger up the tree until you hit node B. It should make sense to you that at this splitting event, one descendant clade eventually gave rise to Taxon 2 and Taxon 3, while the other became the ancestor of taxa 4–6. The polytomy at node C occurs because taxa 4–6 split from one another so quickly that it is not possible to tell which split off earlier or later. You should be able to do the same exercise with the phylogeny of humans shown in Figure 26.2.

Figure 26.3 (page 558) presents a different challenge. Five of the six trees shown in this diagram are identical in terms of the evolutionary relationships they represent. One differs. If you understand how to read a tree, you should be able to determine which one of the six trees is unlike the others.

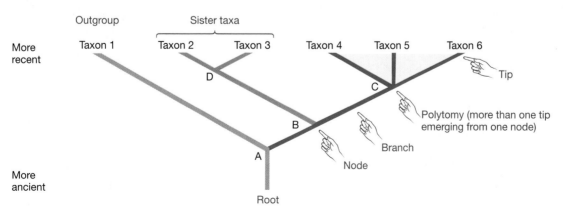

FIGURE 26.1 Reading a Phylogenetic Tree
The parts of a phylogenetic tree, labeled.

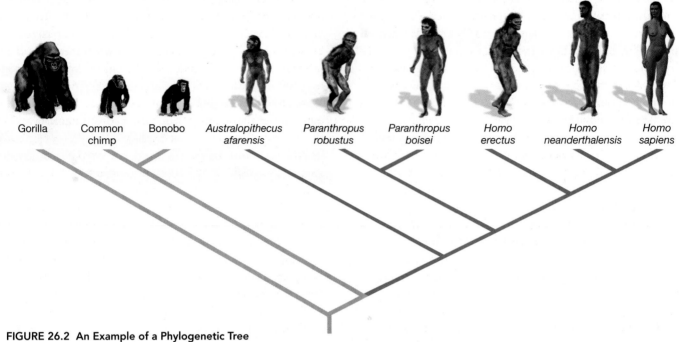

FIGURE 26.2 An Example of a Phylogenetic Tree
A phylogenetic tree showing the relationships among some of the great apes. Chimps and gorillas walk on all four legs, but all other species on this tree walk on two legs and are considered hominins. **EXERCISE** Add a mark on the phylogeny, and label it "origin of walking on two legs." Circle and label a pair of sister species. Circle and label the monophyletic group called hominins. Label one of the outgroups to the hominins.

How Do Researchers Estimate Phylogenies? Like any other pattern or measurement in nature, from the average height of a person in a particular human population to the speed of a passing airplane, the genealogical relationships among species cannot be known with absolute certainty. Instead, the relationships depicted in an evolutionary tree are estimated from data.

To infer the historical relationships among species, researchers analyze the species' morphological or genetic charac-

teristics. For example, to reconstruct relationships among fossil species of humans, scientists analyze aspects of tooth, jaw, and skull structure. To reconstruct relationships among contemporary human populations, investigators usually compare the sequences of bases in a particular gene.

Once the characteristics of different populations have been measured, researchers have several general strategies available for analyzing the data and inferring which species or populations are more closely related or more distantly related. The

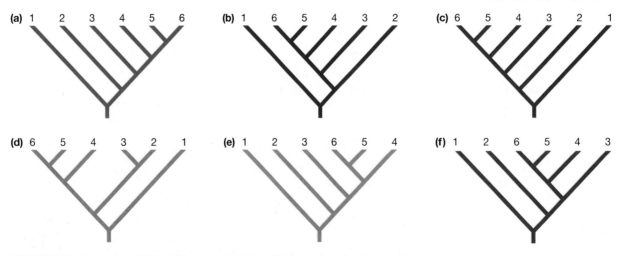

FIGURE 26.3 Alternative Ways of Drawing the Same Tree
QUESTION Five of these six trees describe exactly the same relationships among taxa 1 through 6. Identify the tree that is different from the other five.

fundamental idea is that closely related species should share many of their characteristics, while distantly related species should share fewer characteristics. But there are two general strategies for quantifying the degree of similarity among species: the phenetic approach and the cladistic approach.

The **phenetic approach** to estimating trees is based on computing a statistic that summarizes the overall similarity among populations, based on the data. For example, researchers might use gene sequences to compute an overall "genetic distance" between two populations. A *genetic distance* summarizes the average percentage of bases in a DNA sequence that differs between two populations. A computer program then compares the similarities among populations and builds a tree that clusters the most similar populations and places more divergent populations on more distant branches.

The **cladistic approach** to inferring trees is based on the realization that relationships among species can be reconstructed by identifying shared derived characters, or **synapomorphies** ("union-forms"), in the species being studied. **Figure 26.4** illustrates the logic behind a cladistic analysis. When the ancestral population at the bottom of the figure splits into two descendant lineages at node A, each descendant group begins evolving independently and acquires unique traits. These traits are derived from their common ancestor via mutation, selection, and genetic drift. The traits might be a particular region of DNA that changes as follows:

AAA GCT ACT	ancestral population
AAC GCT ACT	a descendant population
AAA GGT ACT	another descendant population

This is a monophyletic group that shares a derived trait (the "C" in the third position)

This is a monophyletic group that shares a derived trait (the "G" in the fifth position)

AAC GCT ACT

AAA GGT ACT

Trait (in this case, a DNA sequence) in the ancestral population

AAA GCT ACT

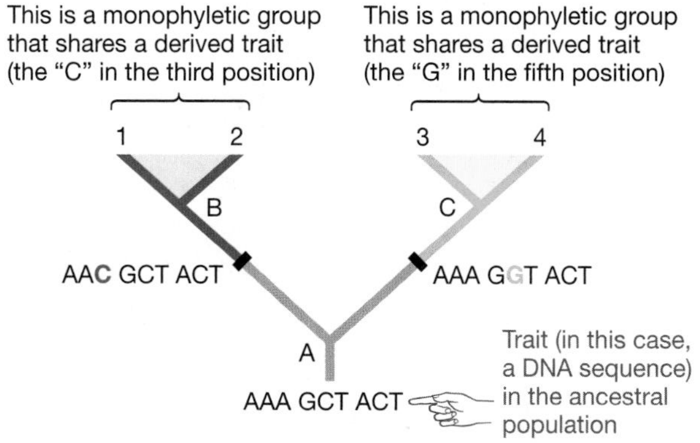

FIGURE 26.4 Synapomorphies Identify Monophyletic Groups Mutation, selection, and drift cause evolutionary changes that are shared in groups derived from the same common ancestor. In this tree, species 1 and 2 can be identified as part of a monophyletic group because base C in the third position of the DNA sequence shown is a synapomorphy. Species 3 and 4 can be identified as part of a different monophyletic group because they have a different synapomorphy—base G in the fifth position of the DNA sequence shown.

When the two lineages themselves split at nodes B and C, the species that result share the derived characteristics. In this way, the ancestor at node B and species 1 and 2 can be recognized as a monophyletic group. Similarly, the ancestor at node C and species 3 and 4 can be recognized as a different monophyletic group. When many such traits have been measured, a computer program can be used to identify which traits are unique to each monophyletic group and then place the groups in a tree in the correct relationship to each other.

Although the logic behind phenetic and cladistic analyses is elegant, problems arise due to convergent evolution. Recall from Chapter 23 that traits in two species may be similar not because they were present in a common ancestor, but because similar traits evolved independently in two distantly related groups. In the example given in Figure 26.4, it is possible that species 2 is not at all closely related to species 1. Its ancestors may have had the sequence AAT GCT AGT, which happened to change to AAC GCT ACT due to mutation, selection, and drift that took place independently of the changes that took place in the ancestors of species 1.

To cope with the existence of convergent evolution, biologists who are using cladistic approaches invoke the logical principle of **parsimony**, which states that the most likely explanation or pattern is the one that implies the least amount of change or the least complexity. For example, a biologist might compare all of the possible branching patterns that could link species 1–4 and count the number of changes in DNA sequences required to produce each pattern. Convergent evolution should be rare compared with similarity due to shared descent, so the tree that implies the fewest overall changes—and thus the least convergence—should be the one that most accurately reflects what really happened during evolution. If the branching pattern in Figure 26.4 is the most parsimonious of all possible trees, then biologists conclude that it is the most accurate.

Whale Evolution: A Case History As an example of how a cladistic approach works, consider the evolutionary relationships of the whales and the group of mammals called the Artiodactyla. Cows, deer, and hippos are artiodactyls. Members of this group have hooves and an even number of toes. They also share another feature: the unusual pulley shape of an ankle bone called the astralagus. The shape of the astralagus is a shared, derived character that identifies the artiodactyls as a monophyletic group. These data support the tree shown in **Figure 26.5a** (page 560).

When researchers began comparing DNA sequences of artiodactyls and other species of mammals, however, the data showed that whales share many similarities with hippos. These results supported the tree shown in **Figure 26.5b**. This tree, supported by the DNA data, conflicts with the morphological data in Figure 26.5a, because it implies that the pulley-shaped astralagus evolved in artiodactyls and then was lost

(a) The astralagus is a synapomorphy that identifies artiodactyls as a monophyletic group.

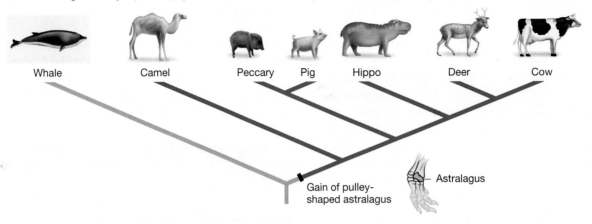

Gain of pulley-
shaped astralagus

Astralagus

(b) If whales are related to hippos, then two evolutionary changes occurred in the astralagus.

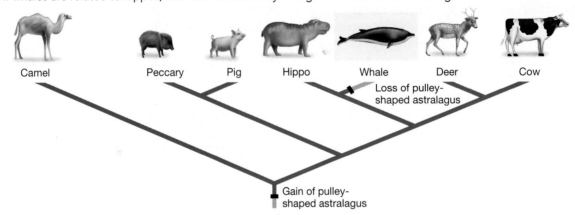

Loss of pulley-
shaped astralagus

Gain of pulley-
shaped astralagus

(c) Data on the presence and absence of SINE genes support the close relationship between whales and hippos.

Locus	1	2	3	4	5	6	7	8	9	10	11	12	13	14	15	16	17	18	19	20
Cow	0	0	0	0	0	0	0	1	1	1	1	1	1	1	1	1	1	1	0	0
Deer	0	0	0	0	0	0	0	1	?	1	1	1	1	1	1	?	1	1	0	0
Whale	1	1	1	1	1	1	1	0	?	1	0	1	1	0	0	0	?	1	0	0
Hippo	0	?	0	1	1	1	1	0	1	1	0	1	1	0	0	0	?	1	0	0
Pig	0	0	0	?	0	0	0	0	?	0	0	0	?	?	0	0	0	1	1	1
Peccary	?	?	?	?	?	?	?	?	?	?	?	?	?	?	?	?	?	?	1	1
Camel	0	0	0	0	0	0	0	0	0	0	0	0	0	0	0	0	0	0	0	0

1 = gene present
0 = gene absent
? = still undetermined

Whales and hippos share four
unique SINE genes (4, 5, 6, and 7)

FIGURE 26.5 Evidence That Whales and Hippos Form a Monophyletic Group
On the basis of a parsimony analysis of the pulley-shaped astralagus, biologists favored **(a)** a tree excluding whales from the Artiodactyla over **(b)** the hypothesis that whales are artiodactyls and are closely related to hippos. **(c)** Data on the presence and absence of SINE genes supports the "whales + hippos" hypothesis. **EXERCISE** The presence of SINE genes 4, 5, 6, 7 identifies hippos and whales as part of a monophyletic group. The presence of SINE genes 8, 11, 14, 15, 17 identifies deer and cows as part of a monophyletic group. SINE genes 10, 12, 13 identify hippos, whales, deer, and cows as part of a monophyletic group. Map the origin of these SINE genes on the tree in part (b). Where did SINE genes 19 and 20 first insert themselves into the genomes of artiodactyls?

during whale evolution. The tree in Figure 26.5b implies two changes in the astralagus, while the tree in Figure 26.5a implies just one. In terms of the evolution of the astralagus, the "whale + hippo" tree is less parsimonious than the tree in Figure 26.5a.

The conflict between the data sets was resolved when researchers analyzed the distribution of the parasitic gene sequences called SINEs (short interspersed nuclear elements), which occasionally insert themselves into the genomes of mammals (see Table 20.1). As the data in **Figure 26.5c** show, whales and hippos share several types of SINES that are not found in other groups. Specifically, whales and hippos share the SINEs numbered 4, 5, 6, and 7. Other SINE genes are present in some artiodactyls but not in others; camels have no SINE genes at all. To explain these data, biologists hypothesize that no SINEs were present in the population that is ancestral to all of the species in the study. Over the course of evolution, however, different SINEs became inserted into the genomes of descendant populations. As a result, the presence of a particular SINE represents a derived character. Because whales and hippos share four of these derived characters, it is logical to conclude that these animals are closely related.

Based on these data, most biologists accepted the phylogeny that is shown in Figure 26.5b as the most accurate estimate of evolutionary history. According to this phylogeny, whales are artiodactyls and share a relatively recent common ancestor with hippos. This observation inspired the hypothesis that both whales and dolphins are descended from a population of artiodactyls that spent most of their time feeding in shallow water, much as hippos do today. Recently this hypothesis was supported in spectacular fashion. In 2001 two teams of researchers announced the independent discoveries of fossil artiodactyls that were clearly related to whales and yet had a pulley-shaped astralagus. One of these fossils is illustrated in Figure 23.3b. The combination of phylogenetic data and data from the fossil record has clarified how a particularly interesting group of mammals evolved.

Over the past two decades, the effort to infer phylogenetic trees has been revolutionized by the availability of DNA sequence data and powerful computers to analyze large data sets. Methods for collecting and analyzing the data used to estimate phylogenies continue to improve. Biological science is currently witnessing an explosion in the quantity and quality of phylogenetic trees.

Using the Fossil Record

Phylogenetic analyses are powerful ways to infer the order in which events occurred during evolution and to understand how particular groups of species are related. But only the fossil record provides direct evidence about what organisms that lived in the past looked like, where they lived, and when they existed. Let's review how fossils form, analyze the strengths and weaknesses of the fossil record, and then survey a few of the major events that have taken place in life's 3.8-billion-year history.

How Do Fossils Form?   Most of the processes that form fossils begin when part or all of an organism is buried in ash, sand, mud, or some other type of sediment. Consider a series of events that begins when a tree falls into a swamp. **Figure 26.6** illustrates the leaves of a tree falling onto a patch of mud, where they are buried by soil and debris before they decay.

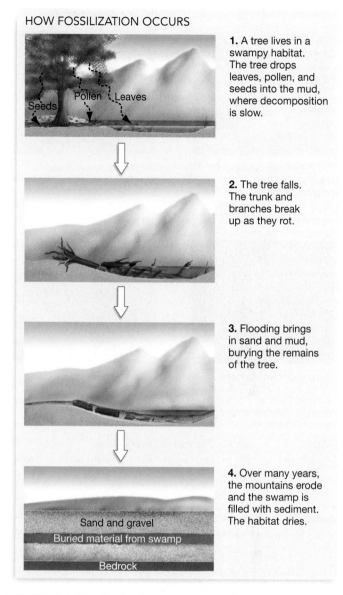

HOW FOSSILIZATION OCCURS

1. A tree lives in a swampy habitat. The tree drops leaves, pollen, and seeds into the mud, where decomposition is slow.

2. The tree falls. The trunk and branches break up as they rot.

3. Flooding brings in sand and mud, burying the remains of the tree.

4. Over many years, the mountains erode and the swamp is filled with sediment. The habitat dries.

Sand and gravel
Buried material from swamp
Bedrock

FIGURE 26.6 Fossilization Preserves Traces of Organisms That Lived in the Past
Fossilization occurs most readily when the remains of an organism are buried in sediments, where decay is slow.

Pollen and seeds settle into the muck at the bottom of the swamp, where decomposition is slow. The stagnant water is too acidic and too oxygen poor to support large populations of bacteria and fungi, so much of this material is buried intact before it decomposes. The trunk and branches that sit above the water line rot fairly quickly, but as pieces break off they, too, sink to the bottom and are buried.

Once burial occurs, several things can happen. If decomposition does not occur, the organic remains can be preserved intact—like the fossil pollen in **Figure 26.7a** (page 562). Alternatively, if sediments accumulate on top of the material and become cemented into rocks such as mudstone or shale, the sediments weight can compress the organic material below into a thin, carbonaceous film. This happened to the leaf in

(a) Intact fossil

The pollen was preserved intact because no decomposition occurred.

(b) Compression fossil

Sediments accumulated on top of the leaf and compressed it into a thin carbon-rich film.

(c) Cast fossil

The branch decomposed after it was buried. This left a hole that filled with dissolved minerals, faithfully creating a cast of the original.

(d) Permineralized fossil

The wood decayed very slowly, allowing dissolved minerals to infiltrate the cells gradually and then harden into stone.

FIGURE 26.7 Fossils Are Formed in Several Ways
Different preservation processes give rise to different types of fossils.

Figure 26.7b. If the remains decompose *after* they are buried—as did the branch in **Figure 26.7c**—the hole that remains can fill with dissolved minerals and faithfully create a **cast** of the remains. If the remains rot extremely slowly, dissolved minerals can gradually infiltrate the interior of the cells and harden into stone, forming a *permineralized* fossil, such as petrified wood (**Figure 26.7d**).

After many centuries have passed, fossils can be exposed at the surface by many mechanisms, including erosion, a road cut, or quarrying. If researchers find a fossil, they can prepare it for study by painstakingly clearing away the surrounding rock (**Figure 26.8**). If the species represented is new, researchers describe its morphology in a scientific publication, name the species, estimate the fossil's age based on dates assigned to nearby rock layers, and add the specimen to a collection so that it is available for study by other researchers. It is now part of the fossil record. This is the information database that supports much of the research reviewed in this chapter.

The scenario just presented is based on conditions that are ideal for fossilization: The tree fell into an environment where decomposition was slow and burial was rapid. In most habitats the opposite situation occurs—decomposition is rapid and burial is slow. In reality, then, fossilization is an extremely rare event. To appreciate this point, consider that there are seven specimens of the first bird to appear in the fossil record,

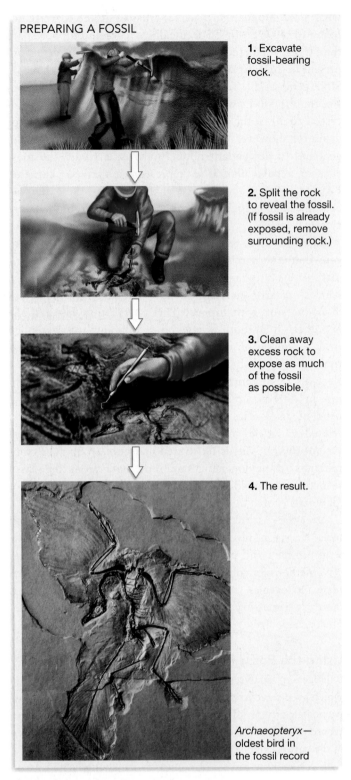

PREPARING A FOSSIL

1. Excavate fossil-bearing rock.

2. Split the rock to reveal the fossil. (If fossil is already exposed, remove surrounding rock.)

3. Clean away excess rock to expose as much of the fossil as possible.

4. The result.

Archaeopteryx— oldest bird in the fossil record

FIGURE 26.8 Preparing a Fossil
EXERCISE Label the head, wings, neck, legs, toes, tail, and ribs on the *Archaeopteryx* specimen. Outline the extent of its feathers.

Archaeopteryx. All were found at the same site in Germany where limestone is quarried for printmaking (the bird's specific name is *lithographica*). If you accept an estimate that crow-sized birds native to wetland habitats in northern Europe would have a population size of around 10,000 and a life span of

10 years, and if you accept the current estimate that the species existed for about 2 million years, then you can calculate that about 2 billion *Archaeopteryx* lived. But as far as researchers currently know, only 1 out of every 286,000,000 individuals fossilized. For this species, the odds of becoming a fossil were almost 40 times worse than your odds are of winning the grand prize in a state lottery.

Limitations of the Fossil Record   Before looking at how the fossil record is used to answer questions about the history of life, it is essential to review the nature of this archive and recognize several features:

- Habitat bias: Because burial in sediments is so crucial to fossilization, there is a strong habitat bias in the database. Organisms that live in areas where sediments are actively being deposited, including beaches, mudflats, and swamps, are much more likely to form fossils than organisms that live in other habitats are. Within these habitats, burrowing organisms such as clams are already underground—pre-buried—at death and are therefore much more likely to fossilize. Organisms that live above ground in dry forests, grasslands, and deserts are much less likely to fossilize.

- Taxonomic bias: Slow decay is almost always essential to fossilization, so organisms with hard parts such as bones or shells are most likely to leave fossil evidence. This requirement introduces a strong taxonomic bias into the record. Clams, snails, and other organisms with hard parts have a much higher tendency to be preserved than do worms. A similar bias exists for tissues within organisms. For instance, pollen grains are encased in a tough outer coat that resists decay, so they fossilize much more readily than do flowers. Teeth are the most common mammalian fossil, simply because they are so hard and decay resistant. Shark teeth are abundant in the fossil record, but shark bones, which are made of cartilage, are extremely rare.

- Temporal bias: Recent fossils are much more common than ancient fossils. To understand why, consider that when two of Earth's tectonic plates converge, the edge of one plate usually sinks beneath the other plate. The rocks composing the edge of the descending plate are either melted or radically altered by the increased heat and pressure they encounter as they move downward into Earth's interior. These alterations obliterate any fossils in the rock. In addition, fossil-bearing rocks on continents are constantly being broken apart and destroyed by wind and water erosion. The older a fossil is, the more likely it is to be demolished.

- Abundance bias: Because fossilization is so improbable, the fossil record is weighted toward common species. Organisms that are abundant, widespread, and present on Earth for long periods of time leave evidence much more often than do species that are rare, local, or ephemeral.

To summarize, the fossil database represents a highly nonrandom sample of the past. **Paleontologists**—scientists who study fossils—recognize that they are limited to asking questions about tiny and scattered segments on the tree of life. Yet, as this chapter shows, the record is a scientific treasure trove. Analyzing fossils is the only way scientists have of examining the physical appearance of extinct forms and inferring how they lived. The ancient Egyptians wrote a Book of the Dead to guide individuals into the afterlife. The fossil record is biology's book of the dead—our guide to past life.

Life's Timeline   A few of the more significant data points in the fossil record are summarized in **Figure 26.9** (page 564). Most of the data on the timelines are "evolutionary firsts" that document important innovations during the history of life. In almost all cases, the dates given were estimated by using the radiometric dating techniques introduced in Chapter 2. When fossils are not available, key events in the history of life may sometimes be dated by using the molecular approaches introduced in **Box 26.1** (page 566). The eras, eons, and periods in the figure do not represent regular time intervals, however, because they were indentified and named long before radiometric and molecular dating techniques became available. Because the fossil record and efforts to date fossils are constantly improving, the timeline in Figure 26.9 is a work in progress.

As you inspect Figure 26.9, notice that the timeline is broken into four segments and that comments on the types of organisms present, the nature of the climate, major geological events, and the positions of the continents accompany most segments:

- Figure 26.9a outlines the interval between the formation of Earth about 4.6 billion years ago and the appearance of most animal groups about 543 million years ago (abbreviated Ma). The entire interval is called the **Precambrian era**, and it is divided into the **Hadean**, **Archaean**, and **Proterozoic eons.** The important things to note about this time are that (1) life was exclusively unicellular for most of Earth's history and (2) oxygen was virtually absent from the oceans and atmosphere for almost 2 billion years after the origin of life.

- Figure 26.9b summarizes the **Paleozoic** ("ancient life") **era**, which begins with the appearance of many animal lineages and ends with the obliteration of almost all multicellular life-forms at the end of the Permian period. The Paleozoic saw the origin and initial diversification of the animals, land plants, and fungi, as well as the appearance of land animals.

- The **Mesozoic** ("middle life") **era** is outlined in Figure 26.9c. This interval is nicknamed the Age of Reptiles. It begins with the end-Permian extinction events and ends with the extinction of the dinosaurs and other groups at the boundary between the Cretaceous period and Tertiary period.

(a) The **Precambrian era** included the origin of life, photosynthesis, and the oxygen atmosphere.

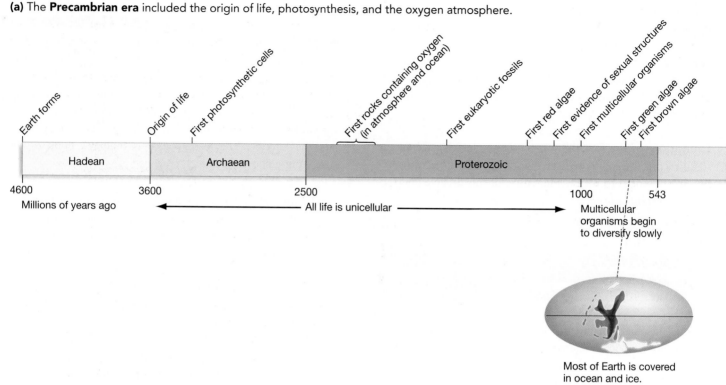

(c) The **Mesozoic era** is sometimes called the Age of Reptiles.

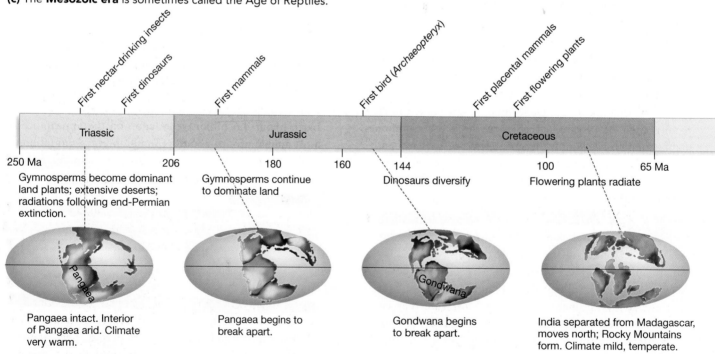

FIGURE 26.9 Life's Timeline
Significant events in Earth's history and the history of life are plotted for the **(a)** Precambrian era, **(b)** Paleozoic era, **(c)** Mesozoic era, and **(d)** Cenozoic era. ("Ma" stands for "million years ago.")

- Figure 26.9d highlights the **Cenozoic** ("recent life") **era**, which is divided into the Tertiary period and the Quaternary period. The Cenozoic is sometimes nicknamed the Age of Mammals, because mammals diversified after the disappearance of the dinosaurs. Events that occur today are considered to be part of the Cenozoic era.

(b) The **Paleozoic era** included the origin and early diversification of animals, land plants, and fungi.

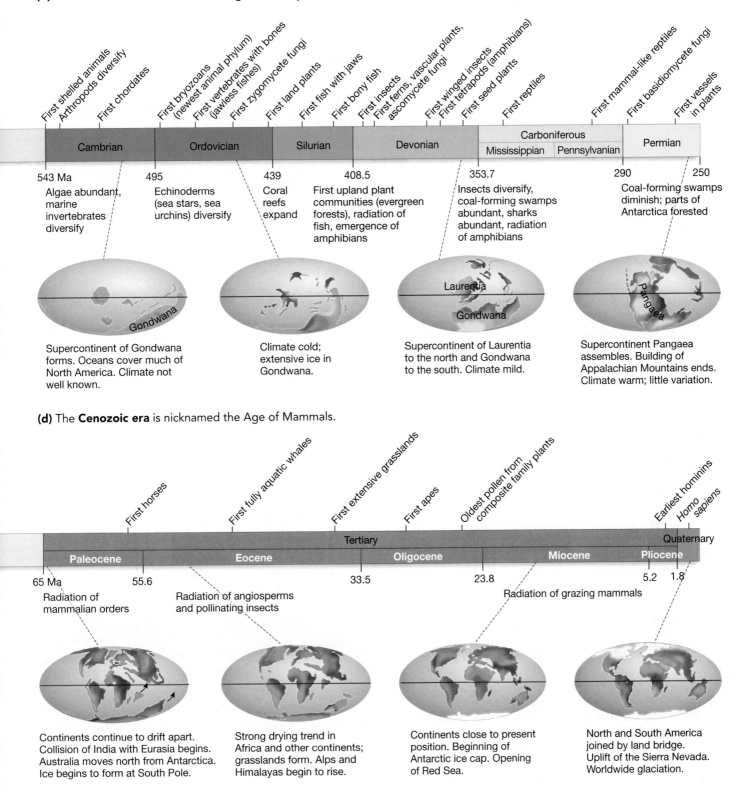

Supercontinent of Gondwana forms. Oceans cover much of North America. Climate not well known.

Algae abundant, marine invertebrates diversify

Echinoderms (sea stars, sea urchins) diversify

Climate cold; extensive ice in Gondwana.

Coral reefs expand

First upland plant communities (evergreen forests), radiation of fish, emergence of amphibians

Supercontinent of Laurentia to the north and Gondwana to the south. Climate mild.

Insects diversify, coal-forming swamps abundant, sharks abundant, radiation of amphibians

Coal-forming swamps diminish; parts of Antarctica forested

Supercontinent Pangaea assembles. Building of Appalachian Mountains ends. Climate warm; little variation.

(d) The **Cenozoic era** is nicknamed the Age of Mammals.

Continents continue to drift apart. Collision of India with Eurasia begins. Australia moves north from Antarctica. Ice begins to form at South Pole.

Radiation of mammalian orders

Radiation of angiosperms and pollinating insects

Strong drying trend in Africa and other continents; grasslands form. Alps and Himalayas begin to rise.

Continents close to present position. Beginning of Antarctic ice cap. Opening of Red Sea.

Radiation of grazing mammals

North and South America joined by land bridge. Uplift of the Sierra Nevada. Worldwide glaciation.

The data in Figure 26.9 show that the nature of life has changed radically since the first cell on Earth appeared. Although the changes that have taken place over the past 3.8 billion years are well documented in the fossil record, the sweep of time involved is difficult for the human mind to comprehend. A semester can seem long to a college student, but biologists who study the fossil record routinely analyze data on events that took place over tens of millions of years. In relation to Earth's history, 100,000 years or even a million years is the blink of an eye.

BOX 26.1 The Molecular Clock

Several researchers have proposed that the fossil record can be supplemented with information on the amino acid and DNA sequences in living species. This hypothesis, called the **molecular clock**, holds that certain types of mutations—such as silent mutations (Chapter 16), or changes in amino acid sequence that do not affect protein function—increase to fixation in populations at a steady rate. For example, consider data on the protein hemoglobin. A biologist compared the amino acid sequence of this molecule in vertebrates ranging from sharks to humans. When he plotted the number of amino acid differences between species against the date that the species diverged, according to the fossil record, the graph shown in **Figure 26.10** resulted. The evolution of hemoglobin seems to proceed at a constant rate over time.

To see how researchers use molecular clocks to supplement the fossil record, let's consider some recent analyses of human evolution. The seven fossil species of *Homo* that have been identified to date, and their estimated interrelationships, are shown in **Figure 26.11**. The lineage leading to both *Homo erectus* and *H. ergaster* and the lineage leading to *H. sapiens* are thought to have split almost

2 million years ago. (The relationship between *H. erectus* and *H. ergaster* is not known.) Two other recent populations of hominins (human-like organisms), called *H. neanderthalensis* and *H. heidelbergensis*, evolved after this split occurred. But it is not yet clear whether these two species represent branches off the *H. erectus* lineage or off the lineage leading to *H. sapiens*. Until more and better-quality fossils can be found and interpreted, researchers are left with a question: When did the most recent common ancestor of our species live? The fossil record indicates that this population existed sometime between the last fossil *H. erectus*, 400,000 years ago, and the first fossil *H. sapiens*, 100,000 years ago.

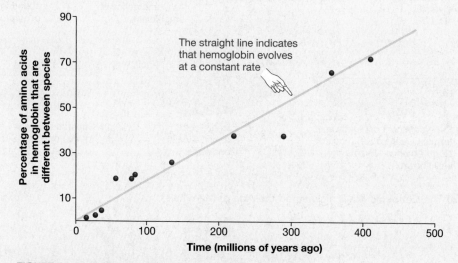

FIGURE 26.10 A Data Set That Supports a Molecular Clock The rate of change in the hemoglobin protein. To interpret this graph, consider that the point at the upper right represents a comparison of hemoglobin proteins in sharks and mammals. According to the fossil record, sharks branched off from other vertebrates about 450 million years ago. QUESTION How fast does this clock tick? That is, what percentage of amino acids in hemoglobin change per 100 million years, on average?

✔ CHECK YOUR UNDERSTANDING

Phylogenies can be estimated by finding synapomorphies that identify monophyletic groups. Using the example of whale evolution, you should be able to explain why the presence and absence of certain SINE genes serve as synapomorphies supporting the hypothesis that hippos and whales recently shared a common ancestor. You should also be able to summarize how whales evolved by synthesizing data from the molecular phylogenies presented in this section with data on the fossil record of whales presented in Chapter 23.

26.2 The Cambrian Explosion

To get a better feel for how biologists use the fossil record to understand the history of life, let's examine recent research on a momentous event—the origin and early diversification of

animals. The first animals appear in the fossil record about 570 million years ago. Their early diversification occurred soon after that, at the start of the Cambrian period. This diversification happened so rapidly that it earned a nickname: the **Cambrian explosion.**

It is impossible to appreciate the drama of this event without recalling that almost all life-forms were unicellular for about 2.5 *billion* years after the origin of life. The exceptions were several lineages of multicellular algae, which show up in the fossil record about 1 billion years ago. At that time, and for tens of millions of years afterward, the organisms that were the ancestors of animals were still unicellular, marine creatures with cilia or flagella—probably not too different from some of the aquatic eukaryotes living today. Then, about 565 million years ago, the first animals—sponges, jellyfish, and perhaps simple worms—appear in the fossil record. Just 40 million years later, virtually every major group of ani-

A team of investigators recently used a molecular clock to answer this question. The group compared the complete mitochondrial genomes sequenced from African, European, and Japanese individuals. These data allowed the team to estimate the number of bases in mitochondrial DNA (mtDNA) that had changed as human populations diverged. To estimate how long it took these genetic differences to develop, the researchers also sequenced the mtDNA of our close relatives, the orangutan and African apes. According to the fossil record, orangutans and African apes diverged 13 million years ago. By dividing the number of bases that differ in the mtDNA of orangutans and African apes by 13 million years, the researchers arrived at an estimate that mtDNA changes at the rate of 7×10^{-8} base per base per year.

The total amount of sequence divergence observed between two species should be equal to 2 × (rate of sequence evolution) × (time since divergence occurred). When the researchers used this equation with their estimate of the number of bases in mtDNA that had changed as human populations diverged, they concluded that the last common ancestor of all living humans lived between 125,000 and 161,000 years ago. Large species of mammals typically last in the fossil record for about 1.5 million years prior to extinction. Humans, then, are a relatively young species.

What did hominin populations look like just before this ancestral population split off and the evolution of *Homo sapiens* began? The molecules cannot answer this question. Only the discovery of new fossils can. Even as you read this, it is virtually certain that researchers are at the fossil-rich localities in Africa and Asia that have yielded evidence of our ancestors, looking for traces of the humans who lived about 150,000 years ago.

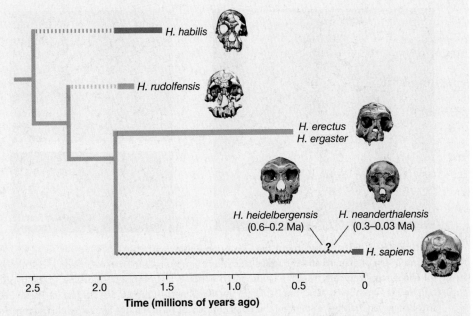

FIGURE 26.11 A Family Tree of Recent Humans
The solid, colored branches on this tree show the range of dates over which fossils of recent hominid species have been found. The dotted lines indicate dates for which researchers expect to find fossils of each species. The line leading to *Homo sapiens* is jagged because it is still not clear where *H. neanderthalensis* and *H. heidelbergensis* fit along this branch.

mals was represented. In a relatively short time, creatures with shells, exoskeletons, legs, heads, tails, and segmented bodies had evolved.

Section 26.3 explores *how* the Cambrian explosion occurred. Here let's focus on *what* happened. By combining evidence from the fossil record and phylogeny reconstruction, researchers have begun to clarify the timing and sequence of key events.

Cambrian Fossils: An Overview

The Cambrian explosion is documented by three major fossil assemblages that record the state of animal life at 570 Ma, at 565 to 544 Ma, and at 525 to 515 Ma. The species collected from each of these intervals are referred to respectively as the Doushantuo fossils (from the Doushantuo formation in China), Ediacaran fossils (from Ediacara Hills, Australia), and Burgess Shale fossils (from British Columbia, Canada), as

Figure 26.12a (page 568) shows. Fossils from the Ediacaran interval and Burgess Shale interval have now been found at localities throughout the world.

Fortunately for biologists, the glimpses of life provided by the three groups of organisms are extraordinarily clear. Each of the assemblages breaks a cardinal rule of fossil preservation. Soft-bodied animals, which usually do not fossilize efficiently, are well represented in all three. By sheer luck, a few habitats happened to exist at each of these time intervals in which burial occurred so rapidly and decomposition occurred so slowly that organisms without shells were able to fossilize.

To appreciate why this extraordinarily efficient preservation is so important, consider the Burgess Shale. In several localities around the world dated to 525–515 million years ago, fossilization processes were typical—only shelled organisms are found. But soft-bodied organisms also fossilized in the atypical conditions of the Burgess Shale. At this locality, over *five times*

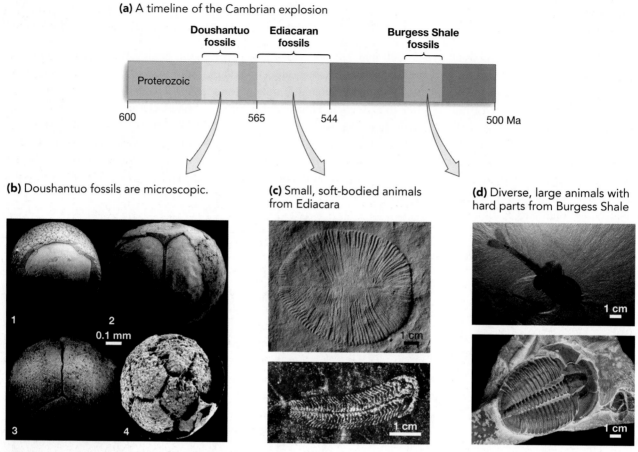

(a) A timeline of the Cambrian explosion

(b) Doushantuo fossils are microscopic.

(c) Small, soft-bodied animals from Ediacara

(d) Diverse, large animals with hard parts from Burgess Shale

FIGURE 26.12 Fossils Document the Cambrian Explosion
(a) The origin of animals and their diversification during the Cambrian explosion is documented by three major fossil assemblages.
(b) These fossilized cells from the Doushantuo deposits are arranged according to the hypothesis that they represent (1) a fertilized egg and then embryos at (2) the 2-cell, (3) the 4-cell, and (4) the 16-cell stages. **(c)** These 560 million-year-old fossils of small, soft-bodied animals were found in the Ediacara Hills of Australia. **(d)** Similar to this shrimplike arthropod (top) and trilobite (bottom), many of the animals fossilized in the Burgess Shale had heads, tails, shells (exoskeletons), and appendages.

as many species are represented. The database is 500 percent better than usual.

The presence of these exceptionally rich deposits before, during, and after the Cambrian explosion makes the fossil record for this event extraordinarily complete. Let's take a detailed look at the **faunas**—the animal types—found in the Doushantuo, Ediacaran, and Burgess Shale fossils.

The Doushantuo Microfossils

Two papers, published within days of each other in 1998, introduced the world to the faunas preserved in the Doushantuo formation of southern China. These phosphate rocks, mined extensively for fertilizer, contain abundant submillimeter-sized fossils. Through careful preparation and microscopy, researchers were able to identify several dozen individual sponges, ranging from 150 μm to 750 μm across, in samples dated at approximately 580 million years ago. Several cell types are present in these fossils, and some cell bodies contain spicules (projections) made of silica-containing compounds. Both of these traits are typical of sponges living today.

In slightly younger deposits found in the formation, dated to about 570 million years ago, a different team of biologists found clusters of cells that they interpreted as animal embryos. This conclusion was based on a simple observation: Their samples contained one-celled, two-celled, four-celled, and eight-celled fossils, along with individuals containing larger cell numbers whose overall size was the same (**Figure 26.12b**). Recall from Chapter 21 that this is exactly the pattern that occurs during cleavage in today's animals. In other words, cell number increases but total volume remains constant. What type of animal did such embryos develop into? The answer is still unknown.

The animals isolated from both samples were scattered among abundant cyanobacterial cells, as well as multicellular algae that may represent early members of what you might know as "seaweeds:" the red, brown, and green algal lineages. These bacteria and algae were undoubtedly photosynthetic. The composite picture, then, is of a shallow-water marine habitat dominated by photosynthetic organisms. Scattered among them were tiny animals that may have made their living by filtering organic debris from the water. They were the first animals on Earth.

The Ediacaran Faunas

In the 1940s paleontologists discovered a variety of animal fossils in the Ediacara Hills of southern Australia. The specimens included the compressed bodies of sponges, jellyfish, and comb jellies, and many burrows, tracks, and other traces from unidentified animal species (**Figure 26.12c**). No animals with shells were present, however. In the decades since the initial discovery, similar faunas that are dated between 565 and 544 million years ago have been found at sites around the world.

Taken together, the fossils from this 20-million-year interval indicate that shallow-water marine habitats contained a diversity of animal species. None of the organisms that fossilized during this period have limbs, however, and none have heads or mouths or feeding appendages. As the scale bars in Figure 26.12c show, these organisms were also tiny compared with today's animals. These observations suggest that Ediacaran animals were small individuals that burrowed in sediments, sat immobile on the sea floor, or floated in the water. There is no evidence that they had structures associated with actively hunting and capturing food. Instead, it is likely that Ediacaran animals simply filtered or absorbed organic material from their surroundings.

The Burgess Shale Faunas

The discovery of fossils in the Burgess Shale formation of British Columbia, Canada, early in the twentieth century ranks among the most sensational additions ever made to the fossil record. Combined with the later unearthing of an extraordinary fossil assemblage in the Chengjiang deposits of China, the Burgess Shale gives researchers a compelling picture of life in the oceans 525–515 million years ago.

Few, if any, species in the Ediacaran faunas are also found in the Burgess Shale–type assemblages 20–40 million years later. New species of sponges, jellyfish, and comb jellies are abundant; but entirely new groups are present as well. Principal among these are the arthropods and mollusks. Today, the arthropods include the spiders, insects, and crustaceans (crabs, shrimp, and lobsters); mollusks include the clams, mussels, squid, and octopi. But echinoderms (sea stars and sea urchins), several types of worm and wormlike creatures, and even a chordate—the group ancestral to today's vertebrates—are found in these fossil faunas. In short, virtually every major animal lineage is documented there.

This tremendous increase in the size and morphological complexity of animals, illustrated in **Figure 26.12d**, was accompanied by diversification in how they made a living. The Cambrian seas were filled with animals that had eyes, mouths, limbs, and shells. They swam, burrowed, walked, ran, slithered, clung, or floated; there were predators, scavengers, filter feeders, and grazers. The diversification of the animals created and filled many of the ecological niches found in shallow-water marine habitats today.

The Cambrian explosion still echoes. Animals that fill today's teeming tide pools, beaches, and mudflats trace their ancestry to species preserved in the Burgess Shale.

26.3 The Genetic Mechanisms of Change

The Doushantuo, Ediacaran, and Burgess Shale faunas document relatively rapid and fundamental changes in the size and complexity of multicellular animals. Now the focus shifts to a new question: *How did the major changes chronicled in the fossil record occur?*

A remarkable coalition of scientists is assembling to answer that question. Paleontologists, comparative anatomists, developmental biologists, and molecular geneticists are all contributing data aimed at clarifying the genetic basis for novel structures such as heads, tails, and limbs. This research program is often called **evo-devo**, because it combines evolutionary and developmental studies. To explore this field, let's consider two important types of mutation that can make major innovations possible: gene duplications and changes in gene expression.

Gene Duplications and the Cambrian Explosion

Chapter 22 emphasizes that the proteins coded by *homeotic genes* are responsible for laying out the three-dimensional pattern of multicellular organisms as they develop. Homeotic genes help specify the location and shape of many key structures. As soon as developmental biologists discovered how important these genes are in specifying the tissues and structures found in animals, they began asking about their role in evolution. Did the origin and elaboration of homeotic genes trigger the origin and elaboration of animal body shapes and appendages that occurred during the Cambrian explosion? Stated another way, was the increase in morphological complexity during the Cambrian caused by an increase in the number and complexity of homeotic genes?

To answer this question, biologists are working to determine the number and identity of homeotic genes found in different animal groups and to compare them with the phylogeny of the same groups. The idea is to look for correlations between the evolutionary history of animal groups, their morphology, and their genetic makeup. Many researchers predicted there would be a strong association between the order in which animal groups appeared during evolutionary history, the number of homeotic genes present in each group, and each group's morphological complexity and body size. Some biologists even suggested that each major animal lineage would have unique homeotic genes associated with its unique body plans and appendages.

The logic behind this "new genes, new bodies" hypothesis was that gene duplication events could have occurred before and during the Cambrian explosion and produced new copies of existing homeotic genes. (Look back at Chapter 20 to review how gene duplication events occur.) These new genes would make possible the new body plans and appendages recorded in the Burgess Shale fauna. Again, the idea was that the number of homeotic genes present would correlate directly with morphological complexity.

To see if the predictions made by this hypothesis hold up, use **Figure 26.13** to analyze the homeotic genes called the *Hox*

genes. To help interpret the data, recall from Chapter 20 that *Hox* genes are always found in clusters, with genes lined up on the chromosome one after the other. Recall, too, that each gene in the *Hox* cluster has a distinct function as an embryo develops. Specifically, each *Hox* gene is involved in a different aspect of pattern formation—the events that organize cells in the space inside an embryo. Finally, note that each *Hox* gene in Figure 26.13 is color coded to indicate homology with other genes. If genes have the same color, it means their DNA sequences are so similar that researchers are confident the genes are related by common descent.

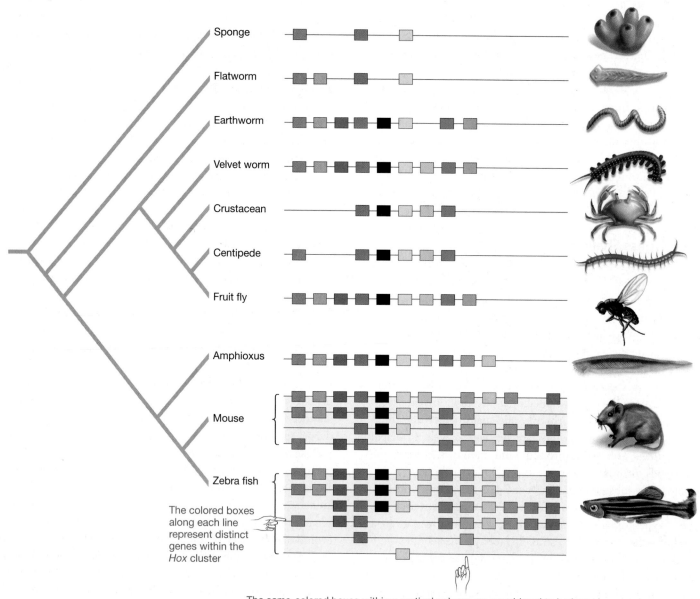

The same-colored boxes within a vertical column are considered to be homologous—
meaning they are related because they are derived from the same ancestral sequence

FIGURE 26.13 *Hox* Genes in Animals

Hox clusters, represented by a horizontal line. (Note that mice have four distinct *Hox* clusters, and zebra fish have six.) **EXERCISE** Next to the illustration of each animal on this evolutionary tree, write the number of *Hox* genes it has.

How do the genes found in different lineages map onto the animal tree? Figure 26.13 illustrates the following points:

- As predicted by the new genes, new bodies hypothesis, the number and identity of *Hox* genes varies widely among animals.

- Groups that branched off early and have small, simple bodies, such as sponges, have fewer *Hox* genes than do groups that branched off later—such as vertebrates. The *Hox* cluster appears to have expanded during the course of evolution (from top to bottom in Figure 26.13). This observation also supports the new genes, new bodies hypothesis.

- It is sensible to argue that the new *Hox* genes were created by gene duplication events, because genes within the cluster are similar in their structure and base sequence. For example, the genes that are colored orange in Figure 26.13, which appear first in flatworms, probably originated when a mutation resulted in a duplication of the gold-colored gene, which is present in sponges. The two important points here are that (1) biologists can use the phylogeny to understand the order in which different genes within the cluster appeared, and (2) gene duplication provides a mechanism for the "new genes, new bodies" hypothesis.

- Vertebrates, represented in Figure 26.13 by the mouse and the zebra fish, have several copies of the entire *Hox* cluster. This observation implies that the complete set of 13 genes was duplicated several times. Because all vertebrates examined to date have the duplicated clusters, the mutations probably occurred close to the origin of the lineage. Vertebrates include some of the largest and most complex animals, so this observation also supports the new genes, new bodies hypothesis. But it is important to note from the figure that in some cases, copies of genes within the duplicated clusters have been lost. Mice, for example, have no pink-colored gene in their second set of *Hox* genes.

The data would certainly appear to support a simple-minded version of the new genes, new bodies hypothesis, except for one crucial observation: Within arthropods (represented by the fruit fly, centipede, and crustacean) and within vertebrates, there is no correspondence at all between the number of *Hox* genes and the complexity of the resulting organisms. For example, fish evolved much earlier than mammals did, and the bodies of mammals are considered more complex than are those of fish; but zebra fish have more *Hox* genes than mice do.

Clearly, the situation is more involved than initially predicted. Duplication of *Hox* genes has undoubtedly been important in making the elaboration of animal body plans possible. But new genes are not the whole story: Changes in the expression and function of existing genes have been equally or even more important.

Changes in Gene Expression: The Origin of the Foot

Recent research on the vertebrate foot provides a compelling example of how changes in gene expression can affect evolution. One of the major innovations during vertebrate evolution was the origin of a limb with feet. This led to the evolution of amphibians, reptiles, and mammals—the **tetrapods** ("four-footed"). As data introduced in Chapter 33 will show, the fossil record indicates that the tetrapod limb evolved from the fins of fish. But tetrapods actually have fewer *Hox* genes than fish have. Might changes in the timing or location of homeotic gene expression be responsible for the fin-to-limb transition?

Biologists have explored this question by comparing the expression of the *hoxd-11* gene and a gene called *Sonic hedgehog* (*Shh*) in the zebra fish and the mouse. In tetrapod embryos, *hoxd-11* is expressed as the limbs begin to bud and grow. The gene product marks locations along the long axis of the limb as it grows out. In contrast, the protein produced by *Shh* marks the front-to-back, or anterior-to-posterior, axis of the developing limb. Both genes are involved in pattern formation—the spatial organization of cells inside an embryo.

Are the timing and location of *hoxd-11* and *Shh* expression identical in fish and tetrapods (**Figure 26.14**, page 572)? To answer this question, researchers treated *limb buds*—the tissues that develop into fins and limbs—with a molecule that hybridizes to the mRNA molecule transcribed from *hoxd-11*. (This technique is called in situ hybridization; see Chapter 21.) They also stained limb buds with a molecule that hybridizes to the gene transcript produced by *Shh*. The two treatments allowed them to identify when and where these genes begin producing their protein product.

When the investigators treated the fish and mammal limb buds early in development, they found no differences in the pattern of gene expression. As part (a) in the "Results" of Figure 26.14 shows, *hoxd-11* gene transcripts appear in the hindmost part of the developing limb in both the fish and the mouse. Similarly, *Shh* transcripts are found in similar amounts and locations in fish and mammal limbs at this stage of development.

When the biologists performed the same experiment late in development, however, a striking difference emerged. In mouse limbs, the location of *hoxd-11* transcripts shifts. Study the late limb bud from a mouse embryo, shown in part (b) in the "Results" of Figure 26.14, and note that the *hoxd-11* product localizes to the part of the structure that is away from the body and faces toward the head. There is also a late expression of *Shh*. But neither of these events occurs in fish. Part (b) of the "Results," for example, shows that *hoxd-11* is expressed in small amounts and only in the hindmost cells late in zebra-fish limb development. The data document a dramatic difference in the timing and location of gene expression in zebra fish versus mice.

The researchers who did the study propose that this difference was an innovation that produced the first hand and foot

Question: Does gene expression differ between the fish fin and the tetrapod limb?

Hypothesis: Genes that help organize developing limbs have different patterns of expression in fins and limbs.

Null hypothesis: Patterns of expression in pattern-formation genes are alike in fins and limbs.

Experimental setup: Stain developing limb buds with molecules that attach to *hoxd-11* gene products.

Prediction: The location or the timing of *hoxd-11* gene expression, or both, will differ when fish and limb buds are compared.

Prediction of null hypothesis: The location and the timing of *hoxd-11* gene expression will be alike in fins and limbs.

Results:

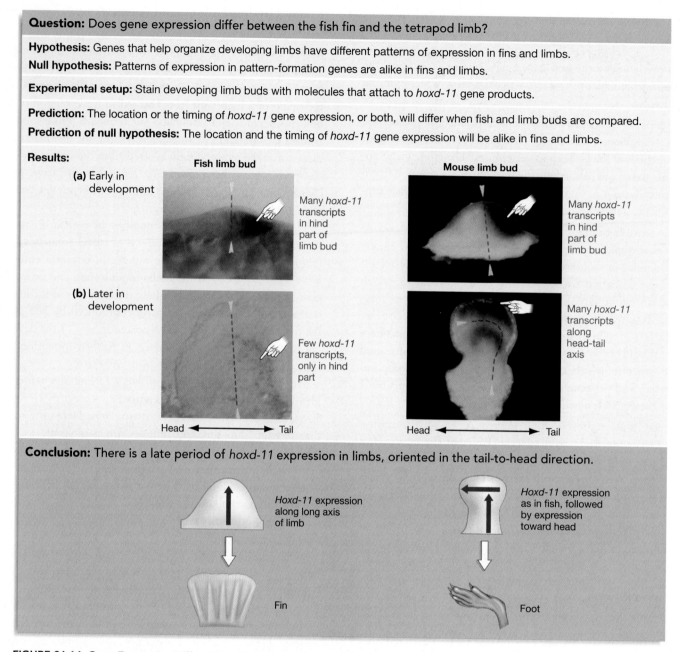

Conclusion: There is a late period of *hoxd-11* expression in limbs, oriented in the tail-to-head direction.

FIGURE 26.14 Gene Expression Differs in the Limb Buds of Fish and Mice
(a) In early development, there is no difference in the pattern of gene expression in the fish and mouse limb buds. **(b)** In later development, there is a striking difference in gene expression. This difference may account for the origin of the hand and foot.

tissues in the history of life. The sketches in the "Conclusion" of Figure 26.14 illustrate their hypothesis that the late and "reoriented" expression of *hoxd-11* and *Shh* added an entirely new element to the limb. More specifically, the researchers suggest that the mutations occurred in air-breathing, swamp-dwelling fish that lived about 400 million years ago. They proposed that the mutations were favored by natural selection, because the new structure made movement on land more efficient for individuals in these populations.

The predictions made by this hypothesis are now being tested. For example, the biologists predict that *hoxd-11* and *Shh*

expression follows the zebra-fish pattern even in fish groups that branched off very early in their radiation. (Zebra fish belong to a group that emerged millions of years after the tetrapods had evolved from fish). Today the fish lineages most closely related to tetrapods are represented by several lungfish and an unusual species called the coelacanth. Some lungfish use their fins to pull themselves along a substrate. As coelacanths swim, they move their fins in a pattern similar to the walking gait of a tetrapod. How are *hoxd-11* and *Shh* expressed as the limbs of lungfish and coelacanths develop? The experiments needed to answer this question are under way.

✓ CHECK YOUR UNDERSTANDING

The emerging field called evo-devo is creating a great deal of excitement among biologists because it offers insight into the genetic changes responsible for major innovations during the history of life. Using the radiation of animals or the origin of the tetrapod foot as an example, you should be able to describe how data from molecular genetics, the fossil record, and phylogenies are being combined to provide a more complete understanding of important evolutionary events.

26.4 Adaptive Radiations

This chapter is about the shape of the tree of life and how and why it has changed through time. Thus far, we have examined the tools biologists use to study the tree of life's past, investigated a major suite of branching events in detail, and probed the genetic mechanisms that underlie major "growth spurts" in the tree of life. Now it's appropriate to ask, If a biologist steps back and looks at the tree as a whole, what broad patterns can be discerned?

When the tree of life is examined from afar, one of the patterns that jumps out is that dense, bushy outgrowths are scattered among the branches. As **Figure 26.15a** shows, this shape results when many large and distinctive groups of organisms branch off from a lineage in a short amount of time. Biologists sometimes call this pattern a *star phylogeny*, because of its starburst shape. Why does this pattern exist? One of the leading causes is a phenomenon known as an adaptive radiation.

An **adaptive radiation** occurs when a single lineage produces many descendant species that live in a wide diversity of habitats and make their living in a variety of ways. When the diversification occurs quickly, it results in a large polytomy or star phylogeny. **Figure 26.15b** shows a few of the Hawaiian honeycreepers—a diverse lineage of songbirds that evolved after a finchlike ancestor happened to colonize the Hawaiian islands. Considering how varied honeycreeper beaks are in size and shape, you should not be surprised that different honeycreepers obtain food by eating insects, sucking nectar, or cracking seeds. **Figure 26.15c** illustrates a few of the Hawaiian silverswords, plants that evolved from a species of tarweed native to California. The silverswords that resulted from the adaptive radiation in Hawaii live in habitats ranging from lush rain forests to austere lava flows, and they range in growth habit from mosslike mat-formers to vines to small trees.

Another classical example of adaptive radiation is the diversification of mammals that took place between 65 and 50 million years ago. During this relatively short interval, the primates (monkeys and apes), bats, carnivores, deer, whales, horses, and rodents originated. The organisms resulting from this rapid divergence represent a remarkable array of adaptive forms. They swim, fly, glide, burrow, swing through trees, walk on four legs, or walk on two legs. They occupy habitats from the open ocean to mountaintops and from rain forests to deserts. They eat fruit, nuts, leaves, twigs, bark, insects, crustaceans, mollusks, fish, and other mammals.

The hallmark of an adaptive radiation is ecological diversification within a single lineage. What makes adaptive radiations occur?

(a) Adaptive radiations produce star phylogenies.

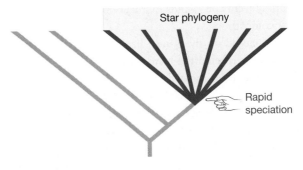

Star phylogeny

Rapid speciation

(b) Hawaiian finches underwent adaptive radiation.

(c) Hawaiian silverswords underwent adaptive radiation.

FIGURE 26.15 Adaptive Radiations Often Produce "Star Phylogenies"
(a) A phylogeny has a starburst shape when rapid speciation has occurred. If speciation is followed by divergence into many different adaptive forms, then an adaptive radiation has taken place. On the Hawaiian islands, adaptive radiations followed the arrival of both a finch from North America and tarweed seeds from California. These radiations produced the Hawaiian **(b)** finches and **(c)** silverswords.

Colonization Events as a Trigger

One of the most consistent themes in adaptive radiations is opportunity. The radiation of mammals, for example, occurred immediately after the extinction of dinosaurs. As mammals diversified, they took over the ecological roles formerly filled by dinosaurs and swimming reptiles. Adaptive radiations often occur when habitats are unoccupied by competitors.

Recently a group of biologists documented this process in detail. They did not study a radiation that followed an extinction event, however. Instead they analyzed radiations triggered by colonization events on islands that had distinct habitats and were free of competitors.

The study focused on the *Anolis* lizards of the Caribbean. Biologists have interpreted the history of this group of lizards as an adaptive radiation for two reasons: The lineage includes 150 species, and there is a strong correspondence between the size and shape of each species and the habitat it occupies. Most *Anolis* species that are twig-dwellers, for example, have relatively short legs and tails, while those that spend most of their time clinging to broad tree trunks or running along the ground tend to have long legs and tails (**Figure 26.16a**). These data suggest that lizard species have diversified in a way that allows them to occupy many different habitats.

Exactly how did the diversification occur? As the first step in answering this question, the biologists estimated the phylogeny of *Anolis* from DNA sequence data. Then they compared the habitats occupied by each species with their relationships on the phylogenetic tree. The results shown in **Figure 26.16b** for species found on two different islands are typical. Notice that the original colonist on each island belonged to a different ecological type. The initial species on Hispaniola, for example, lived on the trunks and crowns of trees, while the original colonist on Jamaica occupied twigs. From different evolutionary starting points, then, an adaptive radiation occurred on both islands. The key point is that, on both islands, the same four ecological types eventually evolved. New species arose on each island independently, but because both islands had similar habitats, each island ended up with a complement of species that was similar in lifestyle and appearance. The same type of convergent evolution occurred repeatedly. The researchers found the same pattern on many other islands. In other words, a series of "miniature adaptive radiations" occurred—one on each island—within the overall *Anolis* radiation. The small-scale and large-scale radiations were triggered by two conditions: opportunity in the form of both available habitat and lack of competitors.

(a) Species of *Anolis* vary in leg length and tail length. Some species are ground dwelling; others live in distinct regions of shrubs or trees.

(b) The same adaptive radiation of *Anolis* has occurred on different islands, starting from different types of colonists.

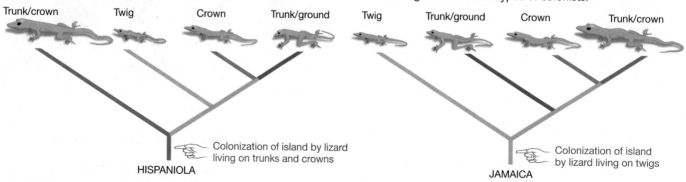

FIGURE 26.16 Adaptive Radiations of *Anolis* Lizards
(a) Short-legged lizard species (left, right) spend most of their time on the twigs of trees and bushes; long-legged species (center) live on tree trunks and the ground. **(b)** The evolutionary relationships among lizard species on the islands of Hispaniola and Jamaica. The initial colonist species was different on these islands; but in terms of how they look and where they live, a similar suite of four species evolved. According to these data, similar adaptive radiations took place independently on the two islands.

(a) Insects have a distinctive body plan.

(b) Flowering plants have a unique reproductive structure.

(c) Cichlids have unique mouths and jaws.

(d) Feathers evolved in the dinosaur ancestors of birds.

FIGURE 26.17 Some Adaptive Radiations Are Associated with Morphological Innovations
(a) The insect body includes several important innovations, including wings, compound eyes, three pairs of legs, and a segmented body organized into three general regions. **(b)** The flower was a morphological innovation that made pollen transfer and reproduction more efficient. **(c)** Cichlid species have distinctive mouthparts, including a pair of specialized jaws located in their throat. [©Don P. Northup www.africancichlidphotos.com] **(d)** The evolution of feathers and flight in dinosaurs triggered the adaptive radiation of birds. Note the long feathers trailing behind the limbs of this theropod dinosaur.

The Role of Morphological Innovation

Other "triggers," besides extinction or colonization events, can ignite adaptive radiations. Foremost among them is morphological innovation—a phenomenon discussed earlier in this chapter. Important new traits such as multicellularity, shells, exoskeletons, and limbs were a driving force behind the adaptive radiation called the Cambrian explosion. The evolution of the limb triggered the diversification of tetrapods. But many of the other important diversification events in the history of life started off with the evolution of a key morphological trait that allowed descendants to live in new areas, exploit new sources of food, or move in new ways:

- The evolution of wings, three pairs of legs, and a protective external skeleton helped make the insects the most diverse lineage on Earth, with perhaps 10 million species today (**Figure 26.17a**).

- Flowers are a unique reproductive structure that triggered the diversification of angiosperms (flowering plants). Today angiosperms are far and away the most species-rich lineage of land plants, with over 250,000 species known (**Figure 26.17b**).

- Cichlids are a lineage of fish that evolved a unique set of jaws in their throat. These second jaws make food process-

ing extremely efficient. Different species have throat jaws specialized for crushing snail shells, shredding tissue from other fish, or mashing bits of algae. Over 300 species of cichlid live in Africa's Lake Victoria alone (**Figure 26.17c**).

- Feathers and wings gave some dinosaurs the ability to fly (**Figure 26.17d**). Today the clade called birds contains about 10,000 species, with representatives that live in virtually every habitat on the planet.

In sum, adaptive radiation, a key pattern in the history of life, is usually associated with a new ecological opportunity or a morphological innovation. After an adaptive radiation, rapid speciation and morphological divergence are tightly coupled.

26.5 Mass Extinctions

Mass extinction events are evolutionary hurricanes. They buffet the tree of life, snapping twigs and breaking branches. They are catastrophic episodes that wipe out huge numbers of species and lineages in a short time, giving the tree of life a drastic pruning. One mass extinction event, about 250 million years ago, nearly uprooted the tree entirely. The end-Permian extinction came close to ending multicellular life on Earth.

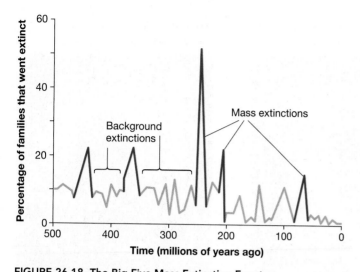

FIGURE 26.18 The Big Five Mass Extinction Events
This graph shows the percentage of families that went extinct over each interval in the fossil record since the Cambrian explosion. The five mass extinction events are drawn in red. EXERCISE Using the data in Figure 26.9, label the graph with the names of the geologic periods in which each of The Big Five mass extinctions occurred.

Mass extinction events need to be distinguished from background extinctions. A **mass extinction** refers to the rapid extinction of a large number of lineages scattered throughout the tree of life. More specifically, a mass extinction occurs when at least 60 percent of the species present are wiped out within 1 million years. **Background extinction** refers to the lower, average rate of extinction observed when a mass extinction is not occurring. Although there is no hard-and-fast rule for distinguishing between the two extinction rates, paleontologists traditionally recognize and study five mass extinction events. **Figure 26.18**, for example, plots the percentage of plant and animal families that died out during each stage in the geologic time scale since the Cambrian explosion. Five spikes in the graph—denoting a large number of extinctions within a short time—are drawn in red. These are referred to as "The Big Five."

How Do Background and Mass Extinctions Differ?

Biologists are interested in distinguishing between background and mass extinctions because these events have contrasting causes and effects. Background extinctions are thought to occur when normal environmental change or competition with other species reduces certain populations to the point at which they die out. Mass extinctions, in contrast, are thought to result from extraordinary, sudden, and temporary changes in the environment. During a mass extinction, species do not die out because individuals are poorly adapted to normal conditions. Rather, species die out from exposure to exceptionally harsh, short-term conditions—such as huge volcanic eruptions or rapid sea-level changes. In a general sense, background extinctions are thought to result primarily from natural selection. Mass extinctions, in con-

trast, function like genetic drift and produce extinctions that are largely random with respect to the fitness of individuals under normal conditions.

To drive these points home, and to see what happens after a mass extinction has occurred, let's examine one of The Big Five in detail. The event we'll analyze, the mass extinction at the end of the Cretaceous period, was not the largest in history. (The mass extinction at the end of the Permian period may have wiped out 90 percent of the multicellular organisms alive at the time.) However, the end-Cretaceous extinction is among the most dramatic: It extinguished the dinosaurs and ushered in the diversification of mammals.

What Killed the Dinosaurs?

The end-Cretaceous extinction of 65 million years ago is as satisfying a murder mystery as you could hope for. Who did it? Decades after biologists recognized the extent of the end-Cretaceous event, the leading hypothesis was that it resulted from climate change. In the early 1970s, however, a researcher offered a dramatic alternative hypothesis. The **impact hypothesis** for the extinction of the dinosaurs claimed that an asteroid struck Earth and caused widespread destruction.

The impact hypothesis for the end-Cretaceous extinction was intensely controversial. As researchers set out to test its predictions, however, support began to grow:

- Worldwide, sedimentary rocks that formed at the Cretaceous-Tertiary (K–T)* boundary were found to contain extremely high quantities of the element iridium. Iridium is extremely rare in Earth rocks, but it is an abundant component of asteroids and meteorites (**Figure 26.19a**).

- Shocked quartz and microtektites are minerals that are found only at documented meteorite impact sites (**Figure 26.19b**). Shocked quartz forms when shock waves from an asteroid impact alter the structure of sand grains. Microtektites form when minerals are melted at an impact site and then cool and resolidify. In Haiti and other locations, both shocked quartz and microtektites have been discovered in abundance in rock layers dated to 65 million years ago.

- A crater the size of Sicily was found just off the northwest coast of Mexico's Yucatán peninsula (**Figure 26.19c**). Microtektites are abundant in sediments from its walls.

Taken together, these data provided conclusive evidence in favor of the impact hypothesis. Researchers now agree that the mystery is solved: An asteroid did it.

Based on currently available data, astronomers and palontologists estimate that the asteroid that struck Earth 65 million years ago was about 10 km across. To get a sense of this

*Geologists use K to abbreviate Cretaceous, because C refers to the Cambrian period.

(a) Iridium is present at high concentration in rocks formed 65 million years ago.

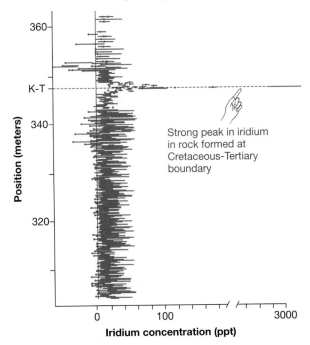

(b) Minerals that form during asteroid impacts

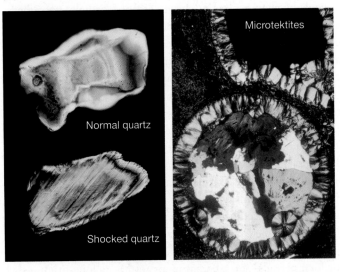

Microtektites

Normal quartz

Shocked quartz

(c) The asteroid left a crater 180 km (112 miles) wide.

FIGURE 26.19 Evidence of an Asteroid Impact 65 Million Years Ago

(a) The concentration of iridium in rocks that formed on either side of the Cretaceous-Tertiary (K-T) boundary, in parts per trillion (ppt). **(b)** Normal quartz grains are markedly different from shocked quartz grains. The striations in the shocked quartz are caused by a sudden increase in pressure. Microtektites (right) are tiny glass particles formed when minerals melt at an impact site and then recrystallize. **(c)** Geologists have identified the walls of a crater off the northwest coast of the Yucatán Peninsula, dated to 65 million years ago.

QUESTION Which of these three pieces of evidence for an asteroid impact do you judge to be the strongest, and why?

scale, consider that Mt. Everest is about 10 km above sea level and that planes cruise at an altitude of about 10 km. Imagine Earth being hit by a rock the size of Mt. Everest, or a rock that would fill the space between you and a jet in the sky.

The distribution of shocked quartz and microtektites dated to 65 Ma indicates that the asteroid hit Earth at an angle and splashed material over much of southeastern North America. To understand the impact's consequences, consider the results of the Tunguska event. On June 30, 1908, a piece of a comet about 30 m across, about 2 megatons in mass, exploded at least 5 km above Earth's surface near the Tunguska River in Siberia. The explosion released about 1000 times as much energy as the atomic bomb that destroyed Hiroshima in World War II. The event incinerated vegetation over hundreds of square kilometers, leveled trees over thousands of square kilometers, and significantly increased dust levels across the Northern Hemisphere. A deafening blast was heard 500 km (300 mi) away; people

standing 60 km away were thrown to the ground or knocked unconscious. These events pale in comparison to what happened 65 million years ago. The energy the K-T asteroid unleashed was 37 *million* times greater than the Tunguska event.

According to both computer models and geologic data, the consequences of the K-T asteroid strike were nothing short of devastating. A tremendous fireball of hot gas would have spread from the impact site; large soot and ash deposits in sediments dated to 65 million years ago testify to catastrophic wildfires, worldwide. The impact site itself is underlain by a sulfate-containing rock called anhydrite. The SO_4^{2-} released by the impact would have reacted with water in the atmosphere to form sulfuric acid (H_2SO_4), triggering extensive acid rain. Massive quantities of dust, ash, and soot would have blocked the Sun for long periods, leading to rapid global cooling and a crash in plant productivity. The upshot? The fossil record suggests that between 60 and 80 percent of all species went extinct.

Selectivity The asteroid impact did not kill indiscriminately. Perhaps by chance, certain lineages escaped virtually unscathed while others vanished. Among vertebrates, for example, the dinosaurs, pterosaurs (flying reptiles), and all of the large-bodied marine reptiles (mosasaurs, ichthyosaurs, and plesiosaurs) expired; mammals, crocodilians, amphibians, and turtles survived.

Why? Answering this question has sparked intense controversy and debate. For years the leading hypothesis was that the K-T extinction event was size selective. The logic here was that the extended darkness and cold would affect large organisms disproportionately, because they require more food than do small organisms. But extensive data on the survival and extinction of marine clams and snails has shown no hint of size selectivity, and small-bodied and juvenile dinosaurs perished along with large-bodied and adult forms. One hypothesis currently being tested is that organisms that were capable of inactivity for long periods—by hibernating or resting as long-lived seeds or spores—were able to survive the catastrophe. But this aspect of the mystery is still unsolved.

Recovery After the K-T extinction, fern fronds and spore dominate the plant fossil record from North America and Australia. These data suggest that extensive stands of ferns replaced diverse assemblages of cone-bearing and flowering plants after the impact and that terrestrial ecosystems were radically simplified. In marine environments, some invertebrate groups do not exhibit normal levels of species diversity in the fossil record until 4–8 million years past the K-T boundary. Recovery was slow.

The organisms present in the Tertiary were markedly different from those of the preceding period. One prominent change was that mammals took the place of the dinosaurs and marine reptiles. The lineage called Mammalia, which had consisted largely of rat-sized predators and scavengers in the heyday of the dinosaurs, exploded after the impact. Within 10–15 million years, all of the major mammalian orders observed today had appeared—from pigs to primates. Why? A major branch on the tree of life had disappeared. With competitors removed, mammals flourished.

The Cretaceous is sometimes called the Age of Reptiles, and the Tertiary the Age of Mammals. The change was not due to a competitive superiority conferred by adaptations such as fur and lactation. Rather, it was due to a chance event: a massive, low-probability impact event.

✓CHECK YOUR UNDERSTANDING

Mass extinctions have occurred repeatedly throughout the history of life. You should be able to evaluate the strength of the evidence supporting the impact hypothesis for the extinction of the dinosaurs, and explain why mass extinctions wipe out species more or less randomly—much the way genetic drift affects changes in allele frequencies.

ESSAY Is a Mass Extinction Event Under Way Now?

As human populations expand, wildlife habitat shrinks. Peter Ward, an expert on the K-T extinction, has reviewed data on accelerating rates of habitat loss around the globe and warned that a "human meteor" is about to strike the planet. The admonition raises a legitimate scientific question: Are humans currently causing a mass extinction event?

Although this is a difficult question to answer, two types of data are relevant. The first dataset involves extinction events over the past 400 years; the second involves very recent changes in the number of threatened species. Both datasets come from a period when human populations have been expanding rapidly (**Figure 26.20**).

Consider data on recently extinct species, tabulated on the right. Although the numbers and percentages appear small at first glance, recall that the time interval involved is just 400 years. If the rates of extinction recorded here were to continue for even a few thousand years, virtually every species in these groups would be wiped out. Based on these data, it appears that a mass extinction event is well under way.

It is important to note, however, that virtually all the extinct species recorded in the table were island forms. On islands, extinctions usually result from overharvesting and from the introduction of nonnative predators and herbivores, such as rats and goats. Island extinctions have already peaked in intensity. In contrast, most current extinctions are occurring in a new place and have a new cause. The extinction threat has shifted to continental habitats, and the causes have shifted to introduced species and habitat destruction—particularly the conversion of tropical wet forests to pastures

Lineage	Species Extinct since 1600	Total Named Species	Percentage of Extinct Species
Mollusks (snails, clams, mussels)	191	100,000	0.20
Crustaceans (shrimp, crabs, lobsters)	4	4000	0.01
Insects	61	1,000,000	0.005
Vertebrates	229	47,000	0.50
Gymnosperms (conifer trees and other plant groups)	2	758	0.30
Dicotyledons (a group of flowering plants)	120	190,000	0.20
Monocotyledons (a group of flowering plants)	462	52,000	0.20

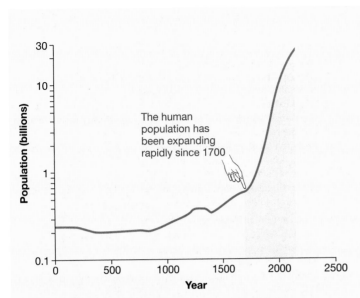

FIGURE 26.20 The Meteoric Rise of Human Populations
The estimated world population size of humans from the year 1 to the year 2150. The line past 2004 is an extrapolation based on the estimate for continued high growth.

and croplands. The destruction of these species-rich habitats is a direct consequence of increasing human populations.

Are humans currently causing a mass extinction event ... ?

How rapidly are humans currently causing extinctions? The only way to answer this question is to extrapolate from very recent events. From 1986 to 1990, for example, 15 vertebrates were known to go extinct. If this rate of extinction continued, it would take only 7000 years to eliminate half of all known vertebrates. Researchers have also attempted to answer the question by assuming that all species currently considered "threatened"—a formal status conferred by international conservation agencies—will go extinct in the next 100 years. If so, extinction rates in plant and animal lineages would vary from 10 to 100 times their background extinction rates.

The message of these admittedly crude analyses is that the human meteor is probably a real, not an imagined, threat. If present rates of human population growth and habitat destruction continue, a mass extinction event will almost undoubtedly occur.

CHAPTER REVIEW

Summary of Key Concepts

■ **Phylogenies and the fossil record are the major tools that biologists use to study the history of life.**

Phylogenetic trees can be estimated by grouping species based on overall similarity in traits or by analyzing shared, derived characters (synapomorphies) that identify monophyletic groups. Phylogenetic trees document the evolutionary relationships among species and identify the order in which events occurred. The fossil record is used in conjunction with phylogenetic analyses because it is the only direct source of data about what extinct organisms looked like and where they lived. Even though the fossil record has strong taxonomic, habitat, and temporal biases, it is an enormously valuable database.

■ **The Cambrian explosion was the rapid morphological and ecological diversification of animals that occurred during the Cambrian period.**

The diversification of animals over a 20- to 40-million-year period, starting about 540 million years ago, is perhaps the best-studied event in the history of life. The animals that lived in marine environments just before and after the Cambrian explosion are documented in the Doushantuo, Ediacaran, and Burgess Shale faunas. During this interval the first heads, tails, appendages, shells, exoskeletons, and segmented bodies evolved.

■ **The new field of "evo-devo" is providing insights into how major events in the history of life occurred, by revealing the genetic mechanisms involved.**

Data on the number and identity of homeotic genes in different animal lineages suggest that at least some of the animal radiation was possible because gene duplication events created new copies of *Hox* genes. But the correlation between morphological complexity and number of homeotic genes is far from perfect. Changes in the timing and location of existing genes were also responsible for many morphological innovations. For example, changes in gene expression may have been involved in the origin of the hand and foot in tetrapods.

■ **Adaptive radiations are a major pattern in the history of life. They are instances of rapid diversification associated with new ecological opportunities and new morphological innovations.**

Adaptive radiations can be triggered by the colonization of a new habitat or the demise of competitors after a mass extinction. Morphological innovations, such as large, complex bodies and limbs with feet and hands, can also initiate adaptive radiations. Speciation events and morphological change occur rapidly during an adaptive radiation, as a single lineage diversifies into a wide variety of ecological roles.

Web Tutorial 26.1   Adaptive Radiation

■ **Mass extinctions have occurred repeatedly throughout the history of life. They rapidly eliminate most of the species alive in a more or less random manner.**

Mass extinctions have altered the course of evolutionary history at least five times. They prune the tree of life more or less randomly and have marked the end of several prominent lineages and the rise of new branches. The Cretaceous-Tertiary extinction killed 60 to 80 percent of existing species and was caused by an asteroid impact. After the devastation of a mass extinction, it can take 10–15 million years for ecosystems to recover their former levels of diversity.

Questions

Content Review

1. Choose the best definition of a fossil.
   a. any trace of an organism that has been converted into rock
   b. a bone, tooth, shell, or other hard part of an organism that has been preserved
   c. any trace of an organism that lived in the past
   d. the process that leads to preservation of any body part from an organism that lived in the past

2. Why are the Doushantuo, Ediacaran, and Burgess Shale fossil deposits unusual?
   a. Soft-bodied animals are preserved in them.
   b. They are easily accessible to researchers.
   c. They are the only fossil-bearing rock deposits from their time period.
   d. They include terrestrial, instead of just marine, species.

3. Which of the following best characterizes an adaptive radiation?
   a. Speciation occurs extremely rapidly, and descendant populations occupy a large geographic area.
   b. A single lineage diversifies rapidly, and descendant populations occupy many habitats and ecological roles.
   c. Natural selection is particularly intense, because disruptive selection occurs.
   d. Species recover after a mass extinction.

4. Which of the following is most accurate?
   a. Mass extinctions are due to asteroid impacts; background extinctions may have a wide variety of causes.
   b. Mass extinctions focus on particularly prominent groups, such as dinosaurs; background extinctions affect species from throughout the tree of life.
   c. Only five mass extinctions have occurred, but many background extinctions have occurred.
   d. Mass extinctions involve at least 60 percent of the species present and extinguish groups rapidly and randomly; background extinctions are slower and result from natural selection.

5. Why do molecular clocks exist?
   a. Natural selection is not important at the molecular level.
   b. Homologous genes have the same structure.
   c. They can be calibrated.
   d. Some DNA sequences, in some lineages, change at a steady rate over time.

6. Why is burial a key step in fossilization?
   a. It slows the process of decay by bacteria and fungi.
   b. It allows tissues to be preserved as casts or molds.
   c. It protects tissues from wind, rain, and other corrosive elements.
   d. All of the above.

Conceptual Review

1. The text claims that the fossil record is biased in several ways. What are these biases? If the database is biased, is it still an effective tool to use in studying the diversification of life? Explain.

2. The initial radiation of animals took place over some 40 million years, at the start of the Cambrian period. Why is the radiation called an "explosion"?

3. What is the "new genes, new bodies" hypothesis? Based on the data presented in this chapter, is this hypothesis correct?

4. Give an example of an adaptive radiation that occurred after a colonization event, after a mass extinction, and after a morphological innovation. In each case, provide a hypothesis to explain why the adaptive radiation occurred.

5. Summarize the evidence that supports the impact hypothesis for the K-T extinction.

6. Why are monophyletic groups identified by shared, derived traits?

Group Discussion Problems

1. Suppose that the dying wish of a famous eccentric was that his remains be fossilized. His family has come to you for expert advice. What steps would you recommend to maximize the chances that his wish will be fulfilled?

2. Summarize the nature of the fossils found in the Doushantuo, Ediacaran, and Burgess Shale deposits. (The three fossil assemblages are listed in order, from most ancient to most recent.) According to the fossil record, what trends or general patterns occurred during the early evolution of animals? For example, did animals tend to get larger or smaller? More complex or less complex?

3. Experiments summarized in Section 24.3 suggest that the tetrapod foot resulted from changes in the timing and location of the expression of certain genes. Based on the information presented in Chapter 18 on the regulation of gene expression in eukaryotes, exactly what sort of mutations would lead to changes in when and where genes are transcribed?

4. The synthesis of results from highly disparate fields, such as developmental genetics and fossil studies, is an exciting recent trend in research on the history of life. Comment on whether you find this synthesis valuable. For example, did reading about the "new genes, new bodies" hypothesis and the experiments on gene expression in tetrapod limbs help you understand how important changes have occurred during the diversification of life? Why or why not?

Answers to Multiple-Choice Questions   1. c; 2. a; 3. b; 4. d; 5. d; 6. d

6

The Diversification of Life

A red trillium in full bloom. In the lineage that this species belongs to, flower parts tend to occur in multiples of three.

27 Bacteria and Archaea

KEY CONCEPTS

■ Bacteria and archaea affect your life. A small percentage of bacteria causes disease. Some species are effective at cleaning up pollution, and photosynthetic bacteria were responsible for the evolution of the oxygen atmosphere. Bacteria and archaea cycle nutrients through both terrestrial and aquatic environments.

■ Bacteria and archaea live in a wide array of habitats and use diverse types of molecules in cellular respiration and fermentation. Although they are small and relatively simple in their overall morphologies, these organisms are extremely sophisticated in the chemistry they can do. Many species are restricted in distribution and have a limited diet.

Although this hot spring looks devoid of life, it is actually teeming with billions of bacterial and archaeal cells from a wide variety of species. As this chapter will show, bacteria and archaea occupy virtually every environment on Earth.

Biologists who study organisms in the domains Bacteria and Archaea are exploring the most wide open frontier in biodiversity. So little is known about the extent of these domains that recent collecting expeditions have turned up an entirely new **kingdom** and numerous **phyla** (singular: **phylum**)—names given to major lineages within each domain. To a biologist, this achievement is equivalent to the sudden discovery of a new group of eukaryotes as distinctive as the ferns, sponges, or jellyfish.

Just how many bacterial and archaeal species are alive today? To date, a mere 5000 species have been formally named and described—most by the morphological species concept introduced in Chapter 25. In reality, it is virtually certain that tens of millions exist. Consider that over 400 species of bacteria are living in your gastrointestinal tract right now. Another 500 species live in your mouth, of which just 300 have been described and named. Norman Pace points out that there may be tens of millions of different insect species but notes that "If we squeeze out any one of these insects and examine its contents under the microscope, we find hundreds or thousands of distinct microbial species." Most of these **microbes** (microscopic organisms) are bacteria or archaea, and virtually all are unnamed and undescribed. If you want to discover and name a new species, then study bacteria or archaea.

Although biologists can only guess at the total number of bacterial and archaeal species living today, the abundance of these groups is well documented:

- The approximately 10^{13} (10 trillion) cells in your body are vastly outnumbered by the bacterial cells that live on and in you. An estimated 10^{12} bacterial cells live on your skin, and an additional 10^{14} cells occupy your stomach and intestines. You are a walking, talking habitat—one that is teeming with bacteria and archaea.

- A mere teaspoon of good-quality soil contains *billions* of microbial cells, most of which are bacteria and archaea.

- In sheer numbers, the bacterium *Prochlorococcus*—found in the plankton (surface-dwelling microbes) of the world's oceans—may be the dominant life-form on Earth. Biologists routinely find this organism at concentrations of 70,000 to 200,000 cells per milliliter of seawater. At these concentrations, a drop of seawater contains a population equivalent to that of a large city. Yet *Prochlorococcus* was first described and named fairly recently—in 1988.

- Biologists estimate the total number of individual bacteria and archaea alive today at 5×10^{30}. There is as much carbon in these cells as there is in all of the plants on Earth. In terms of the total volume of living material on our planet, bacteria and archaea are dominant life-forms.

In addition to being diverse and abundant, bacteria and archaea are found almost everywhere. They live in environments as unusual as oxygen-free mud, hot springs, and salt flats. They have been discovered living in bedrock to a depth of 1500 meters below Earth's surface. In the ocean they are found from the surface to depths of 10,000 m and at temperatures ranging from near 0°C in Antarctic sea ice to over 121°C near submarine volcanoes.

The physical world may be mapped and explored, and many of the larger plants and animals are named. But in **microbiology**—the study of organisms that can be seen only with the aid of a microscope—this is an age of exploration and discovery.

27.1 Why Do Biologists Study Bacteria and Archaea?

Bacteria and Archaea form two of the three largest branches on the tree of life (**Figure 27.1**). The third major branch or domain consists of eukaryotes and is called the Eukarya. Virtually all members of the Bacteria and Archaea are unicellular, and all are prokaryotic—meaning they lack a membrane-bound nucleus. These organisms are distinguished by several important features: Species in the domain **Bacteria** have cell walls made primarily of peptidoglycan (see Chapter 7), plasma membranes similar to those found in eukaryotes, and ribosomes and RNA polymerase that are distinct in structure and function from the homologous structures in the Archaea and Eukarya. Species in the domain **Archaea** have cell walls made of polysaccharides not found in bacterial or eukaryotic cell walls, unique plasma membranes (described in detail

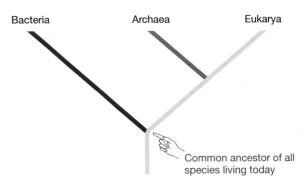

FIGURE 27.1 Bacteria, Archaea, and Eukarya Are the Three Domains of Life
Archaea are more closely related to eukaryotes than they are to bacteria. **EXERCISE** Circle the prokaryotic lineages on the tree of life. **QUESTION** Was the common ancestor of all species living today prokaryotic or eukaryotic? Explain your reasoning.

later), and ribosomes and RNA polymerase similar to those found in eukaryotes.

In general, prokaryotic cells are much smaller than eukaryotic cells and are simpler in overall form, or morphology. But in addition to being much more abundant than members of the Eukarya, Bacteria and Archaea are ancestral to eukaryotes and are much more diverse in the habitats they occupy and the molecules they can use as food. These qualities make them exceptionally interesting to study and important agents in medicine, industry, and quality of the environment. Biologists study these organisms not only to understand the diversity of life on Earth but also to improve human health and welfare.

Bacterial Diseases

No archaea are known to cause disease in humans. But of the hundreds or thousands of bacterial species that live in and on your body, a tiny fraction can disrupt normal body functions enough to cause illness. Bacteria that cause disease are said to be **pathogenic** ("disease-producing"). Pathogenic bacteria have been responsible for some of the most devastating epidemics in human history.

Robert Koch was the first biologist to establish a link between a particular species of bacterium and a specific disease. When Koch began his work on the nature of disease in the late 1800s, microscopists had confirmed the existence of the particle-like organisms we now call bacteria, and Louis Pasteur had shown that bacteria and other microorganisms are responsible for spoiling milk, wine, broth, and other foods. Koch hypothesized that bacteria might also be responsible for causing **infectious diseases**, which spread by being passed from an infected individual to an uninfected individual.

Koch set out to test this hypothesis by identifying the organism that causes *anthrax*. Anthrax is a disease of cattle and other grazing mammals that can result in fatal blood poisoning. The disease also occurs infrequently in humans and mice.

To establish a causative link between a specific microbe and a specific disease, Koch proposed that four criteria had to be met:

1. *The microbe must be present in individuals suffering from the disease and absent from healthy individuals.* By careful microscopy, Koch was able to show that the bacterium *Bacillus anthracis* was always present in the blood of cattle suffering from anthrax, but absent from asymptomatic individuals.

2. *The organism must be isolated and grown in a pure culture away from the host organism.* Koch was able to grow pure colonies of *B. anthracis* in glass dishes on a nutrient medium, using gelatin as a substrate.

3. *If organisms from the pure culture are injected into a healthy experimental animal, the disease symptoms should appear.* Koch demonstrated this crucial causative link in mice. The symptoms of anthrax infection appeared, and then the infected mice died.

4. *The organism should be isolated from the diseased experimental animal, again grown in pure culture, and demonstrated by its size, shape, and color to be the same as the original organism.* Koch did this by purifying *B. anthracis* from the blood of diseased experimental mice.

These criteria, now called **Koch's postulates**, are still used to confirm a causative link between new diseases and a suspected infectious agent. Koch's experimental results also became the basis for the **germ theory of disease**. This theory, which laid the foundation for modern medicine, holds that infectious diseases are caused by bacteria and viruses. Some of the bacteria that can cause illness in human beings are listed in **Table 27.1**. The data in the table indicate that pathogenic forms come from many lineages in the domain Bacteria and that pathogenic bacteria tend to affect tissues at the entry points to the body, such as wounds or pores in the skin, the respiratory and gastrointestinal tracts, and the urogenital canal.

The discovery of antibiotics in the late 1920s and their development over subsequent decades gave physicians effective tools to combat most bacterial infections. **Antibiotics** are molecules that kill bacteria. As the essay at the end of this chapter indicates, however, widespread use of antibiotics in the late twentieth century led to the evolution of drug-resistant strains of bacteria. Most of the bacterial species listed in Table 27.1 now include strains that are resistant to one or more of the commonly prescribed antibiotics. Coping with antibiotic resistance in pathogenic bacteria has become a great challenge of modern medicine.

Although biologists began studying bacteria because of the role these prokaryotes play in disease, the scope of research has broadened enormously. Let's consider some topics that are currently inspiring studies of bacteria and archaea.

Bioremediation

Throughout the industrialized world, some of the most serious pollutants in soils, rivers, and ponds consist of organic compounds that were originally used as solvents or fuels but leaked or were spilled into the environment. Most of these compounds

TABLE 27.1 Some Diseases Caused by Bacteria

Bacterium	Lineage	Tissues Affected	Disease
Borrelia burgdorferi	Spirochaetes	Skin and nerves	Lyme disease
Chlamydia trachomatis	Planctomyces	Urogenital canal	Genital tract infection
Clostridium botulinum	Low-GC Gram positives	Gastrointestinal tract, nervous system	Food poisoning (botulism)
Clostridium tetani	Low-GC Gram positives	Wounds, nervous system	Tetanus
Haemophilus influenzae	Proteobacteria (γ group)	Ear canal, nervous system	Ear infections, meningitis
Helicobacter pylori	Proteobacteria (ε group)	Stomach	Ulcer
Mycobacterium leprae	High-GC Gram positives	Skin and nerves	Leprosy
Mycobacterium tuberculosis	High-GC Gram positives	Respiratory tract	Tuberculosis
Neisseria gonorrhoeae	Proteobacteria (β group)	Urogenital canal	Gonorrhea
Propionibacterium acnes	High-GC Gram positives	Skin	Acne
Pseudomonas aeruginosa	Proteobacteria (β group)	Urogenital canal, eyes, ear canal	Infections of eye, ear, urinary tract
Salmonella enteritidis	Proteobacteria (γ group)	Gastrointestinal tract	Food poisoning
Staphylococcus aureus	Low-GC Gram positives	Skin, urogenital canal	Acne, boils, impetigo, toxic shock syndrome
Streptococcus pneumoniae	Low-GC Gram positives	Respiratory tract	Pneumonia
Streptococcus pyogenes	Low-GC Gram positives	Respiratory tract	Strep throat, scarlet fever
Treponema pallidum	Spirochetes	Urogenital canal	Syphilis
Vibrio parahaemolyticus	Proteobacteria (γ group)	Gastrointestinal tract	Food poisoning
Yersinia pestis	Proteobacteria (γ group)	Lymph and blood	Plague

are highly hydrophobic. Because they do not dissolve in water, they tend to accumulate in sediments. If the compounds are subsequently ingested by burrowing worms or clams or other organisms, they can be passed along to fish, insects, humans, birds, and other species. Most of these compounds are toxic to eukaryotes in moderate to high concentrations. Petroleum from oil spills and compounds that contain ring structures and chlorine atoms, such as the family of compounds called dioxins, are particularly notorious because of their toxicity to humans.

It is not unusual for the sediments where these types of compounds accumulate to become **anoxic**—meaning that they lack oxygen. To understand why this is so, recall from Chapter 9 that many organisms use oxygen as an electron acceptor in cellular respiration. Oxygen-using organisms (aerobes) are found among the decomposers that live in soil, river bottoms, and lake sediments. **Decomposers** ingest sugars and other compounds present in dead bodies or other types of organic debris and then use these compounds as electron donors (food) during cellular respiration. If water currents, wave action, or other types of forces are not strong enough to mix sediments frequently, then decomposers that are aerobes use up all of the available oxygen. When sediments become anoxic, the overall rate of decomposition slows dramatically, because organisms must use electron acceptors that are not as effective as oxygen.

Biologists who are responsible for cleaning up sites polluted with organic solvents and fuels face two challenges: (1) The polluted sediments are usually anoxic, so the overall decomposition rate is slow, and (2) at least some of the toxic compounds present are highly resistant to decomposition. Instead of being broken down into harmless compounds, certain toxic molecules tend to just "sit there." These compounds pose a long-term threat to nearby fish, birds, people, and other organisms.

To clean up sites like these, researchers have begun to explore more extensive use of **bioremediation**, the use of bacteria and archaea to degrade pollutants. Bioremediation is often based on complementary strategies:

- *Fertilizing contaminated sites* to encourage the growth of existing bacteria that degrade toxic compounds. After several recent oil spills, researchers fertilized affected sites with nitrogen but left nearby beaches untreated, as controls. In at least some cases, the fertilized sediments cleaned up much faster than the unfertilized sites (**Figure 27.2**). Dramatic increases occurred in the growth of bacteria that use hydrocarbons as an electron donor, probably because the added nitrogen was used to synthesize nucleic acids and amino acids.

- *"Seeding," or adding, specific species of bacteria* to contaminated sites shows promise of alleviating pollution in some situations. For example, researchers have recently discovered bacteria that are able to render certain chlorinated, ring-containing compounds harmless. Instead of being poisoned by the pollutants, these bacteria use ring-containing, chlorinated compounds as electron acceptors during cellular respiration. In

FIGURE 27.2 Bacteria and Archaea Can Play a Role in Cleaning Up Pollution
On the left is a rocky coast that was polluted by an oil spill but fertilized to promote the growth of oil-eating bacteria. The portion of the beach on the right was untreated.

at least some cases, the by-product is dechlorinated and nontoxic to humans and other eukaryotes. Because these bacteria do not use oxygen as an electron acceptor, they thrive in anoxic conditions. To follow up on these discoveries, researchers are now growing the bacteria in quantity and testing them in the field, to test the hypothesis that seeding can speed the rate of decomposition in contaminated sediments. Initial reports suggest that seeding may help clean up at least some polluted sites.

Extremophiles

Bacteria or archaea that live in high-salt, high-temperature, low-temperature, or high-pressure habitats are called **extremophiles** ("extreme-lovers"). As an example of these habitats, consider hot springs at the bottom of the ocean, where water as hot as 300°C emerges and mixes with 4°C seawater. At such locations, archaea are abundant forms of life. Researchers recently discovered an archaean that grows so close to these hot springs that its surroundings are at 121°C—a record for life at high temperature. This organism can live and grow in water that is heated past its boiling point (100°C) and at pressures that would instantly destroy a human being.

Extremophiles are attracting a great deal of attention from biologists. The genomes of a wide array of extremophiles have been sequenced, and expeditions regularly seek to characterize new species. Why? Part of the answer is an intrinsic fascination with these organisms. How is it possible for a cell wall and plasma membrane to withstand the water pressure that exists in the deep ocean or the osmotic pressure that exists in a salt pond? What structural changes make it possible for enzymes to function in such extreme cold or heat? In addition, biologists hope that understanding extremophiles will provide insights into how life began. Based on models of conditions that prevailed early in Earth's history, it appears likely that the first forms of life were extremophiles.

There are practical reasons for understanding extremophiles as well. Because enzymes that function at low temperature or high temperature may be useful in industrial processes, extremophiles are of commercial interest. Chapter 19 introduced *Taq* polymerase, which is a DNA polymerase that is stable up to 95°C. Recall that *Taq* polymerase is used to run the polymerase chain reaction (PCR) in research and commercial settings and that its heat stability allowed the PCR procedure to be automated. This enzyme was isolated from a bacterium called *Thermus aquaticus* (literally, "hot water"), which was discovered in hot springs in Yellowstone National Park. In addition, **astrobiologists** ("space-biologists") are using extremophiles as model organisms in the search for extraterrestrial life. The idea is that if bacteria and archaea can thrive in extreme habitats on Earth, it is likely that cells might be found in similar environments on other planets or moons of planets.

Global Change

Bacteria and archaea can live in extreme environments and use a wide variety of compounds as food because they produce extremely sophisticated enzymes. This fact, combined with their numerical abundance, has made them potent forces for global change throughout Earth's history. Bacteria and archaea have altered the chemistry of the oceans, atmosphere, and terrestrial environments for billions of years and continue to do so today.

The Oxygen Revolution   Today, oxygen represents almost 21 percent of the molecules in Earth's atmosphere. But climatologists who study the composition of the atmosphere are virtually certain that no free molecular oxygen (O_2) existed for the first 2.3 billion years after the planet formed. This conclusion is based on two observations: (1) There was no plausible source of oxygen at the time the planet cooled to a solid state; and (2) the oldest Earth rocks indicate that, for many years afterward, any oxygen that formed reacted immediately with iron atoms to produce iron oxides, such as hematite (Fe_2O_3) and magnetite (Fe_3O_4). Early in Earth's history, the atmosphere was dominated by nitrogen and carbon dioxide. Where did the oxygen we breathe come from? The answer is cyanobacteria.

Cyanobacteria are a lineage of photosynthetic bacteria (**Figure 27.3**). According to the fossil record, species of cyanobacteria first became numerous in the oceans about 2.7 billion years ago. Their appearance was momentous, because cyanobacteria were the first organisms to perform **oxygenic** ("oxygen-producing") **photosynthesis**. Oxygenic photosynthesis depends on the proteins and pigments in photosystem II. Recall from Chapter 10 that photosystem II includes enzymes capable of stripping electrons from water molecules. The reaction that "splits" water results in the production of oxygen as well as electrons. Although the electrons are required for photosynthesis to continue, the molecules of oxygen are simply released as a waste product.

The fossil record and geological record indicate that oxygen concentrations in the oceans and atmosphere began to increase to significant levels 2.4–2.2 billion years ago. Once oxygen was

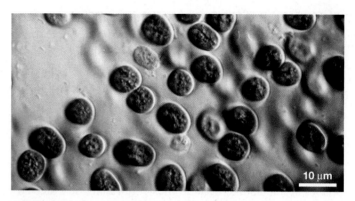

FIGURE 27.3 Cyanobacteria Were the First Organisms to Perform Oxygenic Photosynthesis
Life as we know it today would not have evolved if cyanobacteria had not begun producing oxygen as a by-product of photosynthesis more than 2.7 billion years ago.

common in the oceans, cells could begin to use it as an electron acceptor. Aerobic respiration was now a possibility. This was a crucial event in the history of life. Because oxygen is extremely electronegative, it is an efficient electron acceptor. Much more energy is released as electrons fall down electron transport chains with oxygen as the ultimate acceptor than is released with other substances as the electron acceptor. Once oxygen was available, much more ATP could be produced for each electron donated by NADH or $FADH_2$. As a result, the rate of energy production and metabolism could rise dramatically.

Coincidentally, 2.1 billion years ago is about the time the first macroscopic algae appear in the fossil record. Biologists hypothesize that a causal link was involved with the availability of oxygen. The claim is that multicellularity and large body size were made possible by the high metabolic rates and rapid growth fueled by aerobic respiration.

To summarize, data indicate that cyanobacteria were responsible for a fundamental change in Earth's atmosphere—from one dominated by nitrogen gas and carbon dioxide to one dominated by nitrogen gas and oxygen. Never before, or since, have any organisms done so much to alter the nature of our planet.

The Nitrogen Cycle   In many environments, fertilizing forests or grasslands with nitrogen results in increased growth. Researchers infer that, in such environments, plant growth and overall productivity are limited by the availability of nitrogen.

Organisms must have nitrogen to synthesize proteins and nucleic acids. Although molecular nitrogen (N_2) is extremely abundant in the atmosphere, most organisms cannot use it. To incorporate nitrogen atoms into amino acids and nucleotides, all eukaryotes and many bacteria and archaea have to obtain N in a form such as ammonia (NH_3) or nitrate (NO_3^-).

The only organisms that are capable of converting molecular nitrogen to ammonia are bacteria. The reactions involved are extremely complex and highly endergonic reduction-oxidation (redox) reactions. The enzymes required to accomplish this feat are found only in selected bacterial lineages.

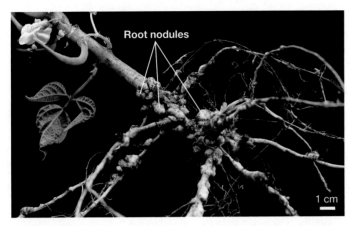

FIGURE 27.4 Some Nitrogen-Fixing Bacteria Live in Association with Plants
Root nodules form a protective structure for bacteria that fix nitrogen.

As a whole, the redox reactions that result in the production of ammonia from N_2 are referred to as **nitrogen fixation**. Certain species of cyanobacteria that live in marine plankton or in association with water plants are capable of fixing nitrogen. In terrestrial environments, nitrogen-fixing bacteria live in close association with plants—often taking up residence in special structures in the roots called *nodules* (**Figure 27.4**).

If bacteria could not fix nitrogen, it is virtually certain that only a tiny fraction of life on Earth would exist today. Large or multicellular organisms would probably be rare to nonexistent, because too little nitrogen would be available to make large quantities of proteins and build a large body.

Nitrate Pollution   Researchers are fascinated by the nitrogen-fixing bacteria because of their ecological significance, the relationships they form with plants, and their practical importance. If bacterial genes for nitrogen-fixing enzymes could be transferred safely to crop plants such as corn and rice, productivity could increase even if fertilizer use were reduced.

At present, though, most farmers must use synthetic fertilizers to add nitrogen to soils and increase crop yields. In parts of the world, massive additions of nitrogen in the form of ammonia are causing serious pollution problems. **Figure 27.5** shows why. When ammonia is added to a corn field—in the midwestern United States, for example—much of it never reaches the growing corn plants. Instead, a significant fraction of the ammonia molecules is used as food by bacteria in the soil. Bacteria that use ammonia as an electron donor to fuel cellular respiration release *nitrite* (NO_2^-) as a waste product. Other bacteria use nitrite as food and release *nitrate* (NO_3^-). Nitrate molecules are extremely soluble in water and tend to be washed out of soils and carried into groundwater or streams. If nitrate enters groundwater, it can contaminate wells that are used for drinking water. If a person drinks large quantities of nitrate, by-products created inside the body may cause cancer or interfere with hemoglobin's ability to carry oxygen. If nitrate enters streams, it can pollute them and, eventually, the ocean.

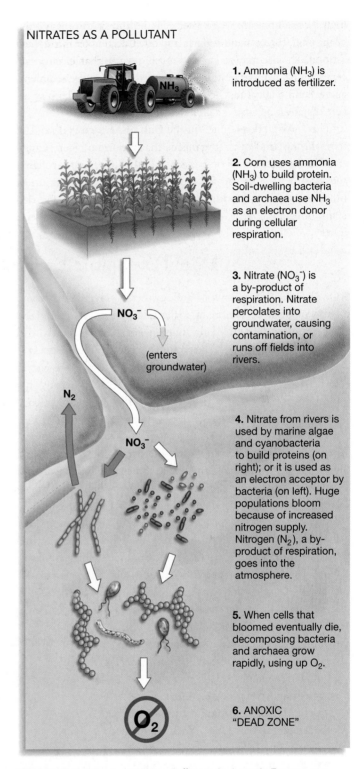

NITRATES AS A POLLUTANT

1. Ammonia (NH_3) is introduced as fertilizer.

2. Corn uses ammonia (NH_3) to build protein. Soil-dwelling bacteria and archaea use NH_3 as an electron donor during cellular respiration.

3. Nitrate (NO_3^-) is a by-product of respiration. Nitrate percolates into groundwater, causing contamination, or runs off fields into rivers.

NO_3^-

(enters groundwater)

N_2

NO_3^-

4. Nitrate from rivers is used by marine algae and cyanobacteria to build proteins (on right); or it is used as an electron acceptor by bacteria (on left). Huge populations bloom because of increased nitrogen supply. Nitrogen (N_2), a by-product of respiration, goes into the atmosphere.

5. When cells that bloomed eventually die, decomposing bacteria and archaea grow rapidly, using up O_2.

6. ANOXIC "DEAD ZONE"

O_2

FIGURE 27.5 Nitrates Act as a Pollutant in Aquatic Ecosystems

To understand why nitrates can pollute the oceans, consider the Gulf of Mexico. Both pollution-related problems and the level of nitrates in the Gulf have increased over the past several decades. Biologists hypothesize that the two events are connected. The hypothesis is that cyanobacteria, along with algae, living in the Gulf of Mexico use as food the nitrates arriving from the Mississippi River. Because the growth of these organisms is limited by nitrogen availability, their numbers explode in response to increased concentrations of nitrate. When these cells die and sink to the bottom of the Gulf, bacteria and archaea and other decomposers use them as food. The decomposers use so much oxygen as an electron acceptor in cellular respiration that oxygen levels in the sediments and even in Gulf waters decline. Nitrate pollution has been so severe that large areas in the Gulf of Mexico are anoxic.

The anoxic "dead zone" in the Gulf of Mexico is devoid of fish, shrimp, and other organisms that require oxygen. Lately the dead zone has encompassed about 18,000 km^2—roughly the size of New Jersey. Similar problems are cropping up in other parts of the world. Virtually every link in the proposed chain of events that leads to nitrate pollution involves bacteria

and archaea. But the hypothesized causative link between nitrate pollution and the development of dead zones is controversial and will probably not be resolved until the experimental addition of nitrates is shown to cause anoxic water to form.

27.2   How Do Biologists Study Bacteria and Archaea?

Biologists who study bacteria and archaea claim that they are investigating some of the most interesting and important organisms on Earth. Bacteria have been important model organisms in biological science for many decades (see **Box 27.1**). It

---

BOX 27.1   A Model Organism: *Escherichia coli*

Of all model organisms in biology, perhaps none has been more important than a common inhabitat of the human gut called *Escherichia coli*. The strain that is most commonly worked on today, called K-12, was originally isolated from a hospital patient in 1922.

Escherichia coli K-12 did not begin to rise to its present prominence as a model organism until 1945, however. In that year, biologists observed that some K-12 cells were capable of conjugation. Recall from Chapter 19 that when conjugation occurs, a copy of a plasmid from one bacterial cell—sometimes joined by one or more genes from the main bacterial chromosome—is transferred to a recipient cell. Conjugation is sometimes referred to as bacterial sex. Its discovery was important, because it gave researchers a mechanism for transferring specific alleles from one cell to another. Conjugation made a wide array of experiments possible. For example, biologists could document the function of the introduced alleles by documenting changes in the recipient cell's phenotype once the transfer process was complete.

During the last half of the twentieth century, key results in molecular biology originated in studies of *E. coli*. These results include the discovery of enzymes such as DNA polymerase, RNA polymerase, DNA repair enzymes, and restriction endonucleases; the elucidation of ribosome structure and function; and the initial characterization of promoters,

regulatory transcription factors, regulatory sites in DNA, and operons. In many cases, initial discoveries made in *E. coli* allowed researchers to confirm that homologous enzymes and processes existed in an array of organisms, often ranging from other bacteria to yeast, mice, and humans. Its success as a model for other species inspired Jacques Monod's claim that "Once we understand the biology of *Escherichia coli*, we will understand the biology of an elephant."

The genome of *E. coli* K-12 was sequenced in 1997 (see Chapter 20), and the strain continues to be a workhorse in studies of gene function, biochemistry, and particularly biotechnology.

In the lab, *E. coli* is usually grown in one of two ways: (1) in suspension

culture or (2) on plates. In **suspension culture**, cells are introduced to a liquid nutrient medium that is rotated continuously to keep the cells suspended and the nutrient solution mixed (**Figure 27.6a**). When cultured on plates, cells are spread onto **agar**, a gelatinous mix of polysaccharides (**Figure 27.6b**). Under optimal growing conditions—meaning before cells begin to get crowded and compete for space and nutrients—a cell takes just 30 minutes on average to grow and divide. At this rate, a single cell can produce a population of over a million descendants in just 10 hours. With the exception of new mutations, all of the descendant cells are genetically identical.

(a) Growth in liquid medium

(b) Growth on solid medium

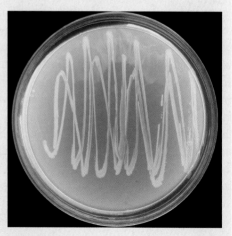

FIGURE 27.6 *E. coli* Is Readily Cultured in the Laboratory
In the lab, *E. coli* is usually grown **(a)** in suspension culture or **(b)** on plates.
QUESTION *Escherichia coli* is grown at a temperature of 37°C. Why?

is legitimate to state, however, that our understanding of the domains Bacteria and Archaea is advancing more rapidly right now than at any time during the past 100 years—and perhaps faster than our understanding of any other lineages on the tree of life.

As an introduction to the domains Bacteria and Archaea, let's examine a few of the techniques that biologists use to answer questions about them. Some of these research strategies have been used since bacteria were first discovered; some were invented less than 10 years ago.

Using Enrichment Cultures

Which species of bacteria and archaea are present at a particular location, and what do they use as food? To answer such questions, biologists rely heavily on the ability to culture organisms in the lab. Of the 5000 species of Bacteria and Archaea that have been described to date, almost all were discovered when they were isolated from natural habitats and grown under controlled conditions in the laboratory.

One classical technique for isolating new types of bacteria and archaea is called **enrichment culture**. Enrichment cultures are based on establishing a specified set of growing conditions—temperature, lighting, substrate, types of available food, and so on. The idea is to sample cells from the environment and grow them under extremely specific conditions. Cells that thrive under the specified conditions will increase in numbers enough to be isolated and studied in detail.

To appreciate how this strategy works in practice, consider research on bacteria that live deep below Earth's surface. One recent study began with samples of rock and fluid from drilling operations in Virginia and Colorado. The samples came from sedimentary rocks at depths ranging from 860 to 2800 meters below the surface, where temperatures are between 42°C and 85°C. The questions posed in the study were simple: Is anything alive down there? If so, what do the organisms use to fuel cellular respiration?

The research team hypothesized that if organisms were living deep below Earth's surface, such organisms might use hydrogen molecules (H_2) as an electron donor and the ferric ion (Fe^{3+}) as an electron acceptor (**Figure 27.7**). (Recall from Chapter 9 that most eukaryotes use sugars as electron donors and use oxygen as an electron acceptor during cellular respiration.) Fe^{3+} is the oxidized form of iron, and it is abundant in the rocks the biologists collected from great depths. It exists at great depths below the surface in the form of ferric oxyhydroxide. The researchers predicted that, if an organism in the samples reduced the ferric ions during cellular respiration, a black, oxidized, and magnetic mineral called magnetite (Fe_3O_4) would start appearing in the cultures as a by-product of cellular respiration.

What did their enrichment cultures produce? In some culture tubes, a black compound began to appear within a week.

Using a variety of tests, the biologists confirmed that the black substance was indeed magnetite. As Figure 27.7 shows, microscopy revealed the organisms themselves—previously undiscovered bacteria. Because they grow only when incubated at

Question: Can bacteria live a mile below Earth's surface?

Hypothesis: Bacteria are capable of cellular respiration deep below Earth's surface by using H_2 as an electron donor and Fe^{3+} as an electron acceptor.

Null hypothesis: Bacteria are not capable of cellular respiration deep below Earth's surface.

Experimental setup:

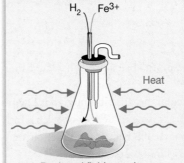

Heat

Rock and fluid samples

Enrichment culture method:

1. Create culture conditions with abundant H_2 and Fe^{3+}; raise temperatures above 45°C.

2. Add rock and fluid samples extracted from drilling operations at depths of about 1000 m below Earth's surface.

Prediction: Black, magnetic grains of magnetite (Fe_3O_4) will accumulate because Fe^{3+} is reduced by growing cells and shed as waste product. Cells will be visible.

Prediction of null hypothesis: No magenetite will appear. No cells will grow.

Results:

Magnetite is detectable, and cells are visible

2 μm

Conclusion: At least one bacterial species that can live deep below Earth's surface grew in this enrichment culture. Different culture conditions might result in the enrichment of different species present in the same sample.

FIGURE 27.7 Enrichment Cultures Isolate Large Populations of Cells That Grow Under Specific Conditions
Bacteria that grow underground can withstand high temperatures. The species enriched in this experiment uses hydrogen (H_2) and ferric ions (Fe^{3+}) as food.

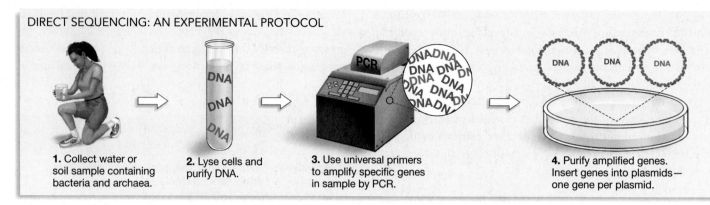

DIRECT SEQUENCING: AN EXPERIMENTAL PROTOCOL

1. Collect water or soil sample containing bacteria and archaea.

2. Lyse cells and purify DNA.

3. Use universal primers to amplify specific genes in sample by PCR.

4. Purify amplified genes. Insert genes into plasmids— one gene per plasmid.

FIGURE 27.8 Direct Sequencing Allows Researchers to Identify Species That Have Never Been Seen
Direct sequencing allows biologists to isolate specific genes from the organisms present in a sample. The polymerase chain reaction is used to generate enough copies of these genes so that the DNA from different species can be sequenced.

between 45°C and 75°C, these organisms are considered **thermophiles** ("heat-lovers").

This discovery was spectacular, because it hinted that Earth's crust may be teeming with organisms to depths of over a mile below the surface. These remarkable bacteria flourish at temperatures that would instantly kill a human being.

Using Direct Sequencing

Researchers estimate that, of all the bacteria and archaea living today, less than 1 percent have been grown in culture. Although enrichment culture is a valid way to answer questions about which bacteria or archaea are present in a particular sample, it is not an efficient way to get a complete picture of bacterial and archaeal diversity. To augment research based on enrichment cultures, researchers are employing a technique called direct sequencing. **Direct sequencing** is a strategy for documenting the presence of bacteria and archaea that represent phylogenetic species but cannot be grown in culture and studied in the laboratory. Recall from Chapter 25 that a phylogenetic species has enough distinctive characteristics that a phylogenetic analysis places it as an independent twig on an evolutionary tree.

Direct sequencing allows biologists to name and characterize organisms that have never been seen. Direct sequencing has revealed huge new branches on the tree of life and produced revolutionary data on the habitats where archaea are found.

Figure 27.8 outlines the steps performed in a direct sequencing study. To begin, researchers collect a sample from a habitat— a few microliters of water or a few micrograms of soil. Next, the bacterial and archaeal cells in the sample are broken open (lysed), and their DNA is purified. Using the sequence of steps described in Figure 27.8 and methods introduced in Chapter 19, researchers can isolate and analyze the DNA sequences of specific genes from species in the sample. By comparing these data with sequences in existing databases, biologists can determine whether the sample contains previously undiscovered archaea or bacteria.

Direct sequencing studies have produced new and sometimes startling results. For two decades after the discovery of the Archaea, for example, researchers thought that these prokaryotes could be conveniently grouped into just four categories: extreme halophiles, sulfate-reducers, methanogens, and extreme thermophiles. Extreme **halophiles** ("salt-lovers") live in salt lakes, salt ponds, and salty soils. **Sulfate reducers** are species that produce hydrogen sulfide (H_2S) as a by-product of cellular respiration. (H_2S may be familiar because it smells like rotten eggs.) **Methanogens** produce methane (natural gas; CH_4) as a by-product of cellular respiration. Extreme thermophiles grow best at temperatures above above 80°C. Based on these early data, researchers thought that archaea were restricted to extreme environments and that the four phenotypes corresponded to separate lineages within the domain Archaea. As a result of direct sequencing studies, however, these generalizations have been discarded. Beginning in the mid-1990s, direct sequencing revealed archaea in habitats as diverse as rice paddies and the Arctic Ocean. Some of these newly discovered organisms appear to belong to an entirely new lineage, tentatively called the **Korarchaeota**. This group's DNA sequences are so distinctive that it might represent a "kingdom" analogous to plants or animals. Yet the Korarchaeota was identified and named before any of its members had been observed.

Biologists usually know nothing about species that are discovered by direct sequencing, except for their habitat and phylogenetic relationships. Still, the technique has revolutionized our understanding of bacterial and archaeal diversity.

Evaluating Molecular Phylogenies

To put data from enrichment culture and direct sequencing studies into context, biologists depend on the accurate placement of species on phylogenetic trees. Recall from Chapter 1

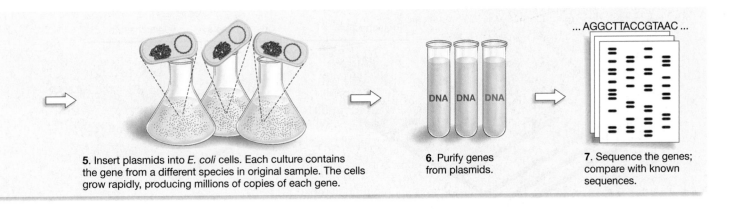

5. Insert plasmids into *E. coli* cells. Each culture contains the gene from a different species in original sample. The cells grow rapidly, producing millions of copies of each gene.

6. Purify genes from plasmids.

7. Sequence the genes; compare with known sequences.

and Chapter 26 that phylogenetic trees illustrate the evolutionary relationships among species and lineages. They are a pictorial summary of which species are more closely or distantly related to others.

Some of the most useful phylogenetic trees for the Bacteria and the Archaea have been based on studies of the RNA molecule found in the small subunit of ribosomes. (See Chapter 16 for more information on the structure and function of ribosomes.) In the late 1960s Carl Woese and colleagues began a massive effort to determine and compare the base sequences of RNA molecules from a wide variety of species. Their goal was to use similarities and differences in the sequence data to infer the evolutionary history of a diversity of organisms. The result of their analysis was shown in Figure 27.1. This diagram came to be known as the *universal tree*, or the *tree of life*.

Woese's tree is now considered a classic result. Prior to its publication, biologists thought that the major division among organisms was between prokaryotes and eukaryotes—between cells that lacked a membrane-bound nucleus and cells that possessed a membrane-bound nucleus (**Figure 27.9**). But based on data from the ribosomal RNA molecule, the major divisions of life-forms are the three groups that Woese named the Bacteria, Archaea, and Eukarya. Follow-up work documented that Bacteria were the first lineage to diverge from the common ancestor of all living organisms. This result means that the Archaea and Eukarya are more closely related to each other than they are to the Bacteria.

Although virtually all biologists accept the three-domains system, it has been very difficult to understand the relationships of major lineages within the Bacteria and Archaea. Analyses of both morphological and molecular characteristics have succeeded in identifying a large series of monophyletic groups within the domains. Recall from Chapter 25 that a **monophyletic group** consists of an ancestral population and all of its descendants. Monophyletic groups can also be called *clades* or *lineages*.

Rapid progress is being made on the question of how the major clades within Bacteria and Archaea are related to each other, and work continues. The phylogenetic tree in **Figure 27.10** (page 592) summarizes recent results but is still considered highly provisional. As more data become available, it is almost certain that at least some of the branches on this tree will change position. In addition, recall that direct sequencing studies have recently led to the discovery of major new groups—meaning that entirely new branches will be added to this tree.

Work on molecular phylogenies continues at a brisk pace. While acknowledging that the tree's resolution will improve as research continues, let's consider some of the innovations that occurred as these groups evolved.

WEB TUTORIAL 27.1
The Tree of Life

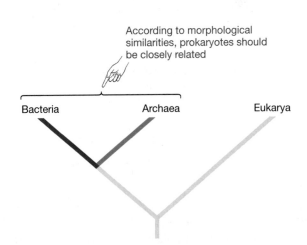

According to morphological similarities, prokaryotes should be closely related

Bacteria   Archaea   Eukarya

FIGURE 27.9 The Tree of Life Based on Morphology Was Incorrect Until recently, the major division among organisms was thought to be between those without a membrane-bound nucleus (the prokaryotes) and those with a membrane-bound nucleus (the eukaryotes). Comparisons of RNA sequences have shown that this tree is not correct. **EXERCISE** Draw the correct tree next to this one. Check your sketch against Figure 27.1. On the correct and incorrect trees, indicate where the nuclear envelope evolved.

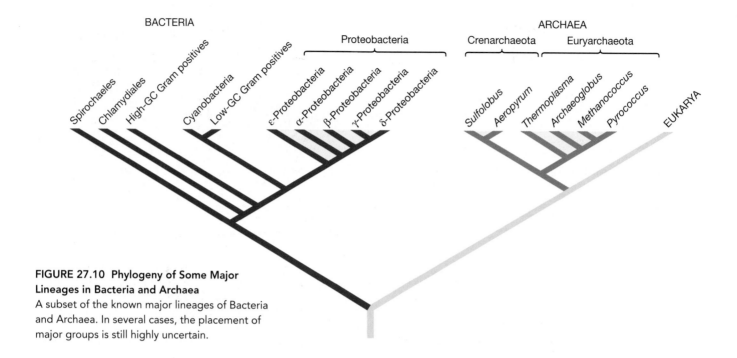

FIGURE 27.10 Phylogeny of Some Major Lineages in Bacteria and Archaea
A subset of the known major lineages of Bacteria and Archaea. In several cases, the placement of major groups is still highly uncertain.

✔ CHECK YOUR UNDERSTANDING

Biologists who study bacterial and archaeal diversity use tools called enrichment cultures and direct sequencing. Enrichment cultures are based on setting up specified conditions in the laboratory and isolating the cells that grow rapidly in response. They create an abundant, pure sample of bacteria that thrive under particular conditions and can be studied further. Direct sequencing is based on isolating DNA from samples taken directly from the environment, purifying and sequencing specific genes, and then analyzing where those DNA sequences are found on the phylogenetic tree of Bacteria and Archaea. You should be able to (1) design an enrichment culture that would isolate species that could be used to clean up oil spills, and (2) outline a study designed to identify the bacterial and archaeal species present in a soil sample near the biology building on your campus.

27.3 What Themes Occur in the Diversification of Bacteria and Archaea?

The lineages in the domains Bacteria and Archaea are ancient, abundant, ubiquitous, and diverse. The oldest fossil cells of any type found to date are 3.4-billion-year-old samples that are virtually indistinguishable from contemporary bacteria. Because eukaryotes do not appear in the fossil record until 1.75 billion years ago, and because archaea probably arose at about the same time as eukaryotes, biologists infer that bacteria may have been the only form of life on Earth for almost 1.7 billion years.

Over the past 3.4 billion years, the Bacteria and Archaea have diversified into hundreds of thousands of distinct species

that live in every conceivable habitat. What overall patterns and themes help biologists make sense of all the diversity observed among these organisms? Let's answer this question by analyzing morphological and metabolic diversity.

Morphological Diversity

The key morphological and biochemical features of bacteria, archaea, and eukaryotes are summarized in **Table 27.2**. The most important point to note is that even though all of the Archaea and almost all of the Bacteria are unicellular, members of these two domains are strikingly different at the molecular level. Bacteria and Archaea are distinguished by the types of molecules that make up their plasma membranes and cell walls, and by their machinery for transcribing DNA and translating messenger RNA into proteins. More specifically, Bacteria have a unique compound called peptidoglycan in their cell walls, and Archaea have unique phospholipids—containing hydrocarbons called isoprenes in their tails—in their plasma membranes. The structures of the RNA polymerases and ribosomes found in Archaea and Eukarya are distinct from those found in Bacteria and similar to each other. For example, antibiotics that poison bacterial ribosomes do not affect the ribosomes of archaea or eukaryotes.

Within the Bacteria and Archaea, morphological diversity is substantial. These organisms range in size from the smallest of all free-living cells—bacteria called mycoplasmas with volumes as small as $0.03 \ \mu m^3$—to the largest bacterium known, *Thiomargarita namibiensis*, with volumes as large as $200 \times 10^6 \ \mu m^3$. Over a billion *Mycoplasma* cells could fit inside an individual *Thiomargarita* (**Figure 27.11a**). Bacteria and archaea exhibit a variety of shapes as well, including filaments, spheres, rods, chains, and spirals (**Figure 27.11b**). Many cells

TABLE 27.2 Characteristics of Bacteria, Archaea, and Eukarya

	Bacteria	**Archaea**	**Eukarya**
Nuclear envelope?	No	No	Yes
Circular chromosome?	Yes (but linear in some species)	Yes	No
DNA associated with histone proteins?	No	Yes	Yes
Organelles present?	No	No	Yes
Unicellular or multicellular?	Almost all unicellular	All unicellular	Many multicellular
Sexual reproduction?	No*	Not known	Common
Structure of lipids in plasma membrane	Glycerol bonded to straight-chain fatty acids via ester linkage	Glycerol bonded to branched fatty acids (synthesized from isoprene subunits) via ether linkage	Glycerol bonded to straight-chain fatty acids via ester linkage
Cell-wall material	Almost all include peptidoglycan, which contains muramic acid	Varies widely among species, but no peptidoglycan and no muramic acid	When present, usually made of cellulose or chitin
Transcription and translation machinery	One relatively simply RNA polymerase; translation begins with formylmethionine; translation poisoned by several antibiotics that do not affect archaea or eukaryotes	Several relatively complex RNA polymerases; translation begins with methionine	Several relatively complex RNA polymerases; translation begins with methionine

*Sexual reproduction begins with meiosis and often involves the exchange of haploid genomes between individuals of the same species. In bacteria, meiosis does not occur. Small numbers of genes can be transferred from one bacterial cell to another, however, and genetic recombination may occur. For more detail, see Chapter 19.

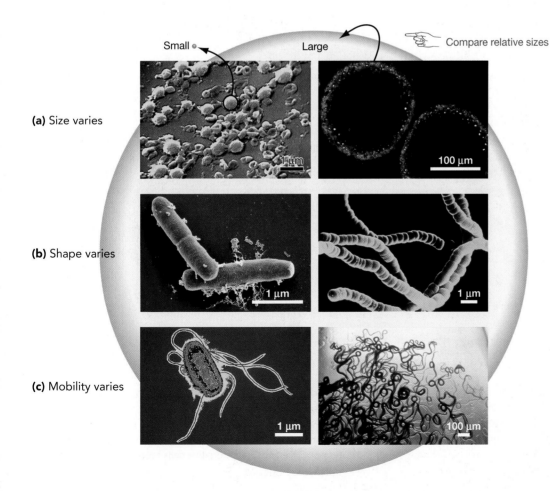

(a) Size varies

(b) Shape varies

(c) Mobility varies

Small ● Large ☞ Compare relative sizes

1 μm   100 μm   1 μm   1 μm   1 μm   100 μm

FIGURE 27.11 Morphological Diversity among Bacteria and Archaea Is Extensive
(a) The sizes of bacteria and archaea vary. (Left) *Mycoplasma* cells are about 0.5 μm in diameter, while (right) *Thiomargarita namibiensis* cells are about 0.15 mm in diameter. **(b)** The shapes of bacteria and archaea vary from (left) rods such as *Bacillus anthracis* and spheres to filaments or spirals such as *Rhodospirillum*. In some species, such as (right) *Streptococcus faecalis*, cells attach to one another and form chains. **(c)** (Left) A wide variety of bacteria and archaea use flagella to power swimming movements. (Right) These cyanobacterial cells move by gliding across a substrate.

are motile, with swimming movements powered by flagella or by gliding over a surface (**Figure 27.11c**). Although gliding movement occurs in several groups, the molecular mechanism responsible for this form of motility is still unknown.

Within bacteria, the overall morphology of the cell wall varies. **Gram-positive bacteria** have a plasma membrane surrounded by a cell wall with extensive peptidoglycan (**Figure 27.12a**). **Gram-negative bacteria**, in contrast, have a plasma membrane surrounded by a cell wall that has two components: a thin gelatinous layer containing peptidoglycan, and an outer phospholipid bilayer (**Figure 27.12b**). When treated with a dye called the **Gram stain**, Gram-positive cells look purple but Gram-negative cells look pink (**Figure 27.12c**). The Gram stain is used routinely to aid in the identification of unknown bacterial cells.

All bacteria and archaea are haploid and reproduce by **fission**—the splitting of a cell into two daughter cells. DNA replication precedes fission, so the two daughter cells are genetically identical. In addition, as noted in Chapter 19, bacterial cells are capable of transferring copies of the extracellular loops of DNA called **plasmids**. When **conjugation** occurs, a copy of a plasmid—and potentially one or more genes from the main bacterial chromosome—moves from one cell to a recipient cell (see Box 19.1). The **conjugation tube** that forms between cells that are transferring and receiving a plasmid is a morphological trait that is unique to bacteria and archaea.

To summarize, members of the Bacteria and the Archaea are remarkably diverse in their overall size, shape, and motility, as well as in the composition of their cell walls and plasma membranes. But when asked to name the innovations that were responsible for the diversification of these two domains, biologists do not point to their morphological diversity. Instead, they point to metabolic diversity—variation in the chemical reactions that go on inside these cells.

Metabolic Diversity

Bacteria and archaea are the masters of metabolism. Taken together, they can use almost anything as food, from hydrogen molecules to crude oil. Bacteria and archaea may be small and relatively simple morphologically, but their biochemical capabilities are dazzling.

(a) Cell walls in Gram-positive bacteria have extensive peptidoglycan.

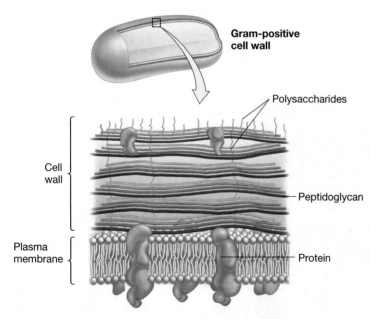

(b) Cell walls in Gram-negative bacteria have some peptidoglycan and an outer membrane.

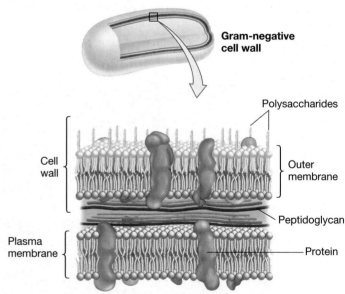

(c) Gram-positive cells retain Gram stain more than Gram-negative cells do.

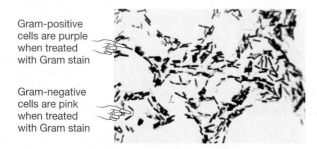

FIGURE 27.12 Gram Staining Distinguishes Two Types of Cell Walls in Bacteria
Both **(a)** the Gram-positive bacteria and **(b)** the Gram-negative bacteria are paraphyletic groups (see Figure 27.10). **(c)** The cell walls of Gram-positive and Gram-negative cells take up the Gram stain differently.

Just how varied are bacteria and archaea when it comes to finding and processing food? To appreciate the answer, recall from Chapters 9 and 10 that organisms have two fundamental nutritional needs: acquiring chemical energy in the form of adenosine triphosphate (ATP) and obtaining carbon in a form that can be used to synthesize fatty acids, proteins, DNA, RNA, and other molecules needed to build cells. Bacteria and archaea produce ATP in three ways:

1. **Phototrophs** ("light-feeders") use light energy to promote electrons to the top of electron transport chains. ATP is produced by photophosphorylation (see Chapter 10).

2. **Organotrophs** oxidize organic molecules with high potential energy, such as sugars. ATP may be produced by cellular respiration—with sugars serving as electron donors—or via fermentation pathways such as those cited in Chapter 9.

3. **Lithotrophs** ("rock-feeders") oxidize inorganic molecules with high potential energy, such as ammonia (NH_3) or methane (CH_4). ATP is produced by cellular respiration, with inorganic compounds serving as the electron donor.

Bacteria and archaea obtain carbon either by processing sources such CO_2 and CH_4 or by absorbing ready-to-use organic compounds from their environment. Organisms that manufacture their own carbon-containing compounds are termed **autotrophs** ("self-feeders"). Organisms that acquire carbon-containing compounds from other organisms are called **heterotrophs** ("other-feeders").

Because there are three distinct ways of producing ATP and two general mechanisms for obtaining carbon, there are a total of six methods for producing ATP and obtaining carbon. **Table 27.3** summarizes these six "feeding strategies"—only two of which are observed in plants, animals, and fungi.

What makes this remarkable diversity possible? Bacteria and archaea have evolved dozens of variations on the most basic themes of metabolism: extracting usable energy from compounds with high potential energy, using light to produce high-energy electrons, and fixing carbon. This metabolic diversity is important because it allows bacteria and archaea to occupy diverse habitats. Let's take a closer look.

Cellular Respiration: Variation in Electron Donors and Electron Acceptors Millions of bacterial, archaeal, and eukaryotic species—including animals and plants—obtain the energy required to make ATP by oxidizing organic compounds such as sugars, starch, or fatty acids. (To review the nature of reduction-oxidation reactions, see Chapter 2.) As shown in Chapter 9, cellular enzymes can strip electrons from organic molecules with high potential energy and transfer them to the electron carriers NADH and $FADH_2$. These compounds, in turn, feed electrons to the molecules that make up electron transport chains. As electrons are stepped down from a high-energy state to a low-energy

TABLE 27.3 Strategies for Obtaining Energy and Carbon: An Overview

Source of Energy (For ATP production)	Source of Carbon (For synthesis of organic compounds)	
	Autotrophs Synthesize their own high-potential-energy organic compounds from CO_2, CH_4, or other inorganic sources.	**Heterotrophs** Use organic compounds with high potential energy produced by other organisms.
Light (phototrophs)	**Photoautotrophs** Cyanobacteria use photosynthesis to produce ATP. They fix CO_2 via the Calvin cycle.	**Photoheterotrophs** Heliobacteria use photosynthesis to produce ATP. They absorb organic molecules with high potential energy from the environment.
Organic Molecules with High Potential Energy (organotrophs)	**(no term)** *Clostridium aceticum* ferments glucose to produce ATP. It uses reactions called the acetyl-CoA pathway to fix CO_2.	**Chemoorganotrophs** *E. coli* uses fermentation or respiration to produce ATP. It absorbs carbon molecules with high potential energy from the environment.
Inorganic Molecules with High Potential Energy (lithotrophs)	**Chemolithotrophs** Nitrifying bacteria, including species of *Nitrosomonas*, produce ATP via respiration, using ammonia (NH_3) as an electron donor. They fix CO_2 via the Calvin cycle.	**Chemolithotrophic Heterotrophs** *Beggiatoa* produces ATP via respiration, using hydrogen sulfide (H_2S) as an electron donor. It absorbs carbon molecules with high potential energy from the environment.

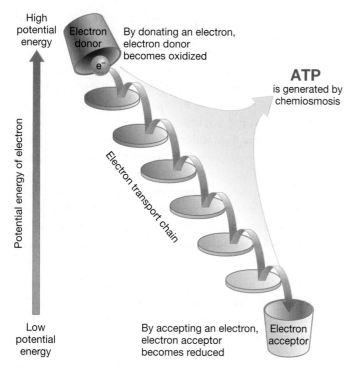

High potential energy

By donating an electron, electron donor becomes oxidized

Electron donor

e^-

Potential energy of electron

Electron transport chain

ATP is generated by chemiosmosis

Low potential energy

By accepting an electron, electron acceptor becomes reduced

Electron acceptor

FIGURE 27.13 Cellular Respiration Requires an Electron Donor and an Electron Acceptor
Bacteria and archaea can exploit a wide variety of electron donors and acceptors. Some of these electron donors are more highly reduced than others, and some electron acceptors are more highly oxidized than others. EXERCISE After studying Table 27.4, add the chemical formula for a specific electron donor, electron acceptor, and reduced by-product to this diagram. Do this for a particular species of bacteria or archaea. Then write in the electron donor, electron acceptor, and reduced by-product observed in humans.

state, the components of the electron transport chain generate a proton gradient across a membrane (**Figure 27.13**). As Chapter 9 pointed out, the resulting flow of protons through the enzyme ATP synthase results in the production of ATP, via the process called chemiosmosis.

The essence of this process, called **cellular respiration**, is that a molecule with high potential energy serves as an original electron donor and is oxidized, while a molecule with low potential energy serves as a final electron acceptor and becomes reduced. Much of the potential energy difference between the electron donor and electron acceptor is transformed into chemical energy in the form of ATP.

In introducing respiration, Chapter 9 focused on the role of organic compounds with high potential energy, such as glucose, as the original electron donor and oxygen as the final electron acceptor (see Figure 9.21). Many bacteria and archaea, as well as all eukaryotes, rely on these molecules. When oxygen acts as the final electron acceptor, water is produced as a by-product of respiration.

Many other bacteria and archaea employ an electron donor other than sugars and an electron acceptor other than oxygen, however, and produce by-products other than water. As **Table 27.4** shows, the electron donors used by bacteria and archaea range from hydrogen molecules (H_2) and hydrogen sulfide (H_2S) to ammonia (NH_3) and methane (CH_4). In addition to using oxygen, some organisms use compounds with low potential energy, such as sulfate (SO_4^{2-}), nitrate (NO_3^-), carbon dioxide (CO_2), or ferric ions (Fe^{3+}), as electron acceptors. It is only a slight exaggeration to claim that researchers have found bacterial and archaeal species that can use almost any compound with relatively high potential energy as an electron donor and almost any compound with relatively low potential energy as an electron acceptor.

Because the electron donors and electron acceptors used by bacteria and archaea are so diverse, one of the first questions that biologists ask about a species is how it accomplishes cellular respiration. The best way to answer this question is through the enrichment culture technique introduced in Section 27.2. Recall that, in an enrichment culture, researchers supply specific electron donors and electron acceptors in the medium and try to isolate cells that can use those compounds to support growth.

TABLE 27.4 Some Electron Donors and Acceptors Used by Bacteria and Archaea

Electron Donor	Electron Acceptor	Product	Category*
Sugars	O_2	H_2O	Organotrophs
H_2 or organic compounds	SO_4^{2-}	H_2S	Sulfate reducers
H_2	CO_2	CH_4	Methanogens
CH_4	O_2	CO_2	Methanotrophs
S or H_2S	O_2	SO_4^{2-}	Sulfur bacteria
Organic compounds	Fe^{3+}	Fe^{2+}	Iron reducers
NH_3	O_2	NO_2^-	Nitrifiers
Organic compounds	NO_3^-	N_2O, NO, or N_2	Denitrifiers (or nitrate reducers)
NO_2^-	O_2	NO_3^-	Nitrosifiers

*The name biologists use to identify species that use a particular metabolic strategy.

The remarkable metabolic diversity of bacteria and archaea is important. First, it explains their ecological diversity. Bacteria and archaea are found almost everywhere because they exploit an almost endless variety of molecules as food. Second, it explains why they play such a key role in cleaning up some types of pollution. Species that use organic solvents or petroleum-based fuels as electron donors or electron acceptors may be effective agents in bioremediation efforts if the waste products are less toxic than the solvents or fuels. Finally, metabolic diversity is what makes bacteria and archaea major players in global change. Nitrogen, phosphorus, sulfur, carbon, and other crucial nutrients cycle from one organism to another because bacteria and archaea can use them in almost any molecular form. The nitrite (NO_2^-) that some bacteria produce as a by-product of respiration does not build up in the environment, because it is used as an electron acceptor by other species and converted (via *nitrification*) to molecular nitrate (NO_3^-). This molecule, in turn, is converted (via *denitrification*) to molecular nitrogen (N_2) by yet another suite of bacterial and archaeal species. Bacteria can convert N_2 in the atmosphere to NH_3 in proteins (by bacterial or archaeal fixation and decomposition). Similar types of interactions occur with molecules that contain phosphorus, sulfur, and carbon. In this way, bacteria and archaea play a key role in the cycling of nutrients (**Figure 27.14**). In total, bacteria and archaea contain as much carbon, and 10 times as much nitrogen and phosphorus, as are in all of the plants on Earth.

Fermentation Chapter 9 introduced **fermentation**, a strategy for making ATP that does not involve electron transport chains. In fermentation, no outside electron acceptor is used; the redox reactions are internally balanced. Because fermentation is a less efficient way to make ATP compared with cellular respiration, in many species it occurs as an alternative metabolic strategy when no electron acceptors are available to make cellular respiration possible. In other species, fermentation does not occur at all; in still other species, fermentation is the only way for cells to make ATP.

Although the presentation in Chapter 9 focused on how glucose is fermented to ethanol or lactic acid, some bacteria and archaea are capable of using other organic compounds as fermentable substrates. For example, the bacterium *Clostridium aceticum* can ferment ethanol, acetate, and fatty acids—as well as glucose. Other species of *Clostridium* ferment complex carbohydrates (including cellulose or starch), proteins, amino acids, or even purines. Species that ferment amino acids produce by-products with names such as cadaverine and putrescine. These molecules are responsible for the odor of rotting flesh. Other bacteria can ferment lactose, a prominent component of milk. This fermentation has two end products: propionic acid and CO_2. Propionic acid is responsible for the taste of Swiss cheese; the CO_2 produced during fermentation creates the holes in cheese.

The diversity of enzymatic pathways observed in bacterial and archaeal fermentations extends the metabolic repertoire of these organisms and supports the claim that as a group, bacteria and archaea can use virtually any molecule with relatively high potential energy as a source of energy. Given this diversity, it is no surprise that bacteria and archaea are found in such widely varying habitats. Different environments offer different energy-rich molecules; various species of bacteria and archaea have evolved the biochemical machinery required to exploit most or all of these food sources.

Photosynthesis Instead of using molecules as a source of high-energy electrons, phototrophs pursue a radically different strategy: **photosynthesis**. That is, they use the energy in light to raise electrons to high-energy states. As these electrons are stepped down to lower energy states by electron transport chains, the energy released is used to generate ATP.

Chapter 10 introduced an important feature of photosynthesis: The process requires a source of electrons. Recall that, in cyanobacteria and plants, the required electrons come from water. When these organisms "split" water molecules apart to obtain electrons, they generate oxygen as a by-product. In contrast, many phototrophic bacteria use a molecule other than water as the source of electrons. In many cases, the electron donor is hydrogen sulfide (H_2S); a few species can use the ion

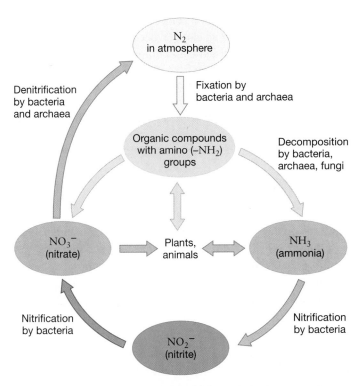

FIGURE 27.14 Nitrogen Atoms Cycle through Environments in Different Molecular Forms

EXERCISE Suppose that bacteria and archaea were no longer capable of fixing nitrogen. Draw an X through the part(s) of the cycle that would be most directly affected.

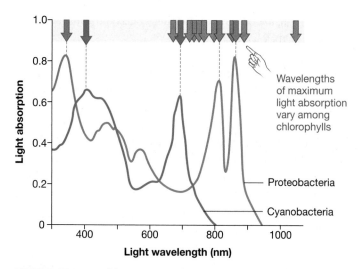

FIGURE 27.15 Highly Diverse Light Wavelengths Are Absorbed by Chlorophylls Found in Photosynthetic Bacteria
The wavelengths of light that are absorbed by the various types of chlorophyll found in different species of photosynthetic bacteria. The complete absorption spectrum is given for the chlorophyll found in cyanobacteria (green curve) and in photosynthetic proteobacteria (blue curve). The red arrows point to absorption peaks observed in chlorophylls isolated from other groups of bacteria.

known as ferrous iron (Fe^{2+}). Instead of producing oxygen as a by-product of photosynthesis, these cells produce elemental sulfur (S) or the ferric ion (Fe^{3+}). They live in habitats where oxygen is completely absent.

Chapter 10 also introduced the photosynthetic pigments found in plants and explored the light-absorbing properties of chlorophylls *a* and *b*. Cyanobacteria have these two pigments. But researchers have isolated seven additional chlorophylls from bacterial phototrophs. As **Figure 27.15** shows, each of these pigments absorbs light best at a different wavelength. Each major group of photosynthetic bacteria has one or more of these distinctive chlorophylls.

Why are bacterial chlorophylls so diverse? The leading hypothesis is that photosynthetic species with different absorption spectra are able to live together without competing for light. If so, then the diversity of photosynthetic pigments observed in bacteria has been an important mechanism for generating species diversity among phototrophs.

Pathways for Fixing Carbon In addition to acquiring energy, organisms must obtain building-block molecules that contain carbon-carbon bonds. Chapters 9 and 10 introduced the two mechanisms that organisms use to procure usable carbon—either making their own or getting it from other organisms. In cyanobacteria and plants, the enzymes of the Calvin cycle transform carbon dioxide (CO_2) to organic molecules that can be used in synthesizing cell material. Animals and fungi, in contrast, obtain carbon from living plants or animals

or by absorbing the organic compounds released as dead tissues decay.

Bacteria and archaea pursue these same two strategies, but with some interesting twists. For example, some species can fix carbon from sources other than CO_2. Certain proteobacteria are called **methanotrophs** ("methane-eaters"), because they use methane (CH_4) as their primary electron donor and carbon source. Among the methanotrophs, CH_4 is assimilated via one of two enzymatic pathways, depending on the species. In addition, some bacteria can use sources of carbon other than CO_2 and CH_4, including carbon monoxide (CO) and methanol (CH_3OH). Further, several groups of bacteria can fix CO_2 using pathways other than the Calvin cycle. To date, researchers have identified three distinct biochemical processes other than the Calvin cycle that result in CO_2 fixation.

What do all these observations mean? Although the enzymes of the Calvin cycle are widespread among bacteria and archaea, several species have evolved different solutions to the problem of fixing carbon. This is yet another example of the metabolic diversity observed within the domains Bacteria and Archaea.

✔CHECK YOUR UNDERSTANDING

Compared with the metabolic capabilities of Bacteria and Archaea, those of plants and animals are very limited. You should be able to describe bacteria or archaea that (1) perform cellular respiration by means of an electron donor other than a sugar and an electron acceptor other than oxygen, (2) perform fermentation from a starting material other than pyruvate derived from glucose and yield a by-product other than lactic acid or ethanol, (3) perform photosynthesis by means of an electron source other than water, and (4) fix carbon to sugar via a pathway other than the Calvin cycle.

27.4   Key Lineages of Bacteria and Archaea

Since the phylogenetic tree identifying the three domains of life was first published, dozens of studies have confirmed the result. It is now well established that all organisms alive today belong to one of the three domains, and that archaea and eukaryotes are more closely related to each other than either group is to bacteria.

Although the relationships among the major lineages within Bacteria and Archaea are still uncertain in some cases (see Figure 27.10), most of the clades themselves have been quite well studied. Let's survey the attributes of species from selected major lineages within the Bacteria and Archaea, with an emphasis on their morphological and metabolic diversity, their impacts on humans, and their importance to other species and to the environment.

Bacteria

The name *bacteria* comes from the Greek root *bacter*, for "rod" or "staff." The name was inspired by the first bacteria to be seen under a microscope, which were rod shaped. But as the following descriptions indicate, bacterial cells come in a wide variety of shapes. If the group were to be named now, biologists might use roots from the Latin words *diversus* or *abundantia*.

Biologists who study bacterial diversity currently recognize at least 14 major lineages, or phyla, within the domain. The lineages reviewed here are just a sampling of bacterial diversity.

Spirochaeles (Spirochetes)

The spirochetes are one of the smaller bacterial phyla. They include 13 genera and a total of 62 species. Recent analyses place spirochetes at the base of the bacterial phylogenetic tree, meaning that they are a particularly ancient group (**Figure 27.16**).

Morphological diversity Spirochetes are distinguished by their unique corkscrew shape and unusual flagella. Instead of extending into the water surrounding the cell, spirochete flagella are contained within a structure called the *outer sheath*, which surrounds the cell. When the flagella beat, the cell lashes back and forth and swims forward.

Metabolic diversity Most spirochetes manufacture ATP via fermentation. The substrate used in fermentation varies among species and may consist of sugars, amino acids, starch, or the pectin found in plant cell walls. A spirochete that lives only in the hindgut of termites can fix nitrogen.

Human and ecological impacts The sexually transmitted disease syphilis is caused by a spirochete. So is Lyme disease, which is transmitted to humans by deer ticks. Spirochetes are extremely common in freshwater and marine habitats; many live only under anaerobic conditions.

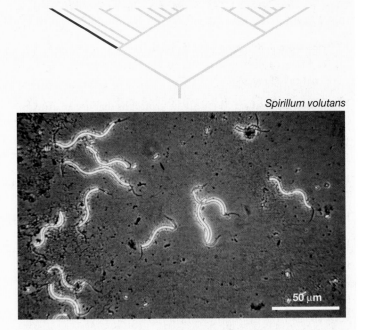

Spirillum volutans

50 µm

FIGURE 27.16 Spirochetes Are Corkscrew-Shaped Cells Inside an Outer Sheath

Chlamydiales

In terms of numbers of species living today, Chlamydiales may be the smallest of all major bacterial lineages. Although chlamydiae are highly distinct phylogenetically, only four species in one genus (*Chlamydia*) are known (**Figure 27.17**).

Morphological diversity Chlamydiae are spherical. They are tiny, even by bacterial standards.

Metabolic diversity All known species live as parasites *inside* host cells and are termed **endosymbionts** ("inside-together-living"). Chlamydiae contain few enzymes of their own and get almost all of their nutrition from their hosts. In Figure 27.17, the chlamydiae are the pink-stained cells, which are living inside blue-stained animal cells.

Human and ecological impacts *Chlamydia trachomatis* infections are the most common cause of blindness in humans. When the same organism is transmitted from person to person via sexual intercourse, it can cause serious urogenital tract infections. One species causes epidemics of a pneumonia-like disease in birds.

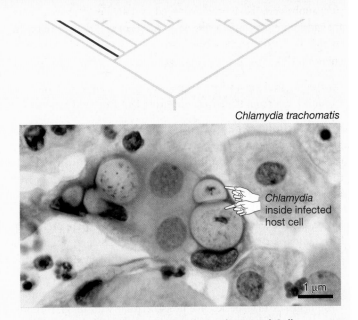

Chlamydia trachomatis

Chlamydia inside infected host cell

1 µm

FIGURE 27.17 Chlamydiae Live Only Inside Animal Cells

High-GC Gram Positives

The name "high-GC Gram positives" is derived from two observations about these bacteria: (1) Their cell-wall material appears purple when treated with the Gram stain, which tests for the absence of a membrane outside the cell wall and is used as an early step in procedures for identifying unknown bacterial cells; and (2) their DNA contains a relatively high percentage of guanine and cytosine. (In some species, G and C represent over 75 percent of the bases present.) Over 1100 species have been described to date (**Figure 27.18**).

Morphological diversity Cell shape varies from rods to filaments. Many of the soil-dwelling species form extensive branched filaments called **mycelia**.

Metabolic diversity Many are heterotrophs that use an array of organic compounds as electron donors and oxygen as an electron acceptor. There are a handful of parasitic species. Like other parasites, they get most of their nutrition from host organisms.

Human and ecological impacts Over 500 distinct antibiotics have been isolated from species in the genus *Streptomyces;* 60 of these—including streptomycin, neomycin, tetracycline, and erythromycin—are now actively prescribed to treat diseases in humans or domestic livestock. Tuberculosis and leprosy are caused by members of this group. One species is critical to the manufacture of Swiss cheese. Species in the genus *Streptomyces* and *Arthrobacter* are abundant in soil and are vital as decomposers of dead plant and animal material.

Some species in these genera live in association with plant roots and fix nitrogen; others can break down toxins such as herbicides, nicotine, and caffeine.

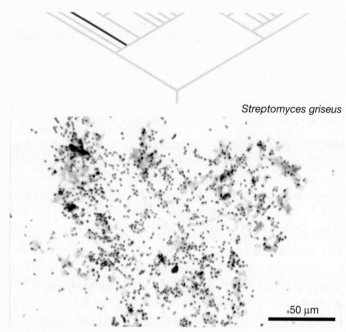

Streptomyces griseus

50 μm

FIGURE 27.18 A *Streptomyces* Species That Produces the Antibiotic Streptomycin

Cyanobacteria

The cyanobacteria were formerly known as the "blue-green algae"—even though algae are eukaryotes. Only about 80 species of cyanobacteria have been described to date, but they are among the most abundant organisms on Earth (**Figure 27.19**). In terms of total mass, cyanobacteria dominate many marine and freshwater environments.

Morphological diversity Cells vary from solitary to colonial. Colonies vary from flat sheets of cells to ball-like aggregations.

Metabolic diversity All perform oxygenic photosynthesis; many can also fix nitrogen. Because cyanobacteria can synthesize virtually every molecule they need, they can be grown in culture media that contain only CO_2, N_2, H_2O, and a few mineral nutrients.

Human and ecological impacts If cyanobacteria are present in high numbers, their waste products can make drinking water smell bad. The compounds are not toxic, however. Cyanobacteria produce much of the oxygen and nitrogen and many of the organic compounds that feed other organisms in freshwater and marine environments. A few species live in association with fungi, forming lichens.

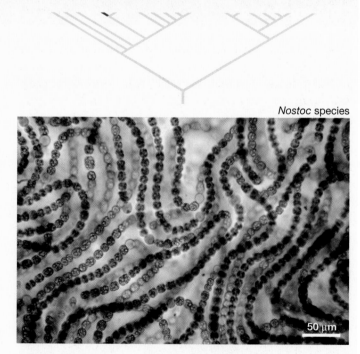

Nostoc species

50 μm

FIGURE 27.19 Cyanobacteria Contain Chlorophyll and Are Green

Low-GC Gram Positives

The low-GC Gram positives react positively to the Gram stain—meaning that they lack a membrane outside their cell wall—but have a relatively low percentage of G and C in their DNA. (In some species, G and C represent less than 25 percent of the bases present.) There are over 1100 species (**Figure 27.20**).

Morphological diversity   Most are rod shaped or spherical. Some of the spherical species form chains or *tetrads* (groups of four cells). A few form a durable resting stage called a **spore**. One subgroup lacks cell walls entirely; another synthesizes a cell wall made of cellulose.

Metabolic diversity   Some species can fix nitrogen; some perform nonoxygenic photosynthesis. Others make all of their ATP via various fermentation pathways; still others perform cellular respiration, using hydrogen gas (H_2) as an electron donor.

Human and ecological impacts   Species in this group cause a variety of diseases, including anthrax, botulism, tetanus, walking pneumonia, boils, gangrene, and strep throat. *Bacillus thuringiensis* produces a toxin that is one of the most important insecticides currently used in farming. Species in the genus *Lactobacillus* are used to ferment milk products into yogurt or cheese. Species in this group are important components of soil.

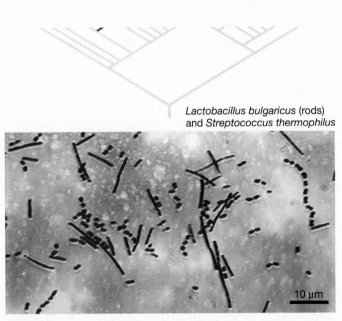

Lactobacillus bulgaricus (rods) and *Streptococcus thermophilus*

10 µm

FIGURE 27.20 Low-GC Gram Positives in Yogurt

Proteobacteria

The approximately 1200 species of proteobacteria form five major subgroups, designated by the Greek letters α (alpha), β (beta), γ (gamma), δ (delta), and ε (epsilon).

Morphological diversity   Proteobacterial cells can be rods, spheres, or spirals. Some form stalks (**Figure 27.21a**). Some are motile. In one group, cells may move together to form colonies, which then transform into the specialized cell aggregate shown in **Figure 27.21b**. This structure is known as a **fruiting body**. Durable spores are produced at the tips of fruiting bodies.

Metabolic diversity   Proteobacteria make a living in virtually every way known to bacteria—except that none perform oxygenic photosynthesis. Various species may perform cellular respiration by using organic compounds, nitrite, methane, hydrogen gas, sulfur, or ammonia as electron donors and oxygen, sulfate, sulfur, or oxygen as an electron acceptor. Some perform nonoxygenic photosynthesis (**Box 27.2**).

Human and ecological impacts   Pathogenic proteobacteria cause Legionnaire's disease, cholera, food poisoning, dysentery, gonorrhea, Rocky Mountain spotted fever, ulcers, and diarrhea. Biologists use *Agrobacterium* cells to transfer new genes into crop plants. Certain acid-loving species of proteobacteria are used in the production of vinegars. Species in the genus *Rhizobium* (α-proteobacteria) live in association with plant roots and fix nitrogen. The bdellovibrios are a group in the δ-proteobacteria that are predators—they drill into bacterial cells and digest them. Pro-

teobacteria are critical players in the cycling of nitrogen atoms through terrestrial and aquatic ecosystems.

(a) Stalked bacterium   **(b)** Fruiting bodies consist of many cells.

Caulobacter crescentus   *Chondromyces crocatus*

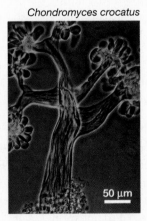

1 µm   50 µm

FIGURE 27.21 Some Proteobacteria Grow on Stalks or Form Fruiting Bodies

BOX 27.2   Lateral Gene Transfer and Metabolic Diversity in Bacteria

If you read the notes on metabolic diversity in Section 27.4 carefully, you will realize that species capable of performing various types of photosynthesis are scattered among many bacterial lineages. The same is true for species that can fix nitrogen.

The "scattered among lineages" pattern is interesting because it is reasonable to predict that extremely complex structures and processes such as photosystem I, photosystem II, and nitrogen fixation evolved just once. If so, then it would be logical to predict that photosynthesizers and nitrogen-fixers would each form monophyletic groups, consisting of an ancestral species and all of its descendants. In contrast, a group that consists of a common ancestor and some *but not all* of its descendants is said to be **paraphyletic** ("beside-group"). If species that perform photosynthesis were monophyletic, then they would be arranged in the pattern shown in **Figure 27.22a**. But instead, the data in **Figure 27.22b** shows that photosynthetic species are paraphyletic. Nitrogen fixers are also paraphyletic.

How could the photosystems and nitrogen fixation evolve just once and yet be paraphyletic in their present phylogenetic distribution? **Lateral gene transfer** is a physical transfer of genes from species in one lineage to species in another lineage. Chapter 20 described several of the mechanisms responsible for lateral gene transfer.

According to the lateral gene transfer hypothesis for the distribution of photosynthesis and nitrogen fixation between lineages, it is correct to claim that

photosystem I, photosystem II, and the nitrogen-fixing enzymes evolved just once. But over the past 3.4 billion years, the genes responsible for these processes have been picked up by organisms from a wide array of bacterial lineages and incorporated into their genomes. Lateral gene transfer is thought to be an important mechanism for generating metabolic diversity in the Bacteria.

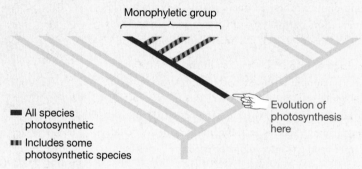

(a) Expected: Monophyletic distribution of photosynthetic groups

Monophyletic group

- ▬ All species photosynthetic
- ▥ Includes some photosynthetic species

Evolution of photosynthesis here

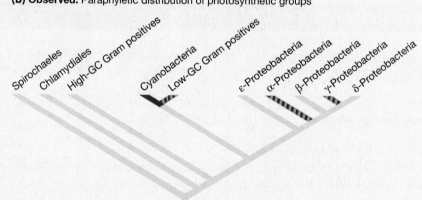

(b) Observed: Paraphyletic distribution of photosynthetic groups

Spirochaeles · Chlamydiales · High-GC Gram positives · Cyanobacteria · Low-GC Gram positives · ε-Proteobacteria · α-Proteobacteria · β-Proteobacteria · γ-Proteobacteria · δ-Proteobacteria

FIGURE 27.22 Photosynthetic Bacteria Are Paraphyletic
(a) If photosynthetic bacteria were monophyletic, their distribution on the phylogenetic tree of Bacteria would look something like this. **(b)** Species that can perform photosynthesis are scattered throughout the phylogenetic tree. Two bacterial lineages not shown—green sulfur bacteria and green nonsulfur bacteria—also perform a type of photosynthesis.

Archaea

The name *archaea* comes from the Greek root *archae*, for "ancient." The name was inspired by the hypothesis that this is a particularly ancient group, which turned out to be incorrect. In addition, archaeans were once thought to be restricted to hot springs, salt ponds, and other extreme habitats. But if the group were to be named now, biologists might start with the Latin root *ubiquit*, for "everywhere."

Archaea live in virtually every habitat known. As far as biologists currently know, there are no parasitic archaea.

The domain Archaea was discovered so recently that major groups are still being discovered and described. Phylogenies based on DNA sequence data have consistently shown that the domain is composed of at least two major phyla, called the **Crenarchaeota** and **Euryarchaeota**. Some data indicate that a third phylum, tentatively called the Korarchaeota, also exists.

Crenarchaeota

The Crenarchaeota got their name because they are considered similar to the oldest archaeans; the word root *cren* refers to a source or fount. Although only 37 species have been named to date, it is virtually certain that thousands are yet to be discovered (**Figure 27.23**).

Morphological diversity   Crenarchaeota cells may be shaped like filaments, rods, discs, or spheres. One species that lives in extremely hot habitats has a cell wall that consists solely of glycoprotein.

Metabolic diversity   Depending on the species, cellular respiration may involve organic compounds, sulfur, hydrogen gas, or Fe^{2+} ions as electron donors and oxygen, nitrate, sulfate, sulfur, carbon dioxide, or Fe^{3+} ions as electron acceptors. Some species make ATP exclusively through fermentation pathways.

Human and ecological impacts   Although no one understands why, no crenarchaeota are known to cause disease. Crenarchaeota have yet to be used in the manufacture of commercial products. In certain extremely hot, high-pressure, cold, or acidic environments, crenarchaeota may be the only life-form present. Acid-loving species thrive in habitats with pH 1–5; some species are found in ocean sediments at depths ranging from 2500 to 4000 m below the surface.

Sulfolobus species

1 μm

FIGURE 27.23 Some Crenarchaetoa Live in Sulfur-Rich Hot Springs

Euryarchaeota

The Euryarchaeota are aptly named, because the word root *eury* means "broad." Members of this phylum live in every conceivable habitat. Some species are adapted to high-salt habitats with pH 11.5—almost as basic as household ammonia. Other species are adapted to acidic conditions with a pH as low as 0. Species in the genus *Methanopyrus* live near hot springs called black smokers that are 2000 m (over 1 mile) below sea level. About 170 species have been identified thus far, and more are being discovered each year (**Figure 27.24**).

Morphological diversity   Euryarchaeota cells may be spherical, filamentous, rod shaped, disc shaped, or spiral. Rod-shaped cells may be short or long or arranged in chains. Spherical cells may be found in ball-like aggregations. Some species have several flagella. Some species lack a cell wall; others have a cell wall composed entirely of glycoproteins.

Metabolic diversity   The group includes a variety of methane-producing species. These methanogens can use up to 11 different organic compounds as electron acceptors during cellular respiration; all produce CH_4 as a by-product of respiration. In other species of Euryarchaeota, cellular respiration may be based on hydrogen gas or Fe^{2+} ions as electron donors and nitrate or sulfate as electron acceptors. Species that live in high-salt environments can use the molecule retinal—which is responsible for light reception in your eyes—to capture light energy and perform photosynthesis.

Human and ecological impacts   Species in the genus *Ferroplasma* (literally, "hot iron") live in piles of waste rock near abandoned mines.

As a by-product of metabolism, they produce acids that drain into streams and pollute them. No species in the Euryarchaeota is known to cause disease in humans. Methanogens live in the soils of swamps and the guts of mammals (including yours). They are responsible for adding about 2 billion tons of methane to the atmosphere each year.

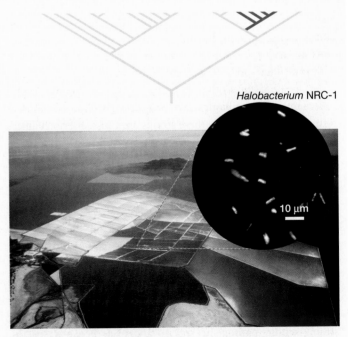

Halobacterium NRC-1

10 μm

FIGURE 27.24 Some Euryarchaeota Live in High-Salt Habitats

ESSAY Antibiotics and the Evolution of Drug Resistance

Antibiotics are molecules that kill bacteria. A wide variety of naturally occurring antibiotics are produced by bacteria or fungi, presumably as a defense against bacterial species that compete with them for space and resources. These molecules work in several ways. Some, such as penicillin, poison enzymes involved in the formation of bacterial cell walls. Others, such as erythromycin, disrupt ribosomes or other parts of the enzymatic machinery involved in protein synthesis. Over 8000 antibiotics are known, and hundreds more are discovered each year. Only a tiny percentage of these is useful in treating human diseases, however.

The first antibiotic that brought important human diseases under chemical control was discovered quite by accident. In 1928 Alexander Fleming was studying the pathogenic bacterium *Staphylococcus* and noticed that cells began dying off in culture plates that had inadvertently been contaminated with a fungus called *Penicillium*. Fleming seized on this observation, confirmed that a molecule from the fungus was responsible for killing the cells, and called it penicillin. In the 1930s a group led by Howard Florey showed that penicillin, administered as a drug, was effective against *Staphylococcus* and *Streptococcus* infections in humans. Florey then led a research consortium that worked out methods for producing the molecule in quantity by the start of World War II.

Penicillin was a "magic bullet"—a drug that killed many different infectious agents without harming the host. The drug saved many millions of lives that would have been cut short by *Staphylococcus* infections or pneumonia. Now, however, the magic is gone. In 1941 physicians could administer 10,000 units of penicillin four times a day for four days and cure pneumonia completely. Now, pneumonia patients who receive 24 *million* units of penicillin a day have a good chance of dying. Similarly, in 1941 all strains of *Staphylococcus aureus* were treated effectively with penicillin. Today, 95 percent of the strains are unaffected by it. Strains of *S. aureus* that are resistant to penicillin and several other important antibiotics have become particularly common in hospitals. In many cases, these strains are causing serious illness in patients who were admitted for other problems. **Figure 27.25** shows how the percentage of patients in intensive care units who are infected with these multiple-drug-resistant strains of *S. aureus* as opposed to nonresistant strains has increased in the United States recently.

The spread of alleles that confer resistance has been favored by natural selection, leading to the evolution of resistant strains. In the case of penicillin and many other antibiotics, however, resistance alleles are found not on the main bacterial chromosome but on plasmids. Because plamids are often transferred from one cell to another, resistance alleles have spread rapidly among strains within pathogenic species and even between species.

Antibiotic-resistance genes are a well-documented example of lateral gene transfer—the phenomenon introduced in Chapter 20 and Box 27.2. Penicillin, for example, acts by binding to the enzymes responsible for creating cross-links in bacterial cell walls. This binding prevents cross-linking and weakens the cell walls, which eventually rupture, causing cell death. But a plasmid-borne gene called β-*lactamase* codes for a protein that cleaves the penicillin molecule. This action renders the drug useless, because it can no longer bind to the cross-linking enzymes. Plasmids that carry the β-*lactamase* gene have been transferred to a wide array of pathogenic species. Lateral transfer of β-*lactamase* has been common.

Drug resistance is now a fact of life. . . .

Drug resistance is now a fact of life for drug companies, physicians, and patients. In response, public health agencies are trying to limit the use of antibiotics in animal feed and to reduce the number of antibiotics prescriptions being written by physicians. In Hungary, for example, a steep reduction in penicillin use in treating pneumonia, encouraged by the National Institute of Public Health, led to a sharp drop in the frequency of penicillin-resistant strains of *Pneumococcus*. If antibiotics are to continue being effective, they will have to be prescribed and taken much more carefully than they have been in the past.

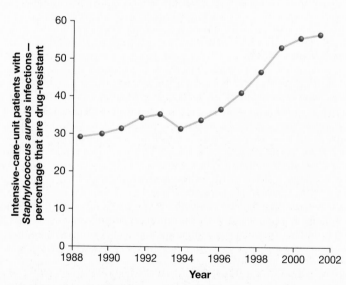

FIGURE 27.25 Drug-Resistant Cells Are Becoming More Common, Particularly in Hospitals
The percentage of *Staphylococcus aureus* infections that are caused in hospital intensive care units by strains resistant to several different antibiotics; from the U.S. Centers for Disease Control and Prevention.

CHAPTER REVIEW

Summary of Key Concepts

■ **Bacteria and archaea affect your life. A small percentage of bacteria causes disease. Some species are effective at cleaning up pollution, and photosynthetic bacteria were responsible for the evolution of the oxygen atmosphere. Bacteria and archaea cycle nutrients through both terrestrial and aquatic environments.**

Bacteria and archaea may be tiny, but they have a huge impact on global ecosystems and human health. Cyanobacteria were responsible for producing Earth's first oxygen-containing atmosphere, and nitrogen-fixing bacteria and archaea keep the global nitrogen cycle running. Bacteria also cause some of the most dangerous diseases in humans, including plague, syphilis, botulism, cholera, and tuberculosis. Disease results when bacteria kill host cells or produce toxins that disrupt normal cell functions.

Enrichment cultures are used to grow large numbers of bacterial or archaeal cells that thrive under specified conditions, such as in the presence of certain electron donors and electron acceptors. To study bacteria and archaea that cannot be cultured, biologists frequently take advantage of direct sequencing. In this research strategy, DNA sequences are extracted directly from organisms in the environment—without first culturing the organisms in the lab. By analyzing where these sequences are placed on the tree of life, biologists can determine whether the branches represent new species. If they do, information on where the original sample was collected can expand knowledge about the types of habitats used by bacteria and archaea.

■ **Bacteria and archaea live in a wide array of habitats and use diverse types of molecules in cellular respiration and fermentation. Although they are small and relatively simple in their overall morphologies, these organisms are extremely sophisticated in the chemistry they can do. Many species are restricted in distribution and have a limited diet.**

Metabolic diversity and complexity are the hallmarks of the bacteria and archaea, just as morphological diversity and complexity are the hallmarks of the eukaryotes. Like eukaryotes, many bacteria and archaea can extract energy from carbon compounds with high potential energy, such as sugars. These molecules are processed through fermentation pathways or by transferring high-energy electrons to electron transport chains with oxygen as the final electron acceptor. But among the bacteria and archaea, many other inorganic or organic compounds with high potential energy serve as electron donors, and a wide variety of inorganic molecules with low potential energy serve as electron acceptors. Dozens of distinct organic compounds are fermented, including proteins, purines, alcohols, and an assortment of carbohydrates.

Photosynthesis is also widespread among bacteria. In cyanobacteria, water is used as a source of electrons during photosynthesis, and oxygen gas is generated as a by-product. But in other species, the electron excited by photon capture comes from a reduced substance such as ferrous iron (Fe^{2+}) or hydrogen sulfide (H_2S) instead of water (H_2O); the oxidized by-product is the ferric ion (Fe^{3+}) or elemental sulfur (S) instead of oxygen (O_2). These organisms also contain chlorophylls not found in plants or cyanobacteria.

To acquire molecules containing carbon-carbon bonds, some species use the enzymes of the Calvin cycle to reduce CO_2. But biologists have also discovered three additional biochemical pathways in bacteria and archaea that transform carbon dioxide (CO_2), methane (CH_4), or other sources of inorganic carbon into organic compounds such as sugars or carbohydrates.

Web Tutorial 27.1   The Tree of Life

Questions

Content Review

1. How do molecules that function as electron donors and those that function as electron acceptors differ?
   a. Electron donors are organic molecules; electron acceptors are inorganic.
   b. Electron donors are inorganic molecules; electron acceptors are organic.
   c. Electron donors have relatively high potential energy; electron acceptors have relatively low potential energy.
   d. Electron donors have relatively low potential energy; electron acceptors have relatively high potential energy.

2. What do some photosynthetic bacteria use as a source of electrons instead of water?
   a. oxygen (O_2)
   b. hydrogen sulfide (H_2S)
   c. organic compounds (e.g., CH_3COO^-)
   d. nitrate (NO_3^-)

3. What is distinctive about the chlorophylls found in different photosynthetic bacteria?
   a. their membranes
   b. their role in acquiring energy
   c. their role in carbon fixation
   d. their absorption spectra

4. What are organisms called that use inorganic compounds as electron donors in cellular respiration?
   a. phototrophs
   b. heterotrophs
   c. organotrophs
   d. lithotrophs

5. How is direct sequencing used?
   a. to grow unknown organisms in laboratory culture
   b. to determine where organisms that cannot be cultured are located on the tree of life
   c. to determine the metabolic capabilities of newly discovered prokaryotes
   d. to improve the accuracy of phylogenetic trees

6. Koch's postulates outline the requirements for which of the following?
   a. showing that an organism is autotrophic
   b. showing that a bacterium's cell wall lacks an outer membrane and consists primarily of peptidoglycan
   c. showing that an organism causes a particular disease
   d. showing that a disease-causing species also has harmless strains

Conceptual Review

1. Biologists often use the term *energy source* as a synonym for "electron donor." Why?

2. The text claims that the tremendous ecological diversity of bacteria and archaea is possible because of their impressive metabolic diversity. Do you agree with this statement? Why or why not?

3. Suppose that universal PCR primers were available for genes involved in electron transport chains or for some of the different types of chlorophyll found in bacteria. Why would it be interesting to use these genes in a direct sequencing study?

4. The text claims that the evolution of an oxygen atmosphere paved the way for increasingly efficient respiration and higher-energy activities by organisms. Explain.

5. Look back at Table 27.4 and note that the by-products of respiration in some organisms are used as electron donors or acceptors by other organisms. In the table, draw lines between the dual-use molecules listed in the "Electron Donor," "Electron Acceptor," and "Product" columns.

6. Explain the statement, "Prokaryotes are a paraphyletic group."

Group Discussion Problems

1. The researchers who observed that magnetite was produced by bacterial cultures from the deep subsurface carried out a follow-up experiment. These biologists treated some of the cultures with a drug that poisons the enzymes involved in electron transport chains. In cultures where the drug was present, no more magnetite was produced. Does this result support or undermine their hypothesis that the bacteria in the cultures perform anaerobic respiration with Fe^{3+} serving as the electron acceptor? Explain your reasoning.

2. *Streptococcus mutans* obtains energy by oxidizing sucrose. This bacterium is abundant in the mouths of Western European and North American children and is a prominent cause of cavities. The organism (and cavities) is virtually absent in children from East Africa. Propose a hypothesis to explain this observation. Outline the design of a study that would test your hypothesis.

3. Suppose that you've been hired by a firm interested in using bacteria to clean up organic solvents found in toxic waste dumps. Your new employer is particularly interested in finding cells that are capable of breaking a molecule called benzene into less toxic compounds. Where would you go to look for bacteria that can metabolize benzene as an energy or carbon source? How would you design an enrichment culture capable of isolating benzene-metabolizing species?

4. Would you predict that disease-causing bacteria, such those listed in Table 27.1, obtain energy from light, organic molecules, or inorganic molecules? When they perform cellular respiration, which substance would you predict that they use as an electron acceptor? Explain your answer.

Answers to Multiple-Choice Questions 1. c; 2. b; 3. d; 4. d; 5. b; 6. c

Protists

28

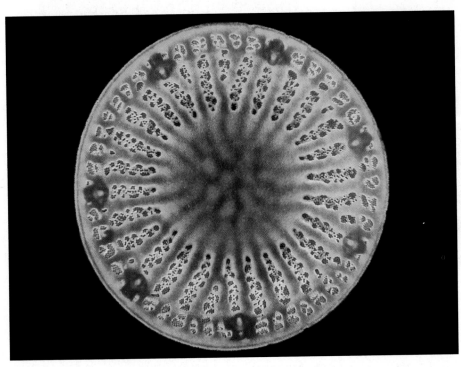

Diatoms are single-celled protists that live inside a glassy case. They may be the most abundant of all eukaryotes found in aquatic environments.

KEY CONCEPTS

■ Protists are a paraphyletic grouping that includes all eukaryotes except the green plants, fungi, and animals. Biologists study protists to understand how eukaryotes evolved, because they are important in marine ecosystems and global warming and because some species cause debilitating diseases in plants, humans, and other organisms.

■ Protists are diverse morphologically. They vary in the types of organelles they contain; they may be unicellular or multicellular, and they may have a cell wall or other external covering, or no such covering.

■ Protists vary widely in terms of how they reproduce and how they find food. Many can reproduce both sexually and asexually. Many species are photosynthetic, while others obtain carbon compounds by ingesting food or parasitizing other organisms.

This chapter introduces the third domain on the tree of life: the **Eukarya**. Eukaryotes range from single-celled organisms that are the size of bacteria to sequoia trees and blue whales. The largest and most morphologically complex organisms on the tree of life—algae, plants, fungi, and animals—are eukaryotes.

Although species in the Eukarya are astonishingly diverse, they share fundamental features. Eukaryotes are defined by the presence of the shared, derived character called the nuclear envelope. All members of the Eukarya possess a nucleus as well as other organelles. All eukaryotes undergo cell division via mitosis, and all have a system of structural and contractile proteins called the cytoskeleton. Like archaea, most eukaryotes have chromosomes where DNA is complexed with proteins called histones.

In introducing the Eukarya, this chapter focuses on an informal grab bag of lineages known as the protists. The term **protist** refers to all eukaryotes that are not green plants, fungi, or animals. Analyzing protist diversity is the key to understanding how eukaryotes came to be and how green plants, fungi, and animals originated.

The first thing to appreciate about protists is their morphological diversity. Many protists are microscopic single cells; others are multicellular organisms up to 60 meters long. Their lifestyles are just as diverse. Some are parasitic, while others are predatory or photosynthetic. They may be stationary all their lives or in virtually constant motion. Their cells can change shape almost continuously or be encased in rigid, glassy shells. The common feature among protists is that they tend to live in

(a) Open ocean:

Surface waters teem with microscopic protists, such as these diatoms.

(b) Shallow coastal waters:

Gigantic protists, such as these kelp, form underwater forests.

(c) Intertidal habitats:

Protists such as these red algae are particularly abundant in tidal habitats.

FIGURE 28.1 Protists Are Particularly Abundant in Aquatic Environments
Protists are abundant in a wide variety of aquatic habitats. In marine environments, they are found in **(a)** the open ocean and **(b)** near-shore and **(c)** intertidal habitats.

moist surroundings such as wet soils, aquatic habitats, or the bodies of other organisms. Protists are crucial in terms of abundance, productivity, and overall impact (**Figure 28.1**).

28.1   Why Do Biologists Study Protists?

Biologists study protists in part because protists are intrinsically interesting, in part because they are so important medically and ecologically, and in part because they are critical to understanding the evolution of plants, fungi, and animals. The remainder of the chapter will focus on why protists are interesting in their own right and how they evolved; here let's consider their impact on human health and the environment.

Impacts on Human Health and Welfare

The most spectacular crop failure in history, the Irish potato famine, was caused by a protist. In 1845 most of the 3 million acres that had been planted to grow potatoes in Ireland became infested with *Phytophthora infestans*—a parasite that belongs to a lineage of protists called Oomycota. Potato tubers that were infected with *P. infestans* rotted in the fields or in storage.

As a result of crop failures in Ireland in 1845 and the following year, an estimated 1 million people out of a population

of less than 9 million died of starvation or starvation-related illnesses. Several million others were driven away. Many people of Irish heritage living in North America, New Zealand, and Australia trace their ancestry to relatives who emigrated from Ireland to evade the famine. As devastating as the potato famine was, however, it does not begin to approach the misery caused by the protist *Plasmodium*.

Malaria   Physicians and public health officials point to three major infectious diseases that are currently afflicting large numbers of people worldwide: tuberculosis, HIV, and malaria. Tuberculosis was introduced in Chapter 23, and HIV is analyzed in Chapter 34. Here we consider malaria.

Malaria ranks as the world's most chronic public health problem. In India alone, over 30 million people each year suffer from debilitating fevers caused by malaria. At least 300 million people worldwide are sickened by it each year, and over 1 million die from the disease annually. The toll is equivalent to eight 747s, loaded with passengers, crashing every day. Most of the dead are children of preschool age.

Four species of the protist *Plasmodium* are capable of infecting humans and causing malaria (**Figure 28.2**). Because each *Plasmodium* species spends part of its life cycle inside mosqui-

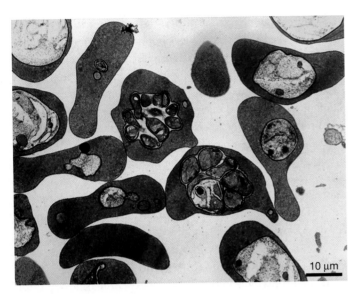

FIGURE 28.2 *Plasmodium* **Causes Malaria**
Over the course of its life cycle, *Plasmodium falciparum* develops into a series of distinct cell types. Each type is specialized for infecting a different host cell in mosquitoes or humans. The yellow- and orange-stained cells in this micrograph are infecting human red blood cells. The infection kills the host cell, contributing to anemia and high fever.

growth in infected people. Efforts to develop a vaccine against *Plasmodium* have also been fruitless to date, in part because the parasite evolves so quickly. (Chapter 34 explains why it is difficult or impossible to develop effective vaccines against rapidly evolving viruses and organisms such as *Plasmodium*, cold and influenza viruses, and HIV.) Although *Plasmodium* is arguably the best studied of all protists, researchers have still not been able to devise effective and sustainable measures to control it.

Biologists study the basic biology of *Plasmodium* in the hopes that a better understanding of the parasite will lead to improved drug therapies or even an effective vaccine. In the meantime, the World Health Organization, the World Bank, and several agencies of the United Nations have joined forces to promote the use of insecticide-treated sleeping nets to prevent bites by *Plasmodium*-infected mosquitoes and to urge the use of "combination therapies"—medications that include at least two different antimalarial drugs. Combination therapies have been shown to slow the evolution of drug resistance.

Unfortunately, malaria is not the only important human disease caused by protists. **Table 28.1** lists protists that have been the cause of human misery and economic losses. Parasitic protists affect hundreds of millions of people every year and are a major concern for physicians worldwide. Harmful protists also worry biologists who manage the fisheries that many people depend on for their food or livelihood.

toes, most antimalaria campaigns have focused on controlling these insects. This strategy has become less and less effective over time, however, because natural selection has favored mosquito strains that are resistant to the insecticides that have been sprayed in their breeding habitats. Further, *Plasmodium* itself has evolved resistance to most of the drugs used to control its

Harmful Algal Blooms When photosynthetic protists experience rapid population growth and reach high densities in marine environments, they are said to "bloom." Unfortunately, a handful of the many species involved in blooms can be

TABLE 28.1 Human Health Problems Caused by Protists

Lineage	Species	Disease
Apicomplexa	Four species of *Plasmodium*, primarily *P. falciparum* and *P. vivax*	Malaria has the potential to affect 40 percent of the world's total population.
	Toxoplasma	Toxoplasmosis may cause eye and brain damage in infants and in AIDS patients.
Dinoflagellates	Many species involved	Toxins released during "red tides" accumulate in clams and mussels and poison people if eaten.
Excavates	*Giardia*	Diarrhea can last for several weeks.
Parabasalids	*Trichomonas*	Trichomoniasis is a reproductive tract infection and one of the most common sexually transmitted diseases. About 2 million young women are infected in the United States each year; some of them become infertile.
Discicristates	*Leishmania*	Leishmaniasis can cause skin sores or affect internal organs—particularly the spleen and liver.
	Trypanosoma gambiense and *T. rhodesiense*	Trypanosomiasis ("sleeping sickness") is a potentially fatal disease transmitted through bites from tsetse flies. Occurs in Africa.
	Trypanosoma cruzi	Chagas disease affects 16–18 million people and causes 50,000 deaths annually, primarily in South and Central America.
Entamoebae	*Entamoeba histolytica*	Amoebic dysentery results from severe infections.
Oomycetes	*Phytophthora infestans*	An outbreak of this protist wiped out potato crops in Ireland in 1845–1847, causing famine.
Microsporidians	*Encephalitozoon cuniculi*	Infections affect the nervous system, respiratory tract, and digestive tract—especially in AIDS patients.

20 μm

FIGURE 28.3 Harmful Algal Blooms Are Sometimes Called Red Tides Dinoflagellates with red coloration have "bloomed" along this beach.

harmful because they release toxins into the surrounding water or, if they are eaten, into a predator. Although the function of these toxins for the protists is not well understood, they may serve to reduce predation by fish.

When a toxin-producing protist reaches high densities in a particular area, a **harmful algal bloom** is said to occur. Toxin-producing protists named dinoflagellates have high concentrations of red accessory pigments called xanthophylls. Hence these outbreaks are known informally as *red tides* (**Figure 28.3**).

Algal blooms can be harmful to people because clams and other shellfish filter photosynthetic protists out of the water as food. During a bloom, high levels of toxins can build up in the flesh of these shellfish. Typically, the shellfish themselves are not harmed. But if a person eats contaminated shellfish, several types of poisoning can result. For example, paralytic shellfish poisoning occurs when people eat shellfish that have fed heavily on protists that synthesize poisons called saxitoxins. In humans, high dosages of saxitoxins cause unpleasant symptoms such as prickling sensations in the mouth or even life-threatening symptoms such as muscle weakness and paralysis.

No antidote exists to the poisons secreted by the protists during harmful blooms. As a result, biologists prevent poisonings by carefully monitoring protist populations in regions where shellfish are harvested for food. If harmful species begin to bloom, the shellfish beds are immediately closed to harvest until they can recover.

Ecological Importance of Protists

As a whole, the protists represent just 10 percent of the total number of named eukaryote species. Although the species diversity of protists may be relatively low, their abundance is extraordinarily high. The number of individual protists found in some habitats is astonishing. One milliliter of pond water can contain well over 500 single-celled protists that swim with the aid of flagella. Under certain conditions, dinoflagellates can reach concentrations of 60 million cells per liter of seawater.

The great abundance of protists is important to an array of ecological events. For example, photosynthetic organisms take in carbon dioxide from the atmosphere and reduce it to form sugars or other organic compounds with high potential energy (Chapter 10). Photosynthesis transforms some of the energy in sunlight into chemical energy that organisms can use to grow and produce offspring. Species that produce chemical energy in this way are called **primary producers**. Diatoms, for example, are photosynthetic protists that rank among the leading primary producers in the oceans, simply because they are so abundant. Production of organic molecules in the world's oceans, in turn, is responsible for almost half of the total carbon that is fixed on Earth. Why is this important?

Protists Play a Key Role in Aquatic Food Chains Photosynthetic protists live in the surface waters of marine and freshwater habitats, where light is available. Small organisms that live near the surface of oceans or lakes and that drift along or swim only short distances are called **plankton**. The organic compounds that are produced by **phytoplankton**—that is, photosynthetic species of plankton—are the basis of food chains in freshwater and marine environments. A **food chain** describes nutritional relationships among organisms. Thus, it also describes how chemical energy flows within ecosystems.

Figure 28.4 shows a food chain for a marine environment. At the bottom of the chain, primary producers—such as photosynthetic protists and photosynthetic bacteria—are eaten by consumers. Primary consumers are just above primary producers in the chain. Scavengers and decomposers occupy the same consumer level in the food chain, but they feed on primary producers that have died. Organisms at this level of the chain are eaten by secondary consumers, which in turn are eaten by tertiary consumers. The fish and shellfish at the middle levels of the food chain are important sources of protein not only for whales, squid, and very large fish (such as tuna) at the uppermost levels but also for people.

Many of the species at the food chain's base are protists. It is not an exaggeration to say that without protists, most food chains in freshwater and marine habitats would collapse.

How Could Protists Help Reduce Global Warming? As noted in Chapter 10, carbon dioxide levels in the atmosphere are increasing rapidly due to human activities such as the burning of fossil fuels and forests. Because carbon dioxide traps heat that is radiating from Earth back out to space, high CO_2 levels in the atmosphere contribute to global warming—arguably today's most pressing environmental problem.

To understand global warming, it is critical to analyze how carbon atoms move among organisms and molecules in both terrestrial and aquatic habitats. The movement of carbon atoms from carbon dioxide molecules in the atmosphere to organisms, in the soil or the ocean, and then back to the atmosphere is called the **global carbon cycle**. To mitigate global warming, researchers are trying to figure out ways to reduce carbon dioxide concentrations in the atmosphere and increase the amount of carbon stored in terrestrial and marine environments.

Recall from Chapter 10 that in terrestrial environments, trees take CO_2 from the atmosphere and use some of the carbon atoms to make wood. Wood is a storage area, or "sink," for carbon, because carbon atoms tend to remain in wood for decades or centuries. As a result, the growth of trees reduces the amount of carbon in the atmosphere and increases the amount of carbon stored in forests. The essay at the end of Chapter 10

suggested that efforts to plant trees in deforested areas might play a role in efforts to reduce atmospheric CO_2 concentrations and slow the rate of global warming.

In the world's oceans, attempts to fertilize photosynthetic protists and bacteria might have a similar effect. To understand why, consider the carbon cycle diagrammed in **Figure 28.5**. The phytoplankton that dominate marine environments have short life spans—they die or are eaten in a matter of days or weeks. As a result, carbon atoms shuttle rapidly among planktonic species. When those organisms die, they are either consumed or sink to the bottom of the ocean. There they may enter one of two long-lived repositories—sedimentary rocks or petroleum:

1. Several lineages of protists have shells made of calcium carbonate ($CaCO_3$). When these shells rain down from the ocean surface and settle in layers at the bottom, the deposits that result are compacted by the weight of the water and by sediments accumulating above them. Eventually they turn into rock. The limestone used to build the pyramids of Egypt consists of protist shells. Limestones and other carbon-containing rocks lock up carbon atoms for tens of millions of years.

2. Although the process of petroleum (oil) formation is not well understood, it begins with accumulations of dead bacteria, archaea, and protists at the bottom of the ocean. CO_2 that passes from the atmosphere to the body of a photosynthetic protist and then to petroleum is removed from the atmosphere for millions of years—unless humans pump the petroleum out of the ground and burn it.

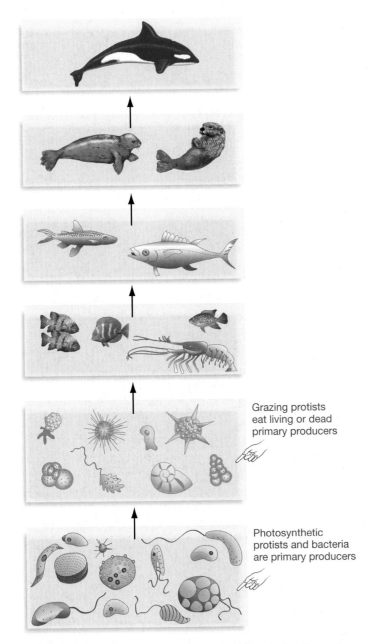

FIGURE 28.4 Protists Are Key Primary Producers in Marine and Freshwater Habitats
In aquatic habitats, photosynthetic protists form the base of the food chain along with photosynthetic bacteria. EXERCISE Indicate the trophic levels at which humans feed.

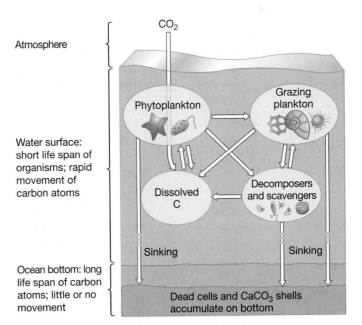

FIGURE 28.5 In the Oceans, Protists Play a Key Role in the Global Carbon Cycle
Carbon atoms tend to shuttle quickly among organisms. But if carbon atoms sink to the bottom of the ocean in the form of shells or dead cells, they may be locked up for long periods in carbon sinks.

Appropriately enough, researchers who study the global carbon cycle refer to long-lived carbon reservoirs as **carbon sinks.**

How do these observations relate to global warming? Recent experiments have shown that when habitats in the middle of the ocean are fertilized with iron, populations of protists and other primary producers increase by a factor of 10. Some researchers hypothesize that, when these blooms occur, the amount of carbon that rains down into carbon sinks in the form of shells and dead cells may increase. If so, then fertilizing the ocean to promote blooms might be an effective way to reduce CO_2 concentrations in the atmosphere. The effectiveness of iron fertilization is hotly debated, however. In addition, some researchers point out that fertilizing the ocean with iron might lead to the formation of anaerobic dead zones (see Chapter 27), toxicity to certain organisms, or other side effects. As a result, many biologists are not convinced that adding iron and inducing massive blooms of protists, bacteria, and archaea would be a good idea. If further research shows that iron fertilization is safe and effective, it could be added to the list of possible approaches for reducing carbon dioxide levels in the atmosphere.

28.2 How Do Biologists Study Protists?

Given that protists are important in medicine and in the world's lakes and oceans, it is not surprising that they have been the focus of intense study. Although great strides have been made in understanding some protist pathogens and the role of protists in the global carbon cycle, it has been extremely difficult for biologists to gain any sort of solid insight into how the group as a whole diversified. The problem is that protists are so diverse that it has been difficult to find any overall patterns in the evolution and diversification of the group.

Recently, researchers have made dramatic progress in understanding protist diversity by combining data on the morphology of key groups and phylogenetic analyses of DNA sequence data. For the first time, a clear picture of how the Eukarya diversified may be within sight. Let's analyze how this work is being done, beginning with classical results on the morphological traits that distinguish major eukaryote groups.

Microscopy: Studying Cell Structure

Using light microscopy, biologists were able to identify and name many of the protist species known today. When transmission electron microscopes became available, a major breakthrough in understanding protist diversity occurred: Detailed studies of cell structure revealed that many protists have a characteristic overall form, organelles with distinctive features, or both. For example, both light microscopy and electron microscopy confirmed that the species that caused the Irish potato famine has reproductive cells with an unusual type of flagellum. Flagella are organelles that project from the cell and whip back and forth to produce swimming movements (Chapter 7). In reproductive cells of *Phytophthora infestans*, one of the two flagella present has tiny, hollow, hairlike projec-

FIGURE 28.6 Species in the Lineage Called Stramenopiles Have a Distinctive Flagellum
The unusual, hollow "hairs" that decorate the flagella of stramenopiles often have three branches at the tip.

tions attached to it. Biologists noted that kelp and other forms of brown algae also have cells with this type of flagellum.

To make sense of these results, researchers interpreted these types of distinctive morphological features as **synapomorphies**—shared, derived traits that distinguish major monophyletic groups (see Chapter 26). Species that have a flagellum with hollow, hairlike projections became known as *stramenopiles* ("straw-hairs"); the hairs typically have three branches at the tip (**Figure 28.6**). In recognizing this group, investigators hypothesized that, because an ancestor had evolved a distinctive flagellum, all or most of its descendants also had this trait. The qualifier "most" is important, because it is not unusual for certain subgroups to lose particular traits over the course of evolution, much as humans are gradually losing the tailbone.

Eventually, eight major groups of eukaryotes came to be identified on the basis of diagnostic morphological characteristics. These groups and the synapomorphies that identify them are listed in **Table 28.2**. Note that the plants, fungi, and animals analyzed in Chapters 29 through 33 represent subgroups within two of the eight major eukaryotic lineages.

Although individual groups of protists were well characterized on the basis or morphology, the relationships among the eight eukaryotic lineages remained almost completely unknown. The next major advance in understanding eukaryotic diversity came when it became possible to obtain and analyze DNA sequence data. In the early 1990s, investigators began using molecular traits to reconstruct the evolutionary history of Eukarya.

Evaluating Molecular Phylogenies

When investigators first began using molecular data to infer the phylogeny of eukaryotes, most studies were based on the gene that codes for the RNA molecule in the small subunit of ribosomes (see Chapter 16). Recall from Chapter 27 that analyses of this gene revealed the nature of the three domains of life. When researchers sequenced this gene from an array of eukaryotes and used the data to infer a phylogeny of the domain Eukarya, they

TABLE 28.2 Major Lineages of Eukaryotes

Lineage	Distinguishing Morphological Feature(s)
Excavates	No mitochondrion is present, although genes derived from mitochondria are found in the nucleus. Cells have a pronounced "feeding groove" where prey or organic debris are ingested.
Discicristates	Cells have mitochondria with distinctive disc-shaped cristae.
Alveolates	Cells have sac-like structures called alveoli that form a continuous layer just under the plasma membrane. Alveoli are thought to provide support.
Stramenopiles	If flagella are present, cells usually have two—one of which is covered with hairlike projections.
Cercozoa	Cells lack cell walls, although some produce an elaborate shell-like covering. When pseudopodia project from the cell to move it, they are slender in shape.
Plants	Cells have chloroplasts with a double membrane.
Amoebozoa	Cells lack cell walls. When pseudopodia project from the cell to move it, they form large lobes.
Opisthokonts	Reproductive cells have a single flagellum at their base. The cristae inside mitochondria are flat, not tube shaped as in other eukaryotes. (This lineage includes protists as well as the fungi and the animals. It is discussed in detail in Chapters 30 through 33.)

found that the eight groups identified on the basis of distinctive morphological characteristics were indeed monophyletic groups. This was important support for the hypothesis that the distinctive morphological features were shared, derived characters that existed in a common ancestor of each lineage.

To understand the relationships among the eight lineages and their subgroups, biologists began analyzing sequence data from other genes and combining these results with data from other morphological traits. Although estimating the phylogeny of Eukarya is still a work in progress, the phylogenetic tree in **Figure 28.7** is the current best estimate of the group's evolutionary history. Understanding where the root or base of the tree lies has been problematic. This tree suggests that a group of single-celled organisms called Excavates, which have a distinctive "feeding groove," forms the most basal, or ancient, lineage of the Eukarya. Groups with specialized types of flagella, unusual mitochondria, or chloroplasts evolved later. The tree also confirms that the lineages grouped under the name *protist* (shown in color in Figure 28.7) are **paraphyletic**—meaning they do not constitute all of the descendants of a single common ancestor (see **Box 28.1**, page 614).

As more and more data become available, our understanding of eukaryote evolution continues to improve. Although a great deal remains to be done to clarify the relationships within Eukarya further, several key events in the history of this group are now well understood. As an example, let's consider how biologists have combined morphological data and phylogenetic analyses to understand better the origin of two key organelles found in eukaryotes: the mitochondrion and the chloroplast.

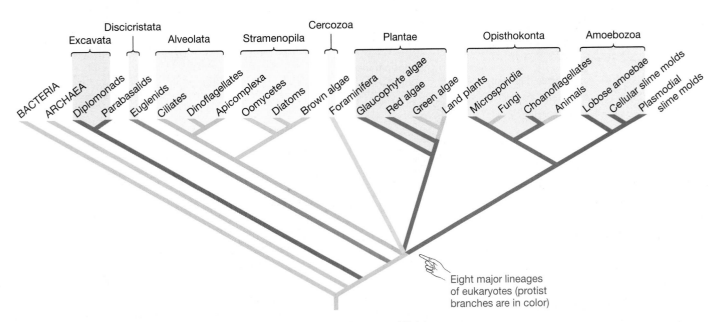

FIGURE 28.7 Phylogenetic Analyses Have Identified Eight Major Lineages of Eukaryotes
Researchers are still evaluating the location of the root on this tree. EXERCISE On the tree, label the node that represents the common ancestor of all eukaryotes. Put a bar and label across the branch where the unusual flagellum of stramenopiles evolved.

BOX 28.1   How Should We Name the Major Branches on the Tree of Life?

Taxonomy is the branch of biology devoted to describing and naming new species and classifying groups of species. Carolus Linnaeus founded this field and published the first work in it in 1735. Linnaeus invented the system of Latin binomials, still being used today, in which each organism is given a unique name consisting of its genus and species. He also invented a hierarchy of more general taxonomic categories that included kingdoms, classes, orders, and families (see Chapter 1). These units were created in an attempt to organize and describe groups of species.

Linnaeus worked long before Darwin had discovered the principle of evolution by natural selection, however. As a result, Linnaeus viewed the groups he described as entities that had been created separately and independently from one another. When biologists came to realize that all organisms are related by common descent, they had to reinterpret the categories that Linnaeus had established. Instead of representing neat "bins" that held increasingly dissimilar species according to a taxonomic (Linnaean) way of thinking (**Figure 28.8a**), biologists recognized that the groupings actually designated twigs, branches, and stems on a tree of life according to a phylogenetic way of thinking (**Figure 28.8b**).

For almost 100 years after Darwin published, however, few reliable techniques existed for reconstructing evolutionary relationships—that is, for estimating phylogenies. As a result, biologists had to continue grouping organisms according to their morphological similarity, much as Linnaeus had done. Biologists were left to hope that these groupings reflected the actual evolutionary relationships on the tree of life.

Over the past 40 years, however, increasingly sophisticated techniques have become available for estimating phylogenies. Chapter 26 introduced these techniques, known as phenetic and cladistic

approaches. In general, the old groupings were remarkably accurate in reflecting actual lineages. But there are some important exceptions. Among the protists, for example, old groups such as "protozoans" and "algae" are now understood to be conglomerations of distantly related, though superficially similar, species. Even the name *protist* will probably be abandoned before long. According to rules that are currently being adopted, most biologists assign names only to monophyletic groups—that is, to branches on the tree of life that include all the descendants of a common ancestor. In violation of this rule, the group that is commonly called protists is paraphyletic. The name refers to some, but not all, descendants of the first eukaryote.

According to a Chinese proverb, "The first step in wisdom is to call things by their right name." In taxonomy, a massive effort is under way to do just that. The initial task is to produce accurate estimates of where each branch occurs on the tree of life. Then monophyletic groups can be named, with confidence that each represents a distinct stem, branch, or twig on the tree.

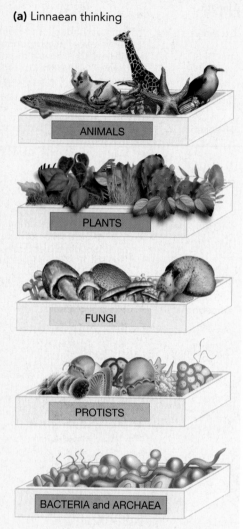

(a) Linnaean thinking

ANIMALS

PLANTS

FUNGI

PROTISTS

BACTERIA and ARCHAEA

(b) Phylogenetic thinking

EUKARYA

BACTERIA   ARCHAEA   Protists   Plants   Fungi   Animals

FIGURE 28.8 Contrasting Approaches to Classification
(a) In Linnaean thinking, organisms are grouped according to morphological similarity. **(b)** In phylogenetic thinking, groups are named if they represent distinct lineages on the tree of life.

Combining Data from Microscopy and Phylogenies: Understanding the Origin of Mitochondria and Chloroplasts

What did the earliest eukaryotes look like? Because the most ancient eukaryotic groups are unicellular, and because virtually all bacteria and all archaeans are also unicellular, biologists infer that the first eukaryote was a single-celled organism. Further, all eukaryotes alive today have a nucleus and a cytoskeleton, so researchers conclude that their common ancestor also had these structures. And because the most ancient eukaryotic lineages lack cell walls, biologists suggest that their common ancestor also lacked this feature. In sum, the earliest eukaryotes were probably single-celled organisms with a nucleus and cytoskeleton but no cell wall. What happened next?

The answer is that one ancestral eukaryote acquired the structure that eventually became a mitochondrion. Recall from Chapter 9 that mitochondria are organelles that have a double membrane and that they generate ATP by using pyruvate as an electron donor and oxygen as the ultimate electron acceptor. All living eukaryotes have mitochondria or at least contain genes that are normally found in mitochondria, so biologists contend that the common ancestor of today's eukaryotes also had mitochondria. Where did these organelles come from?

The Endosymbiosis Theory In 1981 Lynn Margulis expanded on a radical hypothesis—first proposed in the nineteenth century—to explain the origin of mitochondria. The **endosymbiosis theory** proposes that mitochondria originated when a bacterial cell took up residence inside a eukaryote about 2 billion years ago. Its name is inspired by the Greek word roots *endo*, *sym*, and *bio* ("inside-together-living"). **Symbiosis** is said to occur when individuals of two different species live in physical contact; **endosymbiosis** occurs when an organism of one species lives inside an organism of another species.

In its current form, the endosymbiosis theory proposes that mitochondria evolved through a series of steps. As **Figure 28.9** shows, the process began when eukaryotic cells started to use their cytoskeletal elements to surround and engulf smaller prey. The theory proposes that instead of being digested, an engulfed bacterium began to live symbiotically within its eukaryotic host. Specifically, the theory maintains that the engulfed cell survived by absorbing carbon molecules with high potential energy from its host and oxidizing them, using oxygen as a final electron acceptor. The host cell, in contrast, is proposed to be a predator capable only of anaerobic fermentation—meaning it could not use oxygen as an electron acceptor in cellular respiration. The relationship between the host and the engulfed cell was presumed to be stable because a mutual advantage existed between them: The host supplied the bacterium with protection and carbon compounds from its other prey, while the bacterium supplied the host with ATP. Recall from Chapter 9 that cells that can use oxygen during cellular respiration are able to produce much more ATP than are cells that cannot.

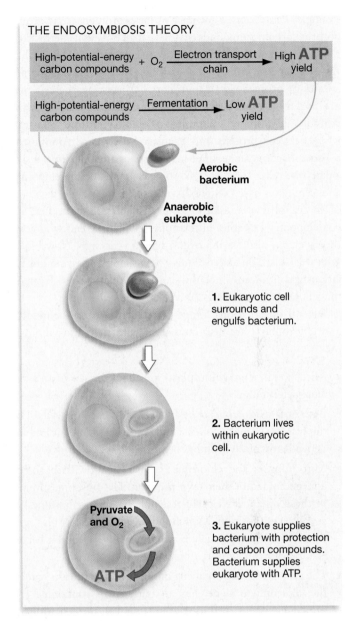

FIGURE 28.9 Proposed Initial Steps in the Evolution of the Mitochondrion
QUESTION According to this hypothesis, how many membranes should surround a mitochondrion?

The endosymbiosis theory contends that chloroplasts originated in an analogous way. In this case, however, the theory maintains that a photosynthetic, endosymbiotic cyanobacterium provided its eukaryotic host with oxygen and glucose in exchange for protection. If the endosymbiosis theory is correct, chloroplasts trace their ancestry to cyanobacteria.

Do the Data Support the Endosymbiosis Theory? When biologists assess a theory, they often begin simply by evaluating its plausibility. For example, is it reasonable to suppose that the steps in the endosymbiosis theory actually happened? Given observations that biologists have made about organisms living today, each step in the endosymbiosis theory passes this test. For example, *Paracoccus*, a member of the α-proteobacteria group, absorbs carbon compounds that have high potential energy from the environment and performs aerobic respiration—much like the presumed ancestral mitochondrion. Dozens of contemporary species of protists obtain nutrients by ingesting bacteria and are anaerobic—much like the presumed ancestral host. More convincingly, many examples of endosymbiotic relationships between protists and bacteria exist today. Among the α-proteobacteria alone, three major groups are found *only* inside eukaryotic cells. There are also some examples of contemporary endosymbioses based on photosynthesis. In such cases, members of the cyanobacteria reside inside protist hosts. In short, there are contemporary examples of relationships that the endosymbiosis theory claims also existed over a billion years ago. Now biologists are focused on understanding how simple endosymbiotic relationships evolved into the complex relationship between mitochondria, chloroplasts, and eukaryotic cells.

Several observations about the structure of mitochondria and chloroplasts are also consistent with the endosymbiosis theory:

- Mitochondria and chloroplasts are about the size of an average bacterium.

- Both organelles replicate by fission, as do bacterial cells. The duplication of mitochondria and chloroplasts takes place independently of division by the host cell.

- Mitochondria and chloroplasts have their own ribosomes and manufacture their own proteins. The ribosomes found in these organelles closely resemble bacterial ribosomes in size and composition. Mitochondrial ribosomes are also poisoned by antibiotics such as streptomycin that inhibit bacterial, but not eukaryotic, ribosomes.

- The photosynthetic organelle of one group of protists, called the glaucophyte algae, has an outer layer containing the same constituent (peptidoglycan) found in the cell walls of cyanobacteria.

- Some cyanobacteria even look like chloroplasts. The bacterium *Prochloron*, for example, contains chlorophylls *a* and *b* and has a system of internal membranes similar to the thylakoids found in chloroplasts (see Chapter 10).

- Both mitochondria and chloroplasts have double membranes, consistent with the engulfing mechanism of origin illustrated in Figure 28.9.

- Mitochondria and chloroplasts have their own genomes, which are organized as circular molecules—much like a bacterial chromosome. Mitochondria and chloroplasts have genes that code for the enzymes needed to replicate and transcribe their own genomes.

Although these data are impressive, they are only consistent with the theory. Stated another way, they do not exclude other explanations. Years after Margulis began to champion this theory, however, data emerged that persuaded virtually all biologists that the endosymbiosis theory was correct. These data were important because they tested predictions made by Margulis's idea that contrasted with predictions made by an alternative theory: that both mitochondria and chloroplasts had evolved within eukaryotic cells, separately from bacteria. According to this "within-eukaryotes" theory, the genes found in both mitochondria and chloroplasts had to have been derived from some of the nuclear genes of ancestral eukaryotes. Margulis's theory, in contrast, proposed that the genes found in mitochondria and chloroplasts were bacterial in origin.

These predictions were tested by studies on the phylogenetic relationships of mitochondrial genes and chloroplast genes (**Figure 28.10**). For example, researchers compared gene sequences isolated from the nuclear DNA of eukaryotes, mitochondrial DNA from eukaryotes, and DNA from several species of bacteria. Exactly as the endosymbiosis theory predicted, the mitochondrial sequences turned out to be much more closely related to the sequences from the α-proteobacteria

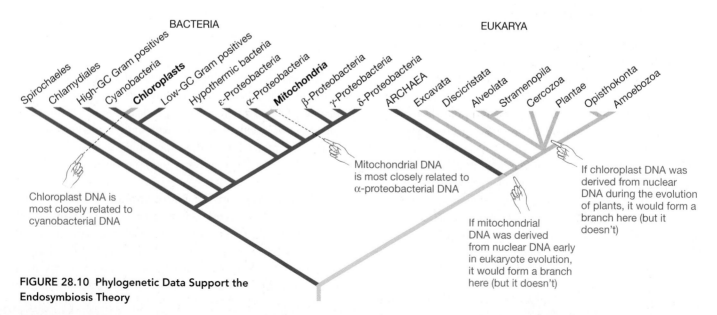

FIGURE 28.10 Phylogenetic Data Support the Endosymbiosis Theory

than to sequences from the nuclear DNA of eukaryotes. The result was considered overwhelming evidence that the mitochondrial genome came from an α-proteobacterium rather than from a eukaryote. A similar analysis showed that chloroplast genes are most closely related to sequences from cyanobacteria, not to those from plants (see Figure 28.10). The endosymbiosis theory was the only reasonable explanation for the data. The results were a stunning vindication of a theory that had once been intensely controversial.

The success of the endosymbiosis theory carries a general message: Protists and other eukaryotes are cut-and-paste jobs. They are chimeras, like the monster from Greek mythology that had the head of a lion, the body of a goat, and the tail of a serpent. The domain diversified after an ancient eukaryote—closely related to archaea—combined with a bacterium. The lineage Eukarya was almost undoubtedly created when two cells began living together. Another major pairing occurred when an ancestor of today's plants engulfed a cyanobacterium.

To put this conclusion into perspective, recall from Chapters 20 and 27 that lateral gene transfer—the transfer of DNA between species that are not closely related—has had a huge impact on the evolution of bacteria and archaea through time and today is playing a large role in the evolution of drug resistance. But even though lateral gene transfer among prokaryotes is impressive, most of the events have involved just one or a handful of genes. These gene movements pale in comparison to the transfer of two entire bacterial genomes to eukaryotes. The evolution of the mitochondrion and the chloroplast resulted in the most massive lateral gene transfers in the history of life.

Discovering New Lineages via Direct Sequencing

What research is currently being done on protist diversity? The answer to this question has several parts. The effort to refine the phylogeny of the Eukarya is continuing, and researchers are developing hypotheses to explain how organelles other than the mitochondria and chloroplasts originated. As **Box 28.2** (page 618) details, protists continue to be used as important model organisms in biology. But of all the research frontiers in eukaryotic diversity, the most exciting may be the one based on the technique called direct sequencing.

As noted in Chapter 27, **direct sequencing** is based on collecting organisms from a habitat and analyzing their DNA without growing larger populations of individuals in laboratory culture. The approach is based on using the polymerase chain reaction (PCR; see Chapter 19) to amplify specific genes in the organisms that have been collected. The resulting DNA sequence data are then used to place the organisms on a phylogenetic tree.

Recall that direct sequencing led to the discovery of previously unknown but major lineages of Archaea. To the amazement of biologists all over the world, the same thing happened when researchers used direct sequencing to survey eukaryotes.

The first direct sequencing studies that focused on eukaryotes were published in 2001 and were motivated by the hypothesis that direct sequencing could detect the presence of previously un-

known species. To test this hypothesis, researchers sampled cells at various depths and locations in the oceans, used PCR primers to amplify genes that encode the RNA component of the ribosome's small subunit, sequenced the genes, and compared the data with the sequences of previously studied eukaryotes. One study sampled organisms at depths from 250 to 3000 m below the surface in waters off Antarctica; another focused on cells at depths of 75 m in the Pacific Ocean, near the equator. Both studies found a wide array of distinctive ribosomal RNA sequences—new species under the phylogenetic species concept introduced in Chapter 25.

Investigators who followed up on these results by examining the samples under the microscope were astonished to find that many of the newly discovered organisms were tiny—from 0.2 μm to 5 μm in diameter. Other direct sequencing studies have now confirmed the existence of eukaryotes that are less than 0.2 μm in diameter. These eukaryotic cells overlap in size with bacteria, which range from 0.5 μm to 2 μm in diameter. The take-home message is that eukaryotic cells are much more variable in size than previously imagined. A whole new world of tiny eukaryotes has just been discovered.

The initial analyses suggest that at least some of the new species may represent important new lineages of dinoflagellates, stramenopiles, or alveolates. As additional direct sequencing studies are done, it is virtually certain that our understanding of eukaryotic diversity will change—perhaps radically—and improve.

✓ CHECK YOUR UNDERSTANDING

Biologists use data from microscopy and DNA sequencing to estimate phylogenetic trees and study the diversity of protists. Combined use of morphological and molecular data has allowed researchers to conclude that the domain Eukarya comprises eight major lineages, members of which have distinctive aspects of cell structure. Recently, direct sequencing has allowed investigators to recognize large numbers of previously undescribed eukaryotes, some of which are extremely small and do not grow in culture. You should be able to (1) compare and contrast the characteristics of at least four of the eight major lineages of eukaryotes, and (2) state a hypothesis for the origin of the chloroplast, list three predictions that follow from that hypothesis, and summarize data that support or contradict those predictions.

28.3 What Themes Occur in the Diversification of Protists?

The protists range in size from bacteria-sized single cells to giant kelp. They live in habitats from the open oceans to dank forest floors. They are almost bewildering in their morphological and ecological diversity. Because they are a paraphyletic group, they do not share derived characteristics that set them apart from all other lineages on the tree of life.

BOX 28.2   A Model Organism: *Dictyostelium discoideum*

Several species of protists have served as model organisms in biology. Among the most important has been the cellular slime mold *Dictyostelium discoideum*. *Dictyostelium* is not always slimy, and it is not a mold—meaning a type of fungus. Instead, it is an amoeba. **Amoeba** is a general term that biologists use to characterize a unicellular protist that lacks a cell wall and is extremely flexible in shape. *Dictyostelium* has long fascinated biologists because it is a social organism. Independent cells sometimes aggregate to form a multicellular structure.

Figure 28.11 shows the *Dictyostelium discoideum* **life cycle**—the sequence of events that occurs over an individual's life span. Note that under most conditions, *D. discoideum* cells are haploid (*n*) and move about in decaying vegetation on forest floors or other habitats. They feed on bacteria by engulfing them whole and reproduce asexually, by mitosis. If food begins to run out, however, the cells begin to aggregate. In many cases, tens of thousands of cells cohere to form a 2-mm-long mass called a **slug**. (This is not the slug that is related to snails. After migrating to a sunlit location, the slug stops and individual cells differentiate according to their position in the slug. Some form a stalk; others form a mass of spores at the tip of the stalk. A *spore* is a single cell that is not formed from gamete fusion and that develops into an adult organism. The entire structure, stalk plus mass of spores, is called a **fruiting body**. Cells that form spores secrete a tough coat and represent a durable resting stage. The fruiting body eventually dries out, and the wind disperses the spores to new locations, where more food might be available. On occasion, *Dictyostelium* may also undergo sexual reproduction. When an aggregation is forming, two cells inside it may undergo syngamy—the fusion of gametes—to form a diploid (2*n*) zygote called a *giant cell*. The giant cell grows by feeding on the haploid amoebae in the aggregation. It then secretes a tough, protective coat. Later the zygote undergoes meiosis to form haploid offspring, which undergo mitosis. Eventually the haploid cells break out of the coating and begin to move about in search of food.

Dictyostelium discoideum has been an important model organism for investigating questions about eukaryotes:

- Cells in a slug are initially identical in morphology but then differentiate into distinctive stalk cells and spores.

Studying this process helped biologists better understand how cells in plant and animal embryos differentiate into distinct cell types.

- The process of slug formation has helped biologists study how animal cells move and how they aggregate as they form specific types of tissues. In addition, the discovery that aggregating *D. discoideum* cells follow trails of cyclic adenosine monophosphate (cAMP) helped investigators better understand the mechanisms responsible for **chemotaxis**—movement in response to a chemical—in other organisms.

- When *D. discoideum* cells aggregate to form a slug, they stick to each other. The discovery of membrane proteins responsible for cell-cell adhesion helped biologists understand some of the general principles in Chapter 8.

- Cells that aggregate to form a slug are not genetically identical. About 20 percent of the cells help other individuals produce offspring by forming the stalk, but die without producing offspring themselves. Why? Researchers are using *D. discoideum* to answer questions about why some individuals sacrifice themselves to help others.

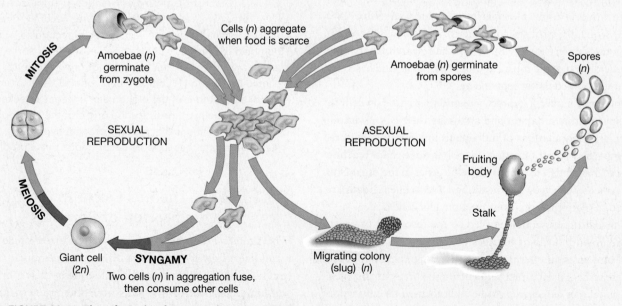

FIGURE 28.11 The Life Cycle of *Dictyostelium discoideum*

Fortunately, several general evolutionary themes tie the diversity of eukaryotes together. The key to understanding the protists is to recognize that a series of important innovations occurred, often repeatedly, as eukaryotes diversified.

Morphological Diversity

Although eukaryotic cells vary widely in size, the domain Eukarya is distinguished by the evolution of the largest cells known. An average-sized eukaryotic cell is 10 times larger in diameter than an average-sized bacterial cell. The evolution of large cells is an important theme in the evolution of the Eukarya.

How is it possible for such large cells to evolve? The answer to this question is not obvious, because large cell size presents an important physical challenge: As cells become larger, their volume increases much more rapidly than does their surface area. This is because a cell's volume increases as the cube of a sphere's diameter (volume $\propto$ diameter^3), while its surface area increases as the square of a sphere's diameter (surface area $\propto$ diameter^2). (The relationship between surface area and volume is explored in more detail in Chapter 41.) As cells get larger, then, the proportion of surface area available gets smaller and smaller relative to the volume present. This creates a problem because food, gases, and waste molecules must diffuse across the cell's surface, while the volume is filled with biosynthetic machinery that requires energy and generates waste. As cell size increases, metabolism in the cell's interior can outstrip the transport and exchange processes that take place along the surface area and that provide the raw materials for metabolism. How do eukaryotes cope?

Organelles Divide a Large Cell into Compartments

Dilemmas regarding surface area and volume are minimized in eukaryotes, because cells are divided into compartments. As an example of how the compartmentalization of the eukaryotic cell works, consider the protist called *Paramecium* (**Figure 28.12**). This organism makes a living by eating bacteria, which it sweeps into an indentation known as the *gullet* and then into the cell mouth. After ingesting a bacterium, a *Paramecium* surrounds it with an internal membrane, forming a compartment called a **food vacuole**. Food vacuoles merge with membrane-bound structures called lysosomes, which hold digestive enzymes, and circulate around the cell. When the food has been digested and nutrients have diffused out of the food vacuole, the vacuole merges with the plasma membrane at a special organelle—the **anal pore**—and expels waste molecules. The food vacuoles provide a large area of internal membrane, allowing nutrients to be delivered efficiently throughout the volume of the cell and waste products to be expelled rapidly.

Although not all eukaryotic cells contain food vacuoles and anal pores, they all have numerous other internal compart-

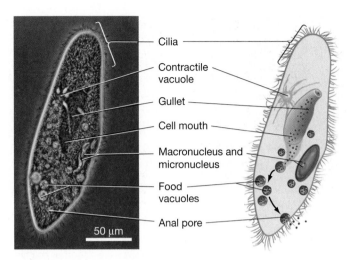

FIGURE 28.12 Protists Have Specialized Intracellular Structures
Paramecium feeds by sweeping bacteria into a subcellular structure called the gullet. **EXERCISE** Next to the drawing of *Paramecium*, list other organelles and intracellular structures that are found in protist cells.

ments. Each of these internal structures—including the nucleus, lysosomes, peroxisomes, mitochondria, chloroplasts, central vacuole, Golgi apparatus, and rough and smooth endoplasmic reticulum (ER) that were introduced in Chapter 7—has a distinct function. Many are membrane bound and are devoted to the synthesis, transport, and distribution of molecules.

In terms of organizing the cell volume, however, the most important feature of eukaryotes may be the cytoskeleton. This system of filaments and tubules, introduced in Chapter 7, provides a scaffolding that supports and organizes the cell and, in many cases, moves it as well. Although some bacteria and archaea have rudimentary cytoskeletal elements, an extensive, highly ordered cytoskeleton is found only in eukaryotes.

In sum, the extensive compartmentalization and differentiation of the eukaryotic cell, along with the ability to generate large amounts of ATP through aerobic respiration, makes large size possible. The evolution of large cell size, specialized organelles, and complex cytoskeletal structures is a major theme in the diversification of protists.

The Evolution of Multicellularity   A few species of bacteria are capable of aggregating and forming fruiting bodies similar to those observed in the protist *Dictyostelium discoideum* (see Box 28.2). Because cells in the fruiting bodies of these bacteria differentiate into specialized stalk cells and spore-forming cells, they are considered multicellular. But the vast majority of multicellular species are members of the Eukarya. The evolution of multicellularity is a prominent theme in the evolution of eukaryotes.

Multicellularity is actually a difficult trait to define. To understand why, consider a group of green algae called the Volvocales.

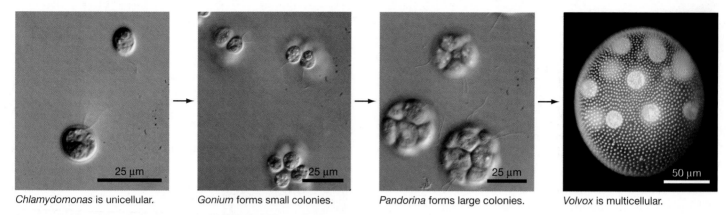

Chlamydomonas is unicellular.	*Gonium* forms small colonies.	*Pandorina* forms large colonies.	*Volvox* is multicellular.

FIGURE 28.13 Multicellularity Evolved in Protists
Members of the green algal family called Volvocales are graded in morphology, from unicellular to colonial to multicellular. The large spheres within the *Volvox* are daughter colonies that are being produced asexually.

As **Figure 28.13** shows, the organisms in this lineage range from single cells such as *Chlamydomonas*, to small clumps of cells such as *Gonium*, to large groupings such as *Pandorina* and the multicellular *Volvox*. In this lineage, there appears to be a smooth continuum between single-celled and multicelled forms. Where do biologists draw the line between the loose aggregations of cells called a **colony** and the more highly structured arrangements of cells that create a body and typify multicellular organisms? For example, why is *Volvox* considered multicellular?

Most biologists would cite two reasons that *Volvox* is multicellular: Individual *Volvox* cells cannot survive on their own, and each cell has a distinct function. Some *Volvox* cells are strictly vegetative, meaning that they perform photosynthesis and supply the organism with ATP and sugar. Sexual reproduction is limited to certain cells that undergo meiosis and form reproductive cells equivalent to sperm and eggs.

The differentiation of cell types is a crucial criterion for multicellulariy. To understand why, recall from Chapter 22 that, when differentiation occurs, only a well-defined subset of each cell's genome is transcribed. Different types of cells express different genes. This feature is used to distinguish **colonial growth**—in which cells aggregate but each performs the same function—from **multicellular growth**. Multicellular organisms have masses of closely interacting cells, distinct cell or tissue types, and large overall size. Multicellularity evolved several times as eukaryotes diversified. This trait arose independently in green plants, red algae, animals, fungi, brown algae, and amoebae.

Structures for Support and Protection Thanks to their cytoskeleton and organelles, all protists have a complex intracellular structure. But in addition, many protists have a rigid internal skeleton or a hard, external structure. Rigid structures inside the plasma membrane, such as the silicon-containing skeletons of protists called heliozoans, support the cell. Many protists have cell walls outside their plasma membrane; others have hard external structures, often called a **test** or a **shell**, that provide support or protection or both.

Shells are found in groups that are scattered about the eukaryotic tree. The structures themselves are equally diverse. As **Figure 28.14** shows, these external skeletons vary from the intricate, chambered calcium carbonate ($CaCO_3$) tests found in foraminifera, to the glass-like, silicon-oxide shells enclosing di-

(a) Foraminiferan

(b) Diatom

(c) Dinoflagellate

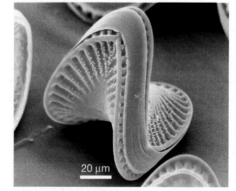

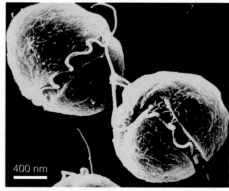

Calcium carbonate test, with chambers

Test made of silicon oxides

Plates made of cellulose

FIGURE 28.14 Hard Outer Coverings in Protists Vary in Composition

(a) Pseudopodia engulf food

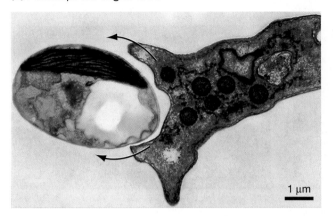

1 μm

(b) Ciliary currents sweep food into gullet

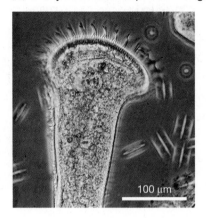

100 μm

FIGURE 28.15 Ingestive Feeding
Methods of prey capture vary widely among ingestive protists. **(a)** Some predators engulf prey with pseudopodia; **(b)** other predators sweep them into their gullets with water currents set up by the beating of cilia.

atoms, to the cellulose plates that surround some dinoflagellates. Some amoebae even cover themselves with tiny pebbles.

To interpret these findings, biologists point out that protists evolved many different ways to support their large cells and protect themselves from predation. Like multicellularity, shells have evolved many times, independently. The evolution of extracellular layers or coatings was an important theme during the evolution of the Eukarya.

How Do Protists Find Food?

Recall from Chapter 27 that bacteria and archaea obtain the chemical energy they need in one of three ways: (1) by absorbing inorganic compounds with high potential energy from the environment; (2) by absorbing high-energy organic compounds, such as sugars, from a host or from decaying organic matter; or (3) through photosynthesis. Among the protists, many species absorb organic compounds for nutrition and many are photosynthetic. Although no protists make a living from inorganic compounds, large cell size and the evolution of the cytoskeleton made a unique feeding strategy possible: Protists can eat bacteria, archaea, or even other protists whole—meaning they can take in packets of food much larger than individual molecules. Thus, eukaryotes feed by (1) ingesting packets of food, (2) absorbing individual molecules, or (3) performing photosynthesis. Let's consider each method.

Ingestive Feeding   Ingestive lifestyles are based on eating live or dead organisms or on scavenging loose bits of organic debris. Protists such as the cellular slime mold *Dictyostelium discoideum* are large enough to engulf bacteria and archaea; many protists are large enough to surround and ingest other protists or microscopic animals. The engulfing process is possible because some protists lack a cell wall and because their flexible membrane and cytoskeleton give these species the ability to surround and "swallow" prey with long, fingerlike projections called **pseudopodia** ("false-feet") (**Figure 28.15**).

Losing a cell wall and gaining the ability to move their plasma membrane around objects was an important innovation during the evolution of protists. Recall from Chapter 27 that bacteria

and archaea are abundant in wet soils and aquatic habitats, where protists also live. Instead of competing with bacteria and archaea for sunlight or food molecules, protists could eat them.

In addition to hunting down prey and engulfing them, many of the ingestive protists are "sit and wait" predators or scavengers. Species that feed in this way often have cilia that create water currents when they beat. In many cases, these protists attach themselves to a substrate and collect food by sweeping particles into their mouths (**Figure 28.15b**). Organisms that filter food out of water in this way are said to be **filter feeders**, which are also known as **suspension feeders**.

Absorptive Feeding   When nutrients are taken up directly from the environment, across the plasma membrane, **absorptive feeding** occurs. Some protists that live by absorptive feeding are decomposers. A **decomposer** is an organism that feeds on dead organic matter, or **detritus**. But many of the protists that absorb their nutrition directly from the environment live inside other organisms. If they damage their host, the absorptive species is called a **parasite**. The protists responsible for the diseases listed in Table 28.1 are parasites (**Figure 28.16**).

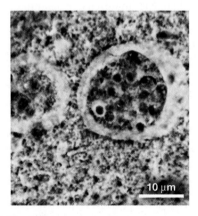

10 μm

FIGURE 28.16 Parasitism
Micrograph of two *Entamoeba histolytica*, which is a common cause of traveler's diarrhea in humans. Parasitic protists get all of their nutrition from their host.

(a) Red: chlorophyll *a* and phycoerythrins

(b) Brown: chlorophyll *a*, chlorophyll *c*, and xanthins

(c) Green: chlorophyll *a* and chlorophyll *b*

(d) Symbiosis

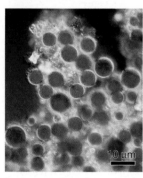

10 µm   100 µm   500 µm   0.5 mm

FIGURE 28.17 Photosynthesis
Many photosynthetic algae are distinguished by their accessory pigments, in addition to chlorophyll *a*. Each accessory pigment intercepts different wavelengths of light. Symbioses are also common in protists. Photosynthetic protists called dinoflagellates live inside the coral on the right and are responsible for its color.

Photosynthesis A wide array of protists are photosynthetic. Plants may be the dominant producers of sugars and other high-energy organic compounds on land, but photosynthetic protists and bacteria are the dominant producers of carbon compounds in oceans and lakes.

When biologists analyze the evolution of photosynthesis in protists, three key themes emerge:

1. The major photosynthetic groups of protists are distinguished by the pigments they contain (**Figure 28.17**). Each of these pigments, in turn, absorbs unique wavelengths of light. As a result, the various groups of photosynthetic protists specialize in harvesting light energy from particular regions of the electromagnetic spectrum. The leading hypothesis to explain this pattern is that photosynthetic species harvest different wavelengths of light to avoid competition. This theme in the diversification of protists echoes the variation in photosynthetic pigments observed in bacteria (see Chapter 27).

2. Many photosynthetic protists live symbiotically with animals or other protists—either inside or outside host cells. Although there are hundreds of examples, perhaps the most important involve the animals called corals. Virtually all reef-building corals harbor single-celled, photosynthetic protists in their skin or gut tissue. The concentrations of endosymbiotic cells can be impressive—up to a million protist cells per square centimeter of coral surface (Figure 28.17).

3. Protists originally obtained chloroplasts by endosymbiosis with a cyanobacterium. But a wide array of photosynthetic protist groups obtained their chloroplasts by ingesting other photosynthetic protists. This process, called **secondary endosymbiosis**, occurs when an organism engulfs a photosynthetic cell and retains its chloroplasts as intracellular symbionts. In groups that obtained chloroplasts via secondary endosymbiosis, the organelle is surrounded by more than two membranes—usually four (**Figure 28.18**). **Figure 28.19**

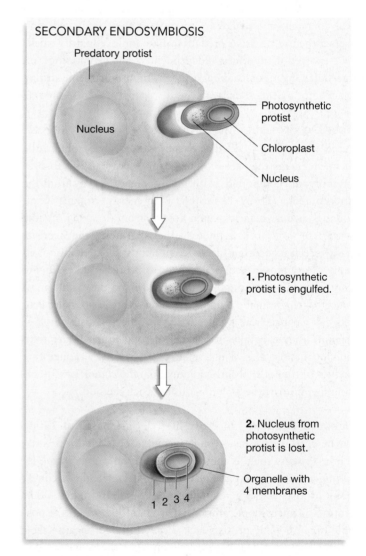

SECONDARY ENDOSYMBIOSIS

Predatory protist

Nucleus

Photosynthetic protist

Chloroplast

Nucleus

1. Photosynthetic protist is engulfed.

2. Nucleus from photosynthetic protist is lost.

Organelle with 4 membranes

1 2 3 4

FIGURE 28.18 Secondary Endosymbiosis Leads to Organelles with Four Membranes
The chloroplasts found in some protists have four membranes and are hypothesized to be derived by secondary endosymbiosis. In species where chloroplasts have three membranes, biologists hypothesize that secondary endosymbiosis was followed by the loss of one membrane.

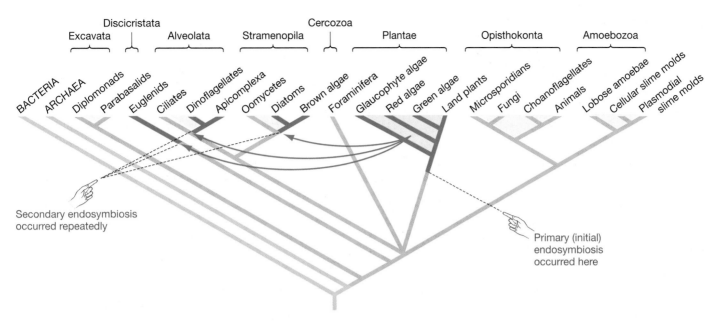

FIGURE 28.19 Photosynthesis Arose in Protists by Primary Endosymbiosis, Then Spread among Lineages via Secondary Endosymbiosis
All of the species above glaucophyte algae on the Eukarya tree have chloroplasts that have two membranes and similar molecular composition, indicating that they arose via primary endosymbiosis. A wide variety of other protist lineages have species with chloroplasts or chloroplast-derived organelles with three or four membranes, indicating that they arose via secondary endosymbiosis.

shows where primary and secondary endosymbiosis occurred on the phylogenetic tree of eukaryotes. The message of this figure is simple: Many lineages of protists acquired the ability to photosynthesize independently of each other.

Diversity in Lifestyles   Before leaving the topic of obtaining food in protists, it is important to recognize that all three lifestyles—ingestive, absorptive, and photosynthetic—occur in many different eukaryote lineages. All three types of food getting can occur within a single clade.

To drive this point home, consider the monophyletic group called the alveolates. There are three major subgroups of this lineage, called dinoflagellates, apicomplexa, and ciliates. Within each subgroup, species vary in how they make a living. About half of the dinoflagellates are photosynthetic, while many others are parasitic. The apicomplexa are parasitic. The ciliates include many species that ingest prey, but some ciliates live in the guts of cattle or the gills of fish and absorb nutrients from their hosts. Other ciliate species make a living by holding algae or other types of photosynthetic symbionts inside their cells.

This diversity in modes of nutrition underscores earlier conclusions. Within each of the eight major lineages of eukaryotes, many different morphologies and ways of feeding evolved.

How Do Protists Move?

Many protists actively move to find food. Predators such as *Dictyostelium discoideum* crawl over a substrate in search of

prey. Most of the unicellular, photosynthetic species are capable of swimming to sunny locations. How are these crawling and swimming movements possible?

Cell crawling is a sliding motion observed in some protists. In the classic mode of cell crawling illustrated in **Figure 28.20a** (page 624), pseudopodia stream forward over a substrate, with the rest of the cytoplasm, organelles, and plasma membrane following. The motion requires ATP and involves interactions between proteins called actin and myosin inside the cytoplasm. The mechanism is related to muscle movement in animals, which is detailed in Chapter 46. But at the level of the whole cell, the precise sequence of events during cell crawling is still somewhat uncertain.

The other major mode of locomotion in protists involves flagella (**Figure 28.20b**) or cilia (**Figure 28.20c**). Flagella and cilia have identical structures. Recall from Chapter 7 that both consist of nine sets of doublet (paired) microtubules arranged around two central, single microtubules. Flagella, however, are long and are usually found alone or in pairs; cilia are short and usually occur in large numbers on any one cell. Flagella and cilia can also be distinguished by the types of structures associated with the basal bodies where they originate. Both structures are markedly different from the flagella found in bacteria and archaea. The cilia and flagella of protists are made up of microtubules, and dynein is the major motor protein. An undulating motion occurs as dynein molecules walk down microtubules. The flagella of bacteria and archaea, in contrast, are composed primarily of a protein called flagellin (see Chapter 7). Instead of undulating, these flagella rotate to produce movement.

(a) Cell crawling via pseudopodia

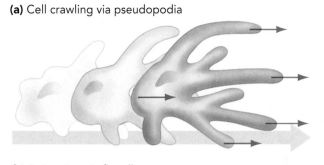

(b) Swimming via flagella

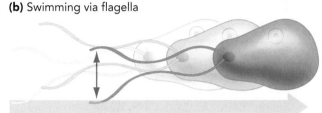

(c) Swimming via cilia

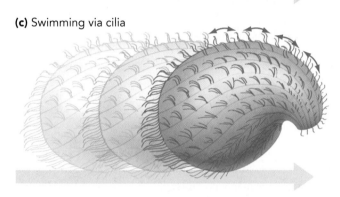

FIGURE 28.20 Modes of Locomotion in Protists Vary
(a) In cell crawling, long pseudopodia stream out from the cell. The rest of the cytoplasm, organelles, and external membrane follow. **(b)** Flagella are long and few in number, and they power swimming movements. **(c)** Cilia are short and numerous. In many cases they are used in swimming.

Even closely related protists can use radically different forms of locomotion. For example, consider again the lineage of protists called the alveolates and the subgroups known as ciliates, dinoflagellates, and apicomplexa. The ciliates swim by beating their cilia. Dinoflagellates swim by whipping their flagella. Apicomplexa move by cell crawling, and they have gametes with flagella. Within the other major lineages of protists, it is common to observe variation in how cells move and common to find species that do not exhibit active movement but instead float passively in water currents. Movement is yet another example of diversification that occurred within each of the eight major eukaryote lineages.

How Do Protists Reproduce?

Sexual reproduction evolved in protists. As pointed out in Chapter 12, sexual reproduction can best be understood in con-

trast to asexual reproduction. Asexual reproduction is based on mitosis in eukaryotic organisms and on fission in bacteria and archaea, and it results in offspring that are genetically identical to the parent. Most protists undergo asexual reproduction routinely. When sexual reproduction occurs, in contrast, offspring are genetically different from their parents and from each other. Sexual reproduction occurs only intermittently in many protists—often at one particular time of year, or when individuals are crowded or food is scarce—and is based on meiosis. The evolution of sexual reproduction ranks among the most significant evolutionary innovations observed in eukaryotes.

Sexual versus Asexual Reproduction Why was the evolution of meiosis important? One hypothesis is that offspring that are genetically variable—meaning their genotypes are different from those of other offspring and from those of their parents—may be able to thrive if the environment changes. For example, offspring with genotypes different from those of their parents may be better able to withstand attacks by parasites that successfully attacked their parents (see Chapter 12). This is a key point, because many types of parasites, including bacteria and viruses, have short generation times and evolve very quickly. Because the genotypes and phenotypes of parasites are constantly changing, natural selection is constantly favoring host individuals with new genotypes. The idea is that new offspring genotypes generated by meiosis may contain combinations of alleles that allow hosts to withstand attack by new strains of parasites. In short, many biologists view sexual reproduction as an adaptation to fight disease.

Variation in Life Cycles A life cycle describes the sequence of events that occurs as individuals grow, mature, and reproduce. The evolution of meiosis not only introduced a new event in the life cycle of protist species, but it also created a distinction between haploid and diploid phases in the life of an individual. Recall from Chapter 12 that diploid individuals have two of each type of chromosome inside each cell, while haploid individuals have just one of each type of chromosome inside each cell. When meiosis occurs in diploid cells, it results in the production of haploid cells.

The life cycle of most bacteria and archaea is extremely simple: A cell divides, feeds, grows, and divides again. Bacteria and archaea do not undergo sexual reproduction, and they are always haploid. In contrast, among protists virtually every aspect of a life cycle is variable—whether meiosis occurs, whether asexual reproduction occurs, and whether the haploid or the diploid phase of the life cycle is the longer and more prominent phase.

Figure 28.21 illustrates some of the variation that occurs in life cycles of protists. Figure 28.21a depicts the haploid-dominated life cycle observed in many unicellular protists. To analyze this life cycle, start with **syngamy**, or fertilization—the fusion of two gametes to form a diploid zygote. Then trace what happens to the zygote. In this case, the diploid zygote undergoes meiosis.

(a) A life cycle dominated by haploid cells

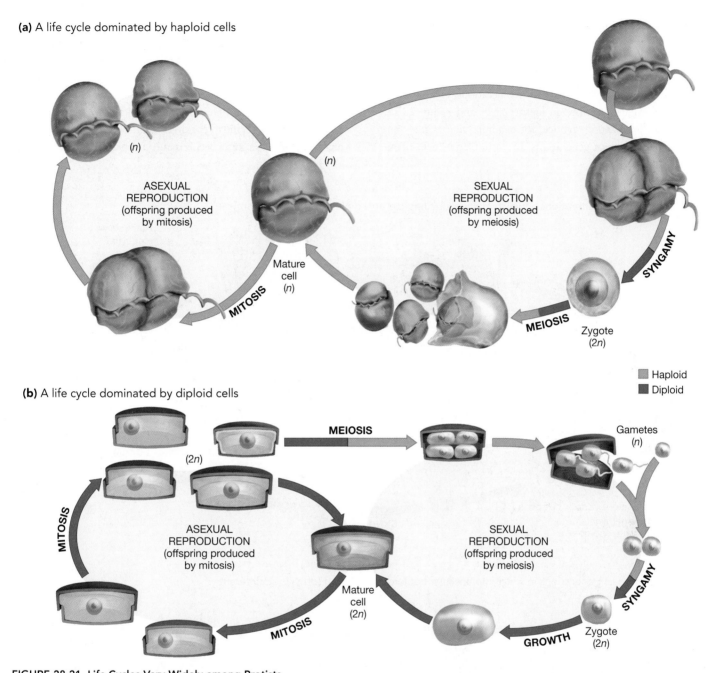

(b) A life cycle dominated by diploid cells

Haploid
Diploid

FIGURE 28.21 Life Cycles Vary Widely among Protists
Many protists can reproduce by both asexual reproduction and sexual reproduction.

The haploid products of meiosis then grow into mature cells that eventually undergo asexual reproduction or produce gametes by mitosis. Contrast that cycle with the diploid-dominated life cycle of Figure 28.21b, in which the diploid zygote develops into a sexually mature adult. In such species, meiosis occurs in the adult rather than in the zygote; gametes are the only haploid cells in the life cycle.

In contrast to the relatively simple life cycle of Figure 28.21, many multicellular protists have one phase in their life cycle that is based on a multicellular haploid form and another phase that is based on a multicellular diploid form. The alternation of multicellular haploid and diploid forms is known as **alternation of generations.** The multicellular haploid form is called a **gametophyte,** because specialized cells in this individual produce gametes by mitosis. The multicellular diploid form is called a **sporophyte,** because it has specialized cells that undergo meiosis to produce haploid cells called spores. Recall that a **spore** is a single cell that is *not* formed by syngamy (the fusion of gametes) and that develops into an adult organism. Gametophytes and sporophytes may be identical in appearance, as in the brown alga illustrated in **Figure 28.22a**. In many cases, however, the gametophyte and sporophyte are different in size or shape or both (**Figure 28.22b**). Among the protists, alternation of generations evolved independently in brown algae and in red algae. To get a better understanding of alternation of generations, compare and contrast the events described in Figure 28.21 and Figure 28.22, starting with fertilization. Notice, for example, that zygotes undergoing alternation of generations do not undergo meiosis im-

mediately. Instead they germinate into diploid individuals: sporophytes. Cells in the sporophyte then undergo meiosis to form haploid spores, which germinate and grow into haploid gametophytes.

Why does so much variation occur in the types of life cycles observed among protists? The answer is not known. Variation in life cycles is a major theme in the diversification of protists. Explaining why that variation exists remains a topic for future research.

(a) Alternation of generations in which multicellular haploid and diploid forms look identical

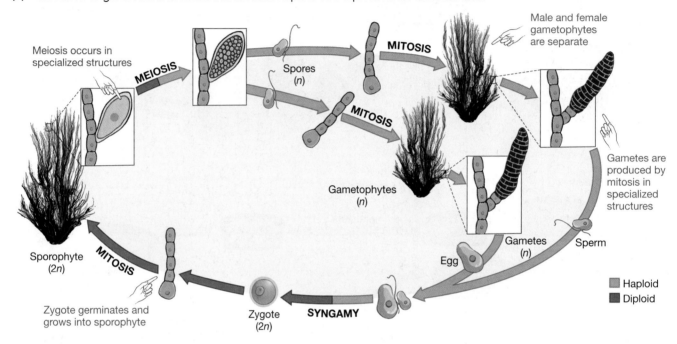

(b) Alternation of generations in which multicellular haploid and diploid forms look different

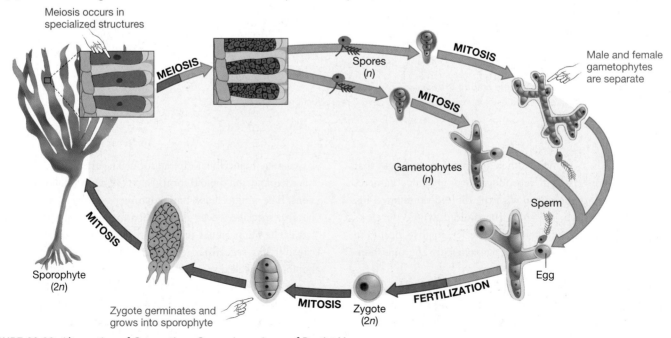

FIGURE 28.22 Alternation of Generations Occurs in an Array of Protist Lineages

Protist diversity can be understood by focusing on key evolutionary themes and innovations. The important innovations in morphological evolution were large cell size, made possible by organelles, a cytoskeleton, and aerobic respiration; multicellularity; and exterior layers or coatings. In terms of finding food, the key methods are ingestive, photosynthetic, or absorptive (often parasitic) lifestyles. Regarding movement, protists employ cell crawling, flagella, or cilia. With respect to reproduction, meiosis and sexual reproduction were a crucial evolutionary innovation. You should be able to (1) explain why organelles made the evolution of large cell size possible, (2) give examples of each of the key evolutionary themes and innovations, (3) provide evidence that traits such as multicellularity and photosynthesis arose repeatedly over the course of eukaryote evolution, and (4) diagram the major types of life cycles observed in protists.

28.4 Key Lineages of Protists

An important generalization jumps out of the Eukarya phylogenetic tree and data on variation among protists in morphology, feeding method, locomotion, and mode of reproduction: Each of the eight major Eukarya lineages has at least one distinctive morphological characteristic. But once an ancestor evolved a set of distinctive characteristics, its descendants diversified into a wide array of lifestyles. For example, parasitic species evolved independently in all eight major lineages. Photosynthetic species exist in most of the eight, and multicellularity evolved independently in four. Similar statements could be made about the evolution of life cycles and modes of locomotion.

In effect, each of the eight lineages represents a similar radiation of species into a wide array of lifestyles. In each case, the radiation began with a morphological innovation—often a change in the structure or function of one or more organelles. Let's take a more detailed look at some representative taxa from seven of the eight major groups. The eighth group—the opisthokonts—includes fungi, animals, and several protist lineages and will be discussed in Chapters 30 through 33. The clade called green plants is analyzed in Chapter 29.

Excavata

The unicellular species that form the Excavata are named for the morphological feature that distinguishes them: an "excavated" feeding groove found on one side of the cell. Because all excavates lack mitochondria, they were once thought to trace their ancestry to eukaryotes that existed *prior* to the origin of mitochondria. But researchers have found that excavates either have nuclear genomes containing genes that are normally found in mitochondria or unusual organelles that appear to be vestigial mitochondria. These observations support the hypothesis that the ancestors of excavates had mitochondria, but that these organelles were lost or reduced over time. Several major lineages are included in the group, two of which are detailed here.

Excavata ▶ Diplomonadida

The first species of diplomonad known was described by the early microscopist Anton van Leeuwenhoek, who found *Giardia intestinalis* while examining samples of his own feces. About 100 species have been named to date. Many live in the guts of animal species without causing harm to their host; other species live in stagnant water habitats. In both types of environment, oxygen availability tends to be very low.

Morphological diversity Each cell has two nuclei, which resemble eyes when viewed under the microscope. Some species lack the organelles called peroxisomes and lysosomes in addition to lacking functional mitochondria; all lack a cell wall.

Feeding and locomotion Some are parasitic, but most ingest bacteria whole. They swim using flagella.

Reproduction Only asexual reproduction occurs; meiosis has yet to be observed in members of this lineage.

Human and ecological impacts Both *Giardia intestinalis* and *G. lamblia* (**Figure 28.23**) are common intestinal parasites in humans. *Hexamida* outbreaks cause heavy losses in turkey farms each year.

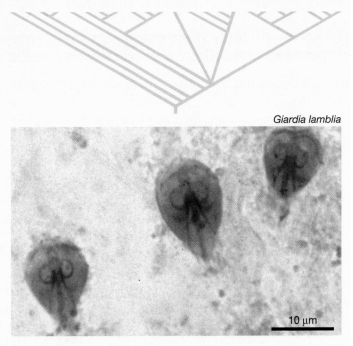

Giardia lamblia

10 μm

FIGURE 28.23 *Giardia* **Causes Intestinal Infections in Humans**

Excavata ▶ Parabasalida

No free-living parabasids are known; all of the species that have been described to date live inside animals. Of the 300 species of parabasalid, several live only in the guts of termites. Termites eat wood but cannot digest it themselves. Instead, the parabasalids produce enzymes that digest the cellulose in wood and release compounds that can be used by the termite host. The relationship between termites and parabasalids is considered mutualistic, because both parties benefit from the symbiosis.

Morphological diversity   Cells lack a cell wall and have a single nucleus.

Feeding and locomotion   Parabasalids feed by engulfing bacteria, archaea, and organic matter. Although the number of flagella present varies widely, four or five is typical.

Reproduction   All reproduce asexually; sexual reproduction has been observed in a few species.

Human and ecological impacts   *Trichomonas* infections can sometimes cause reproductive tract problems in humans (**Figure 28.24**), although members of this genus may also live in the gut or mouth of humans without causing harm. *Histomonas meleagridis* is a serious threat to chicken and turkey farms.

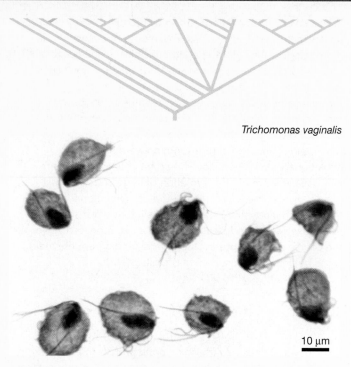

Trichomonas vaginalis

10 μm

FIGURE 28.24 *Trichomonas* **Causes the Sexually Transmitted Disease Trichomoniasis**

Discicristata

The discicristates are all unicellular and were named for the distinctive disc shape of the cristae within their mitochondria. Several major subgroups have been identified, such as amoeboid forms and lineages that include the species responsible for the diseases leishmaniasis and trypanosomiasis in humans (see Table 28.1). Because they are extremely common in ponds and lakes, the euglenids are the best-studied discicristates.

Discicristata ▶ Euglenida

There are about 1000 known species of euglenid. Although most live in freshwater, a few are found in marine habitats. Fossil euglenids have been found in rocks that are over 410 million years old.

Morphological diversity   Most euglenids lack an external wall but have a system of interlocking protein lying under the plasma membrane.

Feeding and locomotion   About one-third of the species have chloroplasts and perform photosynthesis, but most ingest bacteria and other small cells or particles. Some cells have a light-sensitive "eyespot" and use flagella to swim toward light (**Figure 28.25**).

Reproduction   Only asexual reproduction is known to occur among euglenids.

Human and ecological impacts   Euglenids are important components of freshwater plankton and food chains.

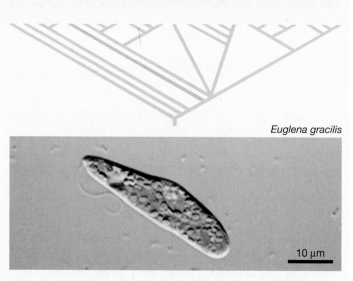

Euglena gracilis

10 μm

FIGURE 28.25 Euglenids Are Common in Ponds and Lakes

Alveolata

Alveolates have small sacs called alveoli that are located just under their plasma membranes. Although all members of this lineage are unicellular, the groups highlighted here—the Ciliata, the Dinoflagellata, and the Apicomplexa—are remarkably diverse in morphology and lifestyle.

Alveolata ▶ Ciliata

Ciliates were named for the cilia that cover them (**Figure 28.26**). Some 12,000 species are known from freshwater habitats, marine environments, and wet soils.

Morphological diversity Cells have two nuclei—a larger *macronucleus* and a smaller *micronucleus*, which is involved only in reproduction. A few ciliates secrete an external skeleton.

Feeding and locomotion Depending on the species, ciliates may be filter feeders, predators, or parasites. They use cilia to swim and have a mouth area where food is ingested.

Reproduction Ciliates divide to produce daughter cells asexually. They also undergo a type of sexual reproduction called **conjugation**. During conjugation, two cells line up side by side and physically connect. Micronuclei exchange between cells and fuse. The resulting nucleus eventually forms a new macronucleus and micronucleus.

Human and ecological impacts Ciliates are abundant in marine plankton and are important consumers. They are common in the digestive tracts of goats, sheep, cattle, and other grazers, where they feed on plant matter and help the host animal digest it.

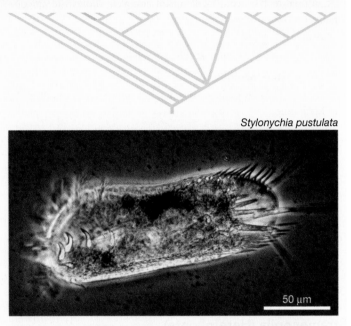

Stylonychia pustulata

50 µm

FIGURE 28.26 Ciliates Are Abundant in Freshwater Plankton

Alveolata ▶ Dinoflagellata

Most of the 4000 species of dinoflagellates are ocean-dwelling plankton, although they are abundant in freshwater as well. Some species are capable of **bioluminescence**, meaning they emit light via an enzyme-catalyzed reaction (**Figure 28.27**).

Morphological diversity Most dinoflagellates are unicellular; some are colonial. Each species has a distinct shape maintained by plates of cellulose inside the cell. Unlike chromosomes of other eukaryotes, chromosomes in dinoflagellates do not contain histones.

Feeding and locomotion About half of the dinoflagellates are photosynthetic. Some of the photosynthetic species live in association with corals or sea anemones, but most are planktonic. The other species are predatory or parasitic. Cells swim in a spinning motion, using two flagella.

Reproduction Both asexual and sexual reproduction occur. Cells that result from sexual reproduction may form tough **cysts**. A cyst is a resistant structure that can remain dormant until environmental conditions improve.

Human and ecological impacts Photosynthetic dinoflagellates are important primary producers in marine ecosystems—second only to diatoms in the amount of carbon they fix per year. A few species are responsible for harmful algal blooms, or red tides.

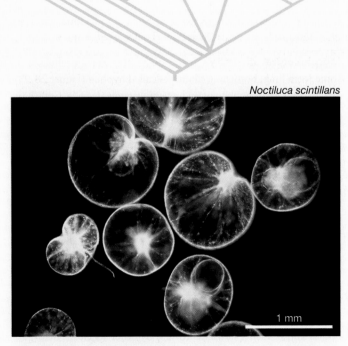

Noctiluca scintillans

1 mm

FIGURE 28.27 Some Dinoflagellates Species Are Bioluminescent

Alveolata ▶ Apicomplexa

All of the 5000 known species of apicomplexa are parasitic.

Morphological diversity Cells have a system of organelles at one end, called the *apical complex*, that is a diagnostic characteristic unique to the group. The apical complex allows the apicomplexan to penetrate the plasma membrane of its host. They have a chloroplast-derived, nonphotosynthetic organelle, indicating that they descended from an algal-like ancestor.

Feeding and locomotion All absorb nutrition directly from their host. They lack cilia or flagella, but some species can move by cell crawling.

Reproduction Reproduction can occur sexually or asexually. In some species, the life cycle involves two distinct hosts, and cells must be transmitted from one host to the next.

Human and ecological impacts Various species in the genus *Plasmodium* infect birds, reptiles, or mammals and cause malaria. *Toxoplasma* is an important pathogen in people infected with HIV (**Figure 28.28**), and chicken farmers estimate that they lose $600 million per year controlling the apicomplexan *Eimeria*. Apicomplexans that parasitize insects have been used to control pests.

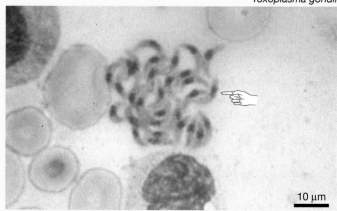

Toxoplasma gondii

10 μm

FIGURE 28.28 *Toxoplasma* **Causes Infections in AIDS patients**

Stramenopila (Heterokonta)

Stramenopiles are sometimes called heterokonts. At some stage of their life cycle, all stramenopiles have flagella that are covered with distinctive hollow "hairs." The structure of these flagella is unique to the stramenopiles. The lineage includes a large number of unicellular forms, although the brown algae are multicellular and include the world's tallest marine organisms, the kelp.

Stramenopila ▶ Oomycota

Based on morphology, oomycetes were thought to be fungi and were given fungus-like names, such as downy mildew and water molds.

Morphological diversity Some oomycetes are unicellular and some form long, branching filaments called **hyphae** (**Figure 28.29**). Species that form hyphae often have multinucleate cells. Cells have walls containing cellulose.

Feeding and locomotion Most species feed on decaying organic material in freshwater environments; a few are parasitic. Mature individuals are sessile.

Reproduction Most species are diploid throughout the majority of their life cycle. In aquatic species, spores that are produced via asexual or sexual reproduction have flagella and swim to find new food sources; in terrestrial species, spores swim in rainwater or are dispersed by the wind in dry conditions.

Human and ecological impacts Oomycetes are extremely important decomposers in aquatic ecosystems. Along with certain bacteria and archaea, they are responsible for breaking down dead bodies and releasing nutrients for use by other species. The organism that caused the Irish potato famine was an oomycete.

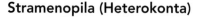

Phytophthora infestans

50 μm

FIGURE 28.29 *Phytophthora infestans* **Infects Potatoes**

Stramenopila ▶ Diatoms

Diatoms are unicellular or form chains of cells. About 10,000 species have been described to date, but some researchers claim that millions of species exist and have yet to be discovered.

Morphological diversity Cells are supported by external, silicon-rich, glassy shells that form a box-and-lid arrangement (**Figure 28.30**).

Feeding and locomotion Diatoms are photosynthetic. Only reproductive cells have flagella and are capable of powered movement. Photosynthetic cells normally just float in the wate, but species that live on a substrate can glide via microtubules that project from gaps in their shells and that move in response to motor proteins.

Reproduction Diatoms divide by mitosis to reproduce asexually or by meiosis to form gametes that fuse to form a new individual. Many species can produce spores that are dormant during unfavorable growing conditions.

Human and ecological impacts In abundance, diatoms dominate the plankton of cold, nutrient-rich waters. They are found in virtually all aquatic habitats and are considered the most important producer of carbon compounds in fresh- and saltwater. Their shells settle into massive accumulations that are mined and sold commercially as diatomaceous earth, which is used in filtering applications and as an ingredient in polishes, paint, cosmetics, and other products.

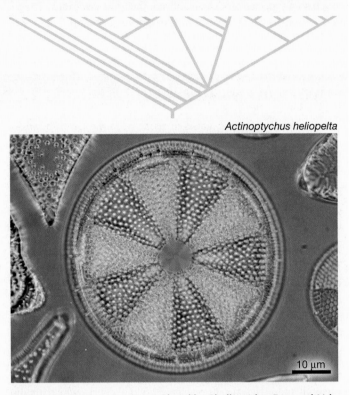

Actinoptychus heliopelta

10 μm

FIGURE 28.30 Diatoms Have Glass-like Shells with a Box-and-Lid Arrangement

Stramenopila ▶ Phaeophyta (Brown Algae)

The color of brown algae is due to their unique suite of photosynthetic pigments. Over 1500 species have been described, most of which live in marine habitats.

Morphological diversity Cells have walls that contain cellulose in addition to other complex polymers. Brown algae are unique among protists, because all species are multicellular. The body typically consists of leaf-like *blades*, a stalk known as a *stipe*, and a root-like *holdfast*, which attaches the individual to a substrate (**Figure 28.31**).

Feeding and locomotion Brown algae are photosynthetic and **sessile**—that is, permanently fixed to a substrate—though reproductive cells may have flagella and be **motile**—capable of locomotion.

Reproduction Sexual reproduction occurs via the production of swimming gametes, which fuse to form a zygote. Most species exhibit alternation of generations. Depending on the species, the gametophyte and sporophyte stages may look similar or different.

Human and ecological impacts In many coastal areas, brown algae form forests or meadows that are important habitats for a wide variety of animals. In the Sargasso Sea, off Bermuda, floating brown algae form extensive rafts that harbor an abundance of animal species. The compound *algin* is purified from kelp and used in the manufacture of cosmetics and paint.

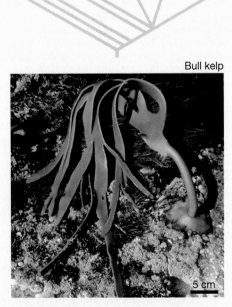

Bull kelp

5 cm

FIGURE 28.31 Many Brown Algae Have a Holdfast, Stalk, and Leaf-like Blades

Cercozoa

The cercozoans are single-celled amoebae that lack cell walls, though some species produce elaborate shell-like coverings. They move by cell crawling and produce long, slender pseudopodia. A number of important subgroups have been identified and named, including the planktonic organisms called actinopods, which synthesize glassy, silicon-rich skeletons. The most prominent group is the Foraminifera.

Cercozoa ▶ Foraminifera

Foraminifera, or forams, got their name from the Latin *foramen*, meaning hole, because produce tests (shells) that have one or two holes through which pseudopodia emerge. Fossil tests of foraminifera are abundant in sediments. There is a continuous record of fossilized forams that dates back 530 million years. Although they are abundant in marine plankton, their biology is relatively poorly known.

Morphological diversity   Cells generally have multiple nuclei. The tests of forams are usually made of organic material stiffened with calcium carbonate, and most species have several chambers (**Figure 28.32**). One species known from fossils was 12 cm long, but most species are much smaller. The size and shape of the test are traits that distinguish foram species.

Feeding and locomotion   Like other cercozoans, forams feed by extending their pseudopodia and using them to capture and engulf bacterial and archaeal cells or bits of organic debris, which are digested in food vacuoles. Some species have symbiotic algae that perform photosynthesis and contribute sugars to their host. Forams simply float in the water.

Reproduction   Asexual reproduction occurs by mitosis; when meiosis occurs, the resulting gametes are released into the open water, where they fuse to form a new individual.

Human and ecological impacts   The tests of dead forams commonly form extensive sediment deposits when they settle out of the water, producing layers that eventually solidify into chalk, limestone, or marble.

Benthic species

0.1 mm

FIGURE 28.32 Forams Are Shelled Amoebae

Plantae

Biologists are beginning to use the name "Plants" to refer to the monophyletic group that includes red algae, green algae, land plants, and glaucophyte algae. All of these lineages are descended from a common ancestor that engulfed a cyanobacterium, beginning the endosymbiosis that led to the evolution of the chloroplast—their distinguishing morphological feature. This initial endosymbiosis probably occurred in an ancestor of today's glaucophyte algae. To support this hypothesis, biologists point to several important similarities between cyanobacterial cells and the chloroplasts that are found in the glaucophytes. Both have cell walls that contain peptidoglycan and that are susceptible to attack by lysozyme and penicillin. The glauco-phyte chloroplast also has a membrane outside its wall that is similar to the membrane found in Gram-negative bacteria (see Chapter 27). Consistent with these observations, phylogenetic analyses of DNA sequence data place the glaucophytes as the most basal, or ancient, group of the Plantae.

The glaucophyte algae are unicellular or colonial and live in plankton or attached to substrates in freshwater environments—particularly in bogs or swamps. Some glaucophyte species have flagella or produce flagellated spores, but sexual reproduction has never been observed in the group. The chloroplasts of glaucophytes have a distinct bright blue-green color. Chapter 29 introduces the green algae and land plants; here we consider just the red algae in detail.

Rhodophyta (Red Algae)

The 6000 species of red algae live primarily in marine habitats. One species lives over 200 m below the surface; another is the only eukaryote capable of living in acidic hot springs. Although their color varies, many species are red because their chloroplasts contain the accessory pigment *phycoerythrin*, which absorbs strongly in the blue and green portions of the visible spectrum.

Morphological diversity   Cells have walls that are comprised of cellulose and other polymers. A few species are unicellular, but most are multicellular. Many of the multicellular species are filamentous, but others grow as thin, hard crusts on rocks or coral (**Figure 28.33**). Some species have erect, leaf-like structures called *thalli*. Some species have cells with many nuclei.

Feeding and locomotion   The vast majority of red algae are photosynthetic, though a few parasitic species have been identified. Red algae are the only type of algae that lack flagella.

Reproduction   Asexual reproduction occurs through production of spores by mitosis. Alternation of generations is common, but the types of life cycles observed in red algae are extremely variable.

Human and ecological impacts   On coral reefs, some red algae become encrusted with calcium carbonate. These species contribute to reef building and help stabilize the entire reef structure.

Cultivation of *Porphyra* (or nori) for sushi and other foods is a billion-dollar-per-year industry in East Asia.

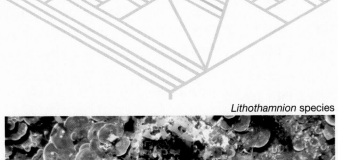

Lithothamnion species

10 cm

FIGURE 28.33 Red Algae Adopt an Array of Growth Forms

Amoebozoa

Species in the Amoebozoa lack cell walls and take in food by engulfing it. They move by means of cell crawling and produce large, lobe-like pseudopodia. The major subgroups in the lineage are lobose amoebae, cellular slime molds, and plasmodial slime molds. The cellular slime mold *Dictyostelium discoideum* was described in detail in Box 28.2. Amoebae are abundant in freshwater habitats and in wet soils; some are parasites of humans and other animals. More details on plasmodial slime molds follow.

Myxogastrida (Plasmodial Slime Molds)

The plasmodial slime molds got their name because individuals form a large, weblike structure that consists of a single cell containing many diploid nuclei (**Figure 28.34**).

Morphological diversity   The huge *supercell* form with many nuclei occurs in few protists other than plasmodial slime molds.

Feeding and locomotion   Cells feed on decaying vegetation and move by cell crawling.

Reproduction   When food becomes scarce, part of the amoeba forms a stalk topped by a ball-like structure in which nuclei undergo meiosis and form spores. The spores are then dispersed to new habitats by the wind or small animals. After spores germinate to form amoebae, two amoebae fuse to form a diploid cell that begins to feed and eventually grows into a supercell.

Human and ecological impacts   Like cellular slime molds, plasmodial slime molds are important decomposers in forests. They help break down leaves, branches, and other dead plant material, releasing nutrients that they and other organisms can use.

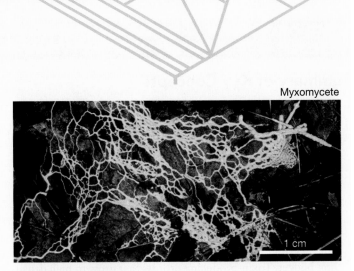

Myxomycete

1 cm

FIGURE 28.34 Plasmodial Slime Molds Are Important Decomposers in Forests

ESSAY Revolutions in Science

In 1962 a philosopher and science historian named Thomas Kuhn published an influential book called *The Structure of Scientific Revolutions*. Most scientific work, Kuhn contends, represents "normal science." In normal science, the questions that scientists ask are inspired by the currently accepted theories and styles of thinking, or what he calls the prevailing paradigm. In Kuhn's view, most scientific results are also "normal" in the sense that they support the prevailing paradigm of their day.

In contrast, Kuhn maintained that entirely new paradigms emerge during "scientific revolutions." Revolutionary science introduces fundamentally new theories and styles of thinking, which eventually displace the older paradigms.

To Kuhn, the Sun-centered theory of the solar system developed by Nicolaus Copernicus represents the classic example of revolutionary science. This theory, published in the early 1500s, maintains that the planets revolve around the Sun rather than the Earth. In addition to being correct, Copernicus's theory defined an entirely new research program in astronomy and physics. It also inspired new ideas in other fields, including religion and philosophy. Copernicus's paradigm inspired new questions, new techniques, and a new style of thinking in science.

This chapter on protists introduced two ideas that could be considered revolutionary. The first is the endosymbiosis theory for the origin of mitochondria and chloroplasts. Prior to this theory's publication, the prevailing paradigm was that the lateral transfer of genes—transfer of genes from one lineage to another—occurred rarely, if ever. The endosymbiosis theory not only inspired a vigorous series of tests, but it also established an entirely new way of thinking about the origin

of the eukaryotes and the importance of lateral gene transfer in the history of life.

The second revolutionary idea is the change from taxonomic to phylogenetic thinking, highlighted in Box 28.1. Prior to

Revolutionary science introduces fundamentally new theories and styles of thinking....

this revolution, the prevailing paradigm was that prokaryotes and eukaryotes represented the two fundamental types of organisms. Traditional taxonomic schemes aimed to organize and categorize the diversity of life around a handful of kingdoms, often defined as the Monera (prokaryotes), Protista, Plantae, Fungi, and Animalia. Biologists analyzed the diversity of life-

forms by placing them into taxonomic categories based on morphological similarity rather than by using morphological and molecular data to infer their history and evolutionary relationships (see Chapter 26).

A change in phylogenetic thinking is currently under way, and it is inspiring the development of a new naming system for the diversity of life. The old five-kingdom system is being replaced by a system of three domains with many kingdoms.

Undoubtedly, 20 years from now you will be able to look back at this book and pinpoint paradigms that have been supplanted by new scientific revolutions. Which of the ideas in this book will be overturned by new insights? What types of data will emerge, showing these ideas to be inadequate? Scientific revolutions are bound to occur. In this sense, they too are "normal."

CHAPTER REVIEW

Summary of Key Concepts

- Protists are a paraphyletic grouping that includes all eukaryotes except the green plants, fungi, and animals. Biologists study protists to understand how eukaryotes evolved, because they are important in marine ecosystems and global warming and because some species cause debilitating diseases in humans and other organisms.

  Protists are often tremendously abundant in marine and freshwater plankton and other habitats. As a result, protists provide food for many organisms in aquatic ecosystems and fix so much carbon that they have a large impact on the global carbon budget. Toxin-producing protists that grow to high densities result in a harmful algal bloom. Parasitic protists cause several important diseases in humans, including malaria.

- Protists are diverse morphologically. They vary in the types of organelles they contain; they may be unicellular or multicellular, and they may have a cell wall or other external covering or no such covering.

  Several morphological features are common to all eukaryotes: a nucleus, a cytoskeleton, and extensive intracellular structuring that includes organelles. In addition to these common features, many aspects of morphology and lifestyle are extremely variable among the protist lineages. The protists include many unicellular organisms as well as multicellular slime molds, red algae, and brown algae. Multicellularity, swimming via the beating of cilia or flagella, cell crawling, and hard outer coverings have evolved in many different protist groups independently.

Eukaryotes also contain mitochondria or have genes indicating that their ancestors once contained mitochondria. Several types of data support the hypothesis that mitochondria originated as endosymbiotic bacteria, possibly engulfed by an early protist closely related to the archaea. The symbiosis is thought to have been successful because the endosymbiotic bacterium provided its host with ATP while the host provided the bacterium with carbon compounds and protection. Similarly, the chloroplast's size, DNA structure, ribosomes, double membrane, and evolutionary relationships are consistent with the hypothesis that this organelle originated as an endosymbiotic cyanobacterium.

■ **Protists vary widely in terms of how they reproduce and how they find food. Many can reproduce both sexually and asexual-** ly. Many species are photosynthetic, while others obtain carbon compounds by ingesting food or parasitizing other organisms.

Protists undergo cell division based on mitosis and reproduce asexually. Many protists also undergo meiosis and sexual reproduction at some phase in their life cycle. Alternation of generations is common in multicellular species—meaning there are separate haploid and diploid forms of the same species. Protists exhibit predatory, parasitic, or photosynthetic lifestyles, which evolved in many groups independently.

Web Tutorial 28.1   Alternation of Generations in a Protist

Questions

Content Review

1. To biologists, the protists comprise all eukaryotes other than which groups?
   a. extinct forms
   b. green plants, fungi, and animals
   c. multicellular forms
   d. those with cells containing a cytoskeleton and nucleus

2. What materials are *not* used by protists to manufacture hard outer coverings?
   a. cellulose
   b. lignin
   c. glass-like compounds that contain silicon
   d. mineral-like compounds such as calcium carbonate ($CaCO_3$)

3. What does cell crawling result from?
   a. interactions among actin, myosin, and ATP
   b. coordinated beats of cilia
   c. the whiplike action of flagella
   d. action by the mitotic spindle

4. According to the endosymbiosis theory, what type of organism is the ancestor of the chloroplast?
   a. a photosynthetic archaean
   b. a cyanobacterium
   c. a primitive photosynthetic eukaryote
   d. a modified mitochondrion

5. Multicellularity is defined in part by the presence of distinctive cell types. At the cellular level, what does this criterion imply?
   a. Individual cells must be extremely large.
   b. The organism must be able to reproduce sexually.
   c. Cells must be able to move.
   d. Different cell types express different genes.

6. Why are protists an important part of the global carbon cycle and marine food chains?
   a. They have high species diversity.
   b. They are numerically abundant.
   c. They have the ability to parasitize humans.
   d. They have the ability to undergo meiosis.

Conceptual Review

1. Why is an advanced cytoskeleton and the lack of a cell wall required for ingestive modes of feeding? How does this mode of feeding relate to increased cell size in protists versus cell size in bacteria and archaea? How does it relate to the acquisition of mitochondria and chloroplasts by endosymbiosis?

2. Compare and contrast sexual reproduction in protists with conjugation in bacteria (see Chapter 19).

3. What is the relationship between meiosis and the alternation of generations?

4. Outline the steps in the endosymbiosis theory for the origin of the mitochondrion. What did each partner provide the other, and what did each receive in return? Answer the same questions for the chloroplast.

5. Provide evidence to support the hypothesis that similar evolutionary innovations occurred within each of the eight major eukaryotic lineages.

6. The text refers to eukaryotes as "chimeras" and draws an analogy with the monster from Greek mythology that was a combination of parts from a lion, a goat, and a snake. Why is this analogy appropriate?

Group Discussion Problems

1. Consider the following:

   • All living eukaryotes have mitochondria or have evidence in their genomes that they once had these organelles. Thus it appears that eukaryotes acquired mitochondria very early in their history.

   • The first eukaryotic cells in the fossil record correlate with the first appearance of rocks formed in an oxygen-rich ocean and atmosphere.

   How are these observations connected? (HINT: Before answering, glance at Chapter 9 and remind yourself what happens in a mitochondrion.)

2. Consider the following:

   • *Plasmodium* has an unusual organelle called an *apicoplast*. Recent research has shown that apicoplasts are derived from chloroplasts via secondary endosymbiosis and have a large number of genes encoded by chloroplast DNA.

   • *Glyphosate* is one of the most widely used herbicides. It works by poisoning an enzyme encoded by a gene in chloroplast DNA.

   • Biologists are testing the hypothesis that glyphosate could be used as an antimalarial drug in humans.

   How are these observations connected?

3. Suppose a friend says that we don't need to worry about global warming. Her claim is that increased temperatures will make planktonic algae grow faster and that carbon dioxide (CO_2) will be removed from the atmosphere faster. According to her, this carbon will be buried at the bottom of the ocean in calcium carbonate tests. As a result, the amount of carbon dioxide in the atmosphere will decrease and global warming will decline. Comment.

4. Biologists are beginning to draw a distinction between "species trees" and "gene trees." A *species tree* is a phylogeny that describes the actual evolutionary history of a lineage. A *gene tree*, in contrast, describes the evolutionary history of one particular gene, such as the gene for chlorophyll *a*. In some cases, species trees and gene trees don't agree with each other. For example, the species tree for green algae indicates that their closest relatives are protists and plants. But the gene tree based on chlorophyll *a* from green algae suggests that this gene's closest relative is a bacterium, not a protist. What's going on? Why do these types of conflicts exist?

Answers to Multiple-Choice Questions   1. b; 2. b; 3. a; 4. b; 5. d; 6. b

Green Plants

29

KEY CONCEPTS

▪ The green plants include both the green algae and the land plants. Biologists study them because humans—along with virtually every other terrestrial organism—rely on plants for food and a healthy environment.

▪ Land plants were the first multicellular organisms to live on land. A series of key adaptations allowed them to survive with their tissues constantly exposed to dry air. In terms of total mass, plants dominate today's terrestrial environments.

▪ Once plants were able to grow on land, a sequence of important evolutionary changes made it possible for them to reproduce efficiently—even in extremely dry environments.

Mosses are common in moist habitats and share many similarities with the earliest land plants. According to data reviewed in this chapter, the earliest land plants evolved from green algae that inhabited ponds, streams, and other freshwater habitats.

The **green plants** consist of the green algae and the land plants. Green algae are important photosynthetic organisms in aquatic habitats—particularly lakes, ponds, and other freshwater settings—while land plants are the key photosynthesizers in terrestrial environments. Before land plants evolved, the only life on the continents consisted of bacteria, archaea, and single-celled protists that thrive in wet soils.

Although green algae have traditionally been considered protists, it is logical to study them along with land plants for two reasons: (1) They are the closest living relatives to land plants, and (2) the transition from aquatic to terrestrial life occurred when land plants evolved from green algae. Land plants were the first organisms that could grow with their tissues exposed to the air instead of being partially or completely submerged. When they evolved, multicellular organisms began to occupy dry terrestrial habitats that had been largely barren of life for over 3 billion years. Land plants made the Earth green. They were pioneers—the first truly terrestrial organisms.

Not long after land plants evolved and began to diversify, fungi and animals also accomplished the feat of moving from aquatic to terrestrial habitats. They could do so, however, only because plants were there first and provided something to eat. Fungi and animals still depend on plants for existence.

By colonizing the continents, plants transformed the nature of life on Earth. In the words of Karl Niklas, the movement of green plants from water to land ranks as "one of the greatest adaptive events in the history of life." How did they do it? Let's delve in and see.

29.1 Why Do Biologists Study the Green Plants?

Biologists study plants because plants are fascinating and because people could not live without them. Along with most other animals and fungi, humans are almost completely dependent on plants for food. People rely on plants for other necessities of life as well—oxygen, fuel, building materials, and the fibers used in making clothing, paper, ropes, and baskets. But we also rely on land plants for important intangible values, such as the aesthetic appeal of landscaping and bouquets. To drive this point home, consider that the sale of cut flowers generates over $1 billion each year in the United States alone.

Based on these observations, it is not surprising that agriculture, forestry, and horticulture are among the most important endeavors supported by biological science. Tens of thousands of biologists are employed in research designed to increase the productivity of plants and to create new ways of using them in ways that benefit people. Research programs also focus on land plants that cause problems for people, either because they are weeds that decrease the productivity of crop plants or newly introduced species that invade and then degrade natural areas. Biologists study plants not only because they are fascinating but also because they keep us alive.

Plants Provide Ecosystem Services

An **ecosystem** consists of all the organisms in a particular area, along with physical components of the environment, such as the atmosphere, precipitation, surface water, sunlight, soil, and nutrients. Plants are said to provide **ecosystem services**, because they add to the quality of the atmosphere, surface water, soil, and other physical components of an ecosystem. Stated another way, plants alter the landscape in ways that benefit other organisms:

- *Plants produce oxygen.* Recall from Chapter 10 that plants perform oxygenic ("oxygen-producing") photosynthesis. In this process, electrons that are removed from water molecules are used to reduce carbon dioxide and produce sugars. In the process of stripping electrons from water, plants release oxygen molecules (O_2) as a by-product. As Chapter 27 noted, oxygenic photosynthesis evolved in cyanobacteria and was responsible for the evolution of the original oxygen atmosphere. The evolutionary success of plants continued this trend by adding huge amounts of oxygen to the atmosphere. Without the green plants, we and other terrestrial animals would be in danger of suffocating for lack of oxygen.

- *Plants build soil.* Leaves and roots and stems that are not eaten when they are alive fall to the ground and provide food for worms, fungi, bacteria, archaea, protists, and other decomposers in the soil. These organisms add organic matter to the soil, which improves soil texture and the ability of soils to hold nutrients and water.

- *Plants hold soil.* The extensive network of fine roots produced by trees, grasses, and other land plants helps hold soil particles in place. And by taking up nutrients in the soil, plants prevent the nutrients from being washed or blown away. When areas are devegetated by grazing, farming, logging, or suburbanization, large quantities of soil and nutrients are lost to erosion by wind and water (**Figure 29.1**).

- *Plants hold water.* Plant tissues themselves take up and retain water. Intact forests, prairies, and wetlands also prevent rain from quickly running off a landscape. There are several reasons why the presence of plants increases the amount of water available in a habitat. Plant leaves soften the physical impact of rainfall on soil; plant roots hold soil particles in place during rainstorms; and plant organic matter builds the soil's water-holding capacity. When areas are devegetated, streams are more prone to flooding and groundwater is not replenished efficiently. It is common to observe streams alternately flooding and then drying up completely when the surrounding area is deforested.

(a) Wind erosion

(b) Water erosion

FIGURE 29.1 Plants Hold Soil in Place
When plants are removed, soils are susceptible to erosion by **(a)** wind and **(b)** water.

- *Plants moderate the local climate.* By providing shade, plants reduce temperatures beneath them and increase relative humidity. They also reduce the impact of winds that dry out landscapes or make them colder. When plants are removed from landscapes to make way for farms or suburbs, habitats become much drier and are subject to more extreme temperature swings.

Perhaps the most important ecosystem service provided by plants, however, involves food. In addition to their beneficial impact on their physical surroundings, land plants are the dominant primary producers in terrestrial ecosystems. (As Chapter 28 indicated, primary producers convert some of the energy in sunlight into chemical energy.) The sugars and oils that land plants produce by photosynthesis provide the base of the food chain in the vast majority of terrestrial habitats. As **Figure 29.2** shows, plants are eaten by **herbivores** ("plant-eaters"), which range in size from insects to elephants. These consumers are eaten by **carnivores** ("meat-eaters"), ranging in size from spiders to lions. Humans are an example of **omnivores** ("all-eaters"), organisms that eat both plants and animals. Omnivores feed at several different levels in the terrestrial food chain. For example, people consume plants, herbivores such as chicken and cattle, and carnivores such as salmon and tuna.

Finally, just as photosynthetic protists and bacteria are the key to the carbon cycle in the oceans, green plants are the key to the carbon cycle on the continents. Plants take CO_2 from the atmosphere and reduce it to make sugars. Although both green algae and land plants produce a great deal of CO_2 as a result of cellular respiration, they fix much more CO_2 than they release. The loss of plant-rich prairies and forests, due to fires or logging or suburbanization, has contributed to increased concentrations of CO_2 in the atmosphere. Higher carbon dioxide levels, in turn, are responsible for the rapid warming that is occurring worldwide.

In many cases, plant growth alters the environment in ways that benefit other organisms as well as themselves.

Agricultural Research: Domestication and Selective Breeding

Until the late 1800s, most people made their living by farming. Hence, it is not surprising that many of the early advances in genetics and developmental biology were motivated by a desire to improve domesticated crops and animal breeds. Agricultural research continues to be a crucial branch of biological science.

Agriculture was a key human activity long before the development of biology, however. Most anthropologists agree that the domestication of crops was the single most important event in the evolution of human culture. But this "single most important event" was not actually a single event: There is compelling evidence that plants were domesticated independently by different people living in widely scattered locations around the world.

Data from archaeological sites have shown that the grains that form the basis of our current food supply were derived from wild species between 10,000 and 2000 years ago. In each case, researchers have observed a distinct set of changes over time as wild forms were brought under cultivation. Samples taken from human habitations show that, over time, maize (corn) kernels and other types of *seeds* (plant embryos plus nutritive tissue) became larger and squash rinds became thicker. What was responsible for these changes? The leading hypothesis is that our ancestors were actively selecting seeds from individual plants that exhibited these traits; they used only the selected seeds to plant the next generation of crops. By repeating these steps year after year, humans gradually changed the characteristics of several wild species, resulting in the characters that are now associated

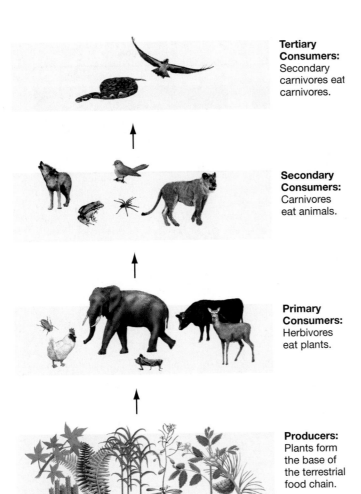

Tertiary Consumers: Secondary carnivores eat carnivores.

Secondary Consumers: Carnivores eat animals.

Primary Consumers: Herbivores eat plants.

Producers: Plants form the base of the terrestrial food chain.

FIGURE 29.2 Plants Are the Basis of Food Chains in Terrestrial Environments
Virtually every organism that lives on land depends on plants for food, either directly or indirectly. EXERCISE Label the levels in this food chain where humans feed.

with domestication (**Figure 29.3a**. This process is called **artificial selection**. It was introduced in Chapter 1 and is analogous to the process of natural selection analyzed in Chapter 23.

Artificial selection was responsible for the original domestication of crop species, and then for a subsequent diversification of crop varieties. For example, researchers have identified 178 distinct varieties of potato being cultivated in a single farming community in the Andes Mountains of South America. Artificial selection also continues today. Over the past 100 years, artificial selection has been responsible for the dramatic increases in the oil content of corn kernels (**Figure 29.3b**). (Corn oil serves as a cooking oil and is used in other products.) Several tree species are now being artificially selected for disease resistance or rapid growth and production of straight trunks.

To summarize, humans have succeeded in domesticating a wide array of plants for food or ornamental use over the past 10,000 years. We eat the seeds of rice, wheat, beans, and corn; the stalks of celery and bok choi; the buds of brussels sprouts; the leaves of spinach, lettuce, and cabbage; the fruits of tomatoes, squash, bell peppers, strawberries, and oranges; and the storage organs of potatoes, onions, carrots, beets, and cassava. Researchers continue to develop new crop plants and select strains of existing crops that have improved productivity or disease resistance. Many of these current research programs rely on the genetic engineering techniques introduced in Chapter 19.

Plant-Based Fuels and Fibers

In addition to relying on plants for food, humans have historically depended on them for cooking and heating fuels and as a source of fibers for clothing and implements such as ropes and nets. For perhaps 100,000 years, wood burning was the primary source of energy used by humans. As

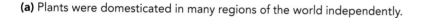

(a) Plants were domesticated in many regions of the world independently.

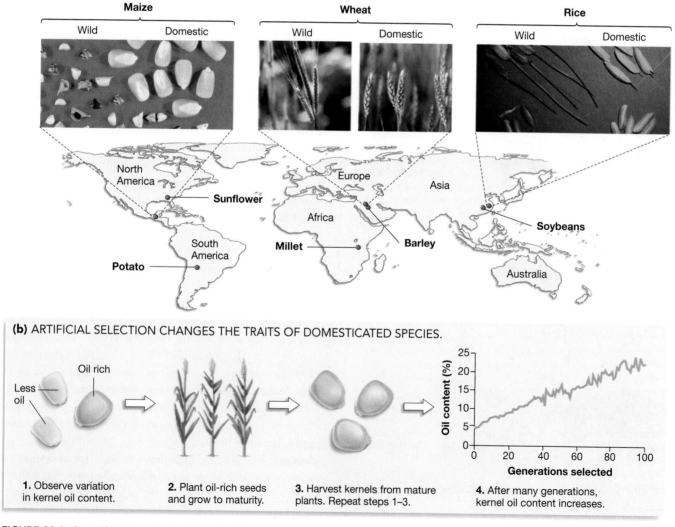

FIGURE 29.3 Crop Plants Are Derived from Wild Species via Artificial Selection
(a) Crop plants were derived from wild relatives in many parts of the world. **(b)** Artificial selection can lead to dramatic changes in plant characteristics.

Figure 29.4a shows, however, wood has been replaced by other sources of energy.

In economies across the globe, the first fuel to replace wood has usually been coal. Coal forms from partially decayed plant material that is compacted over time by overlying sediments and hardened into rock (**Figure 29.4b**). Starting in the mid-1800s, people in England, Germany, and the United States began to mine coal deposits that originally

(a) Plant-based fuels

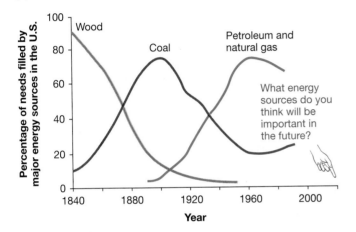

(b) COAL FORMATION

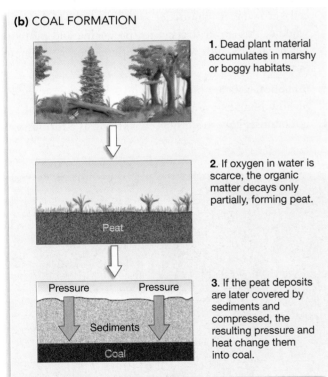

1. Dead plant material accumulates in marshy or boggy habitats.

2. If oxygen in water is scarce, the organic matter decays only partially, forming peat.

3. If the peat deposits are later covered by sediments and compressed, the resulting pressure and heat change them into coal.

FIGURE 29.4 Humans Have Relied on Plant-Based Fuels
(a) The percentage of energy needs in the United States filled by various sources since 1840. Wood was replaced by coal, which in turn was replaced by oil and natural gas. Wood is still the primary cooking and heating fuel in many areas of the world. (b) Coal is formed by terrestrial deposits of partially decayed organisms.

formed during the Carboniferous period some 350–275 million years ago. The coal fueled blast furnaces that smelted vast quantities of iron ore into steel and powered the steam engines that sent trains streaking across Europe and North America. It is no exaggeration to claim that the sugars synthesized by plants about 300 million years ago laid the groundwork for the Industrial Revolution. Coal still supplies about 20 percent of the energy used in Japan and Western Europe and 80 percent of the energy used in China.

The transition from wood burning to the use of fossil fuels was important for another reason. When people burn wood, the carbon that is released to the atmosphere is then removed from the atmosphere as forests regrow over the course of a few decades or centuries. As a result, economies that were based on wood burning added little to the concentration of CO_2 in the atmosphere. But carbon that is released to the atmosphere via burning of coal or petroleum cannot be removed from the atmosphere by the formation of coal and petroleum, because the formation processes are so slow. This logic suggests that the current increase in atmospheric CO_2 concentration (and subsequent global warming) traces its origin to the wood-to-coal transition documented in Figure 29.4a.

Will there be yet another generation of plant-based fuels? Although researchers are working on methods to convert corn to liquid fuels such as ethanol or methanol, the answer is probably no. The currently dominant combustible material, petroleum, is derived from the partially decayed remains of marine organisms—primarily photosynthetic bacteria and protists. Oil is also the source material for nylon and polyester fibers, which have supplanted cotton and other types of plant fibers used in traditional clothing and implements. The energy sources that are predicted to be important in the post-petroleum economy are solar power and wind power, combustion of hydrogen, and perhaps nuclear fusion. None of these sources are based on the carbon compounds found in plants. It is doubtful whether land plants could reduce enough carbon dioxide to meet current world energy demands.

Today, the primary commercial interest in woody plants is for building materials and the fibers used in papermaking. Wood excels in both capacities. Relative to its density, wood is a stiffer and stronger building material than concrete, cast iron, aluminum alloys, or steel. The cellulose fibers refined from trees or bamboo and used in paper manufacturing are stronger under tension (pulling) than nylon, silk, chitin, collagen, tendon, or bone—even though cellulose is 25 percent less dense.

Current research in forestry focuses on maintaining the productivity and diversity of forests that are managed for wood and pulp production. The vast majority of old-growth forests in the Northern Hemisphere have now been cut, and mature forests in the Southern Hemisphere are disappearing rapidly due to logging and burning. The current challenge to foresters is twofold: (1) reducing demand for wood products through recycling and more efficient use of existing timber, and (2) sustaining the productivity of forests after they have been logged repeatedly.

Bioprospecting

Because land plants play such a central role in human health and welfare, biologists have started a research program aimed at discovering new uses for plants. The effort to find naturally occurring compounds that can be used as drugs, fragrances, insecticides, herbicides, or fungicides has been called **bioprospecting** or **chemical prospecting.**

Plants have traditionally been a rich source of useful molecules. For example, the prominent anticancer drugs vincristine and vinblastine were first isolated from a plant native to Madagascar (now found throughout the tropics) called the rosy periwinkle. The anticancer drug taxol was discovered in extracts from the Pacific yew tree, which lives only in northwest North America. **Table 29.1** lists some of the more familiar drugs derived from land plants. Overall, it has been estimated that about 25 percent of the prescriptions written in the United States each year include at least one molecule derived from plants.

In most cases, plants synthesize these compounds in order to harm and thus repel leaf-eating beetles, deer, or other types of herbivores. Stated another way, many of the plant molecules that we use for medicine or other purposes are actually compounds that defend plants against their enemies. For example, experiments have confirmed that the presence of morphine, cocaine, nicotine, caffeine, and other compounds that are psychoactive in humans—meaning that they affect the human mind—is an effective deterrent to insect or mammalian consumers. The natural function of tetrahydrocannabinol (THC), the active ingredient in marijuana, is undoubtedly in defense. Later chapters explore the nature and role of plant defense compounds in more detail.

Researchers are pursuing two paths in the effort to discover new plant compounds that might have medicinal or commercial uses. One path begins in the lab; the other begins in the field:

1. Biologists have been able to stimulate plants that are growing *hydroponically*—that is, in liquid culture instead of in soil—to churn out large quantities of defense chemicals. The researchers expose root cells to a fragment of a bacterial cell wall or a toxin from a fungus, and then collect the defense compounds that are produced in response to the simulated attack. Using this technique, investigators have been able to collect 5000 samples of materials expelled from the roots of 700 plant species. Some of the extracts have been found to kill various types of cancer cells growing in culture. It is still unknown, however, whether this novel strategy for producing and purifying defense compounds will result in the discovery of important new drugs.

2. **Ethnobotanists** are biologists who specialize in the study of how humans use plants (**Figure 29.5**). Teams of ethnobotanists have recently been working at a furious pace

TABLE 29.1 Some Drugs Derived from Land Plants

Compound	Source	Use
Atropine	Belladonna plant	Dilating pupils during eye exams
Codeine	Opium poppy	Pain relief, cough suppressant
Digitalin	Foxglove	Heart medication
Ipecac	Ipecac	Treating amoebic dysentery, poison control
Menthol	Peppermint tree	Cough suppressant, relief of stuffy nose
Morphine	Opium poppy	Pain relief
Papain	Papaya	Reduce inflammation, treat wounds
Quinine	Quinine tree	Malaria prevention
Quinidine	Quinine tree	Heart medication
Salicin	Aspen, willow trees	Pain relief (aspirin)
Steroids	Wild yams	Precursor compounds for manufacture of birth control pills and cortisone (to treat inflammation)
Taxol	Pacific yew	Ovarian cancer
Tubocurarine	Curare vine	Muscle relaxant used in surgery
Vinblastine, vincristine	Rosy periwinkle	Leukemia (cancer of blood)

FIGURE 29.5 Ethnobotanists Research New Uses for Plants
Traditional healers make most of the medicines they administer from local plants. University-trained ethnobotanists are collaborating with traditional healers to research plant products for possible use in Western medicine.

in rain forests around the world, collecting plants from areas that are threatened with destruction and interviewing traditional healers who make their own medicines from local plants. This work is a race against time, for in many cases both the forests and the indigenous people are being overrun by settlers. If present trends continue, it is likely that within a generation, knowledge about traditional uses of rain forest plants—and many of the plant species themselves—will be lost as ecosystems and cultures change.

The use of plant-derived compounds in medicines underscores a key point: Understanding and preserving plant species diversity can be a life-or-death matter. For humans and other animals, plants are the most important of all organisms. Given the importance of plants to the planet in general and humans in particular, it is not surprising that understanding plant diversity is an important component of biological science. Let's consider how biologists go about analyzing the diversity of green plants; then we'll explore the evolutionary innovations that made green plant diversification possible.

29.2 How Do Biologists Study Green Plants?

To understand the genetics and developmental biology of plants, researchers use thale cress, a weedy mustard, as a model organism (**Box 29.1**). To understand how green plants originated and diversified, biologists use three tools: They (1) compare the fundamental morphological features of various green algae and green plants; (2) analyze the fossil record of the lineage; and (3) assess similarities and differences in molecular traits such as the DNA sequences from selected genes. These data are then used to estimate phylogenies. The three approaches are complementary. As a result, it could be argued that biologists have a better understanding of how green plants diversified than they do for any other group on the tree of life. Let's consider each of these research strategies.

Analyzing Morphological Traits

By comparing and contrasting the morphological traits of green plants, biologists have identified several distinct groups of green algae and 12 major lineages, or **phyla** (singular: **phylum**), of land plants.

BOX 29.1   A Model Organism: *Arabidopsis thaliana*

In the early days of biology, the best-studied plants were agricultural varieties such as maize (corn), rice, and garden peas. When biologists began to unravel the mechanisms responsible for oxygenic photosynthesis in the early to mid-1900s, they relied on green algae—often the unicellular species *Chlamydomonas reinhardii*—as an experimental subject. Although crop plants and green algae continue to be the subject of considerable research, a new model organism emerged in the 1980s and now serves as the preeminent experimental subject in plant biology. That organism is *Arabidopsis thaliana*, commonly known as thale cress or wall cress (**Figure 29.6**).

Arabidopsis is a member of the mustard family, or Brassicaceae, so it is closely related to radishes and broccoli. In nature it is a **weed**—meaning a species that is adapted to thrive in habitats where soils have been disturbed. In Europe, for example, *Arabidopsis* is common along roadsides and the edges of agricultural fields. It is also an **annual** plant, which means that individuals do not live from year to year but overwinter as seeds. (Plants that survive from one year to the next are said to be **perennial**.) Indeed, one of the most attractive aspects of working with *A. thaliana* is that individuals can grow from a seed into a mature, seed-producing plant in just four to six weeks. Several other attributes make it an effective subject for study: It has just five chromosomes, has a relatively small genome with limited numbers of repetitive sequences, can self-fertilize as well as undergo cross-fertilization, can be grown in a relatively small amount of space and with a minimum of care in the greenhouse, and produces up to 10,000 seeds per individual per generation.

Arabidopsis has been instrumental in a variety of studies in plant molecular genetics and development, and is increasingly popular in ecological and evolutionary studies. In addition, the entire genome of the species has now been sequenced, and studies have benefited from the development of an international "*Arabidopsis* community"—a combination of informal and formal associations of investigators who work on *Arabidopsis* and use regular meetings, e-mail, and the Internet to share data, techniques, and seed stocks.

FIGURE 29.6 *Arabidopsis thaliana* Is the Most Important Model Organism in Plant Biology
Like fruit flies and *Escherichia coli*, *A. thaliana* is small and short lived enough to grow easily in the laboratory, yet it is complex enough to be an interesting organism for study.

(a) Green algae are strictly aquatic.

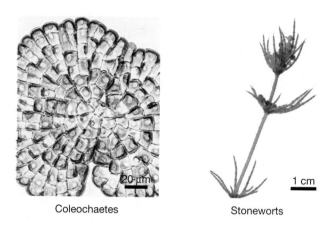

Coleochaetes   Stoneworts

(b) Nonvascular plants do not have vascular tissue to conduct water and provide support.

Liverworts

Hornworts

Mosses

FIGURE 29.7 Morphological Diversity in Green Plants

The green algae have long been hypothesized to be closely related to land plants on the basis of several key morphological traits. Both green algae and land plants have chloroplasts that contain the photosynthetic pigments chlorophyll *a* and *b* and the accessory pigment β-carotene, as well as similar stacks of the internal, membrane-bound sacs called thylakoids (see Chapter 10). The cell walls of green algae and land plants are almost identical in composition, both groups synthesize starch as a storage product in their chloroplasts, and their sperm and peroxisomes are similar in structure and composition. (Recall from Chapter 7 that peroxisomes are organelles in which specialized oxidation reactions take place.)

The green algae include species that are unicellular, colonial, or multicellular and that live in marine or freshwater habitats. Of all the green algal groups, the two that are most similar to land plants are the Coleochaetales and Charales (**Figure 29.7a**). Because the species that make up these groups are multicellular and live in ponds and other types of freshwater environments, biologists hypothesize that the land plants evolved from multicellular green algae that lived in freshwater habitats.

Although some land plants live in ponds or lakes or rivers, the vast majority of species in this lineage live on land. Based on morphology, the 12 most important phyla are traditionally clustered into three broad categories:

1. **Nonvascular plants**, which include the groups called liverworts, hornworts, and mosses. Liverworts and hornworts lack **vascular tissue**—meaning specialized groups of cells that conduct water or dissolved nutrients from one part of the plant body to another. Some moss species have specialized tissues that conduct water and food, but the cells that make up these tissues do not have the reinforced cell walls that define true vascular tissue. Nonvascular plants are extremely abundant in certain habitats but are usually small and grow close to the ground (**Figure 29.7b**).

2. **Seedless vascular plants**, which have well-developed vascular tissue but do not make seeds. A **seed** consists of an embryo and a store of nutritive tissue, surrounded by a tough protective layer. Horsetails, ferns, lycophytes, and whisk ferns are the major groups of seedless vascular plants (**Figure 29.7c**). Although most of the living representatives of these lineages are relatively small, some fossil and some other living representatives of the seedless vascular plants are tree sized.

3. **Seed plants**, which have vascular tissue and make seeds. Although the members of this group vary a great deal in size and shape, seed plants encompass some of the world's largest organisms. There are five major lineages in the group: cycads, ginkgoes, conifers, gnetophytes, and angiosperms (**Figure 29.7d**). The gnetophytes, cycads, ginkgoes, and conifers are collectively known as **gymnosperms** ("naked-seeds"), because their seeds do not develop in an enclosed structure. In the **flowering plants**, or **angiosperms** ("encased-seeds"), seeds develop inside a protective structure called a **carpel**. Today, angiosperms are far and away the most important lineage of land plants in terms of species diversity. Almost 90 percent of the land plant species alive today, and virtually all of the domesticated forms, are angiosperms.

(c) Seedless vascular plants have vascular tissue but do not make seeds.

Lycophytes

Whisk ferns

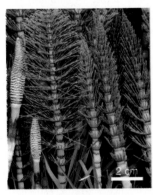

Horsetails

Ferns

FIGURE 29.7 (continued)

(d) Seed plants have vascular tissue and make seeds.

Cycads

Ginkgoes

Conifers

Gnetophytes

Angiosperms

This quick overview of morphological diversity in land plants raises a number of questions:

- Does other evidence support the hypothesis that green algae are ancestral to land plants, and that Coleochaetales and Charales are the closet living relatives to land plants?

- Did the green algae evolve first, followed by nonvascular plants, then seedless vascular plants, and finally seed plants?

- Are both green algae and land plants monophyletic—meaning each is a distinct lineage tracing back to a single common ancestor?

- Are the nonvascular plants, seedless vascular plants, and seed plants monophyletic?

Biologists are answering these questions by analyzing the fossil record of the green plants, analyzing the morphology and DNA sequences of living species, and then using these characteristics to infer the group's phylogeny.

Using the Fossil Record

The first green plants that appear in the fossil record are green algae in rocks that formed 700–725 million years ago. The first land plant fossils are much younger—they are found in rocks that are about 475 million years old. Because green algae appear long before land plants, the fossil record supports the hypothesis that land plants are derived from green algae.

The appearance and early diversification of green algae about 700 million years ago is also significant because, at roughly the same time, the oceans and atmosphere were starting to become oxygen-rich, as never before in Earth's history. Based on this time correlation, it is reasonable to hypothesize that the evolution of green algae contributed to the rise of oxygen levels on Earth. The origin of the oxygen atmosphere occurred not long before the appearance of animals in the fossil record, and may have played a role in the origin and early diversification of animals.

The fossil record of the land plants themselves is massive. In an attempt to organize and synthesize the database, **Figure 29.8** breaks it into five time intervals. Each of these time periods encompasses a major event in the diversification of land plants.

The oldest interval begins 475 million years ago, spans 65 million years, and documents the origin of the group. Most of the fossils dating from this period are microscopic. They consist of the reproductive cells called **spores** and sheets of a waxy coating called **cuticle**. Several observations support the hypothesis that these fossils came from green plants that were growing on land. First, cuticle is a watertight barrier that coats today's land plants and helps them resist drying. Second, the fossilized spores are surrounded by a sheetlike coating. Under the electron microscope, the coating material appears almost identical in structure to a watertight material called **sporopollenin**, which encases spores and pollen from modern land plants and helps them resist drying. Third, fossilized spores that are 475 million years old have recently been found in association with spore-producing structures called **sporangia** (singular: **sporangium**). The fossilized sporangia are similar in appearance to the sporangia observed in some of today's liverworts.

The second major interval in the fossil record of land plants is sometimes called the "Silurian-Devonian explosion." In rocks dated between 410 and 360 million years ago, biologists find large fossils from most of the major plant lineages. Virtually all of the adaptations that allow plants to occupy dry, terrestrial habitats are present, including water-conducting cells and tissues and roots.

The third interval in the fossil history of plants is aptly named the Carboniferous period. In sediments dated from about 350 to 290 million years ago, biologists find extensive deposits of coal. Coal is a carbon-rich rock packed with fossil spores, branches, leaves, and tree trunks. These fossils are frequently derived from lycophytes. Although the only living lyco-

phytes are small, forest-dwelling plants, during the Carboniferous this group was species rich and included a wide array of woody, tree-sized forms. Because coal formation is thought to start only in the presence of water, the Carboniferous fossils indicate the presence of extensive forested swamps.

The fourth interval in land plant history is characterized by gymnosperms. Recall that gymnosperms include the gnetophytes, cycads, ginkgoes, and conifers (such as pines and spruces). Because gymnosperms grow readily in dry habitats, biologists infer that both wet and dry environments on the continents became blanketed with green plants for the first time during this interval, which lasted from 275 to 145 million years ago.

The fifth interval in the history of land plants is still under way. This is the age of flowering plants, the angiosperms. The first flowering plants in the fossil record appear about 145 million years ago. The plants that produced the first flowers are the ancestors of today's grasses, orchids, daisies, oaks, maples, and roses.

According to the fossil record, then, the green algae appear first, followed by the nonvascular plants, seedless vascular plants, and seed plants. Organisms that appear late in the fossil record are much less dependent on moist habitats than are groups that appear earlier. For example, the sperm cells of mosses and ferns swim to accomplish fertilization, while gymnosperms and angiosperms produce pollen grains that are transported via wind or insects and that then produce sperm. These observations support the hypotheses that green plants evolved from green algae and that in terms of habitat use, the evolution of green plants occurred in a wet-to-dry trend.

To test the validity of these observations, biologists analyze data sets that are independent of the fossil record. Foremost among these are the DNA sequences and morphological data that are used to infer phylogenetic trees. Does the phylogeny of land plants confirm or contradict the patterns in the fossil record?

Fragment of plant cuticle

Seed fern leaves

Cones from *Araucaria mirabilis*, an early gymnosperm

Archaefructus, an early angiosperm

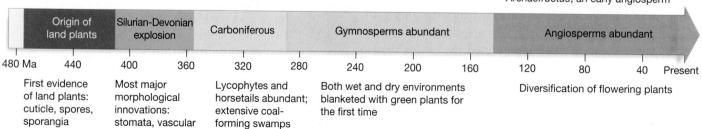

| Origin of land plants | Silurian-Devonian explosion | Carboniferous | Gymnosperms abundant | Angiosperms abundant |

480 Ma   440   400   360   320   280   240   200   160   120   80   40   Present

First evidence of land plants: cuticle, spores, sporangia

Most major morphological innovations: stomata, vascular tissue, roots, leaves

Lycophytes and horsetails abundant; extensive coal-forming swamps

Both wet and dry environments blanketed with green plants for the first time

Diversification of flowering plants

FIGURE 29.8 The Fossil Record of Land Plants Can Be Broken into Five Major Intervals
EXERCISE Add the following to the timeline: (1) first terrestrial vertebrate animals (370 Ma); (2) first mammals (195 Ma).

Evaluating Molecular Phylogenies

Understanding the phylogeny of green plants is a monumental challenge. Fortunately, the challenge is being met by the biologists who are using both morphological traits and DNA sequence data to infer the evolutionary relationships among green plants. In some cases various laboratories are cooperating in analyzing lineages from various parts of the green plant tree, so that in combination a comprehensive picture of green plant evolution can emerge.

The phylogenetic tree in **Figure 29.9** is a recent version of results coming out of laboratories focused on estimating the phylogeny of land plants. There are several important points to note about the relationships implied by this tree:

- The oldest branches on the tree, near the root, lead to various groups of green algae. This result supports the hypothesis that land plants evolved from green algae.

- The green algal group called Charales is the **sister group** to land plants—meaning that Charales are their closest living relative. Because living members of the Charales are multicellular and dwell in freshwater, the data support the hypothesis that land plants evolved from a multicellular ancestor that lived in ponds or lakes or other freshwater habitats.

- The group called "green algae" is paraphyletic. Stated another way, the green algae include some but not all of the descendants of a single common ancestor. The green plants, in contrast, are monophyletic. A single common ancestor gave rise to all of the green algae and land plants.

- The land plants are monophyletic. This result supports the hypothesis that the transition from freshwater environments to land occurred just once. According to the phylogenetic tree, the population that made the transition then diversified into the array of land plants observed today.

- The nonvascular plants—that is, liverworts, hornworts, and mosses—are the earliest-branching, or most basal, groups among land plants. They are not monophyletic, however. Instead, the nonvascular plants represent what biologists call a **grade**—a sequence of lineages that are not monophyletic.

- Mosses have simple water-conducting tissues, and liverworts and hornworts have none at all. This finding suggests that the earliest land plants lacked water-conducting cells and vascular tissue, and that these traits evolved later. The phylogenetic tree supports the hypothesis that water-conducting cells and tissues evolved in a stepwise fashion, with simpler structures preceding the evolution of more complex structures.

- The lycophytes are the sister group to all seedless vascular plants. The whisk ferns, horsetails, and ferns form a monophyletic group, but as a whole the seedless vascular plants form a grade. In contrast, the vascular plants are monophyletic—meaning that vascular tissue evolved once during the diversification of land plants.

- Because they lack roots and leaves, the whisk ferns were traditionally thought to be a basal group in the land plant radiation. Molecular phylogenies challenge this hypothesis and support an alternative hypothesis: The morphological simplicity of whisk ferns is a derived trait—meaning that complex structures have been lost in this lineage.

- The seed plants and the angiosperms are monophyletic, meaning that seeds and flowers evolved once. Although the data available at present suggest that gymnosperms form a monophyletic group, this result has yet to be confirmed.

- The fossil record and the phylogenetic tree agree on the order in which groups appeared. Land plant evolution began with nonvascular plants, proceeded to seedless vascular plants, and continued with the evolution of seed plants.

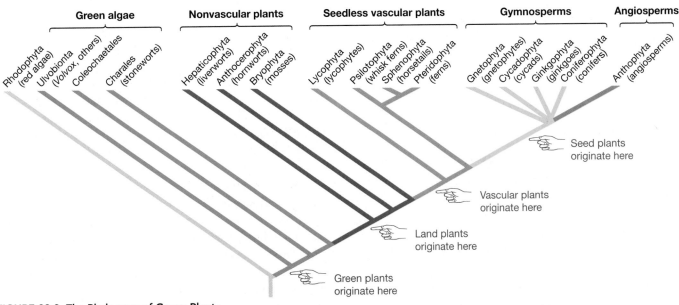

FIGURE 29.9 The Phylogeny of Green Plants
EXERCISE Circle and label the monophyletic groups called (1) land plants, (2) seed plants, and (3) vascular plants. On the figure, identify why the groups called green algae, nonvascular plants, and seedless vascular plants are paraphyletic.

The tree in Figure 29.9 will undoubtedly change and improve as additional data accumulate. Perhaps the most urgent need now is to clarify the relationships among the gymnosperms—the gnetophytes, cycads, ginkgoes, and conifers. Species from these lineages dominated terrestrial environments for over 200 million years, and one of the groups is the sister lineage to the angiosperms—today's most prominent land plant group. If the relationships among gymnosperms can be resolved, the results should improve our understanding of how the flower evolved.

✓CHECK YOUR UNDERSTANDING

Biologists use the fossil record and phylogenetic analyses to study how green plants diversified. Both types of data support the hypotheses that green plants are monophyletic and that land plants evolved from multicellular green algae that inhabited freshwater. The fossil record and molecular analyses agree that the nonvascular plants evolved first, followed by the seedless vascular plants and the seed plants. You should be able to (1) describe three ways that the presence of land plants affects abiotic components of terrestrial habitats; and (2) draw a phylogenetic tree that shows the relationships between the green algae and the three major morphological groups of land plants, and that identifies the monophyletic groups called green plants, land plants, and vascular plants.

29.3   What Themes Occur in the Diversification of Green Plants?

Land plants have evolved from algae that grew on the muddy shores of ponds 475 million years ago to organisms that enrich the soil, produce much of the oxygen you breathe and most of the food you eat, and serve as symbols of health, love, and beauty. How did this happen?

Answering this question begins with recognizing the most striking trend in the phylogeny and fossil record of green plants: The most ancient groups in the lineage are dependent on wet habitats, while more recently evolved groups are tolerant of dry—or even desert—conditions. The story of land plants is the story of adaptations that allowed photosynthetic organisms to move from aquatic to terrestrial environments. Let's first consider adaptations that allowed plants to grow in dry conditions without drying out and dying, and then analyze the evolution of traits that allowed plants to reproduce efficiently when not surrounded by water. This section closes with a brief look at the radiation of flowering plants, which are the most important plants in today's terrestrial environments.

The Transition to Land, I: How Did Plants Adapt to Dry Conditions?

For aquatic life-forms such as the green algae, terrestrial environments are deadly. Compared with a habitat in which the entire organism is bathed in fluid, in terrestrial environments only a portion, if any, of the plant's tissues are bathed in fluid. Tissues that are exposed to air tend to dry out and die.

Once green plants made the transition to survive out of water, though, growth on land offered a bonanza of resources. Take light, for example. The water in ponds, lakes, and oceans absorbs and reflects light. As a result, the amount of light available to drive photosynthesis is drastically reduced even a few feet below the water surface. In addition, the most important nutrient required by photosynthetic organisms, carbon dioxide, is much more readily available in air than it is in water. Not only is it more abundant in the atmosphere, but it diffuses more readily there than it does in water.

Natural selection favored early land plants with adaptations that solved the water problem. These adaptations arose in two steps: (1) prevention of water loss from cells, which kept the cells from drying out and dying; and (2) transportation of water from tissues with direct access to water to tissues without access. Let's examine both of these steps in turn.

Preventing Water Loss: Cuticle and Stomata   Section 29.2 pointed out that sheets of the waxy substance called cuticle are present early in the fossil record of land plants, along with encased spores. This observation is significant because the presence of cuticle in fossils is a diagnostic indicator of land plants. Cuticle is a waxy, watertight sealant that covers the aboveground parts of plants and gives them the ability to survive in dry environments (**Figure 29.10a**). If biologists had to point to one innovation that made the transition to land possible, it would be the production of cuticle.

Covering surfaces with wax creates a problem, however, in terms of the exchange of gases across those surfaces. Plants need to take in carbon dioxide (CO_2) from the atmosphere in order to perform photosynthesis. But cuticle is almost as impervious to carbon dioxide as it is to water. Most modern plants solve this problem with a structure called a **stoma** (plural: **stomata**), consisting of an opening surrounded by specialized **guard cells** (**Figure 29.10b**). The opening, called a **pore**, opens or closes as the guard cells change shape. When guard cells become soft, they close the stomata. Pores are closed in this way to limit water loss from the plant. When guard cells become taut, in contrast, they open the pore, allowing CO_2 to diffuse into the interior of leaves and stems where cells are actively photosynthesizing, and allowing excess O_2 to diffuse out. (The mechanism behind guard-cell movement is explored in Chapter 38.)

Stomata are present in living members of two of the earliest-branching groups on the land plant phylogeny: the hornworts and mosses. Some species in the most basal, or earliest-branching, lineage—the liverworts—have pores but no guard cells. These data suggest that the earliest land plants evolved pores that allowed gas exchange to occur at breaks in the cuticle-covered surface. Later, the evolution of guard cells gave land plants the ability to regulate gas exchange—and control water loss—by opening and closing their pores.

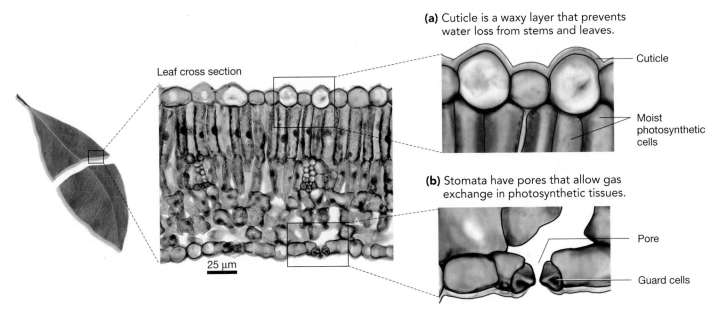

(a) Cuticle is a waxy layer that prevents water loss from stems and leaves.

Leaf cross section

Cuticle

Moist photosynthetic cells

25 μm

(b) Stomata have pores that allow gas exchange in photosynthetic tissues.

Pore

Guard cells

FIGURE 29.10 Cuticle and Stomata Are the Most Fundamental Plant Adaptations to Life on Land
Leaf cells that have been stained blue to make their structure more visible. **(a)** The interior of plant leaves and stems is extremely moist; cuticle prevents water from evaporating away. **(b)** Stomata have pores that open to allow CO₂ to diffuse into the interior of leaves and stems where cells are actively photosynthesizing, and to allow excess O₂ to diffuse out. QUESTION Why was the evolution of guard cells important?

Transporting Water: Vascular Tissue and Upright Growth

Once cuticle and stomata had evolved, plants could keep photosynthesizing while exposed to air. Cuticle and stomata allowed plants to grow in the saturated soils of lake or pond edges. The next challenge? Defying gravity.

Multicellular green algae can grow erect because they float. They float because the density of their cells is similar to water's density. But outside of water, the body of a multicellular green alga collapses. The water that fills its cells is 1000 times denser than air. Although the cell walls of green algae are strengthened by the presence of cellulose, their bodies lack the structural support to withstand the force of gravity and to keep an individual erect in air.

Based on these observations, biologists hypothesize that the first land plants had a low, sprawling growth habit. In addition to lacking rigidity, the early land plants would have had to obtain water through pores or through a few cells that lacked cuticle— meaning they would have had to grow in a way that kept many or most of their tissues in direct contact with moist soil.

The sprawling-growth hypothesis is supported by the observation that the most basal groups of land plants living today (liverworts, hornworts, and mosses) are all low-growing forms. If this hypothesis is correct, then competition for space and light would have become intense soon after the first plants began growing on land. To escape competition, plants would have to grow upright. Plants with adaptations that allowed some tissues to remain in contact with wet soil while other tissues grew erect would have much better access to sunlight compared with individuals that were incapable of growing erect.

For plants to adopt erect growth habits on land, though, two problems had to be overcome. The first is transporting water from tissues that are in contact with wet soil to tissues that are in contact with dry air, against the force of gravity. The second is becoming rigid enough to avoid falling over in response to gravity and wind. As it turns out, vascular tissue solved both problems.

Paul Kenrick and Peter Crane explored the origin of water-conducting cells and erect growth in plants by examining the extraordinary fossils found in a rock formation in Scotland called the Rhynie Chert. These rocks formed about 400 million years ago and contain some of the first large plant specimens in the fossil record (as opposed to the microscopic spores and cuticle found in older rocks). The Rhynie Chert also contains numerous plants that fossilized in an upright position. This indicates that many or most of the Rhynie plants grew erect. How did they stay vertical?

By examining fossils with the electron microscope, Kenrick and Crane established that species from the Rhynie Chert contained elongated cells that were organized into tissues along the length of the plant. Based on these data, the biologists hypothesized that the elongated cells were part of water-conducting tissue and that water could move from the base of the plants upward to erect portions through these specialized water-conducting cells. In addition, some of the water-conducting cells present in these early plants had cell walls with thickened rings containing a molecule called lignin. **Lignin** is a polymer built from six-carbon rings, and it is extraordinarily strong for its weight. It is particularly effective in resisting compressing forces such as gravity. These observations inspired the following hypothesis: The evolution of lignin rings gave stem tissues the strength to remain erect in the face of wind and gravity. The presence of lignin in the cell walls of conducting cells is the defining feature of vascular tissue. In this way, the evolution of vascular tissue allowed early plants both to support erect stems and to transport water to them.

Once simple water-conducting tissues evolved, evolution by natural selection elaborated them. By about 380 million years ago, the fossil record contains the advanced water-conducting cells called tracheids. As **Figure 29.11a** shows, **tracheids** (from the Latin *trachea*, or "windpipe") are elongated cells that are closely packed and organized into a specialized water-conducting tissue. The presence of tracheids marked the evolution of the first vascular plants. Because they die after maturing, tracheids do not contain cytoplasm. These features increased the efficiency of water transport from the base of the plant through vascular tissue to aboveground tissues.

Tracheids that appear later in the fossil record have thickened secondary cell walls with extensive deposits of lignin,

in addition to a primary cell wall made of cellulose (see Chapter 8). The evolution of a secondary cell wall gave tracheids the ability to provide better structural support in addition to increased water-conducting capacity.

To summarize, erect growth was possible due to the evolution of water-conducting cells that contained lignin and thus strengthened stems. This innovation was followed by the evolution of the specialized cells called tracheids and the first vascular tissue. Later, tracheids with secondary cell walls, which added strength, appeared.

Finally, in fossils dated to 250–270 million years ago, biologists have documented the most advanced type of water-conducting cells observed in plants, known as vessels. Vessels are shorter and wider than tracheids. The upper and lower ends of vessels have gaps that lack any sort of cell-wall material—making the transport of water extremely efficient (**Figure 29.11b**). Taken together, the data indicate that vascular tissue evolved in a series of gradual steps that provided an increasing level of structural support (**Figure 29.11c**).

(a) Tracheids have gaps in their secondary cell walls.

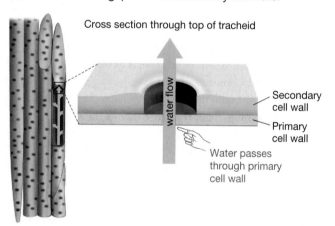

(b) Vessels have gaps in their primary and secondary cell walls.

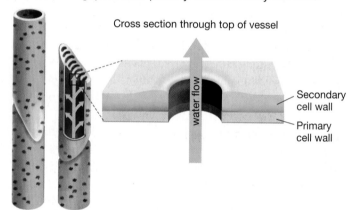

(c) Evolutionary sequence observed in water-conducting cells

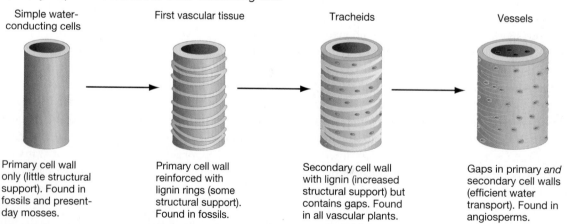

FIGURE 29.11 Vascular Tissue Is Composed of Tracheids or Vessels
(a) Tracheids are water-conducting cells that are dead and empty at maturity. They have secondary cell walls reinforced with lignin. **(b)** In vessels, there are gaps that lack any cell-wall material. **(c)** According to the fossil record and the phylogeny of green plants, the extent of the secondary cell wall increased as vascular tissue evolved. QUESTION Biologists claim that vessels are more efficient than tracheids at transporting water because vessels are shorter and wider than tracheids. Why does this claim make sense?

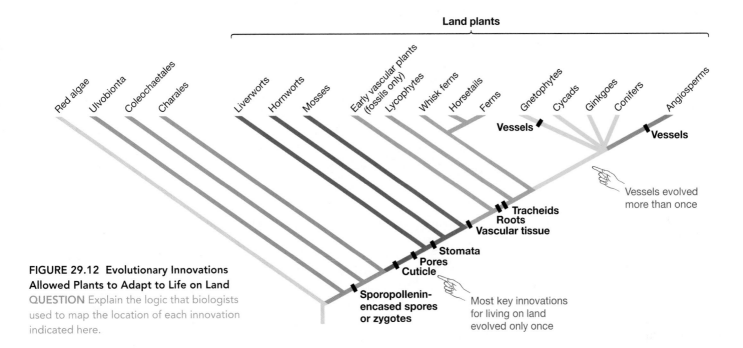

FIGURE 29.12 Evolutionary Innovations Allowed Plants to Adapt to Life on Land QUESTION Explain the logic that biologists used to map the location of each innovation indicated here.

Figure 29.12 summarizes how land plants adapted to dry conditions by mapping where major innovations occurred as the group diversified. The bars across the branches of the phylogenetic tree show when each of the key innovations occurred, based on the fossil record and their occurrence in species living today. Fundamentally important adaptations to dry conditions, such as the cuticle, pores, stomata, and vascular tissue, evolved just once. In addition, current analyses suggest that vessels may have evolved independently in gnetophytes and angiosperms. The evolutionary history of vessels is controversial, however, and research continues. Leaves also evolved more than once. Lycopods have small leaves called **microphylls**, while ferns, horsetails, and all vascular plants have typically larger leaves known as **megaphylls**, which develop in a different way and have a different structure than microphylls have.

The evolution of the cuticle, stomata, and vascular tissue made it possible for plants to avoid drying out and to grow upright, while moving water from the base of the plant to its apex. But for plants to become free of any dependence on wet habitats, one final challenge remained. The nonvascular plants and the most ancient vascular plant lineages, such as ferns, have male gametes that must swim to the egg to perform fertilization. The land plants made their final break with their aquatic origins when they evolved ways to move their gametes without the aid of water.

The Transition to Land, II: How Do Plants Reproduce in Dry Conditions?

The key to understanding how land plants lost their reliance on swimming gametes is to analyze their life cycles. All land plants undergo the phenomenon known as **alternation of generations**, introduced in Chapter 28. Recall that when alternation of generations occurs, individuals have a multicellular haploid phase and a multicellular diploid phase. The multicellular haploid stage is called the **gametophyte**; the multicellular diploid stage is called the **sporophyte**. The two phases of the life cycle are connected by the distinct types of reproductive cells: gametes and spores.

Although alternation of generations is observed in a wide array of protist lineages and in some groups of green algae, it does not occur in the Coleochaetales or Charales, the algal groups most closely related to land plants. In the Coleochaetales and Charales, the multicellular form is haploid (**Figure 29.13**). Only the zygote is diploid. These data suggest that alternation of generations evolved early in the history of the land plants—soon after they split off from green algae.

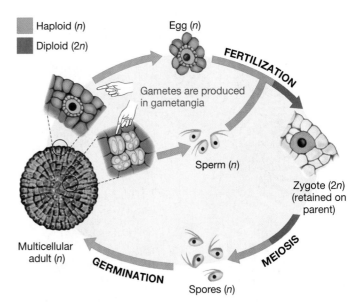

FIGURE 29.13 The Green Algae That Are Closely Related to Land Plants Have Simple Life Cycles The Coleochaetales and Charales do not have alternation of generations. The multicellular stage is haploid.

(a) Alternation of generations always involves the same sequence of events.

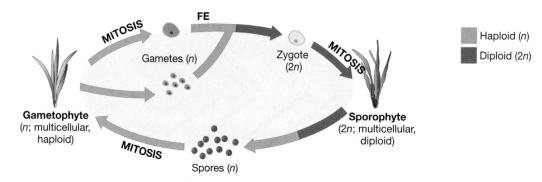

Haploid (n)
Diploid (2n)

(b) **Mosses:** Gametophyte is large and long lived; sporophyte depends on gametophyte for nutrition.

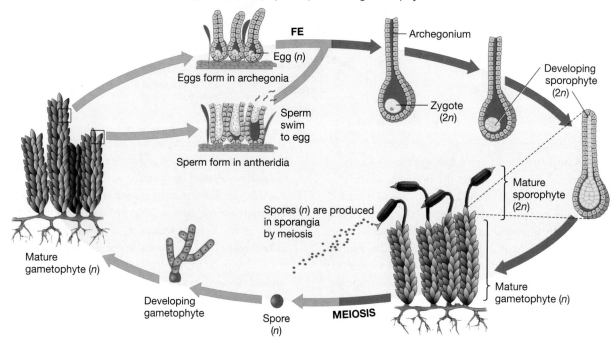

(c) **Ferns:** Sporophyte is large and long lived but, when young, depends on gametophyte for nutrition.

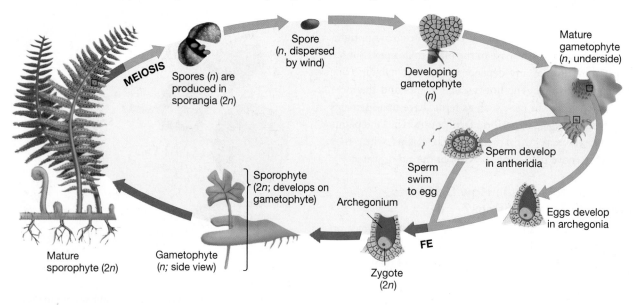

FIGURE 29.14 The Life Cycles of Land Plants
EXERCISE Study part (a). Find each phase of this generalized life cycle in parts (b) through (e).

Alternation of generations always involves the same basic sequence of events (**Figure 29.14a**):

1. Gametophytes produce gametes by mitosis. Both the gametophyte and the gametes are haploid.

2. Two gametes unite to form a diploid zygote.

3. The zygote develops into a multicellular, diploid sporophyte.

4. The sporophyte produces haploid spores by meiosis.

5. Spores develop into a gametophyte.

(d) Conifers: Sporophyte is dominant; gametophyte depends on sporophyte for nutrition.

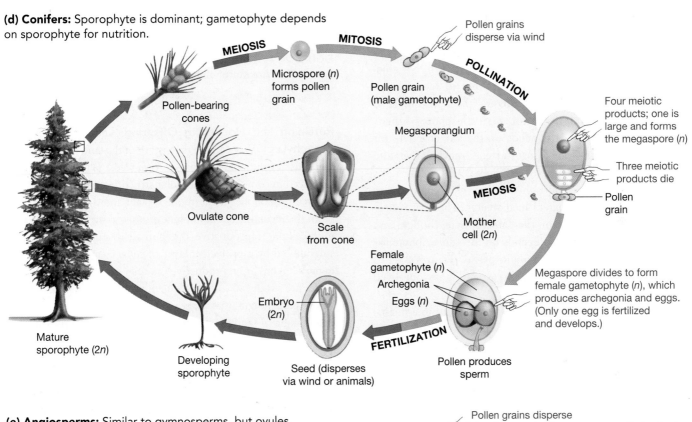

(e) Angiosperms: Similar to gymnosperms, but ovules and seeds form in enclosures called ovaries.

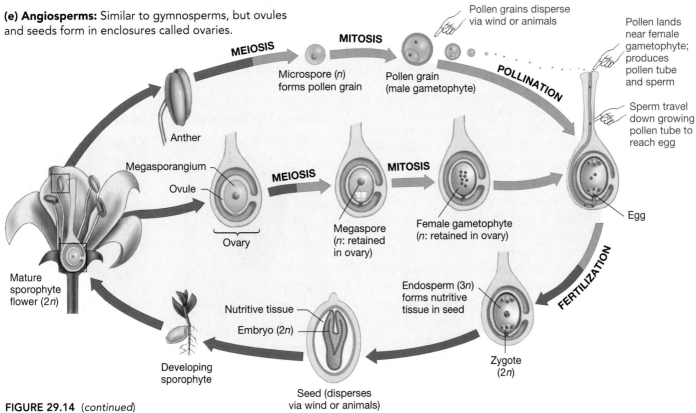

FIGURE 29.14 (continued)

Recall from Chapter 28 that a spore is a cell that grows directly into a multicellular individual without fusing with another cell. In nonvascular plants and seedless vascular plants, spores are encased in a tough, watertight coat made of sporopollenin. As a result, these spores resist drying and are an effective way to disperse offspring to new habitats by wind. It's helpful to recognize the distinction between a spore and a zygote: A zygote forms from the fusion of two cells, such as a sperm and an egg, while spores do not.

Although these five steps are common to all species with alternation of generations, the relationship between the gametophyte and sporophyte is highly variable in land plants. In nonvascular plants such as mosses, the sporophyte is small, short lived, and does not perform photosynthesis (**Figure 29.14b**). Instead, it depends on the gametophyte for nutrition. But in ferns, the sporophyte is much larger and longer lived than the gametophyte (**Figure 29.14c**). The ferns you are familiar with are sporophytes; their gametophytes are typically just a few millimeters in diameter. The fern sporophyte still develops directly on the gametophyte, though; and when young, the sporophyte depends on the gametophyte for nutrition. In gymnosperms and angiosperms (**Figures 29.14d** and **29.14e**), the sporophyte is the largest and most long-lived phase in the life cycle.

The male gametophyte of seed plants is reduced to a microscopic structure called a **pollen grain**. Similarly, the female gametophyte is a small structure that is retained in the sporophyte's female reproductive organs and that produces an egg. In gymnosperms and angiosperms, the gametophyte does not perform photosynthesis at all and has to obtain all of its nutrition from the sporophyte. This is the exact opposite of the situation in liverworts and mosses, in which the sporophyte is nourished by the gametophyte.

Another important innovation found in seed plants is **heterospory**—the production of two distinct types of spore-producing structures and thus two distinct types of spores. The two types of structures are often found on the same plant. **Microsporangia** are spore-producing structures that produce microspores. **Microspores** develop into male gametophytes, which produce the small gametes called sperm. **Megasporangia** are spore-producing structures that produce megaspores. **Megaspores** develop into female gametophytes, which produce the large gametes called eggs (**Figure 29.15a**). Thus, seed plant gametophytes are either male or female, but never both. Some lycophytes and a few ferns are also heterosporous, but most of the seedless vascular plants are **homosporous**—meaning that they produce a single type of spore. Homosporous species produce spores that develop into bisexual gametophytes. Bisexual gametophytes produce both eggs and sperm (**Figure 29.15b**).

Now, what does all this variation in life cycles have to do with plants losing their reliance on swimming gametes? The answer involves a series of remarkable evolutionary innovations that occurred as land plants originated and diversified.

(a) Seed plants are heterosporous.

Microsporangia → Microspores → Male gametophyte → Sperm

Megasporangia → Megaspores → Female gametophyte → Eggs

(b) Most other plants are homosporous.

Sporangium → Spores → Bisexual gametophyte → Sperm / Eggs

FIGURE 29.15 Heterosporous Plants Produce Male and Female Spores That Are Morphologically Distinct

Retaining and Nourishing Offspring: Land Plants as Embryophytes Very early in the history of the land plants, two momentous evolutionary changes occurred: (1) Gametes were produced in complex, multicellular structures; and (2) the embryo was retained on the parent (mother) plant and was nourished by it.

In the most ancient land plant groups, gametophytes produce multicellular reproductive organs called gametangia that are surrounded by a protective layer of cells. **Gametangia** (singular: **gametangium**) are reproductive organs—structures in which gametes are produced. Although members of the Charales also develop gametangia that have a protective layer of cells, the gametangia of land plants are larger and more complex. In both groups, the male and female reproductive organs are separate. The sperm-producing structure is called an **antheridium** (plural: **antheridia**; **Figure 29.16a**), and the egg-producing structure is called an **archegonium** (plural: **archegonia**; **Figure 29.16b**). The antheridium and archegonium have the same function as the testes and ovaries of animals. The elaboration of the multicellular gametangium was an adaptation that helped protect the gametes of land plants and prevent them from drying out.

(a) Sperm form in antheridia.   **(b)** Eggs form in archegonia.

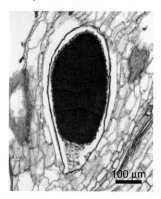 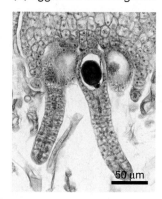

FIGURE 29.16 In All Land Plant Groups but Angiosperms, Gametes Are Produced in Gametangia
Gametangia are complex, multicellular structures that protect developing gametes. **(a)** Sperm-forming gametangia are called antheridia; **(b)** egg-producing gametangia are called archegonia.

(a) Young sporophytes are nourished by the parent plant.

Sporophyte

Gametophyte

1 cm

(b) Transfer cells convey nutrients to the embryo.

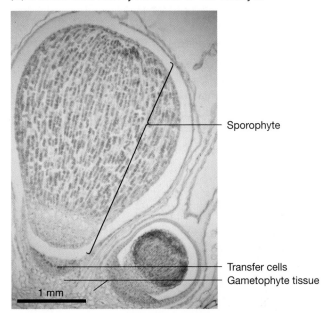

Sporophyte

Transfer cells
Gametophyte tissue

1 mm

FIGURE 29.17 Land Plants Are Also Known as Embryophytes Because Parents Nourish Their Young
(a) Land plants are known as embryophytes because eggs, zygotes, and embryos are retained on the parent plant. Fertilization and early development take place on the parent plant. **(b)** As land plant embryos develop, transfer cells transfer nutrients from the mother to the offspring.

The second innovation that occurred early in land plant evolution involved the eggs that formed inside archegonia. Instead of shedding their eggs into the water or soil, land plants retain them. Eggs are also retained in the green algal lineages that are most closely related to land plants: In Charales and other closely related groups, sperm swim to the egg, fertilization occurs, and the resulting zygote stays attached to the parent. Either before or after fertilization, the egg or zygote receives nutrients from the mother plant. But because these algae live in northern latitudes, the parent plant dies each autumn as the temperature drops. The zygote remains on the dead tissue from the parent, settles to the bottom of the lake or pond, and overwinters. In spring, meiosis occurs, and the resulting spores develop into haploid adult plants.

In land plants, the zygote is also retained on the gametophyte after fertilization. But in contrast to the zygotes of green algae, the zygotes of land plants begin to develop on the parent plant while it is still alive, forming a multicellular embryo. As a result, the developing embryo can be nourished by the parent (**Figure 29.17a**). This trait is important because it means that land plant embryos do not have to manufacture their own food early in life. Instead, they receive most or all of their nutrients from the parent plant. Based on this innovation, biologists call the land plants **embryophytes**—literally, the "embryo-plants." The retention of the embryo is analogous to pregnancy in mammals, where offspring are retained by the mother and nourished through the initial stages of growth. Land plant embryos even have specialized **transfer cells**, which

make physical contact with parental cells and facilitate the transfer of nutrients (**Figure 29.17b**), much like the placenta that develops in a pregnant mammal.

Although the retention of embryos ensures that offspring are well nourished, it has a potential disadvantage: In ferns and horsetails, sporophytes have to live in the same physical location as that of the parent gametophyte. But this limitation is overcome in seed plants, because the embryo is portable and can disperse to a new location.

The Evolution of Pollen During the evolution of land plants, there was clearly a strong trend from gametophyte-dominated life cycles of nonvascular plants to the sporophyte-dominated life cycles of gymnosperms and angiosperms. Why is this trend important? Recall that in the sporophyte-dominated life cycles of gymnosperms and angiosperms, the male sperm-producing structure—the gametophyte—is reduced to a pollen grain. Pollen grains are male gametophytes that are surrounded by a tough, desiccation-resistant coat of sporopollenin. Pollen can be exposed to the air for long periods of time without dying from dehydration. Pollen grains are also tiny enough to be carried to female gametophytes by wind or insects. Upon landing near the egg, the male gametophyte releases the sperm cells that accomplish fertilization.

When pollen evolved, then, the seed plants lost their dependence on water to accomplish fertilization. Instead of swimming to the egg as a naked sperm cell, their tiny gametophytes took to the skies.

The Evolution of the Seed   The evolution of large, complex gametangia protected the eggs and sperm of land plants from drying. Embryo retention allowed offspring to be nourished directly by their parent, and pollen enabled fertilization to occur in the absence of water. Seeds were a fourth innovation that allowed plants to reproduce efficiently on land.

A seed is a structure that encloses and protects a developing embryo (**Figure 29.18a**). Seeds allow embryos to be dispersed to a new habitat, away from the parent plant. Like a bird's egg, the seed provides a protective case for the embryo and a store of nutrients provided by the mother. In addition, seeds are often attached to structures that aid in dispersal by wind, water, or animals. As **Figure 29.18b** shows, seed dispersal is aided by structures such as the wings on a milkweed seed.

By the time seeds had evolved about 360 million years ago, land animals were diverse and abundant enough to play a lead role in seed dispersal. The most important connection between seeds and animals occurred during the evolution of angiosperms, however. Angiosperms are the only organisms that make fruits. A **fruit** is a structure that is derived from the **ovary**, a female reproductive organ found in angiosperms. The angiosperms' ovary encloses the **ovules**, structures where female gametophytes are formed and where eggs are produced. When a fruit is mature, it encloses one or more seeds (**Figure 29.19a**). Tissues derived from the ovary are often nutritious and brightly colored. Fruits such as these attract animals that eat the fruits, digest the nutritious tissue around the seeds, and disperse seeds in their feces (**Figure 29.19b**). Animals were also involved in the other major reproductive innovation in angiosperms—the flower.

(a) Seeds package an embryo with a food supply.

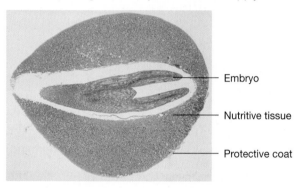

Embryo

Nutritive tissue

Protective coat

(b) Seeds are dispersed by wind, water, and animals.

Wind

Water

Animals

FIGURE 29.18 Seeds Contain an Embryo and a Food Supply and Can Be Dispersed

(a) Seeds consist of an embryo, a store of carbohydrate or oil-rich nutritive tissue, and a protective coat. **(b)** Most seeds have structures that facilitate dispersal by wind, water, or animals.

(a) Fruits are derived from ovaries and contain seeds.

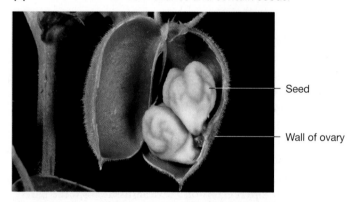

Seed

Wall of ovary

(b) Many fruits are dispersed by animals.

FIGURE 29.19 Fruits Are Derived from Ovaries Found in Angiosperms

(a) A pea pod is one of the simplest types of fruit. **(b)** The ovary wall often becomes thick, fleshy, and nutritious enough to attract animals that disperse the seeds inside.

The Evolution of the Flower Flowering plants, or angiosperms, are the most diverse land plants living today. About 250,000 species have been described, and more are discovered each year. Their success in terms of geographical distribution, number of individuals, and number of species revolves around a reproductive organ: the **flower**. Angiosperms do not produce gametangia. Their gametes are produced by male and female gametophytes, which are derived from microsporangia and megasporangia that develop inside the flower.

As **Figure 29.20** shows, flowers are wonderfully diverse in size, shape, and coloration. They also produce a wide range of scents. Why does all of this variation exist? To answer this question, biologists hypothesize that flowers are adaptations that increase the probability that an animal will perform **pollination**—meaning the transfer of pollen to a plant's female reproductive organ, where eggs are produced and fertilization takes place.

Instead of leaving pollination to an undirected agent such as wind, the hypothesis is that in some circumstances natural selection favored structures that reward an animal—usually an insect—for carrying pollen directly from one flower to another. Under the directed-pollination hypothesis, natural selection has favored flower colors and shapes and scents that are successful in attracting particular types of pollinators. A *pollinator* is an animal that disperses pollen. Pollinators are attracted to flowers because flowers provide the animals with food in the form of protein-rich pollen or a sugar-rich fluid known as **nectar**. In this way, the relationship between flowering plants and their pollinators is mutually beneficial. The pollinator gets food; the plant gets sex.

What evidence supports the hypothesis that flowers vary in size, structure, scent, and color in order to attract different pollinators? The first type of evidence is correlational in nature. In general, the characteristics of a flower correlate closely with the characteristics of its pollinator. A few examples will help drive this point home:

- The carrion flower in Figure 29.20a produces molecules that smell like rotting flesh. The scent attracts carrion flies, which normally lay their eggs in animal carcasses. In effect, the plant tricks the flies. While looking for a place to lay their eggs on a flower, the flies become dusted with pollen. If the flies are already carrying pollen from a visit to a different carrion flower, they are likely to deposit pollen grains near the plant's female gametophyte. In this way, the carrion flies pollinate the plant.

- Flowers that are pollinated by hummingbirds typically have petals that form a long, tubelike structure that corresponds to the size and shape of a hummingbird's beak (Figure 29.20b). Nectar-producing cells are located at the base of the tube. When hummingbirds visit the flower, they insert their beaks and harvest the nectar. In the process, they transfer pollen grains attached to their throats or faces.

- Hummingbird-pollinated flowers also tend to have red petals (Figure 29.20b), while bee-pollinated flowers tend to be purple or yellow (Figure 29.20c). Hummingbirds are attracted to red; bees have excellent vision in the purple and ultraviolet end of the spectrum.

In addition to correlational evidence, there is strong experimental support for the hypothesis that natural selection favors a correspondence between the shapes of flowers and their pollinators. Consider, for example, recent work on two South African populations of the orchid *Disa draconis*. The length of the long tube, or *spur*, located at the back of this flower is very different in the two populations. As predicted by the directed-pollination hypothesis, each population is pollinated by a different insect. Short-spurred plants that grow in mountain habitats are pollinated by horseflies, which have a relatively

(a) Carrion flowers smell like rotting flesh and attract carrion flies.

(b) Hummingbird-pollinated flowers are red and have long tubes with nectar at the base.

(c) Bumble-bee-pollinated flowers are often bright purple.

FIGURE 29.20 Flowers with Different Shapes, Colors, and Fragrances Attract Different Pollinators

(a) Horseflies have a short proboscis and pollinate short-spurred orchids.

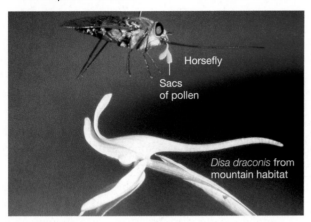

(b) Tanglewing flies have a long proboscis and pollinate long-spurred orchids.

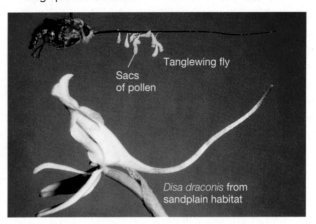

FIGURE 29.21 Populations of the Same Orchid Species Have Different Flower Shapes and Different Pollinators
A *Disa draconis* flower **(a)** from a mountainous habitat with its horsefly pollinator and **(b)** from a lower-elevation, sandplain habitat with its tanglewing fly pollinator. Note the correlation between the length of each flower's spur and the length of each pollinator's proboscis.

short proboscis (**Figure 29.21a**). (A *proboscis* is a specialized mouthpart found in some insects. When extended, it functions like a straw in sucking nectar or other fluids.) Long-spurred orchids that grow on low-lying sandplain habitats are pollinated by tanglewing flies, which have a particularly long proboscis (**Figure 29.21b**). Horseflies are common in mountain habitats but rare at lower elevations. Conversely, tanglewing flies are abundant on low-lying sandplains but absent at high elevations.

The directed-pollination hypothesis makes two key predictions about pollination in these populations. First, it predicts that the ability to attract pollinators is an important component of fitness in *D. draconis*. To test this prediction, biologists pollinated a large number of flowers by hand in each population. Hand pollination ensures that large amounts of pollen are transferred. Then they compared the number of seeds that were produced by the hand-pollinated flowers versus normal, insect-pollinated flowers. The data showed that hand-pollinated individuals produced many more seeds than fly-pollinated individuals did. This is a key observation, because it shows that the fitness of these orchids is limited by their ability to attract pollinators. In this species, traits that increase the likelihood of being pollinated should confer a huge fitness advantage.

The second prediction of the directed-pollination hypothesis is that spurs of a particular length are adaptations that increase the probability of being pollinated by horseflies or tanglewing flies. To test this prediction, researchers artificially shortened the spurs of individuals in the long-spurred population (**Figure 29.22**). The researchers did this by tying off the spurs of randomly selected flowers with a piece of yarn, so that flies could not reach the end of the spur. The idea was that flies would not make contact with the flower's reproductive organs when they inserted their proboscis into the shortened tube. The biologists left nearby flowers alone but tied a piece

of yarn near the spur. The presence of the yarn near the spur controlled for the possibility that the yarn itself—rather than spur length—affected pollinator behavior. The data collected indicate that flowers with short spurs received much less pollen and set much less seed than did flowers with normal-length spurs. These data strongly support the hypothesis that spur length is an adaptation that increases the frequency of pollination in a particular habitat.

Such results support the hypothesis that plants gain large fitness benefits by attracting pollinators. An even more general implication is that the spectacular diversity of angiosperms resulted, at least in part, from natural selection exerted by the equally spectacular diversity of insect, mammal, and bird pollinators. Recall from Section 29.2 that almost 90 percent of the land plants living today reproduce by making flowers. Once green plants could grow and reproduce in terrestrial environments, the story of their diversification revolved around systems for transferring gametes efficiently and producing the largest number of seeds possible. **Figure 29.23** summarizes the adaptations that enabled plants to reproduce effectively on land.

The Angiosperm Radiation

Angiosperms represent one of the great adaptive radiations in the history of life. As Chapter 26 noted, an **adaptive radiation** occurs when a single lineage produces a large number of descendant species that are adapted to a wide variety of habitats. The diversification of angiosperms is associated with three key adaptations: (1) vessels, (2) flowers, and (3) fruits. In combination, these traits allow angiosperms to transport water, pollen, and seeds efficiently. Based on these observations, it is not surprising that most land plants living today are angiosperms.

On the basis of morphological traits, the 235,000 species of angiosperms identified to date have traditionally been grouped

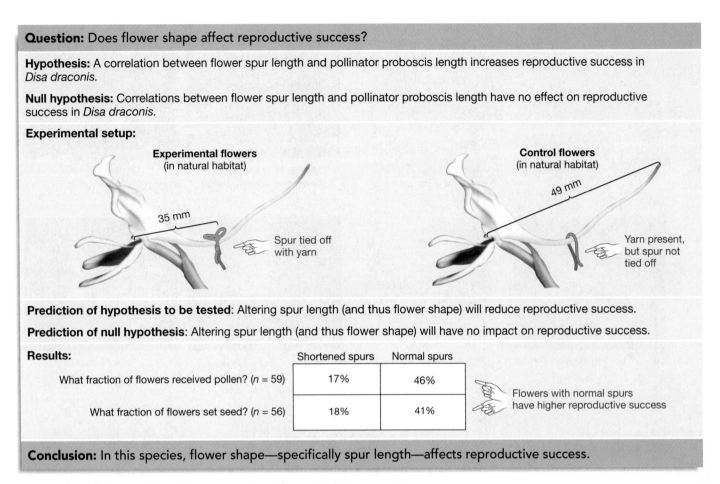

Question: Does flower shape affect reproductive success?

Hypothesis: A correlation between flower spur length and pollinator proboscis length increases reproductive success in *Disa draconis*.

Null hypothesis: Correlations between flower spur length and pollinator proboscis length have no effect on reproductive success in *Disa draconis*.

Experimental setup:

Experimental flowers (in natural habitat)

35 mm

Spur tied off with yarn

Control flowers (in natural habitat)

49 mm

Yarn present, but spur not tied off

Prediction of hypothesis to be tested: Altering spur length (and thus flower shape) will reduce reproductive success.

Prediction of null hypothesis: Altering spur length (and thus flower shape) will have no impact on reproductive success.

Results:

	Shortened spurs	Normal spurs
What fraction of flowers received pollen? (*n* = 59)	17%	46%
What fraction of flowers set seed? (*n* = 56)	18%	41%

Flowers with normal spurs have higher reproductive success

Conclusion: In this species, flower shape—specifically spur length—affects reproductive success.

FIGURE 29.22 The Adaptive Significance of Flower Shape: An Experimental Test
The spurs of experimental *D. draconis* flowers were tied off with yarn 35 mm from the opening; the spurs of control flowers had a piece of yarn loosely tied around them, too, but had normal spurs averaging 49 mm. QUESTION If spur length has no effect on reproductive success, what would the data in this table look like?

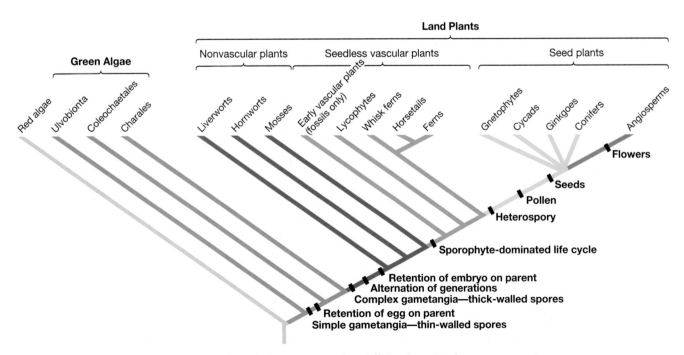

FIGURE 29.23 Evolutionary Innovations Allowed Plants to Reproduce Efficiently on Land
EXERCISE Circle the names of groups (at the tips of the branches) that have flagellated, swimming sperm.

MONOCOTS

One cotyledon (inside seed)

Cotyledon

Vascular tissue scattered throughout stem

Parallel veins in leaves (bundles of vascular tissue)

Flower petals in multiples of 3

DICOTS

Two cotyledons

Vascular tissue in circular arrangement in stem

Branching veins in leaves

Flower petals in multiples of 4 or 5

FIGURE 29.24 Four Morphological Differences between Monocots and Dicots

into two major lineages: the *monocotyledons*, or **monocots**, and the *dicotyledons*, or **dicots**. Some familiar monocots include the grasses, orchids, palms, and lilies; familiar dicots include the roses, buttercups, daisies, oaks, and maples. The names of the two groups were inspired by differences in a structure called the cotyledon. A **cotyledon** is the first leaf that is formed in an embryonic plant. As **Figure 29.24** shows, monocots have a single cotyledon (visible inside the seed) while dicots have two cotyledons (visible in newly germinated plants). The figure also highlights other major morphological differences observed in monocots and dicots, concerning the arrangement of vascular tissue and flower characteristics.

It would be misleading, however, to think that all species of flowering plants fall into one of these two groups—either monocots or dicots. Recent work has shown that dicots do not form a natural group consisting of a common ancestor and all of its descendants. To drive this point home, consider the phylogeny illustrated in **Figure 29.25**. These relationships were estimated by comparing the sequences of several genes that are shared by all angiosperms. Notice that species with dicot-like characters are scattered around the angiosperm phylogenetic tree. Based on this analysis, biologists have concluded that dicots are not a monophyletic group. Instead, they are paraphyletic.

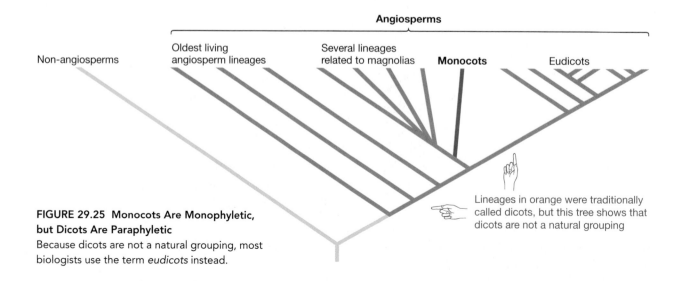

Angiosperms

Non-angiosperms

Oldest living angiosperm lineages

Several lineages related to magnolias

Monocots

Eudicots

Lineages in orange were traditionally called dicots, but this tree shows that dicots are not a natural grouping

FIGURE 29.25 Monocots Are Monophyletic, but Dicots Are Paraphyletic
Because dicots are not a natural grouping, most biologists use the term *eudicots* instead.

Biologists have adjusted the names assigned to angiosperm lineages to reflect this new knowledge of phylogeny. Now they formally name a series of monophyletic groups in the tree, including the **magnoliid complex** (made up of magnolias and laurels and their relatives), the monocots, and the **eudicots** ("true dicots")—a lineage that includes roses, daisies, and maples. The most basal clades consist of the genus *Amborella*, the water lilies, and the star anise and relatives. Plant systematists continue to work toward understanding relationships at the base of the angiosperm phylogenetic tree, and there will undoubtedly be more name changes as our knowledge grows.

> ## ✔ CHECK YOUR UNDERSTANDING
>
> Land plants were able to make the transition to growing in terrestrial environments, where sunlight and carbon dioxide are abundant, based on the evolution of cuticle, pores, stomata, water-conducting tissue, and vascular tissue. Reproduction in dry environments became increasingly efficient as land plants diversified, based on a series of evolutionary innovations: gametangia and the retention of embryos, pollen, seeds, and flowers. You should be able to (1) explain why each of these adaptations was important in survival or reproduction, and (2) map where each of these evolutionary innovations occurred on the phylogenetic tree of green plants.

29.4 Key Lineages of Green Plants

The evolution of cuticle, pores, stomata, and water-conducting tissues allowed green plants to grow on land, where resources for photosynthesis are abundant. Once the green plants were on land, the evolution of gametangia, retained embryos, pollen, seeds, and flowers enabled them to reproduce efficiently even in very dry environments. The adaptations reviewed in Section 29.3 allowed the land plants to make the most important water-to-land transition in the history of life.

To explore green plant diversity in more detail, let's take a closer look at some major groups of green algae. Then we'll discuss the 12 major phyla of land plants in the context of their broad morphological groupings: nonvascular plants, seedless vascular plants, and seed plants.

Green Algae

The green algae are a paraphyletic group that totals about 7000 species. Their bright green chloroplasts are similar to those found in land plants, with a double membrane and chlorophylls *a* and *b* but relatively few accessory pigments. And like land plants, green algae synthesize starch in the chloroplast as a storage product of photosynthesis and have a cell wall composed primarily of cellulose. Green algae are important primary producers in nearshore ocean environments and in all types of freshwater habitats. They are also found in several types of more exotic environments, including snowfields at high elevations, pack ice, and ice floes. These habitats are often colored pink or green by large concentrations of unicellular green algae (**Figure 29.26a**). Although these cells live at near-freezing temperatures, they make all their own food via photosynthesis.

In addition, green algae live in close association with an array of other organisms:

- **Lichens** are stable associations between green algae and fungi or between cyanobacteria and fungi. Lichens are found in terrestrial environments—often in locations such as tree bark or bare rock that lack soil (**Figure 29.26b**). The algae or cyanobacteria do not dry out, because they are protected by the fungus. Of the 17,000 species of lichens that have been described to date, about 85 percent involve a unicellular or filamentous green alga. Lichens are explored in more detail in Chapter 30.

- Unicellular green algae are common endosymbionts in planktonic protists that live in lakes and ponds (**Figure 29.26c**). The association is considered mutually beneficial: The algae supply the protists with food; the protists provide protection to the algae.

Green algae are a large and fascinating group of organisms. Let's take a closer look at just three of the many lineages.

(a) Green algae are responsible for pink snow.

(b) Most lichens are an association between fungi and green algae.

(c) Many unicellullar protists harbor green algae.

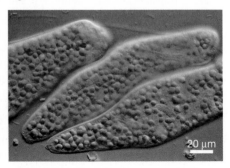

2 cm

20 μm

FIGURE 29.26 Some Green Algae Live in Unusual Environments

Ulvobionta

The Ulvobionta are a monophyletic group composed of several diverse and important subgroups, with a total of about 4000 species. Members of this lineage range from unicellular to multicellular. Most of the green algae you have seen at the ocean or in lakes or streams belong to this group, as do *Volvox* and related species introduced in Chapter 28.

Many of the large green algae in coastal and intertidal habitats along ocean coastlines are members of the Ulvobionta. *Ulva*, the sea lettuce (**Figure 29.27**), is a representative marine species. But there are also large numbers of unicellular or small multicellular species that inhabit the plankton of freshwater lakes and streams.

Reproduction   Most reproduce both asexually and sexually. Asexual reproduction often involves production of spores that swim with the aid of flagella. Sexual reproduction usually results in production of a resting stage—a cell that is dormant in winter. In many species the gametes are not called eggs and sperm, because they are the same size and shape. In most species gametes are shed into the water, so fertilization takes place away from the parent plants.

Life cycle   Many unicellular forms are diploid only as zygotes. Alternation of generations occurs in multicellular species. When alternation of generation occurs, gametophytes and sporophytes may look identical or different.

Human and ecological impacts   Ulvobionta are important primary producers in freshwater environments and in coastal areas of the oceans.

Ulva lactuca

FIGURE 29.27   Green Algae Are Important Primary Producers in Aquatic Environments

Coleochaetales

There are 19 species in this group. Most coleochaetes are barely visible to the unaided eye and grow as flat sheets of cells (**Figure 29.28**). They are considered multicellular, because they have specialized photosynthetic and reproductive cells.

The coleochaetes are strictly freshwater algae. They grow attached to aquatic plants such as water lilies or over submerged rocks in lakes and ponds. When they grow near beaches, they are often exposed to air when water levels drop in late summer.

Reproduction   Asexual reproduction is common and involves production of flagellated spores. During sexual reproduction, eggs are retained on the parent and are nourished after fertilization with the aid of transfer cells—a situation very similar to that observed in land plants. In some species certain individuals are male and produce only sperm, while other individuals are female and produce only eggs.

Life cycle   Alternation of generations does not occur. Multicellular individuals are haploid; the only diploid stage in the life cycle is the zygote.

Human and ecological impacts   Because they are very closely related to land plants, coleochaetes are studied intensively by researchers interested in how land plants made the water-to-land transition.

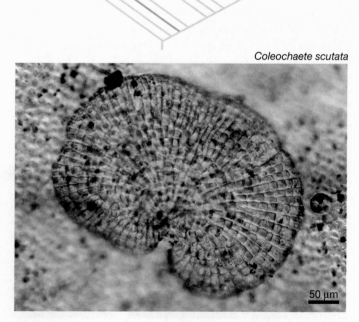

Coleochaete scutata

FIGURE 29.28   Coleochaetes Are Thin Sheets of Cells

Charales (Stoneworts)

There are several hundred species in this group. They are collectively known as stoneworts, because they commonly accumulate crusts of calcium carbonate $(CaCO_3)$ over their surfaces. All are multicellular. Some species can be a meter or more in length.

The stoneworts are freshwater algae. Certain species are specialized for growing in relatively deep waters, though most live in shallow water near lake beaches or pond edges.

Reproduction Primarily sexual reproduction occurs, with production of prominent, multicellular gametangia similar to those observed in early land plants. Eggs are retained on the parent plant, which supplies eggs with nutrients prior to fertilization.

Life cycle Alternation of generations does not occur. Multicellular individuals are haploid; the only diploid stage in the life cycle is the zygote.

Human and ecological impacts Some species form extensive beds in lake bottoms or ponds and provide food for ducks and geese as well as food and shelter for fish (**Figure 29.29**).

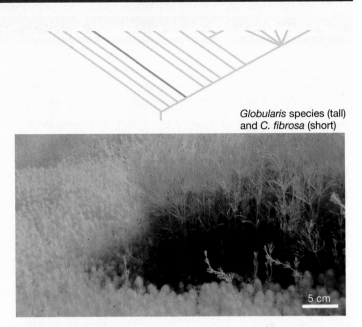

Globularis species (tall) and *C. fibrosa* (short)

5 cm

FIGURE 29.29 Stoneworts Can Form Beds in Lake Bottoms

Nonvascular Plants ("Bryophytes")

The most basal lineages of land plants are collectively known as nonvascular plants, or **bryophytes**. The three lineages with living representatives (liverworts, hornworts, and mosses) do not form a monophyletic group, but instead represent an evolutionary grade. All of the species present today have a low, sprawling growth habit. In fact, it is unusual to find bryophytes that are more than 5 to 10 centimeters (2 to 4 inches) tall. Individuals are anchored to soil, rocks, or tree bark by specialized structures called **rhizoids**. No bryophyptes have true vascular tissue where cells have reinforced walls. In the lineages present today, simple water-conducting cells and tissues are found only in some mosses. All of the bryophytes have flagellated sperm that swim to eggs through raindrops or small puddles on the plant surface. Spores are dispersed by wind.

Hepaticophyta (Liverworts)

Liverworts got their name because some species native to Europe have liver-shaped leaves. According to the medieval *Doctrine of Signatures*, God indicated how certain plants should be used by giving them a distinctive appearance. Thus, liverwort teas were hypothesized to be beneficial for liver ailments. (They are not.) About 6500 species are known. They are commonly found growing on damp forest floors or riverbanks, often in dense mats (**Figure 29.30**), or on the trunks or branches of tropical trees.

Adaptations to land Liverworts are covered with cuticle. Some species have pores that allow gas exchange; in species that lack pores, the cuticle is very thin.

Reproduction Asexual reproduction occurs when fragments of a plant are broken off and begin growing independently. Some species also produce small structures called **gemmae** asexually, during the gametophyte phase. Mature gemmae are knocked off the parent plant by rain and grow into independent gametophytes. During sexual reproduction, sperm and eggs are produced in gametangia.

Life cycle The gametophyte is the largest and longest-lived phase in the life cycle. Sporophytes are small, grow directly from the gametophyte, and depend on the gametophyte for nutrition. Spores are shed from the sporophyte and are carried away by wind or rain.

Human and ecological impacts When liverworts grow on bare rock or tree bark, their dead and decaying body parts contribute to the initial stages of soil formation.

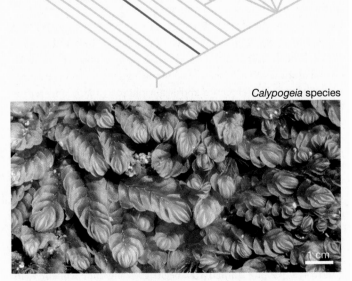

Calypogeia species

1 cm

FIGURE 29.30 Liverworts Thrive in Moist Habitats

Anthocerophyta (Hornworts)

Hornworts got their name because their sporophytes have a horn-like appearance (**Figure 29.31**) and because *wort* is the Anglo-Saxon word for *plant*. About 100 species have been described to date.

Adaptations to land   Hornwort sporophytes have stomata. Research is under way to determine if they can open their pores to allow gas exchange or close their pores to avoid water loss during dry intervals.

Reproduction   Depending on the species, gametophytes may contain only egg-producing archegonia or only sperm-producing antheridia, or both. Stated another way, individuals of some species are either female or male, while in other species each individual has both types of reproductive organs.

Life cycle   The gametophyte is the longest-lived phase in the life cycle. Although sporophytes grow directly from the gametophyte, they are green because their cells contain chloroplasts. Sporophytes manufacture some of their own food but also get nutrition from the gametophyte. Spores disperse from the parent plant via wind or rain.

Human and ecological impacts   Some species harbor symbiotic cyanobacteria that fix nitrogen.

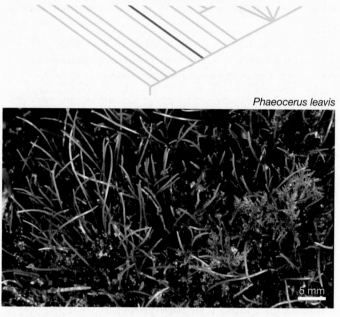

Phaeocerus leavis

5 mm

FIGURE 29.31 Hornworts Have Horn-Shaped Sporophytes

Bryophyta (Mosses)

Over 12,000 species of mosses have been named and described to date, and more are being discovered every year—particularly in the tropics. Mosses are informally grouped with other "bryophytes" (liverworts and hornworts) but are formally classified in their own monophyletic group: the phylum Bryophyta.

Although mosses are common inhabitants of moist forests all over the world, plants in this phylum can also be abundant in more extreme environments, such as deserts and windy, treeless habitats in the Arctic, Antarctic, or mountaintops. In these severe conditions, mosses are able to thrive because their bodies can become extremely dry without dying. When the weather makes photosynthesis difficult, individuals dry out and become dormant or inactive (**Figure 29.32a**). Then when rains arrive or temperatures warm, the plants rehydrate and begin photosynthesis and reproduction (**Figure 29.32b**).

Adaptations to land   One subgroup of mosses contains simple conducting tissues consisting of cells that are specialized for the transport of water or food. But because these cells do not have walls that are reinforced by lignin, they are not considered true vascular tissue. As a result, most mosses are not able to grow much taller than a few centimeters.

Reproduction   Asexual reproduction often occurs by fragmentation, meaning that pieces of gametophytes that are broken off by wind or a passing animal can begin growing independently. In many species, sexual reproduction cannot involve self-fertilization

because the sexes are separate—meaning that an individual plant produces only eggs in archegonia or only sperm in antheridia. A typical sporophyte produces up to 50 million tiny spores. Spores are usually distributed by wind.

(a) Moss in dry weather   **(b)** Moss in wet weather

1 cm   1 cm

FIGURE 29.32 Many Moss Species Can Become Dormant When Conditions Are Dry

(Continued on next page)

Bryophyta (Mosses) *(continued)*

Life cycle The moss life cycle is similar to that of liverworts and hornworts: The sporophyte is retained on the much larger and longer-lived gametophyte and gets most of its nutrition from the gametophyte.

Human and ecological impacts Species in the genus *Sphagnum* are the most abundant plant in wet habitats of northern environments. Because *Sphagnum*-rich environments account for 1 percent of Earth's total land area, equivalent to half the area of the United States, *Sphagnum* species are among the most abundant plants in the world. *Sphagnum*-rich habitats have an exceptionally short growing season, however, so the decomposition of dead mosses and other plants is slow. As a result, large deposits of semi-decayed organic matter, known as **peat**, accumulate in these regions. Researchers estimate that the world's peatlands store about 400 billion metric tons of carbon, which makes these peat deposits among the most important carbon sinks in terrestrial environments. If peatlands begin to burn or decay rapidly due to global warming, the CO_2 released will further exacerbate global warming.

Peat is harvested as a traditional heating and cooking fuel in Ireland and other northern countries. It is also widely used as a soil additive in gardening, because *Sphagnum* can absorb up to 20 times its dry weight in water. This high water-holding capacity is due to the presence of large numbers of dead cells in the leaves of these mosses, which readily fill with water via pores in their walls. The same property made dried sphagnum an effective material in wartime for use as bandages.

Seedless Vascular Plants

The seedless vascular plants are a paraphyletic group that forms a grade between the nonvascular plants and the seed plants. All species of seedless vascular plants have conducting tissues with cells that are reinforced with lignin, forming vascular tissue. Tree-sized club mosses and horsetails are abundant in the fossil record, and tree ferns are still common inhabitants of mountain slopes in the tropics. Tree ferns do not make wood but are supported by fibrous roots.

The sporophyte is the larger and longer-lived phase of the life cycle in all three phyla of seedless vascular plants. The gametophyte is physically independent of the sporophyte, however. Eggs are retained on the gametophyte, and sperm swim to the egg with the aid of flagella. Thus, seedless vascular plants depend on the presence of water for reproduction—they need enough water to form a continuous layer that "connects" gametophytes and allows sperm to swim to eggs. Sporophytes develop on the gametophyte and are nourished by the gametophyte when they are small.

Lycophyta (Lycophytes, or Club Mosses)

Although the fossil record documents lycophytes that were 2 m wide and 40 m tall, the 1000 species of lycophytes living today are all small in stature (**Figure 29.33**). Most live on the forest floor or on the branches or trunks of tropical trees. They are often called "ground pines" or club mosses, even though they are neither pines nor mosses.

Adaptations to land Lycophytes are the most ancient land plant lineage with **roots**—a belowground system of tissues and organs that anchors the plant and is responsible for absorbing water and mineral nutrients. Roots differ from the rhizoids observed in bryophytes, because roots contain vascular tissue and thus are capable of conducting water and nutrients from below ground to the upper reaches of the plant. Small leaves called microphylls extend from the stems.

Reproduction Asexual reproduction can occur by fragmentation or gemmae. During sexual reproduction, spores give rise to bisexual gametophytes in most species—meaning that each gametophyte produces both eggs and sperm. Self-fertilization is extremely rare in most of these species, however. Some club mosses have separate male and female gametophytes.

Life cycle The gametophytes of some species live entirely underground and get their nutrition from symbiotic fungi. In certain species, gametophytes live 6–15 years and give rise to a large number of sporophytes over time.

Human and ecological impacts Tree-sized lycophytes were abundant in the coal-forming forests of the Carboniferous period. In coal-fired power plants today, electricity is being generated by burning fossilized lycophyte trunks and leaves.

Lycopodium species

FIGURE 29.33 Lycopods Living Today Are Small in Stature

Psilotophyta (Whisk Ferns)

Only two genera of whisk ferns are living today and perhaps six distinct species. Whisk ferns are restricted to tropical regions and have no fossil record. They are extremely simple morphologically, with aboveground parts consisting of branching stems that have tiny, scale-like outgrowths instead of leaves (**Figure 29.34**).

Adaptations to land Whisk ferns lack roots. Some species gain most of their nutrition from fungi (see Chapter 30) that grow in association with the whisk ferns' extensive underground stems called **rhizomes**. Other species grow in rock crevices or are **epiphytes** ("upon-plants"), meaning that they grow on the trunks or branches of other plants—in this case, in the branches of tree ferns.

Reproduction Asexual reproduction is common in sporophytes via the extension of rhizomes and the production of new above-ground stems. When spores mature, they are dispersed by wind and germinate into gametophytes that contain both archegonia and antheridia.

Life cycle Sporophytes may be up to 30 cm tall, but gameto-phytes are less than 2 mm long and live under the soil surface. Ga-metophytes absorb nutrients directly from the surrounding soil and from symbiotic fungi. Fertilization takes place inside the archego-nium, and the sporophyte develops directly on the gametophyte.

Human and ecological impacts Some whisk fern species are popular landscaping plants, particularly in Japan. The same species can be a serious pest in greenhouses.

Tmesipteris species

FIGURE 29.34 Psilotophytes Are Extremely Simple Morphologically

Sphenophyta (or Equisetophyta) (Horsetails)

Although horsetails are prominent in the fossil record of land plants, just 15 species are known today. All 15 are in the genus *Equisetum*. Translated literally, *Equisetum* means "horse-bristle." Both the scientific name and the common name, horsetail, come from the brushy appearance of the stems and branches in some species (**Figure 29.35**). Horsetails may be locally abundant in wet habitats such as stream banks or marsh edges.

Adaptations to land Horsetails have an interesting adaptation that allows them to flourish in waterlogged, oxygen-poor soils. Horsetail stems are hollow, so oxygen readily diffuses down the stem to reach roots that cannot obtain oxygen from the surround-ing soil.

Reproduction Asexual reproduction is common in sporophytes and occurs via fragmentation or the extension of rhizomes. From these rhizomes, two types of erect, specialized stems may grow: stems that contain tiny leaves and chloroplast-rich branches and that are specialized for photosynthesis, or stems that bear clusters of sporangia and produce huge numbers of spores by mitosis.

Life cycle Gametophytes perform photosynthesis but are small and short-lived. They normally produce both antheridia and archegonia, but in most cases the sperm-producing structure ma-tures first. This pattern is thought to be an adaptation that mini-mizes self-fertilization and maximizes cross-fertilization.

Human and ecological impacts Horsetail stems are rich in sili-ca granules. The glass-like deposits not only strengthen the stem but also make these plants useful for scouring pots and pans—hence these plants are often called "scouring rushes."

Equisetum arvense

FIGURE 29.35 Horsetails Have Separate Reproductive and Vegetative Stalks

Pteridophyta (Ferns)

With 12,000 species, ferns are by far the most species-rich group of seedless vascular plants. They are particularly abundant in the tropics. About a third of the tropical species are epiphytes, usually growing on the trunks or branches of trees. Species that can grow epiphytically live high above the forest floor, where competition for light is reduced, without making wood and growing tall themselves. The growth habits of ferns are highly variable among species, however, and ferns range in size from rosettes the size of your little fingernail to 20-meter-tall trees (**Figure 29.36a**).

Adaptations to land Ferns are the only seedless vascular plants that have large, well-developed leaves—commonly called **fronds**. Leaves give the plant a large surface area with which to capture sunlight for photosynthesis.

Reproduction In a few species, gametophytes reproduce asexually via production of gemmae. Typically, species that can reproduce via gemmae never produce gametes or sporophytes. In most species, however, sexual reproduction is the norm.

Life cycle Although fern gametophytes contain chloroplasts and are photosynthetic, the sporophyte is typically the larger and longer-lived phase of the life cycle. In mature sporophytes, sporangia are usually found in clusters called **sori** on the undersides of leaves (**Figure 29.36b**).

Human and ecological impacts In many parts of the world, people gather the young, unfolding fronds, or "fiddleheads," of ferns in spring as food. Ferns are also widely used as ornamental plants in landscaping.

FIGURE 29.36 Ferns Are the Most Species-Rich Group of Seedless Vascular Plants

(a) Ferns range from small to tree sized.

Gonocormus minutus

Dicksonia antarctica

(b) Fern sporangia are often located on the underside of fronds.

Polypodium vulgare

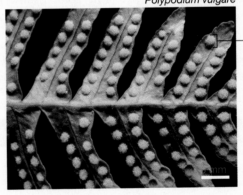

Collection of sporangia

Seed Plants

The seed plants are a monophyletic group that consists of the gymnosperms—gnetophytes, cycads, ginkgoes, and conifers—and the angiosperms. Seed plants are defined by (1) the production of seeds and (2) the production of pollen grains. Recall from Section 29.2 that seeds are a specialized structure for dispersing embryonic sporophytes to new locations. Seeds are the mature form of an ovule, the female reproductive structure that encloses the female gametophyte and egg cell. Also recall from Section 29.3 that pollen grains are tiny, sperm-producing gametophytes that are easily dispersed through air as opposed to water. The structure and function of seed plants is the focus of Chapters 35 through 40.

Seed plants are found in virtually every type of habitat, and they adopt every growth habit known in land plants. Their forms range from mosslike mats to shrubs and vines to 100-meter-tall trees. Seed plants can be annual or perennial, with life spans ranging from a few weeks to almost five thousand years.

Of the several major questions remaining about the evolution of seed plants, perhaps the most prominent is whether the gymnosperms are monophyletic or paraphyletic. Although recent analyses of gene sequence data suggest that gymnosperm ancestry may indeed be traced back to a single common ancestor, the issue is still in doubt. As a result, the phylogenetic trees presented in this chapter show the four gymnosperm groups radiating from a single point. If continued research supports the hypothesis that gymnosperms are monophyletic, then the group would be represented as their own, distinct branch on the tree—meaning a major split occurred between gymnosperms and angiosperms. In the meantime, let's consider each of the five phyla of seed plants in more detail.

Gnetophyta (Gnetophytes)

The gnetophytes comprise about 70 species in three genera. One genus consists of vines and trees from the tropics. A second is made up of desert-dwelling shrubs, including what may be the most familiar gnetophyte—the shrub called Mormon tea, which is common in the deserts of southwestern North America (see Figure 29.6d). The third genus contains a single species that probably qualifies as the world's most bizarre plant (**Figure 29.37**): *Welwitschia mirabilis*, which is native to the deserts of southwest Africa. Although it has large belowground structures, the part that is above ground consists of just two strap-like leaves, which grow continuously from the base and die at the tips. They also split lengthwise as they grow and age.

Adaptations to land   All of the living species make wood as a support structure.

Reproduction   Pollen is transferred by the wind or by insects. Like other gymnosperms, seeds do not form inside an encapsulated structure.

Life cycle   The microsporangia and megasporangia are arranged in clusters at the end of stalks, similar to the way flowers are clustered in some angiosperms.

Human and ecological impacts   The drug ephedrine was originally isolated from species in the gnetophyte genus *Ephredra*, which is native to northern China and Mongolia. Ephedrine is used in the treatment of hay fever, colds, and asthma.

Welwitschia mirabilis

20 cm

FIGURE 29.37 Welwitschia Is a Bizarre Plant

Cycadophyta (Cycads)

The cycads are so similar in overall appearance to palm trees, which are angiosperms, that cycads are sometimes called "sago palms." Although cycads were extremely abundant when dinosaurs were present on Earth 150–65 million years ago, only about 140 species are living today. Most are found in the tropics (**Figure 29.38**).

Adaptations to land   Cycads are woody; many are tree sized.

Reproduction   Pollen is carried by insects (usually beetles or weevils) or, in some species, wind. Cycad seeds are large and are often brightly colored. The colors attract birds and mammals, both of which eat and disperse the seeds.

Life cycle   Cycads are heterosporous. Each sporophyte individual bears either microsporangia or megasporangia, but not both.

Human and ecological impacts   Cycads harbor large numbers of symbiotic cyanobacteria in specialized, above-ground root structures. The cyanobacteria are photosynthetic and fix nitrogen. The nitrogen acts as an important nutrient for nearby plants as well as the cycads themselves. Cycads are popular landscaping plants in some parts of the world.

Cycas revoluta

20 cm

FIGURE 29.38 Cycads Resemble Palms but Are Not Closely Related to Them

Ginkgophyta (Ginkgoes)

Although ginkgoes have an extensive fossil record, just one species is alive today. Leaves from the ginkgo, or maidenhair, tree are virtually identical in size and shape to those observed in fossil ginkgoes that are 150 million years old (**Figure 29.39**).

Adaptations to land   Unlike most gymnosperms, the ginkgo is **deciduous**—meaning that it loses its leaves each autumn. This adaptation allows plants to be dormant during the winter or the dry season, when photosynthesis and growth are difficult.

Reproduction   Pollen is transported by wind. Sperm have flagella, however. Once pollen grains land near the female gametophyte and mature, the sperm cells leave the pollen grain and swim to the egg cells.

Life cycle   Sexes are separate—individuals are either male or female.

Human and ecological impacts   Although today's ginkgo trees are native to east Asia, they are planted widely as an ornamental all over the world. The inside of the seed is eaten as a delicacy in some countries.

(a) Fossil ginkgo
Ginkgo huttoni

(b) Living ginkgo
Ginkgo biloba

FIGURE 29.39 The Ginkgo Tree Is a "Living Fossil"

Coniferophyta (Conifers)

The conifers include the familiar pines, cedars, spruces, and junipers. The group is named for a unique reproductive structure called the **cone**, in which microsporangia and megasporangia are produced (**Figure 29.40**). The 600 species of conifers living today include the world's largest and longest-lived plants. One of the bristlecone pines native to southwestern North America is at least 4900 years old, and redwood trees growing along the Pacific Coast of North America can reach heights of up to 115 m (375 ft) and trunk diameters of over 11 m (36 ft).

Adaptations to land   All conifers are trees or large shrubs; many have narrow, needle-like leaves. Conifer leaves have a small amount of surface area, which is not optimal for capturing sunlight and performing photosynthesis. But because the small surface area reduces water loss from leaves, conifers grow in dry habitats or in cold environments where water is often frozen. Each needle lives from two to four years before being replaced.

Reproduction   Conifers are wind pollinated. As in all seed plants, the female gametophyte is retained on the parent. Thus, fertilization and seed development take place in the female cone. Depending on the species, conifer seeds are dispersed by wind or by seed-eating birds or mammals.

Life cycle   In most species, each individual has both male and female cones. Male cones bear pollen-producing microsporangia. Female cones contain megasporangia, which produce female gametophytes. In pines, pollination and fertilization are separated in time. Once a pollen grain lands on a female cone, it can take over a year for the male and female gametophytes to mature and for fertilization to take place. Seeds mature about 4 months after fertil-

ization occurs. Thus, seeds may be produced almost 18 months after the pollination event.

Human and ecological impacts   Conifers dominate forests that grow at high latitudes and high elevations. Their seeds are key food sources for a variety of birds, squirrels, and mice, and their wood is the basis of the building products and paper industries in many parts of the world. The paper in this book was made from the trunk of a conifer.

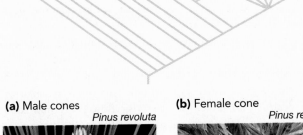

(a) Male cones
Pinus revoluta

(b) Female cone
Pinus revoluta

FIGURE 29.40 Male Cones Produce Microsporangia; Female Cones Produce Megasporangia

Anthophyta (Angiosperms)

The flowering plants, or angiosperms, are far and away the most species-rich lineage of land plants. Over 235,000 species have already been described. They range in size from *Lemna gibba*—a floating, aquatic species that is less than half a millimeter wide—to massive oak trees. Angiosperms thrive in desert to freshwater to rain forest environments and are found in virtually every habitat except the deep oceans. They are the most common and abundant plants in terrestrial environments other than northern and high-elevation coniferous forests.

The defining adaptation of angiosperms is the flower. **Flowers** are reproductive structures that hold either pollen-producing microsporangia or the megasporangia that produce megaspores and eggs, or both. In most cases, an individual flower contains both male and female sporangia surrounded by brightly colored **petals**, which are modified leaves. When a flower is first developing, the entire structure is enclosed by leaflike structures known as **sepals**, which protect the growing flower. Nectar-producing cells are often present at the base of the flower, and the color of petals helps to attract insects, birds, or bats that carry pollen from one flower to another. Some angiosperms are pollinated by wind, however. Wind-pollinated flowers lack both colorful petals and nectar-producing cells (**Figure 29.41**).

Adaptations to land In addition to flowers, angiosperms evolved vessels, the conducting cells that make water transport particularly efficient. Vessels are hollow tubes with empty ends, not unlike the pipes that conduct water in human habitations. Most angiosperms contain both tracheids and vessels.

Reproduction Unlike gymnosperms, angiosperms have a carpel, a structure within the flower that contains and encloses the ovule, which in turn encloses the female gametophyte. If the egg produced by the female gametophyte is fertilized, the ovule develops into a seed. The presence of enclosed ovules inspired the name angiosperm ("encased-seed"). When the ovary—the bottom portion of a carpel—matures, it forms fruit, which contains the seed or seeds. (The structures of flowers, seeds, and fruits are explored in more detail in Chapter 40.) In most cases, male gametophytes are carried to eggs by pollinators that are inadvertently dusted with pollen as they visit flowers to find food. Depending on the angiosperm species, self-fertilization may be common or absent. When fertilization does occur, it involves two sperm nuclei—one of which fuses with the egg to form the zygote, and a second that fuses with two nuclei in the female gametophyte to form a triploid ($3n$) nutritive tissue called **endosperm**. The involvement of two sperm nuclei is unique to angiosperms and is called **double fertilization**.

Life cycle Male and female gametophytes are small and contain a cell that divides by mitosis to form sperm or eggs. Like other seed plants, angiosperms retain their eggs on the parent plant but then release their embryos with a supply of nutrients in their seeds.

Human and ecological impacts It is almost impossible to overstate the importance of angiosperms to humans and other organisms. In most terrestrial habitats today, angiosperms supply the food that supports virtually every other species. For example, many insects eat flowering plants. Historically, the diversification of angiosperms correlated closely with the diversification of insects, which are by far the most species-rich lineage on the tree of life. It is not unusual for a single tropical tree to support dozens or even hundreds of insect species. Angiosperm seeds and fruits have also supplied the staple foods of virtually every human culture that has ever existed.

(a) Male flower

Acer negundo

(b) Female flower

Acer negundo

FIGURE 29.41 Wind-Pollinated Flowers Lack Colorful Petals and Nectar

ESSAY Genetic Diversity in Crop Plants

The spread of mechanized agriculture throughout the globe has produced some spectacular success stories. Due to increased use of fertilizers and new, high-yielding strains of wheat and rice, India became self-sufficient in grain production during the 1970s after decades of importing food. In India and many other countries, the use of high-yielding crop varieties has been instrumental in allowing food production to keep pace with rapid growth in the human population.

This so-called *green revolution* in agriculture has a downside, however. Because planting programs in numerous countries have relied on just a tiny number of source strains, the genetic diversity of crop systems is extremely low. In the United States, for example, virtually all soybeans that are grown trace their ancestry back to just 12 strains imported from northeast China. Similarly, most winter wheat being planted today is descended from two stocks that originated in Poland and Russia.

Low genetic diversity in crop species can lead to two problems: (1) an inability to improve strains further by selective breeding and (2) susceptibility to catastrophic diseases. Recall from Chapter 23 that for natural or artificial selection to occur, genetic variation must be present. Breeders find it difficult to improve crop populations with low genetic variability, because few alleles are present for selection to act on. Populations with low genetic variability are also more susceptible to disease, because disease resistance in plants depends on the presence of particular alleles. If a disease-causing agent evolves the ability to grow rapidly in a crop strain with low genetic variability, it is unlikely that any individuals of that strain will have alleles that allow them to fend off disease. The Irish potato famine of the mid-1800s (introduced in Chapter 28) was particularly devastating because most Irish farmers were planting the same potato variety, which happened to be particularly susceptible to the parasite that had infested the potato crops. To justify concerns about the lack of genetic diversity in crop plants, agricultural scientists also point to more recent events:

- In 1970 a corn blight wiped out 15 percent of the U.S. corn crop, sending prices skyrocketing and numerous farmers out of business.

- In 1980 over a million tons of sugarcane were destroyed by a single fungal disease in Cuba.

- Banana production in Central America was halved during the 1980s by a fungal disease called black sigatoka.

> *In response to scientists' concerns, governments ... have established seed banks to serve as repositories for genetic diversity.*

In response to scientists' concerns, governments from around the world have established seed banks to serve as repositories for genetic diversity. These are labs where seeds from dozens or hundreds of different crop plant strains are stored. Worldwide, there are now over 700 seed collections with some 2.5 million entries. The United States alone spends $20 million each year to maintain and expand its seed banks.

At present, however, the banks represent a largely untapped resource. Researchers have only begun to use the banks in breeding programs designed to improve the disease resistance or productivity of commercial strains. The wild relatives of species such as tomato and rice have been particularly important in recent breeding programs, however. Using sophisticated maps of the tomato genome to identify the locations of particular alleles, researchers have designed crosses between commercial strains of tomato and wild species that dramatically improved fruit characteristics. With traditional breeding techniques, annual improvement in tomato traits such as fruit yield and percentage of soluble solids (important in the production of tomato paste) has been averaging less than 1 percent. After carefully designed crosses with a wild relative, the same traits improved 29 percent and 22 percent, respectively.

With accelerating human populations and continued loss of agricultural lands to urbanization, this research becomes increasingly urgent. Seed banks are an important way to preserve the biodiversity that exists within species that are particularly important to humans.

CHAPTER REVIEW

Summary of Key Concepts

■ The green plants include both the green algae and the land plants. Biologists study them because humans—along with virtually every other terrestrial organism—rely on plants for food and a healthy environment.

Plants hold soil and water in place, build soil, moderate extreme temperatures and winds, and provide food. In addition to these ecosystem services, humans depend on plants for food, fiber, and fuel. This dependence became extreme after the domestication of plant species beginning about 10,000 years ago and with the development of agricultural economies beginning about 8000 years ago. Artificial selection techniques have produced huge increases in the yields and dramatic changes in the characteristics of domesticated plants. Lack of genetic diversity in crop populations is a concern that biologists are trying to address.

■ Land plants were the first multicellular organisms to live on land. A series of key adaptations allowed them to survive with their tissues constantly exposed to dry air. In terms of total mass, plants dominate today's terrestrial environments.

The land plants evolved from green algae and were the first multicellular organisms to grow in terrestrial habitats. The evolution of cuticle allowed their tissues to be exposed to air without dying. The evolution of pores provided breaks in the cuticle and facilitated gas exchange, with CO_2 diffusing into leaves and O_2 diffusing out. Later, the evolution of guard cells allowed plants to control the opening and closing of pores in a way that maximizes gas exchange and minimizes water loss.

Vascular tissue evolved in a series of steps, beginning with simple water-conducting cells and tissues such as those observed in today's mosses. True vascular tissue has cells that are dead at maturity and that have secondary cell walls reinforced with lignin. As a result, vascular tissue conducts water *and* provides structural support that makes erect growth possible. Erect growth is important because it reduces competition for light. Tracheids are water-conducting cells found in all vascular plants; in addition, angiosperms have water-conducting cells called vessels.

Web Tutorial 29.1   *Plant Evolution and the Phylogenetic Tree*

■ Once plants were able to grow on land, a sequence of important evolutionary changes made it possible for them to reproduce efficiently—even in extremely dry environments.

All land plants are embryophytes, meaning that eggs and embryos are retained on the parent plant. Consequently, the developing embryo can be nourished by its mother. Seed plant embryos are then dispersed from the parent plant to a new location, encased in a protective housing, and supplied with a store of nutrients.

All land plants have alternation of generations. Over the course of land plant evolution, the gametophyte phase became reduced in size and life span and the sporophyte phase became more prominent. In seed plants, male gametophytes are reduced to pollen grains and female gametophytes are reduced to tiny structures that produce an egg. The evolution of pollen was an important breakthrough in the history of life, because sperm no longer needed to swim to the egg—tiny male gametophytes could be transported through the air via wind or insects.

Questions

Content Review

1. Which of the following groups is monophyletic?
   a. nonvascular plants
   b. green algae
   c. green plants
   d. gymnosperms

2. What is the difference between tracheids and vessels?
   a. Tracheids are dead at maturity; vessels are alive and are filled with cytoplasm.
   b. Tracheids have a thickened secondary cell wall; vessels also have gaps that lack any cell wall.
   c. Only tracheids have a thick secondary cell wall containing lignin.
   d. Only vessels have a thick secondary cell wall containing lignin.

3. Which of the following statement(s) are true?
   a. Phylogenies estimated from both morphological traits and DNA sequences indicate that the ancestor of the land plants was a green alga in the lineage called Charales.
   b. "Bryophytes" is a name given to the land plant lineages that do not have vascular tissue.
   c. The horsetails and the ferns form a distinct clade, or lineage. They have vascular tissue but reproduce via spores, not seeds.

   d. According to the fossil record and phylogenetic analyses, angiosperms evolved before the gymnosperms. Angiosperms have a unique reproductive structure called the flower. They have vessels but not tracheids.

4. The appearance of cuticle and stomata correlated with what event in the evolution of land plants?
   a. the first erect growth forms
   b. the first woody tissues
   c. growth on land
   d. the evolution of the first water-conducting tissues

5. What do seeds contain?
   a. male gametophyte and nutritive tissue
   b. female gametophyte and nutritive tissue
   c. embryo and nutritive tissue
   d. mature sporophyte and nutritive tissue

6. What is a pollen grain?
   a. male gametophyte
   b. female gametophyte
   c. male sporophyte
   d. sperm

Conceptual Review

1. Why are biologists convinced that the spores, cuticle, and sporangia that appear in the fossil record beginning about 475 million years ago represent the remains of the first land plants?

2. The introduction to this chapter claims that the land plants were the first truly terrestrial organisms. But most biologists contend that saturated soils on the continents were undoubtedly teeming with bacteria, archaea, and protists long before land plants evolved. In light of this, what does the phrase "truly terrestrial" mean? After answering this question, list the physiological problems presented by growth on land. Beside each entry in this list, note the adaptation(s) that allowed land plants to overcome the problem.

3. In the experiments with orchid pollination reviewed in Section 29.3, why did the researchers put yarn around the flowers of control individuals?

4. Land plants may have reproductive structures that (1) protect gametes as they develop; (2) allow fertilization to occur in the absence of water, (3) nourish developing embryos, (4) provide a protective coat so that offspring can be dispersed away from the parent plant, and (5) provide nutritious tissue around seeds that facilitates dispersal by animals. Name each of these five structures, and state which land plant group or groups have each structure.

5. Explain the difference between homosporous and heterosporous plants. Some animal species are *hermaphroditic*—meaning that each individual contains both male and female reproductive organs—while other animal species have separate sexes. Which type of animal is most similar to a homosporous plant?

6. Using what you already know, compare and contrast gametangia and the retention of embryos with analogous structures and processes observed in mammalian reproduction.

Group Discussion Problems

1. What is the significance of the observation that some members of the Charales synthesize sporopollenin and lignin?

2. In many vascular plants, the end walls of tracheids form an extreme diagonal (see Figure 29.11a). Regarding the cell's ability to transport water, why does this make sense? (Hint: Consider how the cell's ratio of surface area to volume affects transport functions.)

3. Angiosperms such as grasses, oaks, and maples are wind pollinated. The ancestors of these subgroups were probably pollinated by insects, however. As an adaptive advantage, why might a species "revert" to wind pollination? Why is it logical to observe that wind-pollinated species usually grow in dense stands containing many individuals of the same species? Why is it logical to observe that in wind-pollinated deciduous trees, flowers form very early in spring—before leaves form?

4. You have been hired as a field assistant for a researcher interested in the evolution of flower shape in orchids. Each of the five species she is working on is pollinated by a different insect. Design an experiment to determine which parts of the five flower types are most important in attracting the pollinators. Assume that you can change the flower's color with a dye and that you can remove petals or nectar stores, add particular scents, add nectar by injection, or switch parts among species by cutting and gluing.

30 Fungi

KEY CONCEPTS

▦ Fungi are important in part because many species live in close association with land plants. They supply plants with key nutrients and decompose dead wood. They are the master recyclers of nutrients in terrestrial environments.

▦ All fungi make their living by absorbing nutrition from living or dead organisms. Fungi secrete enzymes so that digestion takes place outside their cells. Their morphology provides a large amount of surface area for efficient absorption.

▦ Many fungi have unusual life cycles. It is common for species to have a long-lived heterokaryotic stage, in which cells contain haploid nuclei from two different individuals. Although most species reproduce sexually, very few species produce eggs and sperm.

Amanita muscaria lives in association with living tree roots. The mushrooms you see are only reproductive structures—the vast majority of this individual exists underground. *Amanita muscaria* mushrooms contain molecules that can induce hallucinations in mammals. This species has been used in religious ceremonies by some native cultures but is considered highly toxic.

Three major lineages of large, multicellular eukaryotes occupy terrestrial environments: the land plants, the animals, and the fungi. When it comes to making a living, the species in these three groups use radically different strategies. Land plants make their own food through photosynthesis. Animals eat plants, protists, fungi, or each other. Fungi absorb their nutrition from other organisms—dead or alive.

Fungi that absorb nutrients from dead organisms are the world's most important decomposers. For example, although a few types of organisms are capable of digesting the cellulose in plant cell walls, fungi and a few species of bacteria are the only organisms capable of completely digesting both the lignin and cellulose that make up wood. Without fungi, Earth's surface would be piled so high with dead tree trunks and branches that there would be almost no room for animals to move or plants to grow.

Other fungi specialize in absorbing nutrients from living organisms. If fungi absorb these nutrients without providing any benefit in return, then they lower the fitness of their host organism and act as parasites. If you've ever had athlete's foot or a vaginal yeast infection, you've hosted a parasitic fungus.

The vast majority of fungi that live in association with other organisms benefit their hosts, however. In these cases, fungi are not parasites but **mutualists**. Consider that the roots of virtually every land plant in the world are colonized by an array of mutualistic fungi. In exchange for sugars that are synthesized by the host plant, the fungi provide the plant with key nutrients such as nitrogen and phosphorus. Without these nutrients, the host plants grow much more slowly or even starve. It is not possible to overstate the importance of these relationships between living land plants and the fungi that cover their roots. In the soils beneath every prairie, forest, and desert, an underground economy is flourishing. Plants are trading the sugar they manufacture for nitrogen or phosphorus atoms that are available from fungi. These plant-fungal associations are the world's most extensive bartering system. The soil around you is alive with an enormous network of fungi that are feeding the plants you see above ground.

In short, fungi are the master traders and recyclers in terrestrial ecosystems. Some fungi release nutrients from dead plants and animals; others transfer these nutrients to living plants. Because they recycle carbon, nitrogen, phosphorus, and other nutrients, fungi have a profound influence on the productivity and biodiversity of terrestrial ecosystems. In terms of nutrient cycling on the continents, fungi make the world go around.

30.1 Why Do Biologists Study Fungi?

Given their importance to life on land and their intricate relationships to other organisms, it's no surprise that fungi are fascinating to biologists. But there are important practical reasons for studying fungi as well. They nourish the plants that nourish us. They affect global warming, because they are critical to the carbon cycle on land. Unfortunately, a handful of species can cause debilitating diseases in humans and crop plants. Let's take a closer look at some of the ways that fungi affect human health and welfare.

Fungi Feed Land Plants

Fungi that live in close association with plant roots are said to be **mycorrhizal** ("fungal-root"). When biologists first discovered how extensive these fungal-plant associations are, they asked an obvious question: Does plant growth suffer if mycorrhizal fungi are absent? **Figure 30.1** shows a typical result from many experiments and provides a convincing answer. In this case, tree seedlings were grown in the presence and absence of the mycorrhizal fungi normally found on their roots. The photographs document that this tree species grows 3 to 4 times faster in the presence of its normal fungal associates than it does without them. For foresters, farmers, and ranchers, the presence of normal mycorrhizal fungi can mean the difference between profit and loss. Biologists study fungi in part because fungi are critical to the productivity of forests, croplands, and rangelands.

Fungi Speed the Carbon Cycle on Land

The introduction to this chapter claims that fungi are master decomposers and recyclers. To back up this claim, consider two particularly dramatic events documented in the fossil record. One episode was based on a lack of fungi, the other on an abundance of fungi:

1. Researchers who examine the fossils present in coal from the Carboniferous period find remarkably few fungi that are capable of degrading dead plant material. Fungi that make their living by digesting dead plant material are called **saprophytes** ("rotten-plants"). Noting the "dip" in

With normal mycorrhizal fungi

Without normal mycorrhizal fungi

FIGURE 30.1 Plants Grow Better in the Presence of Mycorrhizal Fungi
Typical experimental results when plants are grown with and without their normal mycorrhizal fungi.

the fossil record of saprophytic fungi, researchers hypothesized that their absence was responsible for the enormous buildup of dead plant material that occurred during the Carboniferous period. Recall from Chapter 29 that deposits of compressed, partially decayed plant material are called *peat* and that coal formed when peat was produced during the Carboniferous period was buried under other sediments and subjected to heat and pressure. Peatlands that exist today are highly acidic; so biologists hypothesize that fungi did not grow in the coal-forming swamps of the Carboniferous period, because the pH was too low. The message is that coal exists today because conditions were too acidic for fungi to do their job.

2. At the end of the Permian period, 250 million years ago, the greatest mass extinction in the history of life occurred. It is estimated that in substantially less than a million years, over 90 percent of multicellular species were wiped out. When the end-Permian extinction was first recognized, biologists thought it had affected primarily marine organisms. But then other researchers documented a huge "fungal spike" in rocks that formed in terrestrial

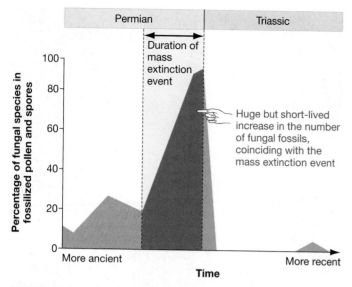

FIGURE 30.2 Fungal Fossils Spike during the End-Permian Extinction
The time interval documents changes in the abundance of fungi before, during, and after the mass extinction event that occurred at the boundary of the Permian and Triassic periods about 250 million years ago.

environments. As **Figure 30.2** shows, the spike is a dramatic but short lived increase in the number of fungal fossils that coincided with the mass extinction event. The explosion of fungal fossils was the first indication that land plants were also devastated during this interval. The hypothesis is that a massive die-off in trees and shrubs produced gigantic quantities of rotting wood and led to the explosion of fungal abundance documented in the spike.

Saprophytic fungi play a key role in today's terrestrial environments as well. To understand why, recall from Chapter 29 that cells in the vascular tissues of land plants have walls containing both lignin and cellulose. Wood forms when stems grow in girth by adding layers of vascular tissue. As mentioned in the introduction to this chapter, the vast majority of organisms that can completely digest wood are fungi. Fungi break down wood into sugars and other small organic compounds that can be used as food by fungi and other organisms.

Figure 30.3 illustrates the consequences of these facts by highlighting the role that fungi play as carbon atoms cycle through today's terrestrial environments. Note that there are two basic components of the **carbon cycle**: (1) the fixation of carbon by land plants—meaning that carbon in atmospheric CO_2 is converted to cellulose, lignin, and other complex organic compounds in the bodies of plants; and (2) the release of CO_2 from plants, animals, and fungi as the result of cellular respiration—meaning the oxidation of glucose and production of the ATP that sustains life. The fundamental point is that, for most carbon atoms, fungi connect the two components. If fungi had not evolved the ability to digest lignin and

cellulose soon after land plants evolved the ability to make these compounds, carbon atoms would have been sequestered in wood for millennia instead of being rapidly recycled into glucose molecules and CO_2. Terrestrial environments would be radically different than they are today and probably much less productive. On land, fungi make the carbon cycle turn much more rapidly than it would without fungi. The nutrients that fungi release feed a host of other organisms.

Fungi Have Important Economic Impacts

In humans, parasitic fungi cause athlete's foot, vaginitis, diaper rash, ringworm, pneumonia, and thrush, among other miseries. But even though these maladies can be serious, in reality only about 30 (out of some 80,000) species of fungi regularly cause illness in humans. Compared with the frequency of diseases caused by bacteria, viruses, and protists, the incidence of fungal infections in humans is low. In addition, soil-dwelling fungi have been the source of many of the most important antibiotics currently being prescribed against bacterial infections (see Chapter 27). On balance, fungi have been much more helpful than harmful in human and veterinary medicine.

The major destructive impact that fungi have on people is through our food supply. Fungi known as rusts, smuts, mildews, wilts, and blights cause annual crop losses computed in the billions of dollars. These fungi are particularly troublesome in wheat, corn, barley, and other grain crops (**Figure 30.4a**). Saprophytic fungi are also responsible for enormous losses due to spoilage—particularly for fruit and vegetable growers (**Figure 30.4b**). But fungi have important positive impacts on the human food supply as well. Domesticated by humans for thousands of years, fungi are essential to the manufacture of bread, soy sauce, cheese, beer, wine, whiskey, and other products. Enzymes derived from fungi are used to improve the characteristics of foods ranging from fruit juice and candy to meat.

In nature, recent epidemics caused by fungi have killed 4 billion chestnut trees and tens of millions of American elm trees in North America. The fungal species responsible for these epidemics were accidentally imported on species of chestnut and elm native to other regions of the world. When the fungi arrived in North America, switched hosts, and began growing in chestnuts and elms native to North America, the results were catastrophic. The local chestnut and elm populations had virtually no genetic resistance to the pathogens and quickly succumbed. The epidemics radically altered the composition of upland and floodplain forests in the eastern United States. Before these fungal epidemics occurred, chestnuts and elms dominated these habitats.

Fungi clearly affect a wide range of species, including humans. How do biologists go about studying them? More specifically, what tools are helping researchers understand the diversity of fungi?

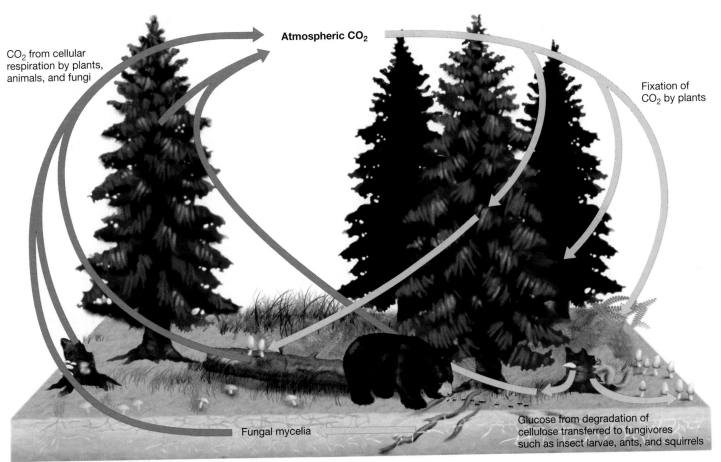

FIGURE 30.3 Fungi Speed Up the Cycling of Carbon Atoms on Land
Carbon atoms cycle through terrestrial ecosystems. If fungi could not degrade lignin to CO_2 and cellulose to glucose, most carbon would eventually be tied up in indigestible woody tissues. As a result, the cycle would slow dramatically.
EXERCISE Draw an X through the arrows that would not exist if fungi could not digest lignin and cellulose.

(a) Parasitic fungi infect corn and other crop plants.

(b) Saprophytic fungi rot fruits and vegetables.

FIGURE 30.4 Fungi Cause Problems with Crop Production and Storage
(a) A wide variety of grain crops are parasitized by fungi. Corn smut is a serious disease in sweet corn, although in Mexico the smut fungus is eaten as a delicacy. **(b)** Fungi decompose fruits and vegetables as well as leaves and tree trunks.

30.2   How Do Biologists Study Fungi?

To date, about 80,000 species of fungi have been described and named; about 1000 more are discovered each year. But the fungi are so poorly studied that the known species are widely regarded as a tiny fraction of the actual total. To predict the actual number of species alive today, David Hawksworth looked at the ratio of vascular plant species to fungal species in the British Isles—the area where the two groups are the most thoroughly studied. According to Hawksworth's analysis, there is an average of six species of fungus for every species of vascular plant on these islands. If this ratio holds worldwide, then the total of 275,000 vascular plant species implies that there are 1.65 million species of fungi.

Although this estimate sounds large, recent data on fungal diversity suggest that it may be an underestimate. Consider what researchers found recently when they analyzed fungi growing on Barro Colorado Island, Panama: Living on the healthy leaves of just two tropical tree species were a total of 418 distinct morphospecies of fungi. (Recall from Chapter 25 that morphospecies are distinguished from each other by some aspect of morphology.) Because over 300 species of trees and shrubs grow on Barro Colorado, the data suggest that tens of thousands of fungi may be native to this island alone. If further work on fungal diversity in the tropics supports these conclusions, there may turn out to be many millions of fungal species.

This viewpoint of fungal diversity was reinforced by a recent analysis of the fungi living in conjunction with the roots of a single species of grass native to Eurasia. In this study, researchers used the direct sequencing approach, introduced in Chapter 27, to analyze the gene that codes for the RNA molecule in the small subunit of fungal ribosomes. The data showed that a total of 49 phylogenetic species were living in conjunction with the grass roots. (The phylogenetic species concept was introduced in Chapter 25.) Most of the species had never before been described, and several represented completely new lineages of fungi. Biologists are only beginning to realize the extent of species diversity in fungi.

Let's consider how biologists are working to make sense of all this diversity, beginning with an overview of fungal morphology.

Analyzing Morphological Traits

Compared with animals and land plants, fungi have very simple bodies. Only two growth forms occur among them:

1. Single-celled forms called **yeasts** (Figure 30.5a), and

2. Multicellular, filamentous structures called **mycelia** (singular: **mycelium**; Figure 30.5b).

Many species of fungus grow as either a yeast or a mycelium, but some regularly adopt both growth forms.

Because yeasts are small and easy to manipulate in the laboratory, they have been extremely important model organisms in cell biology and genetics (**Box 30.1**). But because most fungi form mycelia and because this body type is so fundamental to the absorptive mode of life, most studies of fungal morphology have focused on them. Let's take a closer look at the structure and function of a fungal mycelium.

The Nature of the Fungal Mycelium   If food sources are plentiful, mycelia can be long lived and grow to be extremely large. Researchers recently discovered an individual mycelium growing across 1300 acres (6.5 km^2) in Oregon. This is an area substantially larger than most college campuses. The biologists estimated the individual's weight at hundreds of tons and its age at thousands of years, making it one of the largest and oldest organisms known.

Although most mycelia are much smaller and shorter lived than the individual in Oregon, all mycelia are dynamic. Mycelia constantly grow in the direction of food sources and die back in areas where food is running out. The body shape of a fungus can change almost continuously throughout its life.

The individual filaments that make up a mycelium are called **hyphae** (singular: **hypha**). Most hyphae are haploid or **heterokaryotic** ("different kernel"), meaning each cell contains two haploid nuclei—one from each parent. Hyphae are also similar to one another in structure. As **Figure 30.6a** shows, they are long, narrow filaments that branch frequently. In most fungi, each filament is broken into cell-like compartments by cross-walls called **septa** (singular: **septum**; Figure 30.6b). Unlike the cell walls and membranes of plants and animals, septa do not close off segments of hyphae completely. Instead, gaps called **pores** enable a wide variety of materials, even organelles and nuclei, to flow from one compartment to the next. Septa may have single large openings or a series of small gaps that give the septum a sieve-like appearance. Because nutrients, mitochondria, and even genes can flow though the entire mycelium—at least to a degree—the fungal mycelium is intermediate between a multicellular land plant or animal and an enormous single-celled organism. Some fungal species are even **coenocytic**—they lack septa and therefore have cytoplasm and many nuclei that are continuous throughout the mycelium.

Perhaps the most important aspect of mycelia and hyphae, however, is their shape. Because hyphae are tubes and because mycelia are composed of complex, branching networks, the body

(a) Single-celled fungi are called yeasts.

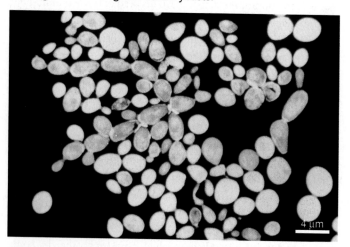

4 μm

(b) Multicellular fungi have weblike bodies called mycelia.

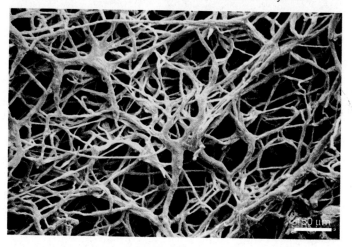

150 μm

FIGURE 30.5 Fungi Have Just Two Growth Forms
Fungi grow **(a)** as single-celled yeasts and/or **(b)** as multicellular mycelia made up of long, thin, highly branched filaments.

BOX 30.1   A Model Organism: *Saccharomyces cerevisiae*

The yeast *Saccharomyces cerevisiae* (see Figure 30.5a) is a favorite experimental subject for researchers interested in the basic cell biology and molecular genetics of eukaryotes. Because it is unicellular and easy to culture and manipulate in the lab, *S. cerevisiae* has become the organism of choice for research on control of the cell cycle and regulation of gene expression in eukaryotes. The morphology of this organism is so simple that it provides an example of a "pure" eukaryotic cell type—one that is suitable for experiments on how cell division occurs and how particular genes are turned on and off. Just as *Escherichia coli* serves as the model bacterial cell, *Saccharomyces*

serves as the model eukaryotic cell. For example, research has confirmed that several genes that control cell division and DNA repair in yeast have homologs in humans that, when mutated, contribute to cancer. Strains of yeast that carry these mutations are now being used to test drugs that might be effective against cancer. *Saccharomyces cerevisiae* was also the first eukaryote whose genome was completely sequenced.

Saccharomyces cerevisiae is a member of the Ascomycota. In nature it lives on the skins of wild grapes and other fruits, which it uses as a source of food. Thousands of years ago, however, humans learned how to use the yeast

Saccharomyces in baking and brewing. It was the first fungus to be cultivated, and domesticated strains of yeast are still vital to food and beverage production. *Saccharomyces* is useful in these industries because it ferments sugars when it grows in the absence of oxygen. The carbon dioxide that forms as a by-product of fermentation by these cells makes bread rise and gives beer and champagne their fizz. Several *Saccharomyces* species are used in brewing, winemaking, and distilling; the ethanol that they produce as a waste product during fermentation acts as the psychoactive ingredient in beer, wine, and spirits.

of a fungus has the highest surface-area-to-volume ratio possible in a multicellular organism. To drive this point home, consider that the hyphae found in any fist-sized ball of rich soil typically have a surface area equivalent to half a page of this book. This surface area is important because it makes absorption extremely efficient, and fungi make their living via absorption.

The extraordinarily high surface area in a mycelium has a downside, however: Fungi are prone to drying out, because the amount of water that evaporates from an organism is a function of its surface area. Due to the high surface area of mycelia,

fungi are most abundant in moist habitats. Fungi often endure dry conditions in the form of tough, watertight spores.

To summarize, mycelia are an adaptation to the absorptive lifestyle of fungi. Thus, it is logical to observe that reproductive organs—not feeding structures—are the only thick, fleshy structures that fungi produce. Mushrooms, puffballs, and other dense, multicellular structures that arise from mycelia do not absorb food. Instead, they function only in reproduction. Typically they are the only part of a fungus that is directly exposed to the air, where drying is a problem. The

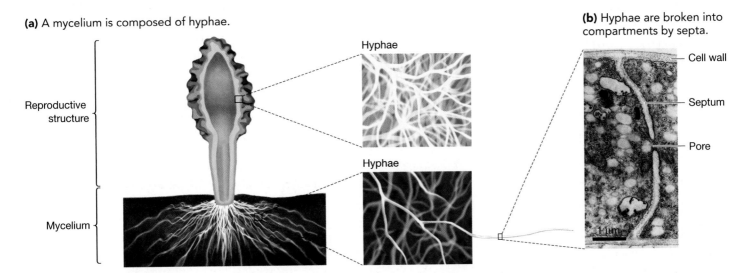

(a) A mycelium is composed of hyphae.

Hyphae

Hyphae

(b) Hyphae are broken into compartments by septa.

Cell wall

Septum

Pore

1 μm

Reproductive structure

Mycelium

FIGURE 30.6 Multicellular Fungi Have Unusual Bodies
(a) The feeding structure of a fungus is a mycelium, which is made up of hyphae. In some species, hyphae come together to form multicellular structures such as mushrooms, brackets, or morels that emerge from the ground. **(b)** Hyphae are often divided into cell-like compartments by partitions called septa, which are broken by pores. As a result, the cytoplasm of different compartments is continuous.

mass of filaments on the inside of mushrooms and other large reproductive structures is protected from drying by the densely packed hyphae forming the surface. However, few species of fungi make the reproductive structures called mushrooms.

Reproductive Structures   When biologists study diversity among land plants, protists, or animals, they normally begin by comparing the morphologies of various species. Because fungal mycelia are so simple and are so similar among species, researchers have focused on the distinctive morphological structures that fungi produce during sexual reproduction. On the basis of these structures, fungi fall into four major groups:

1. The **Chytridiomycota (chytrids)** live primarily in water, and they are the only fungi that have motile cells. The spores that chytrids produce during asexual reproduction have flagella, as do the gametes produced by members of this group during sexual reproduction. These swimming cells are reproductive structures that distinguish chytrids from other fungi (**Figure 30.7a**). Structurally, chytrid flagella are similar to the flagella in the sperm cells of animals.

2. The hyphae of **Zygomycota** are haploid and come in several **mating types**. Instead of having morphologically distinct males and females that produce sperm and eggs, hyphae of different mating types look identical but will not combine unless the individuals have different alleles of one or more genes involved in mating. If chemical messengers released by two hyphae indicate that they are of different mating types, the individuals may become yoked together as shown in **Figure 30.7b**. (The Greek root *zygos* means to be yoked together like oxen. Translated literally, *Zygomycota* means "yoked-together fungi.") Cells from the yoked hyphae fuse to form a spore-producing structure called a **zygosporangium**. Yoked hyphae that form a zygosporangium are the reproductive structure unique to this group.

3. Mushrooms, bracket fungi, and puffballs are among the complex reproductive structures produced by members of the **Basidiomycota**, or **club fungi**. Inside these structures, specialized cells called **basidia** ("little-pedestals") form at the ends of hyphae and produce spores (**Figure 30.7c**). Only members of the Basidiomycota produce basidia.

4. Members of the **Ascomycota**, also called **sac fungi**, produce complex reproductive structures—the largest of which are often cup shaped. The tips of hyphae inside these structures produce distinctive sac-like cells called **asci** (singular: **ascus**; **Figure 30.7d**). An ascus is a spore-producing structure found only in Ascomycota.

In sum, morphological studies allowed biologists to describe and interpret the growth habit of mycelia as an adaptation that makes absorption extremely efficient. Careful analyses of morphological features also allowed researchers to identify four major lineages of fungi, based on the nature of their reproductive structures. How are members of these four major groups related? And how are fungi related to other eukaryotes?

Evaluating Molecular Phylogenies

Biologists who observed that fungi had four distinct types of reproductive structures hypothesized that each type of structure identified a monophyletic group. As a result, the Chytridiomycota, Zygomycota, Ascomycota, and Basidiomycota have traditionally been recognized as separate phyla. Based on the assumption that fungi evolved from aquatic protists, researchers hypothesized that chytrids were the most basal group. As **Figure 30.8a** shows, it was also hypothesized that ascomycetes and basidiomycetes—with their more-complex reproductive structures—were the most recent to evolve.

What Are the Relationships among Major Fungal Groups?   To test the hypotheses that chytrids were the most basal group and that the four phyla of fungi are monophyletic, biologists have sequenced the gene that codes for the RNA found in the ribosome's small subunit from an array of fungal species. The researchers used these data to estimate the phylogeny of fungi. The results supported the hypothesis of monophyly for the ascomycetes and basidiomycetes, but conflicted with the claim of monophyly for chytrids and zygomycetes. Although many aspects of the tree in **Figure 30.8b** are still uncertain, several important conclusions can be drawn:

- It is likely that a lineage of chytrids forms the most basal group of fungi. This result is consistent with the hypothesis

(a) Chytridiomycota make swimming gametes and spores.

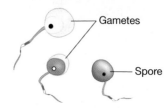

(b) Zygomycota hyphae yoke together and form zygotes.

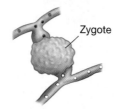

(c) Basidiomycota form spores on basidia (little pedestals).

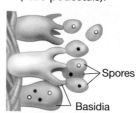

(d) Ascomycota form spores in asci (sacs).

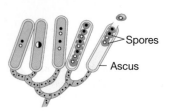

FIGURE 30.7 Four Distinct Reproductive Structures Are Observed in Fungi

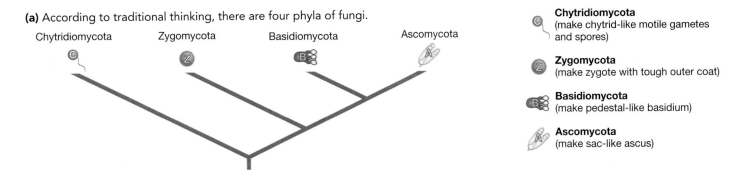

(a) According to traditional thinking, there are four phyla of fungi.

Chytridiomycota   Zygomycota   Basidiomycota   Ascomycota

Chytridiomycota
(make chytrid-like motile gametes and spores)

Zygomycota
(make zygote with tough outer coat)

Basidiomycota
(make pedestal-like basidium)

Ascomycota
(make sac-like ascus)

(b) DNA sequence data have revealed that Glomeromycota, Basidiomycota, and Ascomycota are monophyletic.

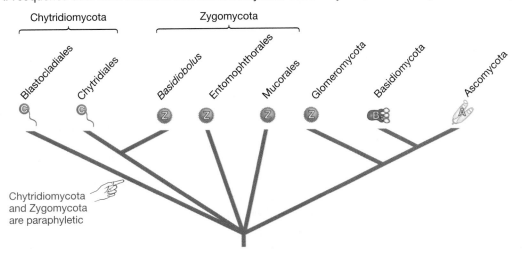

Chytridiomycota   Zygomycota

Blastocladiales   Chytridiales   *Basidiobolus*   Entomophthorales   Mucorales   Glomeromycota   Basidiomycota   Ascomycota

Chytridiomycota and Zygomycota are paraphyletic

FIGURE 30.8 A Phylogeny of the Fungi
(a) A traditional phylogenetic tree based on the four reproductive structures observed in fungi. **(b)** An updated phylogenetic tree based on comparisons of morphological traits and analyses of DNA sequence data. The key indicates the types of sexual reproductive structures in each major lineage.

that fungi evolved from aquatic ancestors. It also suggests that fungi made the transition to land fairly early in their evolution. But according to the data on hand, it is still not clear whether fungi moved from water to land just once or multiple times.

- The Chytridiomycota and Zygomycota appear to be paraphyletic. That is, these groups do not include all descendants of a single common ancestor. Instead, fungal groups with swimming gametes and yoked hyphae are interspersed throughout the basal parts of the tree. If the relationships implied by the molecular data are correct, then either swimming gametes and yolked hyphae evolved more than once, or both were present in a common ancestor but then were lost in certain lineages.

- As predicted by the morphological analyses, the Basidiomycota and Ascomycota are monophyletic and appeared late during the diversification of fungi.

The phylogenetic tree of Fungi is a work in progress. Efforts to understand the relationships of subgroups within the Basidiomycota and Ascomycota are in their infancy, and researchers are still struggling to understand the relationships among the

dozens of subgroups of Chytridiomycota and Zygomycota. Fortunately, a coalition of researchers has recently formed to undertake a series of new and more extensive analyses of DNA sequence data, with the goal of working out the details of the fungal phylogeny. The collaboration is important, because a comprehensive phylogeny of the Fungi should clarify exactly how the masters of the absorptive lifestyle diversified and how the diversification of fungi relates to the diversification of land plants.

In the meantime, it is interesting to note that parasitic, saprophytic, and mutualistic species exist within both the Ascomycetes and the Basidiomycetes. This observation supports the hypothesis that the entire range of fungal lifestyles evolved independently in these two lineages. This conclusion is similar to the situation in protists. Recall from Chapter 28 that species that photosynthesize, ingest food, or absorb food are found within each of the eight major protist lineages and that each of these eight monophyletic groups is distinguished by an aspect of cell structure. In both fungi and protists, an array of feeding strategies evolved in lineages that are distinguished by some fundamental morphological feature.

Where Do Fungi Fit on the Tree of Life?   Although understanding the relationships among the four major groups of fungi and their subgroups has been difficult, researchers have made rapid progress recently on the question of where fungi fit on the overall tree of life. Based on analyses of similarities and differences in the DNA sequences of a number of genes, researchers are increasingly confident that the phylogenetic tree in **Figure 30.9** is an accurate depiction of how fungi relate to other eukaryotes. The tree supports two conclusions about fungi:

1. Their closest living relatives are unusual, single-celled parasites called **microsporidians**. This was an important result, because microsporidians cause serious disease in bee colonies, silkworm colonies, and people with AIDS. Based on the close relationship between fungi and microsporidians, researchers are testing the hypothesis that **fungicides**—meaning molecules that are lethal to fungi—may prove to be effective in combating microsporidian infections as well.

2. Fungi are much more closely related to animals than they are to land plants. In addition to the DNA sequence data, several morphological traits link animals and fungi. For example, most animals and fungi synthesize the tough structural material called chitin (see Chapter 7). **Chitin** is a prominent component of the cell walls of fungi. Also, the flagella that develop in chytrid spores and in chytrid gametes are very similar to those observed in animals: The flagella are single, are located at the back of the body, and move in a whiplash manner. Further, both animals and fungi store food by synthesizing the polysaccharide glycogen. Green plants, in contrast, synthesize starch as their storage product.

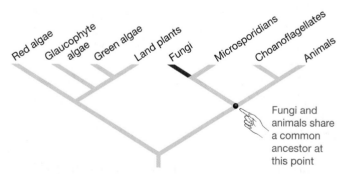

FIGURE 30.9 Fungi Are Closely Related to Animals—More So than to Land Plants
Phylogeny showing the evolutionary relationships among the green plants, animals, fungi, and some groups of protists. (Choanoflagellates are solitary or colonial protists found in freshwater; they are introduced in Chapter 31.) EXERCISE Using the relationships shown here, go back to Figure 30.8 and add microsporidians to the tree in part (b).

Experimental Studies of Mutualism

It is estimated that 90 percent of land plants live in close physical association with certain species of fungi. Stated another way, fungi and land plants often have a **symbiotic** ("together-living") relationship. Although some species of fungi live in association with an array of different land plant species, most documented fungal-plant associations are quite specific. Most fungi live in only a particular type of tissue, in one plant species. Based on this observation, biologists hypothesize that the evolution of symbiotic associations has played a large role in the diversification of fungi. The question is, are these symbiotic relationships *mutualistic*, meaning they benefit both species; *parasitic*, meaning one species benefits at the expense of the other; or *commensal*, meaning one species benefits while the other is unaffected?

Researchers frequently begin studies of plant-fungal associations simply by cataloging the species of fungi present in or on a particular plant tissue. This can be done by analyzing plant tissues directly via light microscopy or electron microscopy. Although growing fungi in the laboratory can be difficult, researchers may also place small samples of plant tissue on a culture plate in the laboratory and then analyze the fungal hyphae that grow out from the tissue into the nutrient medium on the plate.

To understand the nature of the association, biologists turn to experimental approaches. Recall from Section 30.1 that, in early experiments, researchers grew trees or other types of land plants with or without their normal mycorrhizal fungi. These types of experiments are still done, with experimental treatments lacking fungal symbionts created by sterilizing soils with heat or by treating soils and seeds with fungicides. Presence-absence experiments have generally shown that plants grow much larger with their normal symbiotic fungi than they do without. Similarly, fungi that are typically symbiotic are usually unable to grow and reproduce if their regular host plant is absent.

To explore the nature of fungi-plant symbioses in more detail, researchers have used isotopes as tracers for specific elements (**Figure 30.10**). For example, to test the hypothesis that fungi obtain food in the form of carbon-containing compounds from their plant associates, biologists have introduced radioactively labeled carbon dioxide into the air surrounding plants that do or do not contain symbiotic fungi. The experimental plants are usually grown in pots inside laboratory growth chambers. The labeled CO_2 molecules are incorporated into the sugars produced during photosynthesis, and the location of the radioactive atoms can then be followed over time by means of a device that detects radioactivity. If plants feed their fungal symbionts, then labeled carbon compounds should be transferred from the plant to the fungi.

To test the hypothesis that plants are receiving nutrients from their symbiotic fungi in return for sugars, researchers have added radioactive phosphorus atoms or the heavy isotope of nitrogen (^{15}N) to potted plants that do or do not contain symbiotic fungi. If fungi facilitate the transfer of nutrients from soil to plants, then plants grown in the presence of their symbiotic

Question: Are mycorrhizal fungi mutualistic?

Hypothesis: Host plants provide mycorrhizal fungi with sugars and other photosynthetic products. Mycorrhizal fungi provide host plants with phosphorus or nitrogen from the soil.

Null hypothesis: No exchange of food or nutrients occurs between plants and mycorrhizal fungi. The relationship is not mutualistic.

Experimental setup:

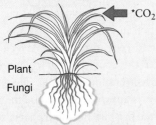 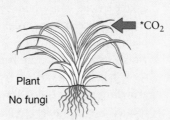 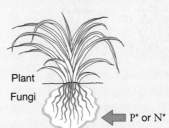 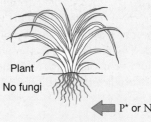

Labeled carbon treatment:
Leaves are exposed to radioactive CO_2. Mycorrhizal fungi present.

Labeled carbon control:
Leaves are exposed to radioactive CO_2. Mycorrhizal fungi absent.

Labeled P or N treatment:
Roots are exposed to radioactive P or heavy isotope of N. Mycorrhizal fungi present.

Labeled P or N control:
Roots are exposed to radioactive P or heavy isotope of N. Mycorrhizal fungi absent.

Prediction for labeled carbon: A large percentage of the labeled carbon taken up by the plant will be transferred to mycorrhizal fungi. In the control, little labeled carbon will be present in the soil surrounding the roots.

Prediction of null hypothesis, labeled carbon: There will be no difference between amounts of labeled carbon in mycorrhizal fungi and in soil when fungi are absent.

Prediction for labeled P or N: A large percentage of the labeled P or N taken up by the fungi will be transferred to the plant. In the control, little or no labeled P or N will be taken up by the plant.

Prediction of null hypothesis, labeled P or N: There will be no difference between amounts of labeled P or N found in plant in presence or absence of fungi.

Results:

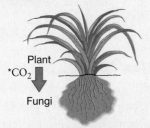  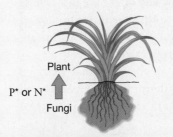 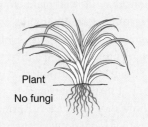

Labeled carbon treatment:
Up to 20% of labeled carbon taken up by plant is transferred to mycorrhizal fungus.

Labeled carbon control:
Little to no labeled carbon is found in soil surrounding plant roots.

Labeled P or N treatment:
Large amount of labeled P or N is found in host plant.

Labeled P or N control:
Little labeled P or N is found in host plant.

Conclusion: The relationship between plants and mycorrhizal fungi is mutualistic. Plants provide mycorrhizal fungi with carbohydrates. Mycorrhizal fungi supply host plants with nutrients.

FIGURE 30.10 Experimental Evidence That Mycorrhizal Fungi and Plants Are Mutualistic
Nutrient transfer experiments indicate that sugars flow from plants to mycorrhizal fungi and that key nutrients flow from mycorrhizal fungi to plants.

fungi should receive much more of the radioactive phosphorus or heavy nitrogen than do plants grown in the absence of fungi.

As the "Results" section of Figure 30.10 shows, experiments with isotopes used as tracers have shown that sugars and other carbon-containing compounds produced by plants via photosynthesis are transferred to their fungal symbionts. In some cases, as much as 20 percent of the sugars produced by a plant end up in

their symbiotic fungi. In exchange, the symbiotic fungi facilitate the transfer of phosphorus or nitrogen or both from soil to the plant. Because phosphorus and nitrogen are in extremely short supply in most environments, the nutrients supplied by symbiotic fungi are critical to the success of the plant. In this way, studies with isotopes have supported the hypothesis that most relationships between fungi and land plants are mutually beneficial.

30.3 What Themes Occur in the Diversification of Fungi?

Why are there so many different species of fungi? This question is particularly puzzling given that fungi share a common attribute: They all make their living by absorbing food directly from their surroundings. In contrast to the diversity of food-getting strategies observed in bacteria, archaea, and protists, all fungi make their living in the same basic way. Fungi are similar to plants in this respect. (Recall from Chapter 29 that virtually all land plants make their own food via photosynthesis.)

Chapter 29 showed that the diversification of land plants was driven not by novel ways of obtaining food, but by adaptations that allowed plants to grow and reproduce in a diverse array of terrestrial habitats. What drove the diversification of fungi? The answer is the evolution of novel methods for absorbing nutrients from a diverse array of food sources.

This section introduces a few of the ways that fungi go about absorbing nutrients from different food sources, as well as how they produce offspring. Let's explore the diversity of ways that fungi do what they do.

Fungi Participate in Several Types of Mutualisms

Not long after associations between fungi and the roots of land plants were discovered and shown to be mutualistic, researchers found that there are two distinct types of mycorrhizal interactions. The two types of mycorrhizae involve different phyla, and each type has a diagnostic morphology, geographic distribution, and function. But mycorrhizae are not the only type of symbiotic fungi found in plants. Researchers have also become interested in fungi that live in close association with the aboveground tissues of land plants—their leaves and stems. Fungi that live in the aboveground parts of plants are said to be **endophytic** ("inside-plants"). Recent research has shown that endophytic fungi are much more common and diverse than previously suspected. Further, data indicate that at least some species of endophytes are mutualistic.

These results support the general realization that most plants are covered with fungi—from the tips of their branches to the base of their roots. Many or even most plants are in-volved in several distinct types of mutualistic relationships with fungi. Let's take a closer look.

Ectomycorrhizal Fungi The type of mycorrhizal fungus illustrated in **Figure 30.11a** is found on virtually all of the tree species that grow in the temperate latitudes of both hemispheres, as well as tree species in northern coniferous forests. In this type of association, mycelia form a dense network of hyphae that covers a plant's roots. As the cross section in Figure 30.11a shows, individual hyphae penetrate between cells in the outer layer of the root, but hyphae do not enter the root cells. Fungi with this growth form are called **ectomycorrhizal fungi (EMF)**. The Greek root *ecto*, which refers to *outer*, is appropriate because the fungi form an outer sheath that is often 0.1 mm thick. Hyphae also extend out from the sheath-like portion of the mycelium into the soil. Most EMF are basidiomycetes; a few are ascomycetes.

How and why do these trees and fungi interact? In the cold, northern habitats where EMF are abundant, the growing season is so short that the decomposition of needles, leaves, twigs, and trunks can be extremely sluggish. As a result, nitrogen atoms tend to remain tied up in amino acids and nucleic acids inside dead tissues instead of being available in the soil. Fortunately, the hyphae of EMF penetrate decaying material and release enzymes (peptidases) that cleave the peptide bonds between amino acids in the dead tissues. The nitrogen released by this reaction is absorbed by the hyphae and then transported to the spaces between the root cells of trees, where it can be absorbed by the plant. EMF are also able to acquire phosphate ions that are bound to soil particles and transfer the ions to host plants. In return, the fungi receive sugars and other complex carbon compounds from the tree.

Researchers have found that, when northern species such as birch tree seedlings are grown with and without their normal EMF in pots filled with forest soil, only the seedlings with EMF are able to acquire significant quantities of nitrogen and phosphorus. Inspired by such data, a biologist has referred to EMF as the "dominant nutrient-gathering organs in most temperate forest ecosystems." The hyphae of EMF are like an army of miners that discover, excavate, and deliver precious nuggets of nitrogen to the trees of northern forests.

Arbuscular Mycorrhizal Fungi (AMF) In contrast to the hyphae of EMF, the hyphae of **arbuscular mycorrhizal fungi (AMF)** grow *into* the cells of root tissue. The name *arbuscular* ("little-tree") was inspired by the bushy, highly branched hyphae, shown in **Figure 30.11b**, that form inside root cells. AMF are also called *endomycorrhizal fungi*, because they penetrate the interior of root cell walls, or *vesicular-arbuscular mycorrhizae* (VAM or VM), because the hyphae of some species form large, balloon-like vesicles inside root cells.

The key point is that the hyphae of AMF penetrate the cell wall and contact the plasma membrane of root cells directly. The highly branched hyphae inside the plant cell wall are thought to be an adaptation increasing the surface area available

(a) Ectomycorrhizal fungi (EMF) form sheaths around roots and penetrate between root cells.

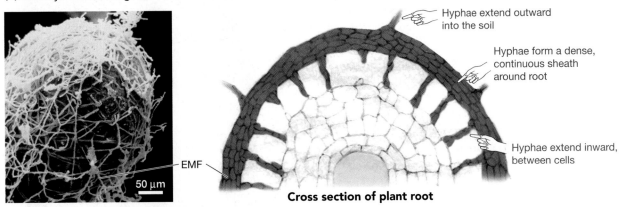

Hyphae extend outward into the soil

Hyphae form a dense, continuous sheath around root

Hyphae extend inward, between cells

EMF

50 μm

Cross section of plant root

(b) Arbuscular mycorrhizal fungi (AMF) contact the plasma membranes of root cells.

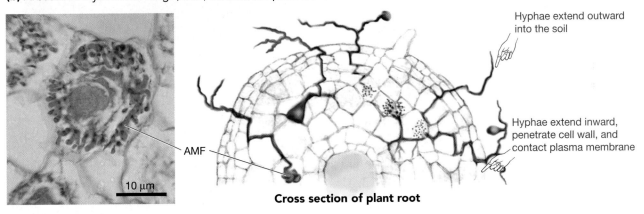

Hyphae extend outward into the soil

Hyphae extend inward, penetrate cell wall, and contact plasma membrane

AMF

10 μm

Cross section of plant root

FIGURE 30.11 Mutualistic Fungi Interact with the Roots of Plants in Two Distinct Ways
(a) Ectomycorrhizal fungi (EMF) form a dense network around the roots of plants. Their hyphae penetrate the intercellular spaces of the root but do not enter the root cells. **(b)** The hyphae of arbuscular mycorrhizal fungi (AMF) penetrate the walls of root cells, where they branch into bushy structures or balloon-like vesicles that contact the root cell's plasma membrane. EXERCISE In the cross sections, add arrows and labels showing the direction of movement of N (nitrogen-containing compounds), P (phosphorus-containing compounds), and C (carbon-containing compounds).

for exchange of molecules via diffusion between the fungus and its host. However, AMF do not form a tight sheath around roots, as do EMF. Instead, they form a pipeline that extends from inside plant cells in the root to the soil well beyond the root.

AMF are zygomycetes, and most belong to the lineage called Glomeromycota (see Figure 30.8b). They are found in a whopping 80 percent of all land plant species and are particularly common in grasslands and in the forests of warm or tropical habitats. Just as EMF are the dominant type of mycorrhizal association in cool, northern environments, AMF are the dominant type in grasslands and in the tropics.

In addition to being extremely common, AMF are ancient. Researchers have found AMF in fossilized root cells that are 400 million years old. This discovery confirms that mycorrhizal associations existed in the most ancient of all land plants. It also supports the hypothesis that plants and fungi colonized terrestrial environments together—meaning that fungi have been nourishing land plants since land plants first evolved.

What do AMF do? In the grasslands and tropical forests where AMF flourish, plant tissues decompose quickly because the growing season is long and warm. As a result, nitrogen is often readily available to plants. In these habitats, phosphorus—not nitrogen—is usually the nutrient that is in greatest demand by plants. Based on these observations, biologists hypothesized that AMF may transfer phosphorus atoms from the soil to the host plant. In most soils, phosphate atoms are found in negatively charged phosphate ions that cling tightly to mineral particles. Experiments with radioactive atoms confirmed the phosphate-transfer hypothesis by showing that, while AMF supply host plants with phosphorus, host plants supply fungi with reduced carbon. EMF mine nitrogen and some phosphorus in the north; AMF mine phosphorus in the south.

Are Endophytes Mutualists? Although endophytic fungi are relatively new to science, they are turning out to be both extremely common and highly diverse. Biologists in Brazil who examine tree leaves for the presence of fungi routinely find several previously undiscovered species of endophytes. Recall from Section 30.2 that a study on fungi in Panama found hundreds of fungal morphospecies living on the leaves of just two tree species. These newly discovered species are endophytes.

Endophytes do not cause disease symptoms in the tissues they invade. Thus, it is clear that they are not parasitic. To test the hypothesis that endophytic fungi are mutualistic, biologists have set up presence-absence experiments with various species of grasses and their endophytes. If endophytic fungi have a mutualistic relationship with their host plants, then host plants should grow more slowly and produce fewer offspring when endophytic fungi are absent. When endophytic fungi are prevented from colonizing grasses, plants that are free of endophytic fungi turn out to be much more susceptible to attack by leaf-eating insects than are plants with a normal array of endophytes. Follow-up work has confirmed that, in grasses, endophytic fungi produce compounds that deter or even kill insect herbivores. The fungus is also hypothesized to benefit by obtaining nutrition from the plant host. Based on these results, biologists have concluded that the symbiotic relationship between endophytes and grasses is mutualistic.

When similar experiments have been done with tree species and their endophytes, however, the results have been much less conclusive. For example, biologists who sprayed spores from endophytic fungi onto the leaves of Emory oak trees confirmed that the treatment dramatically increased the abundance of endophytes. However, the researchers observed no impact on the survival or growth rate of insect larvae that eat the leaves of these trees compared with the growth rate of insects that were eating the leaves of oaks with a normal abundance of endophytes. From these data, the researchers concluded that an increased number of endophytes did not deter insect herbivores in oaks.

Based on such results, the current consensus is that at least some endophytic fungi may be **commensals**—meaning the fungi and the plants simply coexist with no observable effect, either deleterious or beneficial, on the host plant. Research continues, however. Biologists are still working to understand the exact nature of the relationship between endophytic fungi and their tree hosts. In the meantime, it is abundantly clear that most land plants are colonized by a wide array of fungi and that fungi have evolved an array of relationships with their plant symbionts.

Adaptations That Make Fungi Effective Decomposers

The saprophytic fungi are master decomposers and recyclers. Fungi that make their living from dead plant material seek out large, complex molecules such as cellulose, lignin, proteins, and nucleic acids and break them down into hundreds or thousands of smaller compounds. Although bacteria and archaea are also important decomposers in terrestrial environments, fungi and a few bacterial species are the only organisms that can digest

FIGURE 30.12 Fungi Recycle Nutrients
The fungi that are decomposing this section of tree trunk are breaking up its proteins, nucleic acids, lignin, and cellulose. In doing so, the fungi release nitrogen, phosphorus, and other nutrients that can be used by other organisms. **EXERCISE** At least two species of basidiomycetes inhabit the area in this photo. Circle their reproductive structures.

wood completely. Given enough time, fungi can turn even the hardest, most massive trees into soft soils that nourish an array of plants (**Figure 30.12**).

As land plants diversified and spread into terrestrial habitats, the ability to digest dead plant materials opened up the same habitats to fungi. As a result, fungi live wherever land plants live. The capacity to degrade plant bodies was an evolutionary breakthrough that helped trigger the diversification of saprophytic fungi. What adaptations made all this possible?

Extracellular Digestion   Large molecules such as starch, lignin, cellulose, proteins, and RNA cannot diffuse across the plasma membranes of hyphae. Only sugars, amino acids, nucleic acids, and other small molecules can enter the cytoplasm. As a result, fungi have to digest their food before they can absorb it.

Instead of digesting food inside a stomach or food vacuole, as most animals and some protists do, respectively, fungi synthesize digestive enzymes and then secrete them *outside* their hyphae, into their food. Fungi perform **extracellular digestion**—digestion that takes place outside the organism. The simple compounds that result from enzymatic action are then absorbed by the hyphae.

Extracellular digestion is an important adaptation in fungi. As a case study in how this process occurs, consider the enzymes responsible for digesting lignin and cellulose. These are the two most abundant organic molecules on Earth. The term **lignin** refers to a family of extremely strong polymers built from monomers that are six-carbon rings. Recall from Chapter 29 that lignin is found in the secondary cell walls of plant vascular tissues, where it furnishes structural support. **Cellulose** is a polymer of glucose and is found in the primary and secondary cell walls of plant cells.

Basidiomycetes can degrade lignin completely—to CO_2—as well as digest cellulose. Let's take a closer look at how they do it.

Lignin Degradation How do fungi digest lignin? In addition to its biological significance, this question has enormous practical importance. To make soft, absorbent paper products, manufacturers must find efficient ways to degrade lignin without using caustic, dangerous chemicals. The same problem faces environmental scientists charged with cleaning up waste from old sawmills and paper mills.

To find out how fungi digest lignin, biologists began analyzing the proteins that are secreted into the extracellular space by lignin-digesting fungi. After purifying these molecules, the investigators tested each protein for the ability to degrade lignin. Using this approach, investigators from two labs independently discovered an enzyme called **lignin peroxidase.**

The researchers who followed up on the discovery of lignin peroxidase found that the enzyme catalyzes the removal of a single electron from an atom in the ring structures of lignin. This oxidation step creates a free radical—an atom with an unpaired electron (see Chapter 2). This extremely unstable electron configuration leads to a series of uncontrolled and unpredictable reactions that end up splitting the polymer into smaller units.

Biologists have referred to this mechanism of lignin degradation as *enzymatic combustion*. The phrase is apt because the uncontrolled oxidation reactions triggered by lignin peroxidase are analogous to the uncontrolled oxidation reactions that occur when gasoline burns in a car engine. This mechanism is remarkable, because virtually all of the other reactions catalyzed by enzymes are extremely specific. The lack of specificity in the lignin degradation reaction actually makes sense, however, given the nature of the compound. Unlike proteins, nucleic acids, and most other polymers, lignin is extremely heterogeneous. Over 10 types of linkages are routinely found between the monomers that make up lignin. But once lignin peroxidase has created a free radical in the aromatic ring, any of these linkages can be broken.

The uncontrolled nature of the reactions has an important consequence. Instead of being stepped down an orderly electron transport chain as described in Chapter 9, the electrons involved in these reactions lose their potential energy in large, unpredictable jumps. As a result, the oxidation of lignin cannot be harnessed to drive the production of ATP and fuel the growth of hyphae. This prediction accords with an experimental observation: Fungi cannot grow with lignin as their sole source of food.

If wood-rotting fungi don't use lignin as food, why do they produce enzymes to digest it? The answer is simple. In wood, lignin forms a dense matrix around long strands of cellulose. Degrading the lignin matrix gives hyphae access to huge supplies of energy-rich cellulose. Saprophytic fungi are like miners. But instead of seeking out rare, gem-like nitrogen or phosphorus atoms as do EMF and AMF, the saprophytes use lignin peroxidase to blast away enormous lignin molecules, exposing rich veins of cellulose that can fuel growth and reproduction.

Cellulose Digestion Once lignin peroxidase has softened wood by stripping away its lignin matrix, the long strands of cellulose that remain can be attacked by enzymes called **cellulases.** Like lignin peroxidase, cellulases are secreted into the extracellular environment by fungi. But unlike lignin peroxidase, cellulases are extremely specific in their action. For example, biologists have purified seven different cellulases from the fungus *Trichoderma reesei*. Two of these enzymes catalyze a critical early step in digestion—they cleave long strands of cellulose into a disaccharide called cellobiose. The other cellulases are equally specific and also catalyze hydrolysis reactions. In combination, the suite of seven enzymes in *T. reesei* transforms long strands of cellulose into a simple monomer (glucose) that can be used by fungi as a source of food.

Now that the enzymes responsible for lignin and cellulose degradation have been characterized, researchers would like to pursue a practical question: Can lignin peroxidase and cellulases be used to make paper production more efficient or perhaps speed efforts to clean up waste from old sawmill and paper mill operations? If the answer turns out to be yes, fungi could be used along with bacteria and archaea in the bioremediation efforts introduced in Chapter 28.

Variation in Life Cycles

As fungi diversified, they evolved an array of symbiotic relationships with living land plants, as well as the ability to digest dead plant tissue completely. But in addition to being able to absorb nutrition in many ways, fungi reproduce in many ways. Recall from Section 30.2 that fungi may produce swimming gametes and spores, yoked hyphae where nuclei from different individuals fuse to form a zygote inside a protective structure, or specialized spore-producing cells called basidia and asci.

The **spore** is, in fact, the most fundamental reproductive cell in fungi. Spores are the dispersal stage in the fungal life cycle and are produced during both asexual and sexual reproduction. (Recall from Chapter 12 that asexual reproduction is based on mitosis, while sexual reproduction is based on meiosis.) Fungi produce spores in such prodigious quantities that it is not unusual for them to outnumber pollen grains in air samples. If a spore falls on a food source and is able to germinate, a mycelium begins to form. As the fungus expands, hyphae grow in the direction in which food is most abundant. If food begins to run out, mycelia respond by making spores, which are dispersed by wind or animals. Why would mycelia reproduce when food is low? To answer this question, biologists hypothe-

size that spore production is favored by natural selection when individuals are under nutritional stress. The logic is that spore production allows starving mycelia to disperse offspring to new habitats where more food might be available.

Unique Aspects of Fungal Life Cycles   Compared with the life cycles of green plants, protists, and animals, the life cycles of fungi have several unique features. For example, only members of the Chytridiomycota produce gametes, and no species of fungi produce gametes that are so different in size that they are called sperm and egg. In addition, fertilization occurs in two dis-

tinct steps in most fungi: (1) fusion of cells and (2) fusion of nuclei from the fused cells. These two steps can be separated by long time spans and even long distances.

In many fungi, the process of sexual reproduction begins when hyphae from two individuals fuse to form a hybrid hypha. When the cytoplasm of two individuals fuses in this way, **plasmogamy** is said to occur (**Figure 30.13a**). But in most cases, the nuclei from the two individuals stay independent and grow into a heterokaryotic mycelium. The two nuclei that are present in heterokaryotic hyphae function independently, even though gene expression must be coordinated in order for growth and development to

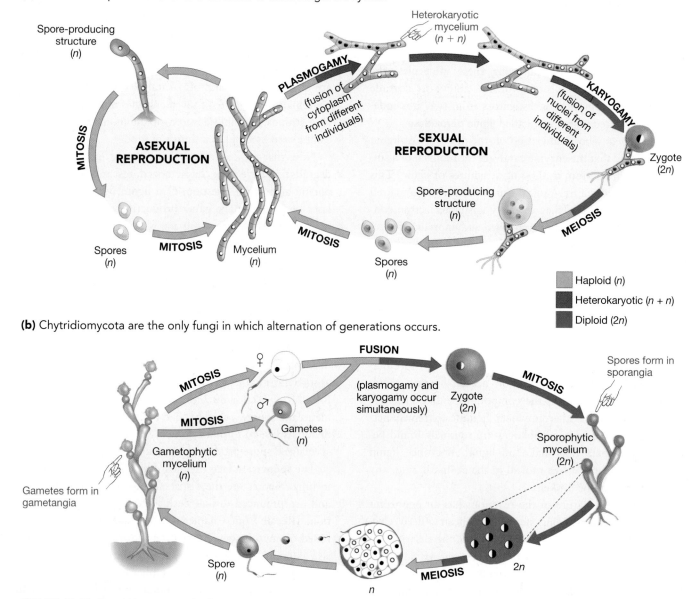

(a) The same sequence of events is common to most fungal life cycles.

(b) Chytridiomycota are the only fungi in which alternation of generations occurs.

FIGURE 30.13 Fungi Have Unusual Life Cycles
(a) A generalized fungal life cycle, showing both asexual and sexual reproduction. The sexual part of the life cycle in the four major groups of fungi: **(b)** Chytridiomycota, **(c)** Zygomycota, **(d)** Basidiomycota, and **(e)** Ascomycota. Asexual reproduction is particularly common in Zygomycota and Ascomycota. **EXERCISE** Fungi spend most of their lives feeding. On each drawing, indicate the longest-lived component of the life cycle. In parts (c) through (e), add a loop labeled "Asexual reproduction" similar to that in part (a). This loop should start at the mycelium and return to it, via the growth of asexually produced spores.

(c) Zygomycota form yoked hyphae that produce a zygote.

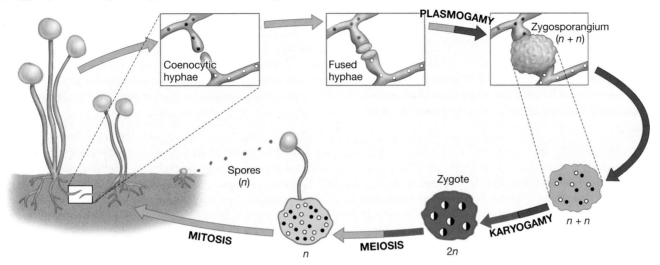

(d) Basidiomycota have reproductive structures with many spore-producing basidia.

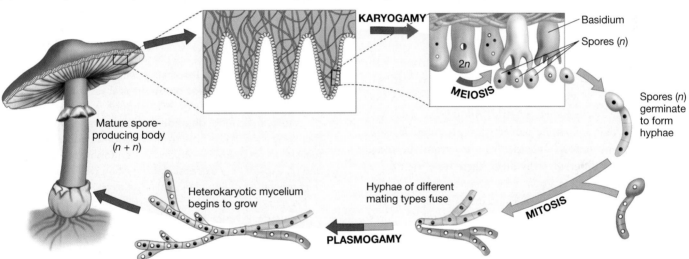

(e) Ascomycota have reproductive structures with many spore-producing asci.

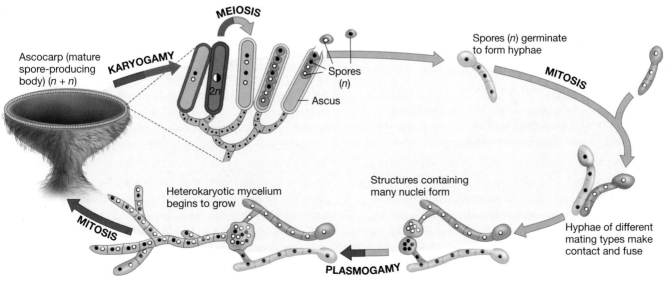

FIGURE 30.13 *(continued)*

occur. For example, the two nuclei divide as the hyphae expand, so each compartment that is divided by a septum contains one of each of the two nuclei. How the activities of the two nuclei are coordinated is almost a complete mystery, and it is an important topic for future research. Eventually, however, one or more pairs of unlike nuclei fuse to form a diploid zygote. The fusion of nuclei is called **karyogamy**. The nuclei that are produced by karyogamy then divide by meiosis to form haploid spores.

Except for some members of the Basidiomycota, most fungal species reproduce asexually as well as sexually. There are, in fact, large numbers of ascomycetes that have never been observed to reproduce sexually. During asexual reproduction, spore-forming structures are produced by a mycelium, and spores are generated by mitosis. As a result, offspring are genetically identical to their parent.

Among the sexually reproducing species of fungi, the nature of the heterokaryotic stage and the morphology of the spore-producing structure vary. Let's take a closer look at each of the four major types of life cycles that have been observed in fungi:

1. The Chytridiomycota are the only fungal group with any species that exhibit alternation of generations. **Figure 30.13b** shows how alternation of generations occurs in the well-studied chytrid *Allomyces*. The key points are that swimming gametes are produced in haploid adults by mitosis and that gametes from the same individual or different individuals then fuse to form a diploid zygote, which grows into a sporophyte. The life cycle is completed when meiosis occurs in the sporophyte, inside a structure called a **sporangium**. The haploid spores produced by meiosis disperse by swimming.

2. In Zygomycota, sexual reproduction starts when hyphae from different individuals fuse, as shown in **Figure 30.13c**. Plasmogamy forms a zygosporangium that develops a tough, resistant coat and is followed by karyogamy. The zygosporangium can persist if conditions become too cold or dry to support growth. When temperature and moisture conditions are again favorable, meiosis occurs. The meiotic products within the sporangium produce haploid spores. When spores are released and germinate, they grow into new mycelia. These mycelia can reproduce asexually by making sporangia, with spores being produced by mitosis and dispersed by wind.

3. Mushrooms, bracket fungi, and puffballs are reproductive structures produced by members of the Basidiomycota (**Figure 30.13d**). Their size, shape, and color vary enormously from species to species, but all members of this group originate from the heterokaryotic hyphae of mated individuals. Inside these reproductive structures, the pedestal-like, spore-producing cells called basidia form at the ends of hyphae. Karyogamy occurs within the basidia.

The diploid nucleus that results undergoes meiosis, and haploid spores mature. Spores are then expelled from the end of the basidia and dispersed by the wind. It is not unusual for a single puffball or mushroom to produce a billion spores.

4. The reproductive structure in Ascomycota is produced by a heterokaryotic mycelium. As **Figure 30.13e** illustrates, the process begins when hyphae from the same ascomycete species but from different mating types fuse, forming a structure containing many independent nuclei. A short heterokaryotic hypha, containing one nucleus from each parent, emerges and eventually grows into a complex reproductive structure whose hyphae have the sac-like, spore-producing structures called asci at their tips. After karyogamy occurs inside each sac, meiosis takes place and haploid spores are produced. When the ascus matures, it splits. The spores may be forcibly ejected and are often picked up by the wind and dispersed.

✓ CHECK YOUR UNDERSTANDING

Most fungi live in close association with land plants. When plants are alive, two types of mycorrhizal associations occur and many fungal species grow as endophytes. When plants die, saprophytic fungi degrade their tissues and release nutrients. You should be able to (1) describe evidence that mutualism occurs in EMF, AMF, and the endophytes of grasses; and (2) explain how extracellular digestion works, using the decomposition of lignin and cellulose as an example.

Instead of being based on the fusion of gametes, sexual reproduction in fungi is usually based on the fusion of hyphae. You should be able to (3) draw a generalized fungal life cycle; (4) explain the difference between plasmogamy and karyogamy, and relate them to the process of fertilization in plants and animals; and (5) explain what a heterokaryotic mycelium is.

30.4   Key Lineages of Fungi

Based on the current estimate for the phylogeny of Fungi, it appears clear that Chytridiomycota and Zygomycota are not monophyletic and thus should not be named as phyla. Until further research is done, however, the relationships among the various subgroups of chytrids and zygomycetes remain too uncertain to name phyla or other major lineages accurately. Stated another way, biologists now know that the traditional fungal phyla are at least partially incorrect, even though a more accurate understanding has yet to emerge. Although the summaries that follow analyze the chytrids and zygomycetes as coherent groups, keep in mind that they are not. At this time, researchers simply lack the data required to identify the major lineages of fungi more accurately.

Chytridiomycota (Chytrids)

Chytrids are largely aquatic and are particularly common in fresh-water environments. Some species live in wet soils, however, and a few have been found in desert soils that are wet only during a short rainy season. Species that are found in dry soils have resting spores that allow them to endure harsh conditions. Resting spores from chytrids have been shown to germinate normally after a resting period of 30 years. Members of this group are the only fungi that produce cells that can move.

Because the animals and protists that are most closely related to fungi are aquatic, it is logical to infer that fungi evolved in water and that chytrids are the most ancient lineage of fungi. The phylogenetic analyses that have been done to date cannot rule out the possibility that a soil-dwelling group of zygomycetes is actually the most basal clade of fungi, however.

Absorptive lifestyle Many species of chytrids have enzymes that allow them to digest cellulose. As a result, these species are important decomposers of plant material in wet soils, ponds, and lakes. Many of the freshwater species are parasitic, and on occasion parasitic chytrids cause disease epidemics in algae or aquatic insects (including mosquitoes). Other species parasitize mosses, ferns, or flowering plants. Mutualistic chytrids are among the most important of the many organisms that live in the guts of deer, cows, elk, and other mammalian herbivores, because the chytrids help these mammals digest their food (**Figure 30.14**). Chytrids produce cellulases that degrade the cell walls of grasses and other plants ingested by the animal, releasing sugars that are used by both the chytrids and their hosts.

Life cycle During asexual reproduction, most chytrid species produce spores that swim to new habitats via a flagellum. A few species reproduce sexually as well as asexually and exhibit alternation of generations. (Recall from Section 30.3 that a few chytrids are the only fungal species to do so.) In species that reproduce sexually, plasmogamy may occur through fusion of gametes, fusion of gamete-forming structures called **gametangia**, or fusion of hyphae.

Human and ecological impacts In addition to their importance as decomposers in aquatic habitats and as mutualists in the guts of large, plant-eating mammals, chytrids have been the focus of intense research. An epidemic of chytrid infections may be responsible for catastrophic declines that recently occurred in frog populations all over the world. For additional details on the amphibian die-off, see the essay at the end of this chapter. Biologists are investigating the possibility of using a chytrid that parasitizes mosquito larvae as a biological control agent, and potato tubers are sometimes invaded and spoiled by a parasitic chytrid that causes black wart disease.

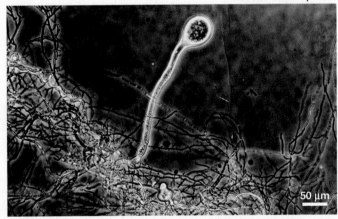
Neocallimastix species

50 μm

FIGURE 30.14 A Chytrid That Lives in the Gut of Cows and Digests the Cellulose in Grass Leaves

Zygomycota ▶ Mucorales and Other Basal Lineages

The Zygomycota ("yoked-fungi") are primarily soil-dwellers. Their hyphae yoke together and fuse during sexual reproduction and then form the durable, thick-walled zygosporangium that is characteristic of the group. Although only a few lineages are highlighted here, over 10 major subgroups have been identified.

In current phylogenetic analyses, most of the major subgroups of Zygomycota cluster together at the base of the fungal tree along with some of the chytrids. These subgroups include Mucorales, Entomophthorales, *Basidiobolus*, and other lineages not discussed here. Because certain chytrids live in wet soils, it is not clear from the phylogeny whether chytrids or zygomycetes were the first fungi

to make the transition from aquatic to terrestrial habitats. In addition, it is not yet known whether fungi made the water-to-land transition once or many times. Understanding the relationships among basal lineages with chytrid and zygomycete characteristics is among the most urgent issues facing biologists working on the phylogeny of fungi.

Absorptive lifestyle Many members of zygomycete lineages are saprophytes and live in plant debris. Some parasitize other fungi, however, or are important parasites of insects and spiders. About 65 species in one subgroup are predatory—primarily on amoebae

(Continued on next page)

Zygomycota ▶ Mucorales and Other Basal Lineages *(continued)*

and other unicellular protists, though some are large enough to capture the microscopic animals called roundworms. The predatory zygomycetes trap their prey by means of sticky substances on their cell walls or catch prey in snares consisting of looped hyphae (**Figure 30.15**). Once a prey individual is captured, hyphae invade its body, digest it, and absorb the nutrients that are released.

Life cycle   Asexual reproduction is extremely common. Ball-like sporangia are produced at the tips of hyphae that form stalks, and mitosis results in the production of spores, which are dispersed by wind. During sexual reproduction, fusion of hyphae occurs only between individuals of different mating types. Fusion of hyphae from the same individual mycelium does not occur.

Human and ecological impacts   The black bread mold *Rhizopus stolonifer* is a common household pest—probably the zygomycete that is most familiar to you. Saprophytic and parasitic members of these lineages are responsible for rotting strawberries and other fruits and vegetables and causing large losses in the fruit and vegetable industries. Some species of *Mucor* are used in the production of steroids for medical use. Species of *Rhizopus* and *Mucor* are used in the commercial production of organic acids, pigments, alcohols, and fermented foods.

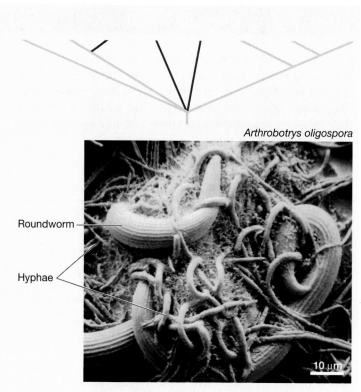

Arthrobotrys oligospora

Roundworm

Hyphae

10 μm

FIGURE 30.15 Some Zygomycetes Are Predatory

Glomeromycota (AMF)

Current phylogenetic analyses place the group of fungi that form arbuscular mycorrhizal associations in their own phylum, indicating that they are a major monophyletic group. Because the Glomeromycota, Basidiomycota, and Ascomycota diverged relatively late in the course of fungal evolution and form the tips of the tree, biologists refer to the three groups as the "crown fungi." Before phylogenies based on molecular characters were available, the lineage now called Glomeromycota was called Glomales and considered part of the Zygomycota.

Absorptive lifestyle   All of the arbuscular mycorrhizal fungi (AMF) are members of the Glomeromycota. Recall from Section 30.3 that these fungi absorb phosphorus-containing ions or molecules in the soil and transfer them into the roots of trees, grasses, and shrubs in grassland and tropical habitats (**Figure 30.16**). In exchange, the host plant provides the symbiotic fungi with sugars and other organic compounds.

Life cycle   Most species form spores underground. Because these species are difficult to grow in the laboratory, their life cycle is not well known.

Human and ecological impacts   Because grasslands and tropical forests are among the most productive habitats on Earth, the AMF are enormously important to both human and natural economies.

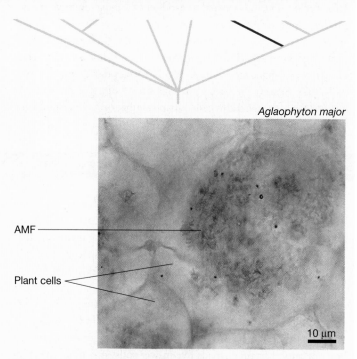

Aglaophyton major

AMF

Plant cells

10 μm

FIGURE 30.16 A Fossilized AMF That Is 398 Million Years Old
[Micrograph courtesy of Hans Kerp and Hagen Hass, Palaeobotanical Research Group, University of Munster, Germany.]

Basidiomycota (Club Fungi)

Although most basidiomycetes form mycelia and produce multicellular reproductive structures, some species have the unicellular growth form. The group is named for basidia, the club-like or pedestal-like cells where meiosis and spore formation occur. About 30,000 species have already been described, and more are being discovered each year.

Absorptive lifestyle Basidiomycetes are important saprophytes. Along with a few soil-dwelling bacteria, they are the only organisms capable of synthesizing lignin peroxidase and completely digesting wood. Other basidiomycetes form ectomycorrhizal associations with trees in temperate and northern forests. One subgroup consists entirely of parasitic forms called *rusts*, including species that cause serious infections in wheat and rye fields. The plant parasites called *smut fungi* are also basidiomycetes. Smuts specialize in infecting grasses; a few infect other fungi.

Life cycle Asexual reproduction through production of spores is common in the Basidiomycota, although not as prevalent as in other groups of fungi. Asexual reproduction also occurs through growth and fragmentation of mycelia in the soil or in rotting wood, resulting in genetically identical individuals that are physically independent. During sexual reproduction, all basidiomycetes—even unicellular ones—produce basidia. In the largest subgroup in this lineage, basidia form in large, aboveground reproductive structures called *mushrooms, brackets, earthstars,* or *puffballs* (**Figure 30.17**).

Human and ecological impacts Because temperate and northern forests provide most of the hardwoods and softwoods used in furniture-making, building construction, and papermaking, the

EMF are enormously important in forest management. Throughout the world, mushrooms are cultivated or collected from the wild as a source of food. Some of the toxins found in poisonous mushrooms are used in biological research; others have hallucinogenic effects on people and are used and traded illegally.

Geastrum saccatum

FIGURE 30.17 Earthstars and Puffballs Can Produce Billions of Spores

Ascomycota (Sac Fungi)

About 75 percent of all known fungi belong to the phylum Ascomycota. This group is named for the sac-like cells, or asci, where meiosis and spore formation take place. The group is extremely large and diverse, however, and the phylogenetic relationships among subgroups are largely unknown. As a result, the discussion of "key lineages" that follows is based on two lifestyles observed in ascomycetes and not on distinct monophyletic groups. Work on the phylogenetic relationships within Ascomycota continues.

Ascomycota ▶ Lichen-Formers

About half of the ascomycetes grow in association with cyanobacteria or single-celled members of the green algae, forming the structures called **lichens** (see Chapter 29). Over 15,000 different lichens have been described to date; in most, the fungus involved is an ascomycete (although a few basidiomycetes participate as well). To name a lichen, biologists usually use the genus and species name assigned to the fungus that participates in the association.

Absorptive lifestyle It is not yet clear whether all of the relationships observed in lichens are mutualistic. In habitats where

lichens are common, neither partner can exist as a free-living organism. The fungus in lichens appears to protect the photosynthetic cells. The fungal hyphae form a dense protective layer that shields the photosynthetic species and reduces water loss. In return, the cyanobacterium or alga provide carbohydrates that the fungus uses as a source of carbon and energy. However, the hyphae of some lichen-forming fungi have been observed to invade algal cells and kill them. This observation suggests a partially parasitic relationship in at least some lichens. The nature of lichen-forming associations is the subject of ongoing research.

(Continued on next page)

Ascomycota ▶ Lichen-Formers *(continued)*

Life cycle As **Figure 30.18a** shows, lichens reproduce asexually via the production of small structures called **soredia** that contain both symbionts. Soredia disperse to a new location via wind or water and then develop into a new, mature individual via the growth of the algal or bacterial and fungal symbionts. In addition, the fungal partner may form asci, which produce spores that then germinate to form a small mycelium. If the growing hyphae encounter enough algal cells, a new lichen can form.

Human and ecological impacts In terms of their abundance and diversity, lichens dominate the Arctic and Antarctic tundras. They are the major food of caribou in these habitats, as well as the most prevalent colonizers of bare rock surfaces throughout the world (**Figure 30.18b**). Rock-dwelling lichens are significant, because they break off mineral particles from the rock surface as they grow—launching the first step in soil formation. About 10,000 tons of lichens are processed annually and used in perfume production.

(a) Cross section of a lichen, showing three layers

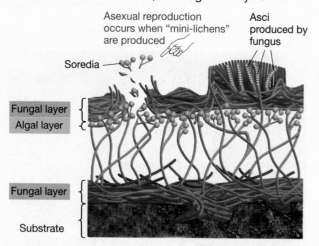

(b) Top view of lichens on a rock

FIGURE 30.18 Lichens Are Associations between a Fungus and a Cyanobacterium or Green Alga
(a) In a lichen, cyanobacteria or green algae are enmeshed in a dense network of fungal hyphae. **(b)** Lichens often colonize surfaces, such as tree bark or bare rock, where other organisms are rare.

Ascomycota ▶ Non-Lichen-Formers

The ascomycetes that do not form lichens are found in virtually every terrestrial habitat, as well as some freshwater and marine environments. Although most ascomycetes form mycelia, the taxonomic group called yeasts is part of the Ascomycota.

Absorptive lifestyle A few members of the Ascomycota form mutualistic ectomycorrhizal associations with tree roots. Ascomycetes are also the most common endophytic fungi on aboveground tissues. Large numbers are saprophytic and are abundant in forest floors and in grassland soils. Parasitic forms are common as well.

Life cycle The aboveground, ascus-bearing reproductive structures of these fungi may be a cup or saucer shape called an **ascocarp** (**Figure 30.19**). For that reason, these ascomycetes are known as **cup fungi**. It is also routine for spores called *conidia* to be produced asexually, at the ends of specialized hyphae called *conidiophores*.

Human and ecological impacts Some saprophytic ascomycetes can grow on jet fuel or paint and are used to help clean up contaminated sites. *Penicillium* is an important source of antibiotics, and *Aspergillus* produces citric acid used to flavor soft drinks and candy. Truffles and morels are so highly prized that they can fetch $400 per pound. A few parasitic ascomycetes cause infections in

humans and other animals. In land plants, parasites from this group cause diseases including Dutch elm disease and chestnut blight.

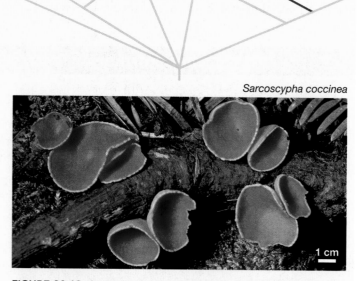

Sarcoscypha coccinea

FIGURE 30.19 Ascomycetes Are Sometimes Called Cup Fungi

ESSAY Why Are Frogs Dying?

Over the past two decades, conservation biologists have become increasingly alarmed about rapid, and in many cases catastrophic, declines in the numbers of frogs and toads around the world. The trend is particularly worrisome because even populations in undisturbed habitats, such as national parks, are being affected. Biologists who have followed the declines closely suspect that several species extinctions have already occurred.

Despite the magnitude of the problem, biologists have had frustratingly few clues as to what has killed the organisms. Researchers would revisit sites where frogs and toads once were abundant and diverse, only to discover that these animals had disappeared. Without dead or dying animals to study, investigators could only speculate about what was happening—until recently.

In the late 1990s an international team of researchers witnessed simultaneous and massive die-offs of amphibians in the rain forests of Australia and Panama. During these events, they collected 120 dead frogs and toads representing 19 species. Although general autopsies showed no obvious tissue abnormalities or lesions, light microscopy and electron microscopy of skin samples were revealing. In every case, the epidermis of the dead frogs was riddled with a parasitic fungus (**Figure 30.20**). An analysis of ribosomal RNA sequences isolated from the fungal cells confirmed what researchers had suspected from the morphology of the spores: The parasite was an undescribed member of the Chytridiomycota. To confirm that this fungus had caused the deaths, the researchers collected spores from the skin of dead frogs, suspended them in water, and added the suspension to aquaria containing six healthy frogs. Within 18 days, all of the experimental frogs had become infected with the fungus and were dead or dying.

This fungus could be the first infectious agent known to cause a species extinction. Biologists who are studying the frog die-off caution, however, that the chytrid is probably not the only agent involved in the worldwide declines. In California, for example, no evidence of fungal infection has been found to date. Instead, pesticides and herbicides are thought to play an important role in the decline of amphibian populations there. It is also entirely possible that well-documented increases in ultraviolet radiation—or other factors—have compromised the immune systems of these organisms, rendering them susceptible to the disease.

This fungus could be the first infectious agent known to cause a species extinction.

Although researchers are convinced that amphibian populations are declining in many parts of the world and that chytrid infections play a role, biologists are far from certain that chytrids are the only causative agent. Research continues.

FIGURE 30.20 Chytridiomycosis Can Be Fatal in Frogs
This frog is heavily infected with a parasitic chytrid. The infected individual is listless and will probably die as a result of the infection.

CHAPTER REVIEW

Summary of Key Concepts

■ **Fungi are important in part because many species live in close association with land plants. They supply plants with key nutrients and decompose dead wood. They are the master recyclers of nutrients in terrestrial environments.**

Living plants are covered with fungi. The roots of grasses and trees that grow in warm or tropical habitats are infiltrated by symbiotic fungi that supply the plant with phosphorus. Similar mycorrhizal associations, based on an exchange of nitrogen for sugars and other products of photosynthesis, are found between basidiomycetes and the roots of trees in cold, northern habitats. Many fungi also live in leaves and stems; in grasses, these endophytic fungi secrete toxins that discourage herbivores.

Although mutualistic associations between fungi and living plants are common, so are parasitic relationships. Parasitic fungi are responsible for devastating blights in crops and other plants.

Once plants die, saprophytic fungi degrade the lignin and cellulose in wood and release nutrients from decaying plant material. Because they free up carbon atoms that would otherwise be locked up in wood, fungi speed up the carbon cycle in terrestrial habitats.

■ **All fungi make their living by absorbing nutrition from living or dead organisms. Fungi secrete enzymes so that digestion takes place outside their cells. Their morphology provides a large amount of surface area for efficient absorption.**

Several adaptations make fungi exceptionally effective at absorbing nutrients from the environment. The two growth habits found among Fungi—single-celled "yeasts" or mycelia composed of long, filamentous hyphae—give fungal cells extremely high surface-area-to-volume ratios. Extracellular digestion, in which enzymes are secreted into food sources, is another important adaptation in fungi. Perhaps the most important reactions catalyzed by these enzymes are the degradations of lignin and cellulose found in wood. Lignin decomposes through a series of uncontrolled oxidation reactions triggered by the enzyme lignin peroxidase. This "enzymatic combustion" contrasts sharply with the breakdown of cellulose, which occurs in a carefully regulated series of steps, each catalyzed by a specific cellulase.

Web Tutorial 30.1   Life Cycle of a Mushroom

■ **Many fungi have unusual life cycles. It is common for species to have a long-lived heterokaryotic stage, in which cells contain haploid nuclei from two different individuals. Although most species reproduce sexually, very few species produce eggs and sperm.**

Many species of fungi have never been observed to reproduce sexually, though most species can produce haploid spores either sexually or asexually. Sexual reproduction usually starts when hyphae from different individuals fuse—an event called plasmogamy. If the fusion of nuclei, or karyogamy, does not occur immediately, a heterokaryotic mycelium forms. Later, heterokaryotic cells produce spore-forming structures where karyogamy and meiosis take place.

The four traditional groupings of fungi are based on distinct reproductive structures: (1) the Chytridiomycota, which are aquatic fungi with motile gametes; (2) the mycorrhizal or saprophytic Zygomycota, which are soil-dwelling fungi with tough sporangia; (3) the mycorrhizal or lignin-decomposing and saprophytic Basidiomycota, which have club-like, spore-forming structures; and (4) the mycorrhizal, saprophytic, or lichen-forming Ascomycota, which have sac-like, spore-forming structures. Recent analyses of DNA sequence data have revealed that only the Basidiomycota and Ascomycota are monophyletic, however. Researchers are still trying to understand how various groups of chytrids and zygomycetes are related.

Questions

Content Review

1. The mycelial growth habit leads to a body with a high surface-area-to-volume ratio. Why is this important?
   a. Mycelia have a large surface area for absorption.
   b. The hyphae that make up mycelia are long, thin tubes.
   c. Hyphae are broken up into compartments by walls called septa.
   d. Hyphae can infiltrate living or dead tissues.

2. What do researchers hypothesize to be the cause of the "fungal spike" at the end-Permian boundary?
   a. unusual soil conditions, which made it more likely that fungal spores and hyphae could fossilize
   b. an adaptive radiation of fungi
   c. a switch to saprophytic lifestyles
   d. a mass extinction event that created a huge quantity of dead plant material

3. The Greek root *ecto* means "outer." Why are ectomycorrhizal fungi, or EMF, aptly named?
   a. Their hyphae form tree-like branching structures inside plant cell walls.
   b. They are mutualistic.
   c. Their hyphae form dense mats that envelop roots but do not penetrate the walls of cells inside the root.
   d. They transfer nitrogen from outside their plant hosts to the interior.

4. The hyphae of AMF form bushy or balloon-like structures after making contact with the plasma membrane of a root cell. Why?
   a. They anchor the fungus inside the root, so the association is more permanent.
   b. They increase the surface area available for the transfer of nutrients.
   c. They produce toxins that protect the plant cells against herbivores.
   d. They appear early in the fossil record of land plants—the first fossil AMF are 390 million years old.

5. Which statement best characterizes a parasitic relationship?
   a. Parasites gain benefits, such as nutrients, and benefit their host in return.
   b. Parasites gain benefits, such as nutrients, and harm their host in the process.
   c. Parasites gain benefits, such as nutrients, without affecting their host.
   d. The participants are not considered symbiotic.

6. Very few organisms besides fungi have a heterokaryotic stage in their life cycle. Which of the following is another usual aspect of the fungal life cycle?
   a. Some fungi exhibit alternation of generations—meaning that there is a multicellular diploid stage and a multicellular haploid stage.
   b. They produce eggs and sperm in approximately equal numbers, instead of many sperm and a few eggs.
   c. Spores have to fuse with each other before they develop into a new mycelium.
   d. Most varieties undergo sexual reproduction without producing eggs or sperm.

Conceptual Review

1. Explain why fungi that degrade dead plant materials are important to the global carbon cycle. Do you accept the text's statement that, without these fungi, "Terrestrial environments would be radically different than they are today and probably much less productive"? Why or why not?

2. Lignin and cellulose provide rigidity to the cell walls of plants. But in most fungi, chitin performs this role. Why is it logical that most fungi don't have lignin or cellulose in their cell walls?

3. Experiments show that AMF contribute phosphate ions to their host plants. Why is this important? Using information from Chapters 3 through 6, include a list of macromolecules found in plants that contain phosphorus.

4. When growing plants in the presence or absence of mycorrhizal or endophytic fungi—to test for the effects of symbiotic fungi on plant growth—researchers sterilize the soil and seeds used in both the presence and absence treatments, and then grow the plants under sterile conditions. Why?

5. Explain why radioactive or heavy isotopes were useful in showing that mycorrhizal fungi have a mutualistic relationship with plants. When researchers introduced radioactive phosphorus or heavy nitrogen atoms into the soil of experimental pots, how could they refute the hypothesis that the nutrients were being taken up by the plant itself—without help from the fungus?

6. Biologists claim that endomycorrhizal fungi are effective mutualists because they have a higher surface area than do plant roots and are more effective at digesting dead organic matter. Provide evidence to back up both claims.

Group Discussion Problems

1. After reading the essay about the amphibian die-off, review the material on Koch's postulates from Chapter 27. These postulates present criteria for implicating an organism as a disease-causing agent. Did the studies on the chytrid found in dead frogs fulfill Koch's postulates? Are you convinced that the fungus is the causative agent? Explain why or why not.

2. Some biologists contend that the ratio of plant species to fungus species worldwide is on the order of 1:6. Explain why you agree or disagree with this claim. In doing so, consider the analyses of endophytic, parasitic, lichen-forming, mycorrhizal, and saprophytic strategies presented in this chapter. Also, consider the diversity of tissues available in plants.

3. Experiments indicate that cellulase genes are transcribed and translated together. If cells are selected to be extremely efficient at digesting cellulose, is this result logical? Would you predict that the gene that codes for lignin peroxidase is transcribed along with the cellulase loci? How would you test your prediction?

4. Many mushrooms are extremely colorful. Why? Fungi do not see, so colorful mushrooms are obviously not communicating with one another. One hypothesis is that the colors serve as a warning to animals that eat mushrooms, much like the bright yellow and black stripes on wasps. Design an experiment capable of testing this hypothesis.

Answers to Multiple-Choice Questions   **1.** a; **2.** d; **3.** c; **4.** b; **5.** b; **6.** d

www.prenhall.com/freeman is your resource for the following: Web Tutorials; Online Quizzes and other Online Study Guide materials; Answers to Conceptual Review Questions; Solutions to Group Discussion Problems; Answers to Figure Caption Questions and Exercises; and Additional Readings and Research.

31 An Introduction to Animals

KEY CONCEPTS

- Animals are a particularly species-rich and morphologically diverse lineage of multicellular organisms on the tree of life.

- Major groups of animals are defined by the design and construction of their basic body plan, which differs in the number of tissues observed in embryos, symmetry, the presence or absence of a body cavity, and the way in which early events in embryonic development proceed.

- Recent phylogenetic analyses of animals have shown that there were three fundamental splits during evolutionary history, resulting in two protostome groups (Lophotrochozoa and Ecdysozoa) and the deuterostomes. The most ancient animal group living today is the sponges. The closest living relatives to animals are choanoflagellates, a group of protists.

- Within major groups of animals, evolutionary diversification was based on innovative ways of feeding and moving. Most animals get nutrients by eating other organisms, and most animals move under their own power at some point in their life cycle.

Jellyfish are among the most ancient of all animals—they appear in the fossil record over 560 million years ago. Compared with most animals living today, they have relatively simple bodies. But like most other animals, they make their living by eating other organisms and are able to move.

As a group, animals are distinguished by two traits: They eat and move. Many unicellular protists also ingest other organisms or dead organic material (detritus) but are small, so they are limited to eating microscopic prey. Animals, in contrast, are multicellular. They are the largest and most abundant predators, herbivores, and detritivores in virtually every ecosystem—from the deep ocean to alpine ice fields and from tropical forests to arctic tundras. Animals find food by tunneling, swimming, filtering, crawling, creeping, slithering, walking, running, or flying. They eat nearly every organism on the tree of life.

Over 1.2 million species of animals have been described and given scientific names to date, and biologists predict that tens of millions more have yet to be discovered. To analyze the almost overwhelming number and diversity of animals, this chapter presents a broad overview of how they diversified. It also provides information on the characteristics of the first groups of animals that evolved. In the next two chapters, we'll follow up with a more detailed exploration of two major phylogenetic groups in animals: protostomes and deuterostomes. Chapter 32 explores the protostomes, which include familiar organisms such as the insects, crustaceans (crabs and shrimp), and mollusks (clams and snails). Chapter 33 features the deuterostomes, which range from the sea stars to the vertebrates, including humans.

31.1   Why Do Biologists Study Animals?

If you ask biologists why they study animals, the first answer they'll give is, "Because they're fascinating." It's hard to argue with this statement. Consider ants. Ants live in colonies that routinely number millions of individuals. But colony-mates co-operate so closely in tasks such as food-getting, colony defense, and rearing young that each ant seems like a cell in a multicel-lular organism instead of an individual. Other species of ant parasitize this cooperative behavior, however. Parasitic ants look and smell like their host species but enslave them, forcing the hosts to rear the young of the parasitic species instead of their own. Ant colonies also vary widely in size and habitat. The smallest ant species forms a colony that would fit inside the brain of the largest ant species. Other species live in trees and protect their host plants by attacking giraffes and other grazing animals a million times their size. There are rancher ants and farmer ants. Rancher ants tend the plant-sucking insects called aphids and eat the sugar-rich honeydew that aphids secrete from their abdomens (**Figure 31.1a**). Farmer ants eat fungi that they carefully plant, fertilize, and cultivate in underground gardens (**Figure 31.1b**). New ant species are discovered every year.

Based on observations like these, most people would agree that ants—and by extension, other animals—are indeed fasci-nating. But beyond pure intellectual interest, there are other compelling reasons that biologists study animals:

- Animals are **heterotrophs**—meaning they obtain the chemi-cal energy and carbon compounds they need from other or-ganisms. Recall from Chapter 28 that photosynthetic protists and bacteria are **primary producers**, which form the base of the food chain in most marine environments. Land plants play the same role in most terrestrial habitats. Het-erotrophs eat producers and other organisms and are called **consumers**. Animals are consumers that occupy the upper levels of food chains in both marine and terrestrial regions. As a result, it is not possible to understand or preserve ecosystems without understanding and preserving animals.

- Animals are a particularly species rich and morphologically diverse lineage of multicellular organisms on the tree of life. Current estimates suggest that there are between 10 million and 50 million species of animals, although only about a million have been formally described and named. Animals range in size and complexity from tiny, sessile (nonmoving) sponges, which contain just a few cell types and no true tis-sues, to blue whales, which migrate tens of thousands of kilometers each year in search of food and contain trillions of cells, dozens of distinct tissues, an elaborate skeleton, and highly sophisticated sensory and nervous systems. A great deal of evolution has gone on in this lineage. To un-derstand the history of life, it is important to understand how animals came to be so diverse.

(a) "Rancher ants" tend aphids and eat their sugary secretions

(b) "Farmer ants" cultivate fungi in gardens.

FIGURE 31.1 Biologists Study Animals because They Are Fascinating
(a) Some species of ant make their living by protecting aphids from predators and then harvesting the aphids' sugary secretion called honeydew. **(b)** Several species of ant cultivate and eat fungi.

- Humans in every country depend on wild and domesticated animals for food. Horses, donkeys, oxen, and other domes-ticated animals also provide most of the transportation and power used in preindustrial societies.

- Efforts to understand human biology depend on advances in animal biology. Most drug testing is done on mice, rats, or primates. Current efforts to understand the human genome are based on analyzing the function of genes in model or-ganisms such as mice, zebrafish, and roundworms.

Given that studying animals is interesting and valuable, let's get started. What makes an animal an animal, and how do biolo-gists go about studying them?

31.2   How Do Biologists Study Animals?

The **animals** are a monophyletic group of multicellular eukaryotes. Most animals move under their own power at some point in their life cycle, and all obtain nutrients by eating other organisms or absorbing nutrients from them. The cells of animals lack walls but have an extensive extracellular matrix, which includes proteins specialized for cell-cell adhesion and communication (see Chapter 8). Animals are the only lineage on the tree of life with species that have muscle tissue and nervous tissue. Although many animals reproduce both sexually and asexually, no animals undergo alternation of generations. During an animal's life cycle, adults of most species are diploid; the only haploid cells are gametes produced during sexual reproduction.

Beyond these shared characteristics, animals are almost overwhelmingly diverse—particularly in morphology. Biologists currently recognize about 34 **phyla**, or major lineages, of animals—including those listed in **Table 31.1**. Each animal phylum has distinct morphological features.

Analyzing Comparative Morphology

In essence, animals are moving and eating machines. A quick glance at the diversity of ways that animals find and capture food, like those illustrated in **Figure 31.2**, should convince you that evolution by natural selection has indeed produced a wide array of ways to move and eat. This diversity is possible because of extensive variation in appendages and in mouthparts or other organs used to capture and process food. Limbs and mouths are specialized structures that make particular ways of moving and eating possible.

In contrast to the spectacular diversity observed among animals in their limbs and mouthparts, the basic architecture of the animal body has been highly conserved throughout evolution. Just as there are only a few basic ways to frame a house—with posts and beams, stud walls, or cement blocks, for example—there are just a handful of ways to design and build an animal body. Once a few different ways of developing a body evolved, an extraordinary radiation of species ensued—based on elaborations of limbs and mouthparts or other structures for moving and capturing food.

Based on this overall pattern of animal evolution, biologists have been able to identify the major lineages of animals by analyzing variation in their core body plan. A **body plan** is an animal's architecture—the major features of its structural and functional design. Four features define the basic elements of an animal's body plan: (1) the number of tissue types found in embryos, (2) the type of body symmetry and degree of cephalization (informally, the formation of a head region), (3) the presence or absence of a fluid-filled cavity, and (4) the way in which the earliest events in the development of an embryo proceed. The origin and early evolution of animals was based on the origin and elaboration of these four features. Let's consider each in detail.

The Evolution of Tissues   Sponges are the only group of animals that lack tissues. Although sponges have several cell types, these cells are not organized into the tightly integrated structural and functional units called **tissues**. Based on this observation, sponges are sometimes referred to as **parazoans** ("beside-animals"). All other animals have tissues; collectively, they are sometimes referred to as **eumetazoans** ("truly-among-animals").

Among the eumetazoans, the number of tissue layers that exist in an embryo is a key trait. Animals whose embryos have two types of tissues are called **diploblasts** ("two-sprouts"); animals whose embryos have three types are called **triploblasts** ("three-sprouts"). By examining developing embryos with the light microscope, biologists documented that embryonic tissues are organized in layers, called **germ layers**. In diploblasts these germ layers are called **ectoderm** and **endoderm**; the third layer in triploblasts is found between these two and is called **mesoderm**. The Greek roots *ecto*, *meso*, and *endo* refer to *outer*, *middle*, and *inner*, respectively; the root *derm* means "skin." All animal embryos except those of sponges have distinct outer and inner layers, or "skins"; most also have a distinct middle layer.

The embryonic tissues found in animals develop into distinct adult tissues, organs, and organ systems. In triploblasts, for example, ectoderm gives rise to skin and the nervous system. Endoderm gives rise to the lining of the digestive tract. The digestive tract is also called the *gut* or *gastrovascular cavity*. The circulatory system, muscle, and

(a) Caterpillar mandibles harvest leaves.

(b) Feather worm tentacles filter debris.

(c) Shark jaws and teeth capture prey.

FIGURE 31.2 Animals Move and Eat in Diverse Ways
Variation in limbs and mouthparts allows animals to move and harvest food in a wide variety of ways.

TABLE 31.1 An Overview of Major Animal Phyla

Group and Phylum	Common Name or Example Taxa	Estimated Number of Species
Protostomes: Lophotrochozoa		
Porifera	Sponges	5500
Cnidaria	Jellyfish, corals, anemones, hydroids, sea fans	10,000
Ctenophora	Comb jellies	100
Acoelomorpha	Acoelomate worms	10
Rotifera	Rotifers	1800
Platyhelminthes	Flatworms	20,000
Nemertea	Ribbon worms	900
Gastrotricha	Gastrotrichs	450
Acanthocephala	Acanthocephalans	1100
Entoprocta	Entoprocts	150
Gnathostomulida	Gnathostomulids	80
Sipuncula	Peanut worms	320
Echiura	Spoon worms	135
Annelida	Segmented worms	16,500
Mollusca	Mollusks (clams, snails, octopuses)	94,000
Phoronida	Phoronids	20
Ectoprocta	Ectoprocts	4500
Brachiopoda	Brachiopods; lamp shells	335
Protostomes: Ecdysozoa		
Nematoda	Roundworms	25,000
Kinorhyncha	Kinorhynchs	150
Nematomorpha	Hair worms	320
Priapula	Priapulans	16
Onychophora	Velvet worms	110
Tardigrada	Water bears	800
Arthropoda	Arthropods (spiders, insects, crustaceans)	1,100,000
Deuterostomes		
Echinodermata	Echinoderms (sea stars, sea urchins, sea cucumbers)	7000
Chaetognatha	Arrow worms	100
Hemichordata	Acorn worms	85
Chordata	Chordates (tunicates, lancelets, sharks, bony fish, frogs, reptiles, mammals)	50,000

(a) Cnidarians and ctenophores are diploblastic.

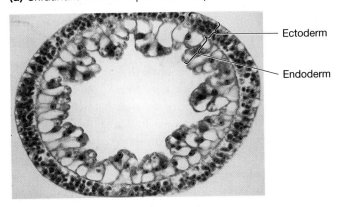

Ectoderm

Endoderm

(b) Cnidaria include hydra, jellyfish, corals, and sea pens (shown).

(c) Ctenophora are the comb jellies.

FIGURE 31.3 Diploblastic Animals Have Bodies Built from Ectoderm and Endoderm

(a) Diploblasts have just two tissue types. **(b)** Like most members of the Cnidaria, this sea pen lives in marine environments. **(c)** Comb jellies, belonging to the Ctenophora, are a major component of planktonic communities in the open ocean. This dark blue comb jelly has just swallowed a whitish comb jelly.

internal structures such as bone and most organs are derived from mesoderm. In general, then, ectoderm produces the covering of the animal and endoderm generates the digestive tract. Mesoderm gives rise to the tissues in between.

Only two groups of diploblastic animals are alive today: the cnidarians and the ctenophorans (**Figure 31.3a**). The Cnidaria (pronounced *ni-DARE-ee-uh*) include jellyfish, corals, hydra, sea pens, and anemones (**Figure 31.3b**). As **Box 31.1** (page 702) indicates, cnidarians have been an important source of model organisms in developmental biology. The Ctenophora

BOX 31.1   A Model Organism: *Hydra*

Cnidarians are of particular interest to biologists, because these diploblasts are the most ancient lineage of animals with tissues. The ectoderm and endoderm present in their embryos gives rise to a number of cell types and tissues in the adults. Cnidarian tissues may be composed of sensory cells that initiate electrical signals in response to environmental stimuli, nerve cells that process those electrical signals and conduct them throughout the body, or muscle cells that contract or relax in response to electrical signals. Cnidarians also have a particularly important type of tissue known as epithelium. **Epithelium** consists of a tightly joined layer of cells that is attached to an extensive extracellular matrix. In animals, epithelium covers the outside of the body and lines the surfaces of internal organs.

To understand why the presence of these tissues is interesting, consider the freshwater cnidarian called hydra (**Figure 31.4**). Most species in the genus

Hydra are about half a centimeter long, live attached to rocks or other firm substrates, and make their living by catching small prey or pieces of organic debris with a cluster of long tentacles. An adult hydra has three major body regions: (1) a *basal disk*, which attaches the individual to a rock; (2) a tubular section that makes up the bulk of the body; and (3) a "head" that contains the mouth and tentacles.

For over 100 years, biologists have been doing experiments based on cutting hydra bodies apart in various ways and studying how missing tissues and body regions **regenerate**—that is, reform. This work has led to a deeper understanding of how nerve cells, muscles, epithelia, and other specialized cells and tissues arise from unspecialized cells called stem cells (see Chapter 22). In particular, hydra experiments provided fundamental insights into how cell-to-cell signals (1) organize cells into tissues and body segments and (2) trigger the differ-

entiation of cells into specialized cell types. Studying how adult hydra regenerate has helped biologists understand how tissues form and specialized cells arise in animal embryos.

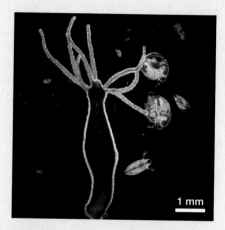

FIGURE 31.4 Hydra Is a Model Organism in Biology
Hydra grow quickly and are relatively easy to maintain in the lab. If an adult is cut into pieces, missing body parts can regenerate in some fragments to form complete adults.

(pronounced *ten-AH-for-ah*) are the comb jellies (**Figure 31.3c**). All other animals, from leeches to humans, are triploblastic.

Symmetry and Cephalization   A basic feature of a multicellular body is the presence or absence of a plane of symmetry. An animal's body is symmetrical if it can be divided by a plane such that the resulting pieces are nearly identical. Animal bodies can have 0, 1, 2, or more planes of symmetry. Most sponges, including the one illustrated in **Figure 31.5a**, are **asymmetrical**—that is, having no planes of symmetry. They cannot be sectioned in a way that produces similar sides.

All other animals exhibit radial ("spoke") symmetry or bilateral ("two-sides") symmetry. Organisms with **radial symmetry** have at least two planes of symmetry. Most of the radially symmetric animals living today either float in water or live attached to a substrate. As **Figure 31.5b** shows, their bodies are often cylinder-like. As a result, they can capture prey or react to predators that approach from more than one direction.

Organisms with **bilateral symmetry**, in contrast, face their environment in one direction. Bilaterally symmetric animals have one plane of symmetry and tend to have a long, narrow

body with a distinct head end (**Figure 31.5c**). The evolution of bilateral symmetry was a critical step in animal evolution, because it triggered a series of associated changes that are collectively known as **cephalization**: the evolution of a **head**, or *anterior* region, where structures for feeding, sensing the environment, and processing information are concentrated. Bilateral symmetry and cephalization made unidirectional movement possible. Feeding and sensory structures on the head face the environment, while *posterior* regions, at the opposite end of the organism, are specialized for locomotion. With the exception of adult forms of species in the phylum Echinodermata, which have radial symmetry, all triploblastic animals have bilateral symmetry. The echinoderms (pronounced *ee-KINE-oh-derms*) include species such as sea stars, sea urchins, feather stars, and brittle stars. Although their larvae are bilaterally symmetric, adult echinoderms are said to have **pentaradial symmetry**—meaning five planes of symmetry.

To explain the pervasiveness of bilateral symmetry, biologists point out that locating and capturing food is particularly efficient when movement is directed by a distinctive head region and powered by a long posterior region. In combination with the origin of mesoderm, which made the evolution of extensive

(a) Asymmetry

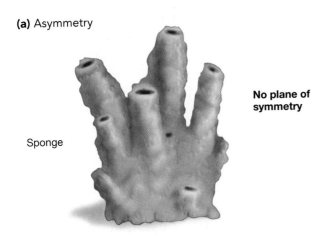

No plane of symmetry

Sponge

(b) Radial symmetry

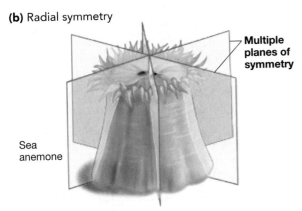

Multiple planes of symmetry

Sea anemone

(c) Bilateral symmetry

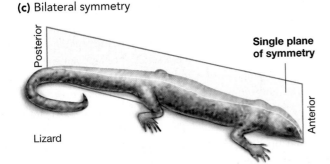

Single plane of symmetry

Posterior

Anterior

Lizard

FIGURE 31.5 There Are Three Types of Body Symmetry in Animals
QUESTION Are the animals in Figures 31.2a, 31.2c, and 31.3 asymmetric, radially symmetric, or bilaterally symmetric? Answer the same question about humans and sea stars.

musculature possible, a bilaterally symmetric body plan enabled rapid, directed movement and hunting. Lineages with a triploblastic, bilaterally symmetric body had the potential to diversify into an array of formidable eating and moving machines.

Evolution of a Body Cavity A third architectural element that distinguishes animal phyla is the presence of an internal fluid-filled cavity called a **coelom** (pronounced *SEE-loam*). Although diploblasts have a central canal that functions in digestion and circulation, they do not have a coelom. The triploblasts

called Platyhelminthes and the Acoelomorpha ("no-cavity-form") also lack a fluid-filled body cavity (**Figure 31.6a**). The remaining triploblasts have a body cavity.

Biologists were able to determine the nature of the body cavity in various phyla through careful observation and dissection of developing embryos and adults. Animals that do not have a coelom are called **acoelomates;** those that possess a coelom are known as **coelomates**. In a few of the coelomate groups, such as the roundworms and rotifers, the enclosed cavity forms between the endoderm and mesoderm layers in the embryo. This design is called a **pseudocoelom,** meaning "false-hollow" (**Figure 31.6b**). The term is unfortunate, because there is nothing false about the fluid-filled cavity—it exists. It simply forms in a different way than a "true" coelom, which forms from within the mesoderm itself and is lined with cells from the mesoderm (**Figure 31.6c**). As a result, muscle and blood vessels can form on either side of the coelomates' body cavity. In this respect, the coelom represents a more effective design than the pseudocoelom and the acoelomate condition in diploblasts.

(a) Acoelomates have no body cavity.

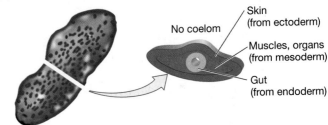

No coelom

Skin (from ectoderm)

Muscles, organs (from mesoderm)

Gut (from endoderm)

(b) Pseudocoelomates have a body cavity partially lined with mesoderm.

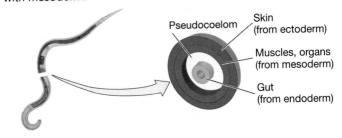

Pseudocoelom

Skin (from ectoderm)

Muscles, organs (from mesoderm)

Gut (from endoderm)

(c) Coelomates have a body cavity completely lined with mesoderm.

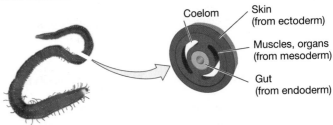

Coelom

Skin (from ectoderm)

Muscles, organs (from mesoderm)

Gut (from endoderm)

FIGURE 31.6 Animals May or May Not Have a Body Cavity

The coelom is important because it creates a container for the circulation of oxygen and nutrients, along with space where internal organs can move independently of each other. In addition, an enclosed, fluid-filled chamber can act as an efficient **hydrostatic skeleton.** Soft-bodied animals with hydrostatic skeletons can move even if they do not have fins or limbs. Movement is possible because the pseudocoelom or coelom of these animals is filled with fluid that is under pressure from the wall of the body cavity—much like a water balloon. The pressurized fluid stiffens the organism, and when muscles in the body wall contract against the pressurized fluid, it moves. When muscles contract, they shorten; when they relax, they lengthen. The shape of the body cavity enclosed by muscles changes in response to muscle movement, because the water inside cannot be compressed. As **Figure 31.7** shows, coordinated muscle contractions and relaxations produce changes in the shape of a hydrostatic skeleton that make writhing or swimming movements possible. By providing a hydrostatic skeleton, the coelom gave bilaterally symmetric organisms the ability to move efficiently in search of food.

(a) Hydrostatic skeleton of a nematode

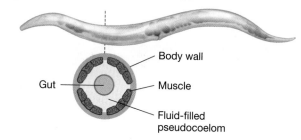

(b) Coordinated muscle contractions result in locomotion.

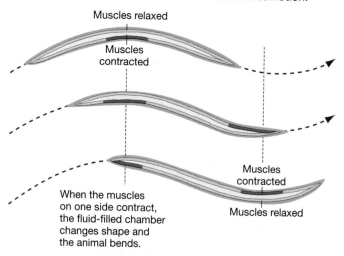

Muscles relaxed

Muscles contracted

Muscles contracted

Muscles relaxed

When the muscles on one side contract, the fluid-filled chamber changes shape and the animal bends.

FIGURE 31.7 Hydrostatic Skeletons Allow Limbless Animals to Move
The nematode (roundworm) moves with the aid of its hydrostatic skeleton. In this case, the hydrostatic skeleton is an enclosed, fluid-filled chamber. QUESTION Suppose muscles on both sides of this nematode contracted at the same time. What would happen?

The Protostome and Deuterostome Patterns of Development

With the exception of adult echinoderms, all of the coelomates—including juvenile forms of echinoderms—are bilaterally symmetric and have three embryonic tissue layers. This huge group of organisms is formally called the **Bilateria,** because they are bilaterally symmetric at some point in their life cycle. The bilaterians, in turn, can be split into two subgroups based on distinctive events that occur early in the development of the embryo. The two groups are the protostomes and the deuterostomes. The vast majority of animal species, including the arthropods (insects, spiders, crustaceans), mollusks, and annelids (segmented worms), are protostomes. Chordates (ascidians, lancelets, fish, frogs, mammals) and echinoderms are deuterostomes.

To understand the differences in how protostome and deuterostome embryos develop, recall from Chapter 21 that the development of an animal embryo begins with cleavage. **Cleavage** is a rapid series of mitotic divisions that occurs in the absence of growth. Cleavage divides the egg cytoplasm and often results in a hollow ball of cells. In many protostomes, these cell divisions take place in a pattern known as **spiral cleavage.** When spiral cleavage occurs, the mitotic spindles of dividing cells orient at an angle to the main axis of the cells and result in a helical arrangement of cells. In many deuterostomes, the mitotic spindles of dividing cells orient parallel or perpendicular to the main axes of the cells, resulting in cells that stack directly on top of each other; this pattern is called **radial cleavage** (**Figure 31.8a**).

After cleavage has created a ball of cells, the process called gastrulation occurs. **Gastrulation** is a series of cell movements that results in the formation of ectoderm, mesoderm, and endoderm—the three embryonic tissue layers. In both protostomes and deuterostomes, gastrulation begins when cells move into the center of the ball of cells. The invagination of cells creates a pore that opens to the outside (**Figure 31.8b**). In **protostomes,** this pore becomes the mouth. The other end of the gut, the **anus,** forms later. In **deuterostomes,** however, this initial pore becomes the anus; the mouth forms later. Translated literally, protostome means "first-mouth" and deuterostome means "second-mouth."

The final difference between the groups arises as gastrulation proceeds and the coelom begins to form. As **Figure 31.8c** indicates, the coelom of protostomes begins to form via a split within a solid block of mesoderm. In deuterostomes, however, layers of mesodermal cells pinch off from the gut to form the coelom.

To summarize, the protostome and deuterostome patterns of development result from differences in three processes: cleavage, gastrulation, and coelom formation. In essence, the protostome and deuterostome patterns of development represent two distinct ways of achieving the same end—the construction of a bilaterally symmetric body that contains a cavity lined with mesoderm.

PROTOSTOMES   **DEUTEROSTOMES**

(a) Cleavage
(zygote undergoes
rapid divisions,
eventually forming
a ball of cells)

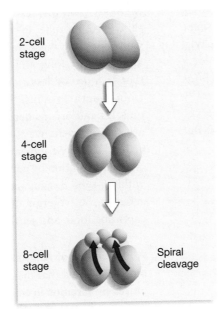
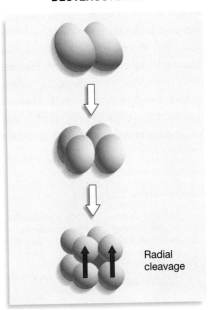

2-cell
stage

4-cell
stage

8-cell
stage   Spiral
            cleavage

Radial
cleavage

(b) Gastrulation
(ball of cells
formed by cleavage
invaginates to form
gut and embryonic
tissue layers)

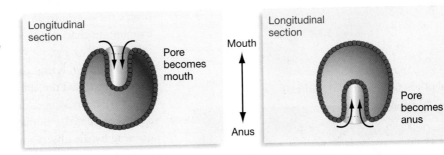

Longitudinal
section

Pore
becomes
mouth

Mouth

Anus

Longitudinal
section

Pore
becomes
anus

(c) Coelom formation
(body cavity lined
with mesoderm
develops)

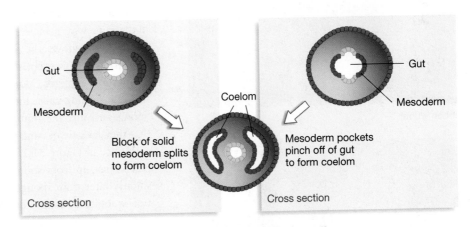

Gut

Mesoderm

Coelom

Block of solid
mesoderm splits
to form coelom

Cross section

Gut

Mesoderm

Mesoderm pockets
pinch off of gut
to form coelom

Cross section

FIGURE 31.8 In Protostomes and Deuterostomes, Three Events in Early Development Differ
The differences between protostomes and deuterostomes show that there is more than one way to build a
bilaterally symmetric, coelomate body plan.

The Tube-within-a-Tube Design   Over 99 percent of the animal species alive today are bilaterally symmetric triploblasts
that have coeloms and follow either the protostome or
deuterostome pattern of development. This combination of features has been a spectacularly successful way to design and
build a moving and eating machine.

Although it might sound complex to call a certain animal a
"bilaterally symmetric, coelomic triploblast with protostome
[or deuterostome] development," the bodies of most animals
are actually extremely simple in form. The basic animal body
is a **tube within a tube.** The inner tube is the individual's
gut, and the outer tube forms the body wall, as illustrated

in **Figure 31.9a.** The mesoderm in between forms muscles and organs. In several animal phyla, individuals have long, thin, tubelike bodies that lack limbs. Animals with this body shape are commonly called **worms**. There are many wormlike phyla, including the nemerteans and sipunculids (**Figure 31.9b**).

What about more complex-looking animals, such as grasshoppers and lobsters and horses? They are bilaterally symmetric, coelomic triploblasts with protostome or deuterostome development, too, and they aren't worms. But

(a) The tube-within-a-tube body plan

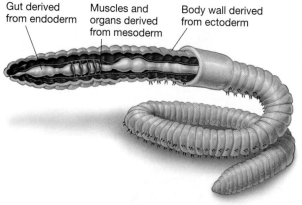

Gut derived from endoderm

Muscles and organs derived from mesoderm

Body wall derived from ectoderm

(b) Many animal phyla have wormlike bodies.

Nemertean (ribbon worm)

Sipunculid (peanut worm)

FIGURE 31.9 The Tube-within-a-Tube Body Plan Is Common in Animals
(a) Many species in the Bilateria have bodies that are variations on the tube-within-a-tube design. **(b)** The Nemertea and Sipunculida are phyla that have wormlike bodies. QUESTION How is the term *worm* similar to the term *yeast* (see Chapter 30)?

a moment's reflection should convince you that, in essence, the body plan of these animals can also be thought of as a tube within a tube, except that the tube is mounted on legs. Consider that most animals with complex-looking bodies are relatively long and thin. They have an outer body wall that is more or less tubelike and an internal gut that runs from mouth to anus. The body cavity itself is filled with muscles and organs derived from mesoderm. Wings and legs are just efficient ways to move a tube-within-a-tube body around the environment.

Once evolution by natural selection produced the basic tube-within-a-tube design, the diversification of animals was triggered by the evolution of novel types of structures for moving, capturing food, and sensing the environment.

A Phylogeny of Animals Based on Morphology When biologists realized that major groups of animals could be characterized by variation in embryonic tissues, body symmetry, type of body cavity, and early development, they used the data to infer the evolutionary relationships shown in **Figure 31.10**. The tree places these groups in a phylogenetic sequence, based on the assumption that complex body plans are derived from simpler forms.

To begin analyzing this tree, note that it identifies a group of protists called the **choanoflagellates** as the closest living relatives of animals and the **Porifera** (sponges) as the most ancient, or basal, animal phylum. These hypotheses were inspired by the observation that choanoflagellates and sponges share several key characteristics. Both are **sessile**, meaning that adults live permanently attached to a substrate. They also feed in the same way, using cells with nearly identical morphology. As **Figure 31.11** shows, the beating of flagella creates water currents that bring organic debris toward the feeding cells of choanoflagellates and sponges. Sponge feeding cells are called **choanocytes**. In these feeding cells, food particles are trapped and ingested. Sponges are also the animals with the simplest body plans. Recall that they lack tissues and that most are asymmetrical.

Notice in Figure 31.10 that the radially symmetric phyla are placed just up from sponges on the tree, meaning they evolved slightly later than sponges. Their placement at the base of the tree is logical, because cnidarians and ctenophorans have just two embryonic tissues and because radial symmetry is thought to be a simpler design than bilateral symmetry. Among the many bilaterally symmetric phyla, the tree predicts that groups evolved in the following order: acoelomates, then pseudocoelomates, and finally coelomates.

What happened *after* the coelomates split into the protostomes and deuterostomes? A close examination of Figure 31.10 suggests that two major events occurred: (1) Radial symmetry evolved as an adult trait in some echinoderms. (2) A type of body architecture called **segmentation** evolved independently in both protostomes and deuterostomes. When a body is divided into a series of repeated structures, such as an earthworm's segments or

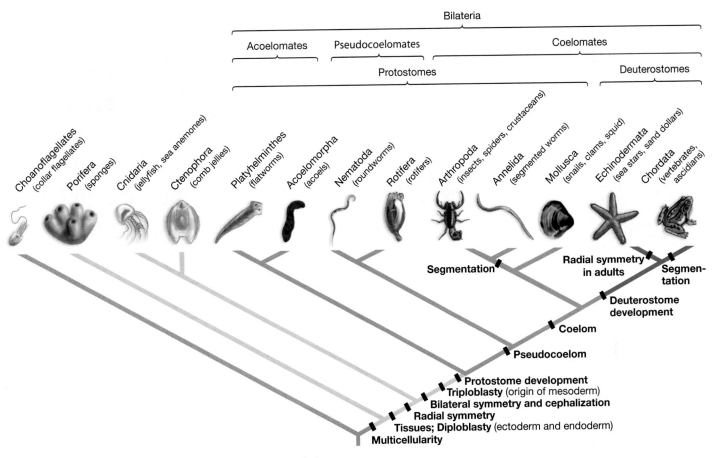

FIGURE 31.10 A Phylogeny of Animal Phyla Based on Morphology
Phylogenetic tree based on similarities and differences in the body plans and developmental sequences of
various animal phyla. The bars along the branches indicate when certain traits originated.

(a) Choanoflagellates are sessile protists; some are colonial.

(b) Sponges are multicellular, sessile animals.

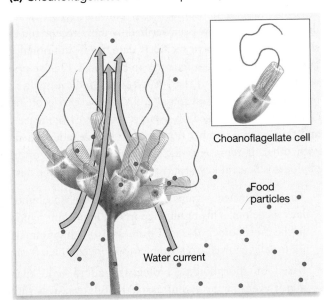

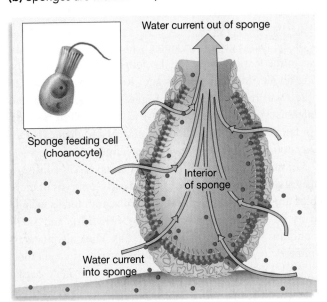

FIGURE 31.11 Choanoflagellates and Sponge Feeding Cells Are Almost Identical in Structure and Function
(a) Choanoflagellates are filter feeders. **(b)** A cross section of a simple sponge. The beating of flagella produces a
water current that brings food into the body of the sponge, where it can be ingested by feeding cells.

(a) Sponges are the first animals in the fossil record.

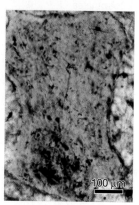

Doushantuo fossils: 580–570 million years old

(b) Cnidarians, ctenophores, and other simple forms appear later.

Ediacaran fossils: 565–544 million years old

(c) Bilaterians appear still later.

Burgess Shale fossils: 525–515 million years old

FIGURE 31.12 The Fossil Record Documents the Origin and Early Evolution of Animals
The fossil record supports the hypothesis that animals evolved in the following order: **(a)** asymmetric species that lacked tissues, **(b)** radially symmetric diploblasts, and **(c)** bilaterally symmetric triploblasts.

a fish's vertebral column and ribs, it is said to be segmented. Segmentation is found in the protostome phylum Annelida and Arthropoda (insects, spiders, crustaceans), as well as in a deuterostome lineage, the vertebrates. **Vertebrates** are a monophyletic lineage defined by the presence of a skull; many vertebrate species also have a backbone. The group called the **invertebrates**, which is defined as all animals that are not vertebrates, is paraphyletic—meaning that they include some, but not all, of the descendants of a common ancestor.

Now, what do other data sets have to say about these relationships? Specifically, do data from the fossil record and from molecular phylogenies agree with the evolutionary tree implied by morphological data, or do they conflict with it?

Using the Fossil Record

Most of the major groups of animals pictured in Figure 31.10 appear in the fossil record over the course of 65 million years, starting about 580 million years ago. Recall from Chapter 26 that the fossil record of animals begins with the Doushantuo microfossils (570 million years ago), continues with the Ediacaran faunas (565–544 million years ago), and then explodes in diversity and complexity with the Burgess Shale deposits (525–515 million years ago). The Doushantuo fossils consist of sponges and what appear to be eggs and early embryos of more complex animals (**Figure 31.12a**). The Ediacaran fossils include an array of sponges, small cnidarians (jellyfish), and small ctenophorans (comb jellies; **Figure 31.12b**). The only bilaterally symmetric organism known from these rocks is a tiny mollusk called *Kimberella*. In contrast, the Burgess Shale fossils include hundreds of bilaterally symmetric and large-bodied species from most major animal phyla. Animals ranging from sponges to chordates are present (**Figure 31.12c**).

In general, then, the fossil record of animal origins is consistent with the overall pattern of evolution described in

Figure 31.10. The first animals to appear were the sponges, followed by the diploblasts and then by the bilaterally symmetric triploblasts. The earliest animals in the fossil record were small, and tens of millions of years passed before large-bodied forms evolved. Do data from phylogenetic analyses of DNA sequence data support these conclusions?

Evaluating Molecular Phylogenies

Perhaps the most influential paper ever published on the phylogeny of animals appeared in 1997. Using sequences from the gene that codes for the RNA molecule in the small subunit of the ribosome, Anna Marie Aguinaldo and colleagues estimated the phylogeny of species from 14 animal phyla. The results were revolutionary. Although they continue to spark intense debate, they have now been verified by more recent and extensive analyses that have included data from additional genes and phyla. The phylogenetic tree in **Figure 31.13** is an updated version of the result of the 1997 study, based on further studies of the genes for ribosomal RNA and several proteins. Because this tree is based on a large amount of sequence data, and thus a large number of traits that evolve independently of each other, it represents the best current estimate of animal phylogeny. Several key observations emerge from the data:

- The most ancient groups of triploblasts, the Acoelomorpha, lack a coelom. This result supports an important hypothesis in the morphological tree (Figure 31.10)—that animal bodies usually evolved from simpler to more complex forms.

- Based on morphology, biologists had thought that the major event in the evolution of the Bilateria was the split between the protostomes and the deuterostomes. The molecular data concur but show that an additional, equally fundamental split occurred within protostomes, forming two major subgroups with protostome develop-

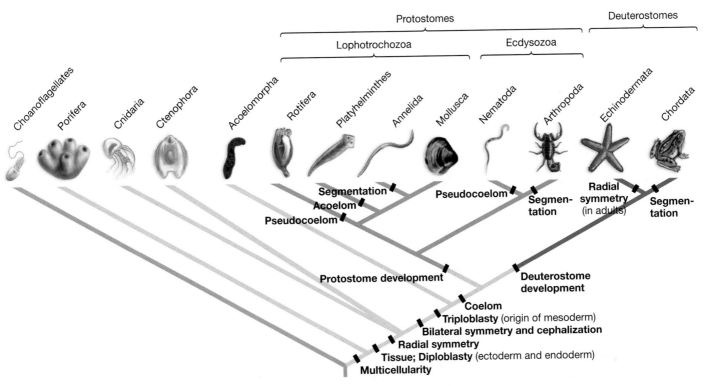

FIGURE 31.13 A Phylogeny of Animal Phyla Based on DNA Sequence Data
Phylogenetic tree based on similarities and differences in the DNA sequences of several genes from various animal phyla. The bars along the branches indicate when certain morphological traits originated. EXERCISE Circle and label at least two branches and two bars that differ from those on the tree in Figure 31.10.

ment: (1) The **Ecdysozoa** (pronounced *eck-die-so-ZOH-ah*) includes the arthropods and the nematodes; (2) the **Lophotrochozoa** (pronounced *low-foe-tro-ko-ZOH-ah*) includes the mollusks and the annelids. Ecdysozoans grow by shedding their external skeletons and expanding their bodies, while lophotrochozoans grow by extending the size of their skeletons. In Chapter 32 we explore the differences between these two lineages in more detail.

- Although both annelids (earthworms and other segmented worms) and arthropods (insect, spiders, and crustaceans) have segmented bodies, the molecular phylogeny shows that segmentation evolved independently in the two lineages as well as in vertebrates. Annelids are members of the Lophotrochozoa; arthropods are ecdysozoans.

- Species in the phylum Platyhelminthes (flatworms) do not have a coelom but are lophotrochozoans. To interpret this result, biologists point out that platyhelminths had to have evolved from an ancestor that had a coelom. Stated another way, the acoelomate condition in these species is a derived condition. It represents the *loss* of a complex trait.

- Twice during the course of evolution, bodies with pseudocoeloms arose from ancestors that had "true" coeloms. A change from coelom to pseudocoelom occurred in the ancestors of today's (1) nematodes (roundworms) and (2) rotifers.

Although biologists are increasingly confident that most or all of these conclusions are correct, the phylogeny of animals is still very much a work in progress. As data sets expand, it is likely that new analyses will not only confirm or challenge these results but also contribute other important insights into how the most species-rich lineage on the tree of life originated and diversified. Stay tuned.

✓ CHECK YOUR UNDERSTANDING

The origin and early diversification of animals was marked by changes in four fundamental features: body symmetry, the number of embryonic tissues present, the evolution of a body cavity, and protostome versus deuterostome patterns of development. You should be able to (1) explain how the evolution of bilateral symmetry is associated with cephalization, why cephalization was important, and why bilateral symmetry in combination with triploblasty and a coelom is responsible for the "tube-within-a-tube" design observed in most animals living today; (2) make a rough sketch of the phylogeny of animals based on molecular sequences, showing choanoflagellates as an outgroup, sponges and jellyfish as basal groups, Acoelomorpha as the most ancient members of the Bilateria, and the splits that produced the Lophotrochozoa, Ecdysozoa, and deuterostomes; (3) on the tree you sketched, mark the origin of multicellularity, triploblasty, protostome development, deuterostome development, and at least one origin of segmentation; and (4) compare and contrast the tree based on molecular data with the phylogeny implied by morphological traits.

31.3 What Themes Occur in the Diversification of Animals?

Within each animal phylum, the basic features of the body plan do not vary from species to species. For example, mollusks are triploblastic, bilaterally symmetric protostomes with a coelom; their body plan features a muscular *foot*, a cavity called the *visceral mass*, and a structure called a *mantle*. But there are over 100,000 species of mollusk. If animal phyla are defined by a particular body plan, what triggered the diversification of species within each phylum?

In most cases, the answer to this question is the evolution of innovative methods for feeding and moving. Recall that most animals get their food by ingesting other organisms. Animals are diverse because there are thousands of ways to find and eat the millions of different organisms that exist. Let's survey the diverse ways that animals feed, move, and reproduce.

Feeding

The feeding tactics observed in animals can be broken into five general types: (1) suspension feeding, (2) deposit feeding, (3) herbivory, (4) predation, and (5) parasitism. Many animals use more than one of these tactics over the course of their lifetime, because they undergo a **metamorphosis** ("between-forms")—a change in form during development—that allows them to exploit different sources of food as a juvenile than as an adult. Metamorphosis and all five feeding tactics are found among animals fossilized in the Burgess Shale and among species in the most diverse and familiar animal phyla living today—the mollusks, arthropods, and chordates. Let's examine each tactic in turn, then analyze how metamorphosis affects feeding.

Suspension (Filter) Feeding Suspension **feeders**, also known as **filter feeders**, capture food by filtering out particles suspended in water or air. The organisms in **Figure 31.14** illustrate a few of the many variations on this theme.

The clam pictured in Figure 31.14a uses a muscular structure called a **foot** to burrow into sediments. While burrowed, clams extend long tubes called **siphons** to maintain contact with the surface. Inside the clam body, cilia on thin structures called **gills** pump water out one siphon and draw it into the other. The incoming water contains food particles that are trapped on the gills and swept toward the mouth by cilia.

Figure 31.14b shows the small marine animals called krill, which suspension feed as they swim. As individuals move forward, their legs wave in and out. Projections on their legs trap food particles that flow past. The food particles are then moved up the body to the mouth, where they are ingested.

Figure 31.14c illustrates how krill are eaten by a group of suspension feeders called the baleen whales. These whales have a series of long plates, made from a horny material called baleen, hanging from their jaws. Baleen whales feed by gulping water that contains krill, squeezing the water out between their baleen plates, and trapping the krill inside their mouths.

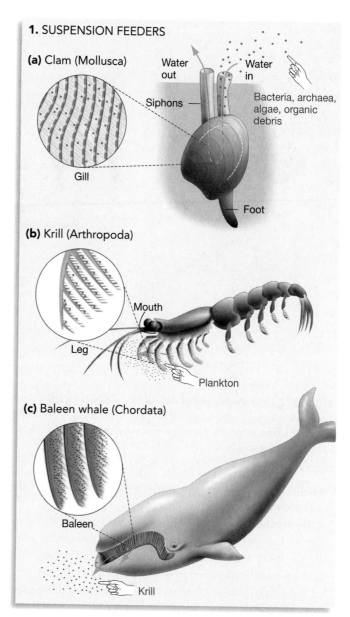

FIGURE 31.14 Suspension Feeders Capture Food by Filtering
(a) Clams, (b) krill, and (c) baleen whales are all suspension feeders. They filter food particles from water, using the trapping structures shown in the close-ups.

It is important to note that suspension feeding is found in a wide variety of animal groups. Clams are mollusks, krill are arthropods, and whales are chordates. A glance at the phylogeny in Figure 31.13 suggests that this strategy has evolved many times, independently. Although the general food-capturing strategy is the same, the mechanism and the type and size of food that is gathered vary from species to species.

Deposit Feeding Deposit **feeders** eat their way through a substrate. Earthworms, for example, are annelids that swallow soil as they tunnel through it. They digest organic matter in the soil and leave behind the mineral material as feces. For these organisms, food consists of soil-dwelling bacteria, protists, fungi, and archaea, along with detritus—the dead and often partially decomposed remains of organisms. Insects that burrow through plant leaves and stems, bore through piles of feces, or mine the carcasses of dead animals or plants can also be considered deposit feeders, because they eat through a substrate. Feeding categories are not rigid, however. Depending on what they eat, deposit feeders can also be considered **herbivores** ("plant-eaters"), parasites, **detritivores** ("detritus-eaters"), or predators. And even though earthworms deposit-feed in soil, they also retrieve and eat dead leaves from the surface.

Unlike suspension feeders, which are diverse in size and shape and use various trapping or filtering systems, deposit feeders are similar in appearance. They usually have simple mouthparts if they eat soft substrates, and a wormlike body shape (**Figure 31.15**). Like suspension feeding, however, deposit feeding occurs in a wide variety of taxonomic groups, including roundworms (Nematoda), segmented worms (Annelida), mollusks (Mollusca), peanut worms (Sipunculida), and chordates such as hagfish.

Herbivory Animals from a diversity of phyla harvest algae or plant tissues. In sharp contrast to suspension feeders and deposit feeders, herbivores have complex mouths with structures

that make biting and chewing or sucking possible. The mouthpart structure called a **radula**, for example, which is found in snails and other mollusks, functions like a rasp or a file. The sharp plates on the radula move back and forth to scrape material away from a plant or alga so that it can be ingested (**Figure 31.16a**). The long, hollow proboscis of a moth (**Figure 31.16b**) is used to suck nectar, while the **mandibles** (chewing mouthparts) of

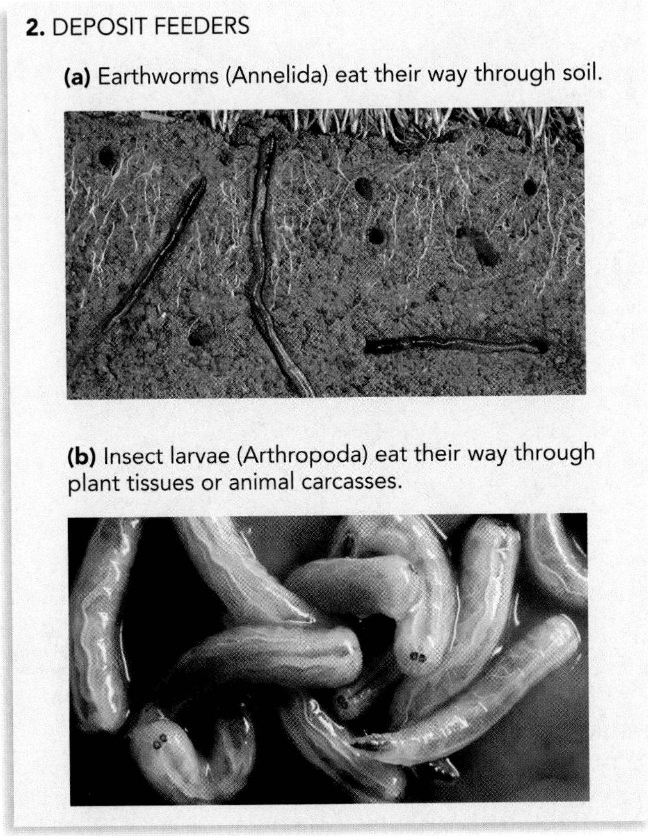

2. DEPOSIT FEEDERS

(a) Earthworms (Annelida) eat their way through soil.

(b) Insect larvae (Arthropoda) eat their way through plant tissues or animal carcasses.

FIGURE 31.15 Deposit Feeders Dig through a Substrate
Deposit feeders, including **(a)** earthworms and **(b)** maggots, generally have long, thin bodies. QUESTION Why is it logical that deposit feeders tend to have tubelike bodies?

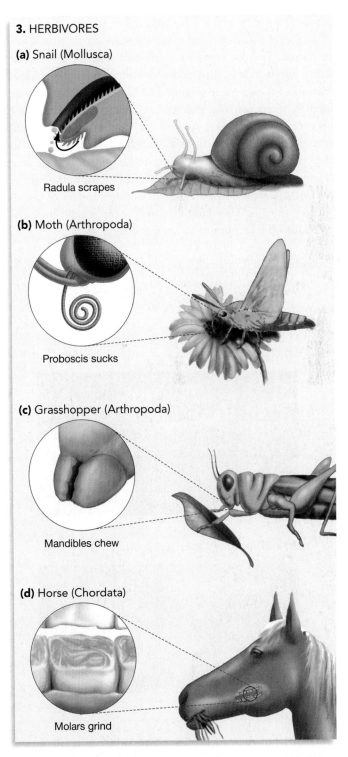

3. HERBIVORES
(a) Snail (Mollusca)
Radula scrapes
(b) Moth (Arthropoda)
Proboscis sucks
(c) Grasshopper (Arthropoda)
Mandibles chew
(d) Horse (Chordata)
Molars grind

FIGURE 31.16 Herbivores Eat Primary Producers
A tiny subset of methods animals use to harvest algae or plant tissues.

grasshoppers (**Figure 31.16c**, page 711) and grinding molars of horses (**Figure 31.16d**) are used to process leaves or stems. Animal mouthparts are a classic example of how the structures found in organisms correlate with their function in harvesting a particular type of tissue.

Predation   Animals use a fascinating variety of strategies and structures to capture and eat other animals. One way to categorize these hunting strategies is to consider whether the **predator** waits for or actively stalks its **prey**.

Frogs are chordates, and many frogs are classic sit-and-wait predators. They sit completely still and wait for an insect or worm to move close, then capture it with a lightening-quick extension of their tongue (**Figure 31.17a**). Stalkers vary in their hunting strategies. For example, wolves and mountain lions prey primarily on members of the deer family. Like other species of dogs, wolves hunt by locating a prey organism and then running it down during an extended, long-distance chase (**Figure 31.17b**). They also live and hunt in family groups called packs. Mountain lions, like nearly all other species of cat, are solitary animals. They hunt by slowly stalking their prey, then pouncing on it or running it down in a short sprint.

Parasitism   It is often difficult to draw a sharp distinction between predation and parasitism. In general, parasites are much smaller than their victims and often harvest nutrients without causing death. Predators, in contrast, are typically larger than their prey or about the same size. Predation almost always leads to the death of the victim.

Like the other feeding methods surveyed here, parasitism is practiced by species from a variety of lineages. The strategies employed also vary widely, but they can be grouped into two broad categories: endoparasitism and ectoparasitism.

Endoparasites live inside their hosts. They are often worm-like and can be very simple morphologically. Tapeworms, found in the intestines of humans and other vertebrates, for example, are platyhelminths with no digestive system. Instead of a mouth, they have hooks or other structures on their head, called a *scolex*, that attach to their host's intestinal wall (**Figure 31.18a**). Instead of digesting food in a gut, they absorb nutrients directly from their surroundings. Most endoparasites ingest their food through a mouth and have a digestive tract, however.

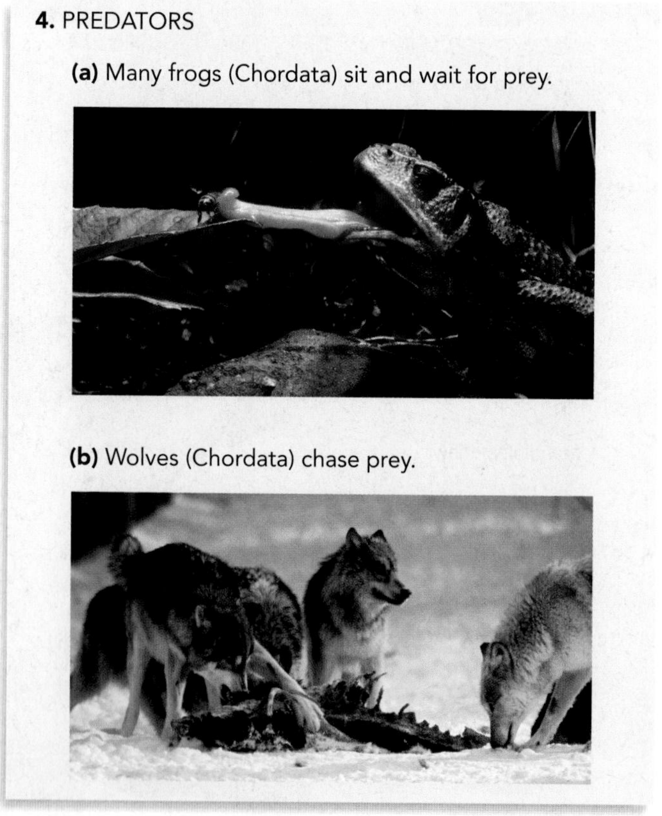

4. PREDATORS

(a) Many frogs (Chordata) sit and wait for prey.

(b) Wolves (Chordata) chase prey.

FIGURE 31.17 Predators Eat Other Animals
(a) Many frogs are sit-and-wait predators. **(b)** Wolf families hunt together by chasing prey.

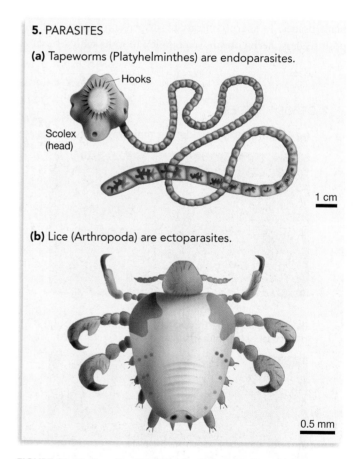

5. PARASITES

(a) Tapeworms (Platyhelminthes) are endoparasites.

Hooks

Scolex (head)

1 cm

(b) Lice (Arthropoda) are ectoparasites.

0.5 mm

FIGURE 31.18 Parasites Take Nutrients from Living Animals
(a) Tapeworms are common intestinal parasites of humans and other vertebrates. They attach to the wall of the digestive tract, using the barbed hooks on their anterior end, and absorb nutrition directly across their body wall. **(b)** Lice are insects that parasitize mammals and birds. This louse, *Phthirus pubis*, uses its clawlike legs to attach to the pubic region of humans. The animal pierces the host's skin with its mouthparts and feeds by sucking body fluids.

Ectoparasites live outside their hosts. They usually have limbs or mouthparts that allow them to grasp the host and mouthparts that allow them to pierce their host's skin and suck the nutrient-rich fluids inside. The louse in **Figure 31.18b** is an example of an insect (Arthropoda) ectoparasite that afflicts humans.

Movement

Many animals are sit-and-wait predators, and some are sessile throughout their adult lives. But the vast majority of animals move under their own power either as a juvenile or as an adult. For example, the eggs of sea anemones hatch into larvae that swim with the aid of cilia (**Figure 31.19a**). Adult sea anemones, however, spend most of their lives attached to a rock or another substrate and make their living by capturing and eating fish or other organisms that pass by (**Figure 31.19b**). In species such as these, larvae function as a dispersal stage. They are a little like the seeds of land plants—a life stage that allows individuals to move to new habitats, where they will not compete with their parents for space and other resources.

In animals that move as adults, locomotion has three functions: (1) finding food, (2) finding mates, and (3) escaping from predators. The ways that animals move in search of food and sex are highly variable; as mentioned in the introduction to this chapter, animals burrow, slither, swim, fly, crawl, walk, or run. The structures that power movement are equally variable—they include cilia, flagella, and muscles that attach to a hard skeleton or compress a hydrostatic skeleton, enabling wriggling movements. The hydrostatic skeleton is an evolutionary innovation unique to animals and is responsible for locomotion in the many animal phyla with wormlike bodies. Another major innovation occurred in animals, however, and made highly controlled, rapid movement possible: the **limb**.

Types of Limbs: Unjointed and Jointed Limbs are a prominent feature of species in many phyla and are particularly important in two major lineages: the ecdysozoans and the vertebrates. Some members of the ecdysozoa, such as onychophorans (velvet worms), have unjointed, sac-like limbs (**Figure 31.20a**); others, such as crabs and other arthropods, have more complex, jointed limbs (**Figure 31.20b**). Jointed

(a) Onychophorans are ecdysozoans with sac-like limbs.

(b) Crabs (Arthropoda) have jointed limbs.

(c) Polychaetes (Annelida) have parapodia.

(d) Sea urchins (Echinodermata) have tube feet.

(a) Motile larval anemone   **(b)** Sessile adult anemone

0.1 mm

5 cm

FIGURE 31.19 If Adults Are Sessile, Then Larvae Disperse
(a) Larval anemones swim under their own power and can disperse to new habitats. **(b)** Anemones are sessile most of their lives. This anemone has captured a blue sea star to eat it.

FIGURE 31.20 Various Animal Appendages Function in Locomotion
Lineages within the Ecdysozoa have **(a)** sac-like legs or **(b)** jointed limbs. **(c)** Some species in the Lophotrochozoa have small projections called parapodia. **(d)** Echinoderms have unusual structures called tube feet.

limbs make fast, precise movements possible and are a prominent type of limb in vertebrates and arthropods.

In essence, the limbs of arthropods and vertebrates work the same way: Limbs move when muscles that are attached to a skeleton contract or relax. The difference between the two groups is that ecdysozoans have an external skeleton, or **exoskeleton** ("outside-skeleton"), while vertebrates have an internal skeleton, or **endoskeleton** ("inside-skeleton"). But in both cases the skeleton has the same function: It is a stiff structure that resists the forces exerted by muscles. The structure and function of muscles and skeletons are detailed in Chapter 46; here the important point is that muscles, limbs, and skeletons are evolutionary innovations observed only in animals.

Are All Animal Appendages Homologous?

Chapter 23 introduced the concept of *homology*, which is defined as similarity in traits due to inheritance from a common ancestor. Traditionally, biologists have hypothesized that appendages used in animal movement evolved independently in a number of groups—meaning that not all animal limbs are homologous. To appreciate the logic behind this hypothesis, it's important to recognize just how diverse animal appendages are. Animals in a wide array of phyla have structures that stick out from the main body wall and function in locomotion. In addition to limbs such as insect and crab legs and the legs and wings of vertebrates, consider the *parapodia* of lophotrochozoans such as polychaetes (**Figure 31.20c**, page 713) and the *tube feet* of echinoderms, such as sea urchins (**Figure 31.20d**). The structure of animal appendages is so diverse that it was logical to maintain that at least some appendages evolved independently of each other. In terms of the concepts introduced in Chapter 23, the low degree of structural homology among the appendages illustrated in Figure 31.20 implies low to nonexistent developmental and genetic homology. Biologists predicted that completely different genes are responsible for each type of appendage.

Recent results have challenged this view, however. The experiments in question involve a gene called *Distal-less*, which was originally discovered in fruit flies. (*Distal* means "away from the body.") *Distal-less*, or *Dll*, is aptly named. In fruit flies that lack this gene's normal protein product, only the most rudimentary limb buds form. The mutant limbs are "distal-less." Based on the morphology of *Dll* mutants, the protein seems to deliver a simple message as a fruit-fly embryo develops: "Grow appendage out this way."

A group of biologists working in Sean Carroll's lab set out to test the hypothesis that *Dll* might be involved in limb or appendage formation in other animals. As **Figure 31.21** shows, they used a fluorescent marker that sticks to the *Dll* gene product to locate tissues where the gene is expressed. When they introduced the fluorescent marker into embryos from annelids, arthropods, echinoderms, chordates, and other phyla, they found that it bound to *Dll* in all of them. More important, the *Dll* gene products were highly localized in cells that form

Question: Is the gene *Dll* involved in limb formation in species other than insects?

Hypothesis: In all animals, *Dll* signals "grow appendage out here."

Null hypothesis: *Dll* is not involved in the development of appendages in species other than insects.

Experimental setup:

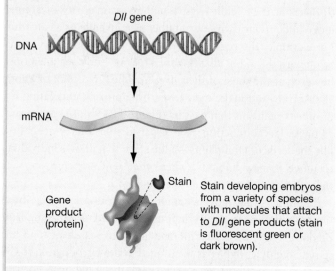

Dll gene

DNA

mRNA

Gene product (protein)

Stain

Stain developing embryos from a variety of species with molecules that attach to *Dll* gene products (stain is fluorescent green or dark brown).

Prediction: In embryos from a wide array of species, stained *Dll* gene products will be localized to areas where appendages are forming.

Prediction of null hypothesis: Stained *Dll* gene products will be localized to areas where appendages are forming only in insects.

Results:

Insect   Onychophoran   Segmented worm

In species representing both Ecdysozoa and Lophotrochozoa, *Dll* is localized in areas of the embryo where appendages are forming.

Conclusion: The gene *Dll* is involved in limb formation in diverse species. The results suggest that all animal appendages may be homologous.

FIGURE 31.21 Experimental Evidence That All Animal Appendages Are Homologous

QUESTION What results would have supported the null hypothesis?

appendages—even in phyla with wormlike bodies that have extremely simple appendages. Other experiments have shown that *Dll* is also involved in limb formation in vertebrates.

Based on these findings, biologists have argued that at least a few of the same genes are involved in the development of *all* appendages observed in animals. To use the vocabulary introduced in Chapter 23, their hypothesis is that all animal appendages have some degree of genetic homology and that they are all derived from appendages that were present in a common ancestor. The idea is that a simple appendage evolved early in the history of the Bilateria and that subsequently, evolution by natural selection produced the diversity of limbs, antennae, and wings observed today. This hypothesis is controversial, however, and research continues at a brisk pace.

Reproduction and Life Cycles

An animal may be efficient at moving and eating, but if it does not reproduce, the alleles responsible for its effective locomotion and feeding will not increase in frequency in the population. As Chapter 23 emphasized, natural selection occurs when individuals with certain alleles produce more offspring than other individuals do. Organisms live to reproduce.

Given the array of habitats and lifestyles pursued by animals, it is not surprising that they exhibit a high degree of variation in how they reproduce. Although animal reproduction will be explored in detail in Chapter 48, a few examples will help drive home just how variable animal reproduction is:

- At least some species in most animal phyla can reproduce asexually, via mitosis, as well as sexually (via meiosis). In the lophotrochozoan phylum Rotifera, a subgroup of species, the bdelloids, reproduces only asexually (**Figure 31.22**). Even certain fish, lizard, and snail species have never been observed to undergo sexual reproduction.

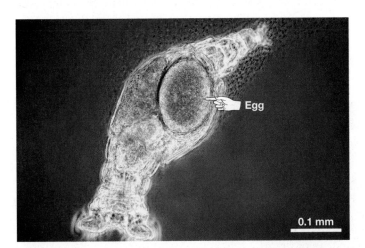

FIGURE 31.22 Bdelloid Rotifers Have Never Been Observed to Reproduce Sexually
Males have never been observed in any of the 370 species of rotifers in the group known as bdelloids. Females produce eggs that hatch into offspring without being fertilized.

(a) Internal fertilization

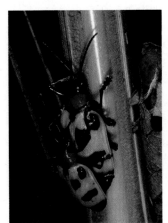

(b) External fertilization

FIGURE 31.23 Fertilization Can Be Internal or External in Animals
(a) Internal fertilization in chrysomelid beetles. **(b)** Male cubera snappers emit sperm that will fertilize a clutch of eggs laid by a female. The water in the photo is cloudy due to a high concentration of snapper sperm.

- When sexual reproduction does occur, fertilization may be internal or external. When internal fertilization takes place, males typically insert a sperm-transfer organ into the body of a female (**Figure 31.23a**). In some cases, males produce sperm in packets, which females then pick up and insert into their own bodies. But in seahorses, females insert eggs into the male's body, where they are fertilized. The male is pregnant for a time and then gives birth to live young. External fertilization is extremely common in aquatic species and involves the fusion of egg and sperm outside the female's body. Females lay eggs onto a substrate or into open water. Males then shed sperm on or near the eggs (**Figure 31.23b**).

- Eggs or embryos may be retained in the female's body during development, or eggs may be laid outside to develop independently of the mother. Species that give birth to live young are said to be **viviparous** ("live-bearing"; **Figure 31.24a**, page 716); species that deposit fertilized eggs are **oviparous** ("egg-bearing"; **Figure 31.24b**); and some species are **ovoviviparous** ("egg-live-bearing"). In ovoviviparous species, the females retain eggs inside their body during early development, but the growing embryos are nourished by yolk inside the egg and not nutrients transferred directly from the mother. In addition to mammals, a few species of sea stars, onychophorans, fish, and lizards are viviparous. The vast majority of animals, however, are oviparous.

In addition to reproducing in a variety of ways, animal life cycles vary widely. Perhaps the most spectacular innovation in animal life cycles involves the phenomenon known as metamorphosis. Recall that metamorphosis is a change from a juvenile to an adult body type.

(a) Viviparity ("live-bearing"): the birth of a shark

(b) Oviparity ("egg-bearing")

FIGURE 31.24 Some Animal Species Give Birth to Live Young, but Most Lay Eggs
(a) A lemon shark is viviparous, meaning its embryos develop for a period of time inside the female's body. **(b)** A corn snake is oviparous, meaning it lays fertilized eggs.

All insects and many other animal species have distinct juvenile and adult stages in their life cycle—often with different body forms and feeding techniques. A juvenile individual is called a **larva** (plural: **larvae**) if it looks substantially different from the adult form or a **nymph** if it looks like a miniature adult. Larvae and nymphs are sexually immature, meaning their reproductive organs are undeveloped and the individual cannot breed. It is important to appreciate just how much larvae can differ from adults in morphology, feeding behavior, and even habitat. For example, the larvae of mosquitoes live in quiet bodies of freshwater, where they suspension-feed on bacteria, algae, and detritus (**Figure 31.25a**). When a larva has grown sufficiently, the individual secretes a protective case. The individual is now known as a **pupa** (plural: **pupae**; **Figure 31.25b**). During **pupation**, the pupa's body is completely remodeled into a new, adult form. In mosquitoes, the adult individual flies and gets its nutrition as a parasite—taking blood meals from mammals and sucking fluids from plants (**Figure 31.25c**).

Fruit flies also undergo **holometabolous** ("whole-change") **metamorphosis**, meaning they undergo a drastic change in form (**Figure 31.26a**). Grasshoppers, in contrast, undergo **hemimetabolous** ("half-change") **metamorphosis**. As a grasshopper grows, it sheds its external skeleton several times and grows. In doing so, it gradually changes from a wingless, sexually immature nymph to a sexually mature adult that is capable of flight (**Figure 31.26b**). Throughout this process, grasshoppers feed on the same food source in the same way: They chew leaves.

In insects, holometabolous metaphorphosis, also known as *complete metamorphosis*, is 10 times more common than hemimetabolous metamorphosis, also called *incomplete metamorphosis*. One hypothesis to explain this observation is based on efficiency in feeding. Because juveniles and adults from holometabolous species feed on different materials in different

(a) Larvae (very different from adult)   **(b)** Pupae   **(c)** Adult

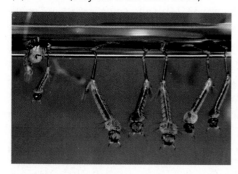      

FIGURE 31.25 During Metamorphosis, Individuals May Change Form Completely
(a) Juvenile forms of organisms that undergo metamorphosis are called larvae. **(b)** During pupation, the juvenile body is remodeled into the adult form. **(c)** Adult forms are sexually mature.

(a) Fruit fly: Complete metamorphosis (holometabolous)

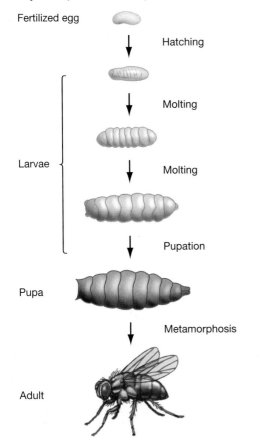

(b) Grasshopper: Incomplete metamorphosis (hemimetabolous)

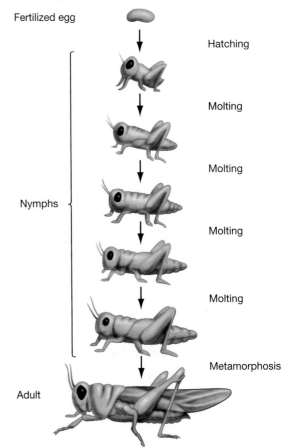

FIGURE 31.26 Incomplete Metamorphosis Does Not Involve a Dramatic Change in Form
(a) In complete metamorphosis, the larval, or sexually immature, form has a different body type than the adult.
(b) In incomplete metamorphosis, nymphs are small versions of sexually mature adults.

ways and sometimes even in different habitats, they do not compete with each other. When complete metamorphosis is part of the life cycle, individuals of the same species but different ages often show dramatic variations in their mode of feeding. An alternative hypothesis to explain the evolutionary success of complete metamorphosis is based on specialization in feeding and mating. In many moths and butterflies, for example, larvae are specialized for feeding, whereas adults are specialized for mating. Larvae are largely sessile, whereas adults are highly mobile. If specialization leads to higher fitness, then complete metamorphosis would be advantageous. Both hypotheses are still being tested, however.

Complete metamorphosis is also extremely common in marine animals. For example, most cnidarians have two distinct body types during their life cycle: (1) A largely sessile form called a **polyp** alternates with (2) a free-floating stage called a **medusa** (plural: **medusae; Figure 31.27**, page 718). Polyps live attached to a substrate, suspension feed on detritus, and fre-

quently form large clusters of individuals called colonies. A **colony** is a group of identical cells that are physically attached but do not perform closely coordinated functions. Medusae, in contrast, float freely in the plankton and feed on fish and other, larger prey. Because polyps and medusae live in different habitats, the two stages of the life cycle exploit different food sources.

✓ CHECK YOUR UNDERSTANDING

The story of animal evolution is based on two themes: (1) the evolution of a small suite of basic body plans (ways to design and build an eating and moving machine) and (2) a diversification of species with the same basic body plan, based on the evolution of innovative structures and methods for capturing food and moving. You should be able to give at least three examples of fundamental variations in the ways that animals find food, move, and reproduce.

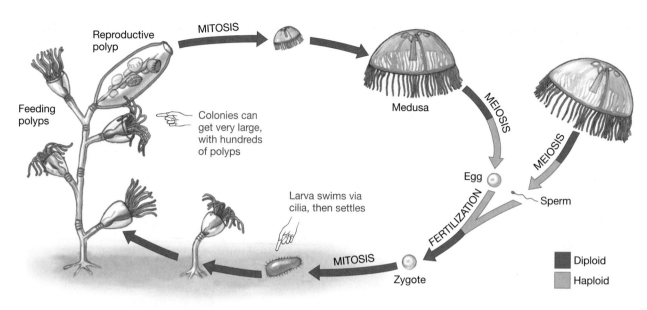

FIGURE 31.27 Cnidarian Life Cycles May Include a Polyp and Medusa Form

31.4 Key Lineages of Animals

The goal of this chapter is to provide a broad overview of how animals diversified into such a morphologically diverse lineage. According to the phylogeny illustrated in Figure 31.13, the closest living relative to the animals are the choanoflagellates, a group of protists. Phylogenetic analyses and the fossil record in-dicate that the phyla called Porifera (sponges), Cnidaria (jellyfish and others), Ctenophora, and Acoelomorpha are the most an-cient of all animal groups. Let's explore the origins of animals by taking a more detailed look at each of these lineages. In Chapters 32 and 33 we'll follow up by examining the Bilateria—the proto-stomes and deuterostomes, respectively—in more detail.

Choanoflagellates (Collar Flagellates)

The microscopic protists known as "collar flagellates" number about 150 species and are found in both freshwater and marine habitats. They are notable because they are the closest living relative of animals and because their cells are virtually identical to the flag-ellated feeding cells, or choanocytes, of sponges (see Figure 31.11a). Choanoflagellates are unicellular or occur in colonies. Recall that a colony is an aggregation of individuals that live in close physical proximity.

Feeding As flagella beat, they create water currents that bring bacteria, archaea, and small pieces of organic debris toward the cell. Food particles are then filtered out of the water by slen-der, hairlike projections of the cell that surround each flagellum (**Figure 31.28**).

Movement Adults attach to a substrate and are sessile. If cells break away from the substrate where they are growing, they can swim via the beating of their flagella.

Reproduction Choanoflagellates reproduce asexually by means of simple fission. Sexual reproduction has never been observed in the group.

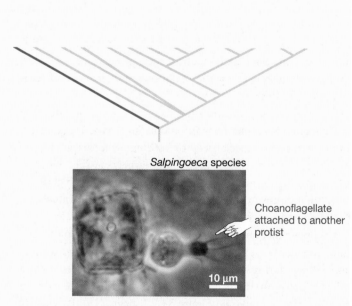

Salpingoeca species

FIGURE 31.28 Choanoflagellates Are Aquatic Suspension Feeders

Porifera (Sponges)

About 5500 species of sponge have been identified to date. Although a few freshwater species are known, most are marine. All sponges are **benthic**, meaning that they live at the bottom of aquatic environments. Sponges are particularly common in rocky, shallow-water habitats of the world's oceans and in coastal areas of Antarctica.

The architecture of sponge bodies is built around a system of tubes and pores that create channels for water currents. Body symmetry varies among sponge species; most are asymmetrical, but some species are radially symmetric (**Figure 31.29**). Sponges are considered multicellular, because they have specialized cell types. They do not, however, have the highly organized, integrated, and specialized groups of cells called tissues. In many species, either flexible collagen fibers or stiff spikes of silica or calcium carbonate ($CaCO_3$) provide structural support for the body. These structures are called **spicules** and serve to stiffen and support the body. One species native to the Caribbean can grow to heights of 2 m.

Sponges have commercial and medical value to humans. The dried bodies of certain sponge species are able to hold large amounts of water and thus are prized for use in bathing and washing. In addition, researchers are increasingly interested in the array of toxins that sponges produce to defend themselves against predators and bacterial parasites. Some of these compounds have been shown to have antibacterial properties or to promote wound healing in humans. The type of bioprospecting being done in plants (see Chapter 29) is also occurring in sponges.

Feeding   All sponges are suspension feeders. Their cells beat in a coordinated way to produce a water current that flows through small pores in the outer body wall, into chambers inside the body, and out through a single larger opening. As water passes by feeding cells, organic debris and bacteria, archaea, and small protists are filtered out of the current and then digested.

Movement   Adult sponges are sessile but may produce larvae that swim with the aid of flagella.

Reproduction   Asexual reproduction occurs in a variety of ways, depending on the species. Sponge cells are *totipotent*, meaning that an isolated adult cell has the capacity to develop into a complete adult organism. Thus, any fragment that breaks off an adult sponge has the potential to grow into a new individual. Although individuals of most species produce both eggs and sperm, self-fertilization is rare because individuals release their male and female gametes at different times. Fertilization usually takes place in the water, but some ovoviviparous species retain their eggs and then release mature, swimming larvae after fertilization and early development have occurred.

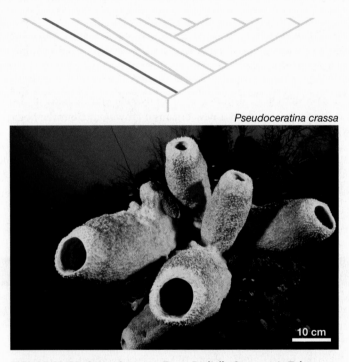

Pseudoceratina crassa

10 cm

FIGURE 31.29 Some Sponges Form Radially Symmetric Tubes

Cnidaria (Jellyfish, Corals, Anemones, Hydroids, Sea Fans)

Although a few species of Cnidaria inhabit freshwater, the vast majority of the 11,000 species are marine. They are found in all of the world's oceans, occupying habitats from the surface to the substrate, and are important predators.

Cnidarians are radially symmetric diploblasts consisting of ectoderm and endoderm layers that sandwich gelatinous material known as **mesoglea**, which contains a few scattered ectodermal cells. The gut is *blind*—meaning that there is only one opening to the environment for both ingestion and elimination of wastes.

Most cnidarians have a life cycle that includes both a sessile polyp form (**Figure 31.30a**, page 720) and a free-floating medusa (**Figure 31.30b**). Anemones, hydra, and coral, however, exist only as polyps—never as medusae. Reef-building corals secrete outer

skeletons of calcium carbonate that create the physical structure of a coral reef—one of the world's most productive habitats (see Chapter 50).

Feeding   The morphological innovation that triggered the diversification of the cnidarians is a specialized cell, called a **cnidocyte**, which is used in prey capture. When cnidocytes brush up against a fish or other type of prey, the cells forcibly eject a barbed, spear-like structure called a *nematocyst* that is coated with toxins. The barbs hold the prey, and the toxins subdue it until it can be brought to the mouth and ingested. Cnidocytes are commonly located near the mouths of cnidarians or on elongated structures called *tentacles*. Cnidarian toxins can be deadly to humans as well as to

(Continued on next page)

Cnidaria (Jellyfish, Corals, Anemones, Hydroids, Sea Fans) *(continued)*

prey organisms; in Australia twice as many people die each year from stings by box jellyfish as from shark attacks. In addition to capturing prey actively, most species of coral and many anemones host photosynthetic dinoflagellates. The relationship is mutually beneficial, because the protists supply the cnidarian host with food in exchange for protection (see Chapter 28).

Movement Both polyps and medusae have simple, muscle-like tissue derived from ectoderm or endoderm. In polyps, the gut cavity acts as a hydrostatic skeleton that works in conjunction with the muscle-like cells to contract or extend the body. Many polyps can also creep along a substrate, using muscle cells at their base. In medusae, the bottom of the bell structure is ringed with muscle-like cells. When these cells contract in a rhythmic fashion, the bell pulses and the medusa moves by **jet propulsion**—meaning a forcible flow of water opposite the direction of movement. Cnidarian larvae swim by means of cilia.

Reproduction Polyps may produce new individuals asexually by (1) budding, in which a new organism grows out from the body wall of an existing individual; (2) fission, in which an existing adult splits lengthwise to form two individuals; or (3) fragmentation, in which parts of an adult regenerate missing pieces to form a complete individual. During sexual reproduction, gametes are usually released from the mouth of a polyp or medusa and fertilization

takes place in the open water. Eggs hatch into larvae that become part of the plankton before settling and developing into a polyp.

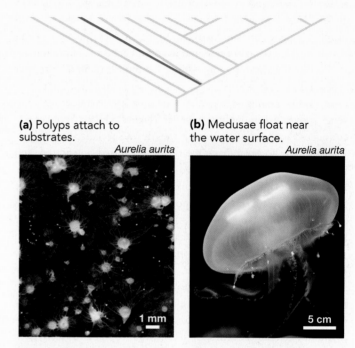

(a) Polyps attach to substrates.
Aurelia aurita

(b) Medusae float near the water surface.
Aurelia aurita

1 mm

5 cm

FIGURE 31.30 Most Cnidarians Have a Polyp Stage and a Medusa Stage

Ctenophora (Comb Jellies)

Ctenophores are transparent, ciliated, gelatinous diploblasts that live in marine habitats (**Figure 31.31**). Although a few species live on the ocean floor, most are planktonic—meaning that they live near the surface. Only about 100 species have been described to date, but some are abundant enough to represent a significant fraction of the total planktonic biomass.

Feeding Ctenophores are predators. Feeding occurs in several ways, depending on the species. Some comb jellies have long tentacles covered with cells that release an adhesive when they contact prey. These tentacles are periodically wiped across the mouth so that captured prey can be ingested. In other species, prey can stick to mucus on the body and be swept toward the mouth by cilia. Still other species ingest large prey whole. Figure 31.3c shows a comb jelly that has just swallowed a second comb jelly almost as large as itself.

Movement Adults move via the beating of cilia, which occur in comblike rows that run the length of the body. Ctenophores are the largest animals known to use cilia for locomotion.

Reproduction Most species have both male and female organs and routinely self-fertilize, though fertilization is external. Larvae are free swimming. The few species that live on the ocean floor un-

dergo internal fertilization and brood their embryos until they hatch into larvae.

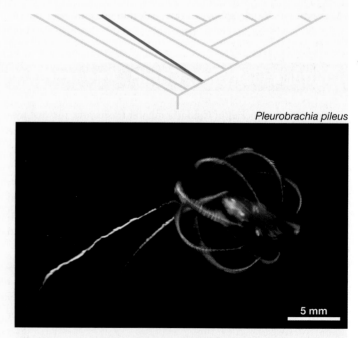

Pleurobrachia pileus

5 mm

FIGURE 31.31 Ctenophores Are Planktonic Predators

Acoelomorpha

As their name implies, the acoelomorphs lack a coelom. They are bilaterally symmetric worms that have distinct anterior and posterior ends and are triploblastic. They have simple guts or no gut at all. In some species with a gut, the mouth is the only opening for the ingestion of food and excretion of waste products. Most acoelomorphs are only a couple of millimeters long and live in mud or sand in marine environments (**Figure 31.32**).

Feeding   Acoelomorphs feed on detritus or prey on small animals or protists that live in mud or sand.

Movement   Acoelomorphs swim, glide along the surface, or burrow through substrates with the aid of cilia that cover either the entire body or the ventral surface.

Reproduction   Adults can reproduce asexually by fission (splitting in two) or by direct growth (budding) of a new individual from the parent's body. Individuals produce both sperm and eggs. Fertilization is internal, and fertilized eggs are laid outside the body.

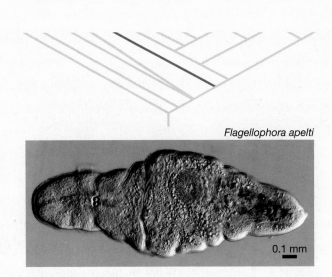

Flagellophora apelti

0.1 mm

FIGURE 31.32 Acoelomorphs Are Small Worms That Live in Mud or Sand

ESSAY   Coral Bleaching

Coral reefs are the most productive habitats in the world. They produce more kilograms of new tissue per square meter per year than does any other habitat on Earth, including tropical wet forests. Reefs may also be the most species rich of all habitats. It has been estimated that as many as 9 million species live in the world's coral reefs.

Coral reefs are found in shallow waters around tropical islands and continental areas. Reefs are built by colonies of cnidarians that secrete calcium carbonate skeletons that protect their bodies (**Figure 31.33a**, page 722). When the individuals die, their hard skeletons remain. Coral skeletons then form a substrate for the growth of additional corals, algae, or other organisms.

The cnidarians that build coral reefs make their living by capturing prey, using cnidocytes arranged on tentacles that project from their calcium carbonate skeletons (**Figure 31.33b**). But reef-building corals also host symbiotic dinoflagellates, which perform photosynthesis and contribute sugars and other products to the host animal. The bright colors of coral reefs are due to the photosynthetic pigments in these symbionts. Because cnidarians build the physical structure of coral reefs and house the most important primary producers as symbionts, corals are the essential component of reef ecosystems worldwide.

The future of the world's coral reefs may be in jeopardy, however, due a phenomenon known as coral bleaching. Corals are said to *bleach* when they expel their photosynthetic symbionts—

meaning that they eject their symbionts into the surrounding water. As a result, the corals turn white.

The frequency of large-scale bleaching events has increased dramatically since 1980, and data from laboratories and monitoring stations around the world have supported the hypothesis that recent bleaching episodes are correlated with periods of elevated water temperature. The molecular mechanism responsible for bleaching is not known, however, and the link between high temperature and bleaching has yet to be confirmed experimentally. Biologists around the world are focused on the problem because, if bleaching continues for an extended period of time, corals begin to die in large numbers. In addition to monitoring the extent and causes of bleaching, researchers hope that gaining a better understanding of the molecular mechanisms involved may suggest methods to mitigate the damage.

The future of the world's coral reefs may be in jeopardy. . . .

Research on coral reefs has become increasingly urgent. If bleaching is a response to temperature stress, then the extent and frequency of bleaching might continue to increase as global warming continues and ocean temperatures rise. It remains to be seen whether bleaching episodes will increase in duration and frequency enough to threaten the long-term stability of the world's coral reefs.

(Continued on next page)

(Essay continued)

(a) Reefs are made from the calcium carbonate skeletons of corals.

(b) Corals have tentacles lined with stinging cells.

FIGURE 31.33 Coral Reefs Are Highly Productive and Species Rich
(a) Coral reefs are found in shallow, tropical waters. The colors are the result of photosynthetic pigments in symbiotic dinoflagellates.
(b) Corals feed by stinging passing prey. QUESTION Why are coral reefs restricted to extremely shallow water habitats?

CHAPTER REVIEW

Summary of Key Concepts

▪ **Animals are a particularly species-rich and morphologically diverse lineage of multicellular organisms on the tree of life.**

The animals consist of about 34 phyla and may number 10 million or more species. Biologists study animals because they are key consumers and because humans depend on them for transportation, power, or food.

▪ **Major groups of animals are defined by the design and construction of their basic body plan, which differs in the number of tissues observed in embryos, symmetry, the presence or absence of a body cavity, and the way in which early events in embryonic development proceed.**

Sponges are the only animals that are asymmetric and lack tissues. The Cnidarians and Ctenophores have radial symmetry and just two embryonic tissues. A handful of species have bilateral symmetry and three embryonic tissues but lack a body cavity, or coelom. The vast majority of animal species have bilateral symmetry, three embryonic tissues, and a coelom. These design features gave rise to a widespread "tube-within-a-tube" body plan. Depending on the species involved, the tube-within-a-tube design is built in one of two fundamental ways—via the protostome pattern or deuterostome pattern of development.

Web Tutorial 31.1   The Architecture of Animals

▪ **Recent phylogenetic analyses of animals have shown that there were three fundamental splits during evolutionary history, resulting in two protostome groups (Lophotrochozoa and Ecdysozoa) and the deuterostomes. The most ancient animal group living today is the sponges. The closest living relatives to animals are choanoflagellates, a group of protists.**

Phylogenetic data support the hypothesis that triploblasty, bilateral symmetry, coeloms, and protostome and deuterostome development all evolved just once. However, phylogenetic data also suggest that coeloms were lost in the phylum Platyhelminthes, that pseudocoeloms evolved independently in the phyla Nematoda and Rotifera, and that segmented body plans arose at least three times independently as animals diversified.

▪ **Within major groups of animals, evolutionary diversification was based on innovative ways of feeding and moving. Most animals get nutrients by eating other organisms, and most animals move under their own power at some point in their life cycle.**

A wide variety of feeding strategies occurs among animals. Suspension feeders filter organic material or small organisms from water; deposit feeders swallow soils or other materials and digest the food particles they contain; herbivores use complex mouthparts to bite, suck, or rasp away plant tissues; predators kill prey by using sit-and-wait or stalking strategies; and parasites can live inside or outside their hosts and take nutrients from the living hosts.

Most animal movement is based on contractions by muscle cells in conjunction with one of three types of skeletons: (1) a hydrostatic skeleton, (2) an exoskeleton, or (3) an endoskeleton. Recent research suggests that, even though the types of appendages used in animal locomotion range from simple saclike limbs to complex lobster legs, all appendages may be homologous.

Questions

Content Review

1. Which of the following is true of *all* animals with bilateral symmetry?
   a. They are triploblastic.
   b. They have a coelom.
   c. They exhibit the protostome pattern of development.
   d. They exhibit the deuterostome pattern of development.

2. Which of the following represents a conflict between phylogenetic trees estimated from morphological data versus DNA sequence data?
   a. Choanoflagellates are the closest living relative of animals.
   b. Protostomes and deuterostomes are a fundamental split within the Bilateria.
   c. Sponges are the most ancient, or basal, lineage of animals.
   d. Body cavities evolved in the following order: acoelomate, pseudocoelomate, coelomate.

3. Which of the following represents a point of agreement between phylogenetic trees estimated from morphological data versus DNA sequence data?
   a. A fundamental split occurred during the evolution of protostomes, creating the Ecdysozoa and the Lophotrochozoa.
   b. Bilateral symmetry evolved once, in conjunction with the evolution of mesoderm.
   c. Segmentation evolved twice, independently, during the evolution of protostomes.
   d. Pseudocoeloms evolved twice, independently, during the evolution of protostomes.

4. Why do some researchers maintain that the limbs of all animals are homologous?
   a. Homologous genes, such as *Dll*, are involved in their development.
   b. Their structure—particularly the number and arrangement of elements inside the limb—is the same.
   c. They all function in the same way—in locomotion.
   d. Animal appendages are too complex to have evolved more than once.

5. In a "tube-within-a-tube" body plan, what is the interior tube?
   a. ectoderm
   b. mesoderm
   c. either the coelom or the pseudocoelom
   d. the gut

6. What is the key difference between choanoflagellates and sponges?
   a. Sponges are multicellular.
   b. Sponges are asymmetrical and do not have tissues.
   c. Choanoflagellates have distinctive flagellated cells that function in suspension feeding.
   d. Choanoflagellates are strictly aquatic.

Conceptual Review

1. Explain the difference between a diploblast and a triploblast. Why was the evolution of a third embryonic tissue layer important?

2. Explain how a hydrostatic skeleton works. How is a hydrostatic skeleton similar to an exoskeleton or endoskeleton? How is it different?

3. Give an example of a suspension feeder that moves and one that is sessile. Describe how these species are able to catch prey.

4. Compare and contrast the types of mouthparts and body shapes you would expect to find in herbivorous insect species that suck plant fluids from stems, bore through stems, bite leaves, and suck nectar from flowers.

5. Explain the differences in the protostome and deuterostome patterns of development. Do you agree with the text's claim that these are just different ways of building a tube-within-a-tube body plan? Explain why or why not.

6. Why was the evolution of cephalization correlated with the evolution of bilateral symmetry? Why were both of these features significant?

Group Discussion Problems

1. Would you expect internal fertilization to be more common in aquatic or terrestrial environments? Explain your answer. What types of differences would you expect to find in the structure of eggs that are laid in aquatic versus terrestrial environments?

2. Ticks are arachnids (along with spiders and mites); mosquitoes are insects. Both ticks and mosquitoes are ectoparasites that make their living by extracting blood meals from mammals. Ticks undergo incomplete metamorphosis, while mosquitoes undergo complete metamorphosis. Based on these observations, would you predict ticks or mosquitoes to be the more successful group in terms of number of species, number of individuals, and geographic distribution? Explain why. How could you test your prediction?

3. Suppose you are walking along an ocean beach at low tide and find an animal that is unlike any you have ever seen before. How would you go about determining how the animal feeds and how it moves? How would you go about determining the major features of its body plan?

4. Suspension feeding is extremely common in aquatic organisms but rare in terrestrial organisms. Generate a hypothesis to explain this observation.

Answers to Multiple-Choice Questions 1. a; 2. d; 3. b; 4. a; 5. d; 6. a

32 Protostome Animals

KEY CONCEPTS

▦ Molecular phylogenies support the hypothesis that protostomes are a monophyletic group divided into two major subgroups: the Lophotrochozoa and the Ecdysozoa.

▦ Although the members of many protostome phyla have limbless, wormlike bodies and live in marine sediments, the most diverse and species-rich lineages—Mollusca and Arthropoda—have body plans with a series of distinctive, complex features.

▦ Key events triggered the diversification of protostomes, including several lineages making the water-to-land transition, a diversification in appendages and mouthparts, and the evolution of metamorphosis in both marine and terrestrial forms.

A small sample of the insects collected from a single tree in the Amazonian wet forest. In numbers of individuals and species richness, protostomes are the most abundant and diverse of all animals.

Protostomes are one of two major monophyletic groups of bilaterally symmetric, coelomate animals. Recall from Chapter 30 that protostomes undergo early development in a dramatically different way than the other major lineage of bilaterians, the deuterostomes, do.

In terms of the total number of species and number of individuals living today, the protostomes are far and away the most successful of all animals. To drive this point home, consider just one lineage in the phylum Arthropoda: the insects. Although biologists can only guess at the total number of insect species, about 925,000 have been formally identified to date. Estimates of the actual number of species that exist range as high as 10 million. Over 33 percent of *all* known animal species on Earth

are in the insect lineage called Coleoptera—the beetles. In addition to being a species-rich clade, beetles are abundant. A single acre of pastureland in England is estimated to contain 17,825,000 individual beetles. In tropical rain forests, beetles and other insects make up 40 percent of the total mass of organisms present. The world population of ants is estimated to be 1 million billion individuals.

It is important to recognize, though, that not all protostome lineages have been as spectacularly successful as the Arthropoda. There are just 135 species in the phylum Echiura (spoon worms) and 150 species in the Kinorhyncha (mud dragons). Protostomes range from the most wildly successful phyla on the tree of life to some of the most obscure.

One of the goals of this chapter is to describe this variation in species richness and abundance. Given that each protostome phylum is defined by aspects of its body plan and by innovations that are involved in feeding and locomotion, why have a few lineages been so inordinately successful?

Before delving into this question, let's consider a fundamental issue. Tens of thousands of biologists have devoted their lives to studying protostomes. Why?

32.1 Why Do Biologists Study Protostomes?

If you asked a group of biologists why it's a good idea to study protostomes, they'd probably reply, "Because they're there." Or better yet, because there are so many of them all over the place. Protostomes include some of the most familiar and abundant organisms on Earth (**Figure 32.1**). The phylum Mollusca, for example, is composed of the snails, clams, chitons, and cephalopods (octopuses and squid)—a total of over 93,000 known species. The phylum Arthropoda contains four major subphyla, each of which is extraordinarily diverse and species-rich. The Arthropoda is made up of the insects, chelicerates (spiders and mites), crustaceans (shrimp, lobster, crabs, barnacles), and myriapods (millipedes, centipedes). A fifth major arthropod lineage, the trilobites, went extinct 250 million years ago. Although no one knows for sure, some biologists estimate that there are over 10 million species of arthropods—most of which are unnamed and undescribed.

Protostomes are organisms that you see every day. They live in just about every habitat that you might explore, and they include some of the most important model organisms in all of biological science (**Box 32.1**). If one of biology's most fundamental goals is to understand the diversity of life on Earth, then protostomes demand our attention. They are also important to human health and welfare: Some species transmit diseases or eat crops; others provide food.

Because of their diversity and abundance, protostomes play key roles in marine and terrestrial ecosystems and in human economic life. Let's take a brief look at how protostomes affect you and the organisms around you—first in the world's oceans, then on the continents.

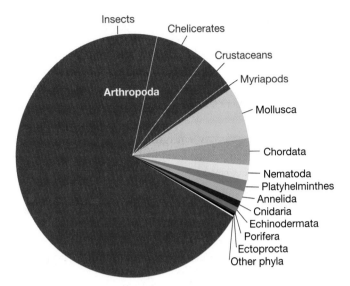

FIGURE 32.1 The Relative Abundances of Animal Lineages About 70 percent of all known species of animals on Earth are insects, most of them beetles. (Humans and other vertebrates are deuterostomes, in the phylum Chordata.) This chapter focuses on the most species-rich lineages of protostomes.

Model Organisms: *Caenorhabditis* and *Drosophila*

Protostomes include two of the most important model organisms in the life sciences: the fruit fly *Drosophila melanogaster* and the roundworm *Caenorhabditis elegans*. (*Caenorhabditis* is pronounced *see-no-rab-DIE-tiss*.) If you walk into a biology building on any university campus around the world, you are almost certain to find at least one laboratory where fruit flies or roundworms are being studied.

Since the early 1900s, *Drosophila melanogaster* has been a key experimental subject in genetics. It was initially chosen as a focus for study by T. H. Morgan because it can be reared in the laboratory easily and inexpensively (**Figure 32.2a**, page 726), matings can be arranged, the life cycle is completed in less than two weeks, and females lay a large number of eggs. These traits made fruit flies valuable subjects for breeding experiments designed to test hypotheses about how traits are transmitted from parents to offspring (see Chapter 13). More recently, *D. melanogaster* has also become a key model organism in the field of developmental biology. The use of flies in developmental studies was inspired in large part by the work of Christianne Nüsslein-Volhard and Eric Wieschaus, who isolated flies with genetic defects in early embryonic development. By

(Continued on next page)

(Box 32.1 continued)

investigating the nature of these defects, researchers have gained valuable insights into how various gene products influence the development of eukaryotes (see Chapter 22).

Caenorhabditis elegans emerged as a model organism in developmental biology in the 1970s, due largely to work by Sydney Brenner and colleagues. This roundworm was chosen for three reasons: (1) Its cuticle (soft outer layer) is transparent, making individual cells relatively easy to observe (**Figure 32.2b**); (2) adults have exactly 959 nonreproductive cells; and, most important, (3) the fate of each cell in an embryo can be predicted because cell fates are invariant among individuals. When researchers examine a 32-cell *C. elegans* embryo, they know exactly which of the 959 cells in the adult will be derived from each of those 32-embryonic cells. In addition, *C. elegans* are small (less than 1 mm long), are able to self-fertilize or cross-fertilize, and undergo early development in just 16 hours. The entire of genome of *C. elegans*, and that of *D. melanogaster*, has now been sequenced (see Chapter 20).

Model organisms do not need to be large or glamorous. They just need to be effective experimental subjects.

(a) Fruit flies (*Drosophila melanogaster*) can be reared in bottles.

(b) *Caenorhabditis elegans* is transparent and can be reared in petri dishes.

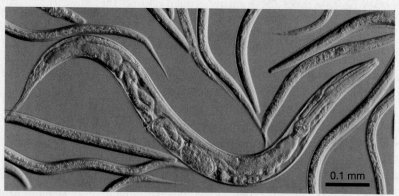

0.1 mm

FIGURE 32.2 Model Organisms Can Be Readily Reared and Studied
(a) Fruit flies (*Drosophila melanogaster*) are small enough that they can be raised on a nutrient medium in bottles. **(b)** It is relatively straightforward to observe changes in cells as *Caenorhabditis elegans* develops, because individuals are transparent.

Crustaceans and Mollusks Are Important Animals in Marine Ecosystems

In human cultures all over the world, crabs, lobster, shrimp, clams, oysters, mussels, and squid are among the highest-priced and most sought-after of all seafood. Although a few of these commercially important marine protostomes are cultivated in pens or other types of semidomesticated situations, the vast majority of the harvest represents animals that are trapped or netted in the wild. The crustacean and mollusk harvest is big business. Regulations focused on achieving sustainable harvests have to be written with input from biologists who understand the life cycles, reproductive systems, and habitat requirements of these animals.

In the bigger picture, crustaceans and mollusks are among the protostomes that act as consumers, predators, and scavengers in many marine food chains—particularly those that form in highly productive coastal habitats outside of the tropics. As **Figure 32.3** illustrates, crustaceans, mollusks, polychaete worms, and other protostomes are particularly important in near-shore habitats with either rocky surfaces or soft substrates such as sand or mud. But because the larvae of many crustaceans, mollusks, and other groups are small and able to swim, protostomes are also important consumers and predators in the plankton of near-shore environments. With the exception of coral reefs and kelp forests, protostomes are key players in virtually every shallow-water marine environment on the planet.

Insects, Spiders, and Mites Are Important Animals in Terrestrial Ecosystems

Each year, insects eat about one-third of the crops planted by farmers around the world. That fact alone makes them the most economically important group of protostomes. Biologists who study insects work with farmers, government agencies, and agribusinesses to devise strategies for controlling the impact of insects on food production and storage. **Entomology**, or the study of insects, is among the most important of all applied fields in biological science.

The goal of **Figure 32.4** is to hint at just how important insects and spiders are in terrestrial ecosystems. Like the crustaceans and mollusks of near-shore marine habitats, insects and spiders are the dominant consumers, scavengers, and predators

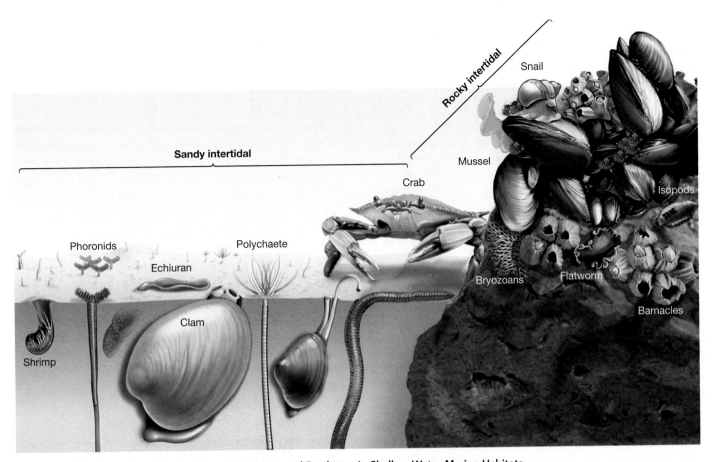

FIGURE 32.3 Protostomes Are Important Consumers and Predators in Shallow-Water Marine Habitats
Along coastlines outside the tropics, protostomes are key consumers of algae and other primary producers. They are also important predators. Protostomes are abundant in habitats with both hard and soft substrates.

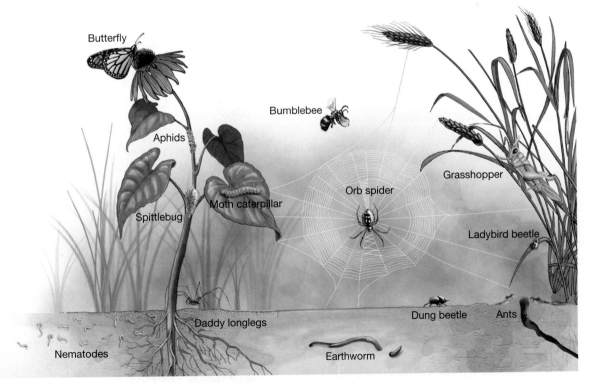

FIGURE 32.4 Protostomes Are Important Consumers and Predators in Terrestrial Habitats
Insects are key consumers of land plants. In terrestrial habitats all over the world, spiders and insects are also the most important predators of insects.

in terrestrial habitats all over the globe. Most flowering plants are pollinated by insects, and many of the leaves, stems, seeds, and roots that they produce are eaten by insects. Insects, in turn, are eaten by other insects and by spiders. Mites, which are closely related to spiders, parasitize or prey on a wide array of insects and other animal species. Insects, spiders, and mites also provide much of the food eaten by birds, lizards, salamanders, sloths, anteaters, and small mammals such as bats and shrews. Although most of the protostomes that live on land are small, their diversity and abundance make their aggregate impact enormous. It is not possible to understand a forest, prairie, or desert fully without understanding the insects, spiders, and mites that live there.

32.2 How Do Biologists Study Protostomes?

Recall from Chapter 31 that analyses of embryonic development, morphology, the fossil record, and phylogenies based on comparisons of DNA sequence data allowed researchers to make several general conclusions about protostomes:

- All protostomes have embryos that develop in a similar way. After an egg is fertilized, the pattern of cell divisions called *spiral cleavage* occurs in many protostome lineages. During gastrulation, the initial invagination that forms in the embryo becomes the mouth. If a **coelom** (a body cavity) forms later in development, it arises when openings form within blocks of mesodermal tissue.

- All protostomes share several common features of their overall body plan. They are bilaterally symmetric organisms that face the environment in one direction. They have a distinctive head region where sensory and information-processing organs are localized and a posterior region that is usually specialized for locomotion. They also have three embryonic tissues. Although most lineages have a coelom, flatworms (Platyhelminthes) are acoelomates and rotifers have a pseudocoelom.

- Certain protostomes have segmented bodies. When **segmentation** occurs, the body has a series of compartments with repeated structures—usually having similar arrangements of muscles, nerves, blood vessels, and gut. Segmentation is a prominent part of the body plan in the Annelida, including earthworms, leeches, and other segmented worms, and the Arthropoda, including insects, myriapods (millipedes and centipedes), crustaceans (crabs and shrimp), and chelicerates (spiders, ticks, and mites).

- In the fossil record, most protostome phyla first appear in the Burgess Shale faunas, dated at about 525–515 million years ago.

- Molecular phylogenies support the hypothesis that protostomes are a monophyletic group. Analyses of DNA sequence data also revealed that a major branching event

occurred within the lineage, creating the groups called **Lophotrochozoa** and **Ecdysozoa**. The lophotrochozoans include the mollusks, annelids, and flatworms (Platyhelminthes; pronounced *pla-tee-hell-MIN-theez*). Members of this group have several distinct features, including growth by incremental additions to their body. The ecdysozoans include the roundworms (Nematoda) and arthropods (Arthropoda). Ecdysozoans grow by molting. They shed an exoskeleton or external covering and then expand their body before a new covering solidifies.

To date about 20 phyla of protostomes have been identified and named. Protostomes range from large, familiar animals such as clams, lobsters, and earthworms to lesser-known groups. For example, the unfamiliar gnathostomulids (pronounced *nath-oh-stoh-MEW-lids*) and gastrotrichs (*GAS-troh-triks*) are less than 2 mm long and live in sands and muds along ocean beaches.

The protostomes are easily the most species-rich and morphologically diverse lineage of animals. What do morphological studies, the fossil record, and molecular phylogenies have to say about how this group diversified?

Analyzing Morphological Traits

Decades of careful study have shown that there is little variation in several aspects of the design and construction of the protostome body. All protostomes are triploblastic and bilaterally symmetric. In embryonic development, the initial invagination during gastrulation always forms the mouth. There is relatively little variation among phyla in how cleavage occurs.

In contrast, radical changes occurred in coelom formation as protostomes diversified. Recall from Chapter 31 that the most dramatic change was a reversion to an **acoelomate** body plan—meaning the lack of a body cavity—in Platyhelminthes. Flukes, tapeworms, and other types of flatworms do not have any sort of body cavity, even though their ancestors did. Data reviewed in Chapter 31 also support the hypothesis that the type of body cavity known as a **pseudocoelom**, which forms from an opening that arises between the ectoderm and mesoderm layers of embryos, arose independently in the protostome phylum Rotifera (rotifers) and the ecdysozoans. The Nematoda (roundworms), for example, have a prominent pseudocoelom.

The other important change that occurred in the formation of body cavities was a drastic reduction of the coelom that occurred in certain phyla. To put this change in context, it's important to note that most protostome phyla have wormlike bodies (**Figure 32.5**) with a basic **tube-within-a-tube** design. The outside tube is the skin, which is derived from ectoderm; the inside tube is the gut, which is derived from endoderm (see Chapter 31). Muscles and organs derived from mesoderm are located between the two tubes. In these animals, the coelom is well developed and forms a hydroskeleton that is the basis of movement. But in the most species-rich and mor-

(a) Priapulids are predators that burrow into the substrate and wait for prey.

(b) Nematomorpha live in freshwater environments.

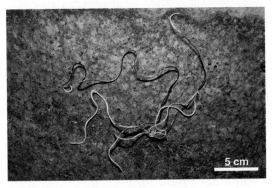

FIGURE 32.5 Members of Most Protostome Phyla Have Wormlike Bodies
(a) Priapulids are also called penis worms, because of their superficial resemblance to the male copulatory organ of mammals.
(b) Nematomorpha are also called hair worms, because they are extremely long and thin. **QUESTION** State a hypothesis for how species with soft, wormlike bodies avoid predation. How do species with wormlike bodies move?

phologically complex protostome phyla—the Arthropoda and the Mollusca (snails, clams, squid)—the coelom is drastically reduced. Arthropods and mollusks have distinct and specialized body plans.

Arthropods are segmented and have an exoskeleton made primarily of the polysaccharide chitin (see Chapter 5); in crustaceans, the exoskeleton is strengthened by the addition of calcium carbonate ($CaCO_3$). As **Figure 32.6a** shows, the arthropod body is segmented and has jointed legs and an exoskeleton. Locomotion is based on muscles that apply force against the exoskeleton to move legs or wings. Based on these observations, biologists suggest that the evolution of the exoskeleton and jointed appendages in arthropods made the coelom and hydrostatic skeleton unnecessary. Arthropods do have a spacious body cavity called the **hemocoel** ("blood-hollow"), however, which provides space for internal organs and circulation of fluids. Developmentally, the hemocoel arises as a cavity between an arthropod embryo's ectoderm and mesoderm layers—meaning it corresponds to a pseudocoelom.

Dramatic changes also occurred in mollusks, although the nature of the body plan that evolved is very different. As **Figure 32.6b** shows, the mollusk body plan is based on three major components: (1) the **foot**, a large muscle that is located at the base of the animal and is usually used in movement; (2) the **visceral mass**, the region that contains most of the main internal organs and external gill; and (3) the **mantle**, a tissue layer that covers the visceral mass and that in many species secretes a shell made of calcium carbonate. Mollusks may have one, two, eight, or no shells. When shells are present, they function to protect the animal. Movement is often based on a hydrostatic skeleton, but not one that is created by a coelom. Instead, the muscular foot itself—much like your tongue—may function as a hydrostatic skeleton. In some species, the fluid enclosed in the visceral mass also functions as a hydrostatic skeleton.

(a) Arthropod body plan (external view)

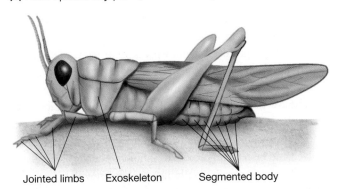

Jointed limbs   Exoskeleton   Segmented body

(b) Mollusk body plan (internal view)

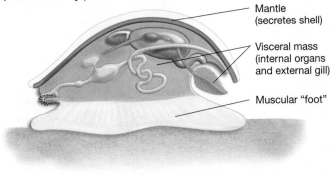

Mantle (secretes shell)

Visceral mass (internal organs and external gill)

Muscular "foot"

FIGURE 32.6 Arthropods and Mollusks Have Specialized Body Plans
(a) Arthropods have segmented bodies and jointed limbs, which enable these animals to move despite their hard outer covering, the exoskeleton. **(b)** The mollusk body plan is based on a foot, a visceral mass, and a mantle.

In addition to functioning in movement, the exoskeletons and shells of arthropods and mollusks are important in defense against predation. To support this claim, biologists point to the secretive behavior of ecdysozoans while individuals are molting. When crabs and other crustaceans have shed an old exoskeleton, their new exoskeleton takes several hours to harden. At this time, individuals hide and do not feed or move about. Experiments have shown that it is much easier for predators to attack and subdue individuals that are not protected by an exoskeleton during this intermolt period.

Using the Fossil Record

The Bryozoa (moss animals) pictured in **Figure 32.7** are the only animal phylum that appears after the Cambrian explosion ended about 515 million years ago. Thus, virtually all of the protostome phyla appear very early in the history of animal evolution. Once these phyla were present, two major events are recorded in the fossil record of protostomes: (1) the extinction of a major group of arthropods called the trilobites and (2) the first appearance of insects.

Trilobites are extremely abundant in marine rocks dated from 550–440 million years ago (**Figure 32.8a**). Like other arthropods, trilobites had a hard exoskeleton, segmented bodies, and jointed limbs. Most species lived on the sea floor and made their living by scavenging dead organic matter or deposit feeding; some may have been predators or suspension feeders. Trilobites gradually decline in numbers and species diversity over time, however, and disappear from the fossil record altogether about 250 million years ago, during the great mass extinction that occurred at the end of the Permian period.

The first insects appear in the fossil record about 400 million years ago. The first winged insects, which are the most prominent dominant group today, are observed in rocks that are about 370 million years old (**Figure 32.8b**). Recall from Chapter 29

(a) Trilobites are extinct. **(b)** Winged insect, 305 Ma

FIGURE 32.8 The Fossil Record of Protostomes Documents Extinctions as Well as the Emergence of Major New Groups
(a) Trilobites appear during the Cambrian explosion but persisted for only about 260 million years. **(b)** The first winged insects appear in the fossil record just 25 million years after the initial diversification of land plants.

that these times coincide with the Silurian-Devonian explosion—the initial, rapid diversification of land plants. Given the close association between land plants and insects today, it is logical to observe that insects initially evolved and diversified soon after land plants did.

Evaluating Molecular Phylogenies

When molecular phylogenies supported the hypothesis that the protostomes are a monophyletic group, biologists were not surprised. But when analyses of DNA sequence data indicated that two major groups of protostomes existed, it was considered a key insight into animal evolution. Analyses of the gene for the small subunit of ribosomal RNA supported the existence of the monophyletic groups Lophotrochozoa and Ecdysozoa. More recent analyses—based on sequence data from the gene for myosin, a protein involved in muscle contraction—also support this hypothesis. Although work and discussion continue, an emerging consensus holds that the divergence of protostomes into two major subgroups was an important event in animal evolution.

Figure 32.9 summarizes recent results on the phylogeny of the protostomes and indicates where major changes occurred in the evolution of protostome body plans. Only a few of the approximately 20 protostome phyla are included, however, because the relationships of most groups within the Lophotrochozoa and Ecdysozoa are still poorly resolved. Stated another way, it is not yet clear how most of the phyla within these two groups are related. As investigators analyze data from additional genes, a more highly resolved and better-supported tree should gradually emerge.

Until such a tree is available, biologists will be unable to make firm conclusions about the sequence of events that occurred as the Lophotrochozoa and Ecdysozoa diverged. Although much remains to be learned about how protostomes diversified, a great deal is understood about what happened

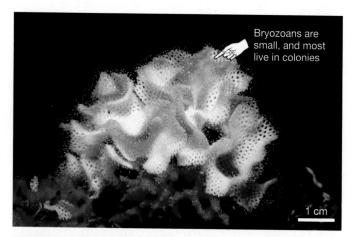

Bryozoans are small, and most live in colonies

1 cm

FIGURE 32.7 Bryozoans Are the Youngest Phylum of Animals
Bryozoans, or moss animals, make their first appearance in the fossil record about 490 million years ago. All other phyla had appeared by about 510 million years ago.

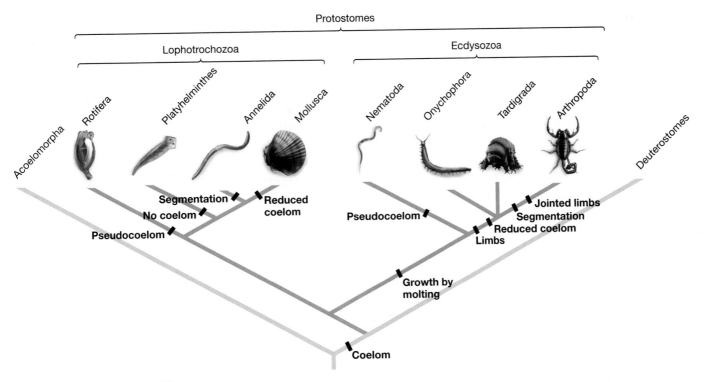

FIGURE 32.9 The Phylogeny of Protostomes
Phylogenetic tree based on analyses of DNA sequences from several genes. The tree does not include protostome phyla that have not yet been included in studies or those with relationships that are still too uncertain.

during their evolution. In addition to evolving the ability to live on land, protostomes underwent dramatic changes in methods for feeding, moving, and reproducing.

32.3 What Themes Occur in the Diversification of Protostomes?

Protostomes are the most abundant animals in the plankton and on the substrates of many marine and freshwater environments. But protostomes are common in virtually every terrestrial setting as well. Like the land plants and fungi, protostomes made the transition from aquatic to terrestrial environments. To help put this achievement in perspective, recall from Chapter 29 that green plants made the move from freshwater to land just once. From Chapter 30, recall that it is not yet clear whether fungi moved from aquatic habitats to land once or several times. Even given present uncertainties in estimating the protostome phylogeny, however, it is clear that a water-to-land transition occurred multiple times as protostomes diversified. The ability to live in terrestrial environments evolved independently in arthropods (at least twice), mollusks, roundworms, and annelids. Repeated water-to-land transitions are a prominent theme in the evolution of protostomes. In each case, the evidence for this claim is based on phylogenetic analyses, which support the hypothesis that the ancestors of the terrestrial lineages in each major subgroup were aquatic.

Why were so many different protostome groups able to move from water to land independently? The short answer is that it was easier for animals to accomplish this than for plants to do

so. For example, land plants had to evolve vascular tissue to transport water and support their stems. Similarly, the first terrestrial animals had to be able to support their body weight on land and move about. But the protostome groups that made the water-to-land transition already had hydrostatic skeletons, exoskeletons, appendages, or other adaptations for support and locomotion that happened to work on land as well as water.

To make the transition to land, two of the most important new problems that animals had to solve were (1) exchanging gases and (2) drying out. As detailed in Chapter 44, land animals exchange gases with the atmosphere readily as long as a moist body surface is exposed to the air. The bigger challenge is to prevent the gas-exchange surface and other parts of the body from drying out. Roundworms and earthworms solve this problem by living in wet soils or other moist environments. Arthropods and many mollusks have a watertight exoskeleton or a shell that minimizes water loss, and their respiratory structures are located inside the body. In insects, the openings to respiratory passages can be closed to minimize water loss. And unlike land plants and fungi, all land animals can move to moister habitats if the area they are in gets too dry.

To summarize, the ability to live in terrestrial environments was a key event in the evolution of several protostome phyla. In addition, the evolution of specialized body plans provided a foundation for diversification in the most species-rich lineages of protostomes—arthropods and mollusks. In large part, the remainder of protostome evolution was driven by innovations in feeding, moving, and reproducing.

Feeding

Protostomes include suspension feeders, deposit feeders, herbivores, predators, and parasites. The diversity of feeding strategies observed among protostomes is reminiscent of that among lineages of protists. Recall from Chapter 28 that each of the major protist lineages has at least one distinct morphological characteristic. Once unique aspects of cell or body structure had evolved, each protist lineage diversified into species that pursue a wide array of lifestyles. In a similar way, the protostome pattern of development represents a unique way of building a bilaterally symmetric animal with a coelom and three embryonic tissue types. Once this method of constructing a body had evolved, protostomes diversified into a wide variety of habitats and lifestyles—primarily as a result of variation in feeding strategies.

A wide diversity in feeding strategies is possible because protostomes have a wide variety of specialized mouthparts. Several of the wormlike phyla, for example, have unique jaws or other mouth and throat structures. Echiurans (spoon worms) suspension feed using an extended structure called a **proboscis**, which forms a gutter leading to the mouth (**Figure 32.10a**). Cells in the gutter of echiurans secrete mucus, which is sticky enough to capture pieces of detritus. The combination of mucus and detritus is then swept into the mouth by cilia on cells in the gutter. The nemerteans (ribbon worms), in contrast, have a proboscis that can extend or retract (**Figure 32.10b**). These worms wrap their extended proboscis around small animals to capture them and then pull them into the mouth. In both phyla, the structure of the mouthparts correlates closely with their function in feeding.

Regarding mouthpart diversity within protostome groups, however, arthropods take the prize. The mouthparts observed in insects and crustaceans alone vary in structure from tubes to pincers and allow them to pierce, suck, grind, bite, mop, chew, engulf, cut, and mash (**Figure 32.10c, d, e**). All insects may have the same basic body plan, but their mouthparts and food sources are highly diverse.

In many species, the jointed limbs of arthropods also play a key role in getting food. The crustaceans called krill use their legs to sweep food toward their mouths as they suspension feed; certain insects, crustaceans, spiders, and mollusks use their appendages to capture prey or hold food as it is being chewed or bitten by the mouthparts.

In most cases, juvenile and adult forms of the same species exploit different food sources. Metamorphosis is the predomi-

(a) Echiurans are suspension feeders with a mucus-lined proboscis.

(b) Nemerteans are predators with an extendable proboscis.

(c) Leaf-cutter ants cut leaves.

(d) Houseflies lap up fluids.

(e) Ladybird beetles are predators.

FIGURE 32.10 Protostome Feeding Structures Are Diverse
(a) Echiurans and **(b)** nemerteans have similar body plans and burrow in soft marine sediments, but they have distinct mouthparts and feed in completely different ways. **(c)** Leaf-cutter ants, **(d)** houseflies, and **(e)** ladybird beetles are all insects. But each species has a different type of mouthpart structure and feeds on a different source of food.

nant mode of development in protostomes and usually results in larvae and adults that live in different habitats and have different overall morphology and mouthparts.

Movement

In protostomes, variation in movement depends on variation in two features: (1) the presence or absence of limbs and (2) the type of skeleton that is present. Wormlike protostomes that lack limbs move with the aid of a hydrostatic skeleton. This type of movement is also observed in caterpillars, grubs, maggots, and other types of insect larvae—even though adults of the same species have a hard exoskeleton and move with the aid of **jointed limbs** (**Figure 32.11a**). Although insect larvae have a highly reduced coelom, they do have an enclosed, fluid-filled body cavity that functions as a hydrostatic skeleton.

The evolution of jointed limbs made rapid, precise swimming and running movements possible and was a key innovation in the evolution of protostomes. The jointed limb is a major reason that arthropods have been so spectacularly successful in terms of species diversity, abundance of individuals, geographic range, and duration in the fossil record.

But the jointed limb is just one of several evolutionary innovations that allowed protostomes to move in unique ways:

- The insect wing is a marvel of engineering. It is also one of the most important adaptations in the history of life. About two-thirds of the multicellular species living today are winged insects. According to data in the fossil record, insects were the first organisms that had wings and could fly. Like most insects today, the earliest insects had two pairs of wings. In most four-winged insects living today, however, the four wings function as two. Beetles fly with only their hind wings; butterflies, moths, bees, and wasps have hooked structures that make their two pairs of wings move together. Flies appear later in the fossil record and have a single pair of wings (**Figure 32.11b**).

(a) Walking, running, and jumping **(b)** Flying **(c)** Gliding and crawling

Jointed legs

Wings (most insects have two pairs of wings)

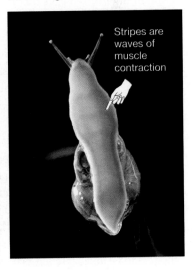

Stripes are waves of muscle contraction

(d) Jet propulsion

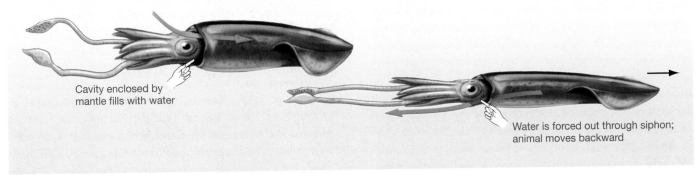

Cavity enclosed by mantle fills with water

Water is forced out through siphon; animal moves backward

FIGURE 32.11 Protostome Locomotion Is Diverse
The evolution of **(a)** jointed limbs and **(b)** wings were key innovations in arthropod movement. **(c)** A wave of muscle contractions, down the length of the foot, allows mollusks to glide along a substrate. **(d)** In jet propulsion, muscular contractions force the water out through a moveable siphon. The siphon can aim outgoing water in a specific direction.

- In mollusks, waves of muscle contractions sweep down the length of the large, muscular foot, allowing individuals to glide along a surface (**Figure 32.11c**). Cells in the foot secrete a layer of mucus, which reduces friction with the substrate and increases the efficiency of gliding.

- Squid have a mantle that is lined with muscle. When the cavity surrounded by the mantle fills with water and the mantle muscles contract, a stream of water is forced out of a tube called a **siphon**. The force of the water propels the squid backward (**Figure 32.11d**). This mechanism, **jet propulsion**, evolved in squid long before human engineers thought of using the same principle to power aircraft.

Reproduction and Life Cycles

When it comes to variation in reproduction and life cycles, protostomes do it all. Asexual reproduction by splitting the body lengthwise or by fragmenting the body is common in many of the wormlike phyla. Many crustacean and insect species reproduce asexually via **parthenogenesis** ("virgin-origin")—the production of unfertilized eggs that develop into offspring. Sexual reproduction is the predominant mode of producing offspring in most protostome groups, however. Sexual reproduction usually begins with external fertilization in sessile forms, such as clams, bryozoans, and brachiopods. It begins with internal fertilization in groups that are capable of movement—such as crustaceans, snails, and insects—because males and females can meet. Females of *ovoviviparous* insects and snails bear live young, although they do so by retaining fully formed eggs inside their bodies and then nourishing them via the nutrient-rich yolk that is inside the egg (see Chapter 31).

Two unique reproductive innovations occurred during protostome diversification: (1) the evolution of metamorphosis and (2) an egg that would not dry out on land. Metamorphosis is common in marine protostomes. In these species, metamorphosis is hypothesized to be an adaptation that allows juveniles to disperse to new habitats by floating or swimming in the plankton. Metamorphosis is also common in terrestrial insects (see Figures 31.17 and 31.18). In these species, metamorphosis is hypothesized to be an adaptation that reduces competition for food between juveniles and adults. With respect to the ability to colonize terrestrial environments, though, the most critical adaptation was an egg that resists drying. Insect eggs have a thick membrane that keeps moisture in, and the eggs of slugs and snails have a thin calcium carbonate shell that helps retain water. Thus, protostomes provide additional evidence that desiccation-resistant eggs evolved repeatedly in populations that made the transition to life on land. As we will see in Chapter 33, a membrane-bound *amniotic egg*, analogous to the membrane-bound eggs of insects, evolved in the terrestrial vertebrates. Recall from Chapter 29 that an analogous adaptation arose in land plants, when the evolution of the gametangia helped protect eggs and embryos from drying.

✓ **CHECK YOUR UNDERSTANDING**

Although all protostomes share fundamental aspects of their body plan, they occupy a wide range of habitats and feed and reproduce in a wide variety of ways. A water-to-land transition occurred multiple times in protostomes. That transition was facilitated by the ability to move away from dry areas and toward moister environments in times of water stress, by the existence of watertight shells in mollusks and exoskeletons in arthropods and by the evolution of desiccation-resistant eggs. You should be able to (1) give examples of variation in mouthpart structure and feeding, structures used in locomotion, and metamorphosis in protostomes; and (2) state a hypothesis to explain why arthropods and mollusks are the most species-rich of all animal phyla.

32.4 Key Lineages: Lophotrochozoans

The Lophotrochozoa is named for the group's two distinctive morphological traits: (1) a feeding structure called a lophophore, which is found in three phyla, and (2) a type of larva called a trochophore, which is common to many of the phyla in the lineage.

As **Figure 32.12a** shows, a **lophophore** ("tuft-bearer") is a specialized structure that rings the mouth and functions in suspension feeding. Lophophores consist of tentacles that have ciliated cells along their surface. The beating of the cilia generates a water current that sweeps protists and other alga, detritus, and animal larvae into the region above the mouth, where they are trapped by the tentacles, swept into the mouth opening, and ingested. Lophophores are found in bryozoans, brachiopods, and phoronids.

Trochophores are a type of larvae common to marine mollusks, annelids that live in the ocean, and several other phyla in the Lophotrochozoa. As **Figure 32.12b** shows, a **trochophore** ("wheel-bearer") larva has a ring of cilia around its middle. These cilia aid in swimming. The beating of the cilia may also produce water currents that sweep food particles into the mouth, though many trochophores feed primarily on yolk provided by the mother. Thus, trochophores may represent a feeding stage as well as a dispersal stage in the life cycle.

Although analyses of extensive DNA sequences and the presence of unique, derived features such as lophophores and trochophore larvae helped biologists identify the Lophotrochozoa, the lineages within this group are highly diverse in morphology. To drive this point home, let's take a closer look at four of the key phyla in the group: (1) **Rotifera**, (2) **Platyhelminthes**, (3) **Annelida**, and (4) **Mollusca**.

(a) Lophophores function in suspension feeding in adults.

0.1 mm

FIGURE 32.12 Distinctive Traits in Lophotrochozoa
(a) Three phyla of lophotrochozoans have the feeding structure called a lophophore. **(b)** Many phyla of lophotrochozoans have the type of larva called a trochophore.

(b) Trochophore larvae swim and feed.

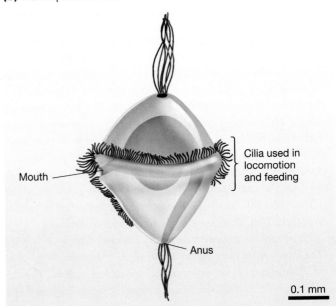

Cilia used in locomotion and feeding

Mouth

Anus

0.1 mm

Rotifera (Rotifers)

The 1800 rotifer species that have been identified thus far live in damp soils as well as marine and freshwater environments. They are important components of the plankton in freshwater and in brackish waters, where rivers flow into the ocean. They have pseudocoeloms, and most species are less than 1 mm long. Although rotifers do not have a lophophore or a trochophore larval stage, extensive similarities in DNA sequence identify them as a member of the lophotrochozoan lineage.

Feeding Rotifers have a cluster of cilia at their anterior end called a **corona** (**Figure 32.13**). In many species, the beating of the cilia in the corona makes suspension feeding possible by creating a current that sweeps microscopic food particles into the mouth. The corona is the signature morphological feature of this group.

Movement Although a few species of rotifers are sessile, most swim via the beating of cilia in the corona.

Reproduction Females produce unfertilized eggs by mitosis, which then hatch into new, asexually produced individuals. Recall from Section 32.3 that the production of offspring via unfertilized eggs is termed *parthenogenesis*. An entire group of rotifers—called the bdelloids—reproduces exclusively asexually; they do so via parthenogenesis. Both sexual reproduction and asexual reproduction are observed in most species, however. Development is direct, meaning that fertilized eggs hatch and grow into adults without going through metamorphosis.

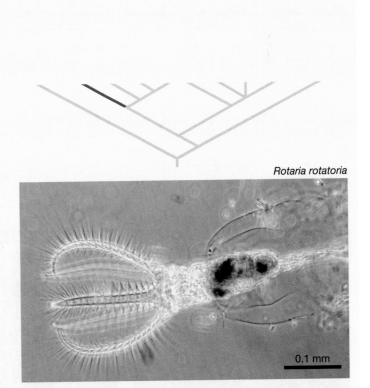

Rotaria rotatoria

0.1 mm

FIGURE 32.13 Rotifers Are Tiny, Freshwater Suspension Feeders

Platyhelminthes (Flatworms)

The flatworms are a large and diverse phylum. More than 20,000 species have been described in three major subgroups. In traditional classification schemes like those described in Chapter 1, each of these lineages was referred to as a *class*. They are (1) the free-living flatworms, (2) the endoparasitic tapeworms, and (3) the endoparasitic or ectoparasitic flukes. The free-living forms are called the Turbellaria (**Figure 32.14a**). Although a few turbellarian species are terrestrial, most live on the substrates of freshwater or marine environments. Tapeworms belong to the subgroup or class called the Cestoda and parasitize fish, mammals, or other vertebrates (**Figure 32.14b**). Most flukes are in a lineage called Trematoda (**Figure 32.14c**). Flukes parasitize vertebrates or mollusks.

Flatworms are named for the broad, flattened shape of their bodies. (The Greek roots *platy* and *helminth* mean "flat-worm.") Species in the Platyhelminthes are unsegmented and lack a coelom. They also lack structures that are specialized for gas exchange. Further, they do not have blood vessels or any other type of system for circulating oxygen and nutrients to their cells. Based on these observations, biologists interpret the flattened bodies of these animals as an adaptation that gives flatworms an extremely high surface-area-to-volume ratio. Because they have so much surface area, a large amount of gas exchange can occur directly across their body wall, with oxygen diffusing into the body from the surrounding water and carbon dioxide diffusing out. Because the volume of the body is so small relative to the surface area available, nutrients and gases can diffuse efficiently to all of the cells inside the animal. This body plan has a downside, however: Because the body surface has to be moist for gas exchange to take place, flatworms are restricted to environments where they are surrounded by fluid.

Feeding Platyhelminthes lack a lophophore and have a digestive tract that is "blind"—meaning it has only one opening for ingestion of food and elimination of wastes. Most turbellarians are hunters that prey on protists or small animals; others scavenge dead animals. Cestodes and trematodes, in contrast, are strictly parasitic and feed on nutrients provided by hosts. Trematodes gulp host tissues and fluids through a mouth and have a blind digestive tract. Cestodes lack both structures and obtain nutrients solely by diffusion through their body wall.

Movement Some turbellarians can swim a little by undulating their bodies, and most can creep along substrates with the aid of cilia on their ventral surface. Cestodes and trematodes are largely or completely sessile; adult cestodes have hooked attachment structures at their anterior end that permanently attach them to the interior of their host.

Reproduction Turbellarians can reproduce asexually by splitting themselves in half. If they are fragmented as a result of a predator attack, the body parts can regenerate into new individuals. Most turbellarians contain both male and female organs and reproduce sexually by aligning with another individual and engaging in mutual and simultaneous fertilization. Trematodes and cestodes also reproduce sexually and either cross-fertilize or self-fertilize. In many cases, the life cycle of trematodes and cestodes involves two or even three distinct host species, with sexual reproduction occurring in the **definitive host** and asexual reproduction occurring in one or more **intermediate hosts**. For example, humans are the definitive host of the blood fluke *Schistosoma mansoni*. Fertilized eggs are shed in a human host's feces and enter aquatic habitats if sanitation systems are poor. The eggs develop into larvae that infect snails. Inside the snail, asexual reproduction results in the production of a different type of larva, which emerges from the snail and burrows into the skin of humans who wade in infected water. Once inside a human, the parasite lives in blood and develops into a sexually mature adult.

(a) Turbellarians are free living.

Pseudoceros ferrugineus

1 cm

(b) Cestodes are endoparasitic.

Taenia species

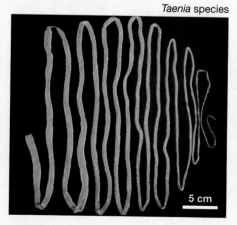

5 cm

(c) Trematodes are endoparasitic.

Dicrocoelium dendriticum

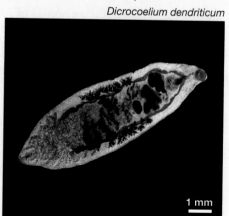

1 mm

FIGURE 32.14 Flatworms Have Simple, Flattened Bodies

Annelida (Segmented Worms)

All annelids have a segmented body and a coelom that functions as a hydrostatic skeleton. The 16,500 species that have been discovered thus far can be divided into two major lineages, traditionally called classes. Those clades are the Polychaeta and Clitellata:

1. The **Polychaeta** (pol-ee-KEE-ta), or polychaetes ("many-bristles"), are named for their numerous, bristle-like extensions called **chaetae**. The chaetae extend from appendages called **parapodia**. Polychaetes are mostly marine, and they range in size from species that are less than 1 mm long to species that grow to lengths of 3.5 m. They include a large number of burrowers in mud or sand, sedentary forms that secrete chitinous tubes, and active, mobile species (**Figure 32.15a**).

2. The **Clitellata** is composed of the oligochaetes (*oh-LIG-oh-keetes*) and leeches. Oligochaetes ("few-bristles") include the earthworms, which burrow in moist soils (**Figure 32.15b**), an array of freshwater species, and a few marine forms. Oligochaetes lack parapodia and, as their name implies, have many fewer chaetae than do polychaetes. The leeches are members of the **Hirudinea**. Their coelom is much reduced in size compared with that of other annelids and consists of a series of connected chambers. Leeches live in freshwater as well as marine habitats (**Figure 32.15c**).

Feeding Polychaetes have a wide variety of methods for feeding: The burrowing forms deposit feed; the sedentary forms suspension feed with the aid of a dense crown of tentacles; and the active forms hunt small animals, which they capture by everting their throats (that is, turning their throats inside out). Virtually all oligochaetes, in contrast, make their living by deposit feeding in soils. About half of the leeches are ectoparasites that attach themselves to fish or other hosts and suck blood and other body fluids. Hosts are usually unaware of the attack, because leech saliva contains an anaesthetic. The host's blood remains liquid as the leech feeds, because the parasite's saliva contains an anticoagulant. Parasitic leeches are still used by physicians to remove blood from particularly large bruises. The nonparasitic leech species are predators or scavengers.

Movement Polychaetes and oligochaetes crawl or burrow with the aid of their hydrostatic skeletons; the parapodia of polychaetes also act as paddles or tiny feet that aid in movement. Leeches can swim by using their hydrostatic skeletons to make undulating motions of the body.

Reproduction Asexual reproduction occurs in polychaetes and oligochaetes by transverse fission or fragmentation—meaning that body parts can regenerate a complete individual. Sexual reproduction in polychaetes may begin with internal or external fertilization, depending on the species. Polychaetes have separate sexes and usually release their eggs directly into the water. Some species produce eggs that hatch into trochophore larvae; others undergo direct development. In oligochaetes and leeches, individuals produce both sperm and eggs and engage in mutual, internal cross-fertilization. Eggs are enclosed in a mucus-rich, cocoon-like structure where they develop directly into miniature versions of their parents.

(a) Most polychaetes are marine.

Alvinella pompejana

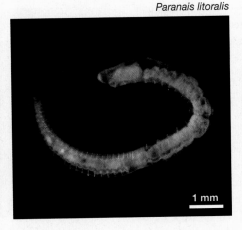

(b) Most oligochaetes are terrestrial.

Paranais litoralis

1 mm

(c) Most leeches live in freshwater.

Helobdella stagnalis

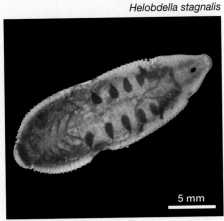

5 mm

1 cm

FIGURE 32.15 Annelids Are Segmented Worms

Mollusca (Mollusks)

The mollusks are far and away the most species-rich and morphologically diverse group in the Lophotrochozoa. They have a specialized body plan based on a muscular foot, a visceral mass, and a mantle that may or may not secrete a calcium carbonate shell. The coelom is much reduced or absent. Over 93,000 species have been described thus far. Although most mollusks live in marine environments, there are some terrestrial and freshwater forms.

Because mollusks are so diverse, let's consider each of the major subgroups or classes in the phylum separately and analyze the traits responsible for their diversification. The four most important lineages of mollusks are (1) **bivalves** (clams and mussels), (2) **gastropods** (slugs and snails), (3) **cephalopods** (squid and octopuses), and (4) **chitons**. The bivalves are suspension feeders; the other three groups of mollusks are herbivores or predators.

Mollusca ▶ Bivalvia (Clams, Mussels, Scallops, Oysters)

The bivalves are so named because they have two separate shells made of calcium carbonate secreted by the mantle. The shells are hinged, and they open and close with the aid of muscles attached to them (**Figure 32.16**). When the shell is closed, it protects the mantle, visceral mass, and foot. The bivalve shell is an adaptation that reduces predation.

Most bivalves live in the ocean, though some freshwater forms are known. Clams burrow into mud, sand, or other soft substrates and are sedentary as adults. Oysters and mussels are largely sessile as adults, but most of them live above the substrate, attached to rocks or other solid surfaces. Scallops are mobile and live on the surface of soft substrates. The smallest bivalves are freshwater clams that are less than 2 mm long; the largest bivalves are giant marine clams that may weigh more than 400 kg. All bivalves can sense gravity, touch, and certain chemicals, and scallops even have eyes.

Because most bivalves live on or under the ocean floor and because they are covered by a hard shell, their bodies are often buried in sediment after death. As a result, bivalves fossilize readily, and the Bivalvia lineage has the most extensive fossil record of any animal, plant, or fungal group. This large database has allowed biologists to conduct thorough studies of the evolutionary history of bivalves.

Bivalves are important commercially. Clams, mussels, scallops, and oysters are farmed or harvested from the wild in many parts of the world and used as food by humans. When sold commercially, they are referred to as "shellfish." Pearl oysters that are cultivated or collected from the wild are the most common source of natural pearls used in jewelry.

Feeding Bivalves are suspension feeders that take in any type of small animal or protist or detritus. Suspension feeding is based on a flow of water across gas-exchange structures called **gills**. The gills are part of the visceral mass and consist of a series of thin membranes where particles are trapped. A water current exists because siphons, which are tubes formed by edges of the mantle, extend from the shell and form a plumbing system. The siphons bring water into the gills and then back out of the body, powered by the beating of cilia on the gills (see Figure 31.12a). Bivalves are the only major group of mollusks that lack a radula.

Movement Clams burrow with the aid of their muscular foot, which functions as a hydrostatic skeleton. But they are otherwise sedentary. Scallops are able to swim by clapping their shells together and forcing water to jet out, pushing them backward. Scallop locomotion is similar to the swimming of cnidarian medusae, which move when muscular contractions force water out of their bell (see Chapter 31). Bivalves produce a swimming trochophore larva that is responsible for dispersing individuals to new locations.

Reproduction Only sexual reproduction occurs. Eggs and sperm are shed into the water, and fertilized eggs develop into trochophore larvae. Trochophores then metamorphose into a distinct type of larva called a **veliger**, which continues to feed and swim before settling to the substrate and metamorphosing into an adult form that secretes a shell.

Lima scabra

FIGURE 32.16 Bivalves Have Two Shells

Mollusca ▶ Gastropoda (Snails, Slugs, Nudibranchs)

The gastropods ("belly-feet") are named for the large, muscular foot on their ventral side. Terrestrial snails can retract their foot and body into a shell when they are attacked or when their tissues begin to dry out. Land slugs and nudibranchs (pronounced *NEW-da-branks*) lack shells but often contain toxins or foul-tasting chemicals to protect them from being eaten. The bright colors of nudibranchs, or sea slugs, are thought to act as a warning to potential predators (**Figure 32.17**). About 70,000 species of gastropods are known.

Feeding Gastropods and other mollusks have a unique structure in their mouths called a radula. In many species the **radula** functions like a rasp to scrape away algae, plant cells, or other types of food. It is usually covered with teeth made of chitin, which vary in size and shape among species. Although most gastropods are herbivores or detritivores, specialized types of teeth allow some gastropods to act as predators. Species called drills, for example, use their radula to bore a hole in the shells of oysters or other mollusks and expose the visceral mass, which they then eat. Cone snails have highly modified, harpoon-like teeth mounted at the tip of an extensible proboscis and armed with poison. When a fish or worm passes by, the proboscis shoots out. The prey is speared by the tooth, subdued by the poison, and consumed.

Movement Waves of contractions down the length of the foot allow gastropods to move by gliding (see Figure 32.11c). The sea butterflies are gastropods that have a reduced or absent shell but a large, winglike foot that flaps and powers swimming movements.

Reproduction Females of some species can reproduce asexually by producing eggs parthenogenetically, but most reproduction is

sexual. Unlike its counterpart in bivalves, sexual reproduction in gastropods begins with internal fertilization. But like bivalves, some marine gastropods produce a trochophore larva that may disperse up to several hundred kilometers from the parent. In most marine species and all terrestrial forms, larvae are not free living. Instead, offspring remain in an egg case while passing through several larval stages, then hatch as miniature versions of the adults.

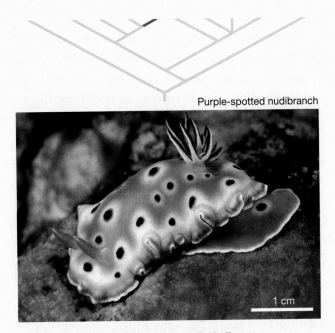

Purple-spotted nudibranch

FIGURE 32.17 Many Gastropods Lack Shells

Mollusca ▶ Polyplacophora (Chitons)

The Greek word roots that inspired the name *Polyplacophora* mean "many-plate-bearing." The name is apt because chitons (pronounced *KITE-uns*) have eight calcium carbonate plates along their dorsal side (**Figure 32.18**). The plates form a protective shell. The approximately 1000 species of chitons are marine. They are usually found on rocky surfaces in the intertidal zone, where rocks are periodically exposed to air at low tides.

Feeding Chitons have a radula and use it to scrape algae and other organic matter off rocks.

Movement Chitons move by gliding on their broad, muscular foot—as gastropods do.

Reproduction Sexes are separate, and fertilization is external. In some species, however, sperm that are shed into the water enter the female's mantle cavity and fertilize eggs inside the body. Depending on the species involved, eggs may be enclosed in a membrane and released or retained until hatching and early development are complete. Most species have trochophore and veliger larvae.

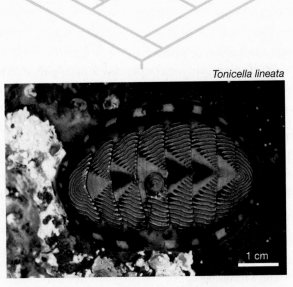

Tonicella lineata

FIGURE 32.18 Chitons Have Eight Shell Plates

Mollusca ▶ Cephalopoda (Squid, Nautilus, Octopuses)

The cephalopods ("head-feet") have a well-developed head and a foot that is modified to form tentacles. **Tentacles** are long, thin, muscular extensions that aid in movement and prey capture (**Figure 32.19**). With the exception of the nautilus, they have either highly reduced shells or none at all. They also have large brains and image-forming eyes with sophisticated lenses.

Feeding   Cephalopods are highly intelligent predators that hunt by sight and use their tentacles to capture prey—usually fish or crustaceans. They have a radula as well as a structure called a **beak**, which can exert powerful biting forces. Squid and octopuses also inject poisons into their prey to subdue them.

Movement   Cephalopods can swim by undulating their bodies and tentacles. They may also move by jet propulsion. They draw water into their mantle cavity and then force it out through a siphon (see Figure 32.11d). Squid are built for speed and hunt small fish by chasing them down. Octopuses, in contrast, crawl along the substrate by using their long, armlike tentacles. They chase down crabs or other crustaceans or pry mussels or clams from the substrate, then use their beaks to crush the exoskeletons of their prey.

Reproduction   Cephalopods have separate sexes, and some species have elaborate courtship rituals that involve color changes and interaction of tentacles. When a male is accepted by a female, he deposits sperm that are encased in a structure called a **spermatophore**. The spermatophore is taken up by the female, and fertilization is internal. Females lay eggs. When they hatch, juveniles develop directly into adults.

Octopus dofleini

10 cm

FIGURE 32.19 Cephalopods Have Highly Modified Bodies

32.5   Key Lineages: Ecdysozoans

The lineage Ecdysozoa is named for a distinctive trait shared by members of the group: They grow by **molting**—that is, shedding an exoskeleton or external covering. The Greek root *ecdysis*, which means "to slip out or escape," is appropriate because during a molt, an individual sheds its outer layer—called a **cuticle** if it is soft or an exoskeleton if it is hard—and slips out of it (**Figure 32.20**). The body expands, and a larger exoskeleton or cuticle then forms. As ecdysozoa grow and mature, they undergo a succession of molts.

The ecdysozoans were first recognized as monophyletic when investigators began using DNA sequence data to estimate the phylogeny of protostomes. There are about eight phyla in the lineage (see Table 31.1), including the Onychophora (*on-ee-KOFF-er-uh*) and the Tardigrada (**Figure 32.21**). They are not species-rich phyla, but both are closely related to arthropods. Onychophorans and tardigrades are similar to arthropods in having limbs and a stiff external covering. Unlike arthropods, their bodies are not segmented and their limbs not jointed. The onychophorans, or velvet worms, are small, caterpillar-like organisms that live in moist leaf litter and prey on small invertebrates. Onychophorans have sac-like appendages and segmented bodies with a hemocoel.

The tardigrades, or water bears, are microscopic animals that live in benthic habitats of marine or freshwater environments. Large numbers of water bears can also be found in the film of water that covers moss or other land plants in moist habitats. Tardigrades have a reduced coelom but a prominent hemocoel, and they walk on their clawed, sac-like legs. Most feed by sucking fluids from plants or animals; others are **detritivores** ("detritus-feeders").

Let's take a closer look at the two most diverse and abundant phyla of the Ecdysozoa: (1) **Nematoda** and (2) **Arthropoda**.

FIGURE 32.20 Ecdysozoans Grow by Molting
Once an ecdysozoan has left its old exoskeleton, hours or days pass before the new exoskeleton is hardened. During this time, the individual is highly susceptible to predation or injury.

(a) Onychophorans have lobe-like limbs.

(b) Tardigrades have lobe-like limbs with claws.

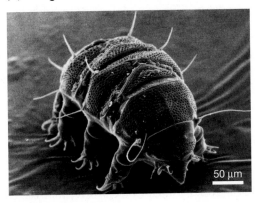

50 μm

FIGURE 32.21 Onychophorans and Tardigrades Have Limbs and Are Closely Related to Arthropods
(a) The Onychophora are called velvet worms because they have wormlike bodies and a velvety appearance.
(b) The Tardigrada are known as water bears because they live in moist habitats and, when viewed under the microscope, resemble bears.

Nematoda (Roundworms)

Species in the phylum Nematoda are commonly called nematodes or roundworms (**Figure 32.22**). Roundworms are unsegmented worms with a pseudocoelom, a tube-within-a-tube body plan, and no appendages. They have no muscles that can change the diameter of the body—their body musculature consists solely of longitudinal muscles that shorten or lengthen the body upon contracting or relaxing, respectively. Although some nematodes can grow to lengths of several meters, the vast majority of species are tiny—most are much less than 1 mm long. They lack specialized systems for exchanging gases and for circulating nutrients and wastes. Instead, gas exchange occurs across the body wall, and nutrients and wastes move by simple diffusion.

Nematodes have been studied intensively because several species are common parasites of humans. Pinworms, for example, infect about 40 million people in the United States alone, and *Onchocerca-volvulus* causes an eye disease that infects around 20 million people in Africa and Latin America. Advanced infections by the roundworm *Wuchereria bancrofti* result in the blockage of lymphatic vessels, causing fluid accumulation and massive swelling—the condition known as *elephantiasis*.

Parasites actually represent a tiny fraction of the 25,000 nematode species that have been described to date, however. The vast majority of nematodes are free living. They are important ecologically, because they are found in virtually every habitat known.

Nematodes are also fabulously abundant. Biologists have found 90,000 roundworms in a single rotting apple and have estimated that rich farm soils contain up to 9 billion roundworms per acre. The simple nematode body plan has been extraordinarily successful.

Feeding   Roundworms feed on a wide variety of materials, including bacteria, fungi, plant roots, small protists or animals, or detritus. In most cases, the structure of their mouthparts is specialized in a way that increases the efficiency of feeding on a particular type of organism or material.

Movement   Roundworms move with the aid of their hydrostatic skeleton (see Figure 31.5). Most roundworms live in soil or inside a host, so when contractions of their longitudinal muscles cause them to wriggle, the movements are resisted by a stiff substrate. As a result, the worm pushes off the substrate, and the body moves.

Reproduction   Sexes are separate in most species, and asexual reproduction is rare or unknown. Sexual reproduction begins with internal fertilization and culminates in egg laying and the direct development of offspring. Individuals go through a series of four molts over the course of their lifetime.

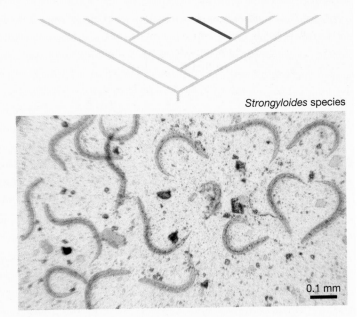

Strongyloides species

0.1 mm

FIGURE 32.22 Most Nematodes Are Free Living, but Some Are Parasites

Arthropoda (Arthropods)

In terms of duration in the fossil record, species diversity, and abundance of individuals, arthropods are easily the most successful lineage of eukaryotes. They appear in the fossil record over 520 million years ago and have long dominated the animals observed in both marine and terrestrial environments. Well over a million living species have been described, and biologists estimate that millions or perhaps even tens of millions of arthropod species have yet to be discovered.

Morphologically, arthropods are distinguished by segmented bodies and a sophisticated, jointed exoskeleton. They have a highly reduced coelom but possess an extensive body cavity called a hemocoel, enclosed by the exoskeleton. The body is organized into distinct **head** and **trunk** regions in all arthropods; in many species there is an additional grouping of segments into trunk regions. The trunk regions are usually called an *abdomen*, and a *thorax* or *tail*.

Metamorphosis is common in arthropods. Their larvae have segmented bodies but may lack a hardened exoskeleton. If metamorphosis occurs, larval and adult forms may grow by molting.

At least some segments along the arthropod body produce paired, jointed appendages. Arthropod appendages have an array of functions: sensing aspects of the environment, exchanging gases, feeding, or locomotion. Arthropod appendages provide the ability to sense environmental stimuli and make sophisticated movements in response. Most species also have sophisticated, image-forming compound eyes. A **compound eye** contains many light-sensing, columnar structures, each with its own lens. (Your eyes are **simple eyes**, meaning they have just one light-sensing structure.) Most arthropods also have a pair of antennae on the head. **Antennae** are long, tentacle-like appendages that are used to touch or smell.

Although the phylogeny of arthropods is still being worked out, most data sets agree that the phylum as a whole is monophyletic; that the myriapods (millipedes and centipedes), chelicerates (spiders), insects, and crustaceans represent four major subphyla within the phylum; and that crustaceans and insects are closely related. The phylum Arthropoda is so large and diverse that a detailed treatment would fill a book this size; space permits just a few notes about the four major lineages.

Arthropoda ▶ Myriapods (Millipedes, Centipedes)

The myriapods have relatively simple bodies, with a head region and a long trunk featuring a series of short segments, each bearing one or two pairs of legs (**Figure 32.23**). If eyes are present, they consist of a few to many simple structures clustered on the sides of the head. The 11,600 species that have been described to date inhabit terrestrial environments all over the world.

Feeding Millipedes and centipedes have mouthparts that can bite and chew. These organisms live in downed, rotting logs and other types of dead plant material that litters the ground in forests and grasslands. Millipedes are detritivores. Centipedes, in contrast, use a pair of poison-containing fangs just behind the mouth to hunt an array of insects. Large centipedes can inject enough poison to debilitate a human.

Movement Myriapods walk or run on their many legs; a few species burrow. Centipedes usually have fewer than 30 segments, with one pair of legs per segment. Some millipedes have over 190 trunk segments, each with two pairs of legs, for a total of over 750 legs.

Reproduction Sexes are separate, and fertilization is internal. Males deposit sperm in packets that are picked up by the female or transferred to her by the male. After females lay eggs in the environment, they hatch into juveniles that develop into adults via a series of molts.

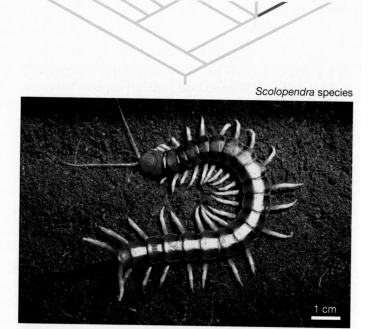

Scolopendra species

1 cm

FIGURE 32.23 Myriapods Have a Pair of Legs on Each Body Segment

Arthropoda ▶ Chelicerata (Spiders, Ticks, Mites, Horseshoe Crabs, Daddy Longlegs, Scorpions)

Most of the 70,000 species of chelicerates are terrestrial, although the horseshoe crabs and sea spiders are marine. The lineage is considered a class in most of the traditional classification schemes and has several subgroups or subclasses. The most prominent of these chelicerate lineages is the arachnids (spiders, scorpions, mites, and ticks).

The chelicerate body consists of anterior and posterior regions. The anterior region lacks antennae for sensing touch or odor but usually contains eyes. The group is named for appendages called **chelicerae** found near the mouth. Depending on the species, the chelicerae are used in feeding, defense, copulation, movement, or sensory reception.

Feeding   Spiders, scorpions, and daddy longlegs capture and sting insects or other prey (**Figure 32.24a**). Although some of these species are active hunters, most spiders are sit-and-wait predators. They create sticky webs to capture prey, which fly or walk into the web and are subsequently trapped. The spider senses the vibrations of the struggling prey, pounces on it, and administers a toxic bite.

Mites and ticks are ectoparasitic and use their piercing mouthparts to feed on host animals (**Figure 32.24b**). Horseshoe crabs eat a variety of protostomes as well as detritus. Most scorpions feed on insects; the largest scorpion species occasionally eat snakes and lizards.

Movement   Like other arthropods, chelicerates move with the aid of muscles attached to an exoskeleton. They walk or crawl on their four pairs of jointed walking legs; some species are also capable of jumping. Horseshoe crabs and some other marine forms can swim slowly. Newly hatched spiders spin long, silken threads that serve as balloons and can carry offspring on the wind more than 400 kilometers from the point of hatching.

Reproduction   Sexual reproduction is the rule in chelicerates, and fertilization is internal in most groups. Courtship displays are extensive in many arthropod groups and may include both visual displays and the release of chemical odorants. In spiders, males use organs that are located on their legs to transfer sperm to females. These organs fit into the female reproductive tract in a "lock-and-key" fashion, and differences in the size and shape of male genitalia are often the only way to tell closely related species of spider apart.

Development is direct, meaning that metamorphosis does not occur. In spiders, males may present a dead insect as a gift that the female eats as they mate; in some species, the male himself is eaten as sperm is being transferred. In scorpions, females retain fertilized eggs. After the eggs hatch, the young climb on the mother's back. They remain there until they are old enough to hunt for themselves.

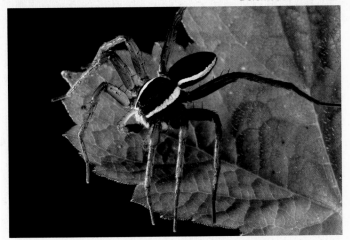

(a) Spider   *Dolomedes fimbriatus*

(b) Mite   *Dermatophagoides* species

50 μm

FIGURE 32.24 Chelicerates Have Head and Thorax Regions

Arthropoda ▶ Insecta (Insects)

About 925,000 species of insects have been named thus far, but it is certain that many more exist. In terms of species diversity and numbers of individuals, insects dominate terrestrial environments. In addition, the larvae of some species are common in freshwater streams, ponds, and lakes. Insects are distinguished by having three body regions—(1) the **head,** (2) **thorax,** and (3) **abdomen** (**Figure 32.25a**). Three pairs of walking legs are located on the ventral surface of the thorax. Most species have one or two pairs of wings, mounted on the dorsal side of the thorax. Typically the head contains three sets of mouthparts, a pair of antennae that are used to touch and smell, and a pair of compound eyes.

Feeding   Because most insect species have four sets of mouthparts (labrum, mandible, maxilla, and labium) that vary greatly in structure among species (see Figure 32.10), insects are able to feed in every conceivable manner and on almost every type of food source available on land. Larvae have wormlike bodies; most are deposit feeders, though some are leaf eaters. Adults are herbivores, predators, or parasites. The diversification of insects was closely correlated with the diversification of land plants, primarily because so many insects make their living by feeding on plant tissues or fluids. Insect predators usually eat other insects; insect parasites usually victimize other arthropods or mammals.

Movement   Insects use their legs to walk, run, or swim, or they use their wings to fly (**Figure 32.25b**). When insects walk or run, the sequence of movements usually results in three of their six legs maintaining contact with the ground at all times.

Reproduction   Sexes are separate. Mating usually takes place through direct copulation, with the male inserting a sperm-transfer organ into the female. Most females lay eggs, but in a few species eggs are retained until hatching. Many species are also capable of reproducing asexually, through the production of unfertilized eggs via mitosis. In the vast majority of species, either incomplete or complete metamorphosis occurs, with complete metamophosis being most common.

(a) Swallowtail butterfly
Pterourus species

(b) Harlequin beetle
Acrocinus longimanus

Head Thorax Abdomen

FIGURE 32.25 Insects Have Distinctive Body Plans and Move Efficiently

Arthropoda ▶ Crustaceans (Shrimp, Lobster, Crabs, Barnacles, Isopods, Copepods)

The 67,000 species of crustaceans that have been identified to date live primarily in marine and freshwater environments. A few species of crab and some isopods are terrestrial, however. (Terrestrial isopods are known as sowbugs, pillbugs, or roly-polies). Crustaceans are common consumers in planktonic habitats and are also important grazers and predators in shallow-water benthic environments.

The segmented body of a crustacean is divided into two distinct regions: (1) the cephalothorax, which combines the head and thorax, and (2) the abdomen. Many crustaceans have a **carapace**—a platelike section of their exoskeleton that covers and protects the cephalothorax (**Figure 32.26a**). They are the only type of arthropod with two pairs of antennae, and they have sophisticated, compound eyes—usually mounted on stalks. Crustaceans and trilobites were the first arthropods to appear in the fossil record.

Feeding   Crustaceans use every type of feeding strategy known. Barnacles (**Figure 32.26b**) and many shrimp are suspension feeders that use feathery structures located on head or body appendages to capture passing prey. Crabs and lobsters are active hunters, herbivores, and scavengers. Typically they have a pair of mouthparts called **mandibles** that bite or chew. Individuals capture and hold their food source with claws or other types of feeding appendages near their mouth, then use their mandibles to shred the food into small bits that can be ingested. As herbivores or detritivores, many species of crustaceans depend on algae for food.

Movement   Crustacean limbs are highly diverse, and it is common for species to have many pairs of limbs and more than one type of limb. Limb structures in crustaceans include paddle-shaped

(Continued on next page)

Mollusca ▶ Crustaceans (Shrimp, Lobster, Crabs, Barnacles, Isopods, Copepods) *(continued)*

forms used in swimming, feathery structures used in capturing food that is suspended in water, and jointed appendages that make sophisticated walking or running movements possible. Barnacles are one of the few types of sessile crustaceans. Adult barnacles cement their heads to a rock or other hard substrate, secrete a protective shell of calcium carbonate, and use their legs to create water currents that bring food to the mouth.

Reproduction Each individual is male or female, and sexual reproduction is the norm. Fertilization is usually internal, and eggs are usually retained by the female until they hatch. Most crus-

taceans pass through several distinct larval stages, and many species include a larval stage called a **nauplius**, which is usually planktonic. A nauplius has a single eye and appendages that develop into the two pairs of antennae and the mouthparts of the adult.

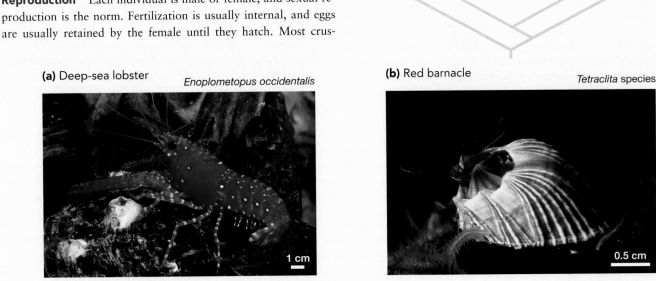

(a) Deep-sea lobster   *Enoplometopus occidentalis*

1 cm

(b) Red barnacle   *Tetraclita* species

0.5 cm

FIGURE 32.26 Crustaceans Have a Carapace That Protects Their Head

ESSAY The Role of Natural History in Biological Science

Science begins with questions about the natural world. Physicists ask questions about matter and forces; chemists ask questions about molecules and reactions; geologists ask questions about landforms. Biologists specialize in questions about organisms. Their job is to figure out what organisms are doing, how they do it, and why.

When biologists ask questions that begin with who, what, when, or where, the work that they are doing is descriptive in nature. Biologists spend a lot of time simply describing what exists in nature, carefully and accurately. Work of this type is called **natural history**. Classically, the core task of a naturalist is to describe organismal diversity—to understand who's who in nature. Naturalists are people who know the names and traits of birds, insects, plants, algae, and other organisms. Much of the information you read in this chapter was assembled by people who would describe themselves as naturalists. But when biologists focus on describing the structure and function of genes, organelles, or cell types, they are essentially doing natural history as well.

In the eighteenth and early nineteenth centuries, virtually all biology was descriptive in nature. During that time, the focus of biological science was simply to observe and describe the diversity of cells and organisms that exist. Today, biology is an experimentally driven science that focuses on questions about how genes, cells, and organisms work and why.

Dragonflies can fly forward or backward at speeds of up to 60 mph.

Even though the nature of biological science has changed dramatically over the years, many biologists would argue that natural history still plays a large role in advancing the science. For example, studying natural history remains an effective way to inspire new types of questions about genes, cells, and organisms. People observe things, wonder what's going on, and start doing experiments. In addition, most biologists got started in the field simply because they've loved organisms ever since they were children.

(Continued on next page)

(Essay continued)

To underscore the importance of natural history in contemporary biology, consider an experience that Tarif Awad had as a graduate student. During a break from lab work at the Marine Biological Laboratories in Woods Hole, Massachusetts, Awad wandered out onto nearby mudflats during low tide. He started digging and found interesting-looking polychaete worms. As he was examining them, he was startled by a woman's voice behind him, asking him what he was doing. When he showed the woman the worms, she responded by clapping her hands and exclaiming, "How wonderful! Someone who loves organisms!" The speaker was Barbara McClintock, Nobel laureate and one of the great geneticists of the twentieth century.

Like many other biologists, McClintock has found the study of natural history to be a lifelong source of emotional and intellectual inspiration. The intellectual inspiration springs from a simple premise: Observing organisms not only arouses new questions but also provides a context for understanding why organisms are the way they are. Natural history is the wellspring of biology and an important science in its own right.

The purely descriptive parts of biological science can be fun. Consider the following observations about some protostome animals:

- Every August, spiny lobsters migrate from shallow-water habitats in the Gulf of Mexico and the Caribbean Sea to deeper ocean channels. They do this by forming long single-file lines and marching for several days.

- Dragonflies can fly forward or backward at speeds of up to 60 mph.

- Spider silk is stronger than steel and has twice the elasticity of nylon. Many spiders eat their old or damaged webs and reuse the silk in constructing a new one.

- Rhinoceros beetles are able to carry 100 times their own body weight over short distances and 30 times their body weight for long periods of time.

- In rocks about 425 million years old, biologists find fossils of giant water scorpions that were over 2 meters long.

- Bolas spiders hunt at night, when *Spodoptera* moths are active. The spiders dangle from a silk thread and release a molecule that is similar in structure to the chemical signal that female moths use to attract male moths as mates. If a male moth mistakenly approaches the spider, the spider shoots it with a thread tipped with a sticky liquid, draws it in, and eats it.

- Snapping shrimp are sit-and-wait predators. When small fish or other prey pass their burrow, they snap a clawlike appendage. The movement produces a shock wave that stuns the prey. The shrimp then pounce on the food and devour it.

Organisms are fascinating. More than one biologist has recharged his or her emotional batteries, or come up with a novel question to answer, simply by going outside and looking at the organisms all around.

CHAPTER REVIEW

Summary of Key Concepts

▪ **Molecular phylogenies support the hypothesis that protostomes are a monophyletic group divided into two major subgroups: the Lophotrochozoa and the Ecdysozoa.**

The protostomes comprise about 20 phyla and were originally identified because their embryos undergo early development in the same way. Only recently did biologists recognize the existence of the lophotrochozoans and ecdysozoans, however. Several phyla in the Lophotrochozoa have characteristic feeding structures called lophophores, and many have a distinctive type of larvae called trochophores. Species in the Ecdysozoa grow by molting—meaning that they shed their old external covering and grow a new, larger one.

Web Tutorial 32.1 Protostome Diversity

▪ **Although the members of many protostome phyla have limbless, wormlike bodies and live in marine sediments, the most diverse** and species-rich lineages—Mollusca and Arthropoda—have body plans with a series of distinct, complex features.

All protostomes are bilaterally symmetric and triploblastic, but the nature of the body cavity is variable. The wormlike phyla of protostomes have well-developed coeloms that provide a hydrostatic skeleton used in locomotion. Flatworms lack a coelom, however, and both nematodes and rotifers evolved a pseudocoelom independently. The coelom is also much reduced or absent in the Mollusca and the Arthropoda lineages, although a body cavity is present. The mollusk's body is made up of a muscular foot, a visceral mass of organs, and a protective mantle. The arthropod body is segmented, divided into specialized regions such as the head, thorax, and abdomen of insects, and protected by an exoskeleton made of chitin.

▪ **Key events triggered the diversification of protostomes, including several lineages making the water-to-land transition, a diversification in appendages and mouthparts, and the evolution of metamorphosis in both marine and terrestrial forms.**

The transition to living on land occurred several times independently during the evolution of protostomes. It was facilitated by watertight shells and exoskeletons and the ability to move to moist locations. Diversification in appendages and mouthparts gave protostomes the ability to move and find food in innovative ways. The evolution of metamorphosis was significant because it allowed juveniles in sessile species to disperse to new habitats and because it made it possible for juveniles and adults to find food in different ways.

Questions

Content Review

1. Why is it logical to observe that mollusks have a drastically reduced coelom or no coelom at all?
   a. They evolved from flatworms, which also lack a coelom.
   b. They are the most advanced of all the protostomes.
   c. Their bodies are encased by a mantle, which may or may not secrete a shell.
   d. They move with the aid of a muscular foot, which functions as a hydrostatic skeleton.

2. What major feature or features distinguish most of the wormlike phyla of protostomes from each other?
   a. Their mouthparts and feeding methods vary.
   b. Their modes of locomotion differ—they burrow, swim, crawl, or walk.
   c. Metamorphosis may be lacking, incomplete, or complete.
   d. They have a well-developed coelom and a tube-within-a-tube body plan.

3. What is the function of the arthropod exoskeleton?
   a. Because hard parts fossilize more readily than do soft tissues, the presence of an exoskeleton has given arthropods a good fossil record.
   b. It has no well-established function. (Trilobites had an exoskeleton, and they went extinct.)
   c. It provides protection and functions in locomotion.
   d. It makes growth by molting possible.

4. Why is it logical to observe that Platyhelminthes have flattened bodies?
   a. They have simple bodies and evolved early in the diversification of protostomes.
   b. They lack a coelom, so their body cannot form a rounded tube-within-a-tube design.
   c. A flat body gives the animals high surface area and low volume—necessary because they lack gas-exchange organs and structures that circulate blood.
   d. They are sit-and-wait predators that hide from passing prey by flattening themselves against the substrate.

5. In number of species, number of individuals, and duration in the fossil record, which of the following phyla of wormlike animals has been the most successful? What feature is hypothesized to be responsible for this success?
   a. Annelida—a segmented body plan
   b. Platyhelminthes—bilateral symmetry but acoelomate condition
   c. Phoronida—a lophophore used as a feeding apparatus
   d. Echiura—a grooved proboscis used in suspension feeding

6. Which protostome phylum is distinguished by having body segments organized into distinct regions?
   a. Mollusca
   b. Arthropoda
   c. Annelida
   d. Nematoda

Conceptual Review

1. Describe the traits that distinguish the Lophotrochozoa and Ecdysozoa. Describe the traits that are common to both groups.

2. Define a segmented body plan. Did segmentation evolve once or multiple times during the evolution of protostomes? Provide evidence to justify your answer.

3. The nature of the coelom changed drastically during the evolution of protostomes. Describe the structure and function of the coelom in flatworms, nematodes, rotifers, the wormlike phyla of protostomes other than flatworms, mollusks, and arthropods.

4. Compare and contrast the role of metamorphosis in the life cycle of a sedentary, marine protostome such as a clam and a terrestrial, mobile protostome such as a butterfly.

5. Explain why the evolution of the exoskeleton and the jointed limb was such an important event in the evolution of arthropods.

6. Explain how squids move by jet propulsion.

Group Discussion Problems

1. Recall that the phylum Platyhelminthes includes three major groups: the free-living turbellarians, the parasitic cestodes, and the parasitic trematodes. Because the parasitic forms are so simple morphologically, researchers suggest that they are derived from more complex, free-living forms. Draw a phylogeny of Platyhelminthes that would support the hypotheses that ancestral flatworms were morphologically complex, free-living organisms and that parasitism is a derived condition.

2. Brachiopoda is a phylum within the Lophotrochozoa. Even though they are not closely related to mollusks, brachiopods look and act like bivalve mollusks (clams or mussels). Specifically, brachiopods suspension feed, live inside two calcium carbonate shells that hinge together in some species, and attach to rocks or other hard surfaces on the ocean floor. How is it possible for brachiopods and bivalves to be so similar if they did not share a recent common ancestor?

3. The Mollusca includes a group of about 370 wormlike species called the Aplacophora. Although aplacophorans do not have a shell, their cuticle secretes calcium carbonate scales or spines. They lack a well-developed foot, and some species have a simple radula.

 • Predict where Aplacophora are found in the phylogenetic tree of Mollusca relative to bivalves, gastropods, chitons, and cephalopods. Explain your reasoning.

 • Predict where Aplacophora live and how they move. Explain your reasoning.

Answers to Multiple-Choice Questions   1. d; 2. a; 3. c; 4. c; 5. a; 6. b

www.prenhall.com/freeman is your resource for the following: Web Tutorials; Online Quizzes and other Online Study Guide materials; Answers to Conceptual Review Questions; Solutions to Group Discussion Problems; Answers to Figure Caption Questions and Exercises; and Additional Readings and Research.

4. Consider the following: (a) The first arthropods to appear in the fossil record are marine crustaceans and trilobites, (b) terrestrial insects appeared relatively late in arthropod evolution, (c) the crustacean group called isopods includes both aquatic and terrestrial forms, (d) myriapods are strictly terrestrial, and (e) analyses of DNA sequence data suggest that crustaceans and insects are closely related. Based on these observations, do you agree or disagree with the hypothesis that arthropods made the water-to-land transition more than once? Explain your reasoning.

Deuterostome Animals

33

KEY CONCEPTS

▨ The most prominent deuterostome lineages are the echinoderms and the vertebrate groups called ray-finned fishes and tetrapods.

▨ Echinoderms and vertebrates have unique body plans. Echinoderms are radially symmetric as adults and have a water vascular system. All vertebrates have a skull, and most have an extensive endoskeleton made of cartilage or bone.

▨ The diversification of echinoderms was triggered by the evolution of appendages called podia; the diversification of vertebrates was driven by the evolution of the jaw and limbs.

▨ Humans are a tiny twig on the tree of life. Chimpanzees and humans diverged from a common ancestor that lived in Africa 6–7 million years ago. Since then, at least 14 humanlike species have existed.

The flukes (tail lobes) of a humpback whale emerge from the water briefly before the animal dives. The presence of a tail is a diagnostic trait of the vertebrates—one of the most prominent lineages of deuterostomes.

The **deuterostomes** include the largest-bodied and most morphologically complex of all animals. They range from the sea stars that cling to dock pilings, to the fish that dart in and out of coral reefs, to the wildebeests that migrate across the Serengeti Plains of East Africa.

Most biologists recognize just three phyla of deuterostomes: the **Echinodermata**, the **Hemichordata**, and the **Chordata**. The echinoderms include the sea stars and sea urchins. The hemichordates, or "acorn worms," are probably unfamiliar—they burrow in marine sands or muds and make their living by deposit feeding or suspension feeding. The chordates include the vertebrates. The vertebrates, in turn, include the sharks, bony fishes, amphibians, reptiles (including birds), and mammals. **Vertebrates** are animals that have a skull. The vast majority of

vertebrates also have a backbone or spinal column. Animals that are not vertebrates are collectively known as **invertebrates.**

The deuterostomes were initially recognized because they all undergo early embryonic development in a similar way. When a humpback whale, sea urchin, or human is just beginning to grow, the gut starts developing from posterior to anterior—with the anus forming first and the mouth second. Although the events that take place early in embryonic development are diagnostic features that unite the deuterstomes, their adult body plans and their feeding methods, modes of locomotion, and means for reproduction are highly diverse. Let's explore who the deuterostome animals are and how they diversified, starting with the question of why so many biologists are drawn to studying them.

33.1 Why Do Biologists Study Deuterostome Animals?

If you were to look at data on the number of biologists who study various lineages on the tree of life, you would probably conclude that a disproportionately large number of people study deuterostomes—particularly vertebrates. The charge of disproportionality is valid because, in terms of numbers of species and numbers of individuals, insects and other protostomes are much more successful than deuterostomes. And regarding their ecological impact—meaning their influence on other species and the physical environment—the bacteria, archaea, protists, fungi, and land plants could be considered far more important than the deuterostomes. And yet, thousands of biologists are drawn to the study of vertebrates and other deuterostomes. Why?

The simplest answer to this question is that, because humans are deuterostomes, other species in our lineage interest us. People identify with vertebrates because they are large and mobile. We interact with them extensively as pets or domesticated livestock. We understand their world more intuitively than we do the world of a bacterium or fungus.

In addition to the natural affinity we have for deuterostomes, they are worthy of study because they play important roles in ecological interactions and human economics:

- In marine environments, deuterostomes are key consumers. Echinoderms and ray-finned fishes are important herbivores, and ray-finned fishes, sharks, and mammals important predators. (Ray-finned fishes are distinct from the lobe-finned fishes and coelocanths introduced later in the chapter.) On land, almost all of the large-bodied herbivores and predators are reptiles and mammals. As **Figure 33.1** shows, the upper trophic levels of food chains in most ecosystems are dominated by deuterostomes. As a result, understanding deuterostomes is critical to understanding how energy and nutrients flow through both marine and terrestrial ecosystems.

- Humans depend on vertebrates for food and, in preindustrial economies, for power. Ray-finned fishes and domesticated livestock are key sources of protein in many cultures. For many thousands of years, agriculture was based on the power generated by oxen, horses, water buffalo, or mules. Today, preindustrial societies continue to rely on horsepower or oxpower for transportation and agricultural work. In the industrialized world, millions of people birdwatch, plan vacations around seeing large mammals in national parks, or keep vertebrates as pets. All over the world, large numbers of people make their living by caring for vertebrates and studying them.

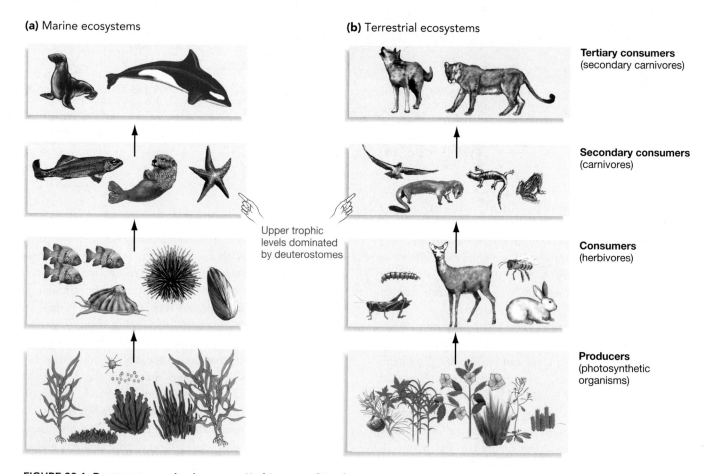

(a) Marine ecosystems **(b)** Terrestrial ecosystems

Tertiary consumers (secondary carnivores)

Secondary consumers (carnivores)

Upper trophic levels dominated by deuterostomes

Consumers (herbivores)

Producers (photosynthetic organisms)

FIGURE 33.1 Deuterostomes Are Important Herbivores and Predators
(a) In marine habitats, echinoderms and ray-finned fishes, both of which are deuterostomes, are the most important large herbivores and predators. **(b)** On land, deuterostomes such as amphibians, reptiles, and mammals play the same role.

(a) Echinoderm larvae are bilaterally symmetric.

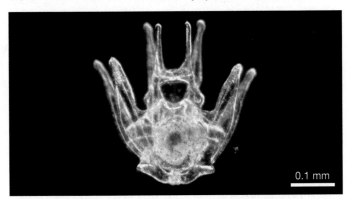

0.1 mm

(b) Adult echinoderms are radially symmetric.

5 cm

FIGURE 33.2 Body Symmetry Differs in Larval and Adult Echinoderms
(a) The larvae and **(b)** adult forms of the sea urchin *Strongylocentrotus franciscanus*. Bilaterally symmetric larvae undergo metamorphosis and emerge as radially symmetric adults.

33.2 How Do Biologists Study Deuterostomes?

To understand the events that produced the diversity of deuterostomes we see today, biologists study the comparative morphology of species from each of the three deuterostome phyla, analyze the fossil record of the group, and compare DNA sequences and other traits to infer their phylogeny. There are three central issues: (1) understanding the diversity of body plans observed in echinoderms, hemichordates, and the chordates; (2) exploring how vertebrates evolved from invertebrates; and (3) grasping how the vertebrates made the transition from living in aquatic environments to living on land.

Analyzing Morphological Traits

All deuterostomes are triploblastic and have a coelom (see Chapter 31). Patterns of early embryonic development are also similar among all members of the lineage. But in terms of the basic components of body plans, one of the most remarkable events in animal evolution occurred during the evolution of echinoderms: There was a reversion to a type of radial symmetry. Adult echinoderms have bodies with five-sided radial symmetry, even though both their larvae and their ancestors are bilaterally symmetric (**Figure 33.2**).

Recall from Chapter 31 that radially symmetric animals do not have well-developed head and posterior regions. As a result, they tend to interact with the environment in all directions at once instead of facing the environment in one direction. If adult echinoderms are capable of movement, they tend to move equally well in all directions instead of only headfirst.

The evolution of radial symmetry in echinoderms was a remarkable event in animal evolution. If echinoderms do not have a head and tail, how is their body organized?

The Water Vascular System of Echinoderms The unique body plan of echinoderms is based on a unique morphological feature. As **Figure 33.3a**, shows, the echinoderm body contains a series of continuous fluid-filled tubes and chambers called the **water vascular system**. One of the tubes is open to the exterior where it meets the body wall, so seawater can flow into and out

(a) Echinoderms have a water vascular system.

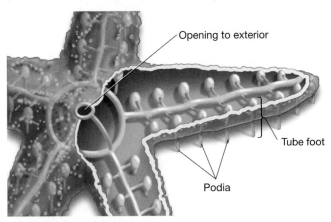

Opening to exterior

Tube foot

Podia

(b) Podia project from the underside of the body.

FIGURE 33.3 Echinoderms Have a Water Vascular System
(a) The water vascular system is a series of tubes and reservoirs that extends throughout the body, forming a sophisticated hydrostatic skeleton. **(b)** Podia aid in movement because they extend from the body and can extend or retract.

(a) Live sand dollar   **(b)** Endoskeleton

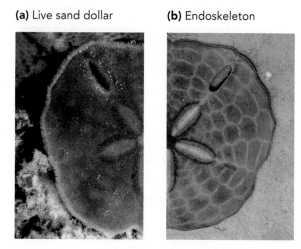

FIGURE 33.4 Echinoderms Have an Endoskeleton
The endoskeleton of an echinoderm is located just under the skin.

of the system. Inside, fluids move via the beating of cilia that line the interior of the tubes and chambers. In effect, the water vascular system forms a sophisticated hydrostatic skeleton.

Figure 33.3a highlights a particularly important part of the system called tube feet. **Tube feet** are elongated, fluid-filled structures. **Podia** ("feet") are sections of the tube feet that project outside the body (**Figure 33.3b**) and make contact with the substrate. As podia extend and contract in a coordinated way along the base of an echinoderm, they alternately grab and release from the substrate. As a result, the individual moves.

The other noteworthy feature of the echinoderm body is its **endoskeleton**, which is a hard, supportive structure inside the body (**Figure 33.4**). As an individual is developing, cells secrete plates of calcium carbonate inside the skin. Depending on the species involved, the plates may remain independent and result in a flexible structure or fuse into a rigid case. This type of endoskeleton is a diagnostic feature of echinoderms, along with radial symmetry and the water vascular system.

The Origin of Chordates The phylum Chordata is defined by the presence of four morphological features: (1) openings into the throat called **pharyngeal gill slits**; (2) a stiff but flexible rod called a **notochord**, which runs the length of the body; (3) a bundle of nerve cells that runs the length of the body and forms a **dorsal hollow nerve cord**; and (4) a muscular **tail**, which extends past the anus.

The hemichordates are not members of the phylum Chordata, because they have just one of these four features. Hemichordates lack a notochord, a dorsal hollow nerve cord, and a tail, although they do have pharyngeal gill slits that function in feeding and gas exchange (**Figure 33.5a**). Hemichordates are sessile suspension feeders and live in the ocean. As the arrows in Figure 33.5a indicate, water enters the mouths of these animals, flows through structures where oxygen and food particles are extracted, and exits through the pharyngeal gill slits.

(a) Hemichordata (acorn worms; a phylum closely related to chordates)

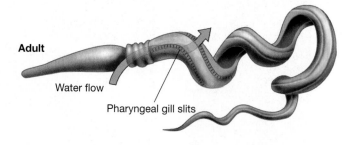

(b) Phylum Chordata consists of three major subgroups.

Urochordata (tunicates)

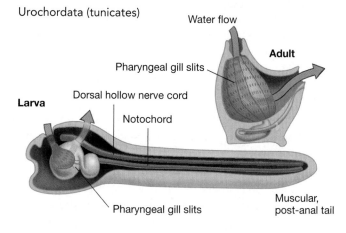

Cephalochordata (lancelets)

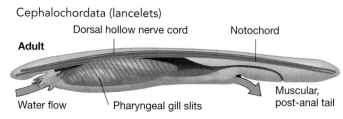

Vertebrata (vertebrates)

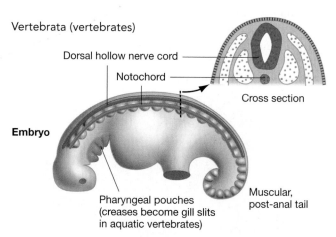

FIGURE 33.5 Four Traits Distinguish the Chordates
(a) Hemichordates, although closely related to echinoderms, have only one of the four traits that distinguish chordates: pharyngeal gill slits. **(b)** In chordates, either larvae or adults or both have notochords and muscular tails in addition to pharyngeal gill slits and dorsal hollow nerve cords. The three major chordate lineages are tunicates, lancelets, and vertebrates.

In contrast, pharyngeal gill slits, notochords, dorsal hollow nerve cords, and tails are found in each of the three major lineages of chordates (**Figure 33.5b**). In the traditional classification schemes introduced in Chapter 1, these groups are considered subphyla. The three groups are (1) urochordates, (2) cephalochordates, and (3) vertebrates. **Urochordates** are also called tunicates or sea squirts; they are small suspension feeders that as adults live attached to hard substrates in the ocean. **Cephalochordates** are also called lancelets or amphioxus; they are small, mobile suspension feeders that look a little like fish. Vertebrates include the sharks, bony fishes, reptiles, and mammals.

Adult urochordates are sessile, ocean-dwelling suspension feeders. As Figure 33.5b shows, pharyngeal gill slits are present in both larvae and adults and function in both feeding and gas exchange, much as they do in hemichordates. The notochord, dorsal hollow nerve cord, and tail are present only in the larvae, however. Because the notochord stiffens the tail, muscular contractions on either side of a larva's tail wag it back and forth and result in swimming movements. As larvae swim or float in the upper water layers of the ocean, they drift to new habitats where food might be more abundant.

Adult cephalochordates live in ocean-bottom habitats, where they burrow in sand and suspension feed with the aid of their pharyngeal gill slits, much like urochordate larvae. The cephalochordates also have a notochord that stiffens their bodies, so that muscle contractions on either side result in fishlike movement when they swim during dispersal or mating.

In vertebrates, the dorsal hollow nerve cord is elaborated into the familiar spinal cord. Structures called *pharyngeal pouches* are present in all vertebrate embryos. In aquatic species, the creases between pouches open into gill slits and develop into part of the main gas-exchange organ—the **gills**. In terrestrial species, however, gill slits do not develop after the pharyngeal pouches form. A notochord also appears in all vertebrate embryos. Instead of functioning in body support and movement, however, it helps organize the body plan. Recall from Chapter 22 that cells in the notochord secrete proteins that help induce the formation of somites. **Somites** are segmented blocks of tissue that form along the length of the body. Although the notochord itself disappears, cells in the somites later differentiate into the vertebrae, ribs, and skeletal muscles of the back, body wall, and limbs. In this way, the notochord is instrumental in the development of the signature adaptation of vertebrates: an endoskeleton made of bone. To understand how bone evolved, let's delve into the fossil record.

Using the Fossil Record

Echinoderms and vertebrates are present in the Burgess Shale deposits that formed during the Cambrian explosion 544–515 million years ago (**Figure 33.6**; see also Chapter 26). The vertebrate fossils in these rocks show that the earliest members of this lineage lived in the ocean about 530 million years ago, had streamlined bodies like fish, and appear to have had a skull, skeletal elements that reinforced the gills, and a notochord made of cartilage. **Cartilage** is a stiff tissue that consists of scattered cells in a gel-like matrix of polysaccharides and protein fibers. The earliest vertebrates had endoskeletons made of cartilage.

Subsequent to the appearance of vertebrates, the fossil record documents a series of key innovations that occurred as this lineage diversified:

- Fossil vertebrates from the early part of the Ordovician period, about 480 million years ago, are the first fossils to have bone. **Bone** is a tissue that consists of cells and blood vessels encased in a matrix made primarily of calcium phosphate ($CaPO_4$), with a small amount of calcium carbonate ($CaCO_3$)

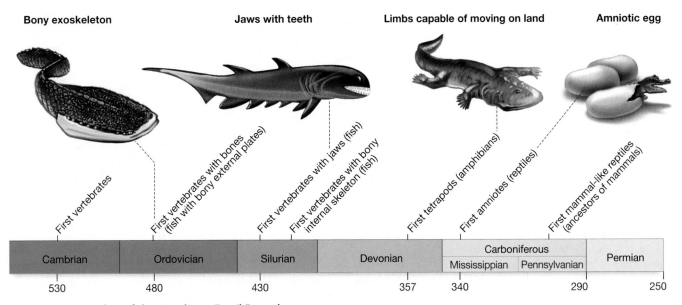

FIGURE 33.6 Timeline of the Vertebrate Fossil Record

and protein fibers. When bone first evolved, it did not occur in the endoskeleton. Instead bone was deposited in scalelike plates that formed an **exoskeleton**—a hard, hollow structure that envelops the body. Based on the fossils' overall morphology, biologists infer that these animals swam with the aid of a notochord, and that they breathed and fed by gulping water and filtering it through their pharyngeal gill slits. Presumably, the bony plates helped provide protection from predators.

- The first vertebrates with **jaws** show up in the fossil record about 430 million years ago. The evolution of jaws was significant because it gave vertebrates the ability to bite, meaning that they were no longer limited to suspension feeding or deposit feeding. Instead, they could make a living as herbivores or predators. Soon after, jawbones with teeth appear

in the fossil record (Figure 33.6). With jaws and teeth, vertebrates became armed and dangerous. The fossil record shows that a spectacular radiation of jawed fishes followed, filling marine and freshwater habitats.

- The next great event in the evolution of vertebrates was the transition to living on land. The first animals that had limbs and were capable of moving on land are dated to about 357 million years ago. These were the first of the **tetrapods** ("four-footed")—animals with four limbs (Figure 33.6).

- About 20 million years after the appearance of tetrapods in the fossil record, the first amniotes are present. The **Amniota** is a lineage of vertebrates that includes all tetrapods other than amphibians. They are named for a signature adaptation: the amniotic egg. An **amniotic egg** is an egg that has a watertight shell or case enclosing a membrane-bound food

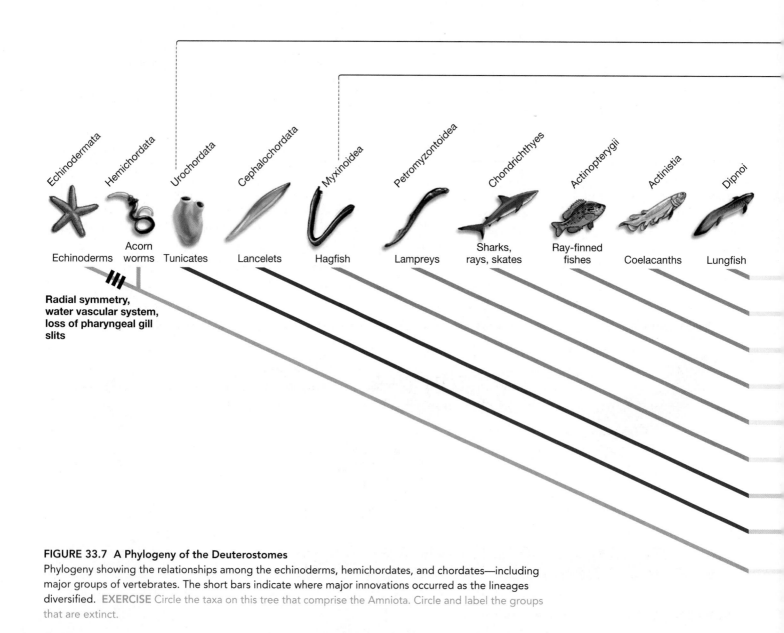

FIGURE 33.7 A Phylogeny of the Deuterostomes
Phylogeny showing the relationships among the echinoderms, hemichordates, and chordates—including major groups of vertebrates. The short bars indicate where major innovations occurred as the lineages diversified. **EXERCISE** Circle the taxa on this tree that comprise the Amniota. Circle and label the groups that are extinct.

supply, water supply, and waste repository. The evolution of the amniotic egg was significant because it gave vertebrates the ability to reproduce far from water. Amniotic eggs resist drying out, so vertebrates that produce amniotic eggs do not have to return to aquatic habitats to lay their eggs.

To summarize, the fossil record indicates that vertebrates evolved through a series of major steps, beginning about 530 million years ago with vertebrates whose endoskeleton consisted of a notochord. The earliest vertebrates gave rise to cartilaginous fishes (sharks and rays) and fish with bony skeletons and jaws. After the tetrapods emerged and amphibians resembling salamanders began to live on land, the evolution of the amniotic egg paved the way for the evolution of the first truly terrestrial vertebrates. The fossil record indicates that a radiation of reptiles followed, along with the animals

that gave rise to mammals. Do phylogenetic trees estimated from analyses of DNA sequence data agree or conflict with these conclusions?

Evaluating Molecular Phylogenies

Figure 33.7 provides a phylogenetic tree that summarizes the relationships among deuterostomes, based on morphological traits and DNA sequence data. The labeled bars on the tree indicate where major innovations occurred during the evolution of deuterostomes. Although the phylogeny of deuterostomes continues to be a topic of intense research, researchers are increasingly confident that the relationships described in Figure 33.7 are accurate.

The overall conclusion from this tree is that the branching sequence inferred from morphological and molecular data

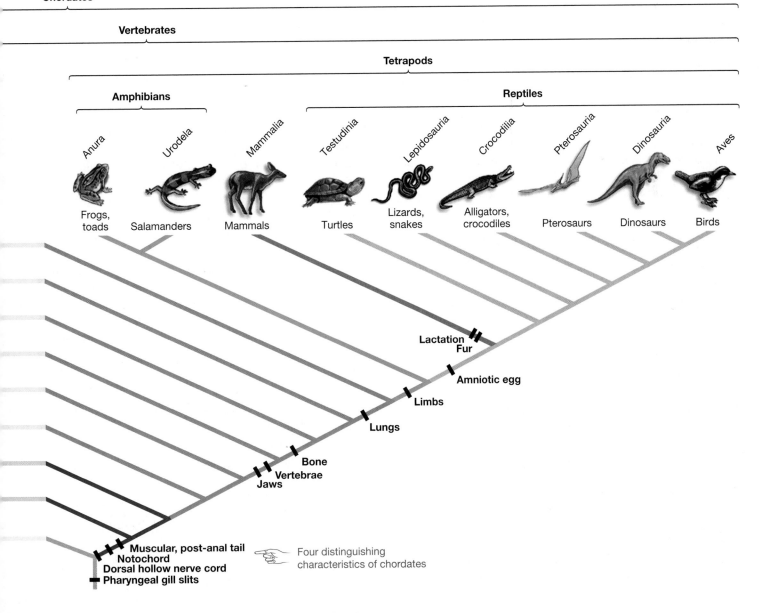

correlates with the sequence of groups in the fossil record. Reading up from the base of the tree, it is clear that the three groups of deuterostomes that have traditionally been recognized as phyla are indeed monophyletic, and that hemichordates and echinoderms are more closely related to each other than they are to chordates. The closest living relatives of the vertebrates are the cephalochordates. The most basal groups of chordates lack the skull and vertebral column that define the vertebrates, and the most ancient lineages of vertebrates lack jaws and bony skeletons. To understand what happened during the subsequent diversification of echinoderms and vertebrates, let's explore some of the major innovations involved in feeding, movement, and reproduction in more detail.

✔CHECK YOUR UNDERSTANDING

Echinoderms are distinguished by their fivefold radial symmetry as adults, water vascular system, and calcium carbonate plates that form an endoskeleton. The water vascular system functions in movement, and the endoskeleton provides protection. You should be able to (1) make a rough sketch of a sea star's body; include labels indicating the structure and function of the water vascular system and endoskeleton.

Chordates are distinguished by the presence of a notochord, a dorsal hollow nerve cord, pharyngeal gill slits, and a muscular tail. You should be able to (2) sketch the body of a tunicate larva or cephalochordate, and add labels indicating the structure and function of the four diagnostic features. On a phylogeny of the vertebrates, you should be able to (3) map the origin of key traits such as the cartilaginous skeleton, bony skeleton, jaws, limbs, and amniotic egg.

33.3 What Themes Occur in the Diversification of Deuterostomes?

Deuterostomes evolved in marine environments, but their diversification carried them into all of the world's major habitat types. Echinoderms, hemichordates, urochordates, and cephalochordates are all strictly marine. The most basal groups of chordates are also ocean-dwelling organisms. Lungfish, certain groups of sharks and rays, and about one-third of the ray-finned fishes live in freshwater. All tetrapods, with the exception of marine reptiles and mammals, live in terrestrial habitats, and all must breathe air.

In terms of numbers of species and range of habitats occupied, the most successful lineages of deuterostomes are the echinoderms and the vertebrates. Echinoderms are both widespread and abundant in marine habitats. In some deepwater environments, echinoderms represent 95 percent of the total mass of organisms. Among vertebrates, the most species-rich and ecologically diverse lineages are the ray-finned fishes and the tetrapods (**Figure 33.8**). Ray-finned fishes occupy habitats ranging from deepwater environments, which are perpetually

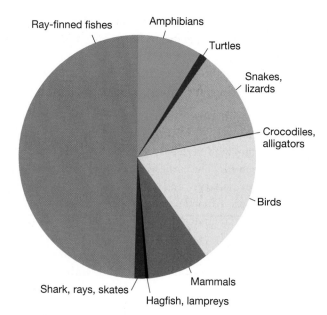

FIGURE 33.8 Relative Species Abundance Among Vertebrates
EXERCISE Circle the taxa that are tetrapods.

dark, to shallow ponds that dry up each year. Tetrapods include the large herbivores and predators in terrestrial environments all over the world.

Today there are about 7000 species of echinoderms, 23,700 species of ray-finned fish, and 25,400 species of tetrapods. To understand why these lineages have been so successful, it's important to recognize that they have unique body plans. Recall that echinoderms are radially symmetric and have a water vascular system, and that ray-finned fishes and tetrapods have a bony endoskeleton. In this light, the situation in deuterostomes appears to be similar to that in protostomes. Recall from Chapter 32 that the most evolutionarily successful protostome lineages are the arthropods and mollusks. A major concept in that chapter was that arthropods and mollusks have body plans that are unique among protostomes. Once their distinctive body plans evolved, evolution by natural selection led to extensive diversification based on novel methods for eating, moving, and reproducing.

Box 33.1 highlights model organisms that biologists have used in experimental work on echinoderms, ray-finned fishes, and tetrapods. To understand how these three groups became so species rich and geographically widespread, let's take a closer look at the adaptations that allow them to eat, move, and produce offspring in diverse ways.

Feeding

Animals eat to live, and it is logical to predict that the 7000 species of echinoderms and nearly 50,000 species of ray-finned fishes and tetrapods eat different things in different ways. It is also logical to predict that echinoderms and vertebrates have traits that make diverse ways of feeding possible.

BOX 33.1   Model Organisms: Sea Urchins, Zebrafish, and Mice

In addition to their evolutionary success, the three most prominent deuterostome lineages—echinoderms, ray-finned fishes, and tetrapods—have furnished key model organisms in biological science. More specifically, sea urchins, zebrafish, and mice (**Figure 33.9**) are used as experimental subjects in labs all over the world. Like other model organisms, part of their appeal is practical: They are relatively inexpensive and easy to maintain in the laboratory, they produce large numbers of offspring, and they develop rapidly. In addition, both sea urchins and zebrafish have transparent eggs and embryos, which facilitates observation of cell movements and other developmental events.

Experiments on urchins, zebrafish, and mice have provided fundamental insights into how gene expression is controlled and how early development occurs. Along with fruit flies, these three species provide the most important model systems in animal genetics and developmental biology. Mice are also closely enough related to humans to provide a tractable study organism for biologists interested in the cellular mechanisms responsible for cancer and other diseases.

(a) Purple sea urchin

(b) Zebrafish

(c) Mouse

FIGURE 33.9 **Model Organisms among the Deuterostomes**

Both predictions are correct. Echinoderms have feeding strategies that are unique among marine animals. Many are based on the use of their water vascular system and podia. Ray-finned fishes and tetrapods, in contrast, depend on their jaws.

Feeding Strategies in Echinoderms Depending on the lineage and species in question, echinoderms make their living by suspension feeding, deposit feeding, or harvesting algae or other animals. In most cases, the animal's podia play a key role in obtaining food. Many sea stars, for example, prey on bivalves. Clams and mussels respond to sea star attacks by contracting muscles that close their shells tight. But by clamping onto each shell with their podia and pulling, sea stars are often able to pry the shells apart a few millimeters (**Figure 33.10a**). Once a gap exists, the sea star extrudes its stomach from its body and forces the stomach through the opening between the

(a) Podia adhere to bivalve shells and pull them apart.

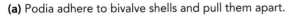

(b) Podia trap particles during suspension feeding.

FIGURE 33.10 **Echinoderms Use Their Podia in Feeding**
(a) Because they have suckers at their tips, sea stars can use their podia to pry open bivalve shells.
(b) Brittle stars extend their podia when they suspension feed.

bivalve's shells. Upon contacting the visceral mass of the bivalve, the stomach of the sea star secretes digestive enzymes. It then begins to absorb the small molecules released by enzyme action. Eventually, only the unhinged shells of the prey remain.

Podia are also involved in echinoderms that suspension feed (**Figure 33.10b**). In most cases, podia are extended out into the water. When food particles contact them, the podia flick the food down to cilia, which sweep the particles toward the animal's mouth. In deposit feeders, podia secrete mucus that is used to sop up food material on the substrate. The podia then roll the food-laden mucus into a ball and move it to the mouth.

The Vertebrate Jaw The most ancient groups of vertebrates have relatively simple mouthparts. For example, hagfish and lampreys lack jaws and cannot bite algae, plants, or animals. They have to make their living as deposit feeders or as ectoparasites.

Vertebrates were not able to harvest food by biting until jaws evolved. The leading hypothesis for the origin of the jaw proposes that natural selection acted on mutations that affected the morphology of **gill arches**, which are curved regions of tissue between the gills. The jawless vertebrates have bars of cartilage that stiffen these gill arches. The gill-arch hypothesis proposes that mutation and natural selection increased the size of an arch and modified its orientation slightly, producing the first working jaw (**Figure 33.11**). Three lines of evidence, drawn from comparative anatomy and embryology, support the gill-arch hypothesis:

1. Both gill arches and jaws consist of flattened bars of bony or cartilaginous tissue that hinges and bends forward.

2. Unlike most other parts of the vertebrate skeleton, both jaws and gill arches are derived from specialized cells in the embryo called neural crest cells.

3. The muscles that move jaws and that move gill arches are derived from the same population of embryonic cells.

Taken together, these data support the hypothesis that jaws evolved from gill arches.

To explain why ray-finned fishes are so diverse in their feeding methods, biologists point to important modifications of the jaw. In ray-finned fishes, for example, the jaw is protrusible—meaning it can be extended to nip or bite out at food. In addition, several particularly species-rich lineages of ray-finned fishes have a set of pharyngeal jaws. The **pharyngeal jaw** consists of modified gill arches that function as a second set of jaws, located in the back of the mouth. Pharyngeal jaws are important because they make food processing particularly efficient. (For more on the structure and function of pharyngeal jaws, see Chapter 43.)

To summarize, the radiation of ray-finned fishes was triggered in large part by the evolution of the jaw, modifications that made it possible to protrude the jaw, and the origin of the pharyngeal jaw. The story of tetrapods is different, however.

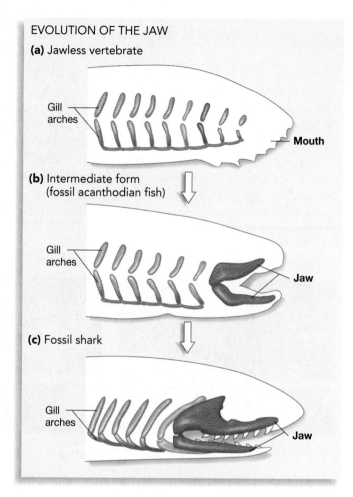

FIGURE 33.11 A Hypothesis for the Evolution of the Jaw
(a) Gill arches support the gills in jawless vertebrates. (b) In the fossil record, jawbones appear first in fossil fish called acanthodians. (c) Fossil sharks that appeared later had more elaborate jaws.
QUESTION The transition from the gill arches to the jaws of acanthodian fishes is complex, and intermediate forms have yet to be found in the fossil record. Would intermediate stages in the evolution of the jaw have any function?

Although jaw structure varies somewhat among tetrapod groups, the adaptation that triggered their initial diversification involved the ability to get to food, not to bite it and process it.

Movement

The signature adaptations of echinoderms and tetrapods involve locomotion. We've already explored the water vascular system and tube feet of echinoderms; here let's focus on the tetrapod limb.

Most tetrapods live on land and use their limbs to move. But for vertebrates to succeed on land, not only did they have to be able to move out of water, but they also had to breathe air and avoid drying out. To understand how this was accomplished, consider the morphology and behavior of their closest living relatives, the lungfish (**Figure 33.12**). Most living species of lungfish inhabit shallow, oxygen-poor water. To supplement the oxygen taken in by their gills, they have lungs and breathe

FIGURE 33.12 Lungfish Have Limb-like Fins
Some species of lungfish can walk or crawl short distances on their fleshy fins.

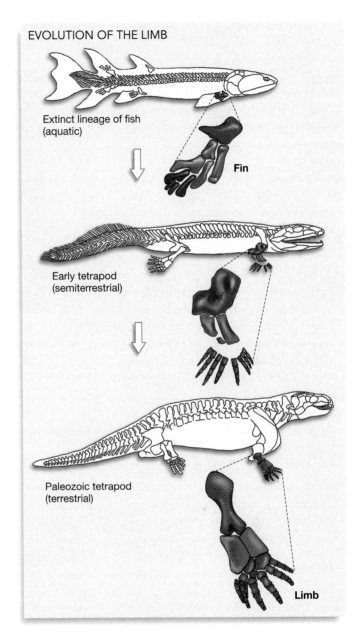

EVOLUTION OF THE LIMB

FIGURE 33.13 Fossil Evidence for a Fin-to-Limb Transition
Both the fossil fish and the early tetrapod are from the Devonian period, about 375 millon years ago. The number and arrangement of bones in the fins and limbs of these two fossil organisms agree with the general form of Paleozoic tetrapods as well as the modern tetrapod limb. The color coding indicates homologous elements.

air. Some also have fleshy fins supported by bones and are capable of walking short distances along watery mudflats or the bottoms of ponds. In addition, some species can survive extended droughts by burrowing in mud.

Fossils provide strong links between lungfish and the earliest land-dwelling vertebrates. **Figure 33.13** shows three of the species involved. The first, an aquatic animal related to today's lung-fishes, is from the Devonian period—about 375 million years ago. The second is one of the oldest tetrapods, or limbed vertebrates, found to date. The third is a more recent tetrapod fossil, dated to about 350 million years ago. The figure highlights the numbers and arrangement of bones in the fossil fish fin and the numbers and arrangement of bones in the limbs of early tetrapods. The color coding emphasizes that each fin or limb has a single bony element that is proximal (closest to the body) and then two bones that are distal (farther from the body) and arranged side by side, followed by a series of distal elements. Because the structures are similar and because no other groups have limb bones in this arrangement, the evidence for homology is strong. Based on the lifestyle of living lungfish, biologists suggest that mutation and natural selection gradually transformed fins into limbs as the first tetrapods became more and more dependent on terrestrial habitats.

The hypothesis that tetrapod limbs evolved from fish fins has also been supported by molecular genetic evidence. Recall from Chapter 26 that several regulatory proteins involved in the development of zebrafish fins and the upper parts of mouse limbs are homologous. Specifically, the proteins produced by *Hox* genes and the homeotic locus *Sonic hedgehog* (*Shh*) are found at the same times and in the same locations in fins and limbs. These data suggest that these appendages are patterned by the same genes. As a result, the data support the hypothesis that tetrapod limbs evolved from fins.

Once the tetrapod limb evolved, natural selection elaborated it into structures that are used for running, gliding, crawling, burrowing, or swimming. In addition, wings and the ability to fly evolved independently in three lineages of tetrapods: the extinct flying reptiles called *pterosaurs* (pronounced *TARE-oh-sors*), birds, and bats. Tetrapods and insects are the only animals that have taken to the skies. **Box 33.2** (page 760) explores how flight evolved in birds.

To summarize, the evolution of the jaw gave tetrapods the potential to capture and process a wide array of foods. With limbs, they could move efficiently on land in search of food. Another challenge remained, however: producing offspring that could survive out of water.

BOX 33.2 The Evolution of Flight in Birds

In 2003 Xing Xu and colleagues published an analysis of a fossilized dinosaur called *Microraptor gui*, which had feathers on both its legs and wings (**Figure 33.14**). This paper was the culmination of a spectacular series of feathered dinosaur species that Xu's group has discovered. Taken together, the newly discovered species answer several key questions about the evolution of birds, feathers, and flight:

- *Did birds evolve from dinosaurs?* On the basis of skeletal characteristics, all of these recently discovered fossil species clearly belong to a lineage of dinosaurs called the *dromaeosaurs*. The fossils definitively link the dromaeosaurs and the earliest known fossil birds.

- *How did feathers evolve?* The fossils support Xu's model that feathers evolved in a series of steps, beginning with simple projections from the skin and culminating with the complex structures observed in today's birds (**Figure 33.15**). *Microraptor gui*, for example, has simple feathers. It is controversial, however, whether the original function of feathers was for courtship or other types of display or for insulation. In today's birds, feathers

FIGURE 33.14 Feathers Evolved in Dinosaurs
An artist's depiction of what *Microraptor gui*, a dinosaur that had feathers on its body and all four limbs, might have looked like in life.

function in display, insulation, and flight.

- *Did birds begin flying from the ground up or from the trees down?* More specifically, did flight evolve with running species that began to jump and glide or make short flights, with the aid of feathers to provide lift? Or did flight evolve from tree-dwelling species that used feathers to glide from tree to tree,

much as flying squirrels do today? Because it is unlikely that *Microraptor gui* could run efficiently with feathered legs, Xu and colleagues propose that flight evolved from tree-dwelling gliders.

Once dinosaurs evolved feathers and took to the air as gliders, the fossil record shows that a series of adaptations made powered, flapping flight

EVOLUTION OF THE FEATHER

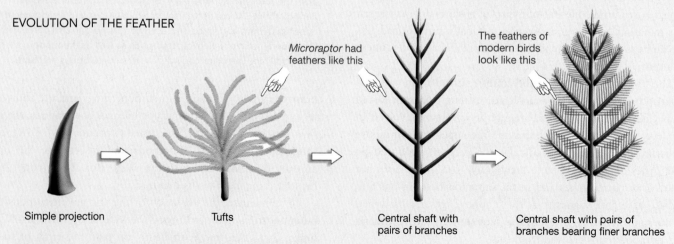

Microraptor had feathers like this

The feathers of modern birds look like this

Simple projection · Tufts · Central shaft with pairs of branches · Central shaft with pairs of branches bearing finer branches

FIGURE 33.15 Feathers Evolved through Intermediate Stages
A model for the evolution of feathers that is supported by the fossil record.

increasingly efficient. As **Figure 33.16** shows, the *sternum* (breastbone) of birds has an elongated projection called the *keel*, which provides a large surface area to which flight muscles attach. (On a chicken or turkey, the flight muscles are called "white meat.") Birds are also extraordinarily light for their size, primarily because they have a drastically reduced number of bones and because their larger bones are thin-walled and hollow—though strengthened by bony "struts" (see Figure 33.16). Birds are also capable of long periods of sustained activity year round, because they are **endothermic**—meaning that they maintain a high body temperature by producing heat in their tissues. From dinosaurs that jumped and glided from tree to tree, birds have evolved into extraordinary flying machines.

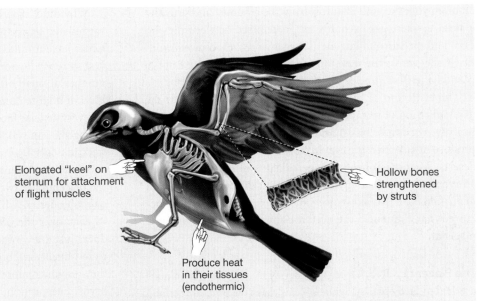

Elongated "keel" on sternum for attachment of flight muscles

Hollow bones strengthened by struts

Produce heat in their tissues (endothermic)

FIGURE 33.16 Birds Have Several Adaptations that Allow for Flight, in Addition to Feathers
The elongated "keel" on the sternum (breastbone), the hollow bones strengthened by crisscrossing struts, and endothermy are all interpreted as adaptations that make bird flight more efficient.

Reproduction

Chapter 31 summarized the diversity of reproductive methods observed in fish. Recall that a few species undergo only asexual reproduction and that when sexual reproduction occurs, it may be based on internal or external fertilization and oviparity, viviparity, or ovoviviparity. In addition, parental care is extensive in some fish species, and often involves guarding eggs from predators and fanning them to keep oxygen levels high. All fish lay their eggs or give birth in water, however. Tetrapods were the first vertebrates that were able to breed on land.

Three major evolutionary innovations gave tetrapods the ability to produce offspring successfully in terrestrial environments: (1) the amniotic egg, (2) the placenta, and (3) elaboration of parental care. Let's explore each of these innovations in turn.

The Amniotic Egg Amniotic eggs have shells that minimize water loss as the embryo develops inside. The first tetrapods, like today's amphibians (frogs, toads, and salamanders), lacked amniotic eggs. Although their eggs were encased by a membrane, the first tetrapods laid eggs that would dry out and die unless they were laid in water. This fact limited the range of habitats that these animals could exploit. Like today's amphibians, the early tetrapods were largely restricted to living in or near marshes, lakes, or ponds.

In contrast, reptiles and the egg-laying mammals produce amniotic eggs. The eggshell is leathery in turtles, crocodiles, and snakes and lizards but hard—due to deposition of the calcium carbonate—in birds and the egg-laying mammals. In addition to having a shell that is largely watertight, an amniotic egg contains a membrane-bound supply of water in a protein-rich solution called **albumen** (Figure 33.17). The embryo is enveloped in a protective inner membrane known as the **amnion**. The **yolk sac** is a membranous pouch that contains nutrients for the growing embryo, and the **allantois** is a membranous pouch that holds waste materials. A middle membrane, the **chorion**, separates the

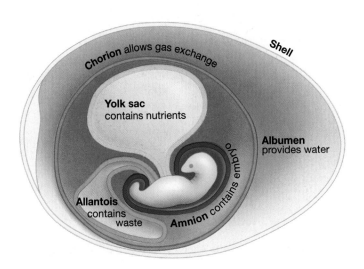

Chorion allows gas exchange

Shell

Yolk sac contains nutrients

Albumen provides water

Allantois contains waste

Amnion contains embryo

FIGURE 33.17 An Amniotic Egg
Amniotic eggs have membrane-bound sacs that hold nutrients, water, and waste and that allow gas exchange.

amnion, yolk sac, and allantois from the albumen and provides a surface where gas exchange can take place between the embryo and the surrounding air. (Oxygen and carbon dioxide diffuse freely across the shell.) Inside an amniotic egg, the embryo is bathed in fluid. The egg itself is highly resistant to drying.

The evolution of the amniotic egg was a key event in the diversification of tetrapods because it allowed turtles, snakes, lizards, crocodiles, birds, and the egg-laying mammals to reproduce in any terrestrial environment—even habitats as dry as deserts. Members of the lineage called Amniota now occupy all types of terrestrial environments. But during the evolution of mammals, a second major innovation in reproduction occurred that eliminated the need for any type of egg laying: the placenta.

The Placenta Recall from Chapter 31 that egg-laying animals are said to be **oviparous**, while species that give birth are termed **viviparous**. In many viviparous animals, females produce an egg that contains a nutrient-rich yolk. Instead of laying the egg, however, the mother retains it inside her body. In these ovoviviparous species, the developing offspring depends on the resources in the egg yolk. In most mammals, however, the eggs that females produce lack yolk. After fertilization occurs and the egg is retained, the mother produces a placenta within her uterus. The **placenta** is an organ that is rich in blood vessels and that facilitates a flow of oxygen and nutrients from the mother to the developing offspring (**Figure 33.18**). After a development period called **gestation**, the embryo emerges from the mother's body.

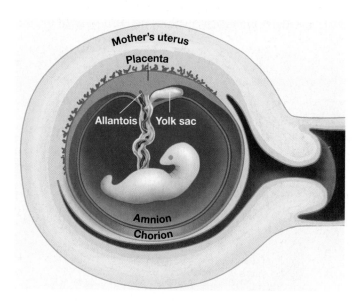

FIGURE 33.18 The Placenta Allows the Mother to Nourish the Fetus Internally

QUESTION Compare the relative positions of the chorion and amnion here with those in the amniotic egg (Figure 33.17). Are they the same or different?

Why did viviparity and the placenta evolve? Biologists have formulated an answer to this question by pointing out that females have a finite amount of time and energy available to invest in reproduction. As a result, a female can produce a large number of small offspring or a small number of large offspring but not a large number of large offspring. Stated more formally, every female faces a trade-off between the quantity of offspring she can produce and their size. In some lineages, natural selection has favored traits that allow females to produce a small number of large, well-developed offspring. Viviparity and the placenta are two such traits. Compared with female insects or echinoderms, which routinely lay thousands or even millions of eggs over the course of a lifetime, a female mammal produces just a few offspring. But because those offspring are protected inside her body and fed until they are well developed, they are much more likely to survive than sea star or insect embryos. Even after birth, many mammals continue to invest time and energy in rearing their young.

Parental Care The term **parental care** encompasses any action by a parent that improves the ability of its offspring to survive, including supplying food, keeping young warm and dry, and protecting them from danger. In some insect and frog species, mothers carry around eggs or newly hatched young; in fishes, parents commonly guard eggs during development and fan them with oxygen-rich water.

The most extensive parental care observed among animals is provided by mammals and birds. In both groups, the mother and often the father continue to feed and care for individuals after birth or hatching—sometimes for many years (**Figure 33.19**). Female mammals also **lactate**—meaning that they produce nutrient-rich fluid called *milk* and use it to feed their offspring after birth. With the combination of the placenta and lactation, placental mammals make the most extensive investment of time and energy in offspring known. The evolution of extensive parental care is hypothesized to be a major reason for the evolutionary success of mammals and birds.

✓CHECK YOUR UNDERSTANDING

The diversification of echinoderms and vertebrates was based on innovations that made it possible for species to feed, move, and reproduce in novel ways. Most echinoderms use their podia to move, but they feed in a wide variety of ways—including using their podia to pry open bivalves, suspension feed, or deposit feed. An array of key innovations occurred during the evolution of vertebrates: Jaws made it possible to bite and process food, limbs allowed tetrapods to move on land, and amniotic eggs could be laid on land. You should be able to (1) explain how the jaw and limb evolved, (2) diagram the structure of an amniotic egg, and (3) explain the role of increased parental care in the evolution of birds and mammals.

(a) Mammal mothers feed and protect newborn young.

(b) Many bird species have extensive parental care.

FIGURE 33.19 Parental Care Is Extensive in Mammals and Birds
(a) Female mammals feed and protect embryos inside their bodies until the young are well developed. Once the offspring is born, the mother feeds it milk until it is able to eat on its own. In some species, parents continue to feed and protect young for many years. **(b)** In birds, one or both parents may incubate the eggs, protect the nest, and feed the young after hatching occurs.

33.4 Key Lineages: Echinodermata

The echinoderms ("spiny-skins") were named for the spines or spikes observed in many species. They are bilaterally symmetric as larvae but undergo metamorphosis and develop into radially symmetric adults. They all have a water vascular system and produce calcium carbonate plates under their skin to form an endoskeleton.

The echinoderms living today make up five major lineages, traditionally recognized as classes: (1) feather stars and sea lilies, (2) brittle stars and basket stars, (3) sea cucumbers, (4) sea stars, and (5) sea urchins and sand dollars. Most feather stars and sea lilies are sessile suspension feeders (**Figure 33.20a**). Brittle stars and basket stars have five long arms that radiate out from a small central disk (**Figure 33.20b**). They use these arms to suspension feed, deposit feed by sopping up material with mucus, or capture small prey animals. Sea cucumbers are sausage-

(a) Feather star

(b) Brittle star (crawling on a sponge)

(c) Sea cucumber

FIGURE 33.20 Three Major Echinoderm Groups: Feather Stars, Brittle Stars, Sea Cucumbers

shaped animals that suspension feed or deposit feed with the aid of tentacles arranged in a whorl around their mouths (**Figure 33.20c**). Sea stars and sea urchins are described in detail in the text that follows.

Echinodermata ▶ Asteroidea (Sea Stars)

The 1700 known species of sea stars have bodies with five or more long arms—in some species up to 300—radiating from a central region that contains the mouth, stomach, and anus (**Figure 33.21**). They range in size from less than 1 cm in diameter when fully grown to 1 m across. They live on hard or soft substrates along the coasts of all the world's oceans. Although the spines that are characteristic of echinoderms are reduced to knobs on the surface of most sea stars, the crown-of-thorns star and a few other species have prominent, upright, movable spines.

Feeding   Sea stars are predators or scavengers. Some species pull bivalves apart with their tube feet and evert their stomach into the prey's visceral mass. Sponges, barnacles, and snails are also common prey. The crown-of-thorns sea star specializes in feeding on corals and is native to the Indian Ocean and western Pacific Ocean. Its population has skyrocketed recently—possibly because people are harvesting their major predator, a large snail called the triton, for its pretty shell. Large crown-of-thorns star populations have led to the destruction of large areas of coral reef.

Movement   Sea stars crawl with the aid of their tube feet. Any one of the five or more arms may be in "front" as the animal moves.

Reproduction   Sexual reproduction predominates, and sexes are separate. At least one sea star arm is filled with reproductive organs that produce massive amounts of gametes—millions of eggs per female, in some species. Species that are native to habitats in the far north, where conditions are particularly harsh, care for their offspring by holding fertilized eggs on their body until the eggs hatch. Most sea stars are capable of regenerating arms that are lost in predator attacks or storms. Some species can reproduce asexually by dividing the body in two, with each of the two individuals then regenerating the missing half.

Pycnopodia helianthoides

FIGURE 33.21 Sea Stars May Have Many Arms

Echinodermata ▶ Echinoidea (Sea Urchins and Sand Dollars)

There are about 800 species of echinoids living today; most are sea urchins or sand dollars. Sea urchins have globe-shaped bodies and long spines and crawl along substrates (**Figure 33.22a**). Sand dollars are flattened and disk-shaped, have short spines, and burrow in soft sediments (**Figure 33.22b**).

Feeding   Sand dollars use their mucus-covered podia to collect food particles in sand or in other soft substrates. Most types of sea urchins are herbivores. In some areas of the world, urchins are extremely important grazers on kelp and other types of algae. In fact, when urchin populations are high, their grazing can prevent the formation of kelp forests. Most echinoids have a unique, jaw-like feeding structure in their mouths that is made up of five calcium carbonate teeth attached to muscles. In many species, this apparatus can extend and retract as the animal feeds.

Movement   Using their podia, sea urchins crawl and sand dollars burrow. Sea urchins can also move their spines to aid in crawling along a substrate.

Reproduction   Sexual reproduction predominates. Fertilization is external, and sexes are separate.

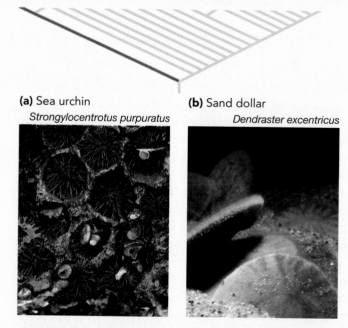

(a) Sea urchin
Strongylocentrotus purpuratus

(b) Sand dollar
Dendraster excentricus

FIGURE 33.22 Sea Urchins and Sand Dollars Are Closely Related

33.5 Key Lineages: Chordata

The chordates comprise three major subgroups or subphyla: (1) the urochordates (also called tunicates or sea squirts), (2) the cephalochordates (lancelets), and (3) the vertebrates. There are about 3000 species of tunicates, 24 species of lancelets, and over 50,000 vertebrates. At some stage in their life cycle, all species in the phylum Chordata have a dorsal hollow nerve cord, a notochord, pharyngeal gill slits, and a muscular tail that extends past the anus.

Tunicates and lancelets were introduced in Section 33.2; here the focus is on the major lineages of Vertebrata. Vertebrates are named for a column of cartilaginous or bony structures, called **vertebrae**, that form along the dorsal sides of most species and that protect the dorsal hollow nerve cord. Not all species of Vertebrata have vertebrae, however. The feature that all species in this lineage have is the **cranium**, or skull—a bony, cartilaginous, or fibrous case that encloses the brain. Biologists once used the term *Craniata* to refer to the vertebrates.

The cranium is thought to be a key innovation in the evolution of vertebrates because it protects the brain along with sensory organs such as eyes. The vertebrate brain develops as an outgrowth of the most anterior end of the dorsal hollow nerve cord and is important to the vertebrate life style. Vertebrates are active predators and herbivores that can make rapid, directed movements with the aid of their endoskeleton. Coordinated movements are possible in part because vertebrates have large brains divided into three distinct parts: (1) a **forebrain**, housing the sense of smell; (2) a **midbrain**, associated with vision; and (3) a **hindbrain**, responsible for balance and hearing. In vertebrate groups that evolved more recently, the forebrain is a large structure called the **cerebrum**, and the hindbrain consists of enlarged regions called the **cerebellum** and **medulla oblongata**. (The structure and functions of the vertebrate brain are analyzed in detail in Chapter 45.) The evolution of a large brain, protected by a hard cranium, was an innovation that defines the vertebrates.

Chordata ▶ Myxinoidea (Hagfish) and Petromyzontoidea (Lampreys)

Although recent phylogenetic analyses indicate that hagfish and lampreys may belong to two independent lineages, some data suggest that they are a single group called the **Agnatha** ("not-jawed"). Because these animals are the only vertebrates that lack jaws, the 110 species in the two groups are still referred to as the jawless fishes. The hagfish and lampreys—several of which are known only from fossils—are the only surviving members of the earliest branches at the base of the vertebrates.

Hagfish and lampreys have long, slender bodies and are aquatic. Most species are less than a meter long when fully grown. Hagfish lack any sort of vertebral column, but lampreys have small pieces of cartilage along the length of their dorsal hollow nerve cord.

Feeding Hagfish are scavengers and predators (**Figure 33.23a**). They deposit feed in the carcasses of dead fish and whales, and some are thought to burrow through ooze at the bottom of the ocean, feeding on polychaetes and other buried prey. Lampreys, in contrast, are ectoparasites. They attach to the sides of fish or other hosts by suction, then use spines in their mouth and tongue to rasp a hole in the side of their victim (**Figure 33.23b**). Once the wound is open, they suck blood and other body fluids.

Movement Hagfish and lampreys have a well-developed notochord and swim by making undulating movements. Lampreys can also move themselves upstream, against the flow of water, by attaching their suckers to rocks and looping the rest of their body forward, like an inchworm. Although lampreys have fins that aid in locomotion, their fins do not occur in pairs as they do in jawed fish.

Reproduction Virtually nothing is known about hagfish mating or embryonic development. Lampreys are **anadromous**, meaning that they spend their adult life in large lakes or the ocean but swim up streams to breed. Fertilization is external, and adults die after breeding once. Lamprey eggs hatch into larvae that look and act like lancelets. The larvae burrow into sediments and suspension feed for several years before metamorphosing into free-swimming adults.

(a) Hagfish

Eptatretus stoutii

(b) Lampreys feeding on fish

Petromyzon marinus

Lampreys

FIGURE 33.23 Hagfish and Lampreys Are Jawless

Chordata ▶ Chondrichthyes (Sharks, Rays, Skates)

The 840 species in this lineage are distinguished by their cartilaginous skeleton (*chondrus* is the Greek word for cartilage), the presence of jaws, and the existence of paired fins. Paired fins were an important evolutionary innovation, because they stabilize the body during rapid swimming—keeping it from pitching up or down, yawing to one side or the other, or rolling.

Most sharks, rays, and skates are marine, though a few species live in freshwater. Sharks have streamlined, torpedo-shaped bodies and an asymmetrical tail—the dorsal portion is longer than the ventral portion (**Figure 33.24a**) In contrast, the dorsal-ventral plane of the body in rays and skates is strongly flattened (**Figure 33.24b**).

Feeding A few species of ray and shark suspension feed on plankton, but most species in this lineage are predators. Skates and rays lie on the ocean floor and ambush passing animals; electric rays capture their prey by stunning them with electric discharges of up to 200 volts. Most sharks, in contrast, are active hunters that chase down prey in open water and bite them. The larger species of shark feed on large fish or marine mammals. Sharks are referred to as the "top predator" in many marine ecosystems, because they are at the top of the food chain—nothing eats them. Yet the largest of all sharks, the whale shark, is a suspension feeder. Whale sharks filter plankton out of water as it passes over their gills.

Movement Rays and skates swim by flapping their greatly enlarged pectoral fins. (*Pectoral fins* are located on the sides of an organism; *dorsal fins* are located on the dorsal surface.) Sharks swim by undulating their bodies and beating their large tail.

Reproduction Sharks have internal fertilization, and fertilized eggs may be shed into the water or retained until the young are hatched and well developed. In some viviparous species, embryos are attached to the mother by specialized tissues in a mammal-like placenta, where the exchange of gases, nutrients, and wastes takes place. Skates are oviparous, but rays are viviparous.

(a) Sharks are torpedo shaped.

Prionace glauca

(b) Skates and rays are flat.

Taeniura melanospila

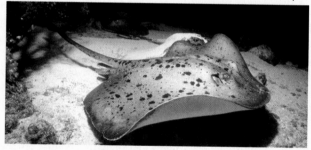

FIGURE 33.24 Sharks and Rays Have Cartilaginous Skeletons

Chordata ▶ Actinopterygii (Ray-Finned Fishes)

Actinopterygii (pronounced *ack-tin-op-teh-RIJ-ee-i*) means "ray-finned." Logically enough, these fish have fins that are supported by long, bony rods arranged in a ray pattern. They are the most ancient living group of vertebrates that have a skeleton made of bone. Their bodies are covered with interlocking scales that provide a stiff but flexible covering, and they avoid sinking in the water with the aid of a gas-filled **swim bladder**. The evolution of the swim bladder was an important innovation, because this structure allowed ray-finned fishes to float. Tissues are heavier than water, so the bodies of aquatic organisms tend to sink. Sharks and rays, for example, have to swim to avoid sinking. But ray-finned fishes have a bladder that changes in volume, depending on the individual's position. Gas is added to the bladder when a ray-finned fish swims down; gas is removed when the fish swims up. In this way, ray-finned fishes maintain neutral buoyancy in water of various depths and thus pressures.

The actinopterygians are the most successful vertebrate lineage based on number of species, duration in the fossil record, and extent of habitats occupied. Almost 24,000 species of ray-finned fishes have been identified. In traditional classifications, Actinopterygii is considered a class.

The most important major lineage of ray-finned fishes is the Teleostei, which includes some 20,000 species. Most of the fish that you are familiar with, such as tuna, trout, cod, and goldfish, are teleosts (**Figure 33.25**).

(Continued on next page)

Chordata ▶ Actinopterygii (Ray-Finned Fishes) *(continued)*

Feeding Teleosts can suck food toward their mouths, grasp it with their protrusible jaws, and then process it with teeth on their jaws and with pharyngeal jaws in their throat. The size and shape of the mouth, the jaw teeth, and the pharyngeal jaw teeth all correlate with the type of food consumed. For example, most predatory teleosts have long, spear-shaped jaws armed with spiky teeth, as well as bladelike teeth on their pharyngeal jaws. In addition to being major predators, ray-finned fishes are the most important large herbivores in both marine and freshwater environments.

Movement Ray-finned fishes swim by alternately contracting muscles on the left and right sides of their bodies from head to tail, resulting in rapid, side-to-side undulations. Their bodies are streamlined to reduce drag in water. Teleosts have a flexible, symmetrical tail, which reduces the need to use their pectoral (side) fins as steering and stabilizing devices during rapid swimming.

Reproduction Most species rely on external fertilization and are oviparous; some species have internal fertilization with external development; still others have internal fertilization and are viviparous. Although it is common for fish eggs to be released in the water and left to develop on their own, parental care occurs in some species. Parents may carry fertilized eggs on their fins, in their mouth, or in specialized pouches to guard them until the eggs hatch. In freshwater teleosts, offspring develop directly, but marine species have larva that are very different from adult forms. As they develop, marine fish larvae undergo a metamorphosis to the juvenile form, which then grows into an adult.

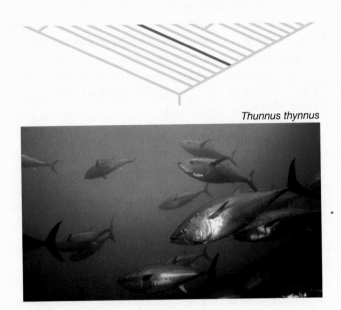
Thunnus thynnus

FIGURE 33.25 Teleosts Are Ray-Finned Fishes that Have a Flexible Tail

Chordata ▶ Actinistia (Coelacanths) and Dipnoi (Lungfish)

Although coelacanths (pronounced *SEEL-uh-kanths*) and lungfish represent independent lineages, they are sometimes grouped together and called **lobe-finned fishes**. Lobe-finned fishes are common and diverse in the fossil record in the Devonian period, about 400 million years ago, but only eight species are living today. They are important, however, because they represent a crucial evolutionary link between the ray-finned fishes and the tetrapods. Instead of having fins supported by rays of bone, their fins are fleshy lobes supported by an array of bones and muscles—similar to those observed in the limbs of tetrapods.

Coelacanths are marine and occupy habitats 150–700 m below the surface (**Figure 33.26**). In contrast, lungfish live in shallow, freshwater ponds (see Figure 33.12). As their name implies, lungfish have lungs and breathe air when oxygen levels in their habitats drop. Some species burrow in mud and enter a quiescent state when their habitat dries up during each year's dry season.

Feeding Coelacanths prey on fish. Lungfish are **omnivorous** ("all-eating"), meaning that they eat algae and plant material as well as animals.

Movement Coelacanths swim by waving their pectoral and pelvic ("hip") fins in the same sequence that tetrapods use in walking with their limbs. Lungfish swim by waving their body, and they can use their fins to walk along pond bottoms.

Reproduction Sexual reproduction is the rule, with fertilization internal in coelacanths and external in lungfish. Coelacanths are ovoviviparous; lungfish lay eggs. Lungfish eggs hatch into larvae that resemble juvenile salamanders.

Latimeria chalumnae

FIGURE 33.26 Coelacanths Are Lobe-Finned Fishes

Chordata ▶ Amphibia (Frogs, Salamanders, Caecilians)

The 4800 species of amphibians living today form three distinct clades traditionally termed orders: (1) frogs and toads, (2) salamanders, and (3) caecilians (pronounced *suh-SILL-ee-uns*). Amphibians are found throughout the world and occupy ponds, lakes, or moist terrestrial environments (**Figure 33.27a**). Translated literally, their name means "both-sides-living." The name is appropriate because adults of most species of amphibian feed on land, but many species lay their eggs in water. When amphibians live on land, gas exchange occurs exclusively or in part across their moist, mucus-covered skin.

Feeding Adult amphibians are carnivores. Most frogs are sit-and-wait predators that use their long, extensible tongues to capture passing prey. Salamanders also have an extensible tongue, which some species use in feeding. Terrestrial caecilians prey on earthworms and other soil-dwelling animals; aquatic forms eat vertebrates and small fish.

Movement Most amphibians have four well-developed limbs. In water, frogs and toads move by kicking their hind legs to swim; on land they kick their hind legs out to jump or hop. Salamanders walk on land; in water they undulate their body to swim like fish. Caecilians lack limbs and eyes; terrestrial forms burrow in moist soils (**Figure 33.27b**).

Reproduction Frogs are oviparous and have external fertilization, but salamanders and caecilians have internal fertilization. Most salamanders are oviparous, but many caecilians are viviparous. In some species of frogs, parents may guard or even carry eggs. In many frogs, young develop in the water and suspension feed on plant or algal material. Salamander larvae are carnivorous. Later the larvae undergo a dramatic metamorphosis into land-dwelling adults. For example, the fishlike tadpoles of frogs and toads develop limbs and replace their gills with lungs.

(a) Frogs and other amphibians lay their eggs in water.   **(b)** Caecilians are legless amphibians.

Bufo periglenes   *Gymnophiona* species

FIGURE 33.27 Amphibians Are the Most Ancient Tetrapods

Chordata: Mammalia (Mammals)

Mammals are easily recognized by the presence of **hair** or **fur**, which serves to insulate the body. Like birds, mammals are endotherms that maintain high body temperatures by oxidizing large amounts of food and generating large amounts of heat. Instead of insulating themselves with feathers, though, mammals retain heat because the body surface is covered with layers of hair or fur. Endothermy evolved independently in birds and mammals. In both groups, endothermy is thought to be an adaptation that enables individuals to maintain high levels of activity—particularly at night or during cold weather.

In addition to being endothermic and having fur, mammals have **mammary glands**—a unique structure that makes lactation possible. The evolution of mammary glands gave mammals the ability to provide their young with particularly extensive parental care. Mammals are also the only vertebrates with facial muscles and lips and the only vertebrates that have jaws formed from a single bone. In traditional classifications, Mammalia is designated as a class.

Mammals evolved when dinosaurs and other reptiles were the dominant large herbivores and predators in terrestrial and aquatic environments. The earliest mammals in the fossil record appear about 195 million years ago; they were small animals that were probably active only at night. Presumably because of their ancestry, most of the 4500 species of mammal living today have good nocturnal vision and a strong sense of smell. The adaptive radiation that gave rise to today's diversity of mammals did not take place until after the dinosaurs went extinct about 65 million years ago. After the dinosaurs were gone, the mammals diversified into lineages of small and large herbivores, small and large predators, or marine hunters—ecological roles that had once been filled by dinosaurs and mosasaurs.

Chordata ▶ Mammalia ▶ Monotremata (Platypuses, Echidnas)

The monotremes are the most ancient lineage of mammals living today, and they are found only in Australia and New Zealand. They lay eggs and have lower metabolic rates than other mammals do. Three species exist: one species of platypus and two species of echidna.

Feeding   Monotremes have a leathery beak or bill. The platypus feeds on insect larvae, mollusks, and other small animals in streams (**Figure 33.28a**). Echidnas feed on ants, termites, and earthworms (**Figure 33.28b**).

Movement   Platypuses swim with the aid of their webbed feet. Echidnas walk on their four legs.

Reproduction   Platypuses lay their eggs in a burrow, while echidnas keep their eggs in a pouch on their belly. The young hatch quickly, and the mother must continue keeping them warm and dry for another four months.

(a) Platypus   *Ornithorhynchus anatinus*

(b) Echidna   *Tachyglossus aculeatus*

FIGURE 33.28 Platypuses and Echidnas Are Egg-Laying Mammals

Chordata ▶ Mammalia ▶ Marsupiala (Marsupials)

The 275 known species of marsupials live in the Australian region and the Americas (**Figure 33.29**). Although females have a placenta that nourishes embryos during development, the young are born after a short embryonic period and are poorly developed. They crawl from the opening of the female's reproductive tract to the female's nipples, where they suck milk. They stay attached to their mother until they grow large enough to move independently.

Feeding   Marsupials are herbivores, omnivores, or carnivores ("meat-eaters"). Many more cases of convergent evolution have occurred than for placental mammals. For example, the recently extinct Tasmanian wolf was a long-legged, social hunter similar to the timber wolves of North America and northern Eurasia. A species of marsupial native to Australia specializes in eating ants and looks and acts much like the South American anteater, which is not a marsupial.

Movement   Marsupials move by crawling, gliding, walking, running, or hopping.

Reproduction   Marsupial young spend more time developing while attached to their mother's nipple than they do inside her body being fed via the placenta.

Didelphis virginiana

FIGURE 33.29 Marsupials Give Birth after a Short Embryonic Period

Chordata ▶ Mammalia ▶ Eutheria (Placental Mammals)

The approximately 4300 species of placental mammals are distributed worldwide. They are far and away the most species-rich and morphologically diverse group of mammals.

Biologists group mammals into 18 lineages called orders. The six most species-rich mammalian orders are the rodents (rats, mice, squirrels; 1814 species), bats (986 species), insectivores (hedgehogs, moles, shrews; 390 species), artiodactyls (pigs, hippos, whales, deer, sheep, cattle; 293 species), carnivores (dogs, bears, cats, weasels, seals; 274 species), and primates (lemurs, monkeys, apes, humans; 235 species).

Feeding   The size and structure of the teeth correlate closely with the diet of placental mammals. Herbivores have large, flat teeth for crushing leaves and other coarse plant material; predators have sharp teeth that are efficient at biting and tearing flesh. Omnivores, such as humans, usually have several distinct types of teeth. The structure of the digestive tract also correlates with the placental mammals' diet. In some plant-eaters, for example, the stomach hosts unicellular organisms that digest cellulose and other complex polysaccharides.

Movement   In placental mammals, the structure of the limb correlates closely with the type of movement performed. Eutherians fly, glide, run, walk, swim, burrow, or swing from trees (**Figure 33.30**). Limbs are reduced or lost in aquatic groups such as whales and dolphins, which swim by undulating their bodies.

Reproduction   In mammals, both fertilization and development are internal. Thus, eutherians are viviparous. An extensive placenta develops from a combination of maternal and fetal tissues, and gestation is relatively long. At birth, young are much better developed than in marsupials—some are able to walk or run minutes after birth. All eutherians feed their offspring milk until the young have grown large enough to process solid food. A prolonged period of parental care, extending beyond the nursing stage, is common as offspring learn how to escape predators and find food.

Hylobates lar

FIGURE 33.30 Eutherians Are the Most Species-Rich and Diverse Group of Mammals

Chordata: Reptilia (Turtles, Snakes and Lizards, Crocodiles, Birds)

The reptiles are a monophyletic group and represent one of the two major living lineages of amniotes—the other lineage consisting of the extinct mammal-like reptiles and today's mammals. The major feature distinguishing the reptilian and mammalian lineages is the number and placement of openings in the skull. These openings are used by jaw muscles that make sophisticated biting and chewing movements possible.

Several features adapt reptiles for life on land. Their skin is made watertight by a layer of scales made of the protein keratin, which is the same protein as that found in mammalian hair. They breathe air through well-developed lungs and lay shelled, amniotic eggs that resist drying out. In turtles, the egg has a leathery shell; in other reptiles, the shell is made of stiff calcium carbonate. Because sperm and egg must fuse before the amniotic membrane and shell form, fertilization in reptiles is internal.

The reptiles include the dinosaurs, pterosaurs (flying reptiles), mosasaurs (marine reptiles), and other extinct lineages that flourished from about 250 million years ago until the mass extinction at the end of the Cretaceous, 65 million years ago. Today, the Reptilia are represented by four major lineages, traditionally recognized as subclasses: (1) turtles, (2) snakes and lizards, (3) crocodiles and alligators, and (4) birds. With the exception of birds, all of these groups are **ectothermic** ("outside-heated")—meaning that individuals do not use internally generated heat to regulate their body temperature. It would be a mistake, however, to conclude that reptiles other than birds do not regulate their body temperature closely. Reptiles bask in sunlight, seek shade, and perform other behaviors to keep their body temperature at a preferred level.

Chordata ▶ Reptilia ▶ Testudinia (Turtles, Tortoises)

The 271 known species of turtles and tortoises inhabit freshwater, marine, and terrestrial environments throughout the world. The testudines are distinguished by a shell composed of bony plates that fuse to the vertebrae and ribs (**Figure 33.31**). Their skulls are highly modified versions of the skulls of other reptiles. Turtles and tortoises lack teeth, but their jawbone and lower skull form a bony beak.

Feeding　Turtles are either herbivorous or carnivorous—feeding on whatever animals they can capture and swallow. They may also scavenge dead material. Most marine turtles are carnivorous. Leatherback turtles, for example, feed primarily on jellyfish, such as the Portuguese man-of-war, and are only mildly affected by the jellyfish's stinging cnidocytes. Tortoises are plant eaters.

Movement　Turtles swim, walk, or burrow. Aquatic species usually have feet that are modified to function as flippers.

Reproduction　All turtles are oviparous. Other than digging a nest prior to depositing eggs, parental care is lacking. The sex of a baby turtle is often not determined by sex chromosomes. Instead, in many species the temperature at which the egg develops determines gender. High temperatures produce mostly males, while low temperatures produce mostly females.

Testudo pardalis

FIGURE 33.31 Turtles Have a Shell Consisting of Bony Plates

Chordata ▶ Reptilia ▶ Lepidosauria (Lizards, Snakes)

Most squamates are small reptiles with elongated bodies and scaly skin. Most lizards have well-developed jointed legs, but snakes are limbless (**Figure 33.32**). The hypothesis that snakes evolved from limbed ancestors is supported by the presence of vestigial hip and leg bones in boas and pythons. There are about 6800 species of lizards and snakes alive now.

Feeding　Small lizards prey on insects. Although most of the larger lizard species are herbivores, the 3-meter-long monitor lizard from the island of Komodo is a predator that kills and eats deer. Snakes are carnivores; some subdue their prey with the aid of poison that is injected via modified teeth called *fangs*. Snakes prey primarily on small mammals, amphibians, and invertebrates, which they swallow whole (usually headfirst).

Movement　Lizards crawl or run on their four limbs. Snakes and limbless lizards burrow through soil, crawl over the ground, or climb trees by undulating their bodies.

Reproduction　Although most squamates lay eggs, many are oviparous. Most squamates are bisexual, but asexual reproduction, via the production of eggs by mitosis, is known to occur in six groups of lizards and one snake lineage.

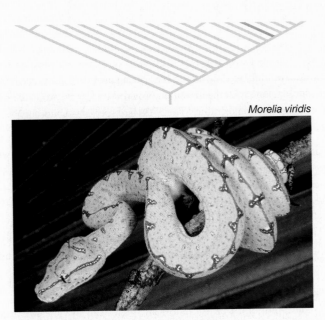

Morelia viridis

FIGURE 33.32 Snakes Are Limbless Predators

Chordata ▶ Reptilia ▶ Crocodilia (Crocodiles, Alligators)

Only 21 species of crocodile and alligator are known. Most are tropical and live in freshwater or marine environments. They have eyes located on the top of their heads and nostrils located at the top of their long snouts—adaptations that allow them to sit semi-submerged in water for long periods of time (**Figure 33.33**).

Feeding Crocodilians are predators. Their jaws are filled with conical teeth that are continually replaced as they fall out during feeding. The usual method of killing small prey is by biting through the body wall. Large prey are usually subdued by drowning. Crocodilians eat amphibians, turtles, fish, birds, and mammals.

Movement Crocodiles and alligators walk or gallop on land. In water they swim with the aid of their large, muscular tails.

Reproduction Although crocodilians are oviparous, parental care is extensive. Eggs are laid in earth-covered nests that are guarded by the parents. When young inside the eggs begin to vocalize, parents dig them up and carry the newly hatched young inside their mouths to nearby water. Crocodilian young can hunt when newly hatched but stay near their mother for up to three years.

Alligator mississippiensis

FIGURE 33.33 Alligators Are Adapted for Aquatic Life

Chordata ▶ Reptilia ▶ Aves (Birds)

The fossil record provides conclusive evidence that birds descended from a lineage of dinosaurs that had a unique trait: **feathers**. In dinosaurs, feathers are hypothesized to have functioned as insulation and in courtship or aggressive displays. In birds, feathers insulate and are used for display but also furnish the lift, power, and steering required for flight. Birds have other adaptations that make flight possible, including large breast muscles used to flap the wings. Bird bodies are lightweight, because they have a reduced number of bones and organs and because their hollow bones are filled with air sacs, linked to the lungs. Instead of teeth, birds have a horny beak. They are **endotherms** ("within-heating")—they have a high metabolic rate and use the heat produced, along with the insulation provided by feathers, to maintain a constant body temperature. The 9700 bird species today occupy virtually every habitat, including the open ocean (**Figure 33.34**).

Feeding Plant-eating birds usually feed on nectar or seeds. Most birds are omnivores, although many are predators that capture insects, small mammals, fish, other birds, lizards, mollusks, or crustaceans. The size and shape of a bird's beak correlates closely with its diet. For example, predatory species such as falcons have sharp, hook-shaped beaks; finches and other seedeaters have short, stocky bills that can crack seeds and nuts; fish-eating species such as the great blue heron have spear-shaped beaks.

Movement Although flightlessness has evolved repeatedly during the evolution of birds, almost all species can fly. The size and shape of birds' wings correlate closely with the type of flying they do. Birds that glide or hover have long, thin wings; species that specialize in explosive takeoffs and short flights have short, stocky wings. Many seabirds are efficient swimmers, using their webbed feet or wings under water. Ground-dwelling birds such as ostrich and pheasants can run long distances at high speed.

Reproduction Birds are oviparous but provide extensive parental care. In most species, one or both parents build a nest and incubate the eggs. After the eggs hatch, parents feed offspring until they are large enough to fly and find food on their own.

Diomedea melanophris

FIGURE 33.34 Birds Are Feathered Descendants of Dinosaurs

33.6 Key Lineages: The Hominin Radiation

Although humans occupy a tiny twig on the tree of life, there has been a tremendous amount of research on human origins. This section introduces the lineage of mammals called the Primates, the fossil record of human ancestors, and data on the relationships among human populations living today.

The Primates

The lineage called Primates consists of two main groups: prosimians and anthropoids. The **prosimians** ("before-monkeys") consist of the lemurs, found in Madagascar, and the tarsiers, pottos, and lorises of Africa and south Asia. Most prosimians live in trees and are active at night (**Figure 33.35a**). The **anthropoids** ("human-like") include the New World monkeys found in Central and South America, the Old World monkeys that live in Africa and tropical regions of Asia, the gibbons of the Asian tropics, and the great apes—orangutans, gorillas, chimpanzees, and humans (Figure 33.35b). The phylogenetic tree in **Figure 33.35c** shows the evolutionary relationships among these groups.

Primates are distinguished by having eyes located on the front of the face. Eyes that look forward provide better depth perception than do eyes on the sides of the face. Primates also tend to have hands and feet that are efficient at grasping, flattened nails instead of claws on the fingers and toes, brains that are large relative to overall body size, complex social behavior, and extensive parental care of offspring. The anthropoids are distinguished from the prosimians by having a fully *opposable thumb*—meaning a thumb that can touch the tips of all the other fingers—which makes grasping particularly efficient.

The lineage in Figure 33.35c that is composed of the **great apes**, including humans, is known as the **hominids**. From extensive comparisons of DNA sequence data, it is now clear that humans are most closely related to the chimpanzees and that our next nearest living relative is the gorilla.

Compared with most types of primate, the great apes are relatively large bodied and have long arms, short legs, and no tail.

(a) Prosimians are small, tree-dwelling primates.

Loris tardigradus

(b) New World monkeys are anthropoids.

Ateles geoffrogi

(c) Humans are apes.

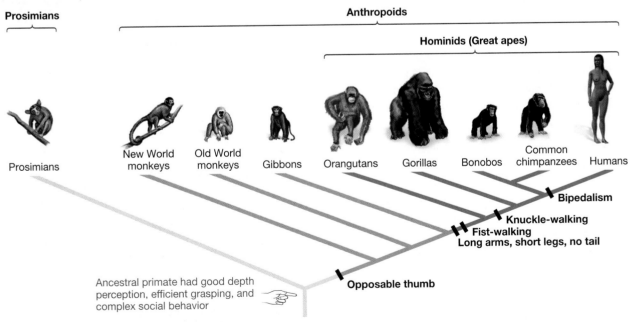

FIGURE 33.35 A Phylogeny of the Monkeys and Great Apes
(a) Prosimians live in Africa, Madagascar, and south Asia. **(b)** Anthropoids include Old World monkeys, New World monkeys, and great apes. **(c)** Phylogenetic tree estimated from extensive DNA sequence data. According to the fossil record, humans and chimps shared a common ancestor 6 to 7 million years ago.

Although all of the great ape species except for the orangutans live primarily on the ground, they have distinct ways of walking. When orangutans do come to the ground, they occasionally walk with their knuckles pressed to the ground. More commonly, though, they "fist walk"—that is, they walk with the backs of their hands pressed to the ground. Gorillas and chimps, in contrast, only "knuckle-walk." They also occasionally rise up on two legs—usually in the context of displaying aggression. Humans are the only great ape that is fully **bipedal** ("two-footed")—meaning they walk upright on two legs. Bipedalism is, in fact, the shared derived character that defines the group called hominins.

Fossil Humans

According to the fossil record, the common ancestor of chimps and humans lived in Africa 6–7 million years ago. As a group, all the species on the branch leading to contemporary humans are called **hominins**. The fossil record of hominins, though not nearly as complete as investigators would like, is rapidly improving. About 14 species have been found, and new fossils that inform the debate over the ancestry of humans are discovered every year. Although naming the hominin species and interpreting their characteristics remain intensely controversial, most researchers agree that they can be organized into four groups:

1. *Australopithecus* Four species of small apes called **gracile australopithecines** have been identified to date (**Figure 33.36a**). The adjective *gracile*, or "slender," is appropriate because these organisms were slightly built. The genus name *Australopithecus* ("southern ape") was inspired by the earliest specimens, which came from South Africa. Several lines of evidence support the hypothesis that the gracile australopithecines were bipedal. For example, the hole in the back of their skulls where the spinal cord connects to the brain is oriented downward, just as it is in our species, *Homo sapiens*. In chimps, gorillas, and other vertebrates that walk on four feet, this hole is oriented backward.

2. *Paranthropus* Three species are grouped in the genus *Paranthropus* ("beside-human"). Like the gracile australopithecines, these **robust australopithecines** were bipedal. But as **Figure 33.36b** illustrates, their skulls were much broader and more robust than those of the gracile forms. All three species had massive cheek teeth and jaws, very large cheekbones, and a *sagittal crest*—a flange of bone at the top of the skull. Because muscles that work the jaw attach to the sagittal crest and cheekbones, researchers conclude that these organisms made their living by crushing large seeds or coarse plant materials. One species is nicknamed "nutcracker man." The name *Paranthropus* was inspired by the hypothesis that the three known species are a monophyletic group

that was a side branch during human evolution—an independent lineage that went extinct.

3. **Early *Homo*** Species in the genus *Homo* are called **humans**. As **Figure 33.36c** shows, species in this genus have flatter

(a) Gracile australopithecines (*Australopithecus africanus*)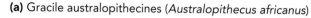

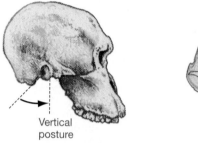

Vertical posture

(b) Robust australopithecines (*Paranthropus robustus*)

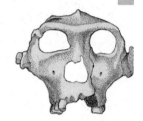

Massive cheek teeth

(c) Early *Homo* (*Homo habilis*)

Flatter face

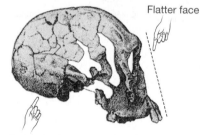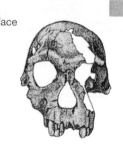

Larger braincase

(d) Recent *Homo* (*Homo sapiens*, Cro-Magnon)

Flattest face

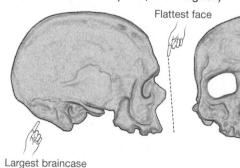

Largest braincase

FIGURE 33.36 African Hominins Comprise Four Major Groups
QUESTION The skulls are arranged as they appear in the fossil record, from most ancient to most recent—(a) to (d). How did the forehead and brow ridge of hominins change through time?

TABLE 33.1 Characteristics of Selected Hominins

Species	Location of Fossils	Estimated Average Braincase Volume (cm³)	Estimated Average Body Size (kg)	Associated with Stone Tools?
Australopithecus afarensis	Africa	450	36	no
A. africanus	Africa	450	36	no
Paranthropus boisei	Africa	510	44	no?
Homo habilis	Africa	550	34	yes
H. ergaster	Africa	850	58	yes
H. erectus	Africa, Asia	1000	57	yes
H. heidelbergensis	Africa, Europe	1200	62	yes
H. neanderthalensis	Middle East, Europe, Asia	1500	76	yes
H. sapiens	Middle East, Europe, Asia	1350	53	yes

and narrower faces, smaller jaws and teeth, and larger braincases than the earlier hominins do. (The **braincase** is the portion of the skull that encloses the brain.) The appearance of early members of the genus *Homo* in the fossil record coincides closely with the appearance of tools made of worked stone—most of which are interpreted as handheld choppers or knives. Although the fossil record does not exclude the possibility that *Paranthropus* made tools, many researchers favor the hypothesis that extensive toolmaking was a diagnostic trait of early *Homo*.

4. **Recent *Homo*** The recent species of *Homo* date from 1.2 million years ago to the present. As **Figure 33.36d** shows, these species have even flatter faces, smaller teeth, and larger braincases than the early *Homo* species do. The 30,000-year-old fossil in the figure, for example, is from a population of *Homo sapiens* called the **Cro-Magnons**. The Cro-Magnons were accomplished painters and sculptors who buried their dead in carefully prepared graves. There is also evidence that **Neanderthal** people (*Homo neanderthalensis*) made art and buried their dead in a ceremonial fashion.

Table 33.1 summarizes data on the geographic range, braincase volume, and body size of selected species within these four groups. **Figure 33.37** provides the time range of each species in the fossil record. Although researchers do not have a solid understanding of the phylogenetic relationships among the hominin species, several points are clear from the available data. First, the shared, derived character that defines the hominins is bipedalism. Second, several species from the lineage were present simultaneously during most of hominin evolution. For example, about 1.8 million years ago there may have been as many as five hominin species living in eastern and southern Africa. Because fossils from more than one species have been found in the same geographic location in rock strata of the same age, it is almost certain that different hominin species lived in physical contact. Finally, compared with the

gracile and robust australopithecines and the great apes, species in the genus *Homo* have extremely large brains relative to their overall body size.

Why did humans evolve such gigantic brains? The leading hypothesis on this question is that early *Homo* began using symbolic spoken language along with initiating extensive tool use. The logic here is that increased toolmaking and language use triggered natural selection for the capacity to reason and communicate, which required a larger brain. To support this

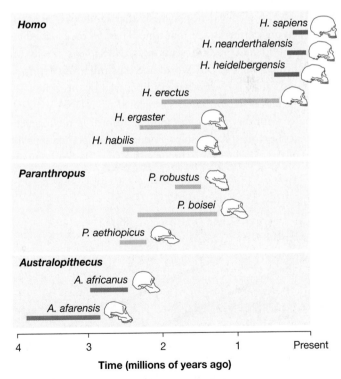

FIGURE 33.37 A Timeline of Human Evolution
Plot of the ages of selected fossil hominins. QUESTION How many species of hominin existed 2.2 million years ago, 1.8 million years ago, and 100,000 years ago?

hypothesis, researchers point out that, relative to the brain areas of other hominins, the brain areas responsible for language were enlarged in the earliest *Homo* species. There is even stronger fossil evidence for extensive use of speech in *Homo neanderthalensis* and early *Homo sapiens*:

- The hyoid bone is a slender bone in the voice box of modern humans that holds muscles used in speech. In Neanderthals and early *Homo sapiens*, the hyoid is vastly different in size and shape from a chimpanzee's hyoid bone. Researchers recently found an intact hyoid bone associated with a 60,000-year-old Neanderthal individual and showed that it is virtually identical to the hyoid of modern humans.

- *Homo sapiens* colonized Australia by boat between 60,000 and 40,000 years ago. An expedition of that type could not be planned and carried out in the absence of symbolic speech.

To summarize, *Homo sapiens* is the sole survivor of an adaptive radiation that took place over the past 3.5 million years. From a common ancestor shared with chimpanzees, hominins evolved the ability to walk upright, make tools, and talk.

The Out-of-Africa Hypothesis

The first fossils of *Homo sapiens* appear in African rocks that date to about 160,000 years ago. For some 100,000 years thereafter, the fossil record indicates that our species occupied Africa while *H. neanderthalensis* resided in Europe and the Middle East. Some evidence suggests that *H. erectus* may still have been present in Asia at that time. Then, in rocks dated between 60,000 and 30,000 years ago, *H. sapiens* fossils are found throughout the Old World and Australia. But *H. neanderthalensis* and *H. erectus* have disappeared by this time.

Phylogenies of *Homo sapiens* estimated from DNA sequence data agree with the pattern in the fossil record. In phylogenetic trees that show the relationships among human populations living today, the first lineages to branch off lead to descendant pop-

ulations that live in Africa today (**Figure 33.38**). Based on this observation, it is logical to infer that the ancestral population of modern humans also lived in Africa. The tree shows that lineages subsequently branched off to form three monophyletic groups. Because the populations within each of these clades live in a distinct area, the three lineages are thought to represent three major waves of migration that occurred as *Homo sapiens* populations dispersed from Africa to (1) southeast Asia and the South Pacific, (2) central Asia and Europe, and (3) northeast Asia and the Americas (**Figure 33.39**). To summarize, the data suggest that modern humans originated in Africa and then spread throughout the world in a series of three major migrations.

What happened to the Neanderthals and to *Homo erectus* as *H. sapiens* expanded its range? This simple question has provoked years of heated controversy. Currently, the debate boils down to two possibilities: (1) Either *H. sapiens* interbred with the other two hominid groups as it moved into Europe and Asia, or (2) it did not. The first possibility is called the **assimilation hypothesis**. It implies that the genetic composition and morphological features of *H. sapiens* are a mixture of ancient traits from Neanderthals and *H. erectus* and recent traits from *H. sapiens*. The second possibility is called the **out-of-Africa hypothesis**. It contends that *H. sapiens* evolved independently of the European and Asian species of *Homo*—meaning there was no interbreeding between *H. sapiens* and Neanderthals or *H. erectus*. Stated another way, the out-of-Africa hypothesis proposes that *H. sapiens* evolved its distinctive traits in Africa, then dispersed throughout the world. The assimilation hypothesis proposes that *Homo sapiens* acquired its distinctive traits at least in part through extensive interbreeding with *H. neanderthalensis* or *H. erectus*, or both.

If the assimilation hypothesis is correct, then modern humans should contain genes descended from *H. neanderthalensis* or *H. erectus*, or both. Recall from Chapter 19 that researchers have tested this prediction by extracting DNA from the fossilized bones of four Neanderthal individuals from three differ-

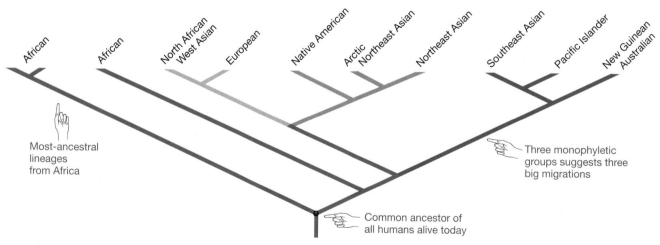

FIGURE 33.38 Phylogeny of Human Populations Living Today
Phylogeny of modern human populations, as estimated from DNA sequence data.

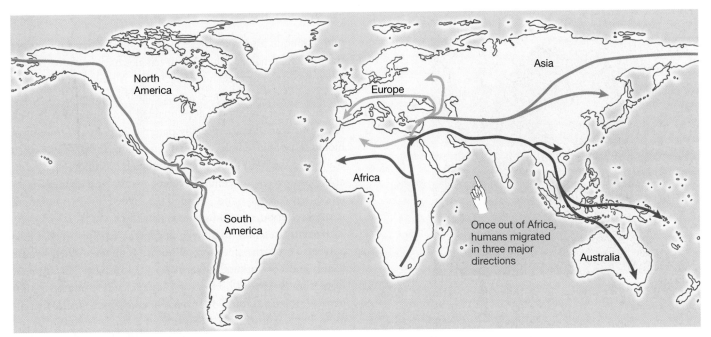

FIGURE 33.39 *Homo sapiens* Originated in Africa and Spread throughout the World
The phylogeny in Figure 33.38 supports the hypothesis that humans originated in Africa and spread out in three major migrations: to southeast Asia and the Pacific Islands, to Europe, and to northeast Asia and the New World.

ent locations in Eurasia. In each case, the researchers ground samples from the intact bone, extracted DNA, and used the polymerase chain reaction to copy a 379-base-pair section of the mitochondrial genome. When they sequenced the Neanderthal DNA and compared it with sequences from about a thousand humans living today, the results were striking: The Neanderthal DNA was extremely different from the DNA of living humans. For example, when sequences from any two randomly chosen *H. sapiens* are compared with each other, an average of 8 bases are unalike. But when randomly chosen *H. sapiens* sequences

are compared to the *H. neanderthalensis* DNA, there is an average of over 25 differences. Further, the sequence differences observed in the Neanderthal DNA are unique—none of the *H. sapiens* sequences contained the nucleotide substitutions found in *H. neanderthalensis*. These results support the hypothesis that *H. sapiens* and *H. neanderthalensis* did not interbreed.

Unfortunately, it has been impossible so far to extract DNA from *H. erectus* fossils and perform the same test. Although the weight of evidence currently tips the scales in favor of the out-of-Africa hypothesis, research continues.

ESSAY So Human an Animal

From a biological point of view, what is a human being? Historically, scientists and philosophers have answered this question by arguing that humans have one or more unique, defining characteristics that set us apart from other organisms. For decades this defining characteristic was thought to be tool use. But then Jane Goodall, who pioneered the study of common chimpanzees in the field, observed that chimps collect and modify twigs to "fish" for termites or ants and eat them (**Figure 33.40,** page 778). Later, Goodall and other biologists reported additional examples of tool use by common chimps, bonobos (also known as pygmy chimps), baboons, sea otters, and various birds.

In response, some observers began to argue that the defining characteristic of humans was not toolmaking but the use of symbolic speech—language based on the abstract symbols we call

Which 1 to 2 percent of the human genome is responsible for the . . . differences between chimps and humans?

words and letters. But researchers have since taught chimps rudimentary aspects of the American Sign Language, and the shape of a recently discovered Neanderthal bone located in the larynx suggests that Neanderthals, too, could speak. These results have a key message: The ability to learn and to use abstract symbols and grammatical rules is not unique to humans.

In short, there may be no single characteristic that "defines" humans. Instead, most contemporary biologists would characterize our

(Continued on next page)

(Essay continued)

FIGURE 33.40 Chimpanzees Teach Their Offspring How to Make and Use Tools
Adult female chimp (bottom) using a stick as a tool to "fish" for ants. Her son, sitting on her back, is also using a stick. Tool use is a culturally transmitted behavior in chimpanzees, meaning that it is learned.

species by listing a suite of traits that are not necessarily unique. Humans are an intensely social, bipedal mammal with drastically reduced fur, a brain that is absolutely huge given our body size, and thumbs that are unusually dextrous. If pressed to come up with a unique trait, a biologist might point out that humans are the only mammals in which adult females have enlarged breasts when they are not pregnant or lactating. A biologist would also point out that humans have a particularly long period of sexual immaturity, or childhood, accompanied by extraordinarily high levels of parental and especially paternal care. In addition, most biologists would emphasize that the changes that have taken place over the past 40,000 years of human evolution have probably been driven by **cultural evolution**—meaning changes fashioned by teaching and learning—as much as by mutation, natural selection, gene flow, and genetic drift.

Recently, research in molecular genetics has furnished a new perspective on the question, Who are humans? When Svante Pääbo and co-workers sequenced a 10,156-base-pair segment from the X chromosomes of humans and common chimpanzees, the team found that the sequences were identical at almost 99 percent of the 10,156 sites.

These findings prompt a question: Which 1 to 2 percent of the human genome is responsible for the morphological and behavioral differences between chimps and humans? Will it be possible to identify the alleles responsible for our bipedal posture, large brains, slow rate of sexual maturation, and huge capacity for learning? Researchers have only begun to tackle these questions. But, with a variety of genome analysis projects now under way, answers may soon be forthcoming.

CHAPTER REVIEW

Summary of Key Concepts

■ **The most prominent deuterostome lineages are the echinoderms and the vertebrate groups called ray-finned fishes and tetrapods.**

Echinoderms, ray-finned fishes, and tetrapods are the most species-rich groups of deuterostomes and are the most important large-bodied predators and herbivores in marine and terrestrial environments.

Web Tutorial 33.1   Deuterostome Diversity

■ **Echinoderms and vertebrates have unique body plans. Echinoderms are radially symmetric as adults and have a water vascular system. All vertebrates have a skull, and most have an extensive endoskeleton made of cartilage or bone.**

Echinoderm larvae are bilaterally symmetric but undergo a metamorphosis into radially symmetric adults. Their water vascular system is composed of fluid-filled tubes and chambers and extends from the body wall in projections called podia. Podia can extend and retract in response to muscle contractions that move fluid inside the water vascular system.

Chordates are distinguished by the presence of a notochord, a dorsal hollow nerve cord, pharyngeal gill slits, and a muscular tail that extends past the anus. Vertebrates are distinguished by the presence of a cranium; most species also have vertebrae. In more recent groups of vertebrates, the body plan features an extensive endoskeleton composed of bone.

■ **The diversification of echinoderms was triggered by the evolution of appendages called podia; the diversification of vertebrates was driven by the evolution of the jaw and limbs.**

Most echinoderms move via their podia, and many species suspension feed, deposit feed, or act as predators with the aid of their podia.

Ray-finned fishes and tetrapods use their jaws to bite food and process it with teeth. Species in both groups move when muscles attached to their endoskeletons contract or relax. Ray-finned fishes called teleosts are efficient swimmers because their flexible, symmetrical tail stabilizes their body during rapid movement. Tetrapods can move on land because their limbs enable walking, running, or flying. The evolution of the amniotic egg allowed tetrapods to lay eggs on land. Parental care was an important adaptation in some groups of ray-finned fishes and tetrapods—particularly mammals.

■ **Humans are a tiny twig on the tree of life. Chimpanzees and humans diverged from a common ancestor that lived in Africa 6–7 million years ago. Since then, at least 14 humanlike species have existed.**

The fossil record of the past 3.5 million years contains at least 14 distinct species of hominins. Several of these organisms lived in Africa at the same time, and some lineages went extinct without leaving descendant populations. Thus, *Homo sapiens* is the sole surviving representative of an adaptive radiation. The phylogeny of living humans, based on comparisons of DNA sequences, agrees with evidence in the fossil record that *H. sapiens* originated in Africa and later spread throughout Europe, Asia, and the Americas. DNA sequences recovered from the fossilized bones of *H. neanderthalensis* suggest that *H. sapiens* replaced this species in Europe without interbreeding.

Questions

Content Review

1. If you found an organism on a beach, what characteristics would allow you to declare that the organism is an echinoderm?
   a. radially symmetric adults, spines, and presence of tube feet
   b. notochord, dorsal hollow nerve cord, pharyngeal gill slits, and muscular tail
   c. exoskeleton and three pairs of appendages; distinct head and body (trunk) regions
   d. mouth forms second (after the anus) during gastrulation

2. What is the diagnostic trait of vertebrates?
   a. cranium
   b. jaws
   c. endoskeleton constructed of bone
   d. endoskeleton constructed of cartilage

3. Why are the pharyngeal jaws found in many ray-finned fishes important?
   a. They allow the main jaw to be protrusible (extendible).
   b. They make it possible for individuals to suck food toward their mouths.
   c. They give rise to teeth that are found on the main jawbones.
   d. They can bite down on food and help process it.

4. Which of the following lineages make up the living Amniota?
   a. reptiles and mammals
   b. viviparous fishes
   c. frogs, salamanders, and caecilians
   d. hagfish, lampreys, and cartilaginous fishes (sharks and rays)

5. Which of the following does *not* occur in either cartilaginous fishes or ray-finned fishes?
   a. internal fertilization and viviparity or ovoviviparity
   b. external fertilization and oviparity
   c. formation of a placenta
   d. feeding of young

6. Researchers agree that modern *Homo sapiens* originated in Africa and then spread throughout Europe, Asia, and eventually the Americas. What do they disagree about?
   a. whether the original African population of *H. sapiens* left any descendants
   b. whether *H. sapiens* members interbred with *H. erectus* members
   c. whether the Neanderthals represent a distinct species from *H. sapiens*
   d. whether *H. neanderthalensis* and *H. erectus* died out concurrently

Conceptual Review

1. Explain how the water vascular system of echinoderms functions as a type of hydrostatic skeleton.

2. List the four morphological traits that distinguish chordates. How are these traits involved in locomotion and feeding in larvae or adults?

3. Describe evidence that supports the hypothesis that jaws evolved from gill arches in fish.

4. Describe evidence that supports the hypothesis that the tetrapod limb evolved from the fins of lobe-finned fishes.

5. The text claims that "*Homo sapiens* is the sole survivor of an adaptive radiation that took place over the past 3.5 million years." Do you agree with this statement? Why or why not?

6. Explain how the evolution of the placenta and lactation in mammals improved the probability that their offspring would survive. Over the course of a lifetime, why are female mammals expected to produce fewer eggs than do female fish?

Group Discussion Problems

1. A growing number of biologists maintain that only monophyletic groups should be given names. According to these researchers, there is no such thing as a fish. Explain this statement.

2. Compare and contrast adaptations that triggered the diversification of the three most species-rich animal lineages: mollusks, arthropods, and vertebrates.

3. Aquatic habitats occupy 73 percent of Earth's surface area. How does this fact relate to the success of ray-finned fishes? How does it relate to the success of coelacanths and other lobe-finned fishes?

4. Mammals and birds are endothermic. Did they inherit this trait from a common ancestor, or did endothermy evolve independently in these two lineages? Provide evidence to support your answer.

Answers to Multiple-Choice Questions   1. a; 2. a; 3. d; 4. a; 5. d; 6. b

34 Viruses

KEY CONCEPTS

▨ Viruses are tiny, noncellular parasites that infect virtually every type of cell known. They cannot perform metabolism on their own—meaning outside a parasitized cell—and are not considered to be alive. Different types of viruses are specialized for infecting particular species and types of cells.

▨ Viruses are highly diverse in overall morphology and in the nature of their genetic material. The genomes of viruses may consist of double-stranded DNA, single-stranded DNA, double-stranded RNA, or one of several types of single-stranded RNA.

▨ The viral infection cycle can be broken down into five steps: (1) entry into a host cell, (2) replication of the viral genome, (3) the production of viral proteins, (4) the assembly of a new generation of virus particles, and (5) exit from the infected cell.

Photomicrograph created by treating seawater with a fluorescing compound that binds to nucleic acids. The smallest, most abundant dots are viruses. The larger, numerous spots are bacteria and archaea. The largest splotches are protists.

Viruses are parasites that afflict every twig on the tree of life. They are not cells and are not made up of cells, so they are not considered organisms. They cannot manufacture their own ATP or carbon-containing compounds, and they cannot make copies of themselves. Viruses enter a **host cell**, take over its biosynthetic machinery, and use that machinery to manufacture a new generation of viruses. Outside of cells, viruses simply exist—they cannot do anything. A **virus** is an obligate, intracellular parasite. The adjective *obligate* is appropriate because viruses cannot replicate unless they enter the inside of a cell.

Because they are not organisms, viruses are referred to as **particles** or **agents** and are not given scientific (genus + species) names. Most biologists would argue that viruses are not even

alive. Yet viruses have a genome, they are superbly adapted to exploit the metabolic capabilities of their host cells, and they evolve. **Table 34.1** summarizes some characteristics of viruses.

The diversity and abundance of viruses almost defy description. Nearly all organisms examined thus far are parasitized by at least one kind of virus. The bacterium *Escherichia coli*, which resides in the human intestine, is afflicted by more than a dozen types of **bacteriophage** (literally, "bacteria-eater"). A bacteriophage is a virus that infects bacteria. The ocean's plankton teems with bacteria and archaea, yet viruses outnumber them in this habitat by a factor of 10 to 1. A wine bottle filled with seawater taken from the ocean's surface contains about 10 *billion* virus particles—almost double the world's population of humans.

TABLE 34.1 Characteristics of Viruses versus Characteristics of Organisms

	Viruses	Organisms
Hereditary material	DNA or RNA; can be single stranded or double stranded	DNA; always double stranded
Plasma membrane present?	No	Yes
Can carry out transcription independently?	No—even if a viral polymerase is present, transcription of viral genome requires use of ATP and nucleotides provided by host cell	Yes
Can carry out translation independently?	No	Yes
Metabolic capabilities	Virtually none	Extensive— synthesis of ATP, reduced carbon compounds, vitamins, lipids, nucleic acids, etc.

34.1 Why Do Biologists Study Viruses?

Any study of life's diversity would be incomplete unless it included a look at the acellular parasites that exploit that diversity. But viruses are also important from a practical standpoint. To health-care workers, agronomists, and foresters, these parasites are a persistent—and sometimes catastrophic—source of misery and economic loss. The nature of viruses has been understood only since the 1940s, but they have been the focus of intense research ever since. Biologists study viruses because they cause illness and death. In the human body, virtually every system, tissue, and cell can be infected by one or more kinds of virus (**Figure 34.1**). Research on viruses is motivated by the desire to minimize the damage they can cause.

In addition, biologists study viruses with the goal of exploiting their ability to enter cells. Recall from Chapter 19 that viruses are being tested as possible therapeutic agents in the treatment of genetic diseases. If viruses can be engineered to carry normal copies of human genes into the cells of patients, it is possible that the gene products could cure symptoms.

Recent Viral Epidemics in Humans

Physicians and researchers use the term **epidemic** ("upon-people") to describe a disease that affects a large number of people at the same time. Viruses have caused the most devastating epidemics in recent human history. During the eighteenth and nineteenth centuries, it was not unusual for Native American tribes to lose 90 percent of their members over the course

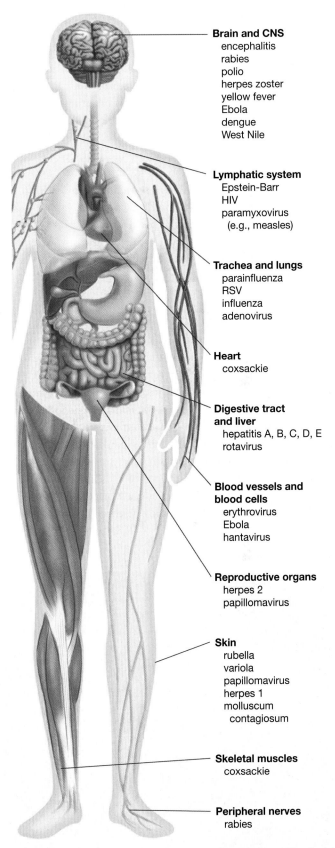

Brain and CNS
encephalitis
rabies
polio
herpes zoster
yellow fever
Ebola
dengue
West Nile

Lymphatic system
Epstein-Barr
HIV
paramyxovirus
(e.g., measles)

Trachea and lungs
parainfluenza
RSV
influenza
adenovirus

Heart
coxsackie

Digestive tract and liver
hepatitis A, B, C, D, E
rotavirus

Blood vessels and blood cells
erythrovirus
Ebola
hantavirus

Reproductive organs
herpes 2
papillomavirus

Skin
rubella
variola
papillomavirus
herpes 1
molluscum
contagiosum

Skeletal muscles
coxsackie

Peripheral nerves
rabies

FIGURE 34.1 Human Organs and Systems That Are Parasitized by Viruses

EXERCISE Choose two viruses that you are familiar with. Next to each, write the symptoms caused by an infection with this virus.

of a few years to measles, smallpox, and other viral diseases spread by contact with European settlers.

In terms of global impact, the influenza outbreak of 1918–1919, called "Spanish flu," qualifies as the most devastating epidemic recorded to date. Influenza is a virus that infects the upper respiratory tract. The strain of influenza virus that emerged in 1918 infected people worldwide and was particularly **virulent**—meaning it tended to cause severe disease. Within hours of showing symptoms, the lungs of previously healthy people often became so heavily infected that affected individuals suffocated to death. Most victims were between the ages of 20 and 40. The viral outbreak occurred just as World War I was drawing to a close and killed far more people than did the conflict itself. For example, ten times as many Americans died of influenza than were killed in combat in the war. Worldwide, the Spanish flu may have killed over 50 million people.

Current Viral Epidemics in Humans: HIV

In terms of the total number of people affected, the measles and smallpox epidemics among native peoples and the 1918 influenza outbreak are almost certain to be surpassed by the incidence of AIDS. **Acquired immune deficiency syndrome (AIDS)** is an affliction caused by the **human immunodeficiency virus (HIV; Figure 34.2)**. Relatively little is known about the strain of influenza that caused the Spanish flu epidemic of 1918–1919, but HIV is far and away the most intensively studied of all viruses. Since the early 1980s, governments and private corporations from around the world have spent hundreds of millions of dollars on HIV research. More biologists

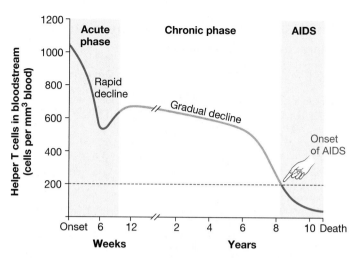

FIGURE 34.3 T-Cell Counts Decline during an HIV Infection
Graph of changes in the number of T cells that are present in the bloodstream over time, based on data from a typical patient infected with HIV. The acute phase of infection occurs immediately after infection and is sometimes associated with symptoms such as fever. Infected people usually show no disease symptoms in the chronic phase, even though their T-cell counts are in slow, steady decline. AIDS typically occurs when T-cell counts dip below 200/mm^3 of blood.

are working on HIV than on any other type of virus. Given this virus's current and projected impact on human populations around the globe, the investment is justified.

How Does HIV Cause Disease?   Like other viruses, HIV parasitizes specific types of cells. The cells most affected by HIV are called *helper T cells* and *macrophages*. These cells are components of the **immune system**, which is the body's defense system against disease. Chapter 49 explains just how crucial helper T cells and macrophages are to the immune system's response to invading bacteria and viruses. If an HIV particle succeeds in infecting one of these cells and reproducing inside, however, the cell dies as hundreds of new particles break out and infect more cells. Although the body continually replaces helper T cells and macrophages, the number produced does not keep pace with the number being destroyed by HIV. As a result, the total number of helper T cells in the bloodstream gradually declines as an HIV infection proceeds (**Figure 34.3**). When the T-cell count drops, the immune system's responses to invading bacteria and viruses become less and less effective. Eventually, too few helper T cells are left to fight off pathogens efficiently, and pathogenic bacteria and viruses begin to multiply unchecked. In almost all cases, one or more of these infections prove fatal. HIV kills people indirectly—by making them susceptible to pneumonia, fungal infections, and unusual types of cancer.

What Is the Scope of the AIDS Epidemic?   Researchers with the United Nation's AIDS program estimate that AIDS

FIGURE 34.2 The Human Immunodeficiency Virus (HIV)
Colorized scanning electron micrograph showing HIV particles (red) emerging from an infected human T cell (green).

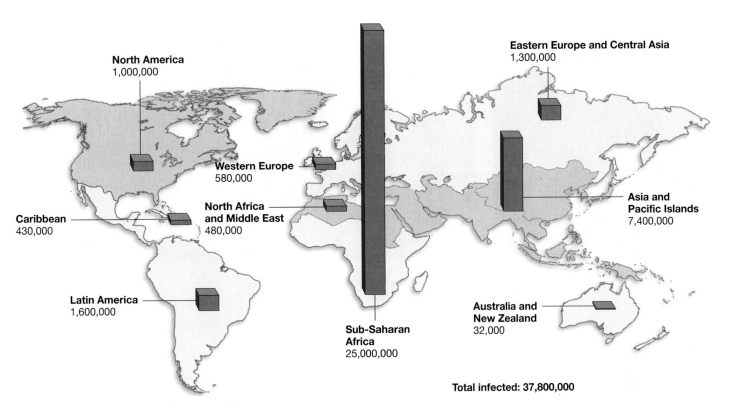

FIGURE 34.4 Geographic Distribution of HIV Infections
Data, compiled by the World Health Organization, showing the numbers of people living with HIV in July 2004, by geographic area.

has already killed 25 million people worldwide. HIV infection rates have been highest in east and central Africa, where one of the greatest public health crises in history is now occurring (**Figure 34.4**). In Botswana and elsewhere, blood-testing programs have confirmed that up to a third of all adults carry HIV. Although there may be a lag of as much as 8–12 years between the initial infection and the onset of illness, virtually all people who become infected with the virus will die of AIDS.

Currently, the UN estimates the total number of HIV-infected people worldwide at about 38 million. An additional 5 million people are infected each year, and the epidemic is growing. Researchers are particularly alarmed because the focus of the epidemic is shifting from its historical center of incidence—central and southern Africa—to south and east Asia. Infection rates are growing as much as 17 percent per year in some of the world's most populous countries—particularly India, China, Russia, Ethiopia, and Nigeria. If present trends continue, 80 million more people will be infected by 2010.

Most viral and bacterial diseases afflict the very young and the very old. But because HIV is a sexually transmitted disease, young adults are most likely to contract the virus and die. People who become infected in their late teens or twenties die of AIDS in their twenties or thirties. Tens of millions of people are being lost in the prime of their lives. Physicians, politicians, educators, and aid workers all use the same word to describe the epidemic's impact: staggering.

34.2 How Do Biologists Study Viruses?

Researchers who study viruses focus on two goals: (1) developing vaccines that allow hosts to fight off disease if they become infected and (2) developing antiviral drugs that prevent a virus from replicating efficiently inside the host. Both types of research begin with attempts to isolate the virus in question.

Isolating viruses takes researchers into the realm of "nanobiology," in which structures are measured in billionths of a meter. (One nanometer, abbreviated nm, is 10^{-9} meter.) Viruses are typically 50 to 100 nm in diameter. They are dwarfed by eukaryotic cells; millions of viruses can fit on the head of a pin.

If virus-infected cells can be grown in culture or harvested from a host individual, researchers can usually isolate the virus by passing the cells through a filter. The filters used to study viruses have pores that are large enough for viruses to pass through but are too small to admit cells. To test the hypothesis that the solution that passes through the filter contains viruses that cause a specific disease, researchers expose uninfected host cells to this filtrate. If the virus-causation hypothesis is correct, then exposing host cells to the filtrate will result in infection. In this way, researchers can isolate a virus and confirm that it is the causative agent of infection. Recall from Chapter 27 that these steps are inspired by Koch's postulates, which established the criteria for linking a specific infectious agent with a specific disease.

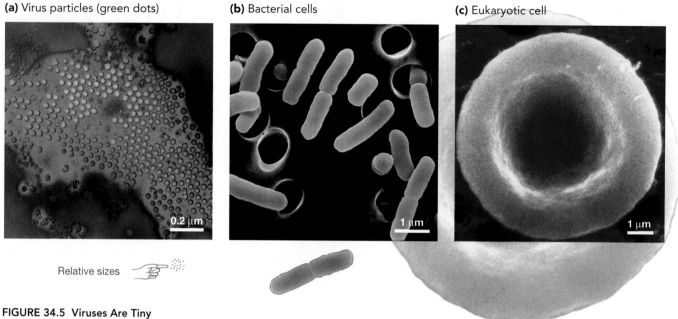

(a) Virus particles (green dots)   **(b)** Bacterial cells   **(c)** Eukaryotic cell

Relative sizes

FIGURE 34.5 Viruses Are Tiny
(a) Viruses are much smaller than **(b)** bacterial cells or **(c)** eukaryotic cells. Notice the scale bars on each micrograph.

Once biologists have isolated a virus, how do they study and characterize it? Let's begin with morphological traits, then consider how viral life cycles vary.

Analyzing Morphological Traits

To see a virus, researchers usually rely on transmission electron microscopy (**Figure 34.5**). Only the very largest viruses, such as the smallpox virus, are visible with a light microscope. Electron microscopy has revealed that viruses come in a wide variety of shapes, and many viruses can be identified by shape alone (**Figure 34.6**). In overall structure, however, they fall into just

two general categories. Viruses can either be (1) enclosed by just a shell of protein called a **capsid** or (2) enclosed by both a capsid and a membrane-like **envelope**. Regarding their morphology, then, the important distinction among viruses is whether they are **nonenveloped** or **enveloped**.

Nonenveloped viruses have an extremely simple structure. They consist of genetic material and possibly one or more enzymes inside a capsid—a protein coat. The nonenveloped virus illustrated in **Figure 34.7a** is an *adenovirus*. You undoubtedly have adenoviruses on your tonsils or in other parts of your upper respiratory passages right now.

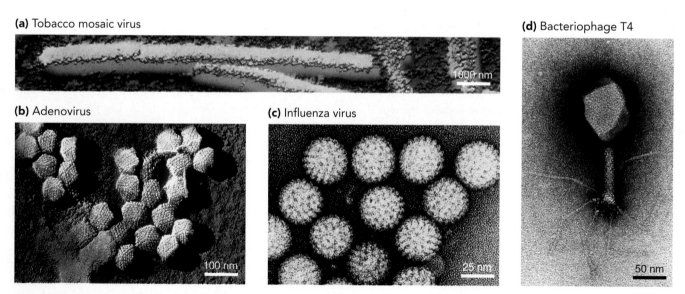

(a) Tobacco mosaic virus   **(d)** Bacteriophage T4

(b) Adenovirus   **(c)** Influenza virus

FIGURE 34.6 Viruses Vary in Size and Shape
Virus shapes include **(a)** rods, **(b)** polyhedrons, **(c)** spheres, and **(d)** complex shapes with heads and tails.

(a) Nonenveloped virus

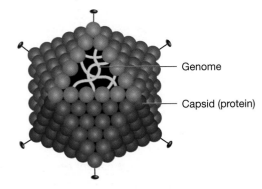

Genome

Capsid (protein)

(b) Enveloped virus

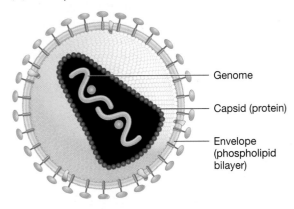

Genome

Capsid (protein)

Envelope
(phospholipid
bilayer)

FIGURE 34.7 Viruses Are Nonenveloped or Enveloped
(a) A protein coat forms the exterior of nonenveloped viruses. **(b)** In enveloped viruses, the exterior is composed of a membranous sphere. Inside this envelope, the hereditary material is enclosed by a protein coat.

Enveloped viruses are slightly more complex. They also have genetic material inside a capsid, but the capsid is surrounded by an envelope. The envelope consists of a phospholipid bilayer with a mixture of viral proteins and proteins derived from the plasma membrane of a host cell—specifically, the host cell in which the virus particle was manufactured. The enveloped virus illustrated in **Figure 34.7b** is HIV.

Once a virus has been isolated and its overall morphology characterized, researchers usually focus on understanding the nature of the virus's replication cycle and on attempts to develop a vaccine. A **vaccine** is a preparation that primes a host's immune system to respond to a specific type of virus. **Box 34.1** explains how vaccines work and why it has not been possible to develop a vaccine for certain viruses—particularly the cold virus or HIV. Here let's focus on variation in virus replication cycles.

Analyzing Variation in Replication Cycles: Lytic and Lysogenic Growth

Although it is likely that millions of types of virus exist, they all infect their host cells in one of two general ways: lytic growth or lysogenic growth. All viruses undergo lytic growth; some can grow lysogenically as well. Both types of viral infection begin when part or all of a virus particle enters the interior of a host cell.

Figure 34.9a shows a **lytic replication cycle**. During lytic growth, the viral genome enters the host cell and viral or host enzymes make copies of it, using nucleotides and ATP provided by the host. The host cell also manufactures viral proteins. When synthesis of the viral genome and viral proteins is complete, a new generation

---

As Chapter 49 will show, one of the primary ways that an individual's immune system responds to parasite attack is through the production of antibodies. An **antibody** ("against-body") is a protein that binds with high specificity to a particular site on another molecule. That is, an antibody will bind only to one particular site on one particular compound. Any molecule that elicits the production of antibodies is considered to be an **antigen** ("against-produce").

Vaccines contain antigens. The antigens are usually components of a virus's exterior—the capsid from a nonenveloped virus or the envelope proteins from an enveloped virus. Exterior proteins are effective antigens because they are the part of the virus that is "seen" by the immune system cells.

To vaccinate a person or a domestic animal, viral antigens are injected or swallowed. Once inside the individual, the antigens stimulate immune system cells called *B cells* (see Chapter 49), which produce antibodies to the antigen (**Figure 34.8**, page 786). Once antigens are coated with antibodies, the antigens are destroyed by other cells and components of the immune system.

Vaccinations are like fire drills or earthquake preparedness drills—they prepare the immune system to respond to a specific type of threat. In effect, they are a faked infection. Immune system cells that are stimulated to produce antibodies by a vaccination remain active in the vaccinated person for a long time—often for life. If a vaccinated person is later exposed to active virus particles, the immune system can respond quickly and effectively enough to stop the infection before it threatens the individual's health.

In many cases, vaccines have been spectacularly successful at curbing or

(Continued on next page)

(Box 34.1 continued)

even eliminating viral diseases in humans and domestic animals. In humans alone, effective vaccines have been developed against smallpox, yellow fever, polio, measles, and some forms of hepatitis. These diseases used to kill or sicken hundreds of thousands of people each year.

There Are Two General Types of Vaccines

Successful vaccines consist of inactivated viruses or attenuated viruses. An **inactivated virus** is not capable of causing an infection, because its genes have been damaged by chemical treatments—often exposure to formaldehyde—or irradiation with ultraviolet light. If you have been vaccinated for hepatitis A or flu, you have received an inactivated virus.

Attenuated viruses are also called **"live" virus vaccines,** because they consist of complete virus particles. The adjective *attenuated* means that they lack virulence. Researchers can make a virus harmless to a host by culturing the virus on cells from species other than that host. In adapting to growth on the atypical cells, viral strains usually lose the ability to grow rapidly in their normal host cells. Although attenuated viruses still provoke a strong immune response, they are not capable of sustaining an infection in a vaccinated individual. The smallpox, polio, and measles vaccines consist of attenuated viruses.

Why Isn't There a Vaccine for Cold Virus, HIV, and the Flu Viruses?

Researchers have not succeeded in developing vaccines against the common cold or HIV and have been only moderately successful in developing a vaccine against flu viruses. The reason is that cold and flu viruses and HIV have exceptionally high mutation rates. HIV actually has the highest mutation rate observed in any organism or virus.

Why does a high mutation rate make vaccine development so difficult? Recall from Chapter 16 that a mutation is de-

fined as a change in DNA. Because the enzymes that copy the genes found in cold and flu viruses and HIV are exceptionally error prone, many of the mature virus particles (virions) that are produced contain mutations in the genes for the virus's envelope proteins. When this DNA is transcribed and translated, the envelope proteins that result are likely to have an altered structure. Due to the high mutation rates of cold and flu viruses and HIV, the envelope proteins in these viruses constantly change through time. Stated another way, new strains of these viruses are constantly evolving. The antibodies

produced against the envelope proteins of earlier strains do not work against strains that appear later, because the envelope proteins of the strains are different. A vaccination that protected an individual against certain strains of cold or flu virus or HIV would not help against other strains.

To date, it has not been possible to design antigens from the cold or flu viruses or HIV that protect vaccinated individuals effectively. In the fight against HIV, drugs that inhibit viral enzymes and thus stop or slow viral replication have been much more successful than vaccination efforts.

HOW VACCINATION WORKS

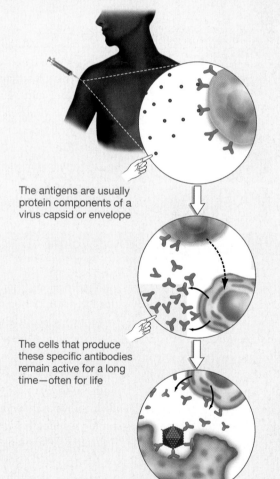

The antigens are usually protein components of a virus capsid or envelope

The cells that produce these specific antibodies remain active for a long time—often for life

1. Viral antigens are introduced into the body.

2. Certain immune system cells recognize antigens.

3. These cells stimulate other immune system cells to produce antibodies to the virus.

4. Later, if the host is exposed to live virus particles, the particles will be coated with antibodies and destroyed by immune system cells.

FIGURE 34.8 Vaccination Induces a Response from the Immune System
The immune system cells that respond to a vaccination are "immortalized," meaning that they stay active for a long time. Thus, the body can respond quickly to any future infections by the same virus.

of virus particles assembles inside the host. A mature virus particle is called a **virion**. The infection ends when the viral agents exit the cell—usually killing the host cell in the process.

Figure 34.9b diagrams a **lysogenic replication cycle**, or **lysogeny**. Only certain types of viruses are capable of lysogenic growth. Although HIV and some other viruses that infect humans are capable of lysogeny, most lysogenic viruses infect bacteria. Recall that viruses that infect bacteria are called bacteriophages. In these viruses, lysogeny is an important variation on the lytic replication cycle. During lysogenic growth, viral DNA becomes incorporated into the host's chromosome. Often this integration occurs without serious damage to the host cell. Once the viral genome is in place, it is replicated by the host's DNA polymerase each time the cell divides. Copies of the viral genome are passed on to daughter cells just like one of the host's own genes. In the lysogenic state, a virus is usually latent, or *quiescent*. This means that no new particles are being produced and no unrelated cells are being infected. The virus is transmitted from one generation to the next along with the host's genes.

In bacteriophages, lysogenic growth is typical when an infected bacterium is growing and dividing rapidly; the viral population then grows along with the bacterial population. But if the infected bacterium is damaged by sunlight or a toxin or if the host begins to starve, then the virus switches from lysogenic to lytic growth. To explain this observation, biologists point to the adaptive value of the switch between lytic and lysogenic replication cycles. If a bacterium is growing and dividing rapidly, then a virus's fitness is maximized through lysogenic growth. But if a bacterium is likely to stop growing or die, then a virus's fitness is maximized through lytic growth.

It is not possible to treat a lysogenic infection with drugs, because the viruses are quiescent—they just sit there. But even lytic infections are notoriously difficult to treat, because viruses use so many of the host cell's enzymes during the lytic replication cycle. Drugs that disrupt these enzymes usually harm the host much more than they harm the virus. To understand viral diversity and how antiviral drugs are developed, let's consider each phase of the lytic cycle in more detail.

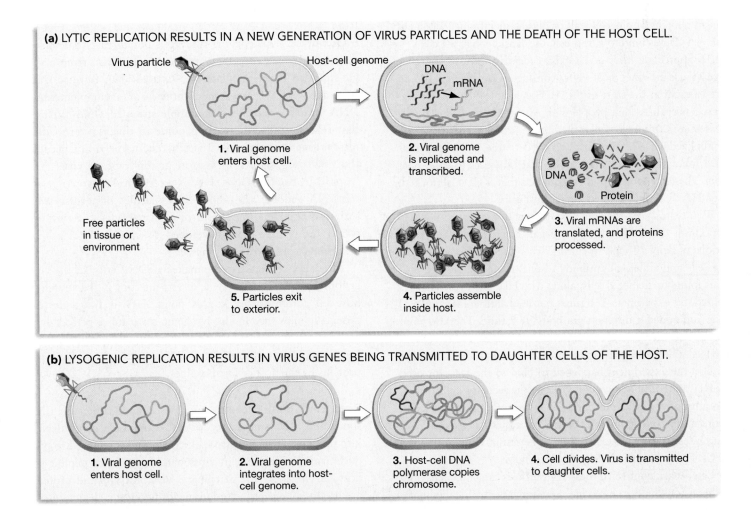

(a) LYTIC REPLICATION RESULTS IN A NEW GENERATION OF VIRUS PARTICLES AND THE DEATH OF THE HOST CELL.

Virus particle

Host-cell genome

DNA
mRNA

1. Viral genome enters host cell.

2. Viral genome is replicated and transcribed.

DNA   Protein

3. Viral mRNAs are translated, and proteins processed.

Free particles in tissue or environment

5. Particles exit to exterior.

4. Particles assemble inside host.

(b) LYSOGENIC REPLICATION RESULTS IN VIRUS GENES BEING TRANSMITTED TO DAUGHTER CELLS OF THE HOST.

1. Viral genome enters host cell.

2. Viral genome integrates into host-cell genome.

3. Host-cell DNA polymerase copies chromosome.

4. Cell divides. Virus is transmitted to daughter cells.

FIGURE 34.9 Viruses Replicate via Lytic or Lysogenic Cycles, or Both
(a) All viruses follow the same general lytic replication cycle. **(b)** Some viruses are also capable of lysogeny, meaning that their genome can become integrated into the host-cell chromosome. **EXERCISE** Compare and contrast a lysogenic virus to the transposable elements introduced in Chapter 20.

Analyzing the Phases of the Lytic Cycle

Five phases are common to lytic growth in virtually all viruses: (1) entry into a host cell, (2) replication and transcription of the viral genome, (3) production and processing of viral proteins, (4) assembly of a new generation of virions, and (5) exit from the infected cell. Each virus has a particular way of entering a host cell and completing the subsequent phases of the lytic cycle. Let's take a closer look.

How Do Viruses Enter a Cell? The replication cycle of a virus begins when a free viral particle enters a target cell. This is no simple task. All cells are protected by a plasma membrane, and many cells also have a cell wall. How do viruses breach these defenses, insert themselves into the cytoplasm inside, and begin an infection?

Most plant viruses enter host cells after a sucking insect, such as an aphid, has disrupted the cell wall with its mouthparts. In contrast, viruses that parasitize bacterial cells or that attack animal cells gain entry by binding to a specific molecule on the cell wall or plasma membrane. In response to this binding event, the genome of a nonenveloped virus enters the host cell, while the capsid remains on the cell wall or plasma membrane. When enveloped viruses bind to a host cell, the capsid enters the cell.

To appreciate how investigators identify the proteins that viruses use to enter host cells, consider research on HIV. In 1981—right at the start of the AIDS epidemic—biomedical researchers realized that people with AIDS had few or no T cells possessing **CD4**, a particular membrane protein. These cells are symbolized CD4$^+$. This discovery led to the hypothesis that CD4 functions as the "doorknob" that HIV uses to enter host cells. The doorknob hypothesis predicts that if CD4 is blocked, then HIV will not be able to enter host cells.

Two teams used the same experimental strategy to test this hypothesis (**Figure 34.10**). They began by growing large populations of helper T cells in culture. Then they took a sample of the cells and added an antibody to one of the cell-surface proteins found on helper T cells, along with HIV particles. They repeated this experiment 160 times but used a different sample of cells and added a different antibody each time. The key point here is that each of the 160 antibodies bound to and effectively blocked a different cell-surface protein. If one of the antibodies used in the experiment happened to bind to the receptor used by HIV, that antibody would cover up the receptor. In this way, the antibody would prevent HIV from entering the cell and protect that cell from infection. This approach led both research teams to reach exactly the same result: Only antibodies to CD4 protected the cells from viral entry.

Later work confirmed that HIV particles can enter cells only if the virions bind to a second membrane protein, called a **co-receptor**, in addition to CD4. In most individuals, proteins called CXCR4 and CCR5 function as co-receptors. The discovery of these co-receptors inspired a search for compounds that would block them and prevent HIV from entering cells. A drug company recently announced the development of a compound that blocks CCR5. Early reports suggest that the new drug is providing some relief for people infected with HIV.

If an HIV particle successfully binds to both CD4 and a co-receptor, however, the lipid bilayers of the particle's envelope and the plasma membrane of the helper T cell fuse. When fusion occurs, HIV has breached the cell boundary. The contents of the virus then enter the cytoplasm, and infection proceeds.

How Do Viruses Copy Their Genomes? Viruses must copy their genes to make a new generation of particles and continue an infection. Many viruses use their own DNA polymerase enzyme to accomplish that crucial step. This protein uses nucleotides provided by the host in synthesizing copies of the viral genome.

In some viruses, however, genes consist of RNA. In most viruses that have an RNA genome, copies of the genome are synthesized by the viral enzyme **RNA replicase**, which is an RNA polymerase. RNA replicase synthesizes RNA from an RNA template, using ribonucleotides provided by the host cell.

In other RNA viruses, however, the genome is transcribed from RNA to DNA by a viral enzyme called **reverse transcriptase**. This enzyme is a DNA polymerase that makes a single-stranded **complementary DNA**, or **cDNA**, from a single-stranded RNA template (see Chapter 19). Reverse transcriptase then catalyzes the synthesis of a complementary DNA strand, resulting in a double-stranded DNA. Viruses that reverse-transcribe their genome in this way are called **retroviruses** ("backward viruses"). The name is apt because the flow of genetic information in this type of virus goes from RNA back to DNA. HIV is a retrovirus. Two copies of the RNA genome and about 50 molecules of reverse transcriptase lie inside the capsid of each particle. The first antiviral drugs that were developed to combat HIV act by inhibiting reverse transcriptase.

After reverse transcriptase makes a cDNA copy of the viral genome, another viral enzyme inserts the copy into a stretch of host-cell chromosome. At this point, HIV is lysogenic. Although it may stay in the lysogenic state for a period, lytic growth is more common. For lytic growth to occur, the viral genes have to be transcribed to mRNA and then translated into proteins by the host cell's ribosomes.

Producing Viral Proteins Viruses cannot make the ribosomes and tRNAs necessary for translating their own mRNAs into proteins. For a virus to make the proteins required to produce a new generation of virus particles, it must exploit the host cell's biosynthetic machinery. Viral mRNAs and proteins are produced and processed in one of two ways, depending on whether the proteins end up in the outer envelope of a particle or in the capsid.

RNAs that code for a virus's envelope proteins follow a route through the cell identical to that of the RNAs of the cell's own

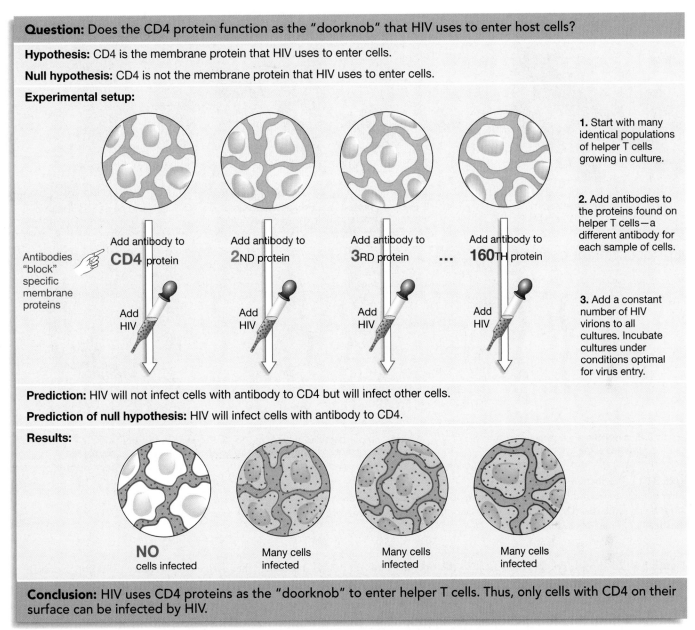

FIGURE 34.10 Experiments Confirmed that CD4 Is the Receptor Used by HIV to Enter Host Cells
In this experiment, the antibodies added to each culture bound to a specific protein found on the surface of helper T cells. Antibody binding blocked the membrane protein, so the protein could not be used by HIV to gain entry to the cells.

membrane proteins. How these viral mRNAs are translated by ribosomes attached to the endoplasmic reticulum (ER) is diagrammed in **Figure 34.11a** (page 790). Afterward the resulting proteins are transported to the Golgi apparatus, where carbohydrate groups are attached, producing glycoproteins. The finished glycoproteins are then inserted into the plasma membrane, where they are ready to be assembled into new virions.

In contrast, a different route is taken by RNAs that code for proteins that make up the capsid or inner core of a virus particle, illustrated in **Figure 34.11b**. These RNAs are translated by ribosomes, but in the cytoplasm, just as non-membrane-bound cellular mRNAs are. The long polypeptide sequences

that result are later cut into functional proteins by a viral enzyme called **protease**. This enzyme cleaves viral polypeptides at specific locations—a critical step in the production of finished viral proteins. The resulting protein fragments are assembled into a new viral core near the host cell's plasma membrane.

The discovery that HIV produces its own protease triggered a search for drugs that would block the enzyme. This search got a huge boost when researchers succeeded in visualizing the three-dimensional structure of HIV's protease, using the X-ray crystallographic techniques introduced in Chapter 4. The molecule has an opening in its interior that is adjacent to the active site,

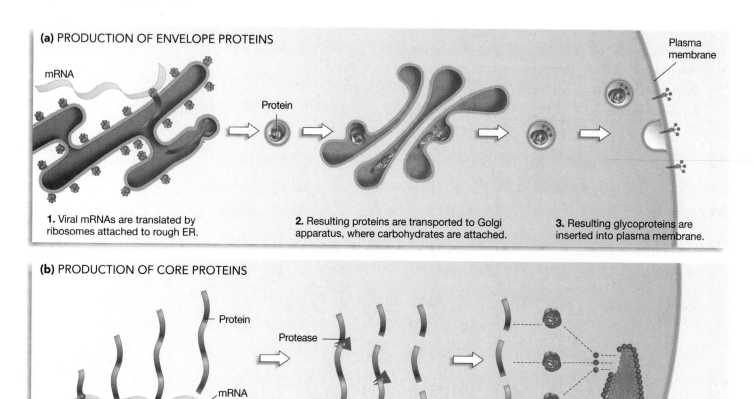

(a) PRODUCTION OF ENVELOPE PROTEINS

mRNA

Protein

Plasma membrane

1. Viral mRNAs are translated by ribosomes attached to rough ER.

2. Resulting proteins are transported to Golgi apparatus, where carbohydrates are attached.

3. Resulting glycoproteins are inserted into plasma membrane.

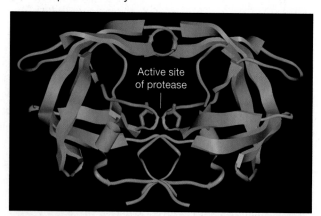

(b) PRODUCTION OF CORE PROTEINS

Protein

Protease

mRNA

Ribosome

Core proteins

1. Viral mRNAs are translated by ribosomes in the cytoplasm.

2. Resulting polypeptides are cut into functional proteins by protease.

3. Resulting core proteins are assembled near plasma membrane.

FIGURE 34.11 Production of Viral Proteins
(a) After being synthesized on the rough ER, envelope proteins are inserted into the plasma membrane.
(b) After being synthesized by ribosomes in the cytoplasm and processed, core proteins assemble near the host cell's plasma membrane.

as **Figure 34.12a** shows. Polypeptides fit into the opening and are cleaved at the active site. Based on these data, researchers immediately began searching for molecules that could fit into the opening and prevent protease from functioning by binding to and blocking the active site (**Figure 34.12b**). Several HIV *protease inhibitors* are now on the market. **Box 34.2** (page 792) explains why they are often prescribed in combination with other antiviral drugs.

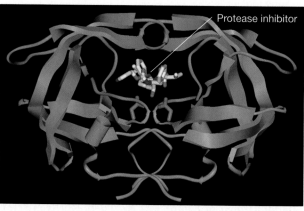

(a) HIV's protease enzyme

Active site of protease

(b) Could a drug block the active site?

Protease inhibitor

FIGURE 34.12 The Three-Dimensional Structure of Protease
(a) Ribbon diagram depicting the three-dimensional shape of HIV's protease enzyme. **(b)** Once protease's structure was solved, researchers began looking for compounds that would fit into the active site and prevent the enzyme from working.

How Are Viruses Transmitted to New Hosts? Viruses leave a host cell in one of two ways: by budding from the plasma membrane or by bursting out of the cell. Viruses that bud from the host cell's plasma membrane take some of that membrane with them. As a result, they incorporate host-cell phospholipids and membrane proteins into their envelope, along with membrane proteins encoded by the viral genome (**Figure 34.13a**). Most budding viruses infect host cells that lack a cell wall. In contrast, viruses that burst erupt from the cell surface, breaking the host cell open in the process, in most cases killing the host cell (**Figure 34.13b**). Most viruses that burst infect host cells that have a cell wall.

Once particles are released, the infection cycle is over. Thousands or millions of newly assembled virions are now in extracellular space. What happens next?

If the host cell is part of a multicellular organism, the new generation of particles begins traveling through the body via the bloodstream or lymph system. There, they may be bound by antibodies produced by the immune system. In vertebrates, this binding marks the particles for destruction. But if a particle contacts an appropriate host cell before it encounters antibodies, then the particle will infect that cell. This starts the replication cycle anew.

What if the virus has infected a unicellular organism, or if the virus leaves a multicellular host entirely? For example, when people cough, sneeze, spit, or wipe a runny nose, they help rid their body of viruses and bacteria. But they also project the pathogens into the environment, sometimes directly onto an uninfected host. From the virus's point of view, this new host represents an unexploited habitat brimming with resources in the

(a) Budding of enveloped viruses

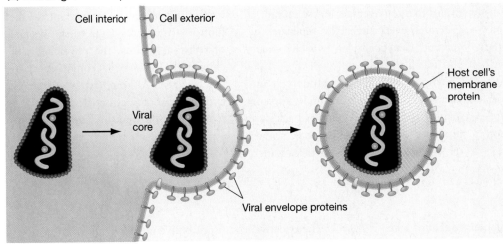
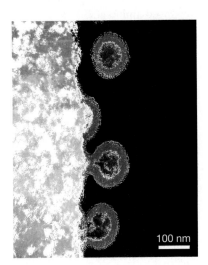

(b) Bursting of nonenveloped viruses

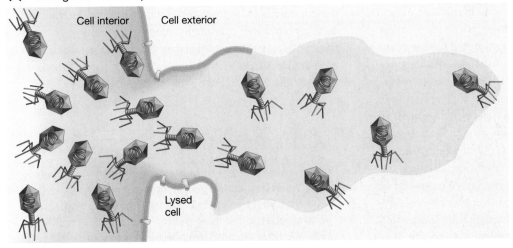
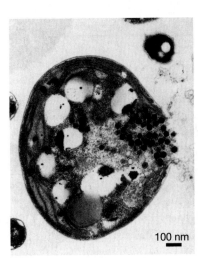

FIGURE 34.13 Viruses Leave Infected Cells by Budding or Bursting
(a) Enveloped viruses bud from a host cell, taking with them a lipid bilayer containing their envelope proteins. **(b)** Nonenveloped viruses burst from a host cell. This event ruptures the plasma membrane (and cell wall, if there is one). **QUESTION** Both budding and bursting kill the host cell. Propose a hypothesis to explain why infection with a budding virus is fatal.

BOX 34.2   HIV Drug Cocktails and the Evolution of Drug Resistance

HIV protease inhibitors were dispensed widely in North America and Europe beginning in the mid-1990s, with spectacular results. After therapy, many patients no longer had detectable levels of HIV in their blood. The drugs knocked HIV populations down.

Within two years, however, HIV levels in many of the patients taking protease inhibitors began to rebound. To investigate why this was happening, researchers sequenced the HIV protease gene in these patients. The researchers found that a series of mutations had occurred in the protease gene. These mutations led to changed amino acid sequences in the enzyme's active site. Because protease inhibitor molecules did not fit as well into the altered version of the active site, the enzyme could function reasonably well even in the presence of the inhibitor molecules. Almost as soon as a new protease inhibitor went into widespread use, researchers had documented the evolution of resistance to the drug.

The scientific literature abounds with examples of bacteria that have evolved resistance to antibiotics and viruses that have evolved resistance to antiviral drugs. But perhaps no organism or virus has evolved resistance to control agents as quickly as HIV has. The leading hypothesis to explain this observation is that HIV's mutation rate is particularly high. Researchers who assay transcripts produced by HIV's reverse transcriptase find that, on average, the wrong base is inserted once every 8000 nucleotides. (In contrast, *E. coli*'s DNA polymerase incorporates the wrong nucleotide about once in every *1 billion* bases.) This means that, on average, a new mutant is generated every time HIV replicates its genome. Genetically, no two HIV particles are alike.

Why is HIV's high mutation rate important? In infected individuals, approximately 10 billion new viral particles are produced daily. It is likely that, among the 10 billion, there are particles with a mutation in the active sites of reverse transcriptase or protease. As a result, HIV populations are almost certain to contain variants that are at least partially resistant to drugs that cripple most other particles in the population.

The message to researchers and physicians is clear: Due to HIV's high mutation rate, the search for drug therapies promises to be an "arms race"—a constant battle between novel drugs and novel, resistant strains of the virus. In attempting to keep ahead of drug-resistant viruses, physicians prescribe **combination therapy**—drug "cocktails" that include a protease inhibitor and one or more reverse transcriptase inhibitors. If patients begin to show signs of resistant strains, the physician will change the dosage or composition of the cocktail. Although it is extremely expensive, combination therapy has extended the life span and improved the quality of life for tens of thousands of AIDS patients.

---

form of target cells. The situation is analogous to that of a multicellular animal dispersing to a new habitat and colonizing it. Viruses that successfully colonize a new host replicate and increase in number. The alleles carried by these successful colonists increase in frequency in the total population. In this way, natural selection favors alleles that allow viruses to do two things: (1) replicate within a host and (2) be transmitted to new hosts.

For physicians and public health officials, reducing the likelihood of transmission is often an effective way to reduce the spread of a virus. HIV particles must be transmitted from person to person via body fluids such as blood, semen, or vaginal secretions. Faced with decades of disappointing results in drug and vaccine development, public health officials are aggressively promoting preventive medicine. Condom use reduces sexual transmission of HIV. Aggressive treatment of venereal diseases may also help; the lesions caused by chlamydia, genital warts, and gonorrhea encourage the transmission of HIV-contaminated blood during sexual intercourse. The most effective forms of preventive medicine are sexual abstinence or monogamy. Where blood supplies are routinely screened, HIV is rarely contracted by means other than unprotected sex with an infected person.

The effectiveness of preventive medicine underscores one of this chapter's fundamental messages: Viruses are a fact of life. Every organism is victimized by viruses; every organism has defenses against them. But the tree of life will never be free of these parasites. Mutation and natural selection guarantee that viral genomes will continually adapt to the defenses offered by their hosts, regardless of whether those defenses are devised by an immune system or by biomedical researchers. Viruses are a constant threat for every organism alive.

✓CHECK YOUR UNDERSTANDING

All organisms and cell types are parasitized by some type of virus. Viruses may undergo lytic or lysogenic growth, or both, when they infect a host cell. You should be able to diagram generalized lytic and lysogenic replication cycles. Your lytic cycle should distinguish five phases: (1) entry into a host cell, (2) replication of the viral genome, (3) production of viral proteins, (4) assembly of a new generation of virions, and (5) exit from the infected cell. You should also be able to give an example of how each phase occurs in a particular virus.

34.3   What Themes Occur in the Diversification of Viruses?

If viruses can infect virtually every type of organism and cell known, how can biologists possibly identify themes that help organize viral diversity? The answer is that, in addition to being identified as enveloped or nonenveloped, viruses can be categorized by the nature of their hereditary material—in essence, the type of molecule their genes are made of. The single most important aspect of viral diversity is the variation that exists in their genetic material.

Two other points are critical to recognize about viral diversity: (1) Biologists do not have a solid understanding of how viruses originate, but (2) it is certain that viruses will continue to diversify. After analyzing diversity in viral genes, let's consider hypotheses to explain where viruses come from and recent data on new or "emerging" viruses.

The Nature of the Viral Genetic Material

DNA is the hereditary material in all cells. As cells synthesize the molecules they need to function, information flows from DNA to mRNA to proteins (Chapter 15). Although all cells follow this pattern, which is called the central dogma of molecular biology, some viruses break it. This conclusion traces back to work done in the 1950s, when biologists were able to separate the protein and nucleic acid components of a particle known as the tobacco mosaic virus, or TMV. Surprisingly, the nucleic acid portion of this virus consisted of RNA, not DNA. Later experiments demonstrated that the RNA of TMV, by itself, could infect plant tissues and cause disease. This was a confusing result, because it showed that, in this virus at least, RNA—not DNA—functions as the genetic material.

Subsequent research revealed an amazing diversity of viral genome types. In some groups of viruses, such as the agents that cause measles and flu, the genome consists of RNA. In others, such as the particles that cause herpes and smallpox, the genome is composed of DNA. Further, the RNA and DNA genomes of viruses can be either single stranded or double stranded. The single-stranded genomes can also be classified as "positive sense" or "negative sense" or "ambisense." In a **positive-sense virus**, the genome contains the same sequences as the mRNA required to produce viral proteins. In a **negative-sense virus**, the base sequences in the genome are complementary to those in viral mRNAs. In an **ambisense virus**, some sections of the genome are positive sense while others are negative sense.

Finally, the number of genes found in viruses varies widely. The tymoviruses that infect plants contain as few as three genes, but the genome of smallpox can code for up to 343 proteins. **Table 34.2** summarizes the diversity of viral genome types.

TABLE 34.2 The Diversity of Viral Genomes

Key: ss = single stranded; ds = double stranded; (+) = positive sense (genome sequence is the same as viral mRNA); (−) = negative sense (genome sequence is complementary to viral mRNA).

Genome		Example(s)	Host	Result of Infection	Notes
(+)ssRNA	+	TMV	Tobacco plants	Tobacco mosaic disease (leaf wilting)	TMV was the first RNA virus to be discovered.
(−)ssRNA	−	Influenza	Many mammal and bird species	Influenza	The negative-sense ssRNA viruses transcribe their genomes to mRNA via RNA replicase.
dsRNA	+ / −	Phytovirus	Rice, corn, and other crop species	Dwarfing	Double-stranded RNA viruses are transmitted from plant to plant by insects. Many can also replicate in their insect hosts.
ssRNA or (+)ssDNA that requires reverse transcription for replication	+ or − or +	Rous sarcoma virus	Chickens	Sarcoma (cancer of connective tissue)	Rous sarcoma virus was identified as a cancer-causing agent in 1911, decades before any virus was seen.
ssDNA—can be (+), (−), or (+) and (−)	+ or −	φ × 174	Bacteria	Death of host cell	The genome for φ × 174 is circular and was the first complete genome ever sequenced.
dsDNA	+ / −	Baculovirus Smallpox Bacteriophage	Insects Humans Bacteria	Death Smallpox Death	These are the largest viruses in terms of genome size and overall size.

Where Did Viruses Come From?

No one knows where viruses originated, but many biologists suggest that they are closely related to the plasmids and transposable elements introduced in Chapter 19 and Chapter 20. Viruses, plasmids, and transposable elements are all acellular, mobile genetic elements that replicate with the aid of a host cell. Simple viruses are actually indistinguishable from plasmids except for one feature: The viruses have a protein coat or membrane-like envelope.

Some biologists hypothesize that simple viruses, plasmids, and transposable elements represent "escaped gene sets." This hypothesis states that mobile genetic elements are descended from clusters of genes that physically escaped from bacterial or eukaryotic chromosomes long ago. According to this hypothesis, the escaped gene sets took on a mobile, parasitic existence because they happened to encode the information needed to replicate themselves—at the expense of the genomes that once held them. In the case of viruses, the hypothesis is that the escaped genes included the instructions for making a protein capsid and possibly envelope proteins. According to the escaped-gene hypothesis, it is likely that each of the distinct types of RNA viruses and single-stranded DNA viruses represents a distinct "escape event."

The same researchers contend that DNA viruses with large genomes originated in a very different manner, however. Here the hypothesis is that the large DNA viruses trace their ancestry back to free-living bacteria that once took up residence inside eukaryotic cells. The idea is that these organisms degenerated into viruses by gradually losing the genes required to synthesize ATP, nucleic acids, amino acids, and other compounds. Although this idea sounds speculative, it cannot be dismissed lightly. Chapter 27 introduced species in the genus *Chlamydia*, which are bacteria that live as parasites inside animal cells. And Chapter 28 provided evidence that the organelles called mitochondria and chloroplasts, which reside inside eukaryotic cells, originated as intracellular symbionts. Investigators contend that, instead of evolving into intracellular symbionts that aid their host cell, DNA viruses became parasites capable of destroying the host.

To date, neither of these hypotheses has been tested rigorously. To support the escaped-genes hypothesis, researchers would probably have to discover a brand new virus that originated in this way, or viruses that had so recently derived from intact bacterial or eukaryotic genes that the viral DNA sequence still strongly resembled the DNA sequence of those genes. To support the degeneration hypothesis, researchers would probably have to find strong genetic homologies between viruses and parasitic species of bacteria that live inside cells. But currently, there is no widely accepted view of where viruses came from.

Emerging Viruses, Emerging Diseases

Although it is not known how the various types of virus originated, it is certain that viruses will continue to diversify. With alarming regularity, the front pages of newspapers carry accounts of deadly viruses that are infecting humans for the first time. In 1993 a hantavirus that normally infects mice suddenly afflicted dozens of people in the southwestern United States. Nearly half of the people who developed hantavirus pulmonary syndrome died. Still higher fatality rates were recorded in 1995 when the Ebola virus, a variant of a monkey virus, caused a wave of infections in the Democratic Republic of Congo. By the time the outbreak subsided, over 200 cases had been reported; 80 percent were fatal. During an Ebola outbreak in 1976, 90 percent of cases were fatal.

Hantavirus pulmonary syndrome and Ebola are examples of **emerging diseases**: new illnesses that suddenly affect significant numbers of individuals in a host population. In these cases, the causative agents were considered **emerging viruses** because they had switched from their traditional host species to a new host—humans. Many investigators consider HIV to be an emerging virus because it originated in chimps and was first transmitted to humans in the early to mid-twentieth century (see **Box 34.3**).

Physicians become alarmed when they see a large number of patients with identical and unusual disease symptoms in the same geographic area over a short period of time. The physicians report these cases to public health officials, who take on two urgent tasks: (1) identifying the agent that is causing the new illness and (2) determining how the disease is being transmitted.

Several strategies can be used to identify a pathogen. In the case of the outbreak of hantavirus pulmonary syndrome, officials recognized strong similarities between symptoms in the U.S. cases and symptoms caused by the Hantaan virus native to northeast Asia. The Hantaan virus rarely causes disease in humans; its normal host is rodents. To determine whether a Hantaan-like virus was responsible for the U.S. outbreak, researchers began capturing mice in the homes and workplaces of afflicted people. About a third of the captured rodents tested positive for the presence of a Hantaan-like virus. (The test that was done is explained in Chapter 49.) DNA sequencing studies confirmed that the virus was a previously undescribed type of hantavirus. Further, the sequences found in the mice matched those found in infected patients. Based on these results, officials were confident that a rodent-borne hantavirus was causing the wave of infections.

The next step in the research program, identifying how the agent is being transmitted, is equally critical. If a virus that normally parasitizes a different species suddenly begins infecting humans, if it can be transmitted efficiently from person to

BOX 34.3 Where Did HIV Originate?

The key to discovering where HIV comes from lies in understanding the evolutionary history, or phylogeny, of the virus. Researchers have reconstructed this history by comparing the composition of HIV genes with sequences from closely related viruses that parasitize chimps, monkeys, and other mammals (**Figure 34.14**). The relationships among these gene sequences support several conclusions:

- HIV belongs to a group of viruses called the lentiviruses, which infect a wide range of mammals, including house cats, horses, goats, and primates. (*Lenti* is a Latin root that means "slow"; here it refers to the long period observed between the start of an infection by these viruses and the onset of the diseases they cause.)

- Many of HIV's closest relatives also have *immunodeficiency* in their name. Like HIV, these agents parasitize cells that are part of the immune system. Several of them cause diseases with symptoms reminiscent of AIDS. Curiously, though, HIV's closest relatives don't appear to cause disease in their hosts. These viruses infect monkeys and chimpanzees and are called **simian immunodeficiency viruses** (**SIVs**).

- There are two distinct types of human immunodeficiency viruses, called HIV-1 and HIV-2. Although both can cause AIDS, HIV-1 is far more virulent and is the better studied of the two. It serves as the focus of this chapter.

- HIV-1's closest relatives are immunodeficiency viruses isolated from chimpanzees that live in central Africa. In contrast, HIV-2's closest relatives are immunodeficiency viruses that parasitize monkeys called sooty mangabeys. In central Africa, where HIV-1 infection rates first reached epidemic proportions, contact between chimpanzees and humans is extensive. Chimps are hunted for food and kept as pets. Similarly, sooty mangabeys are hunted and kept as pets in west Africa, where HIV-2 infection rates are highest. To make sense of these observations, biologists suggest that HIV-1 is a descendant of viruses that infect chimps, and that HIV-2 is a descendant of viruses that infect sooty mangabeys. The "jumps" between host species probably occurred when a human cut up a monkey or chimpanzee for use as food.

- Several strains of HIV-1 exist. A virus **strain** consists of populations that have distinct characteristics. The most important HIV-1 strains are called O for *outgroup* (meaning, the most basal group relative to other strains), N for *new*, and M for *main*. The existence of these distinct strains suggests that HIV-1 has jumped from chimps to humans several times and may do so again.

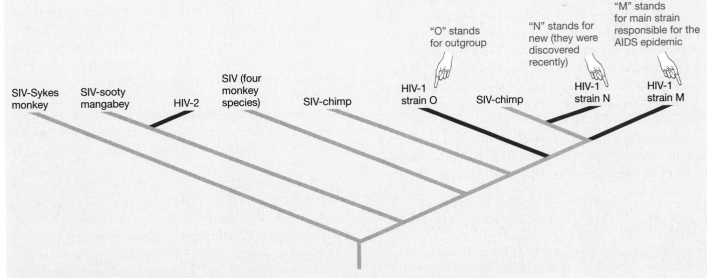

FIGURE 34.14 Phylogeny of HIV Strains and Types
Phylogenetic tree showing the evolutionary relationships among some of the immunodeficiency viruses that infect primates—including chimpanzees, humans, and several species of monkeys. **QUESTION** Based on this tree, researchers conclude that an SIV has "jumped" from chimps to humans at least twice. Do you agree? Explain why or why not.

person, and if it causes serious illness, then the outbreak has the potential to become an epidemic. But if transmission takes place only between the normal host and humans, as is the case with rabies, then the number of cases will probably remain low.

Determining how a virus is transmitted takes old-fashioned detective work. By interviewing patients about their activities, researchers decide whether each patient could have acquired the virus independently. Were the individuals infected with hantavirus in contact with mice? If so, were they bitten? Did they handle rodent feces or urine during routine cleaning chores or come into contact with contaminated food? Was the illness showing up in health-care workers, implying that it was being transmitted from human to human?

In the hantavirus outbreak, public health officials concluded that no human-to-human transmission was taking place. Health-care workers did not become ill; because patients had not had extensive contact with one another, it was likely that each had acquired the virus independently. The outbreak also coincided with a short-term, weather-related explosion in the local mouse population. The most likely scenario was that people had acquired the pathogen by inhaling dust or handling food that contained remnants of mouse feces or urine. In short, hantavirus did not have the potential to cause an epidemic. The best medicine was preventive: Homeowners were advised to trap mice out of living areas, wear dust masks while cleaning any area where mice might have lived, and store food in covered jars.

The Ebola virus, in contrast, was clearly transmitted from person to person. Many doctors and nurses were stricken after tending to patients with the virus. Infections that originate in hospitals usually spread when carried from patient to patient on the hands of caregivers. But by carefully observing the procedures that were being followed by hospital staff, researchers concluded that the Ebola virus was being transmitted only through direct contact with body fluids (blood, urine, feces, or sputum). The outbreak was brought under control when hospital workers raised their standards of hygiene and insisted on the immediate disposal or disinfection of all contaminated bedding, utensils, and equipment. Had the Ebola virus been transmitted by casual contact, such as touch or inhalation—as is the common cold virus—a massive epidemic could have ensued.

✓ CHECK YOUR UNDERSTANDING

Among viruses, several different types of molecules serve as the genetic material. You should be able to (1) describe at least three types of viral genomes other than double-strand-ed DNA, (2) diagram how the information in viral genes is used to produce a new generation of virions during a lytic replication cycle for one of these types of genomes, and (3) explain why a mutation that allowed Ebola virus to spread via airborne particles coughed out by infected individuals would make the virus more or less dangerous to humans.

34.4 Key Lineages of Viruses

Because scientists are almost certain that viruses originated multiple times throughout the history of life, there is no such thing as the phylogeny of all viruses. Stated another way, there is no single phylogenetic tree that represents the evolutionary history of viruses, as there is for the organisms discussed in previous chapters. Instead, researchers focus on comparing base sequences in the genetic material of small, closely related groups of viruses and using these data to reconstruct the phylogenies of particular lineages. Phylogenetic trees for viruses are usually inferred from comparisons of nucleic acid sequence data, using techniques introduced in Chapter 26.

The phylogenetic tree of the simian immunodeficiency viruses and human immunodeficiency viruses in Figure 34.14 is a good example of how researchers construct and interpret the phylogenies of particular groups of viruses. These types of phylogenies have been important in understanding viral diversity and the sources of emerging viruses. For example, phylogenetic analyses allowed biologists to recognize that there are two distinct types of HIV. The phylogenetic data diagrammed in Figure 34.14 indicate that HIV-1 and HIV-2 are distinct types of virus that originated from different host species (see Box 34.3). These data correlate with the observation that HIV-1 and HIV-2 differ in important ways. For instance, although both types of HIV infect immune system cells and are usually transmitted through sexual contact, HIV-2 is much less virulent and much less easily transmitted from person to person than is HIV-1.

To organize the diversity of viruses on a larger scale, researchers group them into eight general categories based on the nature of their genetic material. Within these broad groupings, biologists also identify a total of about 70 virus families that are distinguished by (1) the structure of the virion (often whether it is enveloped or nonenveloped), and (2) the nature of the host species.

Although they do not have formal scientific names, viruses within families are grouped into distinct genera. Within genera, biologists identify and name types of virus, such as HIV, the measles virus, and smallpox. Within each of these viral types, populations with distinct characteristics may be identified and named as strains. The strain is the lowest, or most specific, level of taxonomy for viruses. The O, N, and M strains of HIV-1, which are highlighted in Figure 34.14, are examples of separate virus strains. In the case of HIV-1, the named strains resulted from independent "jumps" of a simian immunodeficiency virus to a new host: humans. Each of these strains has distinguishing characteristics. The M strain, for example, was named *main*, because it is responsible for most of the HIV infections known to date.

To get a sense of viral diversity, let's survey a few of the major groups that can be identified by the nature of their genetic material.

Double-Stranded DNA (dsDNA) Viruses

The double-stranded DNA viruses are a large group, composed of some 21 families and 65 genera. Smallpox (**Figure 34.15**) is perhaps the most familiar of these viruses. Although smallpox had been responsible for millions of deaths throughout human history, it was eradicated by vaccination programs.

Genetic material   As their name implies, the genes of these viruses consist of a single molecule of double-stranded DNA. The molecule may be linear or circular.

Host species   These viruses parasitize hosts from throughout the tree of life, with the notable exception of land plants. They include virus families called the T-even and λ bacteriophages, some of which infect *E. coli*. In addition, the pox viruses, herpesviruses, and adenoviruses—some species of which parasitize humans—have genomes that consist of double-stranded DNA.

Replication cycle   In most double-stranded DNA viruses that infect eukaryotes, viral genes have to enter the nucleus to be replicated. For that to be true, these viral genes must replicate only during S phase, when the host cell's chromosomes are being replicated. As a result, these types of viruses can sustain an infection only in cells that are actively dividing, such as the cells that line the respiratory tract or urogenital canal.

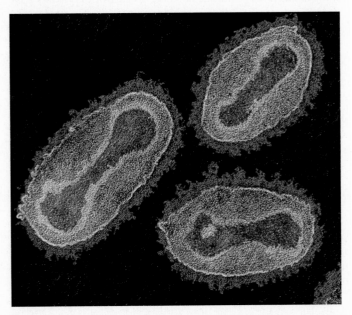

FIGURE 34.15 Smallpox Is a Double-Stranded DNA Virus

RNA Reverse-Transcribing Viruses (Retroviruses)

The genomes of the RNA reverse-transcribing viruses are composed of single-stranded RNA. There is only one family, called the retroviruses.

Genetic material   Virus particles have two copies of their single-stranded RNA genome, so they are diploid.

Host species   Species in this group are known to parasitize only vertebrates—specifically birds, fish, or mammals. HIV is the most familiar virus in this group. The Rous sarcoma virus, the mouse mammary tumor virus, and the murine (mouse) leukemia virus are other retroviruses that have also been studied intensively. Rous sarcoma virus was the first virus shown to be associated with the development of cancer (in chickens); the mouse viruses were the first viruses known to increase the risk of cancer in mammals (**Figure 34.16**).

Replication cycle   Retroviruses contain the enzyme reverse transcriptase inside their capsid. (A typical HIV particle contains about 50 reverse transcriptase molecules.) When the virus's RNA genome and reverse transcriptase enter a host cell's cytoplasm, the enzyme catalyzes the synthesis of a single-stranded cDNA from the original RNA. Reverse transcriptase then makes this cDNA double strand-ed. The double-stranded DNA version of the genome enters the nucleus with a viral protein called *integrase*. Integrase catalyzes the integration of the viral genes into a host chromosome. The virus may remain lysogenic for a period, but eventually the genes are transcribed to RNA to begin lytic growth and the production of a new generation of virus particles.

FIGURE 34.16 Some Retroviruses, Such as the Mouse Mammary Tumor Virus, Are Associated with Cancer

Double-Stranded RNA (dsRNA) Viruses

There are 7 families of double-stranded RNA viruses and a total of 22 genera. Most of the viruses in this group are nonenveloped.

Genetic material   In some families, virus particles typically have a genome consisting of ten to twelve double-stranded RNA molecules; in other families, the genome is composed of just one to three RNA molecules.

Host species   A wide variety of hosts, including fungi, land plants, insects, vertebrates, and bacteria, are victimized by viruses with double-stranded RNA genomes. For humans, the most important viruses in this group are those that cause disease in rice, corn, sugar cane, and other crops. **Figure 34.17** shows rice plants that are being attacked by a double-stranded RNA virus. Infections are also common in *Penicillium*, a filamentous fungus that produces the antibiotic penicillin.

Replication cycle   Once inside the cytoplasm of a host cell, the double-stranded genome of these viruses serves as a template for the synthesis of single-stranded RNAs, which are then translated into viral proteins. The proteins form the capsids for a new generation of virus particles. Copies of the genome are created when a viral enzyme makes the original single-stranded RNA double stranded.

FIGURE 34.17 Double-Stranded RNA Viruses, Such as the Ragged Stunt Virus, Parasitize a Wide Array of Organisms—Here, Rice

Negative-Sense Single-Stranded RNA ([−]ssRNA) Viruses

There are 7 families and 30 genera in this group. Most members of this group are enveloped, but some negative-sense single-stranded RNA viruses lack an envelope.

Genetic material   The sequence of bases in a negative-sense RNA virus is opposite in polarity to the sequence in a viral mRNA. Stated another way, the single-stranded virus genome is complementary to the viral mRNA. Depending on the family, the genome may consist of a single RNA molecule or up to eight separate RNA molecules.

Host species   A wide variety of plants and animals are parasitized by viruses that have negative-sense single-stranded RNA genomes. If you have ever suffered from the flu, the mumps, or the measles, then you are painfully familiar with these viruses (**Figure 34.18**). The Ebola, Hantaan, and rabies viruses also belong to this group.

Replication cycle   When the genome of a negative-sense single-stranded RNA virus enters a host cell, a viral RNA polymerase uses that genome as a template to make new viral mRNA. These viral mRNAs are then translated to form viral proteins. The viral mRNAs also serve as a template for the synthesis of new copies of the negative-sense single-stranded RNA genome.

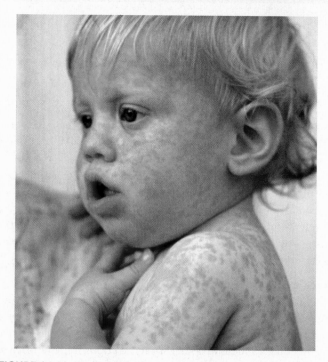

FIGURE 34.18 Negative-Sense Single-Stranded RNA Viruses, Such as the Measles Virus, Cause Some Common Childhood Diseases

Positive-Sense Single-Stranded RNA ([+]ssRNA) Viruses

This is the largest group of viruses known, with 81 genera organized into 21 families.

Genetic material The sequence of bases in a positive-sense RNA virus is the same as that of an mRNA. Stated another way, the genome does not need to be transcribed in order for proteins to be produced. Depending on the species, the genome consists of one to three RNA molecules.

Host species Most of the commercially important plant viruses belong to this group. Because they kill groups of cells in the host plant and turn patches of leaf or stem white, they are often named mottle viruses, spotted viruses, chlorotic (meaning, lacking chlorophyll) viruses, necrotic (meaning, killed cells surrounded by intact tissue) viruses, or mosaic viruses. **Figure 34.19** shows a healthy cowpea leaf and a cowpea leaf that has been attacked by a positive-sense single-stranded RNA virus. Some species in this group of viruses specialize in parasitizing bacteria, fungi, or animals, however. A variety of human maladies, including the common cold, polio, and hepatitis A, C, and E, are caused by positive-sense RNA viruses.

Replication cycle When the genome of these viruses enters a host cell, the single-stranded RNA is immediately translated into viral proteins. These proteins include enzymes that make copies of the genome. The new generation of virus particles is assembled in a complex structure that is associated with the host cell's plasma membrane.

Healthy leaf   Leaf infected with virus

FIGURE 34.19 Positive-Sense Single-Stranded RNA Viruses, Such as the Cowpea Mosaic Virus, Cause Important Plant Diseases

ESSAY The SARS Outbreak

In November 2002, physicians in the southern Chinese province of Guangdong began treating people for an atypical form of **pneumonia,** or inflammation of the lungs. The illness began with a day or two of high fever (over 38°C, or 100.4°F) and progressed to pneumonia and dry coughing, sometimes accompanied by low oxygen levels in the blood. A number of the cases were fatal.

In February 2003 the illness showed up in Hong Kong. At that point, international health officials were alerted. Researchers immediately began to track cases and monitor the spread of the disease. The condition was named **severe acute respiratory syndrome,** or **SARS.** In medicine, a **syndrome** is a group of symptoms that occur together and have the same cause. The syndrome was *severe* because it made people sick enough to require hospitalization and *acute* because it progressed rapidly.

Soon after the disease appeared in Hong Kong, people with the condition began showing up in other countries. By the time SARS was brought under control several months later, a total of 8098 people in 30 countries had fallen ill. Of these, 774—almost 10 percent—had died. This is an extremely high fatality rate. In contrast, the 1918–1919 influenza was fatal in only about 2.5 percent of cases.

To fight the outbreak, public health officials not only quarantined individuals who had been exposed but also restricted access to particularly hard-hit cities or neighborhoods. In addition, many people stopped going to work in affected cities, for fear that they would contract the virus. The Asian Development Bank estimates that this disruption cost the economies of southeast and east Asia $18 billion in gross domestic product, and that an additional $60 billion was lost due to reduced travel and lowered demand for goods and services in the affected regions. These figures work out to an economic loss of about $2 million for each person with SARS.

Early in the outbreak, researchers succeeded in isolating the virus, sequencing its genome, and analyzing its place in a

> *The SARS outbreak did not become an epidemic, because the virus is not easily transmitted from person to person.*

(Continued on next page)

(Essay continued)

phylogeny of closely related viruses. The results showed that SARS was caused by a previously undescribed type of *coronavirus*—a positive-sense single-stranded RNA virus. To determine where the virus originated, researchers followed up on the observation that several of the earliest SARS cases involved workers from restaurants in the city of Shenzhen that specialized in serving exotic meats. Testing showed that the SARS coronavirus was present in several species of wild animals sold in the local exotic meat market, as well as in several of the human vendors. Based on these data, researchers concluded that the SARS coronavirus jumped from a wild animal in the market to an animal vendor or restaurant workers. It is still not known what species in nature is the normal host or "reservoir" of this virus.

The SARS outbreak highlights several key points about virus outbreaks and public health. First, emerging viruses will continue to emerge. Human populations will continually be exposed to novel pathogens, from Ebola virus to Hantaan virus to HIV and SARS coronavirus to viruses that have yet to be identified or even evolve. As this book goes to press, international public health officials continue to monitor pneumonia cases in east Asia for a "reemergence" of SARS. Second, the nature of today's global economy and access to air travel means that emerging diseases can spread worldwide in a matter of weeks or even days. During the SARS outbreak, officials were able to trace the virus's movement to new countries via infected individuals on specific flights of specific airlines. Third, the SARS outbreak did not become an epidemic, because the virus is not easily transmitted from person to person. In the vast majority of cases, people got the virus through extended and direct contact with droplets from an infected person's cough or sneeze. If transmission had occurred through airborne particles instead of water droplets, the virus could have caused a devastating epidemic.

CHAPTER REVIEW

Summary of Key Concepts

■ Viruses are tiny, noncellular parasites that infect virtually every type of cell known. They cannot perform metabolism on their own—meaning outside a parasitized cell—and are not considered to be alive. Different types of viruses are specialized for infecting particular species and types of cells.

Viruses cause illness and death in plants, fungi, bacteria, and archaea, as well as in humans and other animals. Vaccination is frequently effective in preventing virus epidemics. Unfortunately, it is difficult or impossible to design a vaccine that can prepare the immune system for viruses that mutate very rapidly, such as HIV and the cold and flu viruses.

■ Viruses are highly diverse in overall morphology and in the nature of their genetic material. The genomes of viruses may consist of double-stranded DNA, single-stranded DNA, double-stranded RNA, or one of several types of single-stranded RNA.

The most important morphological groups in viruses are the enveloped and nonenveloped viruses. Enveloped viruses contain a capsid consisting of protein and are surrounded by a membranous envelope; nonenveloped viruses are surrounded only by a capsid.

Viral genomes are small and can consist of either RNA or DNA, but not both. Viral genomes do not code for ribosomes, and most do not code for the enzymes needed to translate their own proteins or to perform other types of biosynthesis. When a virus infects a host cell, it uses that cell's enzymes, nutrients, and ribosomes to manufacture a new generation of virus particles. This observation explains why viral diseases are difficult to treat with drugs: Molecules that incapacitate enzymes needed by the virus are likely to damage host cells as well.

■ The viral infection cycle can be broken down into five steps: (1) entry into a host cell, (2) replication of the viral genome, (3) the production of viral proteins, (4) the assembly of a new generation of virus particles, and (5) exit from the infected cell.

A viral infection begins when the contents of a virus particle enter a host cell. For example, when HIV binds to a transmembrane protein called CD4 and a co-receptor, the virus's membrane-like envelope fuses with the host cell's plasma membrane, and the viral genome and proteins enter the cytoplasm. The viral genome is replicated in the second phase of the replication cycle. In HIV, for example, the virus's RNA genome is reverse-transcribed to DNA and integrated into a chromosome of the host cell. In the third phase of the replication cycle, viral proteins are produced. While they are still inside the cell, the viral proteins that result assemble into complete particles (the fourth phase of the infection cycle). In the final phase, the new generation of complete particles buds or bursts from the cell. Viruses that bud from the host cell acquire a lipid bilayer, containing host proteins as well as viral proteins, during the process. Viruses that burst from the host cell do not have this membrane-like envelope. Instead, they are protected only by a protein coat. These coat proteins are coded for by viral genes and assemble inside the host cell's cytoplasm. When a new generation of virus particles buds or bursts from a host cell, the infected cell dies.

Once they are released from the host cell, particles can infect more cells in the same multicellular organism or be transmitted to a new host.

Web Tutorial 34.1   The HIV Life Cycle

Questions

Content Review

1. How do viruses that infect animals enter an animal's cells?
   a. The viruses pass through a wound.
   b. The viruses bind to a membrane protein.
   c. The viruses puncture the cell wall.
   d. The viruses bind to the lipid bilayer.

2. What does reverse transcriptase do?
   a. It synthesizes proteins from mRNA.
   b. It synthesizes tRNAs from DNA.
   c. It synthesizes DNA from RNA.
   d. It synthesizes RNA from DNA.

3. What do host cells provide for viruses?
   a. nucleotides and amino acids
   b. ribosomes
   c. ATP
   d. all of the above

4. When do virus particles acquire a membrane-like envelope, including a lipid bilayer?
   a. during entry into the host cell
   b. during budding from the host cell
   c. as they burst from the host cell
   d. as they integrate into the host cell's chromosome

5. What reaction does protease catalyze?
   a. polymerization of amino acids into peptides
   b. cutting of long peptide chains into functional proteins
   c. folding of long peptide chains into functional proteins
   d. assembly of viral particles

6. Why is it difficult to design a vaccine for viruses with high mutation rates, such as HIV and the cold and flu viruses?
   a. The vaccines tend to be unstable and deteriorate over time.
   b. So many protein fragments are presented by these viruses that the immune system overreacts.
   c. They have no protein fragments that can be recognized by a host cell.
   d. New mutations constantly change viral proteins.

Conceptual Review

1. The outer surface of a virus consists of either a membrane-like envelope or a protein capsid. Which type of outer surface does HIV have? Which type do bacteriophages have? How does the outer surface of a virus correlate with its mode of exiting a host cell, and why?

2. Compare the morphological complexity of HIV with that of bacteriophage T4. Which virus would you predict has the larger genome? Explain the logic behind your hypothesis.

3. Compare and contrast lytic growth and lysogenic growth. When would you expect HIV and bacteriophage T4 to grow lytically versus lysogenically, and why?

4. Explain why viral diseases are more difficult to treat than diseases caused by bacteria.

5. Draw the lytic cycle of a nonenveloped virus with a positive-sense single-stranded RNA genome that infects cells in the roots of rice plants. Describe the modes of action of two drugs that could be developed to treat plants infected with this virus.

6. What type of data convinced researchers that HIV originated when a simian immunodeficiency virus "jumped" to humans? Do you agree with this conclusion? Why or why not?

Group Discussion Problems

1. Suppose you could isolate a virus that parasitizes the pathogenic bacterium *Staphylococcus aureus*. This bacterium causes acne, boils, and a variety of other afflictions in humans. How could you test whether this virus might serve as a safe and effective antibiotic?

2. If you were in charge of the government's budget devoted to stemming the AIDS epidemic, would you devote most of the resources to drug development, vaccine development, or preventive medicine? Defend your answer.

3. Bacteria fight viral infections with restriction endonucleases (bacterial enzymes; introduced in Chapter 19). Restriction endonucleases cut up, or break, viral DNA at specific sequences. The enzymes do not cut a bacterium's own DNA, because the bases in the bacterial genome are protected from the enzyme by methylation (the addition of a CH_3 group). Generate a hypothesis to explain why members of the Eukarya do not have restriction endonucleases.

4. Consider these two contrasting definitions of life:
   a. An entity is alive if it is capable of replicating itself via the directed chemical transformation of its environment.
   b. An entity is alive if it is an integrated system for the storage, maintenance, replication, and use of genetic information.
   According to these definitions, are viruses alive? Explain.

Answers to Multiple-Choice Questions 1. b; 2. d; 3. d; 4. b; 5. b; 6. d

Ecology

Bald eagles make their living by preying on fish and small mammals and by scavenging carcasses. Although bald eagles were once in danger of going extinct, most populations have made a strong recovery.

50 An Introduction to Ecology

KEY CONCEPTS

▨ Ecology focuses on how organisms interact with their environment. Because its goal is to understand the distribution and abundance of organisms, ecology provides a scientific foundation for the conservation of species and natural areas.

▨ An organism's environment consists of both abiotic, or physical, factors and biotic factors, meaning other organisms.

▨ On land, environments are defined by climate—specifically, both the average value and annual variation in temperature and in moisture. In aquatic habitats, environments are defined by physical structure—particularly water depth.

▨ A species' distribution is constrained by a combination of historical, abiotic, and biotic factors.

In addition to providing habitat for a frog, the puddle of water inside this plant provides an environment for an array of microscopic species. The plant itself is growing in the branch of a rain forest tree and is just one of hundreds of plant and insect species that live in association with the tree. Ecology explores how all of these organisms interact with each other and with their environment.

Environments are currently changing faster than they have changed in the past 3.5 billion years—except for the episode 65 million years ago when a mountain-sized asteroid struck Earth (see Chapter 26). As human impacts on the environment have accelerated, ecology has become an increasingly important field in biological science. **Ecology** is the study of how organisms interact with their environment. Efforts to maintain human health and welfare depend on our ability to understand and predict the consequences of the rapid environmental changes that are occurring now.

One of ecology's central goals is to understand the distribution and abundance of organisms. In many cases an ecologist's job is to identify factors that dictate why certain species live where they do and how many individuals live there. Some biologists ask why orangutans are restricted to forests in Borneo and why their numbers are declining rapidly. Other biologists create mathematical models to predict how quickly the human immunodeficiency virus will increase in India and how long it will take for the current human population of 6.35 billion to double. Ecologists seek to recognize and explain patterns in nature—patterns that involve where particular organisms live and in what numbers.

This chapter has two goals: to explore how biologists go about studying ecology and to introduce the major types of environments that organisms occupy. It serves as a springboard to subsequent chapters, which analyze types of interactions between organisms and their envoronment and ask how environmental changes affect humans and other species. In terms of understanding the biological problems that humans are facing now, this may be the most important unit in the book.

50.1   Areas of Ecological Study

To understand why organisms live where they do and in what numbers, biologists break ecology into several levels of analysis. This is a common strategy in biological science. For example, cell biologists study how cells work at increasingly complex levels of organization—from individual molecules to multimolecular machines, organelles, whole cells, and cell-to-cell interactions. Physiologists may analyze processes at the level of ions and molecules as well as what is happening to whole cells, tissues and organs, or a complete system. In ecology, researchers work at four main levels: (1) organisms, (2) populations, (3) communities, and (4) ecosystems. Let's examine each.

Organismal Ecology

At the finest level of organization, ecology focuses on how individuals interact with their environment. Researchers who study organismal ecology explore the morphological, physiological, and behavioral adaptations that allow individuals to live successfully in a particular area. The study of behavior is an important aspect of organismal ecology; it focuses on how organisms respond to particular stimuli from their environment (see Chapter 51). The environmental stimuli that trigger behavior may consist of changes in temperature or moisture, an escape response from prey, or a rival challenging for a mate. Organismal ecology also focuses on the physiological adaptations that allow individuals to thrive in heat, drought, cold, or other demanding physical conditions.

As an example, consider the sockeye salmon (**Figure 50.1a**). After spending four or five years feeding and growing in the ocean, these salmon travel hundreds or thousands of kilometers to return to the stream where they hatched. Females create nests in the gravel stream bottom and lay eggs. Nearby males compete for the chance to fertilize eggs as they are laid. When breeding is finished, all of the adults die.

At the level of organismal ecology, biologists want to know how these individuals interact with their physical surroundings and with other organisms in and around the stream. Which females get the best nesting sites and lay the most eggs, and which males are most successful in fertilizing eggs? How do individuals cope with the transition from living in salt water to living in freshwater?

Population Ecology

A **population** is a group of individuals of the same species that lives in the same area at the same time. When biologists study population ecology, they focus on how the numbers of individuals in a population change over time. Some of the theory that population ecologists use to analyze and predict changes in population size is now being used to evaluate the fate of endangered species. For example, mathematical models of population growth have been used to predict the future of particular salmon populations (**Figure 50.1b**). This work is important. Many

(a) Organismal ecology

How do individuals interact with each other and their physical environment?

Male salmon fight over females during the breeding season

(b) Population biology

How and why does population size change over time?

Each female salmon produces thousands of eggs. Only a few will survive to adulthood. On average, only two will return to the stream of their birth to breed

(c) Community ecology

How do species interact, and what are the consequences?

Salmon are prey as well as predators

(d) Ecosystem ecology

How do energy and nutrients cycle through the environment?

When salmon die and decompose, the nutrients that are released are used by bacteria, archaea, plants, protists, young salmon, and other organisms

FIGURE 50.1 Biologists Study Ecology at Four Main Levels

salmon populations have declined precipitously as their habitats have become dammed or polluted, and salmon are an important source of food for both people and wildlife. If the factors that affect population size can be described accurately enough, mathematical models can assess the impact of proposed dams, changes in weather patterns, altered harvest levels, or specific types of protection efforts.

Community Ecology

A biological **community** consists of the species that interact with each other within a particular area. Researchers who study community ecology ask questions about the nature of the interactions between species and the consequences of those interactions. The work might concentrate on predation, parasitism, competition, or other types of contact and impact. At this level of organization in ecology, biologists also analyze how groups of species respond to disturbances such as fires, floods, and volcanic eruptions. Because human activities are driving species to extinction and disturbing communities on a massive scale, community ecologists are being called on to generate hypotheses and data on how human impacts can be lessened.

As an example of the types of questions asked at the level of community ecology, consider the interactions among salmon and other species in the marine and stream communities where they live. When they are at sea, salmon eat smaller fish and are themselves hunted and eaten by orcas, sea lions, humans, and other mammals. When the salmon return to freshwater to breed, they are preyed on by bears and bald eagles (**Figure 50.1c**). In both marine and freshwater habitats, salmon are subject to parasitism and disease. They are also heavily affected by disturbances—particularly changes in their food supply due to overfishing and the damming and degradation of breeding streams.

Ecosystem Ecology

Ecosystem ecology is an extension of community ecology. An **ecosystem** consists of all the organisms in a particular region along with nonliving components. These nonliving, or **abiotic** ("not-living"), components include air, water, and soil. At the ecosystem level, biologists analyze how chemical elements that act as nutrients cycle through ecosystems and how energy flows through the organisms in a region. More specifically, biologists study how nutrients and energy move between organisms and the surrounding atmosphere and soil or water. Because humans are adding massive amounts of nutrients to ecosystems all over the world and affecting energy flows and climate through global warming, this work has direct public policy implications. Ecosystem ecologists are responsible for assessing the impact of pollution and increased temperature on the distribution and abundance of species—in fact, on Earth's ability to support life.

Salmon are interesting to study at the ecosystem level, because they form a link between marine and freshwater ecosystems. They harvest nutrients in the ocean and then, when they

die and decompose, transport those molecules to streams (**Figure 50.1d**). In this way, salmon transport chemical energy and nutrients from one habitat to another. Because salmon are sensitive to pollution and to changes in water temperature, human-induced changes in marine and freshwater ecosystems may have a large impact on their populations.

How Do Ecology and Conservation Efforts Interact?

The four levels of ecological study are synthesized and applied in conservation biology. **Conservation biology** is the effort to study, preserve, and restore threatened populations, communities, and ecosystems. Ecologists study how interactions between organisms and their environments result in a particular species being found in a particular area and population size; conservation biologists collect the same types of data and apply it by advising efforts to preserve species and restore environments. This is a common theme in biological science. Physicians, dentists, and veterinarians use basic results from molecular biology and physiology to improve the health of people and domesticated animals. Agricultural scientists and farmers apply information from genetics and physiology to improve the quality and quantity of the food we eat. Conservation biologists do the same for natural systems. Ecologists provide theory and data—the knowledge base for understanding how ecosystems work. Conservation biologists prescribe remedies for threatened species and manage land to produce a diversity of species, clean air, pure water, and healthy soils.

50.2   The Nature of the Environment

If ecology is the study of how species interact with their environment, what makes up the environment? The short answer is that an organism's environment has both physical and biological components. The abiotic components include temperature, precipitation, wind, and sunlight. In addition, terrestrial and aquatic environments have unique abiotic components—the chemical content and particle size of soil in terrestrial environments and the chemical composition of water in aquatic habitats. The **biotic** ("living") components, in contrast, consist of other members of the organism's own species as well as individuals of other species. Every organism that interacts with an individual is part of its biotic environment. Biotic factors include interactions with offspring, mates, competitors, predators, prey, and parasites.

It is important to understand the nature of the environments that organisms face, because humans are changing those environments rapidly. People alter the abiotic environment through air and water pollution, global warming, and soil erosion. Human impacts on the biotic environment are also nothing short of massive, through reductions in populations of other species and even outright extinction of species, removal of predators, and the introduction of **exotics**—organisms that are introduced into a region where they do not naturally live.

Because all environments have both abiotic and biotic components, it is not possible to understand ecology without considering both nonliving and living factors. But because subsequent chapters focus on biological interactions, the emphasis here is on the nature of the physical environments that organisms face. What components of the physical world have the most impact on organisms?

Climate

Climate is defined as the prevailing long-term weather conditions in a particular region. **Weather** consists of the specific short-term atmospheric conditions of temperature, moisture, sunlight, and wind. All of these factors are important to organisms.

Temperature is critical because the enzymes that make life possible work at optimal efficiency only in a narrow range of temperatures (see Chapter 3). Most organisms cannot regulate their own body temperatures, so their ability to maintain high levels of enzyme activity, metabolism, and growth or movement depends on the surrounding temperature. Vertebrate animals are an exception to this rule; many vertebrates can maintain a fairly constant body temperature by manufacturing their own heat or changing their position or activity to gain or lose heat. Temperature is also important, because it affects the availability of moisture: Water is frozen at low temperatures and evaporates rapidly at high temperatures.

Moisture is significant to terrestrial organisms because they constantly lose water to the environment through evaporation or transpiration. As a result, they must not only reduce water loss but also replace lost water. Aquatic organisms are surrounded by water and are exquisitely sensitive to changes in its ionic composition (see Chapter 42).

Sunlight is essential to organisms because it is required for photosynthesis and because photosynthesis provides the chemical energy that virtually all organisms depend on to stay alive. In terrestrial habitats the amount of sunlight available to any one organism depends on the extent of shading; in aquatic habitats the critical factor is depth below the water surface—because water absorbs and scatters light. In most habitats the availability of sunlight also varies with the time of year.

Wind is important because it exacerbates the effects of temperature and moisture. Wind increases heat loss due to evaporation and convection, and it increases water loss due to evaporation and transpiration. It can also have a direct physical impact on organisms—particularly birds, flying insects, and plants.

Of the four components of climate, temperature and moisture are far and away the most important to plants. Photosynthesis and plant growth are maximized when temperatures are warm and conditions are wet; conversely, photosynthesis cannot occur efficiently in cold temperatures or under drought stress. Because terrestrial animals depend on plants for food, it follows that temperature and moisture are the most important climatic elements for them as well.

To get a better idea of how ecologists analyze the impact of climate, let's consider a series of specific questions about why climate varies around the globe. Why are some parts of the world warmer and wetter than others, and why do seasons exist?

Why Are the Tropics Warm and the Poles Cold?   In general, areas of the world are warm if they receive a large amount of sunlight per unit area; they are cold if they receive a small amount of sunlight per unit area. The shape of Earth dictates that regions at or near the equator receive a great deal of sunlight per unit area relative to regions that are closer to the poles. At the equator, the Sun is often directly overhead. As a result, sunlight strikes Earth at or close to a 90° angle. At these angles, Earth receives a maximum amount of solar radiation per unit area (**Figure 50.2**). But because Earth's surface slopes away from the equator, the Sun strikes the surface at lower and lower angles toward the poles. When sunlight arrives at a low angle, much less energy is received per unit area. The pattern of decreasing average temperature with increasing **latitude**, or distance from the equator, is a result of Earth's largely spherical shape.

Why Are the Tropics Wet?   One of the most striking weather patterns on Earth involves precipitation. When average annual rainfall is mapped for regions around the globe, it is clear that areas along the equator receive the most moisture, while locations about 30 degrees latitude north and south of the

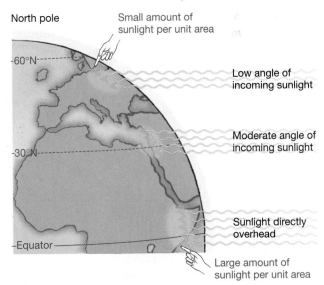

FIGURE 50.2 Solar Radiation per Unit Area Declines with Increasing Latitude
Over the course of a year, the Sun is frequently almost directly overhead at the equator. As a result, equatorial regions receive a large amount of solar radiation per unit area. At latitudes greater than 23.5°, the Sun is never directly overhead. As a result, high-latitude regions receive less solar radiation per unit area than do low-latitude regions.

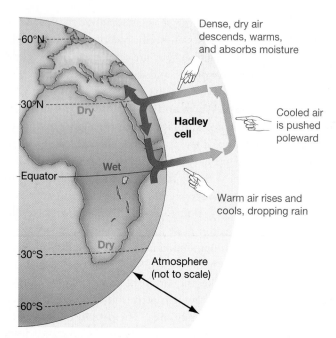

FIGURE 50.3 Global Air Circulation Patterns Affect Rainfall
Hadley cells explain why the tropics are much wetter than regions near 30° latitude north and south. Regions with rising air tend to be wetter than regions with descending air. There are three major air circulation cells on each side of the equator. (Only one of the six is shown.)

equator are among the driest on the planet. Most of the world's great deserts—including the Sahara, Gobi, Sonoran, and Australian outback—lie on or about 30°N or 30°S latitude. Why does this pattern occur?

A major cycle in global air circulation, called a **Hadley cell**, is responsible for making the Amazon River basin wet and the Sahara dry. Hadley cells are named after George Hadley, who in 1735 conceived of the idea of enormous air circulation patterns. As **Figure 50.3** indicates, air that is heated by the strong sunlight along the equator expands and rises. Warm air can hold a great deal of moisture, because warm water molecules tend to stay in vapor form instead of condensing into droplets. As the air rises, however, it radiates heat to space and begins to cool. As it cools, its ability to hold water declines. When water vapor cools, water condenses. The result? High levels of precipitation occur along the equator.

As more air is heated along the equator, the cooler, "older" air above Earth's surface is pushed poleward. When the air mass has cooled enough that its density increases, it begins to sink. As it sinks, it absorbs more and more solar radiation reflected from Earth's surface and begins to warm. As the air warms, it also gains water-holding capacity. Thus, little rain occurs where that air returns to Earth's surface, and the area is bathed in warm, dry air. The result is a band of deserts in the vicinity 30° latitude north and south.

What Causes Seasonality in Weather? The striking global patterns in temperature and moisture that we've just reviewed are complicated by the phenomenon of **seasons**: regular, annual fluctuations in temperature, precipitation, or both. As **Figure 50.4** shows, seasonality occurs because Earth is tilted on its axis by 23.5 degrees. As a result of this incline, the Northern Hemisphere is tilted toward the Sun in June and is the region that faces the Sun most directly. Thus the Northern Hemisphere receives the largest amount of solar radiation per unit area in June. But in December, the Southern Hemisphere is tilted toward the Sun, faces the Sun most directly, and receives the largest quantity of solar radiation per unit area. As a result, it is summer in the Northern Hemisphere in June but summer in the Southern Hemisphere in December. In March and September the equator faces the Sun most directly, so the tropics receive the most solar radiation then. If Earth did not tilt on its axis, there would be no seasons.

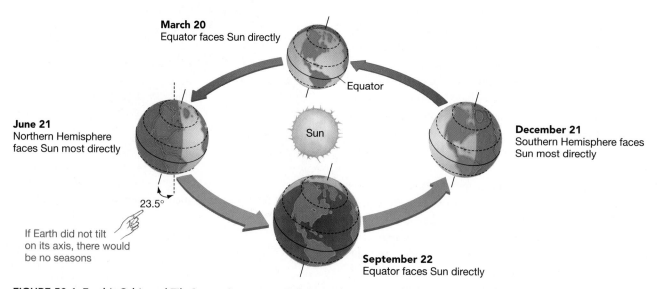

FIGURE 50.4 Earth's Orbit and Tilt Create Seasons at High Latitudes
Seasons occur due to annual variation in the amount of solar radiation that different parts of the Earth receive.

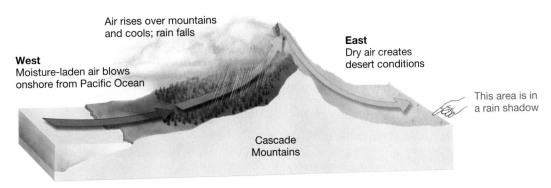

West
Moisture-laden air blows
onshore from Pacific Ocean

Air rises over mountains
and cools; rain falls

East
Dry air creates
desert conditions

This area is in
a rain shadow

Cascade
Mountains

FIGURE 50.5 Mountain Ranges Create Rain Shadows
When moisture-laden air rises over a mountain range, the air cools enough to condense water vapor. As a result, moisture falls to the ground as precipitation. On the other side of the mountain, little water is left to provide precipitation.

Mountains and Oceans: Regional Effects on Climate

The broad patterns of climate that are dictated by global heating patterns, Hadley cells, and seasonality are overlain by regional effects. The most important of these are due to the presence of mountain ranges and proximity to an ocean.

Figure 50.5 shows how mountain ranges affect regional climate, using the Cascade Mountain range along the Pacific Coast of North America as an example. In this area of the world, the prevailing winds are from the west. These winds bring moisture-laden air from the Pacific Ocean onto the continent. As the air masses begin to rise over the mountains, the air cools and releases large volumes of water as rain. As a result, the area along the Pacific Coast from northern California to southeast Alaska hosts some of the only high-latitude rain forests in the world. One area along the Pacific Coast of Washington State gets an average of 340 cm (134 in.) of precipitation each year. (Mountain ranges also occur along the Pacific Coast of Southern California and Mexico; but these areas are arid because they are dominated by dry air that is part of the Hadley circulation.)

Once cooled air has passed the crest of a mountain range, the air is relatively dry, because much of its moisture content has already been released. Areas that receive this dry air are said to be in a **rain shadow**. One of the only high-latitude deserts in the world exists to the east of the Cascade Mountains. It is created by a rain shadow and averages less than 25 cm (10 in.) of moisture annually. Similar rain shadows create the Great Basin desert of western North America and the dry conditions of the Tibetan Plateau.

While the presence of mountain ranges tends to produce extremes in precipitation, the presence of an ocean has a moderating influence on temperature. To understand why this is so, recall from Chapter 2 that water has an extremely high **specific heat**, meaning that it has a large capacity for storing heat energy. Because water molecules form hydrogen bonds with each other, it takes a great deal of heat energy to boil water or melt ice—in fact, to change the temperature of water at all. The hydrogen bonds have to be broken before the water molecules

themselves can begin to absorb heat and move faster, which we measure as increased temperature.

Because water has a high specific heat, it can absorb a great deal of heat from the atmosphere in summer, when the water temperature is cooler than the air temperature. As a result, the ocean moderates summer temperatures on nearby landmasses. Similarly, the ocean releases heat to the atmosphere in winter, when the water temperature is warmer than the air temperature. As a result, islands and coastal areas have much more moderate climates than do nearby areas inland.

50.3 Types of Terrestrial and Aquatic Ecosystems

Climate varies dramatically around the globe, and the type of vegetation present varies just as dramatically. If you could walk from the equator in South America to the North Pole, you would notice startling changes in the organisms around you. Lush tropical forests would give way to seasonally dry forests and then to deserts. The deserts would yield to the vast grasslands of central North America, which end at the boreal forests of the subarctic. If you pressed on, you would reach the end of the trees and the beginning of the most northerly community—the arctic tundra.

Biologists have documented that similar types of changes occur with increased elevation, because of similar changes in climate. In the Siskiyou Mountains of Oregon, for example, researchers have documented that Douglas fir and other drought-tolerant tree species dominate the forests at low elevations, where the climate is warm and dry. At higher elevations, where the climate is cooler and moister, a few white fir trees join the Douglas fir. Douglas fir is absent on higher and cooler slopes, leaving just white fir. The highest-elevation forests, where the climate is cold and wet, are dominated by noble fir and mountain hemlock. Above timberline—meaning elevations higher than trees can grow—the landscape is covered with alpine tundra.

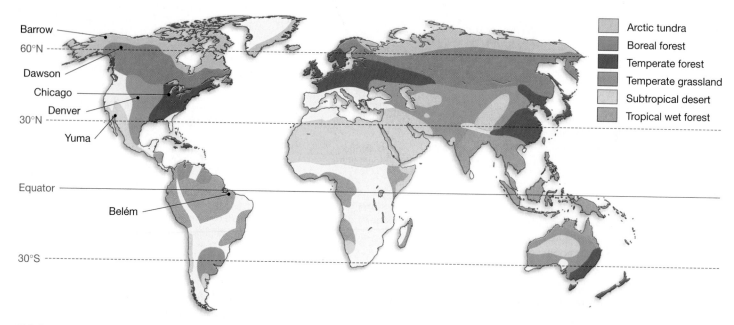

FIGURE 50.6 Distinct Biomes Are Found throughout the World
The distribution of the community types discussed in Section 50.2, and the locations of the specific sites discussed.

Terrestrial Biomes

A journey from the equator to the poles or from sea level to a mountaintop has a simple message: As climate changes, so does the predominant type of vegetation present. To capture this point, biologists identify and analyze ecological units called biomes. A **biome** is a type of terrestrial ecosystem that is unique to a given region and is characterized by a distinct type of vegetation. Grasslands, evergreen forests, and deserts are all types of biomes. The key observation is that each biome is associated with a distinctive combination of temperature and moisture conditions. Biomes represent the major types of environments available to terrestrial organisms.

How do biologists identify climatic regimes and characterize biomes? The most common approach categorizes climate types according to two major features: (1) average annual temperature and precipitation, and (2) annual variation in temperature and precipitation. The idea is that biomes correspond to a particular set of abiotic factors and that temperature and moisture are the most important of these factors—specifically their average values and their annual variations.

Classification schemes that are based on temperature and precipitation data have identified about 24 distinct climate regions around the world. Here we survey six of the most extensive terrestrial biomes, ranging from the wet tropics to the arctic. **Figure 50.6** shows the global distribution of these biomes, while **Figure 50.7** graphs the average temperature and precipitation of regions where these biomes occur in North America.

Tropical Wet Forests   Tropical wet forests, or rain forests, are found in equatorial regions where temperatures and rain-

fall are high, variation in temperature is low, and enough rainfall is available to support year-round growth. Although plants shed older leaves throughout the year, there is no complete, seasonal loss of leaves as in temperate climates and many other regions. In tropical wet forests, plants are broad-leaved evergreens.

As an example of the tropical wet forest biome, consider Belém, Brazil, which is near the equator. Even in the driest month of the year, November, this region receives over 5 cm

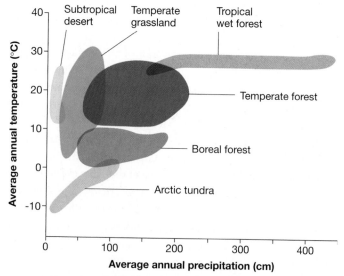

FIGURE 50.7 Biomes Occupy Regions with Distinct Average Annual Temperature and Precipitation
Data showing the distinct combination of temperature and moisture in which North American biomes are found.

(2 in.) of rainfall (**Figure 50.8a**), considerably more than the *annual* rainfall of many deserts. Mean monthly temperatures never drop below 25°C and never exceed 30°C. Tropical wet forests show much less seasonal variation than other biomes do.

As **Figure 50.8b** shows, the favorable year-round growing conditions in tropical wet forests produce riotous growth, leading to extremely high productivity and aboveground biomass. **Productivity** is the total amount of carbon fixed by photosyn-

thesis per unit area per year; **aboveground biomass** is the total mass of living plants, excluding roots. Aboveground biomass reflects **net primary productivity**, which is the total amount of carbon that is fixed per year minus the amount of fixed carbon oxidized during cellular respiration. Fixed carbon that is consumed in cellular respiration provides energy for the organism but is not used for growth—that is, production of biomass.

Tropical wet forests are also renowned for their species diversity. In this biome, researchers routinely find over 200 tree species in a single 10 m × 100 m study plot. Based on counts of the insects and spiders collected from single trees, some biologists contend that the world's tropical wet forests may hold up to 30 million species of arthropods alone.

Further, the diversity of plant sizes and growth forms in wet forest communities produces extraordinary structural diversity. In tropical wet forests, a **canopy** of trees (the uppermost layers of branches) is intermingled with vines, **epiphytes** (plants that grow entirely on other plants), shrubs, and herbs. The diversity of growth forms presents a wide array of habitat types for animals, which in turn pursue an impressive array of feeding strategies. In tropical wet forests, animals that eat fruit, insects, plant tissues, small animals, and dead material all coexist.

Subtropical Deserts Tropical wet forests are found along or near the equator. Because the Hadley cells generated at the equator deliver warm, dry air to areas about 30° latitude north and south of the equator, there is a corresponding band of subtropical deserts in these regions.

To appreciate the climatic regime in subtropical deserts, look at **Figure 50.9a** (page 1152), which provides temperature and precipitation data for Yuma, Arizona, in the Sonoran Desert of southwestern North America. Mean monthly temperatures vary more than in tropical wet climates, but in Yuma temperatures never fall below freezing. (Freezing nighttime temperatures are common elsewhere in the Sonora and in other subtropical deserts.) The most striking feature of the climate of these regions, though, is the low precipitation. The average annual precipitation in Yuma is just 7.5 cm (3 in.).

The scarcity of water in deserts has profound implications. Because conditions are rarely good enough to support photosynthesis, the productivity of desert communities is a tiny fraction of average values for tropical forest communities. Further, as **Figure 50.9b** shows, individual plants are widely spaced—a pattern that may reflect intense competition for water.

Desert species adapt to the extreme temperatures and aridity in one of two ways: They grow at a low rate year-round or break dormancy and grow rapidly in response to any rainfall. Cacti are examples of desert plants that grow year-round. They have small leaves, a waxy coating on the leaves, and the CAM pathway for photosynthesis, introduced in Chapter 10. These adaptations allow cacti to tolerate hot, dry conditions and grow at an extremely low rate throughout the year. Other desert species, though, have traits that allow them to escape drought.

(a) Climate characteristics of tropical wet forest

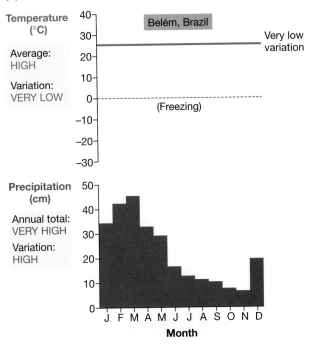

(b) Appearance of tropical wet forest

FIGURE 50.8 Tropical Wet Forest
(a) Graph illustrating the climate of coastal Brazil, typical of a wet tropical climate. Climate graphs have months on the horizontal axis and precipitation and temperature on the vertical axes. The graph shows the average monthly temperature and precipitation for an area.
(b) Tropical wet forests are extremely rich in species.

(a) Climate characteristics of subtropical desert

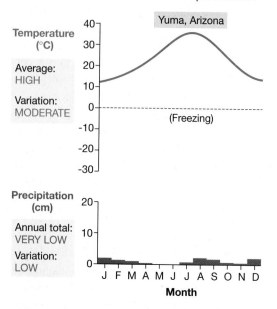

(b) Appearance of subtropical desert

FIGURE 50.9 Subtropical Desert
(a) The climate of Yuma, Arizona, is typical of a subtropical desert.
(b) Saguaro cacti are a prominent feature of the Sonoran Desert in the southwestern United States and northern Mexico.

Some spend much of the year in a dormant state, resuming growth only after rain falls. The shrub called ocotillo, for example, sprouts leaves within days of a rainfall, then drops them a week or two later when soils dry out again. Each ocotillo individual produces and sheds leaves in this way several times a year. Similarly, desert amphibians such as the spadefoot toad spend much of the year dormant in underground burrows. They respond to rain by emerging for a brief time to feed and mate.

In short, deserts provide a stark contrast to tropical forests. Although both biomes feature high temperatures that rarely or never drop below freezing, the absence of water in subtropical deserts leads to low species diversity and productivity.

Temperate Grasslands Figure 50.10a presents climate data for a typical grassland community—in this case, Denver, Colorado. Precipitation in Denver is over four times greater than that in Yuma, Arizona. Nonetheless, conditions are still quite dry; no month has more than 5 cm (2 in.) of precipitation. In comparison with regions to the south and north, temperatures in the area are moderate. Biologists use the term **temperate** to describe regions with temperatures that are moderate relative to the tropics and polar regions, and where summers are typically long and warm and winters short and cold. Grasses are the dominant life-form in temperate regions with relatively low rainfall. Temperate grassland communities

(a) Climate characteristics of temperate grassland

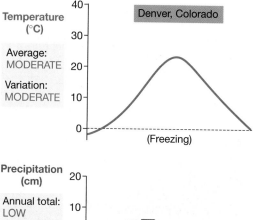

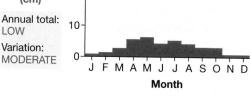

(b) Appearance of temperate grassland

FIGURE 50.10 Temperate Grassland
(a) Denver, Colorado, has a climate typical of a midlatitude steppe.
(b) Grasses are the dominant life-form in prairies and steppes.

are found throughout central North America and the heartland of Eurasia (see Figure 50.6).

Temperature profiles in these regions are also highly seasonal. In Denver, the mean monthly temperature exceeds 20°C in the summer but drops below freezing in the winter. Temperature variation is important because it dictates a well-defined growing season. In the temperate zone, plant growth is possible only in spring, summer, and fall months when moisture and warmth are adequate.

Temperate grasslands, also called **prairies** or **steppes**, usually exist because conditions are too dry to support tree growth but too cold, wet, and seasonal for drought-adapted desert species (**Figure 50.10b**). In temperate areas where rainfall is high enough to support forests, grasslands may develop if recurring fires burn out encroaching trees . Prairie fires, which begin naturally as a result of lightning strikes, were also set intentionally by native people. The plants that dominate prairie communities tolerate fire, because, unlike many plants, they have meristematic tissue, which produces new growth, at the base of their stems. As a result, they resprout quickly after burning.

Although the productivity of temperate grasslands is generally lower than that of forest communities, grassland soils are often highly fertile. The subsurface is packed with roots, which add organic material to the soil as they die and decay. Further, grassland soils retain nutrients, because rainfall is low enough to keep key ions from dissolving and leaching out of the soil. It is no accident, then, that the grasslands of North America and Eurasia are the breadbaskets of those continents. The conditions that give rise to natural grasslands are ideal for growing wheat, corn, and other cultivated grasses.

Temperate Forests In temperate areas with relatively high precipitation, grasslands give way to forests. A typical climate graph for an area near a grassland-forest border is shown in **Figure 50.11a**. Most of the area around Chicago, Illinois, was forested prior to the arrival of European settlers. Temperate forests are found in eastern North America, most of Europe, Chile, and New Zealand (see Figure 50.6).

Temperate forests experience a period in which mean monthly temperatures fall below freezing and plant growth stops. Unlike grassland climates, however, precipitation is moderately high and relatively constant throughout the year. Chicago, for example, has an annual precipitation of 85 cm (34 in.); most months record more than 5 cm (2 in.).

The abundance of moisture allows trees to dominate the landscape (**Figure 50.11b**). Unlike plants in some regions in the tropics, however, plants in temperate forests experience a dormant period. In North America and Europe, temperate forests are dominated by deciduous species, which drop their leaves in the autumn and grow new leaves in the spring. Evergreens, which also enter dormancy each fall, are common as well. In the temperate forests of New Zealand and Chile, however, broad-leaved evergreens predominate.

(a) Climate characteristics of temperate forest

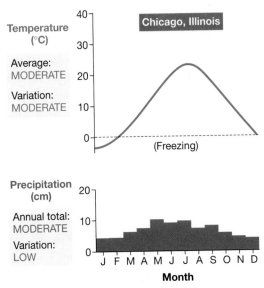

(b) Appearance of temperate forest

FIGURE 50.11 Temperate Forest
(a) Climate graph for Chicago, Illinois, typical of a midlatitude cold and humid climate. **(b)** Temperate forests are dominated by broad-leaved deciduous trees.

Although some of the warmer temperate forests rival or exceed the productivity of tropical forests, most have productivity levels that are lower than those of tropical forests yet higher than those of deserts or grasslands. The level of diversity is also moderate. A temperate forest in southeastern North America may have more than 20 tree species; similar forests to the north may have fewer than 10.

Boreal Forests The boreal forest, or **taiga**, stretches across most of Canada, Alaska, Russia, and northern Europe. Because these regions are just south of the Arctic Circle, they are referred to as subarctic. Their climate is characterized by very cold winters; cool, short summers; and extraordinarily high annual variation in temperature. In the course of a year, subarctic areas may be subject to temperature ranges of more than 70°C.

Figure 50.12a shows a climate graph from Dawson, in the Yukon Territory of Canada. Annual precipitation, at 36 cm (14 in.), is virtually identical to that of the temperate grassland climate of western North America. Regions occupied by boreal forests are so cold, however, that evaporation is minimal; moisture is usually abundant enough to support tree growth as a result. The climate graph also documents the extreme temperature swings typical of these areas. Summer temperatures hover around 10°C, but hot spells in which temperatures exceed 20°C are not uncommon. Winters are cold, with mean monthly temperatures dropping well below freezing for nearly half the year.

As **Figure 50.12b** shows, boreal forests are dominated by highly cold-tolerant conifers, including spruce, pine, fir, and larch trees. With the exception of larch, these species are evergreen. Two hypotheses have been offered to explain why evergreens predominate in cold environments, even though they do not photosynthesize in winter. The first is that they can begin photosynthesizing early in the spring, even before the snow melts, when sunshine is intense enough to warm their needles. The second hypothesis is based on the observation that soils tend to be acidic and contain little available nitrogen. Because leaves are nitrogen rich, species that must produce an entirely new set of leaves each year might be at a disadvantage. To date, however, these hypotheses have not been tested rigorously.

Based on these observations, it is not surprising that the productivity of boreal forests is low. Aboveground biomass is high, however, because slow-growing tree species may be long-lived and gradually accumulate large standing biomass. Boreal forests also have exceptionally low species diversity. The boreal forests of Alaska, for example, typically contain seven or fewer tree species.

Arctic Tundra   Lying poleward from the subarctic is the arctic tundra. This region has climatic conditions similar to those shown in **Figure 50.13a** for Barrow, on the northern coast of Alaska. The growing season is 6–8 weeks long at most; otherwise, temperatures are below freezing. Precipitation is also low. The annual precipitation in Barrow is actually less than that in the Sonoran Desert of southwestern North America. Because of the extremely low evaporation rates, however, arctic soils are saturated year-round.

As **Figure 50.13b** shows, the arctic tundra is treeless. The leading hypothesis to explain the lack of trees is that the growing season is too short and cool to support the production of large amounts of non-photosynthetic tissue. Also, tall plants that poke above the snow in winter might experience substantial damage from wind-driven snow and ice crystals. Woody shrubs, such as willows, birch, and blueberries, are common but rarely exceed the height of a child. Most arctic tundra species hug the ground.

Arctic tundra has low species diversity, low productivity, and low aboveground biomass. Most tundra soils are in the perennially frozen state known as **permafrost**, and the cold temperatures inhibit both the release of nutrients from decaying organic matter and the uptake of nutrients into live roots. Unlike plants in the desert, however, plants in many tundra communities may completely cover the ground. Animal diversity also tends to be low in the arctic tundra, although insect abundance—particularly of biting flies—can be staggeringly high.

(a) Climate characteristics of boreal forest

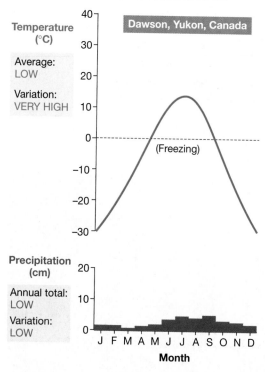

(b) Appearance of boreal forest

FIGURE 50.12 Boreal Forest
(a) Climate graph for Dawson, Yukon Territory, Canada, in the heart of the boreal forest. **(b)** The boreal forest is dominated by needle-leaved evergreens, such as spruce and fir.

(a) Climate characteristics of artic tundra

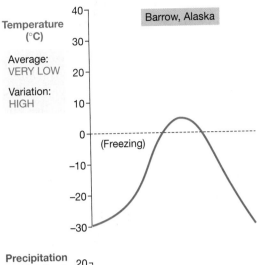
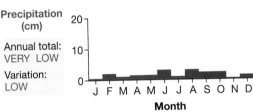

(b) Appearance of artic tundra

FIGURE 50.13 Arctic Tundra
(a) Climate graph for Barrow, Alaska, on the north slope of Alaska.
(b) Arctic tundra is treeless. Very cold-tolerant shrubs and herbaceous plants dominate plant communities.

Aquatic Environments

Different types of aquatic habitats, like different types of terrestrial biomes, are distinguished by abiotic factors. But instead of being defined by different types of moisture and temperature regimes, aquatic environments are distinguished by depth of water and the rate of water movement. Depth of water is particularly important, because it dictates how much light reaches the organisms that live in a particular region. Light, in turn, is

required for photosynthesis and the production of organic matter that other organisms can eat.

The types of wavelengths and the amount of light available to organisms changes dramatically as water depth increases. Water absorbs and scatters light. As **Figure 50.14a** shows, ocean water near the coast specifically removes light in the red and blue regions of the visible spectrum. This is an important observation, because wavelengths in the red and blue regions are required for photosynthesis (see Chapter 10). The total amount of light available to organisms also diminishes rapidly with increasing depth. In pure seawater, the total amount of light available at a depth of 10 m is less than 40 percent of what it is at the surface. Virtually no light reaches depths greater than 40 m in pure seawater (**Figure 50.14b**). In seawater that contains organisms or debris, light penetration is dramatically less than that observed in pure seawater.

(a) Only certain wavelengths of light are available under water.

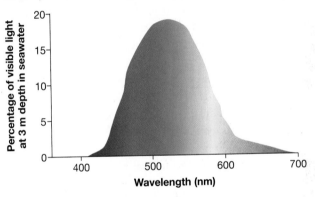

(b) Intensity of light declines with water depth.

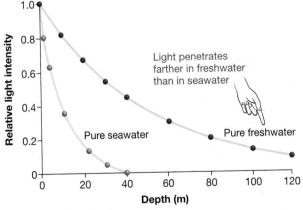

FIGURE 50.14 Availability of Light Changes Dramatically with Increasing Water Depth
(a) Graph showing the wavelengths of light available at a depth of 3 meters in seawater near an ocean coastline. **(b)** Graphs showing how rapidly the amount of light declines with depth of pure freshwater and pure seawater. When water contains organisms and organic debris, light availability declines even faster. EXERCISE
Photosynthesis is driven most efficiently by wavelengths of about 425 nm and 680 nm. Mark and label these wavelengths in part (a).

In addition to water depth, the type and amount of water movement or flow are major influences in aquatic environments. For example, organisms that live in fast-flowing streams have to cope with the physical force of the water, which constantly threatens to move them downstream. Marine organisms that live in intertidal regions are exposed to the air periodically each day as well as to violent wave action during storms.

Different water depths and types of water movement define the array of aquatic environments that are available to organisms, just as different types of climate define the terrestrial biomes. Let's examine the various types of freshwater environments that organisms occupy, and then analyze the array of marine habitats and the challenges that they pose for organisms.

Lakes, Ponds, and Wetlands Bodies of standing freshwater are classified as lakes, ponds, or wetlands. Lakes and ponds are distinguished from each other by size—ponds are small, while lakes are large enough that the water in them can be mixed by wind and wave action—and from wetlands by water depth. **Wetlands** are shallow-water habitats where the soil is saturated with water for at least part of the year. Wetlands are distinct from lakes and ponds for two reasons: They have shallow water and **emergent vegetation**—meaning plants that grow above the surface of the water. Most lakes, ponds, and wetlands form in depressions that were created by the scouring action of glaciers thousands of years ago.

Figure 50.15 diagrams the major regions or zones found within lakes and ponds. The **littoral** ("seashore") **zone** consists of the shallow waters along the shore, where flowering plants are rooted. The littoral zone is shallow enough for sunlight to penetrate and support photosynthesis. As a result, it

hosts abundant **planktonic** organisms, which float or swim near the surface and for the most part are microscopic. In deeper waters, rooted plants are absent and the upper, sunlit portions of the water are dominated by cyanobacteria and unicellular algae. Open water that receives enough light to support photosynthesis is referred to as the **limnetic** ("lake") **zone**. Although too little light penetrates to the bottom of most lakes and ponds to support photosynthesis, the substrate provides habitat for burrowing animals and bacteria and is called the **benthic** ("depths") **zone**. Animals that consume dead organic matter, or **detritus**, are common in the benthic zone, because dead organisms rain down from the littoral and limnetic zones to form sediments. Regions of the littoral, limnetic, and benthic zones that receive sunlight are part of the **photic zone**; regions that do not receive light are termed the **aphotic zone**.

The major message of Figure 50.14 is that water depth is the key abiotic factor in lake and pond environments, primarily because it dictates the extent of light penetration and photosynthesis. Water depth also affects temperature, however. The littoral and limnetic zones are typically much warmer than the benthic zone, simply because they receive so much more solar radiation. **Box 50.1** details how differences in temperature at various depths in a lake, along with changes in the temperature of the surface water throughout the year, promote the mixing of water from different depths. The mixing effect of wind and temperature changes allows both well-oxygenated water from the surface to reach the benthic zone and nutrient-rich water from the benthic zone to enter the littoral and limnetic zones.

Biologists identify three major types of wetlands, as shown in **Figure 50.17**: marshes, swamps and bogs. **Marshes** lack trees and typically exhibit a slow but steady rate of water flow. Marshes are typically connected to lake or stream systems. **Swamps** are similar to marshes but are dominated by trees and shrubs. Because marshes and swamps offer ample supplies of water and sunlight, they are extraordinarily productive habitats. Bogs, in contrast, are remarkably unproductive. **Bogs** develop in depressions where water flow is low or nonexistent. The water in bogs is largely stagnant, so oxygen is used up during the decomposition of dead organic matter faster than it enters via diffusion from the atmosphere. As a result, bog water is oxygen poor or even anoxic (see Chapter 37). Once the oxygen in the water is depleted, decomposition slows. Organic acids and other acids build up, lowering the pH of the water. At low pH, nitrogen becomes unavailable to plants. The combination of acidity, lack of available nitrogen, and anoxic conditions makes bogs a demanding environment. As a consequence, productivity in bogs is low.

Bogs are an exception to the rule that productivity in wetlands is limited by light, but they conform to the rule that the productivity of aquatic ecosystems is limited by nutrients. What features of streams affect productivity?

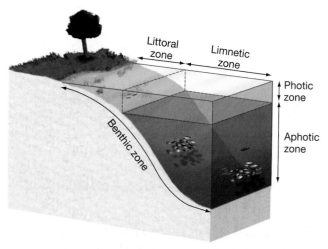

FIGURE 50.15 Lakes Have Distinctive Zones Defined by Water Depth and Distance from Shore
The littoral, limnetic, and benthic zones in a lake present discrete environments to organisms.

BOX 50.1   Thermoclines and Lake Turnover

The northern and temperate regions of the world host many lakes. Each year, these bodies of water undergo remarkable changes known as the spring and fall **turnovers**.

The spring and fall lake turnovers occur in response to changes in air temperature. To see how this happens, examine **Figure 50.16**, which shows the seasonal temperature profiles of a lake. In winter, water at the surface is locked up in ice at a temperature of 0°C. The water just under the ice is slightly warmer, and the water at the bottom of the lake is at 4–5°C. A gradient in tem-

perature such as this is called a **thermocline** ("heat-slope"); thermal stratification is said to occur.

When spring arrives, the ice begins to melt. The temperature of the water rises until it reaches 4°C. This is important, because the density of liquid water is highest at 4°C. The water at the surface of the lake is now heavier than the water below it. As a result, it sinks. The water at the bottom of the lake is displaced and comes to the surface, completing the spring turnover. During the spring turnover, water at the bottom of the lake carries sediments and nutrients from the

benthic zone up to the limnetic zone. This flush of nutrients triggers a rapid increase in the growth of algae and bacteria that biologists call the *spring bloom*.

When temperatures cool in the fall, the water at the surface reaches a temperature of 4°C and sinks, displacing water at the bottom and creating the fall turnover. The fall turnover is important because it brings oxygen-rich water from the surface down to the benthic layer, and because it again brings nutrients from the benthic zone up to the limnetic zone.

FIGURE 50.16 In Temperature Regions, Lakes Turn Over Each Spring and Fall

(a) Marshes have non-woody plants.

(b) Swamps have trees and shrubs.

(c) Bogs are stagnant and acidic.

FIGURE 50.17 Wetland Types Are Distinguished by Water Flow and Vegetation
Water flows through **(a)** marshes and **(b)** swamps, but the water in **(c)** bogs is stagnant.

Streams   Streams are bodies of water that move constantly in one direction. Creeks are small streams, and rivers are large. As **Figure 50.18** shows, the structure of a stream usually varies along its length. Where it originates at a mountain glacier or a spring, a stream tends to be cold, narrow, and fast. As it descends toward a lake or an ocean, a stream accepts water from tributaries and becomes larger, warmer, and slower.

In terms of the environment that is available to organisms, the major physical variables in streams are the speed of the current and the availability of oxygen and nutrients. In small, fast-moving streams, it is rare to find photosynthetic organisms; most of the organic matter present consists of leaves, branches, and other materials that have fallen into the water from outside the stream. Although nutrient levels tend to be low in fast-moving streams, oxygen levels tend to be high because water droplets are exposed to the atmosphere when rapidly moving water splashes over rocks or

FIGURE 50.19 Estuaries Are Highly Productive Environments
Estuaries are productive because they are sunlit and nutrient rich. The salinity of estuary water is intermediate between that of freshwater and that of seawater, so organisms that thrive in estuaries may not do well in either freshwater or marine environments.

other obstacles. Oxygen diffuses into the droplets, contributing to the oxygen-rich quality of fast-moving streams (see Chapter 44).

As creeks and rivers widen and slow down, conditions become more favorable for the growth of algae and plants, and the amount of organic matter and nutrients in the water increases. Without riffles or rapids to oxygenate the water, however, slow-moving streams tend to become relatively oxygen poor. Based on these observations, it is not surprising that a stream often contains completely different types of organisms near its source and near its end, or mouth.

An **estuary** is the environment that forms where rivers meet the ocean (**Figure 50.19**). In these habitats, freshwater mixes with salt water. The salinity of the water in an estuary varies with proximity to the ocean. Salinity and water depth also fluctuate in estuaries in response to tides, storms, and floods. Because the water is shallow and sunlit, and because nutrients are constantly replenished by incoming river water, estuaries are among the most productive environments on Earth.

Marine Environments   As in lakes and ponds, the major physical factor that distinguishes marine environments is water depth. Water depth is a major determining factor in the amount of available light, as Figure 50.14 shows. Like lakes and ponds, oceans have a **photic zone**, which is the area near the surface that receives enough sunlight to support photosynthesis. In addition, oceans have a **benthic zone**, which is the area at the bottom. There may also be an **aphotic zone**, a middle area that is too dark to support photosynthesis (**Figure 50.20**).

Near source, water is fast, cold, nutrient poor, high in O_2

In the middle reaches, water slows, warms, gains nutrients, loses O_2

Near mouth, water is slow, warm, nutrient rich, low in O_2

FIGURE 50.18 Distinctive Environments Appear Along a Stream's Length
The speed, temperature, nutrient level, and oxygen level of stream water usually vary along the stream's length.

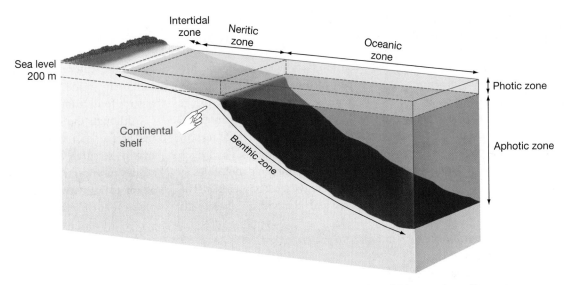

FIGURE 50.20 Oceans Have Distinctive Zones Defined by Water Depth and Distance from Shore
The intertidal, neritic, and oceanic zones present distinctive environments to organisms, as do the photic, aphotic, and benthic zones.

Water depth itself and distance from shore are used to distinguish the three major types of environment found in the ocean: (1) intertidal, (2) neritic, and (3) oceanic zones (Figure 50.20). Within each of these environments are photic and benthic zones and perhaps also an aphotic zone.

The **intertidal** ("between tides") **zone** is along the shore. The intertidal zone consists of a rocky, sandy, or muddy beach that is exposed to the air at low tide but submerged at high tide. Although organisms that live in the intertidal must be able to withstand physical pounding from waves and desiccation at low tide, productivity is high due to the ready availability of nutrients and sunlight. Nutrient levels are relatively high in the intertidal, because estuaries and rivers provide a fairly constant supply and because currents tend to sweep nutrient-laden sediments from offshore areas into the shallower waters near shore.

The **neritic zone** extends from the intertidal zone out toward the open ocean to depths of about 200 m. The edge of the neritic zone is defined by the end of the **continental shelf**—the gently sloping, submerged portion of a continental plate. Beyond the continental shelf, the bottom of the ocean changes from rock of continental origin to the rock of an oceanic plate. Instead of continuing to slope gently, the oceanic plate drops off rapidly in depth. Although it is not as productive as the intertidal zone, the neritic zone is still relatively prolific, because nutrients are available from currents that well up from the deep ocean off the continental shelf. Almost all of the world's major marine fisheries exploit organisms that live in the neritic zone.

In the tropics, shallow portions of the neritic zone may host **coral reefs**. Because the water is warm and sunlight penetrates to the ocean floor in these habitats, coral reefs are among the most productive environments in the world. Recall from Chapter 31 that coral reefs are built by species in the phylum Cnidaria that host symbiotic, photosynthetic dinoflagellates.

In contrast to the neritic zone in general and coral reefs in particular, the **oceanic zone**, or open ocean, ranks as the most unproductive environment on Earth. Although sunlight is abundant in the photic zone of the open ocean, where light penetrates, nutrients are extremely scarce. Iron, which is required for enzymes involved in cellular respiration, is particularly rare. Recall from Chapter 28 that when iron is added to the open ocean experimentally, the number of photosynthetic cells present and the overall productivity of planktonic organisms explode. This zone is nutrient poor simply because there is no mechanism for nutrients to be replenished. When the photosynthetic organisms and the animals that feed on them die, their bodies drift downward out of the photic zone and are lost. In the open ocean, there is no mechanism for bringing nutrients back up from the bottom, as there is in most lakes (see Box 50.1). The aphotic zone of the open ocean is also unproductive, because light is absent—meaning that photosynthesis is impossible. Most organisms present in the aphotic zone survive on the rain of dead bodies from the photic zone. If coral reefs are the rain forests of the ocean, then the oceanic zone is the desert. As **Box 50.2** (page 1160) shows, though, the life-forms that are present include some of the most bizarre organisms known.

The bottom of the ocean comprises the **benthic zone**. Although the benthic zone may be sunlit in the intertidal zone and parts of the neritic zone, it is more usual for light to be completely absent. In most cases, life-forms depend on a

BOX 50.2   Deep-Sea Life

The deep sea has been called the "last frontier" on Earth, with some justification. Using specialized submarines and remote-controlled video cameras that are tethered to research ships, biologists have begun to explore the ocean's coldest, darkest regions, where water is under enormous pressure. The organisms the investigators have found are nothing short of remarkable.

In the vast volume of ocean that lies between the open ocean's photic zone and benthic zone, an array of small fish, shrimp, and other organisms feed on the dead material drifting down from the surface. These organisms are preyed on by small but ferocious-looking predators such as anglerfish and fangtooth (**Figure 50.21a**). Most predators that live in the deep sea have enormous mouths and teeth relative to their overall body size, and many have tissues that emit light—possibly as a lure for unsuspecting prey.

Animals that live at the very bottom of the deep ocean may not look as bizarre as the deep-sea predators, but their adaptations to life in the abyss are just as remarkable. **Figure 50.21b**, for example, shows tubeworms that can be found 2.5 km (1.5 miles) under the surface. Tubeworms, certain clams, and other organisms live near the hot springs called black smokers (see Chapter 2). Bacteria and archaea grow there in profusion, using hydrogen gas, hydrogen sulfide, and other inorganic molecules as a source of energy (see Chapter 27). The tubeworms and clams filter these cells out of the water and digest them or protect them inside their bodies and then absorb food from them.

Although the deep ocean still qualifies as the least well-studied habitat on Earth, research has made promising strides in understanding how organisms adapt to an extraordinarily demanding physical environment.

(a) Predators from the aphotic zone of the deep sea

(b) Animals that live near black smokers

FIGURE 50.21 Deep-Sea Habitats Are Being Explored Intensively
(a) Predatory fish found in the aphotic zone of the open ocean tend to have extremely large mouths and teeth for their size. **(b)** At hot springs within deep-ocean habitats, tubeworms occur with clams and a variety of other animals.

---

rain of organic debris from above, which settles into the benthos and forms sediments. The exceptions to this rule are organisms that live in and around the hot springs called black smokers—organisms that are also featured in Box 50.2.

✓ CHECK YOUR UNDERSTANDING

Global climate patterns are dictated by differential heating of Earth's surface, Hadley cells, and seasonality and are modified at a regional scale by the presence of mountains and oceans. Distinct patterns of temperature and moisture levels and variation are associated with regions of distinct vegetation types, or biomes. Aquatic environments are distinguished by the depth of water and the rate at which it moves. You should be able to explain how variation in the availability of light, moisture, and nutrients causes productivity to vary dramatically among terrestrial biomes and among aquatic environments.

50.4   Biogeography

The diversity of terrestrial and aquatic environments on Earth is impressive. Now the question is, how does the array of abiotic conditions that exist—along with biotic interactions—affect the distributions of organisms? Understanding why organisms are found where they are is one of ecology's most fundamental tasks, and it is a requirement for designing nature preserves in a rational and effective way. The study of how organisms are distributed geographically is called **biogeography**. In Chapter 25 we explored how continental drift has contributed to the current distributions of species; here, let's focus on how interactions with the abiotic and biotic environments affect where a particular species is able to live.

The most fundamental observation about the *range*, or geographic distribution, of species is simple: No one species can survive the full range of environmental conditions present on Earth. *Thermus aquaticus* cells thrive in hot springs with temperatures above 90°C. These archaea can live in this environment because they have enzymes that do not denature at near-boiling temperatures. The same cells would die instantly, however, if they were transplanted to the frigid waters near polar ice. But vast numbers of other archaea are present in cold seawater. These cells have enzymes that can work efficiently at near-freezing temperatures. No organism, however, lives in both hot springs and cold ocean water.

The point here is that organisms tend to be adapted to a limited set of physical conditions. Although some species have much broader geographic ranges than other species do—and humans and the bacteria that live with them may have the widest distribution of any organisms at present—no organism can live everywhere.

To understand a species' distribution fully, however, it is essential to examine historical and biotic factors in addition to the physical conditions present. For example, the temperature and moisture regimes of large areas of western North America are almost identical to those found in the Mediterranean region of Europe and in central Eurasia. Even though the physical conditions at these sites are extremely similar, the species present are almost completely different. Why?

The Role of History

To explain why certain species occur in a particular region while others don't, the first and most obvious factor to consider is history—specifically, the history of dispersal. **Dispersal** refers to the movement of an individual from the place of its birth, hatching, or origin to the location where it lives and breeds as an adult. If a particular species is missing from an area, a physical barrier to dispersal—such as a river, a glacier, a mountain range, or an ocean—may be present. For example, the obvious cause of differences in the plant species found in the grasslands and shrublands of western North America versus those of Eurasia is that the Pacific and Atlantic Oceans are effective barriers against the dispersal of seeds.

In many cases, human activity is circumventing physical barriers to dispersal. For example, public health officials closely monitor annual flu outbreaks in south China because they recognize that new strains disperse all over the planet in a matter of weeks or even days, in the respiratory passages of airplane passengers. Prior to the invention of air travel, disease epidemics in humans were much more likely to be confined to relatively small geographic areas.

Similarly, humans have transported thousands of seeds, birds, insects, and other species across physical barriers to new locations—sometimes purposefully and sometimes by accident. One accidental introduction has had disastrous consequences for the arid shrublands and grasslands of the western United States. In 1889 a native of Eurasia called cheatgrass was accidentally introduced to western North America in a shipment of crop seeds. Cheatgrass is an annual plant with seeds that germinate over winter; plants grow and

FIGURE 50.22 Cheatgrass Has Invaded Extensive Areas of North America
Cheatgrass is an annual plant with seeds that germinate over winter. Plants grow and set seed in early spring.

set seed in early spring. Within 30 years, cheatgrass had taken over tens of thousands of square miles of habitat (Figure 50.22).

If an exotic organism is introduced into a new area, spreads rapidly, and eliminates native species, it is said to be an **invasive species.** In North America alone, dozens of invasive species have had devastating effects on native plants and animals. North American invasives include kudzu, purple loosestrife, reed canary grass, garlic mustard, Russian thistle (also known as tumbleweed), European starlings, African honeybees, and zebra mussels. Similar lists could be compiled for other regions of the world.

Regarding the distribution of organisms, the story of invasive species is simple: They and other introduced species did not exist in certain parts of the world until recently, simply because they had never been dispersed there. But data show that only about 10 percent of the species introduced to an area actually become common, and a smaller percentage than that increases in population enough to be considered invasive. If a species is introduced into an area but does not thrive, what is the cause? Conversely, why are invasives so successful?

Biotic Factors

The distribution of a species is often limited by biotic factors—meaning interactions with other organisms. As an example, recall the data presented in Chapter 25 on the current and former distributions of hermit warblers and Townsend's warblers along the Pacific coast of North America. Both hermit warblers and Townsend's warblers live in evergreen forests, but their ranges do not overlap. Experiments have shown that male Townsend's warblers directly attack male hermit warblers and evict them from breeding territories, and historical data indicate that the geographic range of

(a) Distribution of cattle is limited by distribution of tsetse flies.

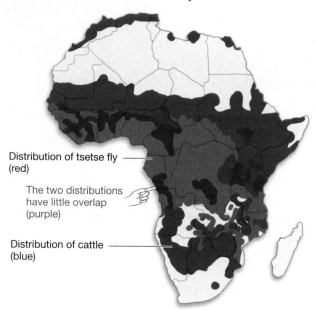

Distribution of tsetse fly
(red)

The two distributions
have little overlap
(purple)

Distribution of cattle
(blue)

(b) Yucca moth distribution is limited by yucca plant distribution.

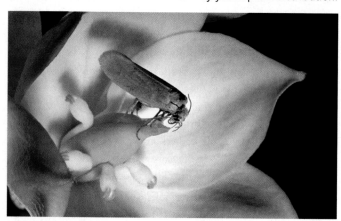

FIGURE 50.23 Distributions Are Limited by Biotic Factors
(a) Because tsetse flies carry a disease that is fatal to cattle, tsetse flies and cattle rarely coexist. **(b)** Because yucca moths lay their eggs only in the flowers of yucca plants, the moths cannot live where yucca plants are absent.

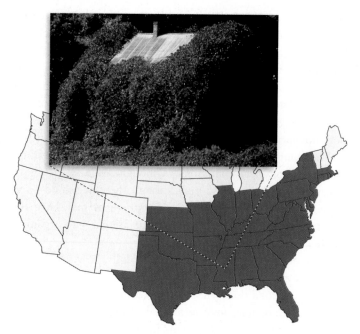

FIGURE 50.24 Distributions Are Limited by Abiotic Factors
Kudzu is an invasive plant that has spread throughout the southeastern United States. This plant's distribution is limited by its inability to grow in soils that freeze to more than a foot in depth.

Townsend's warblers has been expanding steadily at the expense of hermit warblers. The results support the hypothesis that a biotic factor—competition with another species—is limiting the range of hermit warblers.

Competition is not the only biotic factor that can limit a species' distribution, however. In Africa, the range of domestic cattle is limited by the distribution of tsetse flies, because tsetse flies transmit a parasite that causes the disease trypanosomiasis, which is fatal in cattle (**Figure 50.23a**). Similarly, most of the songbirds native to Hawaii are limited to alpine habitats—above the range of the mosquitoes that transmit avian malaria. And the moth pictured in **Figure 50.23b** lays its eggs only in the flowers of yucca plants. As a result, this species does not exist outside the range of yucca plants.

Abiotic Factors

In many cases, introduced species may not become established in a region because of abiotic factors—particularly temperature and moisture. An area may simply be too cold or too dry for an organism to survive. For example, even though the invasive plant kudzu grows prolifically throughout the southeastern United States (**Figure 50.24**), it is killed when the soil freezes to a depth of a foot or more. As a result, it has not expanded its range farther north and west.

It is often difficult to separate the effects of biotic and abiotic factors on a species' range, however. To drive this point home, consider how cheatgrass became established in North America. Cheatgrass does not grow in wet grasslands, because it does not compete well against the tall species that thrive there. But it has been able to invade two important types of biomes in North America: dry temperate grasslands and the arid, shrub-dominated habitats known as **sage-steppe**. In each case, the reasons for the success of cheatgrass were different.

Cheatgrass has been able to invade sage-steppe habitats because it is not affected by fire. Fire is an abiotic factor that can kill the shrubs that dominate sage-steppe (**Figure 50.25**). In fact, because cheatgrass grows in dense beds and dies back in early spring, it actually leads to an increase in the frequency and intensity of wildfires. Normally, fires are rare in sage-

FIGURE 50.25 Sage-Steppe Is an Arid Biome Dominated by Shrubs
Large areas of western North America are covered with sage-steppe vegetation.

steppe biomes, because plants are widely spaced. This distribution usually prevents fires from spreading and affecting a large area. But when this biome begins to dry in late summer, beds of dead cheatgrass offer a continuous, extremely flammable source of fuel. If a fire does come through, it does not affect the cheatgrass for two reasons: (1) Cheatgrass is an annual, so there is no aboveground tissue once the growing season is over, and (2) cheatgrass seeds sprout readily in soils that have been depleted of organic matter by fire. Because sagebrush and other shrubs are perennials, they are killed by hot fires. By increasing the frequency and intensity of fire in sage-steppe biomes, the presence of cheatgrass creates an environment that favors the spread of cheatgrass.

How has this species been able to invade dry grassland biomes? Prior to the arrival of cheatgrass, grasslands in arid areas of the American west were dominated by the types of bunchgrasses shown in **Figure 50.26**. Bunchgrasses do not form sod; instead, they grow in compact bunches. The spaces between the bunches are occupied by what biologists call "black-soil crust." This material is a community of soil-dwelling cyanobacteria, archaea, and other microorganisms that grow when moisture is available. Many of these species also fix nitrogen (see Chapter 37). The nitrogen becomes available to the surrounding plants when the microbial cells die and decompose.

When European settlers arrived and began grazing cattle in this biome, however, both the black soils and bunchgrasses were affected. Cattle grazed on the bunchgrasses until the grasses died, and their hooves compacted and disrupted the black soils. Neither component of the ecosystem was allowed to recover, in part because intensive grazing continued and in

part because cheatgrass arrived. Cheatgrass shades out black-soil crust organisms. When black-soil crust is gone, the amount of nitrogen available to bunchgrasses is reduced. Cheatgrass also competes successfully with bunchgrasses for water and nutrients, because it germinates during the winter and completes its growth early in the spring—before the native bunchgrasses have broken dormancy and begun to grow. And as in sage-steppe habitats, the expansion of cheatgrass increased the frequency and severity of fires, which further degraded the integrity of black-soil crust communities.

To summarize, changes in abiotic and biotic factors allowed cheatgrass to invade native biomes in North America and thrive as an invasive species. Currently, managing cheatgrass is the major problem facing conservation biologists and rangeland managers in the western United States.

Although the story of cheatgrass is particularly dramatic, it is important to remember that the range of every species on Earth is limited by a combination of historical, abiotic, and biotic factors. The range of a particular species depends on dispersal ability, the capacity to survive climatic conditions, and the ability to find food and avoid being eaten. Climate and other abiotic factors were the focus of this chapter; the next several chapters focus on the biotic aspects of ecology. How do interactions among organisms of the same species and interactions among other species affect geographic range and population size?

FIGURE 50.26 Arid Grasslands Feature Bunchgrasses and Black-Soil Crust
As their name implies, bunchgrasses grow in compact bunches instead of forming a sod. The black-soil crust that forms between bunchgrasses is composed of cyanobacteria, archaea, and other microscopic organisms that add organic matter and nutrients to the soil.

ESSAY Battling Invasive Species

Conservation groups now rank invasive species as the second most serious danger to threatened and endangered species in the United States—second only to direct destruction of habitats. Globally, invasive organisms represent one of the gravest threats to native species and the integrity of biomes. Some introduced plants, such as cheatgrass and purple loosestrife—which have taken over extensive areas of marshland in North America—are outcompeting native species. The brown tree snake increased to densities of up to 13,000 individuals per square mile after being introduced to the island of Guam and has virtually wiped out all of the native birds, leading to at least 12 species extinctions. Introduced diseases have dramatically reduced populations of American chestnut and American elm as well as other forest trees. What can be done?

In conservation biology, as in medicine, an ounce of prevention is worth a pound of cure. Although species introductions were once encouraged, most governments now actively inspect imported plants, produce, and other living materials to prevent the accidental introduction of unwanted organisms. Government agencies and private conservation groups have also designed rapid response networks, so that quick action can be taken to eliminate invasive species if they are detected in a new area.

In areas where invasive species are already established, researchers are trying to devise strategies to manage their spread or reduce their numbers. Management strategies to battle invasives range from manually pulling purple loosestrife plants from marshes to lighting early spring fires in stands of cheatgrass, so that these plants will be burned when they are growing quickly and are most vulnerable.

In several cases, conservation biologists are using plant-eating insects or parasites as control agents. The hypothesis here is that invasive organisms are successful in exotic habitats because their natural predators and diseases are absent. By introducing these predators and diseases, researchers hope to control the spread of the invasive species. For example, weevils that feed on the roots or flower buds of purple loosestrife and beetles that eat purple loosestrife leaves were field tested to determine whether they control purple loosestrife effectively and have little to no impact on native vegetation. The results were promising enough that larger-scale cultivation and release of these insects are now being implemented. Similarly, researchers are actively testing the ability of a parasitic fungus to infect and control cheatgrass.

Globally, invasive organisms represent one of the gravest threats to native species. . . .

Although promising strides have been made in the management of invasive species recently, continued and expanded management and research programs are critical. As transcontinental travel and commerce continue to grow, the threat of accidental or malicious introductions continues. Keeping invasive species at bay—and native biomes intact—demands constant vigilance.

CHAPTER REVIEW

Summary of Key Concepts

■ Ecology focuses on how organisms interact with their environment. Because its goal is to understand the distribution and abundance of organisms, ecology provides a scientific foundation for the conservation of species and natural areas.

Biologists study ecology at four main levels: (1) organisms, (2) populations, (3) communities, and (4) ecosystems. The goal of organismal biology is to understand how individuals respond to stimuli from the environment, including members of their own species and other species. Population ecology focuses on how and why populations grow or decline. Community ecology is the study of how two or more species interact, and ecosystem ecology analyzes interactions between communities of organisms and their abiotic environment—particularly the flow of energy and nutrients. In efforts to preserve endangered species and communities, conservation biologists apply analytical methods and results from all four levels.

■ An organism's environment consists of both abiotic, or physical, factors and biotic factors, meaning other organisms.

Abiotic factors that affect organisms include the chemistry of water in aquatic habitats, the nature of soils in terrestrial habitats, and climate, which is a combination of temperature, moisture, sunlight, and wind. Biotic factors include interactions with offspring, mates, competitors, predators, prey, and parasites.

Climate varies around the globe because sunlight is distributed asymmetrically, with equatorial regions receiving more solar radiation than do regions toward the poles on average. Hadley cells create bands of wet and dry habitats, and the tilt of Earth's axis causes seasonality in the amount of sunlight that non-equatorial regions receive. The presence of mountains can create local areas of wet or dry habitats, and proximity to an ocean moderates temperatures in nearby terrestrial habitats.

Web Tutorial 50.1 Tropical Atmospheric Circulation

On land, environments are defined by climate—specifically, both the average value and annual variation in temperature and in moisture. In aquatic habitats, environments are defined by physical structure—particularly water depth.

Distinct climatic regimes are associated with distinct types of terrestrial vegetation, or biomes. Biomes represent the major types of environments available to terrestrial organisms. Because photosynthesis is most efficient when temperatures are warm and water supplies are ample, the productivity and degree of seasonality in biomes varies with temperature and moisture.

Water depth is a crucial factor in aquatic ecosystems because the wavelengths of sunlight that are required for photosynthesis do not penetrate water efficiently. Thus, only certain zones in a lake or ocean can support photosynthesis, and organisms in other aquatic zones must depend on food that rains out of the sunlit areas near the surface.

A species' distribution is constrained by a combination of historical, abiotic, and biotic factors.

The range, or geographic distribution, of a particular species depends on dispersal ability, the capacity to survive climatic conditions, and the ability to find food and avoid being eaten. For each species, a unique combination of historical, abiotic factors, and biotic factors dictates where individuals live and the size of the population.

Questions

Content Review

1. What is a rain shadow?
   a. the part of a mountain that receives prevailing winds and heavy rain
   b. the region beyond a mountain range that receives dry air masses
   c. the region along the equator where precipitation is abundant
   d. the region near 30°N and 30°S latitude that receives hot, dry air masses

2. A region receives less than 5 cm (2 in.) of precipitation annually and has temperatures that never drop below freezing. Which type of biome is present?
   a. subtropical desert   b. temperate grassland
   c. boreal forest   d. arctic tundra

3. What is the predominant type of vegetation in a tropical wet forest?
   a. shrubs and bunchgrasses
   b. herbs, grasses, and vines
   c. broad-leaved deciduous trees
   d. broad-leaved evergreen trees

4. The littoral zone of a lake is most similar to which of the following environments?
   a. oceanic zone   b. intertidal zone
   c. neritic zone   d. marsh

5. Typically, where are oxygen levels highest and nutrient levels lowest in a stream?
   a. near its source
   b. near its mouth, or end
   c. where it flows through a swamp or marsh
   d. where it forms an estuary

6. Which of the following is most important in limiting the distribution of an invasive species?
   a. historical factors
   b. abiotic factors
   c. biotic factors
   d. historical, abiotic, and biotic factors are equally important for invasive species

Conceptual Review

1. Name the four main levels of study in ecology. Write a question about the ecology of humans, salmon, or oak trees that is relevant to each level.

2. In June, does the Northern Hemisphere receive more or less solar radiation than the equator does? Explain your answer.

3. Diagram the Hadley cell that exists between the equator and 30° latitude north and south, indicating the direction of air flow. Similar air circulation patterns exist between 30° and 60° and between 60° and the poles. Diagram these cells as well.

4. Contrast the productivities of the intertidal, neritic, and oceanic zones of marine environments. Explain why large differences in productivity exist.

5. Compare and contrast the biomes found at increasing elevation on a mountain with the biomes found at increasing latitude in Figure 50.6.

6. Explain why productivity is much lower in bogs than it is in marshes or lakes that receive the same amount of solar radiation.

Group Discussion Problems

1. Mars has an even more pronounced tilt than Earth does. Explain why Mars does or does not have seasons.

2. Mountaintops are closer to the Sun than the surrounding area is. Why are mountaintops cold?

3. The southwest corners of some Caribbean islands are extremely dry, even though other areas of the islands receive substantial rainfall. Explain this observation. Base your answer on a hypothesis (one that you propose) for the direction of the prevailing winds in this region of the world.

4. Edinburgh, Scotland, and Moscow, Russia, are at the same latitude, yet these two cities have very different climates—Moscow has much colder winters. Explain why.

Answers to Multiple-Choice Questions **1.** b; **2.** a; **3.** d; **4.** b; **5.** a; **6.** c

51

Behavior

KEY CONCEPTS

▫ After describing a behavior, biologists seek to explain both its proximate and ultimate causes—meaning, how it happens at the genetic and physiological levels and how it affects the individual's fitness.

▫ In a single species, behavior may range from highly stereotyped, invariable responses to highly flexible, conditional responses and from unlearned to learned responses.

▫ The types of learning that individuals do, the way that they communicate, and the way that they orient and navigate correlate closely with their habitat and with the challenges they face in trying to survive and reproduce.

▫ When individuals behave altruistically, they are usually helping close relatives or individuals that help them in return.

Peregrine falcons hunt ducks and other large birds. When diving in pursuit of prey, peregrines can reach speeds of up to 320 km/hr (200 mph).

To biologists, **behavior** is action—specifically, the response to a stimulus. Pond-dwelling bacteria swim toward drops of blood that fall into the water. Time-lapse movies of sunflowers growing in a field document the steady movement of their flowering heads throughout the day, as they turn to face the shifting Sun. Filaments of the fungus *Arthrobotrys* form loops, then release a molecule that attracts roundworms. When a roundworm touches one of these loops, the fungal filaments swell—ensnaring the worm. *Arthrobotrys* cells then proceed to grow into the worm's body and digest it.

Among animals, action may be frequent and even spectacular. Peregrine falcons dive at breathtaking speeds in pursuit of ducks or other flying prey. Some Arctic terns migrate over

32,000 km (20,000 miles) each year. Deep inside a hive, in complete darkness, honeybees can communicate the position and nature of a food source a mile away.

Biologists who study behavior ask questions about what organisms do in response to short-term changes in the environment, how they do it, and why. Although all organisms respond in some way to signals from their environment, most behavioral research is performed on vertebrates, arthropods, or mollusks. With their sophisticated nervous systems and skeletal-muscular systems, these animals can sense, process, and respond rapidly to a wide array of environmental stimuli.

Although most behavioral studies start by describing what animals do in response to a particular stimulus, the eventual goal is to explain the proximate and ultimate causes of behavior. **Proximate**, or mechanistic, **causation** explains *how* actions occur in terms of the neurological, hormonal, and skeletal-muscular mechanisms involved. **Ultimate**, or evolutionary, **causation** explains *why* actions occur—based on their evolutionary consequences and history. Is a particular behavior adaptive, meaning that it increases an individual's fitness? If so, how does it help individuals produce more offspring in a particular environment?

To illustrate the proximate and ultimate levels of causation, consider the spiny lobster shown in **Figure 51.1**. These animals spend the day hiding in cracks or holes in coral reefs. At night they emerge and wander away from the reef in search of clams, mussels, crabs, or other sources of food. Before dawn, they return to one of several dens they use on a regular basis. How do they find their way back? This question addresses how an individual does what it does—representing the proximate level of causation. Although research is continuing, recent experiments have provided strong support for the hypothesis that spiny lobsters can detect Earth's magnetic field and that they navigate by using information on how the orientation and strength of the field changes as they move about. But at the ultimate level of causation, biologists want to know why the behavior occurs. The leading hypothesis to explain homing behavior is that spiny lobsters with the ability to navigate home can search for food over a wide area under cover of darkness, then return to a safe refuge before sharks and other large predators that hunt by sight can find them.

It is important to recognize that analyzing behavior at the proximate and ultimate levels is complementary. To understand what an organism is doing, biologists want to know both how the behavior happens and why. Accomplishing this task requires them to draw on techniques and results from genetics, physiology, and evolutionary biology.

Because behavioral biologists work at the interface of so many fields, behavior is among the most dynamic of all disciplines in biological science. Even so, behavioral biology is essentially ecological in nature. Ecology is the study of how organisms interact with their physical and biological environments, and behavioral biology is the study of how organisms respond to particular stimuli from those environments. What does an *Anolis* lizard do when it gets too hot? How should that same individual respond when a member of the same species approaches it? Or when a house cat approaches?

The goals of this chapter are to explore some key topics in behavioral biology and illustrate the spirit of the field. To begin, let's consider one way to categorize and organize the enormous variety of behavioral responses that is observed in animals.

51.1 Types of Behavior—An Overview

Some types of behavior are performed nearly the same way every time. When people yawn, the action is so similar that every instance takes about 6 seconds. Other types of behavior are highly flexible. When people smile, the action is variable. The same person may smile in dozens of ways, and different people tend to have their own individual style and frequency of smiling. In addition, some types of behavior are readily modified by learning, while other types of behavior are not. Formally, **learning** is defined as a change in behavior that results from a specific experience in the life of an individual. Human infants learn their native language through imitation and trial and error, but they begin to yawn and smile in a normal way before they have a chance to learn these actions from their parents.

Figure 51.2 (page 1168), offers one way of summarizing these general observations about behavior. Notice that the graph has two axes. The horizontal axis maps how flexible a behavior is in terms of its expression within and between individuals of the same species. This axis forms a continuum ranging from fixed, invariant, or highly stereotyped behaviors to highly flexible and variable responses. The vertical axis plots how much learning is involved in the nature of the response. This axis ranges from behaviors that require little or no learning for normal expression to responses that are highly dependent on learning. In humans, yawning would map in the upper left-hand corner of this graph while language acquisition would be plotted in the lower right-hand corner.

When researchers first began to explore animal behavior experimentally, they focused on responses that are highly stereotyped and that are largely unlearned. This general research

FIGURE 51.1 Spiny Lobsters Can Find Their Way Back to Their Home Burrow
Spiny lobsters wander widely in search of food at night, then return to their hiding places in coral reefs before dawn.

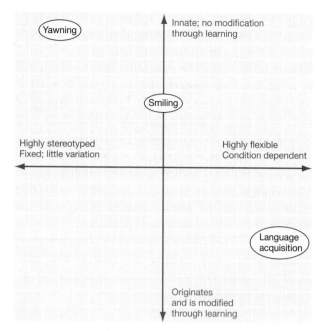

Yawning

Innate; no modification through learning

Smiling

Highly stereotyped
Fixed; little variation

Highly flexible
Condition dependent

Language acquisition

Originates
and is modified
through learning

FIGURE 51.2 Behavioral Traits Vary in Flexibility and Dependence on Learning
These axes represent one way to organize the diversity of behavioral traits observed within and between species. Each type of behavior observed in a species could be plotted at a particular point on the graph.

strategy is common in biological science. When biologists start to analyze a question—ranging from how DNA is transcribed to how plasma membranes work—they start with the simplest possible system. Once simple systems are understood, investigators can delve into more complex situations and questions. Behavioral biology started the same way. Early work focused

on behavioral responses that map to the upper left-hand corner of the graph in Figure 51.2. Let's look at some conclusions from these pioneering studies of stereotyped responses, then consider examples of work on more flexible types of behavior.

Fixed Action Patterns

Highly stereotypical behavior patterns are called fixed action patterns, or FAPs. FAPs have three key characteristics:

1. There is almost no variation in how they are performed. For example, every "headthrow" display given by a male common goldeneye duck involves moving the head back to the tail and then swinging it forward (**Figure 51.3a**). The sequence of movements is so stereotyped that the display takes almost exactly 1.29 seconds every time.

2. They are species specific. Each species of spider builds a distinct type of web (**Figure 51.3b**). Even if a spider never sees a web built by a member of its own species, it still builds a web of its species-specific type.

3. Once the sequence of actions begins, it typically continues to completion. If you put an egg outside a goose's nest, the goose will extend its neck until the egg is just behind its bill. Then the goose retrieves the egg with a side-to-side movement of the head (**Figure 51.3c**). If researchers attach a string to an egg and pull it away after the goose has begun to retrieve it, the goose completes the retrieval movement even without the egg there.

FAPs are examples of what biologists call **innate behavior**—types of behavior that are inherited and show little variation based on learning or the individual's condition. (**Box 51.1** explores the role of genetics in behavior in more detail.) FAPs are

FIGURE 51.3 A Fixed Action Pattern Is a Stereotyped, Inflexible Sequence of Movements
(a) The headthrow display of the male common goldeneye. (b) Spiders of the same species make the same type of web. (c) Geese will continue the motion of retrieving eggs even if the egg has been pulled away.

BOX 51.1   A Closer Look at Behavior Genetics

Biologists explore the genetic bases of behavior to understand the proximate mechanisms responsible for specific behavioral traits. Are certain alleles associated with certain types of behavior? If so, how do the products of these alleles change what an individual does in response to certain stimuli?

As an introduction to the field of behavior genetics, let's consider two strategies that biologists use to explore the genetic basis of behavior. The first involves efforts to identify single genes associated with particular behavioral traits; the second focuses on the more common situation in which many genes influence a behavioral response.

The *foraging* (*for*) gene in fruit flies, discovered by Marla Sokolowski, is a particularly good example of a single gene that influences behavior. As an undergraduate research assistant, Sokolowski noticed that some of the fruit-fly larvae she was studying tended to move after feeding at a particular location, while others tended to stay put. By rearing these "rovers" and "sitters" to adulthood, breeding them, and studying the offspring, she was able to determine that the trait is inherited and that the gene responsible is located at the same location as a previously identified gene called *dg2*. (Recall from Chapter 20 that the entire fruit-fly genome has been sequenced; Chapter 19 introduced the techniques that Sokolowski used to map the *for* gene.) The protein product of *dg2* is involved in the types of signal transduction cascades introduced in Chapters 8 and 47—meaning it is involved in the response to a signal. Rovers and sitters behave differently because they have different alleles of the *for/dg2* gene.

Follow-up experiments in the lab showed that the rover allele is favored when population density is high but that the sitter allele reaches high frequency in low-density populations. These results are considered exciting because they link a proximate mechanism with an ultimate outcome—in this case, the presence of certain alleles with a difference in fitness in specific types of habitats. Sokolowski's current goals are to find out which signaling pathway uses the *for/dg2* gene product and why different alleles lead to different types of feeding behavior.

Although the *for/dg2* gene presents a compelling story, biologists recognize that it is rare for behavioral traits to be so heavily affected by a single gene product. In reality, most aspects of behavior are influenced by the action of many genes. (Think of the number of genes that are likely to be involved in the formation of the human brain and thus in aspects of human personality and decision making.) To use a term that was introduced in Chapter 13, most behavioral traits are *polygenic*. Instead of representing discrete states such as roving and sitting, polygenic traits exhibit quantitative variation—meaning that individuals differ by degree and that many different phenotypes are observed when the population is considered as a whole.

Unfortunately, it is extremely difficult to identify the specific genes and alleles responsible for polygenic traits. To organize their thinking about the genetics of polygenic traits, biologists measure how much the trait varies among individuals in a population and then recognize that every trait in an individual results from the alleles it has and the environment that it has experienced. More formally, biologists distinguish **phenotypic variation** (V_P), **genetic variation** (V_G), and **environmental variation** (V_E). These three quantities have a simple relationship: $V_P = V_G + V_E$. This equation says that the observed variation of a behavioral trait within a population is the sum of variation among individuals in the alleles they possess and the variation in the environments that they have experienced.

This simple formulation helps explain why some types of behavior show little variation among individuals while other types show a great deal of variation. For example, traits show little variability among individuals—meaning that V_P is small—because there is low genetic variation (V_G is small) and/or little environmental influence (V_E small). At the other end of a continuum of behavior, some traits are highly variable among individuals in a population, meaning that V_P is large. In traits such as these, there is a high degree of genetic variation (V_G is large) and/or a great deal of environmental influence on the trait (V_E is large).

The equation also gives biologists a key tool for exploring behavior genetics, even when the specific genes involved are not known. For example, by rearing a large group of individuals in an identical environment (so that V_e is low or zero) and then measuring variation in a particular behavioral trait, researchers can estimate the proportion of the observed variation in the population that is due to genetic differences among individuals. If most phenotypic variation is due to genetic variation, then the trait is likely to respond quickly to natural selection.

Finally, advances in behavior genetics promise to accelerate quickly due to the availability of genome sequences from an array of animals. For example, recent work on the genomes of dogs and chimpanzees is motivated in part by the goal of identifying at least some of the genes involved in complex, polygenic behavioral traits such as aggression and learning. Behavior genetics is an exciting frontier in behavioral biology.

analogous to the types of muscular reflexes introduced in Chapter 46. Recall that if a person's knee is struck with a rubber hammer, the lower leg jerks forward. Like a FAP, a muscular reflex is similar each time it occurs and goes to completion once it begins.

One of the first conclusions made about these simple, highly stereotyped behaviors is that they are responses to very simple stimuli. The stimuli that evoke FAPs are called **releasers** or **sign stimuli**. The presence of a female common goldeneye can "release" the headthrow display of the male common goldeneye. Seeing an egg outside the nest releases the egg-retrieval FAP in geese.

Some of the most classic experiments in the history of behavioral research have focused on identifying releasers. As an example of this work, consider a study that David Lack did on territorial behavior in European robins. Breeding pairs of male and female European robins establish a **territory**—an area of habitat that they defend against other members of their own species. If another robin enters the territory, the owners call, chase the intruder, or even assault it. To explore how this territory defense behavior works, Lack placed a taxidermic mount, or "stuffed" model, of a robin on an established territory. Both territory owners attacked the "intruder." This observation showed that the territorial defense response can be released by the color or shape of an intruder, in the absence of movement or vocalizations.

Lack repeated the experiment to make sure that the same response occurred in other breeding pairs. As he did, he got lucky. One particularly feisty female knocked the head off the stuffed specimen. To Lack's surprise, her attacks continued unabated. Eventually the unfortunate mount was reduced to a puff of feathers. Lack noticed that as long as the mount retained the orange breast feathers found on European robins, the attacks continued.

Based on these observations, Lack proposed a hypothesis: Territorial defense in European robins is a FAP that is released by the sight of orange feathers (**Figure 51.4**). To test the null hypothesis that orange feathers have nothing to do with the territorial response, Lack took a normal mounted specimen of a European robin and painted the orange breast feathers brown, to match the individual's back. When he presented this altered mount to breeding pairs—at the same time of day and year and under the same conditions as he had when presenting a normal mount—the territory owners barely responded. This result supported his hypothesis that orange feathers are required to release territory defense. The result also supports a more general point that releasers consist of small amounts of simple information.

What Is the Adaptive Significance of FAPs?   Geese that retrieve an egg that isn't there and robins that attack puffs of orange feathers appear to be acting in a nonadaptive way. How can alleles that contribute to behavior like this help individuals

Question: What characteristic of a robin intruder releases territorial behavior in breeding European robins?

Hypothesis: Orange feathers act as a releaser for territorial behavior in European robins.

Null hypothesis: Orange feathers have no effect on territorial behavior in European robins.

Experimental setup:

In territory of a breeding pair of European robins, place a full robin mount with normally orange breast feathers painted brown.

In territory of a breeding pair of European robins, place a normal robin mount.

Prediction: The bird mount lacking orange feathers will be ignored. The bird mount with orange feathers will be attacked.

Prediction of null hypothesis: Orange feathers will be ignored.

Results:

Mount that lacks orange feathers is not attacked.

Mount with orange feathers is attacked.

Conclusion: Orange feathers release territorial behavior in European robins.

FIGURE 51.4 Experimental Evidence of a Releaser
FAPs are triggered by simple stimuli. A stimulus is considered a releaser if it alone can trigger a FAP. **EXERCISE** Design an experiment to test the hypothesis that observing someone yawn acts as a releaser for yawning in humans.

produce more offspring and thus increase in frequency in a population? The answer has two parts.

First, natural selection cannot produce adaptations to unusual or "trick" situations, such as an egg being pulled away from a goose or a robin being presented with a puff of orange feathers. Second, even though the sign stimuli that release FAPs are extremely simple, they are exceptionally important for the individual's fitness—its ability to survive and produce offspring. To drive this point home, consider what happens in a darkened environment when a kangaroo rat hears the sound of a rattlesnake's rattle or when you hear a piercing scream nearby. In both cases, a simple sign stimulus releases a FAP—a rapid jumping movement backward, away from the direction of the stimulus. In both cases, the FAP may result in survival. In this light, it is logical to observe that FAPs show little variation among individuals and continue to completion once they have started. Yet the "jump-back" FAP appears nonadaptive if the stimulus is actually a tape recording of a rattlesnake or a friend who is hiding behind a door.

Conditional Strategies

Research on FAPs and releasers dominated behavioral biology through the 1950s. But today investigators focus more on behavioral traits that are triggered by complex information and that are variable and flexible in their expression. As an example of flexible behavior, let's examine a particularly startling phenomenon: female fish that change their sex and become male.

Many species of coral-reef fish have a distinctive mating system (**Figure 51.5**). Males defend territories that contain nesting sites and feeding areas. A group of females lives inside the boundaries of this territory. When these females lay eggs, the male fertilizes the eggs. Thus, a single male monopolizes all the matings in that territory. Invariably, this male is the largest individual in the group. This is logical because the male guards the territory, and because fights between fish are usually won by the biggest contestant.

When the territory-holding male dies, however, something unusual happens. The largest female in the group changes sex. Her reproductive organs change from egg production to sperm production. She becomes the dominant male and begins fertilizing all of the eggs laid in the territory.

Why does she do this? What does the sex-changing female gain by making this switch? These questions are at the ultimate level of explanation and have been addressed by an idea called the size-advantage hypothesis. This hypothesis states that if a group of fish are living in a territory dominated by a single male, females should switch from female to male when they become very large. The hypothesis is based on the observation that fish have *indeterminate growth*—meaning they continue growing throughout their lives.

To understand the logic behind the size-advantage hypothesis, suppose that a small female fish can lay 10 eggs a year but a large female of the same species can lay 20 eggs a year. If six small females and two large females live in a harem, the male that owns the territory fertilizes 100 eggs each year. If the male dies, the largest female can increase the number of offspring she produces each year from 20 to 80 by changing sex and taking over the role of dominant male. Alleles that allow females to do this will increase rapidly in the population. There is no fitness advantage for smaller females to switch sex, though, because they would be defeated in fights and have 0 offspring per year instead of 10.

Biologists refer to this type of behavior as a conditional strategy. In contrast to a FAP, a **conditional strategy** is a behavioral response that depends on conditions—in this case, on the size and sex of other individuals in the same social group. The name is apt because the individuals involved are capable of behaving in several ways. It's important to note, though, that the word *strategy* does not denote any sort of conscious choice. The critical point is that individuals are observed to adopt the behavior that allows them to produce the most surviving offspring.

What other types of factors lead to flexible behavior that varies among individuals in a population? In many species, learning is crucial. Chapter 45 indicated that, at the proximate level, learning occurs because individual neurons and connections between neurons are modified in response to experience. For example, recall that in the sea slug *Aplysia*, the production of the neurotransmitter serotonin changes in response to experience. Changes in serotonin release lead to changes in the behavior of sea slugs in response to attacks by predators. Learning is usually based on changes in neurons.

Learning is a fundamentally important phenomenon in animal behavior. Let's take a closer look at learning from both proximate and ultimate perspectives. How does learning occur, and why?

FIGURE 51.5 Many Species of Coral-Reef Fish Can Change Sex, Depending on Conditions
Bluehead wrasses: a group of females (yellow) surround a male (with the blue head). Bluehead wrasses start life as females but can change sex and become male.

✓CHECK YOUR UNDERSTANDING

Within a given species, behavioral traits may range from highly stereotyped to highly flexible and from innate to learned. Fixed action patterns (FAPs) are highly stereotyped, innate responses released by simple stimuli. In contrast, sex change in fish is an example of a highly flexible response stimulated by complex environmental changes. You should be able to (1) think of an example of a FAP and a condition-dependent behavior in a species that you are familiar with and (2) pose hypotheses to explain why each type of behavior is adaptive.

51.2 Learning

Learning occurs when behavior changes in response to specific life experiences. Learning is particularly important in species that have large brains and a lifestyle that is dominated by complex social interactions—including humans, dolphins, chimps, and crows. In species such as these, FAPs and other types of inflexible, stereotyped behaviors are relatively rare. Instead, each individual is capable of a wide range of behavior. What an individual actually does is condition dependent or learned.

Research on learning is both extensive and fascinating. Here it is possible to touch on only a few highlights, to provide examples of what is known about learning and how biologists go about studying the phenomenon.

Simple Types of Learning: Classical Conditioning and Imprinting

Learning can be quite simple. In response to the stimulus of food, for example, dogs respond by salivating. Ivan Pavlov discovered that if he rang a bell each time food was presented to a dog, the dog would learn to associate the sound of the bell with food. When stimulated by a bell in the absence of food, the dogs Pavlov was training would begin to salivate. This type of learning is called classical conditioning. In **classical conditioning**, individuals are trained by experience to give the same response to more than one stimulus—even a stimulus that has nothing to do with the normal response.

Another type of simple learning takes place in newly hatched ducks, geese, and certain other species of birds. Upon hatching, ducklings and goslings adopt as their mother the first moving thing they see. This type of learning is called **imprinting**. In nature, the "thing" that creates the imprint is the mother duck or goose. But Konrad Lorenz found that if he incubated eggs artificially until they hatched, he could be the first thing that the offspring saw. In response, young greylag geese would imprint on whatever boots he was wearing at the time (**Figure 51.6a**). In contrast, mallard ducklings would imprint on him only if he crawled on all fours and quacked continuously.

Why does such a highly constrained type of learning exist, in response to such simple stimuli? At the ultimate level of causation, the leading hypothesis to explain imprinting is that offspring must quickly learn to recognize and respond to their mother in order to survive. This hypothesis is particularly plausible in species such as ducks and geese, which nest on the ground. Ducklings and goslings hatch synchronously—meaning all of the eggs hatch within a few hours of each other—and are relatively easy for foxes, raccoons, and other predators to find. As a result, mother ducks and geese lead the offspring away from the nest to the safety of nearby water almost immediately. In these bird species, the ability to recognize a moving object and follow it is thought to increase the fitness of offspring.

Lorenz was also able to establish that imprinting occurs only in the early life of the animal, during a short period called the **critical period**, or **sensitive period**. In addition, his research showed that imprinting lasts for life and may establish not only the identity of the offspring's mother but also its species identity. As a result, imprinting has caused problems for conservation biologists who are attempting to raise certain species of endangered birds in captivity. For example, whooping cranes that have been raised by humans not only follow them as chicks but also respond exclusively to humans as sexually mature adults. To prevent captive-reared birds from imprinting on humans, biologists are now careful to raise offspring in the absence of visual or auditory contact with humans and instead to provide recorded sounds of a natural mother in the presence of birdlike puppets or other types of appropriate visual stimulation (**Figure 51.6b**). Offspring raised in this manner imprint normally, behave normally as both chicks and adults, and can be released into the wild.

The key characteristics of imprinting—that it is fast and irreversible and occurs during a critical period—are not typical of most types of learning. Yet imprinting may be a more general type of learning than was originally thought. For example, recent research has shown that the ability of humans to learn a particular language shares at least some of the characteristics of imprinting. You may have personal experience with how quickly children learn a foreign language relative to adults. This observation suggests that there might be something like a critical period for language acquisition. The commonplace observation that babies learn the language of their parents is also important. The babbling noises that babies make are thought to be imitations of words and sounds from their parents. Children who are born deaf cannot hear these sounds. Learning a spoken language may be almost impossible for these individuals, even with intensive training later in life. In contrast, people who become deaf as adults experience relatively slight changes in their speech. These data suggest that auditory feedback is important during the learning process and that language acquisition should shift to the left in Figure 51.2. Let's explore the topic of language learning further by considering research on how birds learn to sing.

(a) Goslings imprint on the first moving object they see.

(b) With care, normal imprinting can occur in captive-reared birds.

FIGURE 51.6 Imprinting Occurs in Some Species of Birds and Lasts for Life
(a) Konrad Lorenz discovered imprinting when captive-reared geese imprinted on him. **(b)** Endangered species raised in captivity are reared with puppets to avoid imprinting on people.

More Complex Types of Learning: Birdsong

Birds sing to attract mates and to mark territories where they find food and raise offspring. In most cases, the songs that a songbird sings are unique to its species. Do young birds learn their species-specific song or suite of songs from their parents, or are songs innate? To answer these questions, biologists perform experiments based on controlling the sound stimuli that birds hear over the course of their lives. The experimental protocols involve taking eggs from the nest, hatching them, and rearing individuals in the laboratory in a controlled auditory environment.

The results of these experiments show that, depending on the species, song-learning behavior falls at various locations on the learning continuum in Figure 51.2. For example, if young chickens or phoebes are raised in isolation from other members of their species and never hear their species-specific song, they still sing a normal song as adults. Song-learning behavior is innate in certain species and may be highly stereotyped.

In contrast, white-crowned sparrows that are raised in isolation do not sing a normal song unless they hear a tape recording of their species-specific song during the early months of life. In this species, singing is learned during a critical period. Yet if researchers play the song of a song sparrow or other closely related species during the critical period for language acquisition, white-crowned sparrow chicks do not learn it. Thus, in this species the ability to learn songs is restricted to songs of their own species (**Figure 51.7**).

In addition, the critical period for song learning in white-crowned sparrows is augmented by practice that occurs when the individual is older. White-crowned sparrows don't begin to sing until they are almost one year old, during the spring. Beginning singers produce a disorganized warbling called *sub-song*.

(a) Young bird reared in silence

(b) Young bird reared with white-crowned sparrow song

(c) Young bird reared with song sparrow song

No song

Normal song

Abnormal song

FIGURE 51.7 White-Crowned Sparrows Have a Critical Period for Song Learning Early in Life
To sing a normal song, white-crowned sparrows have to hear the song of their own species while they are chicks. They also must hear themselves practice their own song when they begin singing at age 1 year.

With practice, this sub-song becomes more and more organized and progressively closer to adult song. Eventually an adult song develops and is "crystallized," meaning that it remains unchanged for the rest of the individual's life. For this improvement to occur, though, the individual must hear itself sing. Individuals that are prevented from hearing their own practice never sing a normal song. Yet if the same individual is deafened after it has developed adult song, it continues to sing a perfectly normal song. These data suggest that individuals have two critical periods for song learning during their lives: one as a nestling, when they must hear the song of their own species, and a second as a 1-year-old, when they must hear themselves practice.

To summarize, singing in white-crowned sparrows is heavily influenced by learning. But in this species, learning is constrained to certain periods of life and occurs only in response to certain types of stimuli.

If chickens are at one end of the learning continuum and if white-crowned sparrows are in the middle, then species such as mynahs, mockingbirds, lyrebirds, and some types of parrots occupy the other end. Individuals of these species continue to learn new songs throughout their lives by imitating sounds in their environment. They have relatively few age-related or species-specific limitations on song learning. "Polly want a cracker" is not a species-specific call in parrots, and the ability to use this call does not depend on hearing it during a critical period early in life.

One of the great challenges that remains for research on song learning is to explain why so much variation exists in song-learning ability and in the size of song repertoires. Why is it adaptive for chickens to sing a few innate songs while mynahs use dozens of learned songs? Research continues.

Can Animals Think?

Research on classical conditioning, imprinting, and song learning has revealed that several types of learning exist and that the ability to learn varies widely among species. It is clear that simple types of learning are widespread in animals and that at least some species can learn a great deal by imitation. But can organisms other than humans think?

Cognition can be defined as the recognition and manipulation of facts about the world, and as the ability to form concepts and gain insights. Unlike the results of classical conditioning, imprinting, and song learning, the results of cognition cannot be observed directly. Instead, to infer that animals are thinking, researchers must design experimental situations that require animals to manipulate facts or information. The experiment must be set up in a way that allows the animal to demonstrate its ability to form novel associations or insights. As an example of work on animal cognition, consider recent experiments on tool use by New Caledonian crows.

New Caledonian crows routinely make and use tools to find food in the wild. Each of these tools is held in the beak and used to remove insect larvae (grubs) or other types of food from plant stems or from crevices. Biologists have documented that the crows manufacture the following general types of tools:

- Straight or curved sticks are broken off and cleaned before use (**Figure 51.8a**). Several other bird species—including one of the Galápagos finches introduced in Chapter 41—are known to modify sticks or plant stems to "fish" for insects. Chimpanzees also make these types of tools (see Chapter 33). Recent laboratory experiments have shown that when New Caledonian crows are presented with an array of straight sticks along with a piece of food in a clear plastic tube, the crows select a stick whose width or length is appropriate for the tube's width and

(a) Straight sticks are used to fish or pry food from crevices.

(b) Hooks are made by breaking parts off leaves.

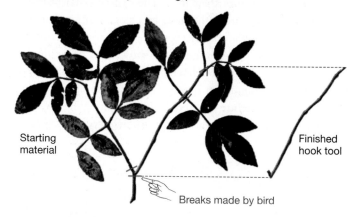

Starting material

Finished hook tool

Breaks made by bird

(c) Several cuts and tears are used to make spearing tools.

Starting material

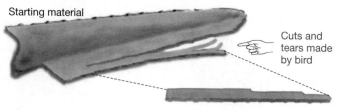

Cuts and tears made by bird

Finished spearing tool

FIGURE 51.8 New Caledonian Crows Make and Use Several Types of Tools
(a) Straight sticks are cleaned and broken off before being used as tools. **(b)** Hooks are made from compound leaves. **(c)** Spears are made from broad, stiff leaves.

the food's distance from the end of the tube. The crows then insert the stick into the tube to fish out the food.

- Hooks are made by breaking complex leaves into pieces (**Figure 51.8b**).
- Spearing tools are constructed from a complex series of cuts and rips, made with the beak, along the edge of leaves from pandanus plants (**Figure 51.8c**).

These observations support the hypothesis that crows can think. More specifically, crows appear to understand facts about the size and shape of raw materials and the location of food. Crows recognize that if they choose or manipulate materials in a certain way, they can use the resulting structure as a tool to acquire the food. The idea is that natural selection has favored the evolution of cognition in this species, because crows with the ability to manufacture and manipulate tools have higher fitness than do crows that lack this ability.

As **Figure 51.9** shows, the crows-can-think hypothesis was supported by recent experiments with two New Caledonian crows in captivity. The experimental setup consisted of a small bucket, containing food, in the bottom of a plexiglas tube that was open at the top, with the handle of the bucket pointing up. In an initial experiment, a male crow and a female crow were given a choice of straight or bent wires to use in retrieving the food from the inside of the tube. The goal of the experiment was to test the hypothesis that the crows can distinguish the two shapes and associate the bent shape with the ability to lift the bucket and acquire the food. During the fifth trial with the apparatus and the choice of two wires, the male happened to remove the hooked wire from the area. Apparently in response, the female picked up the straight wire in her beak, bent it, and successfully used it to lift the bucket and obtain the food.

To confirm the validity of this observation, the researchers gave the pair a straight wire and let them use it until they either successfully retrieved the food or dropped the wire into the tube and were unable to retrieve it. In a total of 17 trials, the male or female dropped the wire into the tube and were unable to retrieve it 7 times. Once the male was able to lift the bucket out and obtain the food using only the unmodified, straight wire. But 9 times, the female bent the straight wire and used it successfully to obtain the food.

These observations are nothing short of astonishing. They imply not only that the female was able to associate the shape of a wire with the ability to lift the bucket successfully, but also that she had the insight to understand that exerting a force on the wire would bend it into the shape required to be successful. In more colloquial terms, she appeared to understand what sort of tool she needed and how to make one. She had no model to imitate when she made her hooks, however, and she had had no prior experience with bending wires. Instead, she appeared to be thinking. Research on animal cognition is an exciting frontier in behavioral research.

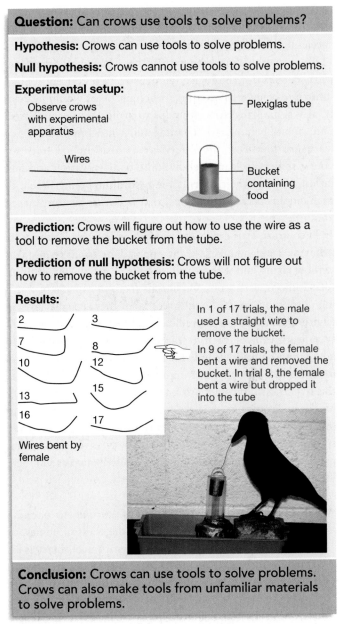

Question: Can crows use tools to solve problems?

Hypothesis: Crows can use tools to solve problems.

Null hypothesis: Crows cannot use tools to solve problems.

Experimental setup:

Observe crows with experimental apparatus

Wires

Plexiglas tube

Bucket containing food

Prediction: Crows will figure out how to use the wire as a tool to remove the bucket from the tube.

Prediction of null hypothesis: Crows will not figure out how to remove the bucket from the tube.

Results:

2   3
7   8
10   12
    15
13
16   17

Wires bent by female

In 1 of 17 trials, the male used a straight wire to remove the bucket.

In 9 of 17 trials, the female bent a wire and removed the bucket. In trial 8, the female bent a wire but dropped it into the tube

Conclusion: Crows can use tools to solve problems. Crows can also make tools from unfamiliar materials to solve problems.

FIGURE 51.9 Experimental Evidence That New Caledonian Crows Understand How to Make Tools
Prior to this experiment, the crows involved had never seen the experimental apparatus or wires of the type available to them. Bending of materials to form hooks has not been observed in the wild.

What Is the Adaptive Significance of Learning?

Research on classical conditioning, imprinting, song learning, and cognition has revealed that several types of learning exist and that the ability to learn varies widely among species. Even within the same species, some types of behavior are innate and inflexible while other types are condition dependent, or learned, and thus highly flexible. Even in people who have devoted their lives to learning, certain simple stimuli act as releasers that trigger FAPs such as yawning and the jump-back response.

These results raise an important question: At the ultimate level of explanation, why does the ability to learn vary among species and among different types of behavior?

To answer this question in the case of FAPs and other non-learned behavior, biologists emphasize two concepts. First, inflexible behavior is adaptive when mistakes would be costly, as when a kangaroo rat jumps to avoid a rattlesnake. Second, innate behavior is advantageous when opportunities to learn are few. For example, there is little opportunity for young spiders to learn web building by observation and imitation, because their parents die before the young hatch. In such circumstances, it is logical that behavior be non-learned and inflexible.

In contrast, learning tends to have an important influence on behavior if individuals have the opportunity to make mistakes without dying and if they have parents or other sources of information to learn from. Organisms that are capable of learning also tend to be long lived, and they usually live in environments that are unpredictable. To explain why the ability to learn evolves, then, researchers hypothesize that it is an adaptation that allows individuals to change their behavior in response to a changing and unpredictable environment. Under this hypothesis, the ability to learn varies because some species live in environments that are much more unpredictable than others. In addition, the type of learning that occurs in a given species should be correlated with the type of environmental unpredictability encountered by that species. Examples will help illustrate these concepts:

- Norway rats are extremely adept at learning to navigate mazes. These animals live in sewers, the walls of homes, or other locations where the ability to learn travel routes is essential for survival (**Figure 51.10a**). Norway rats are also exceptionally good at learning to avoid foods that contain poisons. In nature they feed on a wide variety of fruits, seeds, and insects. The availability of these foods changes with the season, and many potential food items are actually toxic. Thus, it is logical to observe that rats will taste only small quantities of novel foods and that they quickly learn to avoid any foods that induce illness.

- Scrub jays cache seeds and other types of food in storage areas and are proficient at remembering where they have stored food (**Figure 51.10b**). In this species, spatial learning and a good memory are essential for survival. In addition, scrub jays learn to modify their storage behavior based on experience. This species lives in social groups, and it is not unusual for some members of the group to steal food that was cached by other jays in their group. Experiments in the laboratory have shown that if an individual that has stolen food in the past caches food while being watched by other jays, the individual will come back later by itself and move

(a) Lab rats and Norway rats learn how to navigate mazes.

(b) Scrub jays remember where they've stored food.

FIGURE 51.10 Different Species Are Adept at Different Types of Learning
There is often a strong correlation between the type of learning that occurs in a particular species and the problems that the species must solve to survive and reproduce. **QUESTION** Domestic dogs have been trained by people for thousands of years. Domestic dogs are adept at learning via classical conditioning. What is the relationship between these observations?

the stored food to a new location. In this species, the ability to learn and modify behavior is interpreted as an adaptation for coping with unpredictable social situations.

To summarize, the ability to learn and the type of learning that occurs vary within and among species. Learning is an adaptation that helps organisms cope with challenges from their environment.

The same species of animal may learn in a variety of ways, ranging from simple forms such as imprinting and classical conditioning to imitation to cognition. It is common to observe that an individual is capable of learning only certain types of things—such as its species-specific song—only at certain times in its life. You should be able to (1) define learning and give examples of at least two distinct types of learning, (2) discuss the adaptive value of cognition, and (3) offer a hypothesis to explain why cognition appears to be rare among animals.

51.3 How Animals Act: Hormonal Control

Most behavior is modified at least slightly by some type of learning, and most animals have behavioral responses that range from highly stereotyped to highly flexible. Previous examples in the chapter have highlighted the adaptive value of learning and various types of behavior. But we have yet to explore a question about proximate causation in detail. To understand how biologists go about addressing questions at the proximate level, consider the following: When behavior is flexible and condition dependent, it means that an individual has the potential to behave in a variety of ways. How are changes in behavior implemented and controlled? To answer this question, let's explore sexual behavior in the lizard *Anolis carolinensis*.

Anolis carolinensis (**Figure 51.11a**) lives in the woodlands of the southeastern United States. After spending the winter under a log or rock, males emerge in January and establish breeding territories. Females become active a month later, and the breed-

ing season begins in April. By May, females are laying an egg every 10–14 days. By the time the breeding season is complete three months later, the eggs produced by a female will total twice her body mass. **Figure 51.11b** maps this series of events, along with the corresponding changes observed in the male and female reproductive systems.

What causes these dramatic seasonal changes in behavior? The proximate answer is sex hormones—testosterone in males and estradiol in females. Testosterone is produced in the testes of males, and estradiol in the ovaries of females. The evidence for these statements is direct. Testosterone injections induce sexual behavior in castrated males, while estradiol injections induce sexual activity in females whose ovaries have been removed.

This result leaves some key questions unanswered, however. *Anolis* lizards are most successful if they reproduce early in the spring. In springtime, their food supplies are increasing and snakes and other predators are not yet hunting lizards to feed their own young. If testosterone and estradiol levels induced sexual activity in *Anolis* lizards at the wrong time of year, it would be a disaster for the individuals involved. In addition, it is critical for all *Anolis* individuals in a population to start their sexual activity at the same time, so that females can find males that are ready to breed, and vice versa. What environmental cues trigger the production of sex hormones in early spring? How do male and female *Anolis* synchronize their sexual behavior so that they are ready to produce gametes at the same time?

To answer these questions, a biologist brought a large group of sexually inactive adult lizards into the laboratory during the winter and divided them into five treatment groups. The physical environment was exactly the same in all treatments. Each individual received identical food, and in all treatments high "daytime" temperatures were followed by slightly lower "nighttime"

(a) Courtship display of a male *Anolis* lizard

(b) Changes in sexual organs through the year

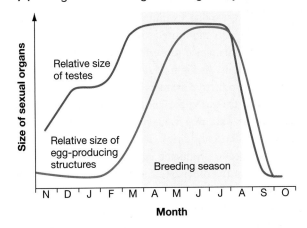

FIGURE 51.11 Sexual Behavior in *Anolis* Lizards
(a) A male *Anolis* lizard courting a nearby female. The male extends a flap of skin called a dewlap and bobs up and down. **(b)** Changes in the sexual organs of *Anolis* lizards through the year: (red curve) the relative size of the follicles inside a female's ovary, which produce eggs; (blue curve) the relative size of a male's testicles.

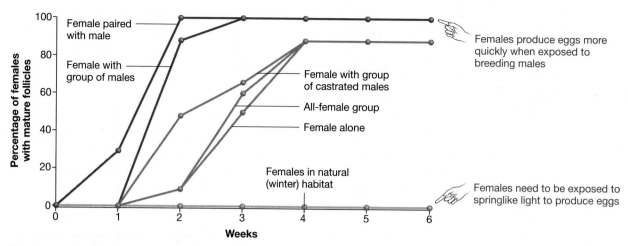

FIGURE 51.12 Exposure to Springlike Conditions and Breeding Males Stimulates Female Lizards to Produce Eggs
The percentage of female lizards with mature follicles, plotted over time. Five treatment groups were exposed to springlike light and temperature; data are also plotted for females left in a natural (winter) habitat. Each data point represents the average value for a group of 6–10 females. QUESTION Some critics contend that the females that remained outside do not represent a legitimate control in this experiment. What conditions would represent a better control?

settings. The biologist also continued to monitor the condition of lizards that remained in natural habitats nearby.

To test the hypothesis that changes in day length signal the arrival of spring and trigger initial changes in sex hormones, the biologist exposed the five treatment groups in the laboratory to artificial lighting that simulated the long days and short nights of spring. To test the hypothesis that social interactions among individuals are responsible for synchronizing sexual behavior, the social setting was varied among treatment groups: (1) single isolated females; (2) groups of females only; (3) single females, each with a single male; (4) single females, each with a group of castrated (nonbreeding) males; or (5) single females, each with a group of uncastrated (breeding) males.

Each week, the researcher examined the ovaries of females in each group. He also monitored the ovaries of females in nearby natural habitats, since those females were not exposed to springlike conditions. As **Figure 51.12** shows, the differences in the animals' reproductive systems were dramatic. Females that were exposed to springlike conditions began producing eggs; females in the field that were not exposed to springlike conditions did not. But in addition, females that were exposed to breeding males began producing eggs much earlier than did the females placed in the other treatment groups. These results support the hypothesis that two types of stimulation are necessary to produce the hormonal changes that lead to sexual behavior. Females need to experience springlike light and temperatures *and* exposure to breeding males.

What aspect of male behavior causes the difference in female egg production? The biologist suspected that visual stimulation from the male's courtship display was important. Specifically, he hypothesized that a flap of skin, called the dewlap, played a

role in stimulating females to produce eggs. To court females, male *Anolis* lizards bob up and down and extend their dewlaps (see the yellow neck pouch in Figure 51.11a). To test this idea, the investigator repeated the previous experiment but added a twist: He placed some females with males that had intact dewlaps, and other females with males whose dewlaps had been surgically removed. The result? Females grouped with dewlap-less males were slow to produce eggs—just as slow, in fact, as the females in the first experiment that had been grouped with castrated males. These latter females had not been courted at all. The result suggests that the dewlap is a key visual signal. The experiments succeeded in identifying the environmental cues that trigger hormone production and the onset of sexual behavior.

To summarize, sexual activity in *Anolis* lizards is a condition-dependent behavior controlled by hormones. This research and similar experiments carry a general message: When hormone levels change in response to environmental changes, they trigger behavior that allows individuals to maximize their fitness in the changed environment. In many cases, flexible behavior is possible because hormone levels are flexible.

51.4 Communication

The bobbing dewlap of a male *Anolis carolinensis*, the song of a bird, and the sentences in this text all have the same overall goal: communication. In biology, **communication** is defined as any process in which a signal from one individual modifies the behavior of a recipient individual. A **signal** is any information-containing behavior. Communication is a crucial component of animal behavior, because it creates a stimulus that elicits a response.

By definition, communication is a social process. For communication to occur, it is not enough that a signal is sent; the signal must be received and acted on. A lizard's bobbing dewlap does not qualify as communication unless another individual sees it and responds to the message. In addition, some biologists maintain that for an event to qualify as communication, the signal that is sent must be intended as a signal. According to these investigators, predicting a person's behavior by watching their body language does not qualify as communication. Similarly, if students unintentionally fall asleep in class and the instructor changes his behavior to wake them up, the exchange does not qualify as communication.

Modes of Communication

The information that organisms communicate may be encoded and delivered in a variety of ways. Communication can be acoustic, as in the song of crickets or birds. It can be visual, as in the color patterns of birds or fish. It can be olfactory, as in scent marking by mice and wolves. It can be tactile, as in communication in some spiders.

One of the most general observations about communication is that the type of signal used by an organism correlates with its habitat. For example, light is quickly absorbed in aquatic habitats, but sound is not. Based on this observation, it is logical to observe that humpback whales rely on songs for long-distance communication. In some cases, humpback whale songs can travel hundreds of kilometers. Groups of whales use acoustic communication to keep together as they move from summer feeding areas to winter breeding grounds, and individual males

sing to attract mates and warn rivals. Correlations between habitat and mode of communication are also observed in visual, olfactory, and tactile communication. Animals that are active during the day and that live in open or treeless habitats tend to rely on visual communication during courtship and territorial displays. Wolves and other animals that are active at night communicate via scent or sound; ants and termites that live underground rely on olfactory and tactile communication.

It is also common to observe several modes of communication being used in conjunction. For example, male red-winged blackbirds establish a breeding territory in the spring by giving a display that combines auditory and visual signals. The display is based on revealing bright red patches of feathers near their shoulders and giving a loud call. To test the hypothesis that both components of the display are important, researchers have manipulated both acoustic and visual elements. In one experiment, biologists used a speaker to play recordings of red-winged blackbird songs in existing territories (**Figure 51.13a**). In response, the territory-owning males flew toward the speaker and appeared to hunt for it. In a second experiment, researchers made a crude model of a male red-winged blackbird by sewing red patches on a black sock stuffed with rags. When the model was placed in existing territories, territory owners approached the model and inspected it (**Figure 51.13b**). But if the model was presented along with a recorded blackbird song, the resident males attacked the model (**Figure 51.13c**). These experiments support the hypothesis that both auditory and visual information are important in territorial displays of red-winged blackbirds. More specifically, the data indicate that while each type of stimulus

(a) Red-winged blackbird call only

(b) Model of male only

(c) Call and model together

Orange lines show pitch and duration of territorial call, from speaker

FIGURE 51.13 Experimental Evidence That Visual and Auditory Stimuli Are Important in Territorial Display
Territory-owning red-winged blackbirds react differently to **(a)** auditory stimuli, **(b)** visual stimuli, and **(c)** auditory plus visual stimuli. **EXERCISE** Researchers have clipped the red feathers from male red-winged blackbirds and released them into territories. Predict how territory owners respond to these individuals.

alone induces a response, the combination of both types of stimuli provokes a much more powerful change in behavior.

Each mode of communication has advantages and disadvantages. Although acoustic communication can be extremely effective in some habitats, songs and calls are short lived. A red-winged blackbird's red feathers remain in place for at least six months, but each call lasts less than 3 seconds. Thus, calls and songs have to be repeated to be effective. Frequent repetition of calls and songs requires a large expenditure of time and energy. In addition, acoustic communication has been shown to attract predators. It is no surprise that when a hawk or falcon approaches a marsh inhabited by red-winged blackbirds, things get very quiet.

Research on animal communication has gone far beyond describing types of communication and documenting correlations between modes of communication and habitat characteristics, however. To appreciate the array of questions that biologists ask about communication and how they go about answering them, let's consider two research programs on animal signaling—one classical and one more recent. The classical work revealed a remarkable form of communication in honeybees; the more recent work has focused on the question of how and why signalers lie.

A Case History: The Honeybee Dance

Honeybees are highly social animals that live in hives. Inside the hive, a queen bee lays eggs that are cared for by individuals called *workers*. In addition to caring for young and building and maintaining the hive, workers obtain food for themselves and other members of the colony by gathering nectar and pollen from flowering plants.

Biologists noticed that bees appear to recruit to food sources, meaning that if a new source is discovered by one or a few individuals, many more bees begin showing up over time. This observation inspired the hypothesis that successful food-finders have some way of communicating the location of food to other individuals. Karl von Frisch suspected that successful food-finders communicate information about food sources when they interact with other workers inside the hive. Because beehives are completely dark, von Frisch hypothesized that communication was tactile in nature.

Von Frisch began studying bee communication by observing bees that built hives inside glass-walled chambers that he constructed. He found that if he placed a feeder containing sugar water near one of these observation hives, a few of the workers began moving in a circular pattern on the vertical, interior walls of the hive. Von Frisch called these movements the "round dance" (**Figure 51.14a**). Other bee workers appeared to follow the progress of the dance by touching the displaying individual as it danced and to respond by flying away from the hive in search of the food source.

To investigate the function of these movements further, von Frisch placed feeders containing sugar water at progressively greater distances from the hive. Using this technique, he was able to get bees to visit feeders at a distance of several kilometers from the hive. By catching bees at the feeders and dabbing them with paint, he could individually mark successful food-finders. Follow-up observations at the hive confirmed that marked foragers danced when they returned to the hive. Follow-up observations at the feeders confirmed that marked foragers returned with greater and greater numbers of unmarked bees. To explain these data, von Frisch proposed that the round dance contained information about the location of food and that workers got information from the dance by touching the dancer and following the dancer's movements.

(a) The round dance

(b) The waggle dance

Other bee workers follow the progress of the dance by touching the displaying individual

FIGURE 51.14 Honeybees Perform Two Types of Dances
(a) During the round dance, successful food-finders move in a circle. **(b)** During the waggle dance, successful foragers move in a circle but then make straight runs through the circle. During the straight part of the dance, the dancer waggles her abdomen.

When von Frisch placed feeders at longer distances from the hive, however, he found that successful food-finders no longer did the round dance. Instead, they performed a new type of display. He named these movements the "waggle dance," because they combined circular movements like those of the round dance with short, straight runs (**Figure 51.14b**). During these runs, the dancer would vigorously move her abdomen from side to side.

These observations supported the hypothesis that both the round dance and waggle dance communicated information about food sources. But a key observation allowed von Frisch to push this result further. He noticed that the orientation of the waggle part of the dance varied and that the variation correlated with the direction of the food source from the hive. In addition, he observed that the length of the straight, "waggling" run was proportional to the distance the foragers had to fly to reach the feeder. By varying the location of the food source and observing the orientation of the waggle dance given by marked workers, von Frisch was able to confirm that dancing bees were communicating the position of the food relative to the current position of the Sun. For example, if he placed food directly away from the Sun's current position, marked bees would give the waggle portion of their dance directly downward. But if the food was 90 degrees to the right of the Sun, bees would waggle 90 degrees to the right of vertical (**Figure 51.15**).

Based on these experiments, von Frisch concluded that the difference between the round dance and the waggle dance was a matter of distance. The round dance is used to indicate the presence of food within 80 to 100 m of the hive. The waggle dance is used to indicate the direction and distance to food that is over 100 m from the hive.

These results are nothing short of astonishing. Honeybees do not have large brains, yet they are capable of symbolic language. In addition, they are able to interpret the angle of the waggle dance performed on a vertical surface and to respond by flying horizontally along the corresponding angle. Further work has confirmed that the dance language of bees includes several modes of communication. In addition to the tactile information in the movements themselves, bees make sounds during the dance and give off scents that indicate the nature of the food source.

Honesty and Deceit in Communication

As far as is known, honeybees are always honest. Stated another way, honeybee dances consistently "tell the truth" about the location of food sources. This observation is logical, because the honeybees that occupy a hive are closely related and cooperate closely in the rearing of offspring. As a result, it is advantageous for an individual to convey information accurately. If a food-finder provided inaccurate or misleading information to

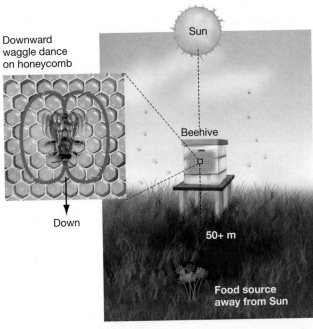

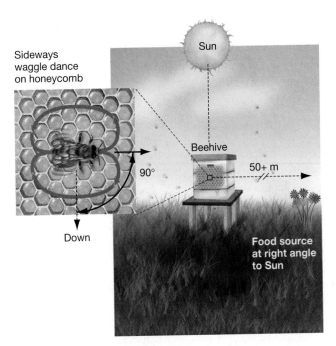

FIGURE 51.15 The Direction of the Waggle Dance Indicates the Location of Food Relative to the Sun
(a) In the language of the honeybee waggle dance, straight runs down the wall of the hive indicate that food is opposite the direction of the Sun. (b) Straight runs to the right indicate that food is 90° to the right of the Sun. Waggle dances are done only when the food is at least 50 m from the hive. **EXERCISE** The length of the straight run in the waggle dance indicates the relative distance of the food. Diagram a waggle dance that indicates food in the direction of the Sun and twice as far away as the food sources indicated by the dances drawn here.

its hivemates, fewer offspring would be reared and thus the food-finder's fitness would be reduced.

In many cases, however, natural selection has favored the evolution of deceitful communication, or what is commonly called lying. Recent research on deceitful communication has highlighted just how complex interactions between signalers and receivers can be. A few examples will help drive this point home.

Deceiving Individuals of Another Species   Figure 51.16 illustrates a few of the hundreds of examples of deceitful communication that have been documented among members of different species. Many of the best-studied instances involve predation:

- The anglerfish in Figure 51.16a has an appendage that looks remarkably like a minnow and that dangles near its mouth. If another fish approaches this "lure" and attempts to eat it, the anglerfish attacks.

- Male and female fireflies flash a species-specific signal to each other during courtship. Predatory *Photinus* fireflies can mimic the pattern of flashes given by females of several other species, however. If a *Photinus* female responds to the courtship flashes made by a male of a different species with the appropriate set of flashes, he waits as she approaches. Instead of copulating with her, however, he is attacked and eaten (Figure 51.16b).

- The butterfly shown in Figure 51.16c is highly palatable to birds and other predators. It is rarely attacked, however, because it strongly resembles butterfly species that contain toxins and are highly unpalatable to predators. Such types of mimicry are discussed in more detail in Chapters 31 and 53.

In each of these examples, individuals increase their fitness by providing inaccurate or misleading information to members of a different species. Deceitful communication is also known in plants, where it functions in pollination. As the essay in Chapter 40 pointed out, several orchid species have flowers that look and smell like female wasps and that accomplish pollination by "fooling" male wasps into attempting copulation.

Deceiving Individuals of the Same Species   In some cases, natural selection has also favored the evolution of lying to members of the same species. When mantis shrimps molt, for example, they lack any external covering and cannot use their large claws. Thus, they are unable to defend the cavities where they live. But if another mantis shrimp approaches with the intent of evicting the individual and taking over the cavity, the molting individual bluffs. It raises its claws in the normal aggressive display and may even lunge at the intruder.

(a) Anglerfish use a "lure" to attract prey.

(b) *Photuris* fireflies flash the courtship signal of another species, then eat males that respond.

(c) This butterfly looks like a bad-tasting species but actually tastes good.

FIGURE 51.16 Deceitful Communication Is Common in Nature
Three examples of species that communicate misleading information to other species.

Perhaps the best-studied type of deceit in nonhuman animals, however, involves the mating system of bluegill sunfish. Male bluegills set up nesting territories in the shallow water along lake edges, fertilize the eggs laid in their nests, care for the developing embryos by fanning them with oxygen-rich water, and protect newly hatched offspring from large predators. Some males cheat on this system, however, by mimicking females. To understand how this happens, examine the female, normal male, and female-mimic male in **Figure 51.17.** Female mimic males look like females but have well-developed testes and produce large volumes of sperm. Mimics also act like females during courtship movements with territory-owning males. They even adopt the usual egg-laying posture. Then when normal females approach the nest and begin courtship, the female-mimic males join in. The territory-owning male tolerates the mimic, apparently thinking that he is successfully courting two females at the same time. But when the actual female begins to lay eggs, the mimic responds by releasing sperm and fertilizing some of them. In this way, the female-mimic male fathers offspring but does not help care for them.

When Does Deception Work? In analyzing deceitful communication in bluegill sunfish and other species, researchers point out that in most cases, lying works only when it is relatively rare. The logic behind this hypothesis is that if deceit becomes extremely common, then natural selection will strongly favor individuals that can detect and avoid or punish liars. But if liars are rare, then natural selection will favor individuals that are occasionally fooled but are more commonly rewarded by responding to signals in a normal way. In mantis

FIGURE 51.17 In Bluegill Sunfish, Some Males Look and Act Like Females
Female-mimic males have fully developed, fully functional testes and produce large numbers of sperm.

shrimps, for example, molting occurs only once or twice a year. Thus, the vast majority of contests over hiding places involve individuals that have an intact exoskeleton and claws that qualify as a deadly weapon. As a result, intruders usually benefit by interpreting threat displays as honest and responding to the stimulus by retreating. But if molting is frequent and bluffing behavior common, then intruders would usually benefit by challenging territory owners, which are likely to be defenseless. Research on deceit and honesty in communication has been extraordinarily productive and continues at a rapid pace.

✓ CHECK YOUR UNDERSTANDING

Communication is an exchange of information between individuals. In most instances, the mode of communication that animals use maximizes the probability that the information will be transferred efficiently. The amount of information that is communicated can be very simple, as in the territorial display of a red-winged blackbird, or extremely complex, as in the dance language of honeybees. Communication can also be honest or deceitful, depending on the intent of the signaler. You should be able to (1) define communication; (2) explain the adaptive significance of different modes of communication and give examples of each; and (3) discuss the mechanisms that humans use to detect when they are receiving misleading or incorrect information from another human, as well as the types of behavior that humans use in response to receiving deceitful communication.

51.5 Orientation, Navigation, and Migration

In response to many stimuli, organisms move. Ants follow odor trails; bees fly toward food sources. A movement that results in a change of position is called **orientation**. The simplest type of orientation is termed **taxis**, and it involves positioning the body, or part of the body, toward or away from a stimulus. A moth attracted to a porch light is an example of a positive **phototaxis**—orientation toward light. A female cricket approaching a calling male is an example of a positive **phonotaxis**—orientation toward sound. A person who retreats from the sound of a siren blaring is an example of negative phonotaxis.

Orientation movements can be extremely sophisticated. Consider bats, which use sound to locate their prey. Hunting bats emit high-pitched sounds and listen for the echo from a flying insect. This process, called **echolocation**, is remarkably acute. If you toss a small stone into the air, a foraging bat will pursue it but then turn away at the last instant. This simple observation suggests that bats are able to tell the

Bats emit high-pitched sounds (shown in blue) and direct their flight toward objects that reflect this sound (shown in purple)

FIGURE 51.18 Bats Hunt by Echolocation
Bats detect the echo from high-pitched sounds that they emit and orient their hunting flights in response.

difference between the echo from an insect and that from a stone (**Figure 51.18**).

Not surprisingly, moths that are preyed on by bats have ears that can detect the sounds emitted by an approaching bat. When these moths hear a bat in the distance, they are able to turn and fly away. But if the bat is very close, moths make a power dive toward the ground. You can make this observation yourself if you ever see moths flying at night during the summer. If you jingle keys at them, the high-frequency sound that results may imitate a bat's hunting call closely enough to elicit a response—moths dropping from the air like stones.

Migration: Why Do Animals Move with a Change of Seasons?

Orienting movements—such as those of a hunting bat or an escaping moth—cover fairly short distances. But in some cases, bats and other species travel thousands of miles in search of places to feed and breed. In ecology, **migration** is defined as the long-distance movement of a population associated with a change of seasons. A few examples will bring home the simple point that migratory movements can be spectacular in their extent and the navigation challenge they pose:

- Arctic terns nest along the Atlantic coast of North America, fly south along the coast of Africa to wintering grounds off Antarctica, and then fly back north along the eastern coast of South America. Chicks follow their parents south but make the return trip on their own. The journey totals over 32,000 km (20,000 miles).

- Many of the monarch butterflies native to North America spend the winter in the mountains of central Mexico or southwest California. Tagged individuals are known to have flown over 3000 km (1870 miles) to get to a wintering area. In spring monarchs begin the return trip north. They do not live to complete the trip, however. Instead, mating takes place along the way, and females lay eggs en route. Although adults that lived through the winter die, offspring continue to head north. After several generations of breeding at the original habitats in northern North America, individuals again migrate south to overwinter. Instead of a single generation making the entire migratory cycle, then, the round trip takes several generations.

- Salmon that hatch in rivers along the Pacific Coast of North America and northern Asia migrate to the ocean when they are a few months to several years old, depending on the species. After spending several years feeding and growing in the North Pacific Ocean, they return to the stream where they hatched. There they mate and die.

In most cases, it is relatively straightforward to generate hypotheses for why migration exists at the ultimate level. Arctic terns feed on fish that are available in different parts of the world at different seasons. Thus, individuals that migrate achieve higher reproductive success than do individuals that do not migrate. Monarch butterflies may achieve higher reproductive success by migrating to wintering areas and new breeding areas than by trying to overwinter in northern North America.

At the proximate level, however, explaining migratory movements is often extremely difficult. How do all of these animals find their way? What cues guide them on these immense journeys? The surprising answer is that in many or even most cases, biologists don't really know.

Navigation: How Do Animals Find Their Way?

To organize research on how animals find their way during migratory movements, biologists distinguish three categories of navigation: (1) **Piloting** is the use of familiar landmarks, (2) **compass orientation** is movement that is oriented in a specific direction, and (3) **true navigation** is the ability to locate a specific place on Earth's surface. A thought experiment will help you understand these categories. Suppose you grew up on the shores of Lake Superior and were transported to the middle of this enormous lake. Because you were blindfolded and made to spin in place en route, you are disoriented and have no idea where you are. But you are given a magnetic compass. The compass tells you where north, south, east, and west are. Unfortunately this information will not help you find home, because you have no idea where you are in relation to home. You are stuck because you have no mechanism of true navigation. Now suppose that a helicopter flying overhead drops you a map. The map has an X marking your current position and an H marking your home. The map also has a scale and a compass symbol. From this information, you determine that home is 100 miles to the west. The map has solved your navigational problem. The compass has solved your orientation problem. Now that you know you live to the west, you can use the compass to find west and paddle home. As you approach the shore, you recognize familiar landmarks. Piloting gets you home.

Organisms other than humans do not have magnetic compasses or printed maps. How do they navigate? Let's answer this question for each of the three categories of navigation.

Piloting A substantial amount of data suggests that at least some migratory animals use piloting to find their way. In some species of migratory birds and mammals, offspring follow their parents south in the fall and north in the spring. Young appear to memorize the route (**Figure 51.19**). Piloting even plays a role in species with more sophisticated navigational abilities. Homing pigeons, for example, can find their way home even when they are released in strange terrain a long distance away. But if these birds are equipped with frosted spectacles so that they see the world as a foggy haze, they return to within a mile or so of their home but do not actually find it. Based on these experiments, researchers have concluded that homing pigeons use piloting to navigate the final stages of a journey home.

Compass Orientation How do animals perform compass orientation? To date, most research on compass orientation has been done on migratory birds. To determine where north is, these animals appear to use the Sun during the day and the stars at night. Let's consider each type of compass orientation separately, because they work differently.

The Sun is difficult to use as a compass reference, because its position changes during the day. It rises in the east, is due south at noon (in the Northern Hemisphere), and sets in the west. To use the Sun as a compass reference, then, an animal must have an internal clock that defines morning, noon, and evening.

FIGURE 51.19 In Some Species, Offspring Learn Migration Routes from Their Parents
In some species of birds and mammals, offspring learn migration routes by following their parents. Here two snow goose parents lead their offspring.

Fortunately, most animals, have such a clock. The **circadian clock** that exists in organisms maintains a 24-hour rhythm of chemical activity. The clock is set by the light–dark transitions of day and night. It tells individuals enough about the time of day that they can use the Sun's position to find magnetic north.

The situation is actually simpler on clear nights, because migratory birds in the Northern Hemisphere can use the North Star to find magnetic north and select a direction for migration. But what if the weather is cloudy and neither the Sun nor stars are visible? Under these conditions, migratory birds appear to orient using Earth's magnetic field. Exactly how they detect magnetism is under debate. One hypothesis contends that the birds can detect magnetism by their visual system, through an unknown molecular mechanism. An alternative hypothesis maintains that individuals have small particles of magnetic iron—the mineral called magnetite—in their bodies. Changes in the positions of magnetic particles, in response to Earth's magnetic field, could then be detected and provide reliable information for compass orientation.

Although research on mechanisms of compass orientation continues, one important point is clear: Birds and perhaps other organisms have multiple mechanisms of finding a compass direction. At least some species can use a Sun compass, a star compass, and a magnetic compass. Which system they use depends on the weather and other circumstances.

51.6 The Evolution of Self-Sacrificing Behavior

The types of behavior reviewed thus far—including FAPs, learned behaviors, honest and deceitful communication, and migration—all have a key common element: They help individuals respond to environmental stimuli in a way that increases their fitness. There is a type of behavior that appears to contradict this pattern, however: altruism. **Altruism** is behavior that has a fitness cost to the individual exhibiting the behavior and a fitness benefit to the recipient of the behavior. It is the formal term for self-sacrificing behavior. Altruism decreases an individual's ability to produce offspring but helps others produce more offspring.

The existence of altruistic behavior appears to be paradoxical, because if certain alleles make an individual more likely to be altruistic, those alleles should be selected against. For example, consider the claim that rodents called lemmings sacrifice themselves for the good of the species. A widely circulated story contends that when lemming populations are high, overgrazing occurs and the habitat is damaged—enough so that most individuals in the population will die of starvation. In response, some individuals throw themselves into the sea and drown. The hypothesis is that suicide is favored because it lowers population density enough to maintain high-quality habitats for the good of the species. Even though individuals suffer, the good-of-the-species hypothesis maintains that the behavior exists because the group benefits.

THE FAR SIDE® BY GARY LARSON

FIGURE 51.20 Altruism Cannot Evolve If "Cheater" Alleles Exist
Most individuals in this population have alleles that lead to self-sacrificing behavior and result in death; the lemming with the inner tube has alleles that prevent self-sacrificing behavior. [*The Far Side*® by Gary Larson ©1980 FarWorks, Inc. All Rights Reserved. Used with permission.]
QUESTION Which alleles will increase in frequency in this population, and why?

Do Organisms Act for the Good of the Species?

The lemming suicide story is false. Although lemmings do disperse from areas of high population density in order to find habitats with higher food availability, they do not throw themselves into the sea. The individual wearing the inner tube in **Figure 51.20** explains why lemmings do not kill themselves for the good of the species. The inner tube represents what biologists called a "cheater" or "selfish" allele. Most of the individuals in the cartoon's population have altruistic alleles that lead them to commit suicide. But some individuals have cheater alleles that lead them to avoid committing suicide. Individuals with self-sacrificing alleles die and do not produce offspring. But individuals with selfish, cheater alleles survive and produce offspring. As a result, selfish alleles increase in frequency while altruistic alleles decrease in frequency. Thus, it is not possible for individuals to do things for the good of the species.

Kin Selection

Even though it is not possible for behavior to evolve for the good of the species, self-sacrificing behavior does occur in nature. For example, black-tailed prairie dogs perform a behavior called alarm calling (**Figure 51.21**). These burrowing mammals live in

FIGURE 51.21 Black-Tailed Prairie Dogs Are Highly Social
Prairie dogs live with their immediate and extended family within large groups called towns. The individual with the upright posture has spotted an intruder and may give an alarm call.

large communities, called *towns*, throughout the Great Plains region of North America. When a badger, coyote, hawk, or other predator approaches a town, some prairie dogs give alarm calls that alert other prairie dogs to run to mounds and scan for the threat. Giving these calls is risky. In several species of ground squirrels and prairie dogs, researchers have shown that alarm-callers draw attention to themselves by calling and are in much greater danger of being attacked than non-callers are.

How can natural selection favor the evolution of self-sacrificing behavior? William Hamilton answered this question by creating a mathematical model to assess how an allele that contributes to altruistic behavior could increase in frequency in a population. To model the fate of altruistic alleles, Hamilton represented the fitness cost of the altruistic act to the actor as C and the fitness benefit to the recipient as B. Both C and B are measured in units of offspring produced. His model showed that the allele could spread if

$$Br > C$$

where r is the coefficient of relatedness. The **coefficient of relatedness** is a measure of how closely the actor and beneficiary are related. Specifically, r measures the fraction of alleles in the actor and beneficiary that are identical by descent—that is, inherited from the same ancestor (see **Box 51.2**).

This result is called **Hamilton's rule.** It is important because it shows that individuals can pass their alleles on to the next generation not only by having their own offspring but also by helping close relatives produce more offspring. Stated another way, an altruist might produce slightly fewer offspring because of the sacrifices the individual makes to help others. But if those sacrifices help close relatives produce many more offspring, then copies of the altruist's alleles will increase in frequency. According to Hamilton's rule, if the fitness benefits of altruistic behavior are high for the recipients, if the recipients are close relatives, and if the fitness costs to the altruist are low, then alleles associated with altruistic behavior will be favored by nat-

BOX 51.2   Calculating the Coefficient of Relatedness

The coefficient of relatedness, r, varies between 0.0 and 1.0. If two individuals have no identical alleles as a result of being inherited from the same ancestor, then their r value is 0.0. Because every allele in pairs of identical twins is identical by descent, their coefficient of relatedness is 1.0.

What about other relationships? **Figure 51.22a** shows how r is calculated between half-siblings. (To review what the boxes, circles, and lines in a pedigree mean, see Figure 13.21.) Half-siblings share one parent. Thus, r represents the probability that half-siblings share alleles as a result of inheriting alleles from their common parent. It is critical to realize that, in each parent-to-offspring link of descent, the probability of any particular allele being transmitted is 1/2. This is so because meiosis distributes alleles from the parent's diploid genome to their haploid gametes randomly. Thus, half the gametes produced by a parent get one of the alleles present at each gene, and half the gametes produced get the other allele. Half-siblings are connected by two such parent-to-offspring links. The overall probability of two half-siblings sharing the same allele by descent is $1/2 \times 1/2 = 1/4$.

To think about this calculation in another way, focus on the red arrows in Figure 51.22a. The left arrow represents the probability that the mother transmits a particular allele to her son. The right arrow represents the probability that the mother transmits the same allele to her daughter. Both probabilities are 1/2. Thus, the probability that the mother transmitted the same allele to both her son and daughter is $1/2 \times 1/2 = 1/4$.

Figure 51.22b shows how r is calculated between full siblings. The challenge here is to calculate the probability of two individuals sharing the same allele as a result of inheriting it through their mother or through their father. The probability that full siblings share alleles as a result of inheriting them from one parent is 1/4. Thus, the probability that full siblings share alleles inherited from either their mother or their father is $1/4 + 1/4 = 1/2$.

(a) What is the r between half-siblings?

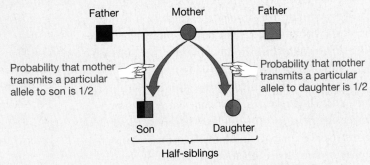

What is the probability that half-siblings inherit the same allele from their common parent?
Answer: r between half-siblings = $1/2 \times 1/2 = 1/4$

(b) What is the r between full siblings?

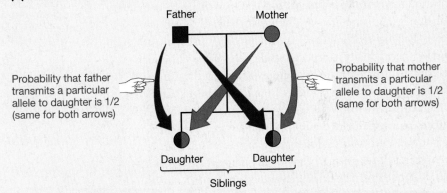

What is the probability that full siblings inherit the same allele from their father or their mother?
Answer: Probability that they inherit same allele from father = $1/2 \times 1/2 = 1/4$
Probability that they inherit same allele from mother = $1/2 \times 1/2 = 1/4$
Overall probability that they share alleles by descent = $1/4 + 1/4 = 1/2$
r between full siblings = $1/2$

FIGURE 51.22 The Coefficient of Relatedness Is Calculated from Information in Pedigrees
QUESTION According to Hamilton's rule, who should be nicer to each other: half-siblings or full siblings? QUESTION What is r between first cousins?

ural selection and will spread throughout the population. The key is that close relatives are very likely to have copies of the altruistic allele. Biologists use the term **kin selection** to refer to natural selection that acts through benefits to relatives.

Does Hamilton's rule work? Do animals really favor relatives when they act altruistically? To test the kin-selection hy-

pothesis, a reseacher studied which of the inhabitants of a black-tailed prairie dog town were most likely to give alarm calls. Within a large prairie dog town, individuals live in small groups called *coteries* that share the same underground burrow. Members of each coterie defend a territory inside the town. By tagging offspring that were born over several generations, the researcher identified the genetic relationships among individuals in the town. More specifically, the researcher knew, within each coterie, to which of the following three categories each prairie dog belonged: (1) an individual with no close genetic relatives in its coterie; (2) an individual with no offspring in the coterie but at least one sibling, cousin, uncle, aunt, niece, or nephew: and (3) an individual with at least one offspring or grandoffspring in the coterie.

The kin-selection hypothesis predicts that individuals who do not have close genetic relatives nearby will rarely give an alarm call. To evaluate this prediction, the biologist recorded the identity of callers during 698 experiments. In these studies, a stuffed badger was dragged through the colony on a sled. Were prairie dogs with close relatives nearby more likely to call, or did kinship have nothing to do with the probability of alarm calling? The data in **Figure 51.23** illustrate that black-tailed prairie dogs are much more likely to call if they live in a coterie that includes close relatives. This same pattern—of preferentially dispensing help to kin—has been observed in many other species of social mammals and birds. Most cases of self-sacrificing behavior that have been analyzed to date are consistent with Hamilton's rule and are hypothesized to be the result of kin selection.

Reciprocal Altruism

During long-term studies of highly social animals, such as lions, chimpanzees, and vampire bats, biologists have observed non-relatives helping each other. Chimps and other primates, for example, occasionally spend considerable time grooming unrelated members of their social group—cleaning their fur and removing ticks and other parasites from their skin. In vampire bats, individuals that have been successful in finding food are known to regurgitate blood meals to both kin and non-kin that have not been successful and that are in danger of starving.

How can self-sacrificing behavior like this evolve if kin selection is not acting? The leading hypothesis to explain altruism among nonrelatives is called **reciprocal altruism**, which is an exchange of fitness benefits that are separated in time. Data that have been collected so far support the reciprocal-altruism hypothesis in at least some instances. Among vervet monkeys, for example, individuals are most likely to groom unrelated individuals that have groomed or helped them in the past. Similarly, vampire bats are most likely to donate blood meals to non-kin that have previously shared food with them. Reciprocal altruism is also widely invoked as an explanation for the helpful and cooperative behavior commonly observed among unrelated humans.

To summarize, altruistic behavior increases the fitness of individuals by favoring kin or by increasing the likelihood of receiving help in the futrue from non-kin. Altruism is a flexible, condition-dependent behavior that occurs in a surprisingly wide array of species that live in social groups.

Question: Do black-tailed prairie dogs prefer to help relatives when they give an alarm call?

Kin-selection hypothesis: Individuals give an alarm call only when close relatives are near.

Null hypothesis: The presence of relatives has no influence on the probability of alarm calling.

Experimental setup:

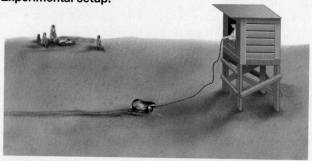

1. Determine relationships among individuals in prairie dog coterie.

2. Drag stuffed badger across territory of coterie.

3. From observation tower, record which members of coterie give an alarm call.

4. Repeat experiment 698 times. Each individual prairie dog coterie is tested 6-9 times over 3-year period.

Prediction of kin-selection hypothesis: Individuals in coteries that contain a close genetic relative are more likely to give an alarm call than are individuals in coteries that do not contain a close genetic relative.

Prediction of null hypothesis: The presence of relatives in coteries will not influence the probability of alarm calling.

Results:

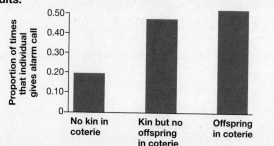

Conclusion: Alarm calling usually benefits relatives.

FIGURE 51.23 Experimental Evidence That Black-Tailed Prairie Dogs Are More Likely to Give Alarm Calls If Relatives Are Nearby

✓ CHECK YOUR UNDERSTANDING

Hamilton's rule states that alleles for altruistic behavior increase in frequency if the fitness cost of the behavior for the actor is low, the fitness benefit to the recipient is high, and the actor and recipient are closely related. Using Hamilton's rule, you should be able to (1) describe situations in which a black-tailed prairie dog is likely and unlikely to give an alarm call and (2) explain why reciprocal altruism has been observed only in species where individuals have good memories and live in small, long-lived social groups.

ESSAY Children at Risk

Some of the concepts discussed in this chapter—including altruism, Hamilton's rule, kin selection, and natural selection—may seem a little abstract at first. Yet Martin Daly and Margo Wilson have shown that they can be enormously important tools for understanding certain aspects of human behavior.

Daly and Wilson study the "dark side" of human behavior—homicide, domestic violence, and adultery—from a biological perspective. They try to determine whether some of these events are consistent with the types of actions favored by natural selection.

For example, let's consider data that Daly and Wilson have collected on child abuse. These researchers hypothesized that natural selection should favor parents who invest resources in biological children but not in stepchildren—with whom the parents have no genetic relationship. Based on this hypothesis, the researchers predicted that child abuse should be much more common in households containing a stepparent than it is in homes containing only biological parents. Further, Daly and Wilson predicted that, if stepparents choose to abuse their stepchildren, infants should be most at risk. The logic behind this prediction is that very young children demand the most time and resources and are least able to defend themselves.

Are the data consistent with these predictions? Daly and Wilson have analyzed the most extreme form of child abuse, which is the killing of children by parents. Using a database on all homicides reported in Canada between 1974 and 1983, the investigators found 341 cases in which a child was killed by a biological parent and 67 in which the perpetrator was a stepparent. Because households containing only biological parents were much more common, however, Daly and Wilson realized that they needed to compare the *rate* of violence in the two types of households, rather than the absolute number of incidents.

Daly and Wilson expressed the data as a rate—the number of children killed per million "child-years" by their parents. (A *child-year* is a year that a child of a particular age lived in a particular family situation.) The histograms in **Figure 51.24** resulted. The pattern is striking. Children who live with a stepparent are at much higher risk of abuse than are children who live with biological parents. Kids who are less than 2 years old are, in fact, 70 times more likely to be killed in a household with a stepparent than in a household with only biological parents.

... the "dark side" of human behavior... from a biological perspective.

The study points up the value of applying "selection thinking" to human behavior. It also has an urgent practical message: Society should be especially alert to indications of violence in households with stepparents where very young stepchildren are present. Several points about the data are worth noting, however. It is *extremely* important to recognize that violence occurs among biological kin as well as non-kin and that the vast majority of stepparents are solicitous and generous with their stepchildren. Further, the behavior is undoubtedly pathological rather than adaptive. Human perpetrators are jailed and treated as abhorrent. It is extremely unlikely, then, that the act of killing a stepchild improves the reproductive success of the murderer.

(a) Rates of child homicide in families with only biological parents

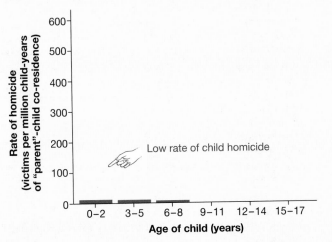

(b) Rates of child homicide in families with a stepparent

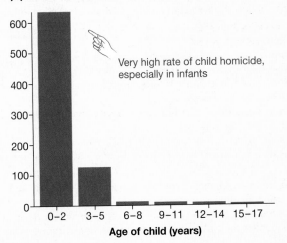

FIGURE 51.24 Rates of Child Homicide Are Higher in Families Containing a Stepparent
Rates of child homicide expressed as victims per "million child-years" in Canadian households containing only **(a)** a biological parent versus **(b)** a stepparent.

CHAPTER REVIEW

Summary of Key Concepts

■ After describing a behavior, biologists seek to explain both its proximate and ultimate causes—meaning how it happens at the genetic and physiological levels and how it affects the individual's fitness.

At the proximate level, experiments and observations focus on understanding how specific gene products, neuron activity, and hormonal signals cause behavior. At the ultimate level, researchers seek to understand the adaptive significance of behavior, or how it enables individuals to survive and reproduce. By combining proximate and ultimate viewpoints and studying the genetic basis of behavior, biologists can seek a comprehensive understanding of how and why animals do what they do.

Web Tutorial 51.1 Homing Behavior in Digger Wasps

■ In a single species, behavior may range from highly stereotyped, invariable responses to highly flexible, conditional responses and from unlearned to learned responses.

FAPs are examples of highly stereotyped, unlearned behaviors that are released by extremely simple stimuli. Language acquisition and tool making are examples of behavior that are highly dependent on learning, cognition, and the environmental conditions experienced by an individual. Most of the behavior observed in animals is condition dependent, meaning an individual has the genetic, neuronal, and hormonal mechanisms in place for behaving in a variety of ways, and the actu-

al behavior observed depends on environmental conditions. Sex change in coral-reef fish and the onset of sexual behavior in *Anolis* lizards are examples of condition-dependent behaviors triggered by complex environmental stimuli.

■ The types of learning that individuals do, the way that they communicate, and the way that they orient and navigate correlate closely with their habitat and with the challenges they face in trying to survive and reproduce.

Animals usually behave in a way that increases their ability to survive and reproduce in their current environment. Deceitful communication can be adaptive—as when anglerfish fool prey into attacking a "lure" attached to the anglerfish's face or when molting mantis shrimps try to bluff their way into retaining ownership of a hiding place.

■ When individuals behave altruistically, they are usually helping close relatives or individuals that help them in return.

Although it is common to observe self-sacrificing behavior in animals, individuals do not do things for the good of the species. Instead, most self-sacrificing behavior is directed toward close relatives. When this is the case, alleles that lead to self-sacrificing can increase in frequency due to kin selection. In addition, some animals that live in close-knit social groups engage in reciprocal altruism—meaning they exchange help over time.

Questions

Content Review

1. What do proximate explanations of behavior focus on?
   a. how displays and other types of behavior have changed through time, or evolved
   b. the functional aspect of a behavior, or its "adaptive significance"
   c. genetic, neurological, and hormonal mechanisms of behavior
   d. psychological interpretations of behavior—especially motivation

2. What evidence suggests that there is a critical period for song learning in some bird species?
   a. In mynahs and mockingbirds, individuals continue to sing new songs throughout their lives.
   b. Individuals do not learn to sing normally unless they are allowed to hear themselves practice.
   c. Individuals that never hear their species-specific song can still sing normally as adults.
   d. Birds that are deafened soon after hatching never sing normally, but birds that are deafened several months after hatching sing normally.

3. What is the difference between orienting behavior and piloting?
   a. Orienting movements are fast responses to visual or auditory stimuli; piloting is way-finding using landmarks.
   b. Orienting is the ability to follow a compass during navigation; piloting is the ability to use map information during navigation.
   c. Orienting is used during long-distance movements such as migration; piloting is used in short-distance movements involved in homing.
   d. Orienting is the ability to use a Sun or star compass; piloting is the ability to use a magnetic compass.

4. Which of the following statements about the waggle dance of the honeybee is not correct?
   a. The length of a waggling run is proportional to the distance from the hive to a food source.
   b. Sounds and scents produced by the dancer provide information about the nature of the food source.
   c. No elements of the round dance are used.
   d. The orientation of the waggling run provides information about the direction of the food from the hive, relative to the Sun's position.

5. Why are biologists convinced that the sex hormone testosterone is required for normal sexual activity in male *Anolis* lizards?
   a. Male *Anolis* lizards with larger testes court females more vigorously than do males with smaller testes.
   b. The molecule is not found in female *Anolis* lizards.
   c. Male *Anolis* lizards whose gonads had been removed did not develop dewlaps.
   d. Male *Anolis* lizards whose gonads had been removed did not court females.

6. What does Hamilton's rule specify?
   a. why animals do things "for the good of the species"
   b. why reciprocal altruism can lead to fitness gains for unrelated individuals
   c. when alleles that favor self-sacrificing acts increase in frequency via kin selection
   d. the conditions under which more complex behaviors evolve from simpler behaviors

Conceptual Review

1. Make a graph like that in Figure 51.2, with axes that plot the degree to which a particular behavior is modified by learning and the degree to which behavior is stereotyped and inflexible. Plot the following behaviors on the graph: tool use in New Caledonian crows; exploration of mazes by rats; singing by chickens, by white-crowned sparrows, and by mynahs; and blushing in humans.

2. Discuss the proximate and ultimate causes of the following behaviors: (a) homing behavior by spiny lobsters, (b) sexual behavior by *Anolis* lizards, and (c) the "jump-back" response of kangaroo rats.

3. Propose a hypothesis to explain the adaptive significance of tool use in New Caledonian crows. Recall that a female crow bent an unfamiliar material into a hooked tool. Explain why this observation supports the hypothesis that individuals of this species can think.

4. What environmental stimuli cause changes in hormone levels that lead to egg laying in *Anolis* lizards? Based on the data presented on sexual behavior in *Anolis*, propose a hypothesis to explain the proximate basis of sex switching in coral-reef fish.

5. For an animal to navigate, it must have a map and a compass. Explain why both types of information are needed. Describe three types of compasses that have been identified in migratory or homing species.

6. Compare and contrast kin selection and reciprocal altruism. Be sure to identify the conditions under which self-sacrificing behavior is expected to evolve under kin selection versus reciprocal altruism.

Group Discussion Problems

1. Mated pairs of cranes give territorial displays called unison calls. Cranes that are raised in isolation from other cranes give unison calls normally, and the display is performed the same way every time. Your friend argues that the unison call of cranes is a FAP, because it is stereotyped and not influenced by learning. Another friend argues that it is a condition-dependent behavior, because only mated pairs do it and because the tendency to give unison calls varies among pairs and with time of year. (Some pairs are more aggressive than others and give unison calls more frequently in territorial contexts.) Who's right?

2. Most tropical habitats are highly seasonal. But instead of alternating warm and cold seasons, there are alternating wet and dry seasons. Most animal species breed during the wet months. If you were studying a species of *Anolis* native to the tropics, what environmental cue would you simulate in the lab to bring them into breeding condition? How would you simulate this cue? Further, think about the sensory organs that lizards use to receive this cue. Are they the same as or different than the receptors that *Anolis carolinensis* uses to recognize that spring has arrived in the southeastern United States?

3. What is the significance of the observation that the product of the *for/dg2* gene in fruit flies is involved in a signal transduction pathway? Propose a hypothesis to explain why fruit-fly larvae with the rover genotype and phenotype have higher fitness when population density is high—meaning individuals are crowded.

4. J. B. S. Haldane once remarked that he'd be willing to lay down his life to save two brothers or eight cousins. Explain what he meant. Based on the theory of kin selection and reciprocal altruism, predict the conditions under which people are expected to be nice to one another.

Answers to Multiple-Choice Questions   **1.** c; **2.** d; **3.** a; **4.** c; **5.** d; **6.** c

www.prenhall.com/freeman is your resource for the following: Web Tutorials; Online Quizzes and other Online Study Guide materials; Answers to Conceptual Review Questions; Solutions to Group Discussion Problems; Answers to Figure Caption Questions and Exercises; and Additional Readings and Research.

52

Population Ecology

KEY CONCEPTS

■ Life tables summarize how likely it is that individuals of each age class in a population will survive and reproduce.

■ The growth rate of a population can be calculated from life-table data or from the direct observation of changes in population size over time.

■ A wide variety of patterns is observed when researchers track changes in population size over time, ranging from no growth, to regular cycles, to continued growth independent of population size.

■ Data from population ecology studies help biologists evaluate prospects for endangered species and design effective management strategies.

A flock of white pelicans in a channel of the Mississippi River. This chapter explores how and why growth rates in populations change through time.

A **population** is a group of individuals from the same species that live in the same area at the same time. In both ecology and evolutionary biology, populations are the basic unit of analysis. Evolutionary biology is the study of how the characteristics of populations change through time; ecology is the study of how populations interact with their environment.

Population ecology is the study of how and why the number of individuals in a population changes over time. Biologists also analyze changes in the ages of individuals in a population, the proportion of males to females, and geographic distribution. With the explosion of human populations across the globe, the massive destruction of natural habitats, and the resulting threats to species throughout the tree of life, population ecolo-

gy has become an extraordinarily vital field in biological science. The mathematical and analytical tools introduced in this chapter help biologists predict changes in population size and design management strategies to save threatened species.

Biologists ask a wide array of questions in population ecology. How are individuals distributed in space? How old are they, and how likely are they to reproduce or die? Is the population growing or declining or staying the same through time, and why?

To answer these questions, let's review some of the basic tools that biologists use to study populations, then follow up with examples of how biologists study changes in population size over time. The chapter concludes by asking how all of these elements fit together in efforts to save endangered species.

52.1 Demography

The number of individuals that are present in a population depends on four processes: birth, death, immigration, and emigration. Populations grow due to births—here meaning any form of reproduction—and **immigration**, which occurs when individuals enter a population by moving from another population. Populations decline due to deaths and **emigration**, which occurs when individuals leave a population to join another population. Analyzing birth rates, death rates, immigration rates, and emigration rates is fundamental to **demography**: the study of factors that determine the size and structure of populations through time.

To make detailed predictions about the future of a population, however, biologists need to understand the makeup of the population in more detail. If a population consists primarily of young individuals with a high survival rate and reproductive rate, the population size should increase. But if a population is comprised chiefly of old individuals with low reproductive rates and low survival rates, then it is almost certain to decline. To predict the future of a population, biologists need to know how many individuals of each age are alive, how likely individuals of different ages are to survive to the following year, how many offspring are produced by females of different ages, and how many individuals of different ages immigrate and emigrate each **generation**—the average time between a mother's first offspring and her daughter's first offspring.

These types of demographic data provide an important tool for biologists charged with designing management programs for endangered species. To understand the nature of these data and how they are used in conservation programs, let's examine a classical tool for describing the demography of a population.

Life Tables

Formal demographic analyses of populations are based on a type of data set called a life table. A **life table** summarizes the probability that an individual will survive and reproduce in any given year over the course of its lifetime. Life tables were invented almost 2000 years ago; in ancient Rome they were used to predict food needs. In modern times, life tables have been the domain of life-insurance companies, which have a strong financial interest in predicting the likelihood of a person dying at a given age. More recently, biologists have employed life tables to study the demographics of endangered species and other non-human populations.

To understand how researchers use life tables, consider the lizard *Lacerta vivipara* (**Figure 52.1**). *Lacerta vivipara* is a common resident of open, grassy habitats in Western Europe. Researchers set out to estimate the life table of a low-elevation population in the Netherlands, with the goal of comparing

FIGURE 52.1 *Lacerta vivipara* **Are Native to Europe**
Translated literally, *vivipara* means "live birth." Females in *L. vivipara* populations from northern Spain and southwest France lay eggs; in other populations, females bear live young.

the results to data that other researchers had collected from *L. vivipara* populations in the mountains of Austria and France and in lowland habitats in Britain and Belgium. The researchers wanted to know if populations that interact with different environments vary in basic demographic features.

To complete the study, the researchers visited their study site daily during the seven months that these lizards are active during the year. Each day the researchers captured and marked as many individuals as possible. Because this program of daily monitoring continued for seven years, the biologists were able to document the number of young produced by each female in each year of its life. If a marked individual was not recaptured in a subsequent year, they assumed that it had died sometime during the previous year. These data allowed them to calculate the number of individuals that survived each year in each particular age group and how many offspring each female produced. What did the data reveal?

Survivorship Survivorship is a key component of a life table and is defined as the proportion of offspring produced that survive, on average, to a particular age. For example, suppose 1000 *L. vivipara* are born in a particular year. These individuals represent a **cohort**—a group of the same age that can be followed through time. How many individuals would survive to age 1, age 2, age 3, and so on?

As **Table 52.1** (page 1194) shows, the biologists calculated that, in the population living in the Netherlands, survivorship from birth to age 1 as 0.424. If 1000 females were born in a particular year in this population, on average 424 would still be alive one year later. Survivorship from birth to age 2 was 0.308, meaning an average of 308 female lizards would survive for two years.

TABLE 52.1 Life Table for *Lacerta vivipara*

Year	Number alive	Survivorship	Fecundity	Survivorship × fecundity = average number of offspring produced per female born
0	1000	1.000	0.00	0.00
1	424	0.424	0.08	0.03
2	308	0.308	2.94	0.91
3	158	0.158	4.13	0.65
4	57	0.057	4.88	0.28
5	10	0.010	6.50	0.07
6	7	0.007	6.50	0.05
7	2	0.002	6.50	0.01

Data are from Strijbosch and Creemers, 1988.

To recognize general patterns in survivorship and make comparisons among populations or species, biologists plot the logarithm of the number of survivors versus age. The resulting graph is called a **survivorship curve**. Studies on a wide variety of species indicate that three general types of survivorship curves exist (**Figure 52.2a**). Humans have what biologists call a type I survivorship curve. In this pattern, survivorship throughout life is high, and most individuals approach the species' maximum life span. Songbirds, in contrast, experience relatively constant mortality throughout their lives once they have left the nest, resulting in a type II survivorship curve. Many plants have type III curves due to extremely high death rates for seeds and seedlings. **Figure 52.2b** provides a graph for you to plot survivorship of *L. vivipara*.

Survivorship curves are important in conservation work, because they pinpoint the stage of life when endangered species have particularly low survivorship. Biologists can then put measures into place to protect individuals during that period. For example, trees or herbaceous plants can be protected from drought when they are very young by germinating them in a nursery and from predation by putting a wire cage over seedlings.

Fecundity The number of female offspring produced by each female in the population is termed **fecundity**. (In most cases, biologists only keep track of females when calculating life-table data, because only females produce offspring.) Researchers documented the reproductive output of the same *L. vivipara* lizard females year after year and thus were able to calculate a quanti-

(a) Three general types of survivorship curves

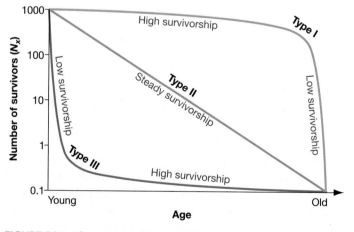

(b) Exercise: Survivorship curve for *Lacerta vivipara*

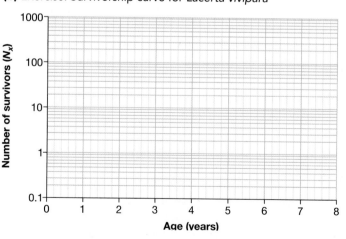

FIGURE 52.2 Three General Types of Survivorship Curves
(a) Generalized graphs showing the number of survivors, on a logarithmic scale, plotted against age.
EXERCISE Fill in the graph in part **(b)** with data on survivorship of *Lacerta vivipara* given in Table 52.1. Compare the shape of the curve to the generalized graphs in part (a).

ty called **age-specific fecundity**, which is defined as the average number of female offspring produced by a female in age class *x*. An **age class** is a group of individuals of a specific age—for instance, all female lizards between 4 and 5 years old. As **Box 52.1** (page 1196) shows, data on survivorship and fecundity allow researchers to calculate the growth rate of a population. How do data on survivorship and fecundity in the Netherlands populations compare with populations in different types of habitats?

The Role of Life History

The life-table data in Table 52.1 are interesting because they contrast with results from other populations of *L. vivipara*. For example, notice that almost no 1-year-old female *L. vivipara* lizards in the Netherlands breed. But in Brittany, France, 50 percent of 1-year-old female lizards breed. Both situations contrast sharply with data on females in the mountains of Austria. In populations that live at high elevation, females do not begin breeding until they are 4 years old. In addition, most females in the Austrian population live much longer than do individuals in either lowland population—in the Netherlands or France. When the three sites are compared, it is clear that fecundity is high but survivorship is low in Brittany. In contrast, fecundity is low but survivorship is high in Austria. The population in the Netherlands is intermediate in both characteristics.

To make sense of this type of variation in fecundity and survivorship, biologists point out that every individual has a restricted amount of energy and resources at its disposal. Resources are limited. As a result, there is a fundamental trade-off between the size of eggs that a female can produce and the number of eggs that she can produce. No female can produce a very large number of very large eggs. Similarly, individuals may allocate resources to reproduction and achieve high fecundity, or they may allocate resources to increasing survivorship. But as the data in **Figure 52.3** indicate, it is not possible to achieve

both high fecundity and high survivorship. To capture this point, biologists say that a trade-off exists between survival and reproduction.

Organisms with high fecundity tend to grow quickly, reach sexual maturity at a young age, and produce many small eggs or seeds. The mustard plant *Arabidopsis thaliana*, for example, germinates and grows to sexual maturity in just four to six weeks. In this species, individuals usually live only a few months but may produce as many as 10,000 tiny seeds. In contrast, organisms with high survivorship tend to grow slowly and invest resources in traits that reduce damage from enemies and increase their own ability to compete for water, sunlight, or food. An oak tree, for example, may take decades to reach sexual maturity but can live for several hundred years and produce offspring each year. Oaks invest resources in making wood, an extensive root system, large seeds (acorns), and bitter-tasting molecules (tannins) that deter predators. These traits increase the survivorship of both seedlings and adults but decrease fecundity.

An organism's **life history** consists of how an individual allocates resources to growth, reproduction, and activities or structures that are related to survival. An *Arabidopsis thaliana* plant and an oak tree represent two ends of a broad continuum that exists in life-history characteristics (**Figure 52.4**). Even within the species *Lacerta vivipara*, analyses of life tables have shown that there is considerable variation in life-history traits.

Research has also documented that, within populations, life-history traits can change if conditions change. For example, suppose that a series of wet years produced an abundance of food for lizards. Females might survive better in response, or they might begin breeding at an earlier age or produce more offspring at a particular age. Such changes might directly affect a population's growth rate. How do biologists analyze growth rates in populations? And what factors make growth rates increase or decrease?

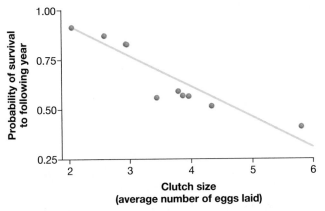

FIGURE 52.3 There Is a Trade-off between Survival and Reproduction
Graph of survivorship versus fecundity. Each data point represents a different species of bird. A statistical analysis called regression was used to create a "line of best fit" to these data points.

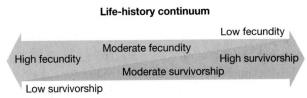

FIGURE 52.4 Life-History Traits Form a Continuum
Every organism can be placed somewhere on this life-history continuum. The placement of a species is most meaningful when it is considered relative to closely related species. **EXERCISE** Place *Arabidopsis thaliana* and an oak tree on the continuuum. For plants, list the types of traits you would expect to see on the left, middle, and right of the continuum. Include growth habit (herbaceous, shrub, or tree) as well as relative disease- and predator-fighting ability, seed size, seed number, body size, and investment in roots versus shoots.

BOX 52.1 Using Life Tables to Calculate Population Growth Rates

If immigration and emigration are ignored, the data in a life table can be used to calculate a population's growth rate. This is logical, because survivorship and fecundity are ways to express death rates and birth rates—the other two factors that influence population size. To see how a population's growth rate can be estimated from life-table data, let's look at each component in a life table more carefully.

Survivorship is symbolized as l_x, where x represents the age class being considered. Survivorship for age class x is calculated by dividing the number of individuals in that age class (N_x) by the number of individuals that existed as offspring (N_0):

$$l_x = \frac{N_x}{N_0} \qquad \text{(Eq. 52.1)}$$

Age-specific fecundity is symbolized m_x, where x again represents the age class being considered. Age-specific fecundity is calculated as the total number of offspring produced by females of a particular age, divided by the total number of females of that age class present. It represents the average number of offspring produced by a female of age x.

Documenting age-specific survivorship and fecundity allows researchers to calculate the net reproductive rate, R_0, of a population:

$$R_0 = \sum_{i=0}^{x} l_x m_x \qquad \text{(Eq. 52.2)}$$

The **net reproductive rate** represents the growth rate of a population per generation. The logic behind the equation for R_0 is that the growth rate of a population per generation is identical to the average number of female offspring that each female produces over the course of her lifetime. A female's average lifetime reproduction, in turn, is a function of survivorship to each age class and fecundity at each age class. In Table 52.1, R_0 is the sum of the survivorship × fecundity values in the rightmost column.

If R_0 is greater than 1, then the population is increasing in size. If R_0 is less than 1, then the population is declining. According to the data in Table 52.1, R_0 in the Netherlands population of *L. vivipara* is 2.0—in other words, an average *L. vivipara* female produces 2.00 female offspring over the course of her lifetime. When this study was being conducted, the population in the Netherlands was growing rapidly.

52.2 Population Growth

The most fundamental questions that biologists ask about populations involve growth or decline in numbers of individuals. For conservationists, the ability to analyze and predict changes in population size is fundamental in efforts to manage threatened species.

Recall that four processes affect a population's size: Births and immigration add individuals to the population; deaths and emigration remove them. It follows that a population's overall growth rate is a function of birth rates, death rates, immigration rates, and emigration rates. In this section we consider only the impact of births and deaths on population growth; Section 52.4 includes an analysis of the movement of individuals into and out of populations.

A population's growth rate is the change in the number of individuals in the population (ΔN) per unit time (Δt). If no immigration or emigration is occurring, then the growth rate is equal to the number of individuals (N) in the population times the difference between the birth rate per individual (b) and death rate per individual (d). The difference between the birth rate and death rate per individual is called the **per-capita rate of increase** and is symbolized r. (*Per capita* means "per individual.") If the per-capita birth rate is greater than the per-capita death rate, then r is positive and the population is growing. But if the per-capita death rate begins to exceed the per-capita birth rate, then r becomes negative and the population declines. Within populations, r varies through time and can be positive, negative, or 0.

When conditions are optimal for a particular species—meaning birth rates per individual are as high as possible and death rates per individual are as low as possible—then r reaches a maximal value called the **intrinsic rate of increase, r_{max}.** When this occurs, a population's growth rate is expressed as

$$\Delta N/\Delta t = r_{max} N$$

Exponential Growth

The graphs in **Figure 52.5** plot changes in population size for various values of the intrinsic rate of increase. In species such as *Arabidopsis* and fruit flies, which breed at a young age and produce many offspring each year, r_{max} is high. In contrast, r_{max} is low in species such as giant pandas and oak trees, which take years to mature and produce few offspring each year. Stated another way, r_{max} is a function of a species' life-history traits.

The sweeping curves in Figure 52.5 illustrate exponential growth. **Exponential growth** occurs when r does not change over time. The key point about exponential growth is that the growth rate does not depend on the number of individuals in the population. When increases in the size of the population do not affect r, biologists say that growth is **density independent**.

It's important to emphasize that exponential growth adds an increasing number of individuals as the total number of individuals N gets larger. As an extreme example, an r of 0.02 per year in a population of 1 billion adds over 20 million individuals per year. The same growth rate in a population of 100 adds just over 2 individuals per year. Even if r is constant, the number of individuals added to a population is a function of N.

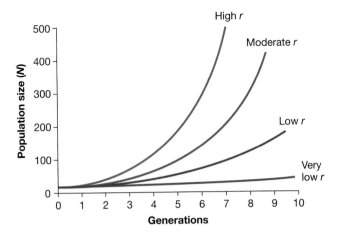

FIGURE 52.5 Exponential Growth Is Independent of Population Size
When the per-capita growth rate r does not change over time, exponential growth occurs. Population size may increase slowly or rapidly, depending on the size of r. Exponential growth occurs when growth is not affected by population density. QUESTION Consider a species on the left-hand side of the life-history continuum in Figure 52.4 (high fecundity, low survivorship) versus a species on the right-hand side (low fecundity, high survivorship). Under ideal conditions, which species has higher r? Explain your reasoning.

In nature, exponential growth is observed when a few individuals found a new population in a new habitat or when a population has been devastated by a storm or some other type of catastrophe and then begins to recover, starting with a few surviving individuals. But it is not possible for exponential growth to continue indefinitely. If *Arabidopsis* populations grew exponentially for a long period of time, they would eventually fill all available breeding habitat. When **population density**—the number of individuals per unit area—gets very high, we would expect the population's per-capita birth rate to decrease and the per-capita death rate to increase, causing r to decline. Stated another way, growth would be **density dependent**.

Logistic Growth

To analyze what happens when growth is density dependent, biologists use a parameter called carrying capacity. **Carrying capacity, K**, is defined as the maximum number of individuals in a population that can be supported in a particular habitat over a sustained period of time. The carrying capacity of a habitat depends on a large number of factors: food, space, water, soil quality, resting or nesting sites, and the intensity of disease and predation. Carrying capacity can change from year to year, depending on conditions.

If a population of size N is below the carrying capacity K, then the population should continue to grow. More specifically, a population's growth rate should be proportional to $(K - N)/K$:

$$\frac{\Delta N}{\Delta t} = r_{max} N \left(\frac{K - N}{K} \right)$$

This expression is called the **logistic growth equation**. In the expression $(K - N)/K$, the numerator defines the number of additional individuals that can accommodated in a habitat with carrying capacity K; dividing $K - N$ by K turns this number of individuals into a proportion. Thus, $(K - N)/K$ describes the proportion of "unused resources and space" in the habitat. It can be thought of as the environment's resistance to growth. When N is small, then $(K - N)/K$ is large and the growth rate should be high. But as N gets larger, $(K - N)/K$ gets smaller. When N is at carrying capacity, then $(K - N)/K$ is equal to 0 and growth should stop. The key idea here is that as a population approaches a habitat's carrying capacity, its growth rate should slow.

The logistic growth equation describes **logistic population growth**, or changes in growth rate that occur as a function of population size. Just as exponential growth is density independent, logistic growth is density dependent. **Box 52.2** explores population growth models and how they are applied in more detail.

WEB TUTORIAL 52.1
Modeling Population Growth

BOX 52.2 Developing and Applying Population Growth Equations: A Closer Look

To explore how biologists model changes in population size in more detail, let's consider data on whooping cranes, which are large, wetland-breeding birds native to North America. Whooping cranes may live more than 20 years in the wild and over 40 years in captivity, but it is common for females to be six or seven years old before they breed for the first time. Female cranes lay just two eggs per year and usually rear just one chick. It is logical to expect, based on these life-history characteristics, that the growth rate of whooping crane populations will be extremely low.

Conservationists monitor this species closely, because hunting and habitat destruction reduced the total number of individuals in the world to about 20 in the mid-1940s. Since then, intensive conservation efforts have resulted in a current total population of about 425. That number includes a group of 189 individuals that breeds in Wood Buffalo National Park in the Northwest Territories of Canada.

In addition to the Wood Buffalo population, two new populations of cranes have been established recently by releasing offspring raised in captivity. One of

(Continued on next page)

(Box 52.2 continued)

these groups lives near the Atlantic coast of Florida year round; the other group migrates from breeding areas in northern Wisconsin to wintering areas along Florida's Gulf coast.

According to the biologists who are in charge of managing whooping crane recovery, the species will no longer be considered endangered when the two newly established populations have at least 25 breeding pairs—meaning there would be about 125 individuals in each—and when neither population needs to be supplemented with captive-bred young in order to be self-sustaining. How long will this take?

Discrete Growth

Whooping cranes breed once per year, so the simplest way to express a crane population's growth rate is to compare the number of individuals at the start of one breeding season to the number at the start of the following year's breeding season. The current total at Wood Buffalo is 189. Suppose that biologists count 198 cranes on the breeding grounds next year. The population growth rate could be calculated as $198/189 = 1.048$. Stated another way, the population will have grown at the rate of 4.8 percent per year.

When populations breed during discrete seasons, their growth rate can be calculated as for whooping cranes. To create a general expression for how these populations grow, biologists use N to symbolize population size. N_0 is the population size at time zero (the starting point), and N_1 the population size one breeding interval later. In equation form, the growth rate is given as

$$N_1/N_0 = \lambda \qquad \text{(Eq. 52.3)}$$

The parameter λ (lambda) is called the **finite rate of increase**. (In mathematics, a *finite rate* refers to an observed rate over a given period of time. A *parameter* is a

variable or constant term that affects the shape of a function but does not affect its general nature.) In the whooping crane example, λ was 1.048. Rearranging the expression in Equation 52.3 gives

$$N_1 = N_0 \lambda \qquad \text{(Eq. 52.4)}$$

Stated more generally, the size of the population at the end of time t will be given by

$$N_t = N_0 \lambda^t \qquad \text{(Eq. 52.5)}$$

This equation summarizes how populations grow when breeding takes place seasonally. The size of the population at time t is equal to the starting size, times the finite rate of increase multiplied by itself t times. In a sense, λ works like the interest rate at a bank. For species that breed once per year, the "interest" on the population is compounded annually. A savings account with a 5 percent annual interest rate increases by a factor of 1.05 per year. If a population is growing, then its λ is greater than one. The population is stable when λ is 1.0 and declining when λ is less than 1.0.

Continuous Growth

The parameters λ and r—the finite rate of increase and the per-capita growth rate—have a simple mathematical relationship. The best way to understand their relationship is to recall that λ expresses a population's growth rate over a discrete interval of time. In contrast, r gives the population's per-capita growth rate at any particular instant. As a result, r is also called the *instantaneous rate of increase*. The relationship between the two parameters is given by

$$\lambda = e^r \qquad \text{(Eq. 52.6)}$$

where e is the base of the natural logarithm, or about 2.72. (In general, the relationship between any finite rate and any instantaneous rate is given by finite rate $= e^{\text{instantaneous rate}}$.)

Substituting Equation 52.6 into Equation 52.5 gives

$$N_t = N_0 e^{rt} \qquad \text{(Eq. 52.7)}$$

This expression summarizes how populations grow when they breed continuously, as do humans and bacteria, instead of at defined intervals. For species that breed continuously, the "interest" on the population is compounded continuously. When the growth rates λ and r are equivalent, however, the differences between discrete and continuous growth are negligible.

Because r represents the growth rate at any given time, and because r and λ are so closely related, biologists routinely calculate r for species that breed seasonally. In the whooping crane example, $\lambda = 1.048 = e^r$. To solve for r, take the natural logarithm of both sides. In this case, $r = 0.0469$.

The instantaneous rate of increase, r, is also directly related to the net reproductive rate, R_0, introduced in Box 52.1. In most cases, r is calculated as $\ln R_0/g$, where g is the generation time. Thus, r can be calculated from life-table data. It is a more useful measure of growth rate than R_0, because r is independent of generation time.

To summarize, biologists have developed several ways of calculating and expressing a population's growth rate. λ has the advantage of being easy to understand, and R_0 has the advantage of being calculated directly from life-table data. Although r is slightly more difficult conceptually, it is the most useful expression for growth rate, because it is independent of generation time and is relevant for species that breed either seasonally or continuously.

Applying the Models

To get a better feel for r and for Equation 52.7, consider the following series of questions about whooping cranes. The key to answering these questions is to realize that Equation 52.7 has just four parameters. Given three of these parameters, you can calculate the fourth:

1. If 20 individuals were alive in 1941 and 425 existed in 2004, what is r? Here $N_t = 425$, $N_0 = 20$, and $t = 63$ years. Substitute these values into Equation 52.7 and solve for r. Then check your answer at the end of this box.[1]

2. In the most recent report issued by the biologists working on the Wood Buffalo crane recovery program, it is estimated that the flock should be able to sustain an r of 0.046 for the foreseeable future. If the flock currently contains 189 individuals, how long will it take that population to double? Here $N_0 = 189$ and $N_t = 2 \times 189 = 378$. In this case,

3. In 2002 a pair of birds in the flock that lives in Florida year-round successfully raised offspring (nine years after the first cranes were introduced there). In 2003 another pair of cranes in this population bred successfully. If the number of breeding pairs continues to double each year, how long will it take to reach the goal of 25 breeding pairs? (Note that $\lambda = 2.0$ if a population is doubling each year.) Given that $N_0 = 2$ and $N_t = 25$, solve for t again, and check your answer.[3]

4. The whooping crane flock that migrates between Wisconsin and Florida was founded in 2001. If its

you solve Equation 52.7 for t. Then check your answer.[2]

development is like that of the resident flock in Florida, the first successful breeding attempt will occur in 2010—nine years after the initial introduction. Suppose that the instantaneous growth rate for the number of breeding pairs in this population will be 0.05. In what year will the breeding population reach 25 pairs? This is the year that whooping cranes should come off the endangered species list. Here $N_0 = 1$, $N_t = 25$, and $r = 0.05$. Solve for t, and add this number of years to 2010; then check your answer.[4]

[1] $r = 0.05$
[2] $t = 15$ years
[3] $t = 3.64$ years
[4] 2074

Figure 52.6a illustrates density-dependent growth for a hypothetical population. Notice that the graph has three sections. Initially, growth is rapid. With time, N increases to the point at which competition for resources or other density-dependent factors begins to occur. As a result, the growth rate begins to decline and eventually reaches 0. When the population is at the habitat's carrying capacity, the graph of population size versus time is flat.

Figure 52.6b shows data on logistic growth from laboratory populations of two species of protists in the group called ciliates (see Chapter 28): *Paramecium aurelia* and *P. caudatum*. In this experiment, an investigator placed 20 individuals from one of the *Paramecium* species into 5 mL of a solution. He set up made many replicates of these 5-mL environments for each species separately. He kept conditions as constant as possible in each replicate by adding the same number of bacterial cells

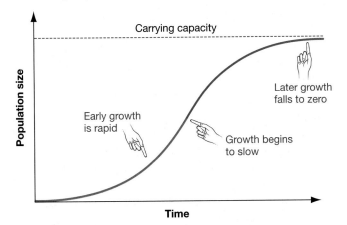

(a) Density dependence: growth rate is a function of population size.

(b) Data from laboratory experiments

FIGURE 52.6 Logistic Growth Is Dependent on Population Size
(a) Hypothetical curve illustrating the pattern predicted by the logistic growth equation. This pattern often occurs when a small number of individuals colonizes an unoccupied habitat. Initially, *r* is high because competition for resources is low to nonexistent. Carrying capacity depends on the quality of the habitat.
(b) Data from laboratory experiments with two species of *Paramecium*. EXERCISE In part (b), draw dashed horizontal lines indicating the carrying capacity for each species.

every day as food, washing the solution every second day to re- move wastes, and maintaining the pH at 8.0. As the graphs show, both species exhibited logistic growth in this environ- ment. The carrying capacity differed in the two species, however. The maximum density of *P. aurelia* averaged 448 individuals per mL, but that for *P. caudatum* averaged just 128 individuals per mL.

To summarize, exponential growth can occur only for a rel- atively short time. Eventually a population's growth rate has to decline. Why? What factors cause growth rates to change?

What Limits Growth Rates and Population Sizes?

Population sizes change as a result of two general types of fac- tors: (1) density-independent and (2) density-dependent factors. **Density-independent factors** change birth rates and death rates irrespective of the number of individuals in the population. Density-independent changes are usually triggered by changes in the abiotic environment—variation in weather patterns or catastrophic events such as cold snaps, hurricanes, volcanic eruptions, or drought. In contrast, **density-dependent factors** are usually biotic and change in intensity as a function of pop- ulation size. For example, predation rates on deer may increase when population sizes are high, or competition for food and starvation may become more common. When trees are crowd- ed, they have less water, nutrients, and sunlight at their dispos- al and make fewer seeds. Death rates may increase and birth rates decrease when populations are at high density.

To get a better understanding of how biologists study factors that affect population size, consider the data in **Figure 52.7**:

- The graph in Figure 52.7a is from an experimental study of a coral-reef fish called the bridled goby. Each data point repre- sents an identical artificial reef, constructed by a researcher from the rubble of real coral reefs. The initial density, plotted along the horizontal axis, is the number of marked bridled go- bies that were released on each artificial reef at the start of the experiment. The proportion surviving, plotted on the vertical axis, is the number of these introduced individuals that were still living at that reef 2.5 months later. This graph shows a strong density-dependent relationship in survivorship.

- The graph in Figure 52.7b is from a long-term study of song sparrows on Mandarte Island, British Columbia. Each data point represents a different year. The density of females, plot- ted along the horizontal axis, is the number of females that bred on the island; the clutch size, plotted on the vertical axis, is the average number of eggs laid by each female. This graph indicates a strong density-dependent relationship in fecundity.

Density-dependent changes in on survivorship and fecundity cause logistic population growth. In this way, density-dependent factors define a particular habitat's carrying capacity. If gobies get crowded, they die or emigrate. If song sparrows are crowded, they cannot produce a normal number of eggs and offspring.

(a) Survival varies as a function of population density.

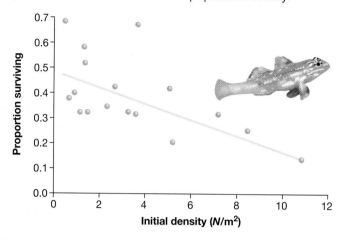

(b) Fecundity varies as a function of population density.

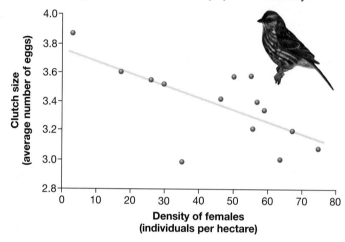

FIGURE 52.7 Density-Dependent Growth Results from Changes in Survivorship and Fecundity
(a) When bridled gobies are introduced at different population densities to artificial reefs, survivorship is highest at low density. **(b)** In the song sparrow population on Mandarte Island, there is a strong negative correlation between population density in a particular year and the average number of eggs (clutch size) produced by females. QUESTION When song sparrow populations are at high density and extra food is provided to females experimentally, average clutch size is much higher than expected from the data in part (b). Based on these data, state a hypothesis to explain why average clutch size declines at high density.

It's important to recognize that *K* varies among species and populations, however, and that this variation affects both growth rates and population sizes. *K* varies because for any par- ticular species, some habitats are better than other habitats due to differences in food availability, predator abundance, and other density-dependent factors. Stated another way, *K* varies in space. It also varies with time, as conditions in some years are better than others. In addition, the same region may have a very different carrying capacity for different species. The same area will tend to support many more individuals of a small-bodied species than of a large-bodied species, simply because large individuals de- mand more space and resources. These simple observations help explain the variation in total population size that exists among species and among populations of the same species.

When populations of the same species are considered, r_{max} may vary as well. To drive this point home, recall that life-table data vary in low-elevation and high-elevation populations of *Lacerta vivipara* lizards. Because female lizards in low-elevation areas mature quickly and begin breeding early in life, they have higher lifetime fecundity than do females in alpine regions, which start breeding much later. Thus, r_{max} is higher in lowland populations. In general, species or populations that have relatively long life spans and large body size tend to have much lower r_{max} than do small-bodied, short-lived species.

✔CHECK YOUR UNDERSTANDING

Populations grow exponentially unless slowed by density dependence. Density-dependent factors that influence population growth rates include predation, disease, and competition for resources such as food and sunlight. You should be able to (1) propose a density-dependent factor that limits growth in *Paramecium aurelia* and design an experiment to test this hypothesis, and (2) propose a hypothesis to explain why the carrying capacity for *P. caudatum* in Figure 52.6b is so much lower than the carrying capacity for *P. aurelia* and design an experiment to test this hypothesis.

52.3 Population Dynamics

The theory introduced in the previous two sections provides a foundation for exploring how biologists study **population dynamics**—changes in populations through time. As an initial example, consider data on the longest-running experiment in the history of biological science: the Park Grass study in Rothamsted, United Kingdom. In 1856 researchers established a series of 0.1–0.2 ha (about 1/3-acre) plots in a hay meadow that had uniform soil characteristics and vegetation. Since then, some of the plots have received regular fertilizer treatments or lime (a mix

of calcium and magnesium oxides), and some have been left untreated as controls. Each year researchers recorded which plant species were present in each plot, as an index of population size.

The weather was the same in all plots, the fertilizing and liming regimes were constant in treated plots, and presumably all plots were exposed to the same predators and diseases. Even so, several dramatic changes in population size have taken place. When the data from 1920–1979 were analyzed, four major patterns emerged among the 43 species present. Ten of the species showed a clear spike in abundance, followed by a decline over the course of the 60-year period (**Figure 52.8a**). Seven species increased in abundance and then reached a plateau (**Figure 52.8b**). Five species declined steadily, and 21 showed no discernible change (**Figure 52.8c**).

In several cases, patterns of change over time appeared to correlate with life-history traits. For example, species that maintained a constant population size throughout the 60-year interval have longer life spans and a lower r_{max} than do other species at the site. To interpret this observation, biologists suggest that these species allocate more energy and resources to competitive ability than to reproduction. As a result, these species are able to maintain abundance over a longer period. In contrast, species that spiked and declined or that increased and maintained high abundance had a higher r_{max}, an earlier age at first reproduction, and shorter life spans than did species that maintained a constant population size. Researchers are still working on the question of why some of these species remained constant over time and why other species showed gradual declines over the course of the study.

Population Cycles: The Case of the Red Grouse

One of the most striking patterns in population dynamics has also been one of the most difficult to understand. **Population cycles** are regular fluctuations in size that some animal populations exhibit. For example, red grouse are native to heather moorlands in Britain. Over three-fourths of red grouse populations in Britain

(a) Some species increased, then declined.

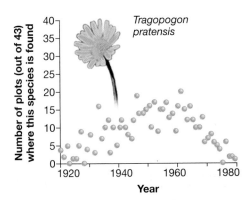

(b) Some species increased and maintained high population size.

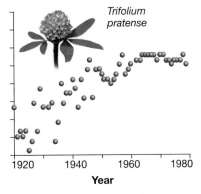

(c) Some species maintained high population size.

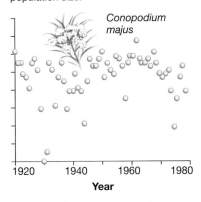

FIGURE 52.8 Distinct Patterns of Long-Term Changes in the Populations of Meadow Species
Graphs showing some of the distinct patterns in population dynamics from 1920 to 1979 among species in the Park Grass experiment, which began in 1856.

rise and fall in regular cycles. Grouse cycles in Britain average between four and eight years in length.

Most hypotheses to explain population cycles hinge on some sort of density-dependent factor. The idea is that predation, disease, or food shortages intensify dramatically at high population density and cause population numbers to crash. It has been extremely difficult, however, to document a causal relationship between population cycles and factors such as predation intensity, disease outbreaks, or food availability.

What factor causes red grouse populations to cycle? A group of biologists suspected that a parasitic roundworm called *Trichostrongylus tenuis* is responsible. They contended that infection rates skyrocket when grouse densities are high, causing populations to drop rapidly. The biologists also hypothesized that infection rates lessen when grouse densities are low, allow-

ing populations to recover. The idea here is that the parasite is transmitted readily at high population densities, and that individuals in dense populations are less well fed and thus less able to fend off an infection.

To test these propositions experimentally, the researchers caught from 1000 to 3000 red grouse in each of several populations during the year when a crash was expected. They treated the individuals with a drug that kills roundworms, released the grouse, and later monitored the number of birds shot by hunters in the ensuing years in each of the treatment populations. Finally, they compared these data to hunting success in two nearby populations where hunting intensity was comparable, and where birds were not treated with the anti-roundworm drug.

Some results of the experiment are shown in **Figure 52.9**. The control populations showed a typical dramatic four-year

Question: What factors cause red grouse populations to cycle?

Hypothesis: Density-dependent infections by a parasitic roundworm cause red grouse populations to cycle.

Null hypothesis: Parasitic roundworm infections do not influence red grouse population cycles.

Experiment setup:

Experimental populations

1. Treat red grouse with anti-roundworm drug during a year when a population crash is expected.

2. Monitor the number of grouse shot by hunters in ensuing years.

Control populations

1. No drug treatment.

2. Monitor the number of grouse shot by hunters in ensuing years.

Prediction: The experimental populations will not crash following the treatments, but the control populations will.

Prediction of null hypothesis: Both the experimental and control populations will crash.

Results:

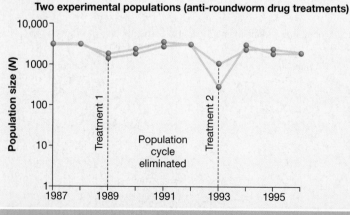

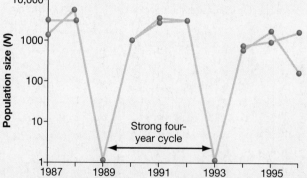

Conclusion: Density-dependent factors such as parasite infection cause red grouse populations to cycle.

FIGURE 52.9 Experimental Evidence That Disease Outbreaks Cause Population Cycles in Red Grouse
The data points on the graphs indicate the number of birds shot by hunters each year out of experimental and control populations. The number killed is interpreted as an index of total population size.

cycle in numbers, but the treated populations maintained high densities. This is strong evidence that red grouse cycles are driven by density-dependent changes in disease rates. The study also illustrates the power of a well-designed experiment to solve long-standing controversies in biology.

Age Structure

As the life-table data presented in Section 52.1 showed, age has a dramatic effect on the probability that an individual will reproduce and survive to the following year. But the previous analysis of life-table data left out a critical point about the ages of individuals in a population: A population's **age structure**—meaning the proportion of individuals that are at each possible age—has a dramatic influence on the population's growth over time. To see how changes in age structure can affect population dynamics, let's consider two case histories for which biologists have documented changes in age structure—one example concerns a flowering plant, and the other focuses on our own species.

Age Structure in a Woodland Herb   The common primrose, pictured in **Figure 52.10a**, grows in the woodlands of Western Europe. Primroses grow on the forest floor and can germinate, mature, and produce offspring only in the relatively high-light environment created when a tall tree falls and opens a gap in the forest canopy. These sunny spaces are short lived, however, because mature trees and saplings in and around the gap grow and fill the space. As they do, light levels in the gap decline and the environment becomes increasingly unsuitable for herbs such as primrose.

Biologists recently set out to document how primrose populations respond to this ephemeral woodland environment. The researchers hypothesized that populations in new gaps and populations in large gaps—meaning those with the highest light availability—would experience high rates of reproduction and an age structure characterized by large numbers of juveniles. The researchers also hypothesized that as light levels declined in a particular gap due to the growth of surrounding trees, primrose reproduction would decline. As a result, the age structure of populations in older gaps should be characterized by a large proportion of individuals in adult stages, with fewer juveniles.

To test these hypotheses, the investigators selected eight common primrose populations growing in a range of light levels, from high to low. The team established 1-square-meter study plots inside each gap and studied approximately 350 individuals per population. Each individual was assigned to one of five ages: seedlings, juveniles, or adults in one of three size classes. As **Figure 52.10b** shows, the data provided strong support for their hypothesis that in gaps with high light levels, populations have a larger proportion of juveniles than do gaps with low light levels. In addition, the proportion of juvenile individuals declined over time in all gaps as light levels declined. These

data have three important messages about the dynamics of primrose populations:

1. Populations that are dominated by juveniles should experience rapid growth. Population size should then decline with time, however, due to a density-independent factor: shading by trees.

2. The long-term trajectory of the overall primrose population in an area may depend primarily on the frequency and severity of windstorms that knock down trees and create sunlit gaps. If so, it means that the dynamics of primrose populations are governed by an abiotic, density-independent factor.

3. In a large tract of forest, the primrose population will consist of a large number of subpopulations, each found in

(a) Common primroses live in sunlit gaps in forests.

(b) Age structure of primrose population varies with age of gap.

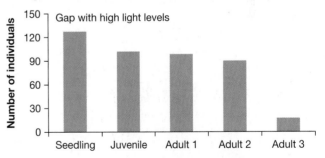

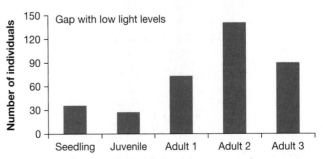

FIGURE 52.10 The Age Structure of Common Primrose Populations Changes over Time
(a) Common primroses are forest herbs. **(b)** "Adult 1," "Adult 2," and "Adult 3" individuals are of increasingly large size and thus age.

canopy gaps. Some of these subpopulations are likely to be very small, and each subpopulation will arise, grow, and then go extinct over time. Biologists say that a group of small, isolated populations such as this represents a **metapopulation**—a population of populations. Metapopulation structure turns out to be extremely common. The history of a metapopulation is driven by the birth and death of populations, just as the dynamics of a single population are driven by the birth and death of individuals. As Section 52.4 will show, this aspect of population structure has important implications for conservation biology.

Age Structure in Human Populations As another example of how age structure affects population dynamics, consider humans. In countries where industrial and technological development occurred many decades ago, survivorship has been high and fecundity low for several decades. As a result, these populations are not growing quickly. When researchers create a graph with horizontal bars representing the numbers of males and females of each age group for such a population, an **age pyramid** like that in **Figure 52.11a** results. The most striking pattern in the data is that there are similar numbers of people in most age classes. The evenness occurs because the same number of infants are being born each year and because most survive to old age. In this way, analyzing an age pyramid can give biologists important information about a population's history.

Studying age distributions can also help researchers predict a population's future. For example, the populations of developed nations are not expected to grow especially quickly, because only modest numbers of individuals will reach reproductive age in the near future. The red and blue lines in Figure 52.11a show the projected age structure of a developed nation (Sweden) in 2050. Modest changes in the age distribution have occurred since 2000 because survivorship has increased while fecundity has remained the same. The projections highlight a major public policy concern in countries such as this: how to care for an increasingly aged population.

In contrast, the age distribution is bottom-heavy in the less-developed nations. As **Figure 52.11b** shows, these populations are dominated by the very young. This type of age distribution occurs when populations have undergone rapid growth—with more children being born than exist in older age classes. Due to dramatic improvements in health care, most people in the younger age classes now survive to reproductive age.

For Figure 52.11b, the projected age distribution in 2050 is based on continued high survivorship and a pattern that has occurred repeatedly around the world: When infant mortality drops in a country, fecundity rates begin to decline within a generation or two. The red and blue lines in Figure 52.11b are relatively even for ages 0–55—meaning that rapid growth has stopped—because fecundity is expected to be relatively low and survivorship high. The red and blue lines also illustrate a major public policy concern in less-developed countries—providing

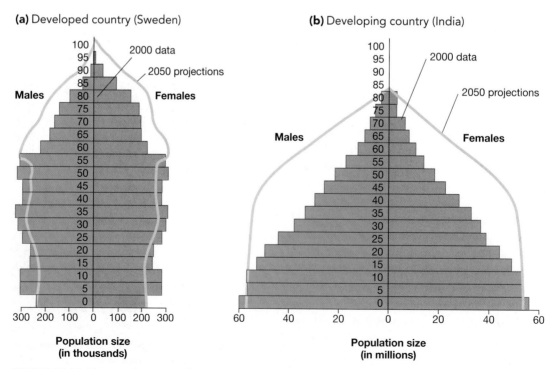

(a) Developed country (Sweden)

(b) Developing country (India)

Population size
(in thousands)

Population size
(in millions)

FIGURE 52.11 The Age Structure of Human Populations Varies Dramatically
Age distributions showing the number of males and females in 5-year age increments from 0 to 100.

education and jobs for an enormous group of young people who will be reaching adulthood. As these examples show, understanding age structure is a key component of analyzing population dynamics in humans and other species.

Analyzing Change in the Growth Rate of Human Populations

Studies on changes in whooping crane and Park Grass populations over 60-year periods have been extraordinarily valuable. But six decades turns out to be short in comparison to data available on human population dynamics.

Archeological and anthropological data have been used to estimate the size of human populations over the past 250 years (**Figure 52.12**). Although the shape of the curve is similar to the examples of exponential growth in Figure 52.5, the growth rate for humans has increased over time since 1750, leading to a very steeply rising curve over the past few decades. The highest growth rates occurred between 1965 and 1970, when populations increased at an average of 2.04 percent per year.

Since 1970, however, the growth rate in human populations has been dropping. Between 1990 and 1995 the overall growth rate in human populations averaged 1.46 percent per year; currently the rate is 1.2 percent annually. In humans, r may be undergoing the first long-term decline in history. What will human population size be at its peak?

To answer this question, the UN Population Division makes regular projections for how population will change between now and 2050. The UN projections are based on three scenarios. These scenarios hinge on different values for fertility rates—the average number of children that each woman has during her lifetime. Currently, the worldwide average is 2.7. (This is a huge reduction from the fertility rates during the 1950s, which averaged 5.0 children per woman.) **Figure 52.13a** shows how total population size is expected to change by 2050 if average fertility continues to decline and averages 2.5, 2.1, or 1.7 children per woman. The middle number, 2.1, is particularly important because it represents the replacement

rate. When fertility is at the **replacement rate**, each woman produces exactly enough offspring to replace herself and her offspring's father. When this fertility rate is sustained for a generation, $r = 0$—meaning there is **zero population growth (ZPG)**.

A glance at Figure 52.13a should convince you that the three scenarios are starkly different. The high projection predicts that world population in 2050 will reach nearly 11 billion—a 75 percent increase over today's population, which is about 6.4 billion—with no signs of peaking. The low projection, in contrast, predicts that the total human population in 2050 will be at 7.4 billion and will already have peaked.

Figure 52.13b makes another important point. The UN had to alter its population projections dramatically between 1992 and 1998, primarily to account for the impact of AIDS. Growth rates are dropping dramatically in areas of the world that are being hit hard by the epidemic.

(a) 2002 Projections based on three average fertility rates

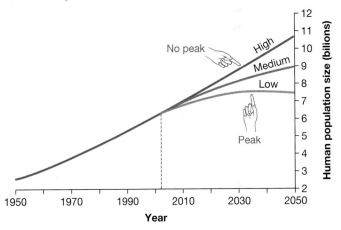

(b) 1992 Projections

Fertility rate	Projected population in 2050
High	12.5 billion
Medium	10.15 billion
Low	7.8 billion

The 1992 projections for 2050 are higher than those from 2002, primarily because the earlier projections did not account for the impact of AIDS.

FIGURE 52.13 Projections for Human Population Growth Depend on Fertility Rates
(a) The UN Population Division's population growth database includes a 50-year projection based on high, medium, and low average fertility rates over the interval. These projections, released in 2002, were based on an average of 2.5 (high), 2.1 (medium), or 1.7 (low) children born per woman. **(b)** Summary of projections made in 1992.

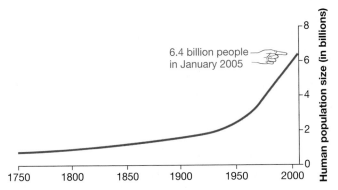

FIGURE 52.12 The Human Population Has Been Growing Rapidly
Graph based on estimates from historical and census data.

To summarize, humans may be approaching the end of a period of rapid growth that lasted well over 500 years. How quickly growth rates decline and how large the population can eventually become will be decided primarily by changes in fertility rates, and to a lesser extent by the future of the AIDS epidemic.

Are density-dependent processes such as disease and resource shortages responsible for these changes? In other words, are humans approaching Earth's carrying capacity? This question is controversial and is explored in more depth in the essay at the end of this chapter. For now, let's turn to questions about how conservation biologists use the concepts and data presented in this chapter.

✔ CHECK YOUR UNDERSTANDING

To explain changes in the size of populations over time, biologists test hypotheses using observational data—such as the life history of species in the Park Grass experiment—or experiments, such as treating red grouse populations with an antiparasite drug. In many cases, understanding population dynamics depends on understanding changes in a population's age structure. You should be able to explain (1) why the age structure of human populations differs between developed and developing nations and (2) why changes in fertility rates have such a dramatic impact on projections for the total human population in 2050.

52.4 How Can Population Ecology Help Endangered Species?

It should come as no surprise to learn that conservationists draw heavily on concepts and techniques from population ecology when designing programs to save species threatened with extinction. As habitat destruction and invasive species push other species from throughout the world into decline, the study of population growth rates and population dynamics has taken on an increasingly practical tone.

To illustrate how biologists use the theory and results introduced in this chapter, let's analyze how an understanding of geographic structure and life-table data can help direct conservation action.

Preserving Metapopulations

If you browse through an identification guide for trees or birds or butterflies, you'll likely find range maps indicating that most species occupy a broad area. In reality, however, the habitat preferences of many species are restricted, and individuals occupy only isolated patches within that broad area. The sunlit patches occupied by common primroses are an example of this pattern, and so are the meadows occupied by Glanville fritillaries—an endangered species of butterfly native to the Åland islands off the coast of Finland (**Figure 52.14a**).

Recall that if species occupy many small patches of habitat, such as the canopy gaps occupied by primroses, they are said to represent a metapopulation, or population of populations. Glanville fritillaries and common primroses naturally exist as metapopulations. But because humans are reducing large, contiguous areas of forest and grasslands to isolated patches or reserves, more and more species are being forced into a metapopulation structure.

Recent research on Glanville fritillaries by Ilkka Hanski and colleagues illustrates the consequences of metapopulation structure for endangered species. The work began with a survey of meadow habitats on the islands. Because fritillary caterpillars feed on just two types of host plant, *Plantago lanceolata* and *Veronica spicata*, the team was able to pinpoint potential butterfly habitats. They estimated the population size within each patch by counting the number of larval webs in each. Of the 1502 meadows that contained the host plants, 536 had Glanville fritillaries. The patches ranged from 6 m² to 3 ha in area. Most had only a single breeding pair of adults and one larval group, but the largest population contained hundreds of pairs of breeding adults and 3450 larvae.

Figure 52.14b illustrates how a metapopulation like this is expected to change over the years. Given enough time, each population within the larger metapopulation is expected to go extinct. The cause could be catastrophic, such as a storm or an oil spill; it could also be a disease outbreak or a sudden influx of predators. Migration from nearby populations can reestablish populations in these empty habitat fragments, however. In this way, there is a balance between extinction and recolonization. Even though subpopulations blink on and off over time, the overall population is maintained at a stable number of individuals.

Does this metapopulation model correctly describe the population structure of fritillaries? To answer this question, Hanski's group conducted a *mark-recapture study*. (**Box 52.3**, page 1208, provides details on how a mark-recapture study is done.) Of the 1731 butterflies that the biologists marked and released, 741 were recaptured over the course of the summer study period. Of the recaptured individuals, 9 percent were found in a previously unoccupied patch. This migration rate is high enough to suggest that patches where a population has gone extinct will eventually be recolonized.

Hanski and co-workers repeated the survey two years after their initial census. Just as the metapopulation model predicts, some populations had gone extinct and others had been created. On average, butterflies were lost from 200 patches each year; 114 unoccupied patches were colonized and newly occupied each year. In addition to confirming that the metapopulation model is valid, the work has important implications for endangered species. A small, isolated population—even one within a nature preserve—is unlikely to survive over the long term. For Glanville fritillaries, the data indicate that subpopu-

(a) Glanville fritillaries live in patches of meadow.

(b) Within metapopulations, subpopulations come and go.

A metapopulation is made up of small, isolated populations.

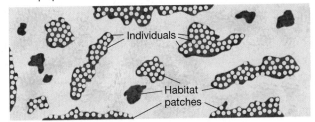

Although some subpopulations go extinct over time...

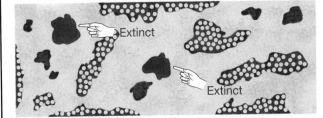

...migration can restore or establish subpopulations.

FIGURE 52.14 Metapopulation Dynamics Depend on Extinction and Recolonization
(a) The Glanville fritillary is extinct in much of its former range; it is now found only on the Åland islands near Finland. **(b)** The overall population size of a metapopulation stays relatively stable even if subpopulations go extinct. These populations may be restored by migration, or unoccupied habitats might be colonized.

lations that were large, that occupied larger areas, and that were closer to neighboring populations (and hence more likely to be colonized) were less likely go extinct. The data also show that fritillary populations with low genetic diversity—probably due to inbreeding—were more likely to go extinct than were populations of the same size that had higher genetic diversity.

Results like these have important messages for conservation biologists:

- Areas that are being protected for threatened species should be substantial enough in area to maintain large populations that are unlikely to go extinct in the near future.

- When it is not possible to preserve large tracts of land, an alternative is to establish systems of smaller tracts that are connected by corridors of habitat, so that migration between patches is possible.

- If the species that is threatened exists as a metapopulation, it is crucial to preserve at least some patches of unoccupied habitat to provide future homes for immigrants.

These results also have a more general message: Although traditional population growth models such as the expressions for exponential and logistic growth are simple and elegant, the factors they ignore—immigration and emigration—are crucial to understanding the dynamics of most populations. Because metapopulation structure is common, biologists must use more sophisticated models to predict the fates of populations.

Using Life-Table Data to Make Population Projections

Demographic data such as age-specific survivorship and fecundity are important for saving endangered species and for other applied problems. To understand why, suppose that you were in charge of reintroducing a population of lizards to a nature reserve, and that research conducted when this species occupied the site previously documented the survivorship and fecundity data. Your initial plan is to take 1000 newly hatched females from a captive breeding center and release them into

BOX 52.3 Mark-Recapture Studies

To estimate the population size of sedentary (largely immobile) organisms, researchers sample appropriate habitats by counting the individuals that occur along transects or inside rectangular plots—called quadrats—set up at random locations in the habitat. These counts can be extrapolated to the entire habitat area to estimate the total population size. In addition, they can be compared to later censuses to document trends in population size.

In contrast, estimating the total population size of organisms that are mobile and that do not congregate into herds or flocks or schools is much more of a challenge. In species such as the Glanville fritillary, the population inside sample quadrats or along transects changes constantly as individuals move in and out. Further, it can be difficult to track whether a particular individual has already been counted.

If individuals can be captured and then tagged in some way, however, the total population size of a mobile species can be estimated by using an approach called mark-recapture. To begin a **mark-recapture study**, researchers catch individuals in live traps and mark them with leg bands, ear tags, or some other methods of identification. After the marked individuals are released, they are allowed to mix with the unmarked animals in the same population for a period of time. Then a second trapping effort is conducted, and the percentage of marked individuals that were captured is recorded.

To estimate the total population from these data, researchers make a key assumption: The percentage of marked and recaptured individuals is equal to the percentage of marked individuals in the entire population. This assumption should be valid when no bias exists regarding which individuals are caught in each sample attempt. It is important that individuals do not learn to avoid traps after being caught once and that they do not emigrate or die as a result of being trapped.

The relationship between marked and unmarked individuals can be concisely expressed algebraically:

$$\frac{m_2}{n_2} = \frac{n_1}{N}$$

In this equation, m_2 is the number of marked animals in the second sample (the recapture), n_2 is the total number of animals (marked and unmarked) in the second sample, n_1 is the number of marked animals in the first sample, and N is the total population size. Having measured m_2, n_2, and n_1, the researcher can estimate N. For example, suppose researchers marked 255 animals and later were able to trap a total of 162 individuals in the population, of which 78 were marked. What is the estimate for total population size? Solve for N and check your answer.[5]

[5] 530

---

the habitat. Is this population likely to become established and grow, or will you need to keep introducing offspring that have been raised in captivity?

To answer this question, you need to use the life table data in **Figure 52.15a** and calculate (1) how many adults will survive to each age class each year and (2) how many juveniles will be produced by each adult age class each year over the course of several years. **Figure 52.15b** starts these calculations for you. The fate of the original 1000 females is indicated in red. In the second year, just 330 of these individuals have survived; 40 are left as 3-year-olds.

How many offspring did this generation produce? After one year, 330 of the original 1000 females remain. As the first purple number in Figure 52.16b indicates, these survivors have an average of 3.0 female offspring apiece, so they contribute 990 new juveniles to the population. In the second year, the 200 females that are left have an average of 4.0 female offspring each and contribute 800 juveniles. In their third year, the 40 surviving females average 5.0 offspring and contribute 200 juveniles.

Figure 52.15c extends the calculations by showing what happens as the offspring of the original females begin to breed.

Their offspring are shown in green. By adding subsequent generations and continuing the analysis, you could predict whether the population will stay the same, decline, or increase over time. In this way, life-table data can be used to predict the future of populations.

Part of the value of a population projection like the one begun in Figure 52.15 is that it allows biologists to alter values for survivorship and fecundity at particular ages and assess the consequences. For example, suppose that a predatory snake began preying on juvenile lizards. According to the model in Figure 52.15, what would be the impact of a change in juvenile mortality rate? Analyses like this allow biologists to determine which aspects of survivorship and fecundity are particularly sensitive for particular species. The studies done to date support some general conclusions:

- Whooping cranes, sea turtles, spotted owls, and many other endangered species have high juvenile mortality, low adult mortality, and low fecundity. In these species, the fate of a population is extremely sensitive to increases in adult mortality. Based on this insight, conservationists have recently begun an intensive campaign to reduce the loss of adult

(a) Life table

Age (x)	Survivorship (l_x)	Fecundity (m_x)
0 (birth)	----	0.0
1	0.33	3.0
2	0.2	4.0
3	0.04	5.0

(b) Fate of first-generation females

Year	0 (newborns)	1-year-olds	2-year-olds	3-year-olds	Total population size (N)
1st	1000 (just introduced)				1000
2nd	990 (= 330 × 3.0)	330 (= 1000 × 0.33)			1320 (= 990 + 330)
3rd	800 (= 200 × 4.0)		200 (= 1000 × 0.20)		
4th	200 (= 40 × 5.0)			40 (= 1000 × 0.04)	
5th					

(c) Fate of first- and second-generation females

Year	0 (newborns)	1-year-olds	2-year-olds	3-year-olds	Total population size (sum across all rows)
1st	1000				1000
2nd	990	330			1320
3rd	800 + 981 (981 = 327 × 3.0)	327 (= 990 × 0.33)	200		2308 (= 800 + 981 + 327 + 200)
4th	200 + 792 (792 = 198 × 4.0)		198 (= 990 × 0.20)	40	
5th	195 (195 = 39 × 5.0)			39 (= 990 × 0.04)	

FIGURE 52.15 Life-Table Data Can Be Used to Project the Future of a Population
(a) Life table providing age-specific survivorship and fecundity for a hypothetical population. **(b)** Predicted fate of 1000 1-year-old females introduced into a habitat just before the breeding season. The number of individuals in this cohort is shown in red; the offspring they produce each year is indicated in purple. **(c)** Extension of the data in part (b), indicating how many of the offspring produced by the original females in their first year survived in subsequent years, shown in purple, and how many offspring they produced in each subsequent year, shown in green. **EXERCISE** Assume that all 4-year-old females die after producing three young. Fill in the fourth and fifth years in the table.

female sea turtles in fishing nets. Previously, most conservation action had focused on protecting eggs and nesting sites.

- In humans and other species with high survivorship in most age classes, rates of population growth are extremely sensitive to changes in age-specific fecundity. Based on this result, programs to control human population growth focus on two

issues: lowering fertility rates through the use of birth control, and delaying the age of first reproduction by improving women's access to education.

In some or even most cases, however, the population projections made from life-table data may be too simplistic to be useful. For example, conservationists may need to expand the basic demographic models to account for occasional disturbances such as fires or storms or disease outbreaks. If the population exists in the fragmented habitats typical of a metapopulation, planners also need to assess the impact of emigration and immigration from nearby populations. Let's close this chapter with a look at how that assessment is done.

Population Viability Analysis

A **population viability analysis**, or PVA, is a model that estimates the likelihood that a population will avoid extinction for a given time period. In most cases, PVA attempts to combine basic demographic models for a species with data on geographic structure and the rate and severity of habitat disturbance. Typically, a population is considered viable if the analysis predicts that it has a 95 percent probability of surviving for at least 100 years. Natural resource managers are currently using PVA to assess the effects of logging, suburbanization, and other land management practices on sensitive species and to evaluate the merits of alternative recovery plans for endangered species.

A recent study—conducted on Leadbeater's possum, an endangered marsupial—illustrates the value of a carefully constructed PVA. Leadbeater's possum (**Figure 52.16a**) inhabits old-growth forests in southeastern Australia and relies on dead trees for nest sites. The goal of the PVA was to assess the effects that logging these habitats might have on the viability of the species. The question was, if logging reduced the ability of possums to migrate to new habitats, what would the impact on the population be?

Life-table data are difficult to obtain for Leadbeater's possum, because this species is rare, lives in trees, and is active at night. Enough data were available from field studies, however, to allow for an estimate of age-specific survival and average fecundity per female per year. Biologists used these data to make population projections while varying the spatial configuration

of habitat patches. The analysis allowed the researchers to assess the impact of migration and to simulate the effects of logging, fires, storms, and other disturbances.

Figure 52.16b illustrates the researchers' results. The four curves on the graph describe the changes in population size predicted to occur over time, based on different assumptions about the migration rates between groups of possums in patches of old-growth forest. When migration is high, the overall population size is predicted to stabilize at approximately 65 individuals. When no migration among fragments occurs, the final population size is predicted to be fewer than 20 individuals. Based on these results, the biologists concluded that extensive timber harvesting would pose a serious threat to the species. The PVA showed that reducing the size and number of remaining old-growth fragments would reduce the possibility for migration and lead to rapid population declines.

Like all analyses based on simulations, though, a PVA makes many assumptions about future events and is only as accurate as the data entered into it. In particular, the usefulness of PVAs has been challenged because data on age-specific fecundity and survivorship and other basic demographic information are lacking or poorly documented in many endangered species.

A group of biologists recently defended the approach, however, by analyzing 21 long-term studies that have been completed on threatened populations of animals. To do their analysis, the researchers split each of the 21 data sets in two. Data from the first half of each study was used to run a PVA on each of the 21 species. After comparing the predictions of the PVAs to the actual data from the second half of each study, the researchers found that the correspondence was extremely close. This result strengthens confidence in PVA as a good predictive method for land managers.

(a) Leadbeater's possum

(b) Population viability analysis (PVA)

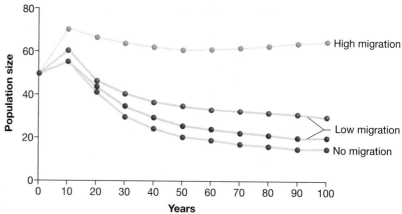

FIGURE 52.16 A PVA Can Predict the Consequences of Habitat Loss and Fragmentation
(a) Leadbeater's possum, a marsupial that breeds in old-growth forests of southeast Australia. **(b)** Results of a PVA: the fate of possum populations for four different rates of migration among habitat patches. The rates range from a case in which patches are isolated and no migration occurs to a case in which 5 percent of the individuals migrate each year. The no-migration scenario simulates what might happen if extensive deforestation occurred and created isolated patches of old-growth forest.

ESSAY What Limits Human Population Growth?

On October 12, 1999, the United Nations Population Division estimated that the human population exceeded six billion (6,000,000,000) for the first time. Why has the growth rate of the human population increased over time? Is our species at or near Earth's carrying capacity?

Population biologists point to several key events in human history that have contributed to the rising growth rate. The first event occurred about 10,000 years ago, when groups of humans began to develop agriculture. The transition from a nomadic existence to a settled, agricultural one was associated with dramatic increases in fecundity. More recently, improvements in medicine and hygiene have led to dramatic declines in death rates—particularly infant mortality. These technological advances are responsible for the dramatic rise in human populations over the past 100 years.

Until recently, most people considered population growth to be a good thing for religious, economic, and political reasons. As the twentieth century progressed and human population size continued to explode, however, a few people expressed concern about the ecological consequences of human population growth. Paul Ehrlich, who began his career studying butterflies, has been one of the most prominent voices arguing that human population growth must cease. His concern is that humans are exceeding Earth's carrying capacity and are causing an ecological collapse that might lead to our own downfall. At the root of Ehrlich's arguments is the concern that exponential growth is shooting us past Earth's carrying capacity for humans and is causing irreparable harm to the environment. What is Earth's carrying capacity, and how do we know if we have surpassed it?

Population demographers who have attempted to calculate Earth's carrying capacity for humans have produced estimates that range from 2 billion people to over 1 trillion people. The low estimate is much less than the current population; the high estimate assumes that people would live in floating cities above the planet's surface. The estimates are variable because carrying capacity depends on technological advancements and on the standard of living that people are willing to accept. Ehrlich's dire predictions have not proven correct to date, because the green revolution of the 1970s and 1980s—dramatic improvements in crop plant productivity, introduced in Chapter 29—produced enough food to keep pace with human population growth. Now, however, the availability of groundwater is declining in areas that depend on irrigated agriculture, and the amount of productive farmland available is beginning to peak. Do these observations mean that Earth's carrying capacity for humans has already been exceeded?

Is our species at or near Earth's carrying capacity?

Joel Cohen argues that the classical concept of carrying capacity has little meaning for humans. Although he acknowledges that density-dependent factors such as competition and disease are already affecting human populations, Cohen emphasizes that our future hinges much more directly on how many children the 6.4 billion of us decide to have and how many resources each of us demands. His point is that human choice is the most important element in defining carrying capacity for our species.

From this point of view, Earth's carrying capacity is what we decide it will be. The human population is already too large for all 6.4 billion of us to live with the same standard of living as that of Americans and Western Europeans. Increased demand for this level of resource use, or fertility rates that continue to exceed the replacement rate, would almost certainly bring about the type of ecological collapse Ehrlich envisioned. Over the course of your lifetime, the decisions made by 6.4 billion people about the number of children they have and the amount of resources they use will determine the future of the human population—and by extension, the future of life on Earth.

CHAPTER REVIEW

Summary of Key Concepts

A population is a group of individuals of the same species that occupy the same area at the same time. Population ecology is the study of how and why the number of individuals in a population change over time. Population size changes in response to changes in birth, death, immigration, and emigration rates. Biologists use a variety of mathematical and analytical tools to study population ecology.

■ **Life tables summarize how likely it is that individuals of each age class in a population will survive and reproduce.**

Life tables are a basic tool in demography—the study of patterns in births and deaths in populations. When life tables of different species or populations are compared, it is clear that individuals may have distinct ways of allocating energy and resources to activities that promote survival versus activities that promote reproduction. Because the resources available to an individual are always limited, any increase in allocation of resources to survival and competitive ability necessitates a decrease in the resources allocated to reproduction. This trade-off between survival and reproduction is the most fundamental aspect of a species' life history.

■ **The growth rate of a population can be calculated from life-table data or from the direct observation of changes in population size over time.**

One of the most basic characteristics of a population is its growth rate. Exponential growth occurs when the per-capita growth rate, r, does not change over time. Eventually, however, growing populations approach the carrying capacity of their environment—resulting in logistic growth.

Laboratory and field studies have confirmed the existence of exponential and logistic growth in species ranging from whooping cranes to *Paramecium*. Field research has documented that as population density increases, survivorship and fecundity may decrease—leading to increased death rates and lowered birth rates and thus a reduction in growth rate. Density dependence in population growth is due to competition for resources, disease, predation, or other factors that increase in intensity when population size is high.

Web Tutorial 52.1   Modeling Population Growth

■ **A wide variety of patterns is observed when researchers track changes in population size over time, ranging from no growth, to regular cycles, to continued growth independent of population size.**

In the Park Grass experiment, species with high fecundity and low survivorship tended to increase over time, although some later declined. Experiments have shown that density-dependent factors drive the regular population cycles observed in certain species. In the case of red grouse, roundworm infections spread rapidly when population density is high and then cause the population to crash.

Changes in population size through time may occur as a result of changes in the age structure of populations—specifically the number of individuals in various age classes. A population with few juveniles and many adults past reproductive age, like the human populations of developed nations, may be declining or stable in size. In contrast, a population with a large proportion of juveniles is likely to increase rapidly in size. Human populations in the developing nations currently have this type of age distribution.

The total human population has been increasing rapidly since about 1750 and is currently about 6.4 billion. In countries where survivorship has increased due to medical advances, fecundity has decreased. Based on various scenarios for average female fertility, the total human population in 2050 is expected to be between 7 and 11 billion.

Web Tutorial 52.2   Human Population Growth and Regulation

■ **Data from population ecology studies help biologists evaluate prospects for endangered species and design effective management strategies.**

Human activities are isolating populations into metapopulations occupying small, fragmented habitats. The history of a metapopulation is driven by the birth and death of populations, just as the dynamics of a population are driven by the birth and death of individuals. Migration among habitat patches is essential for the stability of a metapopulation, so conservationists are trying to preserve unoccupied patches of habitat and establish corridors that link habitat fragments.

Demographic data are the basis of population projections that are fundamental to population viability analysis, or PVA. Most PVAs attempt to model the effects that different management strategies might have on populations of endangered species. A PVA estimates the probability that a population will persist for a certain number of years under a prescribed set of demographic and habitat conditions.

Questions

Content Review

1. What is the defining feature of exponential growth?
   a. It lasts indefinitely.
   b. The growth rate is constant.
   c. The growth rate increases rapidly over time.
   d. The growth rate is very high.

2. What four factors define population growth?
   a. age-specific birth rates, age-specific death rates, age structure, and metapopulation structure
   b. survivorship, age-specific mortality, fecundity, death rate
   c. mark-recapture, census, quadrat sampling, transects
   d. births, deaths, immigration, emigration

3. In what populations does exponential growth tend to occur?
   a. in populations that colonize new habitats
   b. in populations that experience intense competition
   c. in populations that experience high rates of predation
   d. in metapopulations

4. Which of the following is not a reason why population growth declines as population size approaches the carrying capacity?
   a. Climate becomes unfavorable.
   b. Competition for resources increases.
   c. Predation rates increase.
   d. Disease rates increase.

5. If most individuals in a population are young, why is the population likely to grow rapidly in the future?
   a. Death rates will be low.
   b. The population has a skewed age distribution.
   c. Immigration and emigration can be ignored.
   d. Many individuals will begin to reproduce soon.

6. Why have population biologists become particularly interested in the dynamics of metapopulations?
   a. because humans exist as a metapopulation
   b. because whooping cranes exist as a metapopulation
   c. because many populations are becoming restricted to small islands of habitat
   d. because metapopulations explain why populations occupying large, contiguous areas are vulnerable to extinction

Conceptual Review

1. Explain Equations 52.3 and 52.5 in words.

2. Draw type I, II, and III survivorship curves on a graph with labeled axes. Explain why the growth rate of species with a type I survivorship curve depends primarily on fertility rates. Explain why the growth rate of species with a type III survivorship curve is extremely sensitive to changes in adult survivorship.

3. Offer a hypothesis to explain why humans have undergone near-exponential growth for over 500 years. Why can't exponential growth continue indefinitely? Give two examples of density-dependent factors that influence population growth in natural populations.

4. Compare and contrast the dynamics of a population that resides in a large contiguous habitat to the dynamics of a metapopulation. Be sure to consider density-dependent and density-independent factors on birth rates and death rates, and the importance of immigration and emigration. Assume that the total amount of area occupied is the same in both populations.

5. Make a rough sketch of the age distribution in developing versus developed countries, and explain the significance of the differences. How is AIDS, which is a sexually transmitted disease, affecting the age distribution in countries hard hit by the epidemic?

6. Compare the pros and cons of using R_0, λ, and r to express growth rates. What is the difference between r (the per-capita rate of increase) and r_{max} (the maximum or intrinsic growth rate)?

Group Discussion Problems

1. When wild plant and animal populations are logged, fished, or hunted, only the oldest or largest individuals tend to be taken. What impact does harvesting have on a population's age structure? How might harvesting affect the population's life table and growth rate?

2. Design a system of nature preserves for an endangered species of beetle whose larvae feed on only one species of sunflower. The sunflowers tend to be found in small patches that are scattered throughout dry grassland habitats. Explain the rationale behind your proposal.

3. Snowshoe hares are preyed on by lynx. Both species show pronounced population cycles. The cycles are roughly synchronized, although the rise and fall of lynx populations (the predator) slightly lags the rise and fall of hare populations (the prey). Design an experiment to test the hypotheses that these cycles are driven by (1) predation of hares by lynx, (2) food shortages when hares are at high density, or (3) a combination of food shortages and predation. Assume that you are able to exclude lynx from certain study areas and add food for hares in other areas.

4. In most species the sex ratio is at or near 1.00, meaning that there is an approximately equal number of males and females. In China, however, there is a strong preference for male children. According to the 2000 census there, the sex ratio of newborns is almost 1.17, meaning that close to 117 boys are born for every 100 girls. Based on these data, researchers project that there will soon be about 50 million more men than women of marriageable age in China. Discuss how this skewed sex ratio might affect the population growth rate in China.

53 Community Ecology

KEY CONCEPTS

▨ Interactions among species, such as competition and consumption, have two main outcomes: (1) They affect the distribution and abundance of the interacting species, and (2) they are agents of natural selection and thus affect the evolution of the interacting species.

▨ The assemblage of species found in a biological community changes over time and is primarily a function of climate and chance historical events.

▨ Species diversity is high in the tropics and lower toward the poles. The mechanism responsible for this pattern is still being investigated.

▨ Within at least some biomes, there is a strong positive relationship between the number of species present and the productivity and stability of the community.

Like many regions of the world, the area near Australia's Great Barrier Reef contains an array of biological communities.

Chapter 52 focused on the dynamics of populations—how and why they grow or decline and how they change over time and through space. That chapter considered each population as an isolated entity. But in reality, populations of different species form complex assemblages called communities. A biological **community** consists of interacting species, usually living within a defined area.

The concept of a biological community was introduced in Chapter 50. Each of the biomes analyzed in that chapter represents a broad type or category of plant community. The goal of that chapter was to describe the general characteristics of selected biomes and analyze broad correlations between a region's climate and the type of biome present.

This chapter focuses on a spatial scale between biomes and populations. The task is to analyze the dynamics of biological communities—how they develop and change over time. Biolo-

gists want to know how communities work. How do species inside communities interact with each other, and what are the long-term consequences? Why is the number of species higher in some areas than others, and why is species diversity important? Let's delve in.

53.1 Species Interactions

The species in a community interact almost constantly. Members of different species eat one another, pollinate each other, exchange nutrients, compete for resources, and provide habitats for each other. As a result, the fate of a particular population may be tightly linked to the other species that share its habitat.

Biologists organize all the possible interactions among species by considering the interactions' effects on the fitness of the individuals involved. Recall from Chapter 23 that fitness is defined as the ability to survive and produce offspring. Does the relationship between two species provide a fitness benefit to members of one species (a "+" interaction) but hurt members of the other species (a "−" interaction)? Or does the association have no effect on the fitness of a participant (a "0" interaction)? Three broad categories of interaction are analyzed in this chapter: the −/− relationship known as *competition*, the +/− interactions called *consumption* and *parasitism*, and the +/+ association termed *mutualism*. A fourth category of interaction, **commensalism**, is defined as a +/0 association. An example is the birds that follow moving army ants in the tropics. As the ants march along the forest floor, they hunt insects and small vertebrates. As they do, birds follow and pick off prey species that fly or jump up out of the way of the ants (**Figure 53.1**). The birds are **commensals** that benefit from the association (+) but have no impact on the ants (0).

In addition to introducing the array of tools that biologists use to study competition, consumption, and mutualism, it is important to focus on three key themes:

1. Species interactions may affect the distribution and abundance of a particular species. Recall from Chapter 52 that most of the density-dependent factors that produce logistic growth are based in species interactions—predation, disease, or competition for space and resources. Changes in species interactions often explain short-term changes in population size and distribution.

2. Species act as agents of natural selection when they interact. Deer are fast and agile in response to natural selection exerted by their major predators, wolves and cougars. The speed and agility of deer, in turn, promote natural selection that favors wolves and cougars that are fast and that have superior eyesight and senses of smell. To capture this point, biologists say that species interactions resemble an arms race. In humans, an arms race is said to occur when one nation develops a new weapon, which prompts a rival country to develop a defensive weapon, which pushes the original country to manufacture a more powerful weapon, and so on. In biology, a **coevolutionary arms race** occurs between predators and prey, parasites and hosts, and other types of interacting species. The adjective *coevolutionary* is appropriate because the pairs of species influence each others' evolution. In this way, changes in species interactions lead to long-term changes in the characteristics of populations—in addition to short-term impacts on population size.

3. The outcome of interactions among species is dynamic and conditional. Consider the relationship between army ants and birds that follow them. If attacks by birds force other insects into the path of the ants, then both ants and birds benefit and the relationship is mutualistic. But if birds begin to steal prey from ants, then the relationship becomes parasitic. The outcome of the interaction may depend on the number and types of prey, birds, and ants present and may change over time.

Competition

Competition is a −/− interaction that occurs when different individuals use the same resources and when those resources are limited—meaning they are in short supply relative to demand. Ever since ecological studies began, researchers have focused on competition as an important interaction within and between species. The attention is justified in part by the central place that competition holds in Darwin's theory of evolution by natural selection. Darwin pointed out that individuals within a population compete for the resources that are required for growth and reproduction. Further, some individuals are more successful in this competition and leave more offspring than others do. If the traits that lead to success are heritable, then the frequency of alleles in the population changes and evolution by natural selection occurs.

Competition that occurs between members of the same species is called **intraspecific** ("within species") **competition**. In addition to creating opportunities for natural selection to occur, intraspecific competition for space, sunlight, food, and other resources intensifies as a population's density increases. As a result, intraspecific competition is a major cause of density-dependent growth (see Chapter 52).

How does competition affect members of different species? Stated another way, what happens when **interspecific** ("between species") **competition** occurs? To answer this question, it is important to recognize that every species has a unique **niche**, or set of habitat requirements. A niche can be thought of as the range of resources that the species is able to use or the range of

FIGURE 53.1 Commensals Gain a Fitness Advantage but Don't Affect the Species They Depend On
Birds that associate with army ants are commensals. They have no measurable fitness effect on the ants but gain from the association by capturing insects that try to fly or climb out of the way of the ants.

(a) One species eats seeds of a certain size range.

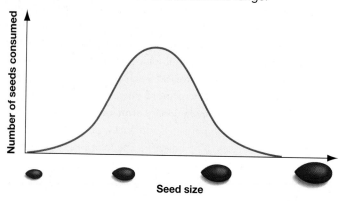

(b) Partial niche overlap: competition for seeds of intermediate size

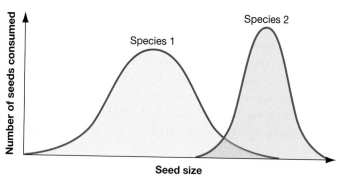

FIGURE 53.2 Niche Overlap Leads to Competition
(a) Graph describing one aspect of a species' fundamental niche, meaning the range of resources that it can use or range of conditions it can tolerate. **(b)** Competition occurs when the niches of different species overlap. In this case, both species use seeds of intermediate size. EXERCISE Mark the horizontal axis to indicate the range of seed sizes where competition occurs.

conditions it can tolerate. G. Evelyn Hutchinson proposed that a species' niche could be envisioned by plotting these habitat requirements along a series of axes. **Figure 53.2a**, for example, represents one niche axis for a hypothetical species. In this case, the habitat requirement plotted is the size of seeds eaten by members of this population, which might be a function of mouth or tooth size. Other niche axes could represent other types of foods used or the temperatures, humidity, and other environmental conditions tolerated by the species.

Interspecific competition occurs when the niches of two species overlap. The two species plotted in **Figure 53.2b**, for example, compete for seeds of intermediate size. When competition occurs, each individual will get fewer seeds on average. In this way, competitive interactions reduce the amount of resources available to each species. This is why competition is considered a – / – interaction.

The Consequences of Competition—Theory According

to G. F. Gause, it is not possible for species with the same niche to coexist. This hypothesis is called the **competitive exclusion**

principle, and it was inspired by a series of competition experiments Gause did with similar species of the unicellular pond-dweller *Paramecium*. When Gause placed small populations of *Paramecium caudatum* and *P. aurelia* in separate laboratory cultures, both species exhibited logistic growth (see Chapter 52). But Gause also showed that when the two species are put in the same culture together, only the *P. aurelia* population exhibits a logistic growth pattern. *Paramecium caudatum*, in contrast, is driven to extinction (**Figure 53.3a**).

(a) Competitive exclusion in two species of *Paramecium*

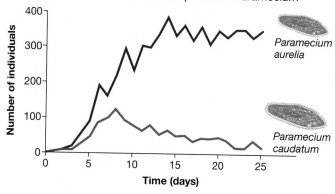

(b) Competitive exclusion occurs when competition is asymmetric ...

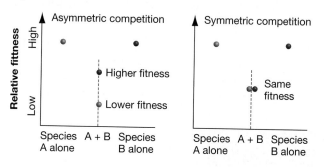

(c) ... and niches overlap completely.

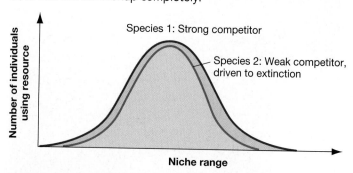

FIGURE 53.3 Competitive Exclusion Occurs when One Species Is a Superior Competitor
(a) In laboratory culture, *Paramecium aurelia* outcompetes *P. caudatum*. **(b)** When competition is asymmetric, one of the interacting species suffers a larger fitness loss than the other does. **(c)** If two species have completely overlapping niches, there is no refuge for the weaker competitor. It may be driven to extinction.

A result such as Gause's is produced by what contemporary investigators call asymmetric competition. When **asymmetric competition** occurs, one species suffers a much greater fitness decline than the other species does (**Figure 53.3b**). Under **symmetric competition**, however, each species experiences a roughly equal decrease in fitness. In Gause's laboratory cultures, competition between *P. aurelia* and *P. caudatum* was asymmetric. Subsequent research has shown that most competition is asymmetric.

Gause's work highlights a central question in the study of competition: If competition is a common interaction, why haven't the superior competitors outcompeted all other species? Is coexistence possible despite competition?

Before considering experimental work on this question, let's consider the predictions made by theory. Imagine a scenario in which asymmetric competition is occurring. If the two species have completely overlapping niches, as diagrammed in **Figure 53.3c**, then the stronger competitor is likely to drive the weaker competitor to extinction, just as Gause predicted.

If the niches do not overlap completely, then the weaker species should be able to retreat to an area of non-overlap. In cases like this, an important distinction arises between a species' **fundamental niche**, which is the combination of resources or areas used or conditions tolerated in the absence of competitors, and its **realized niche**, which is the portion of resources or areas used or conditions tolerated when competition occurs (**Figure 53.4a**). When competition does not result in the complete exclusion of one species, it may result in **niche differentiation**, in which competing species evolve traits that allow them to exploit different resources or live in different areas (**Figure 53.4b**).

It's important to recognize that some species may be stronger competitors in some habitats or environments than others and that the types of competition that occur vary widely. The actual mechanism of competition depends on the species and resource involved. As **Figure 53.5** (page 1218) shows, mechanisms of competition include the following:

1. *Consumptive competition*, in which two species consume the same resources (Figure 53.5a).

2. *Preemptive competition*, in which one species makes space unavailable to other species (Figure 53.5b).

3. *Overgrowth competition*, in which one species grows above another (Figure 53.5c).

4. *Chemical competition*, in which one species produces toxins that negatively affect another species (Figure 53.5d).

5. *Territorial competition*, in which a mobile species protects its feeding or breeding territory against other species (Figure 53.5e).

6. *Encounter competition*, in which two species interfere directly for access to specific resources (Figure 53.5f).

To summarize, competition occurs when niches overlap, and it can result in extinction, changes in niche use, and niche dif-

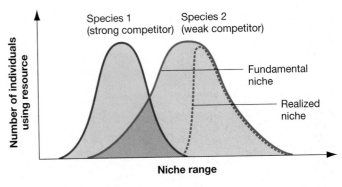

(a) If niches do not overlap completely, weaker competitors use nonoverlapping resources.

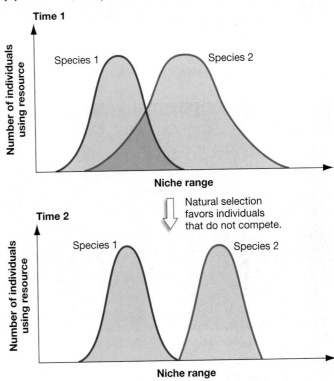

(b) Over time, competition can cause niche differentiation.

Natural selection favors individuals that do not compete.

FIGURE 53.4 Partial Niche Overlap Has Several Consequences **(a)** When the niches of competing species overlap, the weaker competitor uses only a subset of its fundamental niche. **(b)** Over time, evolution can lead to changes in niche breadth—a process called niche differentiation.

ferentiation over time. Competition theory makes predictions that are clear, logical, and compelling. Is there evidence that niche differentiation actually occurs in nature? What types of experiments can reveal the existence of competition and identify its consequences?

The Consequences of Competition—Experimental Studies

Joseph Connell initiated a classic study of competition after observing an interesting pattern in an intertidal rocky shore in Scotland. He noticed that there were two species of barnacles with

(a) Consumptive competition: Organisms consume the same resources.

These trees are competing for water and nutrients.

(b) Preemptive competition: Individuals occupy space, preventing access to resources.

Space preempted by these barnacles is unavailable to competitors.

(c) Overgrowth competition: One organism grows over another.

The large fern has overgrown other individuals and is shading them.

(d) Chemical competition: One species produces toxins that negatively affect another.

Few plants are growing under these *Salvia* shrubs.

(e) Territorial competition: Mobile organisms protect a feeding or breeding territory.

Grizzly bears drive off black bears.

(f) Encounter competition: Organisms interfere directly for access to specific resources.

Spotted hyenas and vultures fight over a kill.

FIGURE 53.5 Many Types of Competition Exist
A few of the major types of competition.

(a) Barnacle species are distributed in distinct zones.

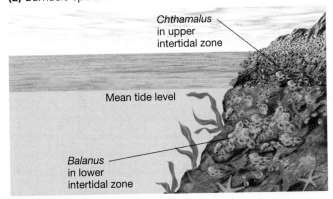

FIGURE 53.6 **Experimental Evidence for Competitive Exclusion**
(a) In natural habitats, adult *Chthamalus* and *Balanus* barnacles do not coexist. **(b)** Testing the hypothesis that *Chthamalus* is competitively excluded from the lower intertidal zone by *Balanus*.
EXERCISE In part (b), label the experimental and control treatments on each rock. Why was it important to carry out the experimental and control treatments on the same rock? Why not use separate rocks?

distinctive distributions. Barnacle larvae are mobile, but adults live attached to rocks. The adults of one species, *Chthamalus stellatus*, occurred in an upper intertidal zone, while the adults of the other species, *Balanus balanoides*, were restricted to a lower intertidal zone (**Figure 53.6a**). The upper intertidal zone is a more severe environment for barnacles, because it is exposed to the air for longer periods each day. The young of both species were found together in the lower intertidal zone, however.

To explain these observations, Connell hypothesized that adult *C. stellatus* were competitively excluded from the lower intertidal zone. His view was that *Chthamalus* larvae occupy the species' fundamental niche but that adults are constrained by competition for attachment space. The alternative hypothesis is that adult *Chthamalus* are absent from the lower intertidal zone because they do not thrive in the physical conditions there.

To test these hypotheses, Connell performed the experiment shown in **Figure 53.6b**. He began by removing a number of rocks from the upper intertidal zone that had been colonized by *Chthamalus* and transplanting them into the lower intertidal zone. He screwed the rocks into place and allowed *Balanus* larvae to colonize them. Once the spring colonization period was over, Connell divided each rock into two treatments. In one half, he removed all *Balanus* that were in contact with or next to a *Chthamalus*.

This experimental design allowed Connell to document *Chthamalus* survival in the absence of competition with *Balanus* and compare it with survival during competition. This is a common experimental strategy in competition studies: One of the competitors is removed, and the response by the remaining species is observed.

Connell's results support the hypothesis of competitive exclusion. In the unmodified areas, *Balanus* killed many of the young *Chthamalus* by growing against them and lifting them off the

(b) Testing the hypothesis that competition occurs

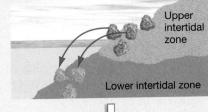

Question: Why is the distribution of adult *Chthamalus* restricted to the upper intertidal zone?

Hypothesis: Adult *Chthamalus* are competitively excluded from the lower intertidal zone.

Alternative hypothesis: Adult *Chthamalus* do not thrive in the physical conditions of the lower intertidal zone.

Experimental setup:

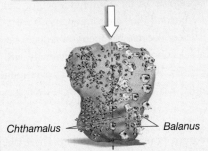

1. Transplant rocks containing young *Chthamalus* to lower intertidal zone.
2. Let *Balanus* colonize the rocks.
3. Remove *Balanus* from half of each rock. Monitor survival of *Chthamalus* on both sides.

Prediction: *Chthamalus* will survive better in the absence of *Balanus*.

Prediction of alternative hypothesis: *Chthamalus* survival will be low and the same in the presence or absence of *Balanus*.

Results:

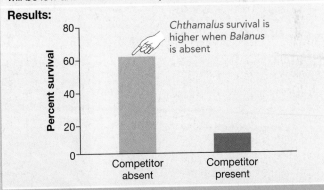

Conclusion: *Balanus* is competitively excluding *Chthamalus* from the lower intertidal zone.

substrate. As the graph in Figure 53.6b shows, *Chthamalus* survival was much higher when all of the *Balanus* were removed.

Connell's experiment was one of the first examples of how competition can modify the distribution of a species. In this case, the poorer competitor (*Chthamalus*) is restricted to the upper intertidal zone, which is a more severe habitat because it is prone to drying and high temperatures. *Chthamalus* is able to grow there, however, because it is more resistant to drying than *Balanus* is. *Chthamalus* has a much broader fundamental niche than *Balanus* has, so there is a region where the niches of the two species do not overlap. As a result, the overall outcome of competition between these species is coexistence by habitat partitioning. If both species had identical fundamental niches, *Chthamalus* would probably be eliminated entirely from the area.

In the area where Connell worked, the fundamental and realized niches of *Balanus* are identical, but the fundamental and realized niches of *Chthamalus* are different. The situation is dynamic, however. If an environmental change were to bring a new parasite into the area and *Chthamalus* were more resistant to it than *Balanus* was, then the outcome of competition and the realized niches might be quite different.

To summarize, competition is a common interaction that has a wide range of ecological outcomes. Most competition is asymmetrical, and in extreme cases competition can lead to the complete exclusion of one species. When the two competitors have slightly different fundamental niches, however, the poorer competitor can take refuge in areas that are beyond the better competitor's tolerance, thus achieving coexistence. Because competition reduces fitness in both species, natural selection should favor individuals with traits that minimize competition. When species interact over long periods of time, then, competition for resources should decrease.

Consumption

Consumption occurs when one organism eats another. There are three major types of consumption:

1. **Herbivory** takes place when **herbivores** ("plant-eaters") consume plant tissues. Bark beetles mine cambium; aphids suck sap; caterpillars chew leaves.

2. **Parasitism** occurs when a **parasite** consumes relatively small amounts of tissue or nutrients from another individual, called the **host**. Parasitism often occurs over a long period of time. It is not necessarily fatal, and parasites are usually small relative to their host. An array of worms and unicellular protists parasitizes humans; ticks parasitize cattle and other large mammals.

3. **Predation** occurs when a **predator** kills and consumes all or most of another individual. The consumed individual is called the **prey**. Woodpeckers eat bark beetles; finches eat seeds; ladybird beetles devour aphids; wasps kill caterpillars.

To illustrate how biologists analyze the impact of consumption, let's consider a series of questions about predators, herbivores, parasites, and their victims.

How Do Prey Defend Themselves? With respect to fitness, predation is costly for the prey species and beneficial for the predator. Prey individuals do not passively give up their lives to increase the fitness of their predators, however. Instead, prey may hide or run, fly, or swim away when they sense the presence of a predator. Analyses have shown that many species find safety in numbers—schooling and flocking behavior is an effective way to reduce the risk of predation, in part because predators become confused when they dive or swim into a mass of prey. Other prey species sequester or spray toxins or employ weaponry such as spines or kicking hooves. Traits such as these are called **standing** or **constitutive defenses**, because they are always present (**Figure 53.7a**).

(a) Constitutive defenses of animals vary.

Camouflage: blending into the background

Schooling: safety in numbers

Weaponry: fighting back

(b) Mimicry can protect both dangerous and harmless species.

Müllerian mimics
Common wasp   Bumblebee   Honeybee

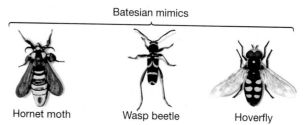
Batesian mimics
Hornet moth   Wasp beetle   Hoverfly

FIGURE 53.7 Constitutive Defenses Are Always Present
(a) Prey have an array of adaptations to reduce the likelihood of predation. **(b)** Müllerian mimicry occurs among dangerous prey species; Batesian mimicry occurs between a dangerous prey species and harmless prey species.

Some of the best-studied constitutive defenses involve the phenomenon called mimicry. **Mimicry** occurs when one species closely resembles another species. **Figure 53.7b** illustrates two major types of mimicry. The wasp on the far left is brightly colored and dangerous, because of its stinger. It looks like other dangerous insects—particularly other wasps and bees. When harmful prey species resemble each other, **Müllerian mimicry** is said to occur. To explain the existence of Müllerian mimics, biologists propose that the existence of similar-looking dangerous prey in the same habitat increases the likelihood that predators will learn to avoid them. In this way, Müllerian mimicry should reduce the likelihood of dangerous individuals being attacked. But wasps also act as a model for harmless species of moth, beetle, and fly that resemble wasps. To explain these mimics, biologists propose that predators avoid the harmless mimics because they mistake them for a dangerous wasp. This resemblance is known as **Batesian mimicry.**

The central point here is that prey have adaptations that reduce the likelihood that they will become victims. These adaptations are responses to natural selection exerted by predators. Constitutive defenses are expensive, however, in terms of the energy and resources that must be devoted to producing and maintaining them. Based on this observation, it should not be surprising to learn that many prey species have **inducible defenses**—meaning defensive traits that are produced only in response to the presence of a predator. Induced defenses then decline if predators leave.

To see how inducible defenses work, consider recent research on blue mussels that live in an estuary along the coast of Maine (**Figure 53.8a**). Biologists had documented that predation on mussels by crabs was high in an area of the estuary with relatively slow tidal currents (a "low-flow" area) but low in an area of the estuary with relatively rapid tidal currents (a "high-flow" area). The researchers hypothesized that if blue mussels possess inducible defenses, then heavily defended prey individuals should occur in the low-flow area, where predation pressure is higher, but not in the high-flow area, where water movement reduces the number of crabs present.

To evaluate this hypothesis, the biologists measured mussel shell characteristics in the two areas. They found that mussels in the high-predation area were more strongly attached to their substrate and had thicker shells than did mussels in the low-predation area (**Figure 53.8b**). These traits make the mussels more difficult to remove from the substrate and harder to crush and so function as effective antipredator defenses.

Like any correlational study, however, the data in Figure 53.7b are open to interpretation. A critic of the inducible defense explanation could offer a reasonable alternative hypothesis—for example, that only constitutive defenses exist and that crabs have eliminated weakly attached mussels with thin shells from the low-flow areas. The observed differences could also be due to differences in sunlight, water temperature, or other abiotic factors and have nothing to do with predation.

(a) Prey and predator

Blue mussels   Crabs

(b) Correlation between predation rate and prey defense

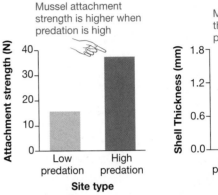 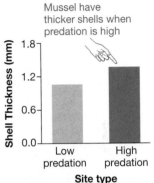

FIGURE 53.8 Inducible Defenses Are Produced Only when Prey Are Threatened
(a) In shallow-water environments in Maine, blue mussels (left) are preyed on by crabs (right). **(b)** To measure how strongly mussels are attached to the substrate, researchers drilled a hole in their shells and measured the force, in newtons (N), required to pull them off. They also measured how large mussel shells were relative to an individual's size by dividing shell weight by soft-tissue weight.

To test the inducible-defense hypothesis more rigorously, biologists did the experiment diagrammed in **Figure 53.9** (page 1222). The tank on the left allowed the researchers to measure shell growth in mussels that were "downstream" from crabs that were fed fish. As predicted by the induced-defense hypothesis, the mussels exposed to a crab in this way developed significantly tougher shells than did mussels that were not exposed to a crab. These results suggest that, even without direct contact, mussels can sense the presence of crabs and increase their investment in defenses. In a similar experiment, the investigators compared mussels that were exposed to water running through broken mussel shells versus intact but empty mussel shells. A significant increase in shell thickness in the tank downstream from the broken shells was measured. This result supports the hypothesis that mussels can detect the presence of predators from molecules released by broken shells.

Results like these underscore several themes of this section: Species interactions are dynamic and result in coevolution—in this case, adaptations and counteradaptations in an evolutionary arms race between predators and prey.

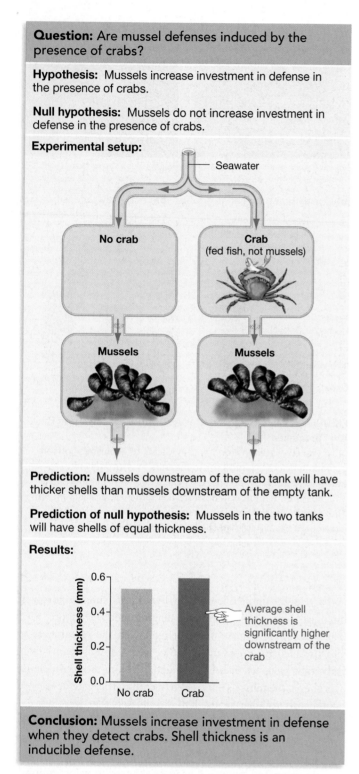

Question: Are mussel defenses induced by the presence of crabs?

Hypothesis: Mussels increase investment in defense in the presence of crabs.

Null hypothesis: Mussels do not increase investment in defense in the presence of crabs.

Experimental setup:

Seawater

No crab

Crab (fed fish, not mussels)

Mussels

Mussels

Prediction: Mussels downstream of the crab tank will have thicker shells than mussels downstream of the empty tank.

Prediction of null hypothesis: Mussels in the two tanks will have shells of equal thickness.

Results:

Average shell thickness is significantly higher downstream of the crab

Conclusion: Mussels increase investment in defense when they detect crabs. Shell thickness is an inducible defense.

FIGURE 53.9 Experimental Evidence for Inducible Defenses Experiment conducted in tanks in a laboratory.

Are Animal Predators Efficient Enough to Reduce Prey Populations?

Research on animal defense systems supports the hypothesis that species interactions have a strong impact on the evolution of predator and prey populations. Can predators also affect the short-term distribution and abundance of prey populations? Prey are typically smaller than predators, have

larger litter or clutch sizes, and tend to begin reproduction at a younger age. As a result, they have a much larger *intrinsic growth rate*, r_{max}—the maximum growth rate that a population can achieve under ideal conditions (see Chapter 52). If prey reproduce rapidly and are also well defended, it is not clear whether predators should be able to kill enough of them to reduce the prey population significantly—particularly if predators tend to take old or sick members of a population.

In several cases, data from predator-removal programs—in which wolves, cougars, coyotes, or other predators are actively killed by human hunters—support the hypothesis that predators actually do reduce the size of prey populations. For example, a wolf control program in Alaska during the 1970s decreased predator abundance to 55–80 percent below pre-control density. Concurrently, the population of moose, on which wolves prey, tripled. This observation suggests that wolf predators had reduced this moose population far below the number that could be supported by the available space and food.

Other types of experiments have also supported the hypothesis that predators play a role in density-dependent growth of prey populations. Recall from Chapter 52 that some populations go through regular cycles and that the regular population cycle of red grouse in England appears to be driven, at least in part, by density-dependent increases in parasitism. **Figure 53.10** shows another classic case of population cycling: snowshoe hare and lynx populations in northern Canada. Hares are herbivores, and lynx prey primarily on hares. Two hypotheses have been proposed to explain the cycles these populations exhibit: (1) Hares use up all their food when their populations reach high density and starve; in response, lynx also starve. (2) Lynx populations reach high density in response to increases in hare density. At high density, lynx eat so many hares that the prey population crashes. Stated another way, either hares control lynx population size or lynx control hare population size.

To test these hypotheses, researchers set up a series of 1-km² study plots in boreal forest habitats that were as identical as possible (**Figure 53.11**). Three plots were left as unmanipulated controls. One plot was ringed with an electrified fence with a mesh that excluded lynx but allowed hares to pass freely. Two plots received additional food for hares year-round. One plot had a predator-exclusion fence and was also supplemented with food for hares year-round. The biologists then monitored the size of the hare and lynx populations over an 11-year period, or enough time for a complete cycle in the two populations. As the data in Figure 53.11 show, plots with predators excluded showed higher hare populations at the peak of the cycle than did control plots. This result supports the hypothesis that predation by lynx reduces hare populations. Plots with supplemental food had even higher peak populations, however, and the plot with supplemental food and predators excluded had a hare population over 15 times as large as the average in control plots. These data support the hypotheses that hare populations are limited by availability of

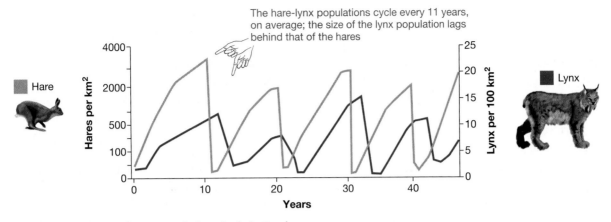

FIGURE 53.10 Hare and Lynx Populations Cycle in Synchrony
To make it easier to read, the left-hand vertical axis, which indicates hare density, is not linear.

Question: What factors control the hare-lynx population cycle?

Hypothesis: Predation, food availability, or a combination of those two factors controls the hare-lynx cycle.

Null hypothesis: Predation, food availability, or a combination of those two factors does *not* control the hare-lynx cycle.

Experimental setup:

Document hare population in seven study plots (similar boreal forest habitats, each 1 km²) over 11 years (duration of one cycle).

3 plots: Unmanipulated controls

1 plot: Electrified fence excludes lynx but allows free access by hares

2 plots: Supply extra food for hares

1 plot: Electrified fence excludes lynx but allows free access by hares; supply extra food for hares

Prediction: Hare populations in at least one type of manipulated plot will be higher than the average population in control plots.

Prediction of null hypothesis: Hare populations in all of the plots will be the same.

Results:

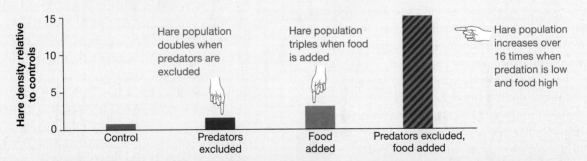

Conclusion: Hare populations are limited by both predation and food availability. When predation and food limitation occur together, they have a greater effect than either factor does independently.

FIGURE 53.11 Experimental Evidence That Predation and Food Availability Drive the Hare-Lynx Cycle
QUESTION State a hypothesis to explain why lack of predation and increased food, in combination, have such a large effect on hare population size.

food as well as by predation, and that food availability and predation intensity interact—meaning the combined effect of food and predation is much larger than their impact in isolation. Research continues to figure out why this is so. In the meantime, it is increasingly clear that, in many instances, predators are efficient enough to reduce prey populations.

Why Don't Herbivores Eat Everything—Why Is the World Green? If predators affect the size of populations in prey that can run or fly or swim away, then consumers should have a devastating impact on plants and on mussels, anemones, sponges, and other sessile (nonmoving) animals. In some cases, this prediction turns out to be correct. For example, consider the results of a recent **meta-analysis**—a study of studies, meaning an analysis of a large number of data sets on a particular question. Biologists who compiled the results of more than 100 studies on herbivory found that the median percentage of mass removed from aquatic algae by herbivores was 79 percent. This figure supports the hypothesis that herbivores eat the vast majority of algal food available in aquatic biomes. The figure dropped to just 30 percent for aquatic plants, however, and only 18 percent for terrestrial plants.

These data raise the question of why herbivores don't eat more of the available plant food. Stated another way, why is the world green? Biologists routinely consider three possible answers to this question. Herbivores could be kept in check by predation or disease; plant tissues could offer poor or incomplete nutrition; or plants could defend themselves effectively against attack:

1. The **top-down control** hypothesis states that herbivore populations are limited by predation or disease. The term *top-down* is inspired by the concept of a food chain, which is explained in detail in Chapter 54; see also Chapter 28. As an example, consider the boreal forest biome of northern Canada. Lynx are top predators, meaning they are not subject to predation. Lynx eat hares, which in turn eat vegetation on the forest floor. The data in Figure 53.11 showed that lynx limit the number of hares and thus increase the amount of vegetation. Biologists can also point to "natural experiments" in top-down control that have taken place throughout the world. For example, after a dozen pairs of European rabbits were introduced to Australia in 1859, the population exploded—exceeding 250,000 individuals by 1865. As a result, vast areas of the continent were virtually stripped of vegetation. Similar consequences have occurred when goats and pigs have escaped from captivity and established large, wild populations in areas that lack their natural predators and diseases.

2. The **poor-nutrition hypothesis** contends that plants are a poor food source in terms of the nutrients they provide

for herbivores. More specifically, plant tissues have less than 10 percent of the nitrogen found in animal tissues, by weight. If the growth and reproduction of herbivores is limited by the availability of nitrogen, then populations will be kept too low to consume more than a small fraction of the available food. But if the nitrogen concentration of plants is increased via fertilization, then herbivores should grow faster and reproduce more. To test this prediction, a recent meta-analysis examined 185 studies of how insect herbivores had responded to plant fertilization. In over half of the cases, herbivores showed a significant increase in growth rate or reproduction when the plants that they feed on were fertilized. Based on this result, the researchers concluded that nitrogen limitation is an important factor in a large fraction of plant-herbivore interactions.

3. The **plant-defense hypothesis** holds that plants defend themselves effectively enough to limit herbivory. Most plant tissues are defended by weapons such as thorns, prickles, or hairs or by potent poisons such as nicotine, caffeine, and cocaine. In addition, no animal species can, without help from protists or bacteria, digest cellulose or lignin, which are components of wood. Manufacturing defensive compounds seems like the perfect solution to the problem of herbivory. In practice, however, plants face a complex challenge in defending themselves. Consider data on interactions between cottonwood trees and two of their herbivores: beavers and leaf beetles. Cottonwoods resprout after they are cut down by a beaver (**Figure 53.12a**), and resprouted trees contain high concentrations of a defensive compound that deters further attack by beavers (**Figure 53.12b**). This is an induced defense. But caterpillars of a particular leaf beetle species eat this defensive compound readily. In fact, the caterpillars store enough of it in their bodies to act as a defensive compound against their own major predator—ants. The data in **Figure 53.12c** show that caterpillars that grow on resprouted cottonwood trees survive longer than do caterpillars that grow on normal cottonwoods when the caterpillars are placed on an ant mound. This is an example of an indirect effect in species interactions. The response by cottonwoods to herbivory by beavers benefits another herbivore—leaf beetles. The net result is that there is no perfect, one-size-fits-all defensive strategy. Natural selection should favor plants that evolve an ever-changing suite of compounds to deter the ever-changing array of herbivores they face.

Taken together, the data reviewed here suggest that there is no single answer to the question of why herbivores don't eat a greater fraction of the available plant food. All three of the hypotheses we have examined are correct. Top-down control, nitrogen limitation, and effective defense are all important factors

(a) Cottonwood tree felled by beavers

(b) Resprouted trees have more defensive compounds.

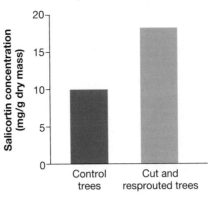

(c) Herbivorous larvae survive longer on resprouted trees.

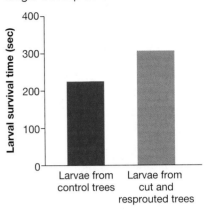

FIGURE 53.12 Defensive Compounds in Cottonwood Trees Discourage Beavers but Favor Beetles
(a) If felled by a beaver, most cottonwood trees respond by sending up new shoots.
(b) Cottonwoods that have resprouted following attack by beavers have significantly higher concentrations of defensive compounds than do cottonwoods that were never attacked by beavers. **(c)** Leaf beetle larvae placed on the mounds of ants (their primary predator) survived significantly longer if the larvae had been reared on cottonwood leaves with high levels of defensive compounds.

in limiting the impact of herbivory, although the particular mix of factors will vary from plant species to plant species and from habitat to habitat.

Adaptation and Arms Races Over the long term, how do species that interact via consumption affect each other's evolution? To answer this question, let's focus on interactions between humans and the most serious of all human parasites—species in the genus *Plasmodium*. Recall from Chapter 28 that *Plasmodium* are unicellular protists that cause malaria in a wide array of vertebrates. Malaria kills at least a million people a year, most of them children of preschool age. Recent data suggest that humans and the causative agent of malaria—four species in the genus *Plasmodium*—are locked in a coevolutionary arms race.

Plasmodium has a complex life cycle. It is a unicellular organism, but it has several distinct cell types. Each cell type corresponds to the host and host tissue being infected. For example, one of these cell types lives in the salivary glands of infected mosquitoes and is transferred to a human host when the infected mosquito bites a person. The *Plasmodium* cells travel to the human's liver, reproduce asexually, and form a large population of a new cell type that is later released into the host's bloodstream. This generation of parasite cells invades the host's red blood cells, reproduces asexually, and eventually causes the infected blood cells to rupture. Loss of red blood cells causes severe anemia in the human host; the infected person's immune system responds so strongly to the release of the parasite cells that debilitating fevers result. The parasite's life

cycle continues when a mosquito bites an infected person and ingests infected red blood cells.

Cells in the human immune system do not sit passively by as *Plasmodium* destroys liver cells and red blood cells. Instead, certain cells (introduced in Chapter 49) seek out and destroy *Plasmodium* cells. In West Africa, for example, there is a strong association between an allele called *HLA-B53* and protection against malaria. In liver cells that are infected by *Plasmodium*, HLA-B53 proteins on the surface of the liver cell display a parasite protein called cp26 (**Figure 53.13a**, page 1226). The display is a signal that immune system cells can read. It means "I'm an infected cell. Kill me before they kill all of us." Immune system cells destroy the liver cell before the parasite cells inside can multiply. In this way, people who have at least one copy of the *HLA-B53* allele are better able to beat back malaria infections before the infection progresses. People who have the *HLA-B53* allele appear to be winning the arms race against malaria.

Unfortunately, follow-up research has shown that the arms race is far from won. *Plasmodium* populations in West Africa now have a variety of alleles for the protein recognized by HLA-B53. Some of these variants bind to HLA-B53 and trigger an immune response in the host, but others escape detection. Furthermore, many people in West Africa are infected with several different strains of *Plasmodium*. In some cases, the recognition step by HLA-B53 breaks down when certain strains are found together (**Figure 53.13b**). To make sense of these observations, researchers suggest that natural selection has favored the evolution of *Plasmodium* strains with weapons that counter HLA-B53.

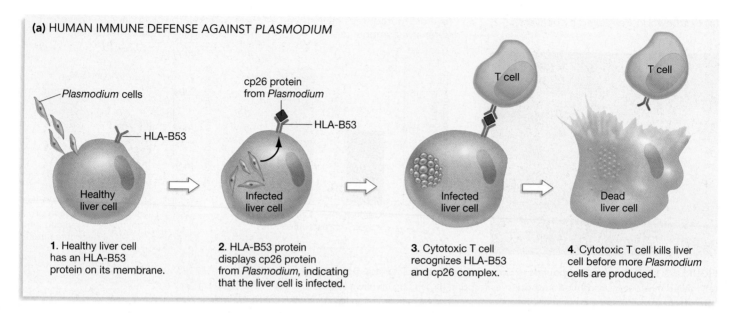

(a) HUMAN IMMUNE DEFENSE AGAINST *PLASMODIUM*

1. Healthy liver cell has an HLA-B53 protein on its membrane.

2. HLA-B53 protein displays cp26 protein from *Plasmodium*, indicating that the liver cell is infected.

3. Cytotoxic T cell recognizes HLA-B53 and cp26 complex.

4. Cytotoxic T cell kills liver cell before more *Plasmodium* cells are produced.

(b) *Plasmodium* strains have different versions of the cp protein. How successful are these strains at infecting people?

Plasmodium strain	Infection rate	Interpretation
cp26	Low	HLA-B53 binds to these proteins. Immune response is effective.
cp29	Low	
cp26 and cp29 strains together	High	Immune response fails when these strains infect the same person.
cp27	High	HLA-B53 does not bind to these proteins. Immune response is not as effective.
cp28	Average	

FIGURE 53.13 Interactions between the Human Immune System and *Plasmodium*

(a) If HLA-B53 binds to a particular *Plasmodium* protein, then infected cells are recognized and destroyed. **(b)** Some strains of *Plasmodium* appear to avoid detection by the immune system better than others do. Also, certain strains defeat the immune response if they infect the same person at the same time.

To summarize, an arms race continues to rage between *Plasmodium* and humans. Certain human proteins act as antimalarial weapons. But as predicted by coevolutionary theory, *Plasmodium* has evolved effective responses and continues to evolve.

Can Parasites Manipulate Their Hosts?

Parasites do not just have to invade tissues and grow while evading defensive responses by their host. They also have to be transmitted to new hosts. To a parasite, an uninfected host represents uncolonized habitat, teeming with resources. What have biologists learned about how parasites are transmitted to new hosts?

To answer this question, consider species of land snails that are parasitized by flatworms—specifically, flukes in the genus *Leucochloridium*. Researchers who studied this association discovered something unusual. When the flukes have matured and are ready to be transmitted to their next host, a bird, they burrow into the snail's tentacles and wriggle. Infected snails also become attracted to light, even though uninfected snails avoid sunlit areas and prefer dark, shady environments (**Figure 53.14**). When infected snails move out of the shade into the open and glide about with wriggling tentacles, they are more easily spot-

Birds that prey on snails are the next host for the parasite

Infected snails move to open sunny areas; tentacles wriggle

Uninfected snails stay in shaded areas; tentacles do not wriggle

FIGURE 53.14 A Parasite That Manipulates Host Behavior
The behavior of snails that are infected with flukes in the genus *Leucochloridium* is dramatically different from the behavior of uninfected snails. **EXERCISE** Design an experiment to test the hypothesis that infected snails are more likely than uninfected snails to be eaten by birds.

ted and consumed by birds. To interpret these observations, biologists suggest that the worms manipulate the behavior of the snail, and that the change in snail behavior makes the parasite more likely to be transmitted to a new host.

Studies of how parasites are transmitted to new hosts reinforce a general message: Extensive coevolution occurs among species that interact via consumption. Experiments on mutualistic interactions carry the same message.

Mutualism

Mutualisms are $+/+$ interactions that involve a wide variety of organisms and rewards. Many species of bees, for example, visit flowers to harvest nectar and pollen. Bees benefit because nectar is used as a food source for adult bees while the pollen is fed to larvae. Flowering plants also benefit because, in the process of visiting flowers, foraging bees carry pollen from one plant to another and accomplish pollination. Chapters 29 and 40 detailed some of the adaptations found in flowering plants that increase the efficiency of pollination. Other chapters highlighted an array of other mutualisms:

- One of the most important of all mutualisms occurs between fungi and plant roots. Chapter 30 reviewed experimental evidence indicating that fungi receive sugars and other carbon-containing compounds in exchange for nitrogen or phosphorus needed by the plant partner.

- Arguably the most critical of all mutualisms involves bacteria that fix nitrogen and certain species of plants. As pointed out in Chapters 27 and 37, the partnership is based on host plants providing sugars and protection and the bacteria supplying nitrogen in return.

- Chapter 31 cited interesting mutualisms between rancher ants that protect aphids in exchange for sugar-rich honeydew and between farmer ants that cultivate fungi for food.

Figure 53.15 illustrates some other interesting mutualisms. Figure 53.15a shows ants in the genus *Crematogaster*, which live in acacia trees native to Africa and the New World tropics. The ants live in bulbs at the base of acacia thorns and feed on small structures that grow from tree branches. These ants protect the tree by attacking and biting herbivores and by cutting vegetation from the ground below the host tree. Figure 53.15b illustrates cleaner shrimp in action. These shrimp pick external parasites from the jaws and gills of fish. In this mutualism, one species receives dinner while the other obtains medical attention.

As these examples show, the rewards from mutualistic interactions range from the transportation of gametes to food, housing, medical help, and protection. It is important to note, however, that even though mutualisms benefit both species, the interaction does not involve individuals from different species being altruistic or "nice" to each other. Expanding on a point that Charles Darwin introduced in 1862, Judith Bronstein recently described mutualisms as "a kind of reciprocal parasitism; that is, each partner is out to do the best it can by obtaining what it needs from its mutualist at the lowest possible cost to itself." Her point is that the benefits received in a mutualism are a by-product of each individual pursuing its own self-interest, by maximizing its ability to survive and reproduce.

In this light, it is not surprising that some species that are closely related to mutualists "cheat" on the system. For

(a) Mutualism between ants and acacia trees

(b) Mutualism between cleaner shrimp and fish

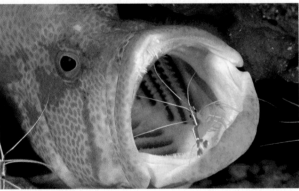

FIGURE 53.15 Mutualisms Take Many Forms
(a) In certain species of acacia tree, ants in the genus *Crematogaster* live in large bulbs at the base of spines and attack herbivores that threaten the tree. The ants eat nutrient-rich tissue produced at the tips of leaves or on branches. **(b)** Cleaner shrimp remove and eat parasites that take up residence on the gills of fish. **EXERCISE** In part (a), label the entrance to the ant colony and the nutrient-rich tissue on which the ants are feeding.

example, Chapter 40 introduced deceit pollination, in which certain species of plants produce a showy flower but no nectar reward. Pollinators have to be deceived to make a visit and carry out pollination. Evolutionary studies show that deceit pollinators evolved from ancestral species that did provide a reward. Over time, a +/+ interaction evolved into a +/− interaction.

A recent experimental study of mutualism provides another good example of the dynamic nature of these interactions. This study focused on ants and treehoppers. Ants are insects that live in colonies; treehoppers are small, herbivorous insects that feed by sucking sugar out of the phloem of plants. Treehoppers excrete the sugary solution honeydew from their posteriors. The honeydew, in turn, is harvested for food by ants.

It is clear that ants benefit from this association. But do the treehoppers? Biologists hypothesized that the ants might protect the treehoppers from their major predator, jumping spiders. These spiders feed heavily on juvenile treehoppers.

To test the hypothesis that ants protect treehoppers, the researchers studied ant-treehopper interactions over a three-year period. As **Figure 53.16** shows, the researchers marked out a 1000-m² study plot. Each year they randomly assigned the treehopper host plants inside the plot to one of two groups. The biologists removed the ants from one group but left the others alone to serve as a control. Then they compared the growth and survival of treehoppers on plants with and without ants. Recall that this is a common research strategy for studying species interactions. To assess the fitness costs or benefits of the interaction, researchers remove one of the participants experimentally and document the effect on the other participant's survival and reproduction, compared with the survival and reproduction of control individuals that experience a normal interaction.

In both the first and third years of the study, the number of treehopper young on host plants increased in the treatment with ants but showed a significant decline in the treatment with no ants (Figure 53.16). This result supports the hypothesis that treehoppers benefit from the interaction with ants because the ants protect the treehoppers from predation by jumping spiders.

In the second year of the study, however, the researchers found a very different pattern. There was no difference in offspring survival, adult survival, or overall population size between treehopper populations with ants and those without ants. Why? The researchers were able to answer this question because they also measured the abundance of spiders in each of the three years. Their census data showed that in the second year of the study, spider populations were very low.

Based on these results, the investigators concluded that the benefits of the ant-treehopper interaction depend entirely on predator abundance. Treehoppers benefit from their interaction with ants in years when predators are abundant but are unaffected in years when predators are scarce. If producing honey-

dew is costly to treehoppers, then the +/+ mutualism changes to a +/− interaction when spiders are rare.

To summarize, mutualism is like parasitism, competition, and other types of species interactions in an important respect: The outcome of the interaction depends on current conditions. Because the costs and benefits of species interactions are fluid, an interaction between the same two species may vary from par-

Question: Is the relationship between ants and treehoppers mutualistic?

Hypothesis: Ants harvest food from treehoppers and protect treehoppers from jumping spiders.

Null hypothesis: Ants harvest food from treehoppers but are not beneficial to treehopper survival.

Experimental setup:

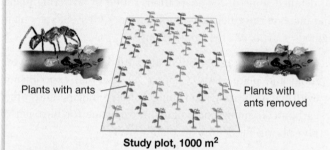

Plants with ants — Plants with ants removed

Study plot, 1000 m²

Prediction: More young treehoppers will be found when ants are present than when ants are absent.

Prediction of null hypothesis: There will be no difference in the number of young treehoppers on the plants.

Results (Year 1):

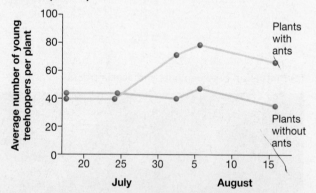

Conclusion: Treehoppers benefit from the interaction with ants, which protect treehoppers from predation by jumping spiders.

FIGURE 53.16 Experimental Evidence That the Treehopper-Ant Interaction Is Mutualistic
An experiment designed to test whether the presence of ants is beneficial for treehoppers. Treehoppers are small herbivorous insects that suck sugar out of the phloem of plants and secrete honeydew. In some years, the number of young treehoppers increased over time when ants were present and declined over time when ants were absent.

TABLE 53.1 Summary of Species Interactions

Type of Interaction	Fitness Effects	Short-Term Impact: Distribution and Abundance	Long-Term Impact: Coevolution
Competition	– / –	Reduces population size of both species; if competition is asymmetric, competitive exclusion reduces range of one species.	Niche differentiation via selection to reduce competition
Consumption	+ / –	Impact on prey population depends on prey density and effectiveness of defenses.	Strong selection on prey for effective defense; strong selection on consumer for traits that overcome defenses
Parasitism	+ / –	Impact on host population depends on parasite density and effectiveness of defenses.	Strong selection on host for effective defense; strong selection on parasite for traits that overcome defenses
Mutualism	+ / +	Population size and range of both species are dependent on each other.	Strong selection on both species to maximize fitness benefits and minimize fitness costs of relationship
Commensalism	+ / 0	Population size and range of commensal may depend on size and distribution of host.	Strong selection on commensal to increase fitness benefits in relationship; no selection on host

asitism to mutualism to competition. **Table 53.1** summarizes the fitness effects, short-term impacts on population size, and long-term evolutionary aspects of species interactions.

✓ CHECK YOUR UNDERSTANDING

Species interactions affect the short-term distribution and abundance of the populations involved and result in coevolution. For example, natural selection favors the evolution of traits that decrease competition, resulting in changes in a species' niche over time. Consumers may reduce the population size of the species on which they feed and often exert strong natural selection for effective defense mechanisms. Mutualisms benefit the species involved and can lead to highly coevolved associations, such as mycorrhizal fungi and symbiotic nitrogen-fixing bacteria. **You should be able to** (1) explain what a coevolutionary arms race is and (2) give a specific example of how competition, consumption, and mutualism affect population size and long-term evolution.

53.2 Community Structure

Biologists have made important progress in understanding the nature of species interactions and their consequences. In terms of understanding the structure and function of biological communities, however, research on species interactions has a limitation: It usually focuses on just two species at a time. But biological communities contain many thousands of species. To understand how communities work, biologists broaden the scope of research and explore how combinations of many species interact.

The question that biologists initially asked about communities concerned structure: Do biological communities have a tightly prescribed organization and composition, or are they merely loose assemblages of species? If communities are highly structured entities, then their makeup should be pre-

dictable. For example, if a community is destroyed by a disturbance and then allowed to recover, the diversity and abundance of species at that site should be identical when recovery is complete. But if communities can be made up of many different combinations of species, depending on which arrive earlier or later, then community composition will be difficult to predict. The diversity and abundance of species found in a region would vary substantially before a disturbance and after recovery.

Clements and Gleason: Two Views of Community Dynamics

Beginning with a paper published in 1936, Frederick Clements promoted the view that biological communities are stable, integrated, and orderly entities with a highly predictable composition. His hypothesis was that species interactions are so extensive and coevolution is so important that the groups of species called communities have become highly integrated and interdependent units in nature. Stated another way, the species within a community cannot live without each other.

To drive his idea home, Clements likened the development of a plant community to the development of an individual organism. He argued that communities develop by passing through a series of predictable stages dictated by extensive interactions among species and that this development culminates in a stable final stage known as a **climax community**. According to Clements, the nature of the climax community is determined by the area's climate and does not change over time. Further, he held that if a fire or other disturbance destroys the climax community, it will reconstitute itself by repeating its predictable developmental stages.

Henry Gleason, in contrast, contended that the community found in a particular area is neither stable nor predictable. He claimed that plant and animal communities are ephemeral associations of species that just happen to share similar climatic requirements. According to Gleason, it is largely a matter of chance whether a similar community develops in the same area after a disturbance occurs. Gleason downplayed the role of biotic factors, such as species interactions, in structuring communities. To him, abiotic factors and history—for example, which seeds and juvenile animals happened to arrive after a disturbance—were the key elements in determining which species are found at a particular location.

Which viewpoint is more accurate? To answer this question, let's consider a series of data sets that are relevant to the issue, then synthesize the results into a general picture of how communities are structured.

Keystone Species   According to Clements, the composition of a community is dictated by climate and structured by biotic interactions. Some of the best data on the role of biotic interactions in communities contradict his view of communities as static, predictable entities, however. Experiments have shown that if a single species of predator or herbivore is removed from a community, the structure of an entire community can change dramatically.

As an example of this work, consider a series of classical experiments that Robert Paine did in intertidal habitats of the Pacific Northwest of North America. In this environment, the sea star *Pisaster ochraceous* is an important predator. When Paine removed *Pisaster* from experimental areas, what had been diverse communities of algae and invertebrates became overgrown with solid stands of the California mussel *Mytilus californianus* (**Figure 53.17**). Although *M. californianus* is a dominant competitor, its populations had been held in check by sea-star predation. When this predator was gone, the species richness and structural complexity of the habitat changed radically.

To capture the effect that a predator such as *Pisaster* can have on a community, Paine coined the term *keystone species*. A **keystone species** is a species that has a much greater impact on the surrounding species than its abundance would suggest. In grassland biomes of central North America, bison act as a keystone species even though they represent a tiny fraction of the total mass of organisms present. Because they graze primarily on grasses, bison herds decrease the abundance of grasses and increase the importance of **forbs**—herbaceous, flowering plants such as daisies and members of the pea family.

Sea stars, bison, and other keystone species may vary in abundance in space or over time. As a result, the composition and structure of their communities will vary. These observations suggest that communities are dynamic entities whose composition is difficult to predict.

Historical Data on Community Structure   Data on the historical composition of plant communities began to accumulate during the 1970s and 1980s. Studies of fossil pollen documented that the distribution of plant species and communities at specific locations throughout North America has changed radically since the end of the last ice age about 11,000 years ago. An important pattern emerged from these data: Contrary to what Clements would predict, groups of species do not come and go in the fossil record in tightly integrated units. Instead, the ranges of individual species tended to change independently of one another. For example, at a study site in Mirror Lake, New Hampshire, the percentage of pollen present from pines and oaks rose and fell in tandem over the past 14,000 years—just as Clements would predict for a tightly integrated, pine-oak forest community. In contrast, the percentage of pollen from spruce, fir, cedar, beech, and other tree species changed independently of each other. In general, studies of fossil pollen suggest that the composition of most plant communities has been dynamic and contingent on historical events rather than static.

Experimental Tests   As an example of experimental work on how communities are structured, consider a recent study of

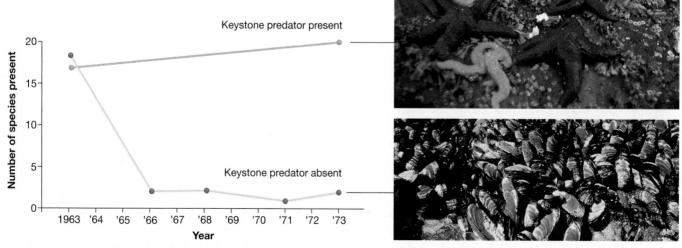

FIGURE 53.17 Keystone Predation Alters Community Structure in a Rocky Intertidal Habitat
When the sea star *Pisaster ochraceous* is present, rocky intertidal habitats have a large diversity of species.
When *Pisaster* is absent, the same habitats are dominated by beds of California mussels.

the planktonic organisms in ponds. To explore how predictable community structure is, biologists constructed 12 identical ponds (**Figure 53.18**). They filled the ponds at the same time with water that contained enough chlorine to kill any preexist-

ing organisms. At the outset, then, the ponds were sterile. If Clements's view of community structure is correct, each pond should develop the same community of species once the chlorine vaporized and made the water habitable. If Gleason's view of community structure is correct, then each pond should develop a different community.

To test this prediction, the researchers sampled water from the ponds repeatedly for one year. They measured temperature, chemical makeup, and other physical characteristics of the water and recorded the diversity and abundance of each species of plankton by examining the samples under the microscope. They found a total of 61 species in all of the ponds but discovered that individual ponds had only 31 to 39 species. This observation is important. Each pond contained just half to two-thirds of the total number of species that lived in the experimental area and that were available for colonization. Most species occurred in most or all of the 12 ponds, but each pond had a unique species assemblage. Why?

To explain their results, the researchers contended that some species are particularly good at dispersing and are likely to colonize all or most of the available habitats. Other species disperse more slowly and tend to reach only one or a few of the available habitats. Further, the investigators proposed that the arrival of certain competitors or predators early in the colonization process greatly affects which species are able to invade successfully later. As a result, the specific details of community assembly and composition are somewhat contingent and difficult to predict. At least to some degree, communities are a product of chance and history.

The overall message of research on keystone species, the history of plant communities, and experiments is that Clements's position was clearly too extreme and that Gleason's view is closer to being correct. Although both biotic interactions and climate are important in determining which species exist at a certain site, chance and history also play a large role.

Disturbance and Change in Ecological Communities

Perhaps the most dramatic contrast between the views held by Clements and contemporary biologists concerns how communities change over time. Instead of viewing communities as the static entities that Clements envisioned, researchers now recognize that communities are dynamic. Community composition and structure may change radically in response to changes in abiotic and biotic conditions. Biologists have become particularly interested in how communities respond to disturbance. A **disturbance** is any event that removes some individuals or biomass from a community. (Recall from Chapter 50 that biomass is the total mass of living organisms.)

Forest fires, windstorms, floods, the fall of a large canopy tree, disease epidemics, and herbivore outbreaks all qualify as disturbances. These events are important because they alter light levels, nutrients, unoccupied space, or some other aspect of resource availability. What biologists have come to realize is that the impact of disturbance is a function of three

Question: Do identical communities develop in identical environments?

"Clementsian" hypothesis: Identical communities will always develop in identical environments.

"Gleasonian" hypothesis: Identical communities will not always develop in identical environments.

Experimental setup:

1. Construct 12 identical ponds. Fill at the same time and sterilize water so that there are no preexisting organisms.

2. Examine water samples from each pond. Identify each plankton species present in each sample.

"Clementsian" prediction: Identical plankton communities will develop in all 12 ponds.

"Gleasonian" prediction: Different plankton communities will develop in different ponds.

Results (after 1 year):

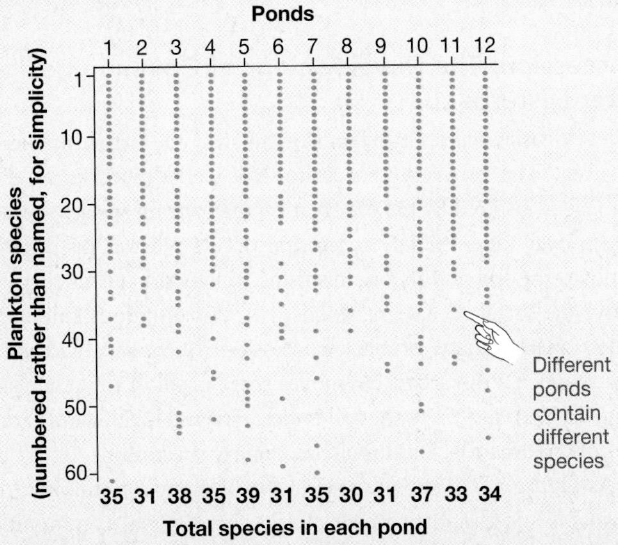

Different ponds contain different species

Conclusion: Although about half of the species present appear in all or most ponds, each pond has a unique composition. Both hypotheses are partially correct.

FIGURE 53.18 Experimental Evidence That Identical Communities Do Not Develop in Identical Habitats

QUESTION How many species are found in all 12 ponds?

factors: (1) the type of disturbance, (2) its frequency, and (3) its severity—for example, the speed and duration of a flood, or the intensity of heat during a fire. Let's take a closer look.

What Disturbances Occur, and How Frequent and Severe Are They?

Most communities experience a characteristic type of disturbance. In most cases, disturbances occur with a predictable frequency and severity. To capture this point, biologists refer to a community's **disturbance regime**. For example, fires kill all or most of the existing trees in a boreal forest every 100 to 300 years, on average. In contrast, numerous small-scale tree falls, usually caused by windstorms, occur in temperate and tropical forests every few years.

How Do Researchers Determine a Community's Disturbance Regime?
Ecologists use two strategies to determine a community's natural pattern of disturbance. The first approach is based on inferring long-term patterns from data obtained in a short-term analysis. For example, an observational study might document that 1 percent of all boreal forest on Earth burns in a given year. Assuming that fires occur randomly, researchers project that any particular piece of boreal forest has a 1 in 100 chance of burning each year. According to this reasoning, fires will recur in that particular area every 100 years, on average.

This extrapolation approach is straightforward to implement, but it has important drawbacks. In boreal forests, for example, fires do not occur randomly in either space or time. They are more likely in some areas than in others, and they tend to occur in particularly dry years. Unless sampling is extensive, it is difficult to avoid errors caused by extrapolating from particularly disturbance-prone or disturbance-free areas or years.

The second approach to determining disturbance regimes is based on reconstructing the history of a particular site. Flooding frequency, for example, can be estimated by analyzing sediments, because floods deposit distinctive groups of sediment particles. Researchers estimate the frequency and impact of storms by finding wind-killed trees and determining their date of death. (Investigators do so by comparing patterns in the growth rings of the dead trees with those of living individuals nearby; see Chapter 35.) The disturbances that have been most extensively studied using historical techniques, however, are forest fires.

Forest fires often leave a layer of burned organic matter and charcoal on the surface of the ground. As a result, researchers can dig a soil pit, find charcoal layers, and use radioisotope dating to establish when the fires occurred. It is also possible to date the death of trees killed by fire by comparing their growth rings with those of living trees. Further, trees that are not killed by fire are often scarred. When a fire burns close enough to kill a patch of cambium tissue, a scar forms. These fire scars occur most often at the tree's base, where dead leaves and twigs accumulate and furnish fuel. Fire scars can be dated by analyzing growth rings.

Why Is It Important to Understand Disturbance Regimes?
To appreciate why biologists are so interested in understanding disturbance regimes, consider a recent study on the fire history of giant sequoia groves in California (**Figure 53.19a**). Giant sequoias grow in small, isolated groves on the western side of the Sierra Nevada range. Individuals live more than a thousand years, and many have been scarred repeatedly by fires. A biologist obtained samples of cross sections through the bases of 90 giant sequoias in five different groves. As **Figure 53.19b** shows, the cross sections contained numerous rings that had been scarred by fire. To determine the date of each disturbance, the researcher counted tree rings back from the present. He found that, in most of the groves, 10 to 53 fires had occurred each century for the past 1530 to 2000 years (**Figure 53.19c**). The data indicated that each tree had been burned an average of 64 times.

This study established that fires are extremely frequent in the community examined. Because not enough time would pass for large amounts of fuel to accumulate between fires, they were probably of low severity. Partly because of this work, the biologists responsible for managing sequoia groves now set controlled fires or let low-intensity natural fires burn instead of suppressing them immediately. Similarly, studies of disturbance regimes along the Colorado River in southwestern North America inspired land managers to release a huge pulse of water from the reservoirs behind dams on the waterway recently. The flood that resulted was designed to mimic a natural disturbance. According to follow-up studies, the artificial flood appears to have benefited the plant and animal communities downstream.

Succession: The Development of Communities after Disturbance
Severe disturbances remove all or most of the organisms from an area. The recovery that follows is called **succession**. The name was inspired by the observation that certain species succeed others over time. **Primary succession** occurs when a disturbance removes the soil and its organisms as well as organisms that live above the surface. Glaciers, floods, volcanic eruptions, and landslides often initiate primary succession. **Secondary succession** occurs when a disturbance removes some or all of the organisms from an area but leaves the soil intact. Fire and logging are examples of disturbances that initiate secondary succession.

As **Figure 53.20** shows, a sequence of plant communities develops as succession proceeds. Early successional communities are dominated by species that are short lived and small and that disperse their seeds over long distances. Late successional communities are dominated by species that tend to be long lived, large, and good competitors for resources such as light and nutrients. The specific sequence of species that appears over time is called a **successional pathway**. What determines the pattern and rate of species replacement during succession at a particular time and place?

(a) Giant sequoias after a fire

(b) Fire scars in the growth rings

(c) Reconstructing history from fire scars

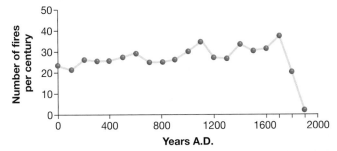

FIGURE 53.19 History of Disturbance in a Fire-Prone Community
(a) A giant sequoia grove. **(b)** Trunk cross section of a giant sequoia; the thick black arcs are fire scars. **(c)** Because trees form one ring (light band/dark band) every year, researchers can count the rings to determine how often fires have occurred during the last 2000 years.

Old field

Disturbance (plowing) ends, site is invaded by short-lived weedy species.

Pioneering species

Weedy species are replaced by longer-lived herbaceous species and grasses.

Early successional community

Shrubs and short-lived trees begin to invade.

Mid-successional community

Short-lived tree species mature; long-lived trees begin to invade.

Late-successional community

Long-lived tree species mature.

Climax community

FIGURE 53.20 Succession in Midlatitude Temperate Forests
Sequence of photos showing how succession leads to the development of a temperate forest from a disturbed state (in this case, an abandoned agricultural field).

Theoretical Considerations Biologists focus on three factors to predict the outcome of succession in a community: (1) the particular traits of the species involved, (2) how species interact, and (3) historical and environmental circumstances such as the size of the area involved and weather conditions. Before going on to consider a detailed case history, we need to explore these factors in more detail.

Species traits, such as dispersal capability and the ability to withstand extreme dryness, are particularly important early in succession. As common sense would predict, recently disturbed sites tend to be colonized by plants and animals with good dispersal ability. When these organisms arrive, however, they often have to endure harsh environmental conditions. These **pioneering species** tend to have "weedy" life histories. (A **weed** is a plant that is adapted for growth in disturbed soils.) Early successional species devote most of their energy to reproduction and little to competitive ability. They have small seeds, rapid growth, and a short life span, and they begin reproducing at an early age. As a result, they have a high reproductive rate. But in addition, they can tolerate severe abiotic conditions, such as high light levels, poor nutrient availability, and drying.

Once colonization is under way, the course of succession tends to depend less on how species cope with aspects of the abiotic environment and more on how they interact with other species. This change occurs because plants that grow early in succession change abiotic conditions in a way that makes the conditions less severe. Because plants provide shade, they reduce temperatures and increase humidity. Their dead bodies also add organic material and nutrients to the soil. As abiotic conditions improve, biotic interactions become more important.

During succession, existing species can have one of three effects on subsequent species: (1) facilitation, (2) tolerance, or (3) inhibition. **Facilitation** takes place when the presence of an early-arriving species makes conditions more favorable for the arrival of certain later species by providing shade or nutrients. **Tolerance** means that existing species do not affect the probability that subsequent species will become established. **Inhibition** occurs when the presence of one species inhibits the establishment of another, as when a plant species that requires high light levels to germinate is inhibited late in succession.

In addition to species traits and species interactions, the pattern and rate of succession depend on the historical and environmental context in which they occur. For example, researchers found that the communities that developed after forest fires disturbed Yellowstone National Park in 1988 depended on the size of the burned patch and how hot the fire had been at that location. Succession is also affected by the particular weather or climate conditions that occur during the process. Variation in weather and climate causes different successional pathways to occur in the same place at different times.

In summary, analyzing species traits, species interactions, and the historical/environmental context provides a useful structure for understanding why particular successional pathways occur. To see this theoretical framework in action, let's examine data on the course of primary succession that has occurred in Glacier Bay, Alaska.

A Case History: Glacier Bay, Alaska An extraordinarily rapid and extensive glacial recession is occurring at Glacier Bay (**Figure 53.21a**). In just 200 years, glaciers that once filled the bay have retreated approximately 100 km, exposing extensive tracts of barren glacial sediments to colonization. Because of this event, Glacier Bay has become an important site for studying succession.

Figure 53.21a shows the plant communities found in the area. The oldest sites are dense forests of Sitka spruce and western hemlock. Areas that have been deglaciated for about 100 years are inhabited by scattered spruce trees and dense thickets of a shrub called Sitka alder. Sites that have been deglaciated for 45 to 80 years are also covered with dense alder thickets, but the emergent trees are primarily cottonwood. Locations that have been ice free for 20 years or less do not have a continuous plant cover. Instead, they host scattered individuals of willow and a small shrub called *Dryas*.

These observations inspired a hypothesis for the pattern of succession in Glacier Bay: With time, the youngest communities of *Dryas* and willow succeed to alder thickets, which subsequently become dense spruce-hemlock forests. A recent study has challenged this hypothesis, however. Researchers who reconstructed the history of each community by studying tree rings found that three distinct successional pathways have occurred (**Figure 53.21b**): (1) In the lower part of the bay, soon after the ice retreated, Sitka spruce began growing and quickly formed dense forests. Western hemlock arrived after spruce and is now common in the understory. (2) At middle-aged sites in the upper part of the bay, alder thickets were dominant for several decades, and spruce is just beginning to become common. These forests will probably never be as dense as the ones in the lower bay, however, and there is no sign that western hemlock has begun to establish itself. (3) In contrast, the youngest sites in the uppermost part of the bay may be following a third pathway. Alder thickets became dominant fairly early, but spruce trees are scarce. Instead, cottonwood trees are abundant.

These data present a challenge: How do species traits, species interactions, and dispersal patterns interact to generate the three observed successional pathways?

Species traits may be especially important in explaining certain details about the successional pathways. Western hemlock, for example, is abundant at older sites but largely absent from young ones. This is logical, because its seeds germinate and grow only in soils containing a substantial amount of organic matter and because the young trees can tolerate deep shade but not bright sunlight. Western hemlock's intolerance of early-successional conditions explains why none of the three pathways began with colonization by this species.

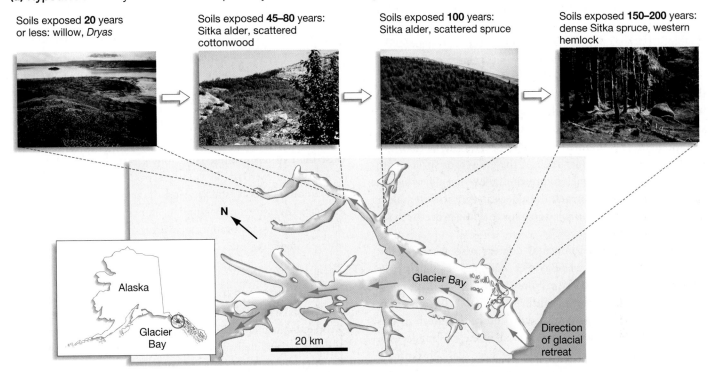

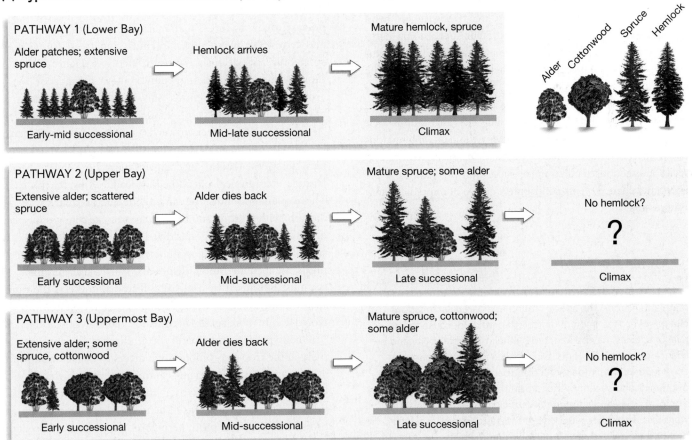

(a) Hypothesis 1: Only one successional pathway occurs in Glacier Bay.

Soils exposed **20** years or less: willow, *Dryas*

Soils exposed **45–80** years: Sitka alder, scattered cottonwood

Soils exposed **100** years: Sitka alder, scattered spruce

Soils exposed **150–200** years: dense Sitka spruce, western hemlock

(b) Hypothesis 2: Three distinct successional pathways occur in Glacier Bay.

FIGURE 53.21 Evidence for Multiple Successional Pathways at Glacier Bay
(a) Several distinct plant communities are found in Glacier Bay, Alaska. The first hypothesis about succession at Glacier Bay proposed that these communities represent stages in the same successional pathway. **(b)** Recent data indicate that three successional pathways exist.

Species interactions have been important in all three pathways. For example, research has shown that alder facilitates the growth of Sitka spruce. The facilitating effect occurs because symbiotic bacteria that live inside nodules on the roots of alder convert atmospheric nitrogen (N_2) to nitrogen-containing molecules that alder can use to build proteins and nucleic acids. Although spruce trees are capable of invading and growing without the presence of alder, they grow faster when alder stands have added nitrogen to the soil.

Competition is another important species interaction. For example, shading by alder reduces the growth of spruce until spruce trees are tall enough to protrude above the alder thicket. Once the spruce trees breach the alder canopy, however, alder dies out, because it is unable to compete with spruce trees for light.

Historical and environmental context also clearly influences succession at Glacier Bay. For example, geologists have found evidence that the ice was more than 1100 m thick in the upper part of Glacier Bay during the mid-1700s. Because forests grow to an elevation of only 700 m or 800 m in this part of Alaska, the glacier eliminated all of the existing forests. In the lower part of the bay, however, the ice was substantially thinner. As a result, some forests remained on the mountain slopes beyond the ice. As the glacier retreated from these areas, the forests on the slopes provided a source of spruce and hemlock seeds and set a dramatically different successional pathway in motion. In this way, environmental context—in this case, distance to existing forests—helped determine how the community developed.

To summarize, successional pathways are determined by an array of factors. These factors include the adaptations that certain species have to their abiotic environment, interactions among species, and the history of the site. Species traits and species interactions tend to make succession predictable, while history and chance events contribute a degree of unpredictability to succession.

✓CHECK YOUR UNDERSTANDING

The composition of a community is primarily a function of climate and history. The species found at a particular location change over time and are difficult to predict, because the arrival of species may be affected by their dispersal ability and by chance or historical events. Disturbance is a normal part of communities. The impact of a disturbance depends on its type, frequency, and severity. After a disturbance occurs, a succession of species and communities replaces the individuals that were lost. The exact sequence of species observed is a function of their traits, their interactions, and the history of the site. You should be able to cite specific observational or experimental evidence that supports each of these statements.

53.3 Species Diversity in Ecological Communities

The diversity of species present is a key feature of biological communities, and it can be quantified in two ways. **Species richness** is a simple count of how many species are present in a given community. **Species diversity**, in contrast, is a weighted measure that incorporates a species' relative abundance as well as its presence or absence (see **Box 53.1**). At a global scale, species abundance data are often not available. In such cases, ecologists sometimes use the two terms interchangeably. Research on richness and diversity has focused on two questions: First, why are some communities more species rich than others? Second, why is species diversity important? As human populations increase and species losses mount, answering these questions becomes increasingly urgent.

Global Patterns in Species Diversity

Biologists who began cataloguing the flora and fauna of the tropics in the mid-1800s immediately recognized that these communities contained many more species than did temperate or subarctic environments. Data compiled in the intervening years have confirmed the existence of a latitudinal gradient in species diversity for communities as a whole as well as for many taxonomic groups. In birds, mammals, fish, reptiles, many aquatic and terrestrial invertebrates, orchids, and trees, for example, species diversity declines as latitude increases (**Figure 53.23**). Although this pattern is not universal—a number of marine groups, as well as shorebirds, show a positive relationship between latitude and diversity—it is widespread. Why does it occur?

To explain why species diversity might decline with increasing latitude, biologists have to consider two fundamental principles. First, the causal mechanism must be abiotic, because latitude is a physical phenomenon produced by Earth's shape. The explanation must be an abiotic factor that varies predictably with latitude and that could produce changes in species diversity. Second, the species diversity of a particular area is the sum of four processes: speciation, extinction, immigration (colonization), and emigration (dispersal). Thus, the latitudinal gradient must be caused by a physical factor that affects the rate of speciation, extinction, immigration, or emigration in a way that would lead to more species in the tropics and fewer near the poles.

One hypothesis to explain the gradient is that high productivity in the tropics promotes high diversity by increasing speciation rates and decreasing extinction rates. (Recall from Chapter 50 that productivity is the total amount of photosynthesis per unit area per year.) The idea is that increased biomass production supports more herbivores and thus more predators and parasites and scavengers. In addition, speciation rates should increase when niche differentiation occurs within populations of herbivores, predators, parasites, and scavengers.

BOX 53.1   Measuring Species Diversity

To measure species diversity, a biologist could simply count the number of species present in a community. The problem is that simple counts provide an incomplete picture of diversity. The relative abundance of species is also an important component of diversity. To grasp this point, consider the composition of the three hypothetical communities shown in **Figure 53.22**. These communities are nearly identical in species composition but differ greatly in the relative abundance of each species.

These data can be used to compare two measures of species diversity. Species richness is simply the number of species found in a community. In this case, communities 1 and 2 have equal species richness and community 3 is lower in richness by one species. It is important to note, however, that communities 2 and 3 have similar relative abundances of each species, or what biologists call high *evenness*. Community 1, in contrast, is highly uneven. Fifty-five percent of the individuals in community 1 belong to species A, and other species are relatively rare. An uneven community

has lower effective diversity than its species richness would indicate.

To take evenness into account, other diversity indices have been developed. A simple example is the **Shannon-Weaver index**, given by the following equation:

$$H' = -\Sigma p_i \log(p_i)$$

In this equation, p_i is the proportion of individuals in the community that belong

to species i. The index is summed over the i species in the study. The Shannon-Weaver index for the three hypothetical communities is shown at the bottom of Figure 53.22. Notice that while communities 1 and 2 have the same species richness, community 2 has higher diversity because of its greater evenness. Community 3 has lower species richness than community 1 but higher diversity.

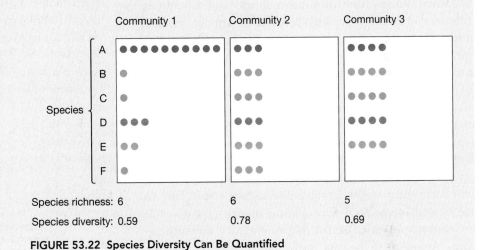

FIGURE 53.22 Species Diversity Can Be Quantified

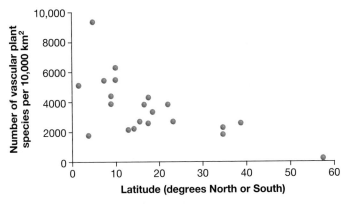

FIGURE 53.23 There Is a Strong Latitudinal Gradient in Species Diversity

One example of a latitudinal gradient in species diversity. The number of plant species in a 10,000-km² area declines from up to 10,000 in equatorial regions to less than 1000 at very high latitudes.

Although this **high-productivity hypothesis** is supported by the global-scale correlation between productivity and diversity, experimental studies challenge it. For example, researchers who add fertilizer to aquatic or terrestrial communities routinely observe significant increases in productivity but decreases in diversity.

Based on results such as these, researchers have concluded that productivity alone is probably not a sufficient explanation for the higher species diversity in the tropics. Research continues, however, and investigators have recently begun to focus on an abiotic factor that is correlated with productivity: temperature. Biologists who analyzed data on gastropods and other marine invertebrates have documented a strong correlation between the temperature of marine waters and species diversity. This observation has inspired the *energy hypothesis*: Temperature causes changes in species diversity by influencing

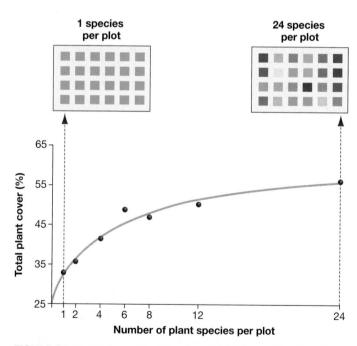

productivity *and* the ability of organisms to tolerate the physical conditions in a region.

A third hypothesis is that tropical regions have had more time for speciation to occur than other regions have. Temperate and arctic latitudes were repeatedly scoured by ice sheets over the last 2 million years, but tropical regions were not. Recent data suggest, however, that tropical forests were dramatically reduced in size by widespread drying trends during the ice ages. Existing forests may be much younger than originally thought. If so, then the contrast in the age of northern and southern habitats may not be enough to explain the dramatic difference in species diversity.

A fourth hypothesis was inspired by the observation that species diversity is much higher in mid-successional communities than in pioneer or mature communities. The **intermediate disturbance hypothesis** holds that regions with a moderate type, frequency, and severity of disturbance should have high species richness and diversity. The logic here is that, with intermediate levels of disturbance, communities will contain pioneering species as well as species better adapted to late-successional conditions. For example, recent studies have confirmed that tree falls and canopy gaps occur regularly in tropical forests and that fires occur in these biomes occasionally. As yet, however, there are no convincing data showing that intermediate levels of disturbance are more likely to occur in the tropics than they are at higher latitudes.

To summarize, there is no simple answer to the question of why some communities are more diverse than others. Each of the factors discussed here may influence diversity, but no single one offers a convincing explanation for the global diversity gradient. According to the data available at present, the presence of high energy inputs seems most responsible for communities with high species diversity.

The Role of Diversity in Ecological Communities

Research on species richness and species diversity has recently taken a practical turn: Does the number of species present affect the productivity and stability of communities? If so, then the rapid loss of diversity that is occurring now is almost certain to have serious consequences for biological communities all over the planet.

Does Species Richness Affect Productivity?
Primary productivity is the total amount of photosynthesis per unit area per year. **Net primary productivity (NPP)** is the amount of primary productivity that ends up in biomass. NPP is a key component of communities, because it represents the amount of food that is available to consumers and **detritivores**—species that eat dead material.

One of the more robust conclusions emerging from recent work in community ecology is that, at least in grassland biomes, productivity increases with increasing species richness.

Consider an experiment in which researchers grew seedlings from 24 species native to North American prairies and planted them into a total of 147 study plots. Each plot received a similar number of individuals but a different number of species: 1, 2, 4, 6, 8, 12, or 24. The research team allowed the plants to grow for a year and weeded the plots to remove any invading species. In the second year, the team measured productivity by quantifying the area of the plot covered by plants. As **Figure 53.24** shows, plant cover was higher in the more species-rich plots. Similar results have been obtained in other types of grassland habitats and when the investigators measure aboveground biomass instead of total cover. The more species in a grassland, the more biomass is produced.

Does Species Richness Affect Community Stability?
When biologists refer to the stability of a community, they mean its ability to (1) withstand disturbance without changing and (2) recover to former levels of productivity or species richness after a disturbance. **Resistance** is a measure of how much a community is affected by a disturbance. **Resilience** is a measure of how quickly a community recovers following a disturbance.

How does species richness affect resistance, resilience, and productivity? Several biologists have proposed that diversity is positively correlated with stability, meaning that highly diverse

FIGURE 53.24 Evidence That Species-Rich Communities Are More Productive
Total plant cover is defined as the percentage of a study plot's area that is covered by plants. It is interpreted as an index of productivity.
QUESTION Extrapolating from these data, what would total plant cover be in a plot that contained 53 plant species?

communities are more resistant and more resilient than are less diverse communities. The logic behind this hypothesis is that when a large number of species are present, some are likely to be redundant in their role in the community. For example, there are likely to be many types of herbivores, carnivores, pollinators, and so on. Because of this redundancy, wiping out one or a few species is unlikely to affect the community as a whole profoundly.

To test the hypothesis that diversity leads to high resistance and resilience, biologists followed up on a natural experiment. In 1987–1988 a severe drought hit sites in Minnesota where a research team had been measuring species richness and other characteristics in a series of permanent study plots. When the drought ended, the team was able to ask whether resistance and resilience correlated with the number of species that existed before the disturbance. Some of their results are shown in **Figure 53.25**. The graph documents the change in total biomass that occurred from the year prior to the drought to the height of the drought. This quantity reflects how resistant the community is to disturbance. A completely resistant community would show no change in biomass. As predicted, drought resistance appeared to be higher in more diverse communities than in less diverse ones. In addition, the group analyzed the change in biomass in each plot four years after the drought versus biomass prior to the drought. This analysis focused on how resilient the community is. A completely resilient community would recover quickly from the disturbance and have the same biomass at both times. The data indicated that most plots containing five or fewer species showed a significant lowering of biomass after the disturbance, indicating that they had not recovered. But in all of the plots that contained more than five species, biomass after the drought was the same as biomass prior to the disturbance.

These experiments support the hypothesis that species richness affects how communities function. In North American grasslands, at least, communities that are more diverse appear to be more productive, more resistant, and more resilient than communities that are less diverse. Why this pattern occurs is not clear, however. More experiments are needed to confirm these results and determine the mechanisms responsible for these patterns.

Question: Are more species-rich communities more stable than less-diverse communities?

Hypothesis: Resistance to disturbance should increase with increasing species richness.

Null hypothesis: No relationship exists between resistance and species richness.

Experimental setup:

Before drought   **During drought**

Compare biomass of experimental plots before drought and at peak of drought. (This was a natural experiment— severe drought just happened to occur during study.)

Prediction: Plots that were more species rich before the drought will be more resistant to change.

Prediction of null hypothesis: All the plots will have similar resistance regardless of species richness before the drought.

Results:

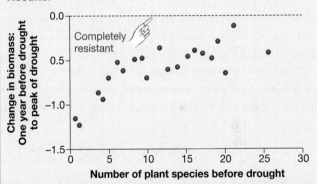

Conclusion: Resistance to disturbance increases with increasing species richness.

FIGURE 53.25 Evidence That Species-Rich Communities Resist Disturbance
Data support the hypothesis that diverse communities are more stable, meaning they change less during a disturbance, than less-diverse communities are. QUESTION The 0.0 values on the vertical graph axes are labeled "Completely resistant." Why?

ESSAY Let-It-Burn Policies

The summer of 1988 was extremely dry in the western United States. In the Greater Yellowstone Area (GYA), which includes Yellowstone National Park, almost no rain fell from mid-July through August. Thunderstorms rolled through with ample lightning but no rain. As a result of the severe conditions, fires ignited. A great deal of fuel (in the form of dry brush) was available in forested areas of the GYA, because park managers had actively suppressed forest fires since the park's founding in 1872. But in 1988 park personnel allowed the fires to burn. Research had shown that the GYA experienced frequent fires

(Continued on next page)

(Essay continued)

before the arrival of European settlers, and that many of the region's biological communities depend on fires to thin out dense stands of trees and allow regeneration. By late July, however, the blazes had grown so large that the National Park Service decided to suppress them. The conflagration continued to grow, however. By September 1 the fire perimeter enclosed approximately 353,800 ha—an area substantially larger than the state of Rhode Island (**Figure 53.26**).

The onset of winter, combined with a $120-million firefighting effort, eventually extinguished the fires. Cold, wet weather did not extinguish a firestorm of controversy, however. Critics of the let-it-burn policy decried the destruction of the National Park Service's crown jewel.

But was the park destroyed? The species traits of trees that are particularly abundant in the GYA suggest that the answer is no. For example, both aspen and lodgepole pine require fire in order to reproduce. Aspen cannot tolerate shade and does not sprout beneath its own canopy. Lodgepole pine is also shade intolerant and has cones that are glued shut with resin. The cones open and shed their seeds only in the heat of a fire.

Since 1988 the regeneration promised by biologists has started to become visible. Tourists have continued to visit the park, and the debate over the fires has quieted somewhat. In other areas where fires have historically been frequent, most land managers continue their policies of letting natural fires burn.

In many regions of western North America, fires have been suppressed for so long that unnaturally large amounts of fuel, in the form of dead and drying trees, have built up. Once fires start in an area like this, they can quickly reach the proportions of the Yellowstone fires. In some of the sequoia preserves introduced earlier in this chapter, for example, enough fuel has built up that an uncontrolled fire would threaten to kill even the largest, oldest trees. To remedy the situation, biologists are attempting to reintroduce fire through controlled burnings. The idea is that setting and managing a series of fires when conditions are moist will gradually reduce the amount of fuel and reduce the danger of a catastrophic fire. In many cases, land managers are also being urged to use logging as a way to reduce the amount of fuel in fire-prone forests quickly. This tactic has been disputed, however. Logging opponents maintain that, for logging operations to be economical, valuable old trees will have to be removed and smaller trees—which have little economic value but pose high fire danger—will be left.

Controversy over how to manage natural disturbances is almost certain to continue.

Management of these forests is also complicated by the dramatic increase in the number of homes that are being built in fire-prone areas. In addition, efforts to fight forest fires aggressively take an enormous cost in money and lives, with sometimes little gain in reducing the actual number of hectares burned. Controversy over how to manage natural disturbances is almost certain to continue.

(a) Extent of 1988 fires in Yellowstone Park

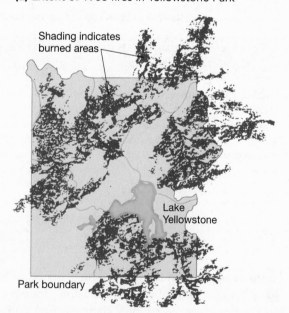

Shading indicates burned areas

Lake Yellowstone

Park boundary

(b) If fuel is abundant, fires burn canopy trees as well as ground cover.

FIGURE 53.26 Extensive Forest Fires in and around Yellowstone Park, 1998
(a) Map of the Greater Yellowstone Area, showing the regions that burned in the 1988 fires. **(b)** One of the blazes in progress.

CHAPTER REVIEW

Summary of Key Concepts

■ **Interactions among species, such as competition and consumption, have two main outcomes: (1) They affect the distribution and abundance of the interacting species, and (2) they are agents of natural selection and thus affect the evolution of the interacting species.**

A community is an assemblage of interacting species.

To categorize the different types of interactions that occur among species, biologists consider whether each participant experiences a net fitness cost or benefit from the interaction. These costs and benefits depend on the conditions that prevail at a particular time and place and may change through time.

Competition is a −/− interaction that occurs when the niches of two species overlap—meaning they use the same resources. Competition may result in the complete exclusion of one species. It may also result in niche differentiation, in which competing species evolve traits that allow them to exploit different resources or live in different areas.

Consumption is a +/− interaction that occurs when consumers eat prey, which resist through standing defenses or inducible defenses. Predators are efficient enough to reduce the size of many prey populations. Levels of herbivory are relatively low in terrestrial ecosystems, however, because predation and disease limit herbivore populations, because plants provide little nitrogen, and because many plants contain toxic compounds or other types of defenses. Parasites generally spend all or part of their life cycle in or on their host (or hosts) and usually have traits that allow them to escape host defenses. In turn, hosts have evolved counteradaptations that help fight off parasites.

Mutualism is a +/+ interaction that provides participating individuals with food, shelter, transport of gametes, or defense against predators. For each species involved, the costs and benefits of a mutualism may vary over time and from place to place.

Web Tutorial 53.1 Life Cycle of a Malaria Parasite

■ **The assemblage of species found in a biological community changes over time and is primarily a function of climate and chance historical events.**

Within at least some biomes, there is a strong positive relationship between the number of species present and the productivity and stability of the community.

Ecologists have debated whether communities are fixed, predictable entities or simply places where the distributions of various species overlap. Historical and experimental evidence support the view that communities are dynamic rather than static and that their composition is neither entirely predictable nor stable over time.

In addition to climate, the composition of a community is influenced by disturbance. In extreme cases, disturbance may remove all organisms and all soil from a large area. Each community has a characteristic disturbance regime—meaning a type, severity, and frequency of disturbance that it experiences. Three types of factors influence the pattern of succession that occurs after a disturbance. First, the historical and environmental context of the site affects which species are available to join the resulting communities. The dispersal ability of different species also affects their availability. Second, a species' physiological traits influence the kinds of abiotic environmental conditions it can tolerate and dictate when it can successfully join a community. Third, interactions among species influence if and when a species appears during succession.

Web Tutorial 53.2 Primary Succession

■ **Species diversity is high in the tropics and lower toward the poles. The mechanism responsible for this pattern is still being investigated.**

One of the most widely studied patterns in community ecology is the latitudinal gradient in species diversity. Within most specific taxonomic groups and within communities as a whole, species diversity declines from the equator to the poles. Recent research suggests that the high species diversity observed in the tropics results from a combination of higher rates of speciation and lower rates of extinction.

■ **Within at least some biomes, there is a strong positive relationship between the number of species present and the productivity and stability of the community.**

Experimental evidence supports the hypothesis that, at least in grassland biomes, the productivity and stability of communities increase as a function of species richness. If further research confirms this conclusion, the rapid loss of species occurring now will have profound impacts on the integrity of communities worldwide.

Questions

Content Review

1. What is competitive exclusion?
   a. the evolution of traits that reduce niche overlap and competition
   b. interactions that allow species to occupy their fundamental niche
   c. the degree to which the niches of two species overlap
   d. the claim that species with the same niche cannot coexist

2. What is niche differentiation?
   a. the evolution of traits that reduce niche overlap and competition
   b. interactions that allow species to occupy their fundamental niche
   c. the degree to which the niches of two species overlap
   d. the claim that species with the same niche cannot coexist

3. Why is the phrase "coevolutionary arms race" an appropriate way to characterize the long-term effects of species interactions?
   a. Both plants and animals have evolved weapons for defense that are so effective that many plants are not eaten and predators cannot reduce prey populations to extinction.
   b. Adaptations that give one species a fitness advantage in an interaction are likely to be countered by adaptations in the other species that eliminate this advantage.
   c. In all species interactions except for mutualism, at least one species loses (suffers decreased fitness).
   d. Even mutualistic interactions can become parasitic if conditions change. As a result, interacting species are always "at war."

4. Why are inducible defenses advantageous?
   a. They are always present—thus, an individual is always able to defend itself.
   b. They make it impossible for a consumer to launch surprise attacks.
   c. They result from a coevolutionary arms race.
   d. They make efficient use of resources, because they are produced only when needed.

5. Which of the following is *not* correlated with species diversity?
   a. latitude
   b. productivity
   c. longitude
   d. resilience

6. What is net primary productivity?
   a. an individual's lifetime reproductive success (lifetime fitness)
   b. an individual's average annual reproductive success
   c. the total amount of photosynthesis that occurs in an area of a given size per year
   d. the amount of energy that is stored in standing biomass per year

Conceptual Review

1. The text claims that species interactions are conditional and dynamic. Do you agree with this statement? Why or why not? Cite specific examples to support your answer.

2. Species interactions affect the distribution and abundance of populations. Summarize experimental evidence that population size for snowshoe hares depends on both predation rates by lynx and competition for food among hares.

3. State three hypotheses that have been proposed to explain the low level of herbivory in terrestrial plant communities. Are these hypotheses mutually exclusive? (In other words, can more than one be correct?) Explain why or why not.

4. State the hypotheses that Clements and Gleason proposed to explain community composition and structure. State the predictions that each of these hypotheses makes with respect to (a) the presence and impact of keystone species, (b) changes in the distribution of the species in a particular community over time, and (c) the communities that should develop at sites where abiotic conditions are identical. Which hypothesis appears to be more accurate?

5. What is a disturbance? List five examples of disturbance. Compare and contrast their effects. For each type of disturbance, compare and contrast the consequences of high-frequency and low-frequency disturbances and high and low severity of disturbances.

6. Describe the latitudinal gradient in species diversity that exists for most taxonomic groups. Discuss the pros and cons of one hypothesis to explain this pattern.

Group Discussion Problems

1. Some insects harvest nectar by chewing through the wall of the structure that holds the nectar As a result, they obtain a nectar reward, but pollination does not occur. Suppose that you observed a certain bee species obtaining nectar in this way from a particular orchid species. Over time, how would you expect the characteristics of the orchid population to change in response to this bee behavior?

2. Using this chapter's information on fire regimes in giant sequoia groves, propose a management plan for Sequoia National Park. Explain the logic behind your plan.

3. Suppose that a two-acre lawn on your college's campus is allowed to undergo succession. Describe how species traits, species interactions, and the site's history might affect the community that develops.

4. Design an experiment to test the hypothesis that increasing species richness increases a community's productivity and ability to resist disturbance and recover from disturbance. Use a biome other than grasslands.

Ecosystems

54

The Qori Kalis glacier in the Andes Mountains of Peru in the year 2000. In 1978 the glacier extended past the edge of the lake in the lower left-hand corner of this photograph. Global warming is causing dramatic reductions in the size of glaciers and ice fields in many locations throughout the world.

KEY CONCEPTS

▦ An ecosystem has four components: (1) the abiotic environment, (2) primary producers, (3) consumers, and (4) decomposers. These components are linked by the movement of energy and nutrients.

▦ The productivity of terrestrial ecosystems is limited by warmth and moisture, while nutrient availability is the key constraint in aquatic ecosystems. As energy flows from producers to consumers and decomposers, much of it is lost.

▦ To analyze nutrient cycles, biologists focus on the nature of the reservoirs where elements reside and on how quickly elements move between reservoirs.

▦ Humans are causing large, global changes in the abiotic environment. The burning of fossil fuels has led to rapid global warming. Extensive fertilization is increasing productivity and causing pollution.

An **ecosystem** consists of the organisms that live in an area together with their physical, or abiotic, environment. With the explosive growth of human populations, dramatic changes are occurring in both the biotic and abiotic components of ecosystems around the globe. Extinctions and other effects on the biotic elements of ecosystems are explored in Chapter 55; this chapter analyzes changes to the chemical and physical characteristics of the environment. Human impacts on the abiotic environment include global warming, acid rain, the development of a hole in the ozone layer over Antarctica, the formation of anaerobic "dead zones" in the oceans due to nitrate pollution, and the development of algal blooms and eutrophication in lakes as a result of phosphate pollution.

This chapter has three goals: to introduce (1) what ecosystem ecology is, (2) how biologists study ecosystems, and (3) how human activities are affecting ecosystems. One of its most fundamental tasks is to explore how organisms interact with the physical environment to produce an ecosystem. The relevant abiotic components include energy and nutrients. Biologists analyze how the energy in sunlight dissipates as it flows through an ecosystem, and how atoms and molecules cycle between the abiotic and biotic worlds. Because human activities are adding massive amounts of energy to ecosystems, disrupting nutrient cycles, and changing the chemistry of lakes, oceans, and the atmosphere, ecosystem ecology has taken on an increasingly applied role in biological science.

How does an ecosystem differ from a community? In most cases, ecosystems are composed of multiple communities along with their chemical and physical environments. For example, biologists who study lakes recognize a number of communities within a lake. There is a distinct community of interacting species along the lake bottom, another community near the surface of the water, and others in other locations (see Chapter 50). Those different communities are studied as a unit, called the lake ecosystem, because energy and matter flow among them.

This chapter introduces ecosystem studies with a look at how energy flows among the components of an ecosystem; it continues by exploring how carbon, nitrogen, and other key elements cycle through organisms, sediments, the oceans, and the atmosphere; and it ends by analyzing how humans are affecting the abiotic environment. Farming, logging, suburbanization, and other human activities are having a massive impact on energy flows and nutrient cycles. Understanding ecosystem ecology is fundamental to managing the future of our planet.

54.1 Energy Flow and Trophic Structure

If an ecosystem is thought of as an economy, then energy is its currency. As **Figure 54.1** shows, ecosystems have four components that are linked by a flow of energy: (1) the abiotic environment, (2) primary producers, (3) consumers, and (4) decomposers.

A **primary producer** is an **autotroph** ("self-feeder"), which is any organism that can synthesize its own food—from inorganic sources (Chapter 10). In most ecosystems, primary producers use solar energy to manufacture their own food via photosynthesis. But in some ecosystems, such as the deep-sea hydrother-

mal vents (introduced in Chapter 2) and the iron-rich rocks deep below Earth's surface (explored in Chapter 27), bacteria use the chemical energy contained in inorganic compounds such as hydrogen (H_2), methane (CH_4), or hydrogen sulfide (H_2S) to make food.

Primary producers form the basis of ecosystems by transforming the energy in sunlight or inorganic compounds into the chemical energy stored in sugars. Primary producers use the chemical energy at their disposal in two ways: Most is used for maintenance or respiratory costs, but some supports growth and reproduction. The energy that is invested in new tissue is called **net primary productivity (NPP)**.

Net primary productivity represents the amount of energy that is available to the second component of an ecosystem: consumers. **Consumers** eat other organisms. **Herbivores** are consumers that eat plants; **carnivores** are consumers that eat animals. **Decomposers**, or *detritivores*, are consumers that obtain energy by feeding on the dead remains of other organisms or waste products. Decomposers form the third component of ecosystems. The fourth and final component is the abiotic environment, which includes the soil, the climate, the atmosphere, and the particulate matter and solutes in water. Energy moves from the Sun or inorganic compounds to consumers, decomposers, and the abiotic environment.

Because the amount of energy available to consumers and decomposers depends on NPP, biologists have focused on that as a key topic of research. Let's consider two of the most fundamental questions about net primary productivity: How does it vary among the world's ecosystems, and what happens to it?

Global Patterns in Productivity

Figure 54.2 summarizes data on NPP from around the globe. To interpret this map, begin by focusing on terrestrial ecosystems. A quick look at the color key should convince you that the terrestrial ecosystems with highest productivity are located in the wet tropics. Notice that, with the exception of the world's major deserts, NPP on land declines from the equator toward the poles. Productivity patterns in marine ecosystems are somewhat different, however. Marine productivity is highest along coastlines, and it can be as high near the poles as it is in tropics. The oceanic zones, introduced in Chapter 50, in contrast, have extremely low NPP. Typically, a square meter of open ocean produces a maximum of 35 gr (1.2 oz) of organic matter each year. In terms of productivity, the open ocean is a desert.

Figure 54.3 presents the same NPP data a different way—organized by biome instead of by geography. Figure 54.3a provides data on average NPP per square meter per year for each biome; Figure 54.3b documents the total area that is covered by each type of ecosystem; and Figure 54.3c presents the percentage of the world's total productivity—the result of multiplying the data in part (a) by the data in part (b). It's

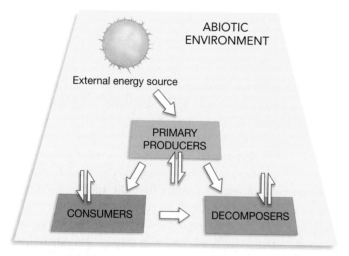

FIGURE 54.1 The Four Components of an Ecosystem Interact
The arrows represent energy. A similar diagram could be drawn to represent the flow of nutrients among the abiotic environment, primary producers, consumers, and decomposers.

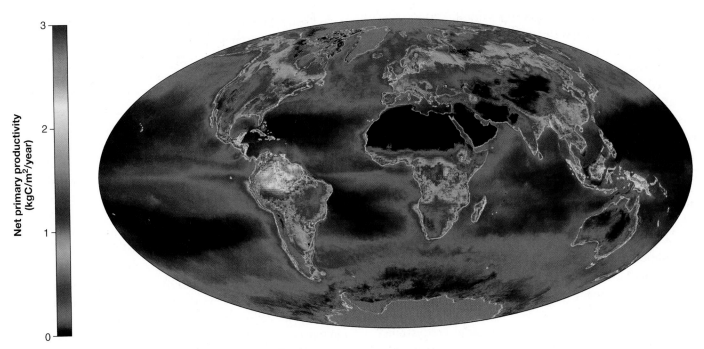

FIGURE 54.2 Net Primary Productivity Varies among Regions
The terrestrial ecosystems with the highest primary productivity are found in the tropics, where warm temperatures and high moisture encourage high rates of photosynthesis. Tundras and deserts have the lowest productivity. The highest productivity in the oceans occurs in nutrient-rich coastal areas.

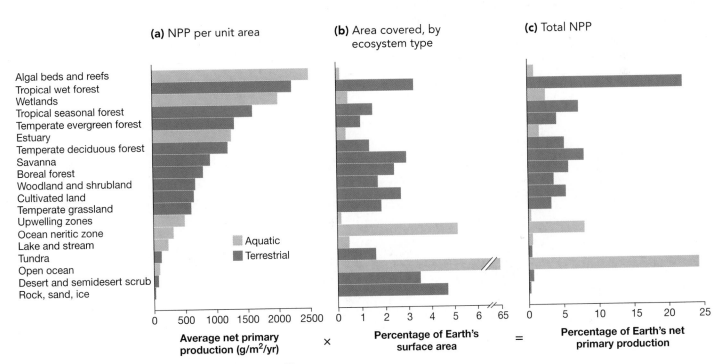

FIGURE 54.3 Net Primary Productivity Varies among Biomes
(a) Among biomes, average annual NPP per square meter varies by over three orders of magnitude. (b) Most of Earth's surface is covered by open ocean. The most common terrestrial habitat consists of unvegetated rock, sand, or ice. (c) Even though NPP per square meter is low, the open ocean is so vast that it is responsible for over 25 percent of Earth's total NPP.

important to note several important patterns. Tropical seasonal forests, which are warm year-round but have a dry season, and tropical wet forests cover less than 5 percent of Earth's surface but together account for over 30 percent of total NPP. Among aquatic ecosystems, the most productive habitats are marshes, estuaries, and algal beds and coral reefs. Most of the total NPP provided by aquatic ecosystems derives from the open ocean, however. Even though NPP per square meter is low in these regions, the open ocean is so extensive that its total production is high.

These patterns raise an interesting question: What limits NPP in terrestrial and marine ecosystems? The short answer is that productivity is limited by any factor that limits the rate of photosynthesis—specifically, temperature and the availabilities of water, sunlight, and nutrients. Different limiting factors prevail in different environments, however. Let's take a closer look.

What Limits the Productivity of Terrestrial Ecosystems?

The data in Figure 54.2 and Figure 54.3 document that terrestrial productivity is lowest in deserts and arctic regions. This observation suggests that the overall productivity of terrestrial ecosystems is limited by a combination of temperature and availability of water and sunlight. The same conclusion has been reinforced by a recent analysis of changes in the productivity of terrestrial ecosystems between 1982 and 1999. When temperatures are averaged over the globe, the interval covered by this study included the warmest years since data collection began in 1861.

Figure 54.4 shows how average annual NPP changed in terrestrial environments during the 1982–1999 warm interval. Notice that NPP increased dramatically in most tropical wet forests, particularly in the Amazon River basin of South America. The leading hypothesis to explain this observation is that increased temperatures led to decreased cloud cover in these regions, and thus higher incident sunlight and higher productivity. Regions that decreased in productivity, in contrast, are thought to have experienced increased cloud cover or drier conditions in response to the elevated temperatures. Even though global terrestrial NPP increased by over 6 percent during the 17-year study interval, variation among regions was high because of complex interactions between temperature and the availability of sunlight and moisture.

What Limits the Productivity of Marine Ecosystems?

To explain why the productivity of marine habitats is so much higher along coastlines than in deepwater regions, biologists focus on nutrient limitation. As pointed out in Chapter 50, the neritic and intertidal zones along coasts receive nutrients from two major sources: (1) rivers that carry and deposit nutrients from terrestrial ecosystems and (2) near-shore ocean currents that bring nutrients that have rained down into the cold, deep water of the oceanic zone back up to the surface. Both of these sources are absent in the surface waters of the open ocean. In addition, nutrients found in organisms near the surface of the open ocean—where sunlight is abundant—constantly rain down to dark, deeper waters in the form of dead cells and are lost.

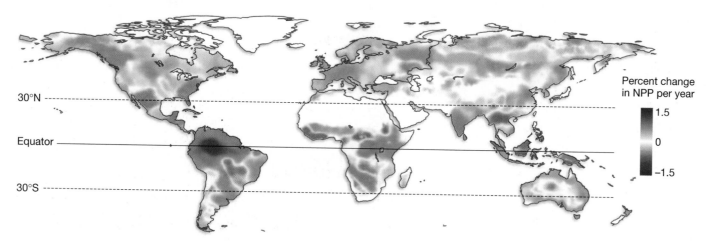

FIGURE 54.4 Changes in NPP during a Warming Trend, 1982–1999
The 1980s and 1990s had the highest average global temperature on record. During this time interval, NPP increased in the green regions and decreased in the red areas, compared with average NPP before this interval.

Analyses of waters that support populations of planktonic organisms have confirmed that trace elements such as zinc, iron, and magnesium are particularly rare in the open ocean. These atoms are important because they are required as enzyme cofactors. Iron, for example, is essential to the proteins that are involved in electron transport chains (see Chapter 9). On the basis of these observations, biologists have proposed that the productivity of open-ocean ecosystems could be increased dramatically by fertilizing them with iron. The results of one recent iron-fertilization experiment are shown in **Figure 54.5**. The data show large increases in the concentration of chlorophyll *a* in surface waters over a two-week interval. Results like these provide strong support for the hypothesis that NPP in marine ecosystems is limited primarily by the availability of nutrients.

How Does Energy Flow through an Ecosystem?

What happens to NPP? Net primary productivity results in biomass—organic material that non-photosynthetic organisms can eat. In every terrestrial and marine environment in the world, the chemical energy in primary producers eventually moves to one of two types of organisms: primary consumers or decomposers.

A **primary consumer** is an herbivore—an organism that eats plants or algae or other photosynthetic cells. Primary consumers are a key link in what biologists call the **grazing food web**: the collection of organisms that eat plants, along with the organisms that eat the herbivores. Consumers that eat herbivores are called **secondary consumers**.

Not all plant tissue is consumed by herbivores, however. Tissues that are not consumed eventually die. When they do, they enter a **decomposer food web**, which is composed of species that eat the dead remains of organisms. In forest ecosystems, for example, dead animals and the dead tissues that accumulate, creating plant *litter*, are collectively known as **detritus**. Detritus is consumed by a variety of **primary decomposers**: bacteria, archaea, fungi, roundworms, earthworms, and millipedes (**Figure 54.6**, page 1248). These primary decomposers are in turn consumed by secondary consumers, such as centipedes, spiders, salamanders, and shrews.

Trophic Levels, Food Chains, and Food Webs

To describe how energy flows from primary producers to primary consumers and secondary consumers, biologists identify distinct feeding levels in an ecosystem. Organisms that obtain their energy from the same type of source are said to occupy the same **trophic** ("feeding") **level**. As an example, consider the organisms that eat the living or dead plants illustrated in Figure 54.6. These primary consumers occupy the same trophic

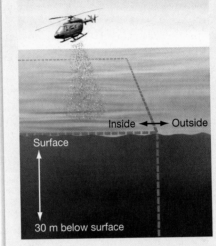

Question: Is net primary productivity (NPP) in the open ocean limited by nutrients?

Hypothesis: NPP in the open ocean is limited by availability of iron.

Null hypothesis: NPP in the open ocean is not limited by availability of iron.

Experimental setup:

1. Add 350 kg iron (as $FeSO_4$) to a patch of ocean 8 km × 10 km.

Inside ← → Outside
Surface
30 m below surface

2. Take water samples for a two-week period outside and inside the patch, at surface and a depth of 30 m, and record amount of chlorophyll *a* present (as indicator of NPP).

Prediction: Amount of chlorophyll *a* near the surface inside the patch will increase relative to amounts outside the patch or at 30 m below the surface.

Prediction of null hypothesis: Amount of chlorophyll *a* will be the same in all measurements.

Results:

Inside, surface
Outside
Inside, 30 m

Conclusion: NPP in the open ocean is limited by the scarcity of nutrients—specifically, iron.

FIGURE 54.5 Fertilization with Iron Increases NPP in the Open Ocean

QUESTION Most of the increased chlorophyll *a* was present in a single species of diatom. Recall from Chapter 28 that diatoms have glassy cases made of silicon. What happens to the biomass present in diatoms after they die?

FIGURE 54.6 The Decomposer Food Web
Dead leaves, sticks, dead animals, and other types of detritus are eaten by an enormous variety of organisms. These decomposers, in turn, are eaten by salamanders, shrews, spiders, centipedes, and other predators.

level, as illustrated in **Figure 54.7**. A **food chain** connects the trophic levels in a particular ecosystem. In doing so, it describes how energy moves from one trophic level to another. Organisms at the top trophic level are not killed for food by any other organisms but enter the decomposer food chain when they die.

Most organisms eat more than one type of food, however, and many species feed at several trophic levels. As a result, food chains are usually embedded in more complex **food webs**. The food web shown in **Figure 54.8** is a more complete description of the trophic relationships among the organisms in an ecosystem. Food chains and webs are among the most basic ways to describe how energy flows through an ecosystem. Let's consider two fundamental questions that biologists ask about food chains and webs.

Why Is Energy Lost at Each Trophic Level? Ecosystems share a characteristic pattern: Total biomass present declines from one trophic level up to the next. Although the biomass and thus the chemical energy present at each trophic level varies widely among ecosystems, the data in **Figure 54.9** are representative of many terrestrial communities. The most striking observation is that as much as 90 percent of the energy present in a given trophic level is lost at the next highest trophic level.

Trophic level	Feeding strategy	Decomposer food chain	Grazing food chain
4	Tertiary consumer	Fox	Cooper's hawk
3	Secondary consumer	Salamander	Robin
2	Primary consumer	Earthworm	Cricket
1	Producer	Dead maple leaves	Maple tree leaves

FIGURE 54.7 Organisms That Share a Trophic Level Obtain Their Energy from the Same Source
Examples from a temperate-forest ecosystem are shown; many other species exist at each level in this ecosystem.

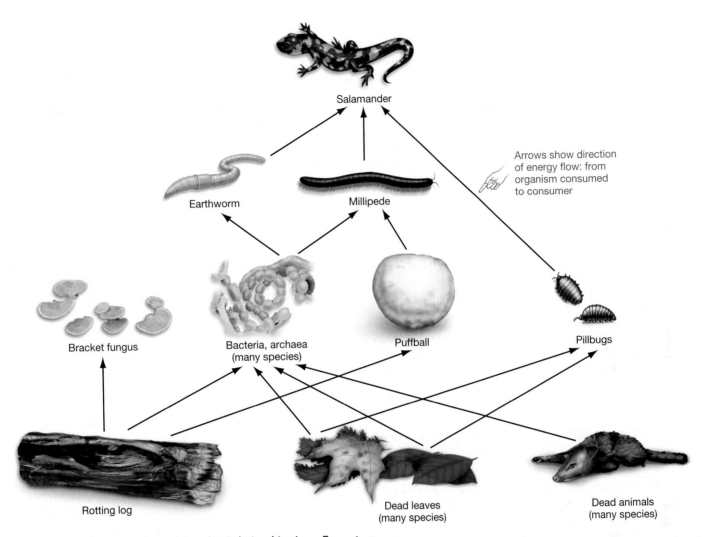

Arrows show direction of energy flow: from organism consumed to consumer

FIGURE 54.8 Food Webs Describe Trophic Relationships in an Ecosystem
Food webs offer a more comprehensive analysis of feeding relationships than food chains do. This food web shows only a fraction of the total feeding relationships and species in the decomposer food web of a temperate deciduous forest.

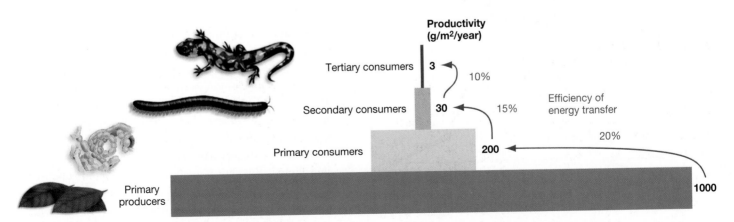

FIGURE 54.9 Productivity Declines at Higher Trophic Levels
In all ecosystems, productivity is highest at the first (bottom) trophic level and declines at higher levels. This pattern is the pyramid of productivity.

To understand why this pattern occurs, consider the interface between primary producers and primary consumers. Much of the net primary productivity that exists is unavailable to herbivores, because it resides in indigestible substances such as the lignin in wood or because it is protected by noxious defensive compounds. NPP that is not eaten by an herbivore enters the decomposer food web. Even if the material is ingested by an herbivore, most of the energy stored in the chemical bonds of carbon compounds is lost as heat as it is metabolized or used to keep the consumer alive—not to grow or reproduce. At the next higher trophic level, many herbivores are never consumed by secondary consumers, because these herbivores defend themselves effectively and die of other causes. Of the energy that is successfully ingested by secondary consumers, some is again lost as heat or used up in staying alive and trying to capture prey.

The general point here is simple: The amount of biomass present at the second trophic level must be less than productivity at the first trophic level; productivity at the third trophic level must be less than that at the second. This pattern holds true for the entire food chain and produces a **pyramid of productivity**. Production of biomass is highest at the lowest trophic level.

What Limits the Length of Food Chains?

It is interesting to note that none of the food chains, food webs, or trophic structures presented in this chapter thus far have more than four trophic levels. When a biologist reviewed the characteristics of food chains that had been documented by researchers working in a wide variety of ecosystems, he found that the maximum number of links ranged from 1 to 6 (**Figure 54.10**). Each link joins two trophic levels. In this study, terrestrial and lake ecosystems had 3.7 links per food chain, on average, while streams had an average of 3.2 links. Why don't ecosystems have 8 or 9 or 10 trophic levels? Why does the overall average number of levels seem to be about 3.5? Several competing hy-

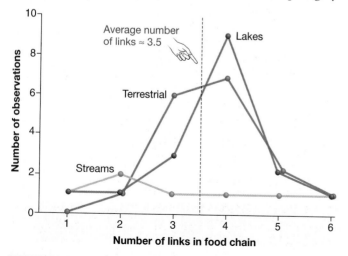

FIGURE 54.10 Most Food Chains Are Short
The vertical axis plots the number of research studies that described food chains with 1 to 6 links in stream, lake, and terrestrial habitats.

potheses have been offered to explain this observation. Let's consider three:

- *Hypothesis 1: Energy Transfer* As energy is transferred up a food chain, a large fraction of that energy is lost. By the time energy reaches the top trophic level, there may not be enough left to support an additional suite of consumers. Suppose that the efficiency of energy transfer between each pair of trophic levels is 10 percent. If the initial trophic level in a hypothetical ecosystem produces 10,000 kcal per day, then the second, third, and fourth levels will produce 1000, 100, and 10 kcal per day, respectively. Can any organism obtain enough energy to survive at a fifth trophic level?

  The hypothesis that food-chain length is limited by productivity leads to a strong prediction: There should be more trophic levels in ecosystems with higher productivity or higher efficiency of energy transfer than in ecosystems with lower productivity or lower efficiency of energy transfer. When Stuart Pimm analyzed data on the productivity of four aquatic and four terrestrial ecosystems, however, he found that low-productivity ecosystems were as likely to contain four trophic levels as high-productivity ones were. To date, research has not supported the prediction that the most productive ecosystems have the longest food chains.

- *Hypothesis 2: Stability* Pimm proposed an alternative hypothesis to the energy-transfer explanation for why food chains are short. His idea was that long food chains are easily disrupted by environmental perturbations and thus tend to be eliminated. Using mathematical models, he demonstrated that long food chains took longer to return to their previous state following a disturbance than did short food chains. He then proposed that long food chains are unlikely to persist in a variable environment.

  Pimm's hypothesis also predicts that the length of food chains should increase with the stability of the environment. A biologist tested this prediction by comparing the animal communities that develop inside tree holes in Australia and Great Britain. After water accumulates in depressions or in the holes in tree branches or roots, leaf litter from the trees falls in and forms the basis of a food web. The researcher found that tree-hole communities in Australia and Great Britain were similar in many respects, except that annual leaf fall was much more variable in the British communities than in the Australian ones. As predicted by the stability hypothesis, the British ecosystems had only two trophic levels while Australian habitats supported three.

  Other researchers have challenged the assumptions behind Pimm's theoretical analysis, however, and support for the stability hypothesis remains very tentative. Experimental and observational tests of this hypothesis are continuing.

- *Hypothesis 3: Environmental Complexity* Biologists have also hypothesized that food-chain length is a function of an ecosystem's physical structure. Specifically, researchers have proposed that tundra, grasslands, lake and sea bottoms, streambeds, and intertidal zones offer a largely flat, or two-dimensional, surface to the organisms living there. In contrast, forests and open-water environments in lakes and rivers offer three-dimensional volumes. This hypothesis predicts that three-dimensional ecosystems should have longer food chains than two-dimensional sites do.

To test this hypothesis, researchers examined 113 publications that described food chains in a wide variety of aquatic and terrestrial environments. These data supported the prediction that food webs are significantly longer in three-dimensional ecosystems. As a result, the investigators concluded that dimensionality does influence food-web structure. The mechanism for this pattern remains to be determined, however. It remains to be seen why three-dimensional ecosystems allow longer food webs to develop.

To summarize, there is unlikely to be a single, simple answer to the question of what limits food-chain length. Inefficiencies of energy transfer, environmental stability, and environmental complexity may all influence the number of trophic levels that can be supported in a given ecosystem. Multiple causation is not unusual in biological science. In genetics, for example, gene expression is rarely triggered by a single regulatory site in DNA and just one regulatory protein. Instead, the amount and timing of gene expression result from interactions among dozens of regulatory DNA sequences and proteins. Recognizing multiple causation is productive, because it can suggest exciting new experiments and analyses. For example, if food-chain length does have multiple causes, then habitats that are particularly productive, stable, and highly complex in structure should have the longest food chains. Research continues on this fundamental question in ecosystem ecology.

Analyzing Energy Flow: A Case History

Research on energy flow in ecosystems has uncovered a series of important general principle. In both terrestrial and marine ecosystems, NPP varies with region and biome. On land, the most important constraints on productivity are warmth and moisture; in aquatic habitats, nutrient availability is key. The chemical energy in biomass flows into consumer or decomposer food webs, where it is passed up a series of trophic levels. Productivity declines with each trophic level, however.

To see these general principles in action, let's analyze energy flow in a specific ecosystem. For almost five decades, a team of researchers has been studying how energy and nutrients flow through a temperate forest ecosystem in the northeastern United States: the Hubbard Brook Experimental Forest in New Hampshire.

As **Figure 54.11** shows, energy flow in this ecosystem begins when plants capture the energy in solar radiation via photosynthesis. At Hubbard Brook, the amount of energy entering the ecosystem from sunlight varies throughout the year and, to a

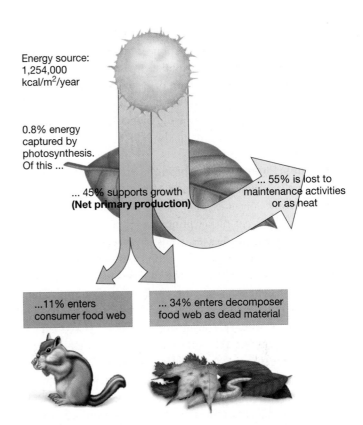

Energy source: 1,254,000 kcal/m²/year

0.8% energy captured by photosynthesis. Of this ...

... 45% supports growth **(Net primary production)**

... 55% is lost to maintenance activities or as heat

...11% enters consumer food web

... 34% enters decomposer food web as dead material

FIGURE 54.11 Energy Flow through the Hubbard Brook Forest Ecosystem
Energy from sunlight is transformed to chemical energy by photosynthesis. Photosynthetic products fuel new plant growth, which eventually enters either the consumer or decomposer food web. **EXERCISE** Calculate gross photosynthetic efficiency and NPP in kcal/m²/year. Calculate the amount of energy that enters the consumer and decomposer food webs each year, in kcal/m².

lesser extent, from year to year. From June 1, 1969, to May 31, 1970, which the research team describes as a typical year, 1,254,000 kilocalories (kcal) of solar radiation per square meter (m²) reached the forest. If this amount of energy were available in the form of electricity, it would easily power two 75-watt lightbulbs that burned continuously for one year.

By documenting rates of photosynthesis in a large sample of forest plants, the biologists calculated that the plants used 10,400 kcal/m² of energy in photosynthesis. This value represents **gross primary productivity**, which is the total amount of photosynthesis in a given area and time period. The team also calculated **gross photosynthetic efficiency**, or the efficiency with which plants use the total amount of energy available to them, as the ratio of gross photosynthesis to solar radiation in kcal/m². At Hubbard Brook, efficiency was $10,400 \div 1,254,000 = 0.8$ percent. This value is typical of other ecosystems as well.

Why is efficiency so low? The answer has several components. Plants cannot photosynthesize in winter or during dry periods of a day or year, although sunlight is available. Even when conditions are ideal, the pigments that drive photosynthesis can respond to only a fraction of the wavelengths available and thus a fraction of the total energy received.

Given that only a tiny fraction of incoming sunlight is converted to chemical energy and that only a fraction of this gross primary productivity is used to build biomass, let's consider the next question: What happens to the NPP at Hubbard Brook?

How Does Energy Flow through the Hubbard Brook Ecosystem?

Between 1969 and 1971 the amount of energy that entered the consumer food web at Hubbard Brook varied from about 1 percent of net primary productivity per year to more than 40 percent. Leaves, seeds, and fruits were the most commonly consumed plant parts; wood was rarely eaten.

Figure 54.12 illustrates what happened to the energy obtained by consumers, using chipmunks as an example. Chipmunks are small rodents that are seed predators. On average, these primary consumers harvested 31 kcal/m^2 of energy each year. Of that total, 17.7 percent was excreted and 80.7 percent was lost to respiration and other processes. Just 1.6 percent went into the production of new chipmunk tissue by growth and reproduction. The production of new tissue by primary consumers is called **secondary production.** Secondary production was much higher in ectothermic consumers such as caterpillars, which transformed about 5.4 percent of the energy they ingested into new tissue. Because ectotherms rely principally on heat gained from the environment and do not oxidize sugars to keep warm, they devote much less energy to cellular respiration than endotherms do. Even so, it is clear that only a tiny fraction of

the available solar radiation is involved in secondary production. These data reflect the general principles that productivity drops sharply at each trophic level.

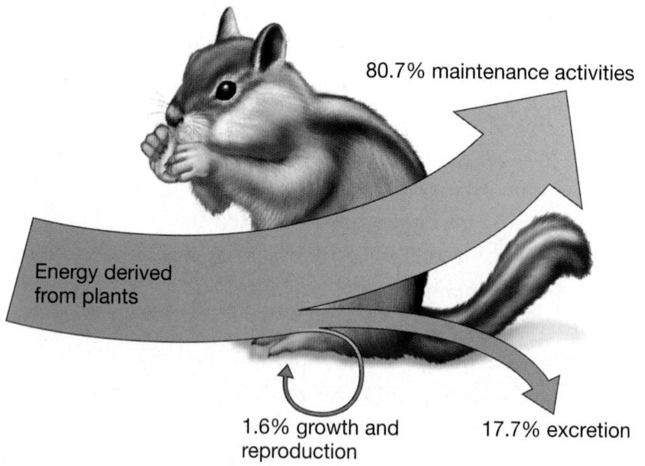

What about biomass that enters the decomposer food web? At Hubbard Brook, about 75 percent of net primary productivity is not eaten while alive. This energy enters the decomposer food web and is passed up subsequent trophic levels, just as it is in the consumer food web. But in addition, the researchers at Hubbard Brook found that large amounts of energy leave the decomposer food web in the form of detritus that washes into streams. This transfer of energy into the aquatic ecosystem is important, because it is a major source of energy for aquatic organisms. At Hubbard Brook, photosynthesis by aquatic plants introduced only about 10 kcal/m^2/yr of energy. In contrast, each year 6039 kilocalories washed into each square meter of streambed from the surrounding forest.

To summarize, long-term studies have documented the flow of energy into, through, and out of a temperate-forest ecosystem. Although the specific numbers found at Hubbard Brook are unique to that site and the time interval of that study, the general patterns have turned out to be typical of ecosystems around the globe.

✓ CHECK YOUR UNDERSTANDING

In an ecosystem, energy flows from sunlight or inorganic compounds with high potential energy to producers, and from there to consumers and decomposers. In most ecosystems, NPP is limited by conditions that limit the rate of photosynthesis: temperature and/or the availability of sunlight, water, and nutrients. Productivity diminishes at each subsequent trophic level in the ecosystem. You should be able to (1) summarize and provide a brief explanation of variation in NPP among ecosystems, (2) explain why productivity diminishes with increasing trophic levels, and (3) evaluate at least one hypothesis to explain why most food chains have only two or three links.

54.2   Biogeochemical Cycles

Energy is not the only quantity that is transferred when one organism eats another. The organisms that are eaten also contain carbon (C), nitrogen (N), phosphorus (P), calcium (Ca), and other elements that act as nutrients. Atoms are constantly reused as they cycle through trophic levels and air, water, and soil. The path that an element takes as it moves from abiotic systems through organisms and back again is referred to as its **biogeochemical** ("life-Earth-chemical") **cycle.**

Because humans are now disturbing biogeochemical cycles on a massive scale, research on this aspect of ecosystem ecology is exploding. To get a basic understanding of how a particular

80.7% maintenance activities

Energy derived from plants

1.6% growth and reproduction   17.7% excretion

FIGURE 54.12 How Do Consumers Use Primary Production?
A small fraction of the energy consumed by chipmunks, which are primary consumers (herbivores), is used for secondary production. Most of the energy is used for cellular respiration.

biogeochemical cycle works, researchers focus on three fundamental questions:

1. What are the nature and size of the **reservoirs**, or pools where elements are stored for a period of time? In the case of carbon, the biomass of living organisms is an important reservoir, along with sediments and soils. Another significant carbon reservoir is buried in the form of coal and oil.

2. How fast does the element move between reservoirs, and what factors influence these rates? The global photosynthetic rate, for example, measures the rate of carbon flow from CO_2 in the atmosphere to living biomass. When fossil fuels burn, carbon that was buried in coal or petroleum moves into the atmosphere as carbon dioxide (CO_2).

3. How does this biogeochemical cycle interact with other cycles? For example, researchers are trying to understand how changes in the nitrogen cycle affect carbon.

Before taking up these issues, though, let's begin with a very general overview of what nutrient cycling is.

Biogeochemical Cycles in Ecosystems

Figure 54.13 shows a simplified version of a generalized terrestrial nutrient cycle. In this case, the cycle starts when nutrients are taken up from the soil by plants and incorporated into plant tissue. If the plant is eaten, the nutrients pass to the animal members of the ecosystem's consumer food web; if the plant dies, the nutrients enter the decomposer food web. Once ingested by an animal, nutrients are excreted in fecal matter or urine, taken up by a parasite or predator, or added to the dead biomass reservoir when the animal dies.

The nutrients in plant litter, animal excretions, and dead animal bodies are broken down by bacteria, archaea, roundworms, fungi, and other organisms. The combination of breakdown products and microscopic decomposers forms the soil organic matter. **Soil organic matter** is a complex mixture of partially and completely decomposed detritus. Completely decayed organic material is called **humus**, because it is rich in a family of carbon-containing molecules called *humic acids*. (Chapter 37 describes other components of the soil.) Eventually, the nutrients in soil organic matter are converted to an inorganic form. For example, cellular respiration by soil-dwelling bacteria and archaea converts the nitrogen present in amino acids that are found in detritus to ammonium (NH_4^+) or nitrate (NO_3^-) ions. Once this step is accomplished, the nutrients are available for uptake by plants.

A key feature of this process is that nutrients are reused. Reuse is not total, however. Nutrients leave the ecosystem whenever plant or animal biomass is removed. For example, plant biomass is removed when an herbivore enters the ecosystem, eats a plant, and leaves the ecosystem before excreting nutrients or dying. Nutrients also leave ecosystems

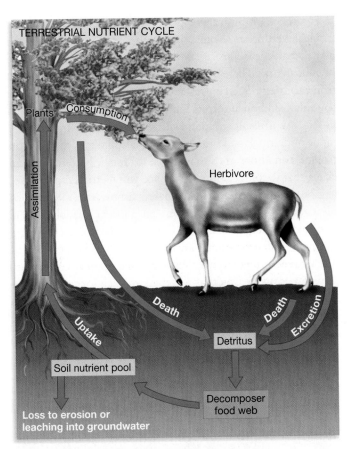

FIGURE 54.13 Generalized Terrestrial Nutrient Cycle
Nutrients cycle from organism to organism in an ecosystem via assimilation, consumption, and decomposition. Nutrients are exported from ecosystems through the migration of organisms out of the area or in flowing water or groundwater.

when flowing water or wind removes particles or inorganic ions. Soil erosion has a huge impact on nutrient cycles: It removes nutrients rapidly and in potentially large quantities.

What Factors Control the Rate of Nutrient Cycling in Ecosystems? Of the many links in a nutrient cycle, the decomposition of detritus most often limits the overall rate at which nutrients move through an ecosystem. The decomposition rate, in turn, is influenced by two types of factors: abiotic conditions such as temperature and precipitation, and the quality of the detritus as a nutrient source for the fungi, bacteria, and archaea that accomplish decomposition.

To appreciate the importance of abiotic conditions on the decomposition rate, consider the difference in detritus accumulation between boreal forests and tropical wet forests. Chapter 50 indicated that boreal forests occur in areas where temperature is low. As a result, soils in these ecosystems are cold and wet. Tropical wet forests, in contrast, occur in areas where temperatures and rainfall are high. Soils tend to remain moist and warm all year long.

Figure 54.14 illustrates typical soils from boreal forests and tropical wet forests. Notice that the uppermost part of the soil in a boreal forest consists of partially decomposed detritus and organic matter. There is no such layer at the top of the soil in a tropical forest. The contrast occurs because the cold and wet conditions in boreal forests limit the metabolic rates of decomposers. As a result, decomposition fails to keep up with the input of detritus, and organic matter accumulates there. In the tropics, conditions are so favorable for fungi, bacteria, and archaea that decomposition keeps pace with detrital inputs.

The quality of detritus also exerts a powerful influence on the decomposition rate and thus on nutrient availability. For example, decomposers can be limited by the presence of large compounds in detritus that are difficult to digest, such as lignin. The presence of lignin, which is one of the primary constituents of wood, is one reason that wood takes much longer to decompose than leaves do. The growth of decomposers is also inhibited if detritus is low in nitrogen.

(a) Boreal forest: Extensive litter accumulation (dead wood and leaves)

(b) Tropical wet forest: Almost no litter accumulation

FIGURE 54.14 Temperature and Moisture Affect Decomposition Rates
(a) In boreal forests, decomposition rates are limited by cold soil temperatures. Organic matter builds up, because the input of detritus into the soil exceeds the decomposition rate. **(b)** In tropical wet forests, warm temperatures allow decomposition to proceed rapidly, so organic matter does not build up.

What Factors Influence the Rate at Which Nutrients Are Exported from Ecosystems? Nutrient availability has a profound effect on productivity. The rate of nutrient loss is thus an important characteristic of an ecosystem. Several of the major impacts that humans have on ecosystems—such as farming, logging, burning, and soil erosion—accelerate nutrient loss.

To test the effect of vegetation removal on nutrient cycling, researchers initiated a large-scale experiment at the Hubbard Brook Experimental Forest. They chose two similar **watersheds**—areas drained by a single stream—for study (**Figure 54.15**). They then removed all vegetation, including the trees, from the forests in one of the two watersheds. In the following two years, this *clear-cut* area was treated with an herbicide to prevent vegetation from regrowing. As a result, the experimental watershed was devegetated. The untreated watershed served as a control.

Before removing the vegetation, the researchers had documented that 90 percent of the nutrients in the ecosystems were in soil organic matter, and an additional 9.5 percent was in plant biomass. After the vegetation was removed, the team monitored the concentrations of nutrients in the streams exiting the two watersheds. The graphs in Figure 54.15 document the amount of dissolved substances that subsequently washed out of the stream in each watershed over the course of four years. Losses were typically 10 times as high in the clear-cut site as in the control site.

Based on these data, the researchers concluded that devegetation has a huge impact on nutrient export. Instead of being held in the ecosystem and recycled, nutrients wash out. The loss occurs because nutrients in the soil are either dissolved in water or attached to small particles of sand or clay. If plant roots no longer hold the soil particles in place and if they no longer take up and recycle nutrients, then the molecules and ions wash out of the soil and are lost to the ecosystem. Consequently, the productivity that an area can support may decline over time if the area is kept in a devegetated state. Long-term devegetation of this type has occurred in formerly forested areas of the Middle East, North Africa, and elsewhere due to intensive farming and grazing.

Global Biogeochemical Cycles

When nutrients leave one ecosystem, they enter another. In this way, the movement of ions and molecules among ecosystems links local biogeochemical cycles into one massive global system. Local and global cycles interact when water, organisms, or wind move nutrients. As an introduction to how these global biogeochemical cycles work, let's consider the global water, carbon, and nitrogen cycles. These cycles have recently been heavily modified by human activities—with serious ecological consequences.

The Global Water Cycle A simplified version of the **global water cycle** appears in **Figure 54.16**. The diagram shows the estimated amount of water that moves between major components of the cycle over the course of a year.

Question: How does the presence of vegetation affect the rate of nutrient export in an ecosystem?

Hypothesis: Presence of vegetation lowers the rate of nutrient export because it increases soil stability and recycling of nutrients.

Null hypothesis: Presence of vegetation has no effect on the rate of nutrient export.

Experimental setup:

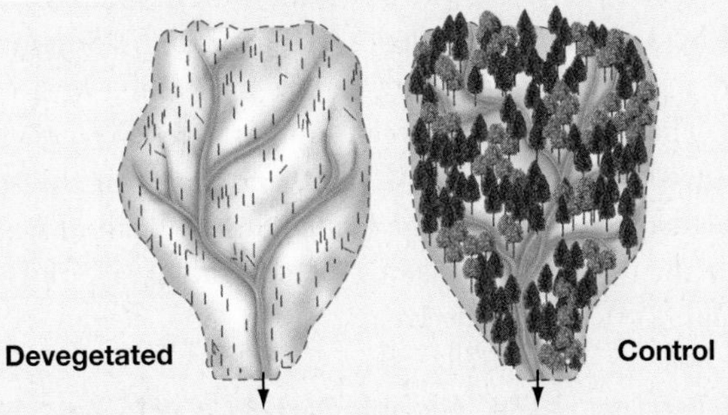

Devegetated   Control

1. Choose two similar watersheds. Document nutrient levels in soil organic matter, plants, and streams.

2. Devegetate one watershed, and leave the other intact.

3. Monitor the amount of dissolved substances in streams.

Prediction: Amount of dissolved substances in stream in devegetated watershed will be much higher than amount of dissolved substances in stream in control watershed.

Prediction of null hypothesis: No difference will be observed in amount of dissolved substances in the two streams.

Results:

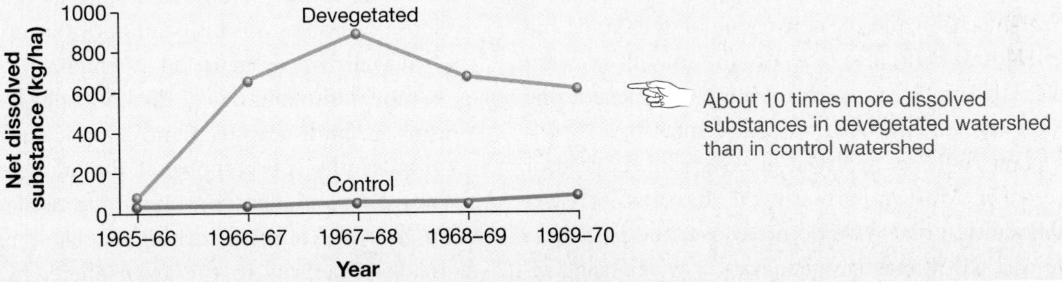

About 10 times more dissolved substances in devegetated watershed than in control watershed

Conclusion: Presence of vegetation limits nutrient loss. Removing vegetation leads to large increases in nutrient export.

FIGURE 54.15 Land-Use Changes Affect the Rate of Nutrient Loss from Ecosystems
QUESTION In effect, this experiment removed one of the arrows in Figure 54.13. Which one?

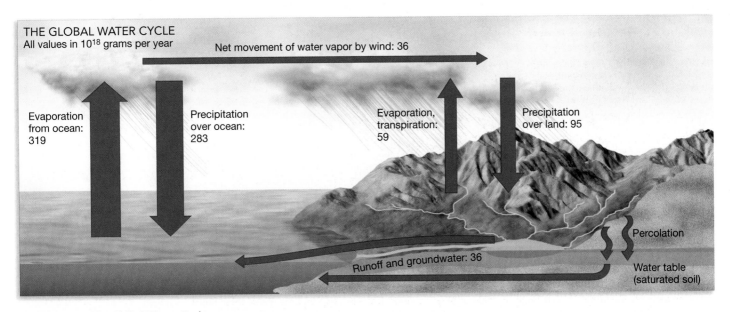

FIGURE 54.16 The Global Water Cycle
QUESTION Predict how the amount of water evaporated from the ocean is changing in response to global warming. Discuss one possible consequence of this change.

To analyze this cycle, begin with evaporation of water out of the ocean and precipitation back into it. The key observation is that, for the marine component of the cycle, evaporation exceeds precipitation—meaning that, over the oceans, there is a net gain of water to the atmosphere. When this water vapor moves over the continents, it is augmented by a small amount of water that evaporates from lakes and streams and a large volume of water transpired by plants. The total volume of water in the atmosphere over land is balanced by the amount of precipitation that occurs on the continents. The cycle is completed by both water that moves from the land to the oceans via streams and **groundwater**—meaning water that is found in soil.

Humans are affecting the water cycle in complex ways. Perhaps the simplest and most direct impacts concern rates of groundwater replenishment. Asphalt and concrete surfaces from suburbanization reduce the amount of precipitation that percolates through the soil from the surface and enters deep soil layers. Surbanization and the conversion of grasslands and forests to agricultural fields also increases the amount of water that runs off Earth's surface into streams instead of penetrating to groundwater layers. This occurs because soft, shaded soils packed with plant roots are replaced by soils with reduced or no vegetative cover. Most importantly, the dramatic increases in irrigated agriculture that have occurred over the past three decades have resulted in massive quantities of water being removed from groundwater storage and brought to the surface.

In combination, these impacts have resulted in alarming drops in groundwater storage. The **water table** is the upper limit below the surface that the ground is saturated with stored water, and it is dropping on every continent. Between 1991 and 1996 the water table north of Beijing, China, experienced drops that averaged 1.5 m (5 ft) per year. Throughout India, water tables are falling at a rate of between 1 m and 3 m per year. Similar rates are being documented in Yemen, parts of Mexico, and states in the southern Great Plains region of the United States. Lack of water is exacerbating political tensions in several areas of the world, including the Middle East.

The Global Carbon Cycle   As **Figure 54.17** shows, the **global carbon cycle** documents the movement of carbon among terrestrial ecosystems, the oceans, and the atmosphere. Of these three reservoirs, the ocean is by far the largest. The atmospheric reservoir is also important despite its relatively small size, because carbon moves into and out of it rapidly.

The arrows in Figure 54.17 emphasize that carbon frequently moves into and out of the atmospheric pool through organisms. In both terrestrial and aquatic ecosystems, for example, photosynthesis is responsible for taking carbon out of the atmosphere and incorporating it into tissue. Cellular respiration, in contrast, releases carbon that has been incorporated into living organisms to the atmosphere, in the form of carbon dioxide.

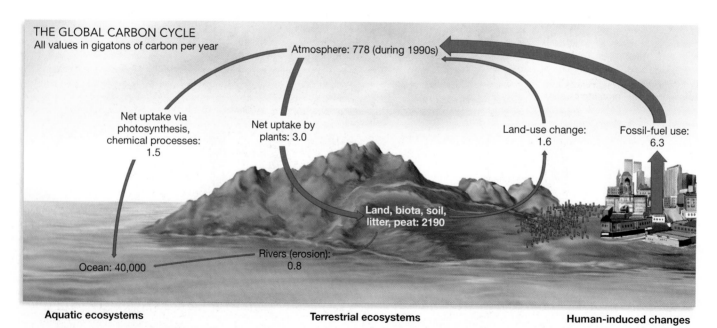

FIGURE 54.17 The Global Carbon Cycle
The arrows indicate how carbon moves into and out of ecosystems. Deforestation and the use of fossil fuels are adding 7.9 gigatons of carbon to the atmosphere each year. Of that amount, 2 gigatons are fixed by photosynthesis in terrestrial ecosystems and 2 gigatons are fixed by physical and chemical processes in the oceans. The remainder—3.9 gigatons—is added to the atmosphere.

How have humans changed the carbon cycle? The fossil fuels found on Earth, which are derived from carbon-rich sediments, are estimated to contain a total of 5000–10,000 gigatons of carbon. (A gigaton, symbolized Gt, is a billion tons). This is one-eighth to one-fourth the size of the oceanic reservoir and 2.5 to 5 times the size of the terrestrial reservoir. In effect, burning fossil fuels moves carbon from an inactive geological reservoir to an active reservoir—the atmosphere.

Land-use changes have also altered the global carbon cycle. Deforestation, for example, reduces an area's net primary productivity. It also releases CO_2 when fire is used for clear-cutting or when dead limbs, twigs, and stumps are left to decompose. Researchers who used data on the amount of land cleared for agriculture and forestry to estimate the amount of carbon released found that, at a global scale, a net movement of carbon to the atmosphere has been occurring from terrestrial ecosystems for at least the past 100 years.

Figure 54.18a shows the amount of carbon released from fossil-fuel burning and land-use changes over the past century; **Figure 54.18b** shows the consequences. In just 45 years, CO_2 in the atmosphere at Mauna Loa Observatory on the island of Hawaii has increased from about 315 to over 375 parts per million (ppm)—meaning milligrams of CO_2 per kilogram of air. The same trend has been observed at sites around the globe. As the data in **Figure 54.18c** show, CO_2 concentrations in the atmosphere have now risen to levels far above the 280 ppm that were typical prior to 1860.

These changes in the global carbon cycle are important because carbon dioxide functions as a **greenhouse gas**: It traps heat that has been radiated from Earth and keeps it from being lost to space, similar to the way the glass of a greenhouse traps heat. More specifically, carbon dioxide is one of several gases in the atmosphere that absorb and reflect the infrared wavelengths radiating from Earth's surface. Increases in amounts of greenhouse gases have the potential to warm Earth's climate by increasing the atmosphere's heat-trapping potential.

The Global Nitrogen Cycle Figure 54.19 (page 1258) illustrates the **global nitrogen cycle**. A key aspect of this biogeochemical cycle is that plants are able to use nitrogen only in the form of ammonium or nitrate ions (NH_4^+ or NO_3^-). As a result, the vast pool of molecular nitrogen (N_2) that is in the air blanketing Earth—N_2 makes up 78 percent of the atmosphere—is unavailable to plants. Nitrogen is added to ecosystems in a usable form only when it is reduced, or "fixed," meaning when it is converted from N_2 to NH_3. Nitrogen fixation results from lightning-driven reactions in the atmosphere and from enzyme-catalyzed reactions in bacteria that live in the soil and oceans. (Chapter 37 explained how bacteria fix nitrogen.)

The nitrogen cycle has been profoundly altered by human activities. The amount of nitrogen fixation from human sources is now approximately equal to the amount of nitrogen

(a) Increases in CO_2 emissions due to fossil-fuel use and forest destruction

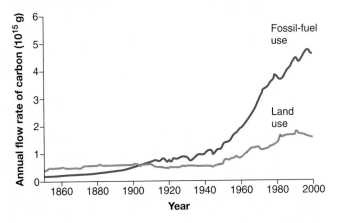

(b) Atmospheric CO_2

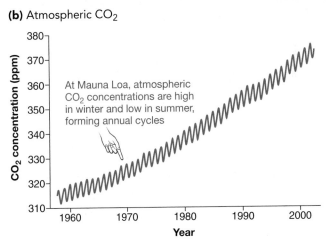

At Mauna Loa, atmospheric CO_2 concentrations are high in winter and low in summer, forming annual cycles

(c) Changes in CO_2 concentration over time

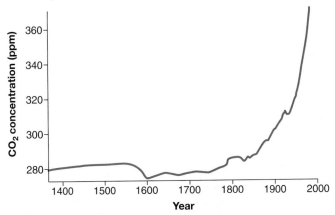

FIGURE 54.18 Humans Are Causing Increases in Atmospheric Carbon Dioxide

(a) Rates of carbon flow from fossil-fuel burning and land-use changes have been increasing since 1860. **(b)** Because the Mauna Loa Observatory in Hawaii is far from large-scale human influences, it should accurately represent the average condition of the atmosphere in the Northern Hemisphere. **(c)** For centuries, average CO_2 concentrations in the atmosphere were about 280 parts per million (ppm). **QUESTION** Why are atmospheric CO_2 concentrations low in the Northern Hemisphere in summer and high in winter? What pattern would you expect in the Southern Hemisphere?

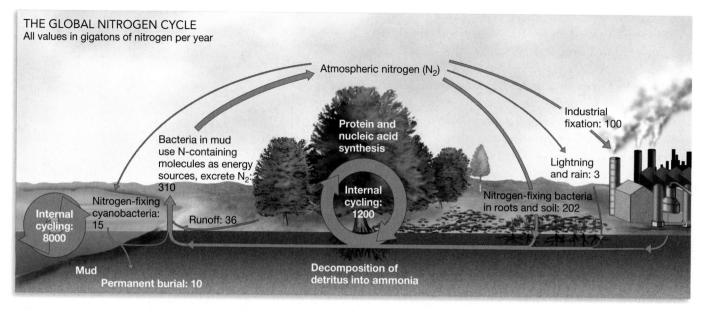

THE GLOBAL NITROGEN CYCLE
All values in gigatons of nitrogen per year

Atmospheric nitrogen (N_2)

Industrial fixation: 100

Protein and nucleic acid synthesis

Bacteria in mud use N-containing molecules as energy sources, excrete N_2: 310

Lightning and rain: 3

Internal cycling: 1200

Nitrogen-fixing cyanobacteria: 15

Nitrogen-fixing bacteria in roots and soil: 202

Internal cycling: 8000

Runoff: 36

Mud

Permanent burial: 10

Decomposition of detritus into ammonia

FIGURE 54.19 The Global Nitrogen Cycle
Nitrogen enters ecosystems as ammonia or nitrate via fixation from atmospheric nitrogen. It is exported in runoff and as nitrogen gas given off by bacteria that use nitrogen-containing compounds as an electron acceptor.

fixation from natural sources, as **Figure 54.20** shows. There are three major sources of this human-fixed nitrogen: (1) industrially produced fertilizers, (2) the cultivation of crops, such as soybeans and peas that harbor nitrogen-fixing bacteria, and (3) the release of nitric oxide during the combustion of fossil fuels.

Adding nitrogen to terrestrial ecosystems usually increases productivity. In some cases, then, a massive increase in nitrogen availability is beneficial. But in other situations, it is not.

Chapter 27, for example, detailed how excessive applications of nitrogen-containing fertilizers on farmlands have produced nitrogen-laced runoff, which has had disastrous impacts on numerous aquatic ecosystems. Researchers have also documented that nitrogen inputs can lead to a significant loss of biodiversity in terrestrial ecosystems. For example, when biologists added nitrogen to plots of native grasslands in midwestern North America, a few competitively dominant species tended to take over. As those species grew rapidly, they displaced other species that did not respond to nitrogen inputs as strongly. In this ecosystem, increased nitrogen boosted productivity but decreased species diversity, by altering the balance of competitive interactions. The same result has occurred in study plots within the Park Grass experiment, introduced in Chapter 52. The plots that have been fertilized with nitrogen since 1856 contain many fewer species than unfertilized plots do.

To summarize, several of the most pressing environmental problems facing our species result from recent and massive alterations in biogeochemical cycles. Even local changes in biogeochemical cycles tend to have large-scale consequences, because nutrients are transported among ecosystems. If you are in your late teens or early twenties as you read this text, you are part of a generation that is projected to experience the most traumatic episode of environmental change in human history. The trauma has two sources: the massive loss of species documented in Chapter 26 and Chapter 55 and the profound changes in global biogeochemistry recorded here. Next we'll take a closer look at the consequences of human activities.

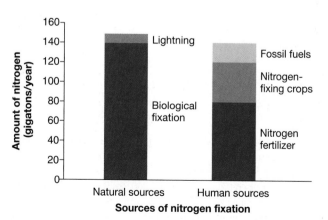

FIGURE 54.20 Humans Are Adding Large Amounts of Nitrogen to Ecosystems
Human activities now fix almost as much nitrogen each year as natural sources do. Thus, human activities have almost doubled the total amount of nitrogen available to organisms.

54.3 Human Impacts on Ecosystems

Two factors are responsible for the human impacts on ecosystems currently being documented by biologists. The first is the rapid increase in human population size, analyzed in Chapter 52. The second is the rapid increase in resources used by people. Figure 54.18a introduced this second factor by showing increases in fossil-fuel consumption and land use that have occurred over about the past 150 years. **Figure 54.21** follows up by providing data on the variation in average annual energy consumption per person around the world. The key point is that residents of industrialized countries, though relatively few in number, have a disproportion-

ately large impact on biogeochemical cycles because they use so much energy, water, food, and other resources. Let's look at two of the consequences of population growth and resource use: global warming and productivity increases.

Global Warming

Recall that carbon dioxide concentration in the atmosphere has been increasing throughout the twentieth century and that most analyses indicate that the increase is due to the burning of fossil fuels and clear-cutting of land, particularly forests, for agriculture. Whether the increase in CO_2 is producing **global warming**—an increase in Earth's surface temperature, averaged over the globe, has been intensely controversial until recently, however.

In 1988 an international group of scientists, called the Intergovernmental Panel on Climate Change (IPCC), was formed to evaluate the consequences of rising CO_2. The group has since produced a series of reports summarizing the state of scientific knowledge on the issue. In 1998 the report concluded that current evidence suggests a "discernible human influence on climate." This was the IPCC's first statement supporting the hypothesis that rising CO_2 concentrations due to human activities are having a measurable impact on climate. The IPCC's most recent report, released in 2001, stated that "There is new

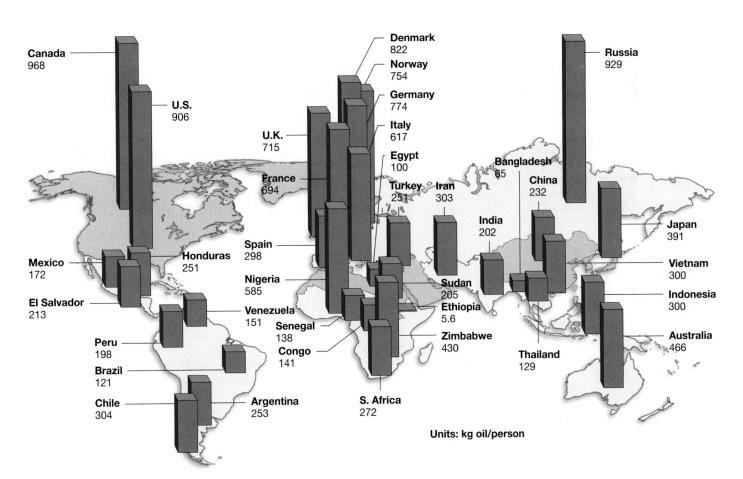

FIGURE 54.21 Energy Consumption Varies among Countries
Per-capita energy use in 1999.

and stronger evidence that most of the warming observed over the last 50 years is attributable to human activities." The panel has now taken the position that rising concentrations of greenhouse gases are at least partially, and perhaps primarily, responsible for global warming.

How much will average temperatures rise in our lifetimes? Predicting the future state of a system as complex and variable as Earth's climate is extremely difficult. Global climate models are the primary tools that scientists use to make these projections. A global climate model is based on a large series of equations that describe how the concentrations of various gases in the atmosphere, solar radiation, transpiration rates, and other parameters interact to affect climate. The models currently being used suggest that average global temperature will undergo additional increases of up to 3.5°C over the next 50 years and 1.4–5.8°C (2.2–10°F) by the year 2100.

How will ecosystems respond to this increase? Answering this question is equally difficult, because ecosystems respond to warming in ways that increase or decrease CO_2 concentrations and thus exacerbate or mitigate warming. Positive feedbacks, for example, occur when warmer and drier climate conditions lead to more fires, which in turn release more CO_2, which leads to even more warming. Researchers recently documented that a form of positive feedback is already occurring in arctic tundras. Traditionally, tundras sequester carbon in the form of soil organic matter, because decomposition rates are extremely low there. During a series of warm summers in the 1980s, however, researchers found that decomposition rates increased sufficiently to release carbon from stored soil organic matter and to produce a net flow of carbon to the atmosphere.

Negative feedbacks, in contrast, arise when warmer conditions lead to increased photosynthetic rates and hence to an increase in the uptake of CO_2. In addition, the growth rates of several tree species and some agricultural crops have been shown to increase in direct response to increasing atmospheric CO_2. Because CO_2 is required for photosynthesis, it can act as a fertilizer.

Will ecosystem responses increase or decrease global warming? Currently, it is not clear whether positive or negative feedbacks will predominate. Researchers are working to answer this question by using computer models, experiments, and analyses of how ecosystems responded to past climate changes.

Biologists have also documented dramatic impacts that warming temperatures are already having on organisms. **Figure 54.22a** shows recent data on the ranges of copepods in the North Atlantic. Copepods are small crustaceans that are often the dominant predator in marine plankton. The maps

(a) Copepods that live in cold water are declining in the North Atlantic.

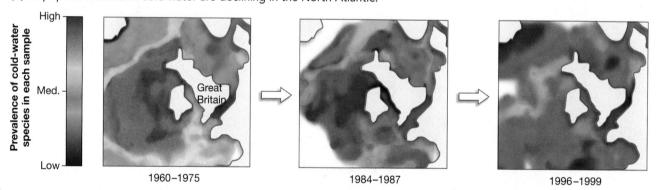

1960–1975   1984–1987   1996–1999

(b) Flowering times for some species in midwestern North America are earlier in the year.

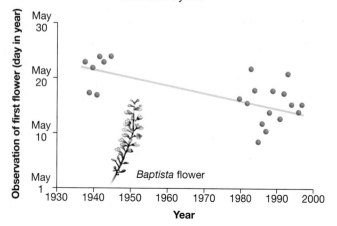

FIGURE 54.22 The Ranges and Behavior of Some Organisms Are Changing in Response to Global Warming

(a) In the North Atlantic, species that live in cold water are declining in frequency; species that live in warm water are increasingly common. **(b)** In a data set from Wisconsin, 19 of the 55 events tracked over a 61-year period showed a significant increase in earliness. Only one event declined in earliness; the rest showed no change.

show changes that occurred in a 40-year period, beginning in 1960, in the number of copepod species typical of marine waters off southern Europe versus the number of copepod species typical of northern European waters. The key observation is that southern species have moved north, while the range of cold-water species has declined. The graph in **Figure 54.22b**, which shows changes in the date of first flowering for a plant native to the midwestern United States, has a similar message. Over the past 60 years, the first day of the year when flowers are produced by certain species has moved up by almost 20 days. To interpret this observation, biologists hypothesize that the climate has warmed enough to promote growth and flowering earlier in the year. Global warming has caused significant changes in the geographic ranges and behavior of organisms, and it promises to cause more.

Productivity Increases

Several of the changes that humans are inducing in biogeochemical cycles have the same effect: They increase productivity. Warming temperatures, the addition of nitrogen and other nutrients, and rising CO_2 levels have all been shown to increase NPP, at least in certain ecosystems. The question is, What are the short-term and long-term consequences of large-scale increases in productivity?

In previous chapters, increased productivity has been viewed as beneficial for plant communities. Chapter 53 pointed out that high productivity resulting from high temperatures is one of the hypotheses to explain why tropical wet forests are so species rich. But several other lines of evidence suggest that increased productivity may have negative effects, at least in certain ecosystems:

- Data introduced in Section 54.2 suggest that if grassland habitats are fertilized with nitrogen, productivity increases but species richness declines over time.

- Nitrate ions derived from fertilizers applied to cornfields in the midwestern United States are washing into the Gulf of Mexico. Increased nitrate concentrations have stimulated the growth of planktonic organisms, whose subsequent decomposition has used up available oxygen and triggered the formation of anaerobic "dead zones" in the gulf (see Chapter 27). In this way, excessive fertilization has caused a severe pollution problem.

- Biologists hypothesize that recent increases in levels of nitrate and other fertilizers of human origin have contributed to the frequency or intensity of harmful algal blooms—large populations of dinoflagellates that release toxins into the surrounding water (see Chapter 28).

- During the 1960s and 1970s it became popular in industrialized nations to use phosphate-based detergents for washing clothes. Much of this phosphate ended up in lakes and streams, where it triggered a massive growth of cyanobacteria—which normally are phosphorus limited in these ecosystems. The response was **eutrophication**: the conversion of a lake to a highly productive ecosystem with rapid decomposition, low oxygen levels, and rapid filling with decomposing organic matter.

Overall, it is not clear whether increased productivity will be beneficial or detrimental to ecosystems. The answer should be forthcoming, however. Because large-scale additions of nitrogen, phosphorus, and carbon dioxide are continuing, humans are implementing a global-scale fertilization experiment.

ESSAY Phosphate Pollution, Acid Rain, and the Ozone Hole: Hope for Ecosystem Recovery?

A central message of this chapter is that energy flows and nutrient cycling are undergoing extraordinary changes in ecosystems throughout the world. Although the outcome of these changes is uncertain, it is important to recognize that humans have already identified and acted on several recent changes in the abiotic environment that clearly had negative consequences. The events and responses took place at the local, regional, and global levels.

The first example of an effective response to an abiotic change took place at a local level and involved a global nutrient cycle. Like many other elements and molecules, phosphorus cycles through ecosystems. As the close of this chapter notes, the use of phosphate-based detergents in the industrialized

Effective responses have occurred ... when biologists documented serious problems in ecosystem ecology.

countries led to a large increase in the concentrations of phosphate in lakes and streams, triggering rapid and widespread eutrophication—particularly in shallow lakes that received outflow from municipal sewage systems. In response, governments in North America and Western Europe encouraged or required the use of phosphate-free detergents, and sewage plants were upgraded to remove more phosphate during treatment. Although

(Continued on next page)

(Essay continued)

phosphate pollution from farm fertilizers remains a serious problem, the crisis conditions of the 1960s and 1970s (**Figure 54.23a**) have largely been alleviated.

The second example involves changes in the pH of rainwater at a regional level. The problem began with sulfur oxides and nitrogen oxides that are pumped into the atmosphere by coal-burning electrical power plants and vehicles that lack catalytic converters. When exposed to sunlight and water vapor, the molecules react to form sulfuric acid (H_2SO_4) and nitric acid (HNO_3). Normal rainwater has a slightly acidic pH, about 5.6. But in areas affected by **acid rain**, precipitation can have a pH as low as 4.2 or 4.4. During the 1980s and 1990s, biologists documented that forests and lakes in eastern North America and northern Europe were being affected by acid rain (**Figure 54.23b**). Tree growth slowed in response to the acidification of soils, and lakes became less productive and less diverse. Once biologists had documented the problem, governments instituted stricter controls on the amounts of sulfur oxides and nitrogen oxides that could be emitted from power plants, cars, and trucks. Over the past decade, the intensity of acid rain has diminished and some of the ecosystems that were being affected have begun to recover.

A third example involves changes in atmospheric chemistry that were global in scale. Widespread use of the compounds called *chlorofluorocarbons* (CFCs) in refrigeration and aerosol products resulted in the release of thousands of tons of CFCs into the atmosphere. When CFCs accumulated in the upper atmosphere, they participated in chemical reactions that released chlorine atoms. These chlorine atoms subsequently reacted with ozone (O_3) molecules, which also accumulate in the upper atmosphere. In some years, the loss of ozone due to these reactions was so severe that an **ozone hole** opened over Antarctica (**Figure 54.23c**). This issue concerned scientists from around the world, because ozone absorbs large amounts of ultraviolet (UV) radiation. When the ozone layer that surrounds Earth thins or is wiped out, an excess of UV radiation can reach Earth's surface and act as a mutagen and carcinogen (see Chapter 18). Fortunately, soon after the problem was documented in the early 1990s, international treaties scheduled and enforced the rapid phasing out of CFC production and use. Scientists have recently been able to document the first signs that the size and duration of the ozone hole may be moderating.

The message of these examples is clear: Effective responses have occurred at the local, regional, and global levels when biologists documented serious problems in ecosystem ecology. It remains to be seen whether the same success can be achieved in response to global warming, nitrate pollution, and other current problems in the abiotic environment.

(a) Phosphate pollution: lake eutrophication

(b) Sulfur oxide and nitrogen oxide pollution: acid rain

(c) CFC pollution: ozone hole (blue region) over Antarctica

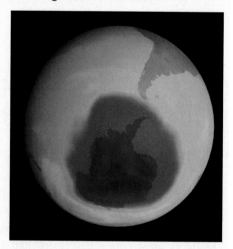

FIGURE 54.23 Some Pollution Problems Are Getting Better
(a) Although eutrophication reached a crisis situation in some lakes in the 1960s, declining phosphate release from sewage systems has alleviated the problem in many areas. **(b)** Acid rain was an increasing problem in some regions of the world until clean air legislation resulted in air-quality improvements in the 1990s. **(c)** The ozone hole has started to show its first signs of shrinking, in response to lowered CFC emissions.

CHAPTER REVIEW

Summary of Key Concepts

■ **An ecosystem has four components: (1) the abiotic environment, (2) primary producers, (3) consumers, and (4) decomposers. These components are linked by the movement of energy and nutrients.**

An ecosystem consists of one or more communities of interacting species and their abiotic environment. As energy flows through ecosystems and as nutrients cycle through them, energy and nutrients are exchanged between biotic and abiotic components of the ecosystem. Energy flows into ecosystems as a result of photosynthesis or the respiration of inorganic molecules with high potential energy. Chemical energy from producers enters the consumer food web or the decomposer food web.

■ **The productivity of terrestrial ecosystems is limited by warmth and moisture, while nutrient availability is the key constraint in aquatic ecosystems. As energy flows from producers to consumers and decomposers, much of it is lost.**

Among terrestrial ecosystems, net primary productivity (NPP) is highest in tropical wet forests and tropical dry forests. Among aquatic ecosystems, productivity is highest in coral reefs, wetlands, and estuaries. Although productivity is extremely low in the oceanic zone, the area covered by this ecosystem is so extensive that the open ocean accounts for the highest percentage of Earth's overall productivity.

Organisms that acquire energy from the same type of source are said to occupy the same trophic level. Most ecosystems have at least three trophic levels: primary producers, herbivores or primary decomposers, and carnivores. Because energy transfer from one trophic level to the next is inefficient, ecosystems have a pyramid of productivity: Biomass is highest at the lowest trophic level and lower at each higher trophic level.

The feeding relationships among species in a particular ecosystem are described by a food chain or food web. Food chains rarely exceed five or six trophic levels. The energy-flow, dynamic-stability, and environmental-complexity hypotheses have been proposed to explain why food webs and food chains are not longer.

■ **To analyze nutrient cycles, biologists focus on the nature of the reservoirs where elements reside and on how quickly elements move between reservoirs.**

Nutrients move through ecosystems in biogeochemical cycles. The rate of nutrient cycling is strongly affected by the rate of decomposition of detritus. The decomposition rate, in turn, is affected by abiotic environmental conditions such as temperature and by the quality of the detritus. Nutrients are also lost from ecosystems. Experiments have shown that the loss of vegetation greatly increases the rate of nutrient loss.

Web Tutorial 54.1   The Global Carbon Cycle

■ **Humans are causing large, global changes in the abiotic environment. The burning of fossil fuels has led to rapid global warming. Extensive fertilization is increasing productivity and causing pollution.**

Average global temperatures are increasing rapidly because land-use changes and burning of fossil fuels have increased the flow of carbon in the form of CO_2 into the atmosphere, and because carbon dioxide acts as a greenhouse gas. Nitrogen fixation from fertilizer production and from the planting of nitrogen-fixing crop species is now approximately equal to the amount of nitrogen fixation from natural sources. These increases have led to increased productivity, but also to pollution and to loss of biodiversity.

Questions

Content Review

1. What is the difference between a community or group of communities and an ecosystem?
   a. A community and the abiotic environment comprise an ecosystem.
   b. An ecosystem is a type of community.
   c. A biome includes only the plant community or communities present in an environment.
   d. An ecosystem includes only the abiotic aspects of a particular environment.

2. Which of the following ecosystems would you expect to have the highest primary production?
   a. subtropical desert
   b. temperate grassland
   c. boreal forest
   d. tropical dry forest

3. Most of the net primary productivity in an ecosystem is used for what purpose?
   a. respiration by primary consumers
   b. respiration by secondary consumers
   c. growth by primary consumers
   d. growth by secondary consumers

4. According to the dynamic-stability hypothesis for food-chain length, food chains will be shorter in which type of environment?
   a. cold
   b. constant
   c. variable
   d. low in nutrient availability

5. Which of the following is normally the longest-lived reservoir for carbon?
   a. atmosphere (CO_2)
   b. marine plankton (primary producers *and* consumers)
   c. petroleum
   d. wood

6. Devegetation has what effect on ecosystem dynamics?
   a. It increases belowground biomass.
   b. It increases nutrient export.
   c. It increases rates of groundwater recharge (penetration of precipitation to the water table).
   d. It increases the pool of soil organic matter.

Conceptual Review

1. Draw a pyramid of productivity for a temperate forest ecosystem, and explain its shape.

2. Explain the difference between gross primary productivity and net primary productivity. Which is larger, and why? Why is total primary production always greater than total secondary production?

3. Explain why decomposition rates are higher in some ecosystems than in others, and give examples. How does the decomposer food web regulate nutrient availability in an ecosystem?

4. Compare and contrast the energy-flow, dynamic-stability, and environmental-complexity hypotheses for food-chain length.

5. Draw a diagram of the global water, carbon, or nitrogen cycle. Label major reservoirs and flows. Compare and contrast the life span of water or an element in each reservoir, and evaluate factors that affect the rate of movement between reservoirs.

6. Why are the open oceans nutrient poor? Why are neritic zones and intertidal habitats relatively nutrient rich?

Group Discussion Problems

1. Suppose you had a small set of experimental ponds at your disposal and an array of pond-dwelling algae, plants, and animals. How could you use radioactive isotopes of carbon or phosphorus to study energy flows or nutrient cycling in these experimental ecosystems?

2. Some researchers are concerned that fertilizing the open oceans with iron would lead to overfertilization of neritic zones, because larger amounts of nutrients would be carried into coastal regions by ocean currents. Outline an experiment and a computer simulation study to test this hypothesis. Evaluate which approach—the experiment or the computer simulation—would likely be more effective in addressing the hypothesis.

3. Suppose that herbivores were removed from a temperate deciduous forest ecosystem. Predict what would happen to the rate of nitrogen cycling. Explain the logic behind your prediction.

4. Explain why human-caused changes to the global carbon cycle are affecting Earth's climate. State why you think that these changes are beneficial or detrimental. List three things that you, local institutions, and state and national governments can do to either augment or mitigate these changes.

Answers to Multiple-Choice Questions 1. a; 2. d; 3. a; 4. c; 5. c; 6. b

Biodiversity and Conservation

55

Lowland wet forest along the Segama River in Borneo, Indonesia. Most of the world's biodiversity is found in tropical wet forests.

KEY CONCEPTS

▨ Biodiversity can be analyzed at the genetic, species, and ecosystem levels.

▨ Humans depend on biodiversity for the products that wild species provide and for ecosystem services that protect the quality of the abiotic environment.

▨ If present trends in human population growth and habitat destruction continue, a mass extinction event is under way.

▨ Effective conservation programs start with sound biological research and involve economists, politicians, and local residents.

In a technologically advanced world filled with dizzying amounts of information, it is easy to assume that biologists have already answered most of the important questions about the natural world. But this assumption could not be further from the truth. No one knows how many species live on Earth. Of the approximately 1.4 million species that have been described to date, only a tiny percentage could be described as well studied. Biologists are only beginning to understand the details of how species interact with one another; we have even more to learn about how groups of species interact with the abiotic world.

One thing is certain, however: Researchers who study biological diversity are doing their work with a newfound urgency. The world population of humans will surpass 6.5 billion people by the time you read this chapter and promises to exceed 9 billion during your lifetime. Human activities are changing ecosystems all over Earth at an ever-increasing rate. For existing species, most of these changes are negative. The global loss of biodiversity and degradation of ecosystems are arguably the greatest challenge our species has ever faced.

Meeting this challenge will require an expanded global community dedicated to the conservation of biodiversity. The role of the biologist in this community is central. While conservation is a complex issue requiring the expertise and cooperation of economists, conservationists, government administrators, educators, business professionals, and community leaders, any effective course of action must be based on accurate scientific information. To design and implement successful conservation measures, we need to understand more about how ecosystems work—particularly the roles of their component species. The data that biologists collect are the starting point for conservation action.

Let's delve into the science behind conservation work by examining what is known about biodiversity, beginning with its often unrecognized significance to our own lives. Then we can review the principal human activities that cause environmental degradation and loss of biodiversity, and close by exploring some of the local and global efforts to conserve Earth's biodiversity.

55.1 What Is Biodiversity?

Perhaps the simplest way to think about biodiversity is in terms of the tree of life that was introduced in Chapter 1. Recall that biologists have used DNA sequence data and morphological traits to estimate the evolutionary relationships among all forms of life. These analyses have supported Charles Darwin's hypothesis that all organisms are related to each other by descent from a common ancestor. The exact relationships among lineages and species are also coming into clearer and clearer focus as biologists compare DNA sequences from more species, measure and interpret additional morphological traits, and discover and analyze new fossil remains. Stated another way, biologists are learning more and more about which species are closely or distantly related, and thus about the size and shape of the tree of life (see Chapters 27 through 34).

Biodiversity Can Be Measured and Analyzed at Several Levels

In a fundamental sense, **biodiversity** can be thought of as the tips of the branches on the tree of life—meaning all distinctive populations and species living today. More formally, biologists recognize and analyze biodiversity on three levels:

1. **Genetic diversity** is the total genetic information contained within all individuals of a species and is measured as the number and relative frequency of all alleles present in a species. Because no two members of the same species are genetically identical, each species is the repository of an immense array of alleles. At this level, biodiversity is everywhere around you, from the variety of apples at the grocery store to the variation in coloration and singing ability of sparrows in your backyard.

2. **Species diversity** represents the variety of life-forms on Earth—the twigs on the tree of life. Recall from Chapter 53 that, in practice, species diversity is measured as the number and relative frequency of species in a particular region. There is an additional aspect of species diversity, however, that biologists call *taxonomic diversity*. When biologists assess taxonomic diversity, they attempt to incorporate the evolutionary relationships among the species present. For example, some lineages on the tree of life are extremely species rich, while others may be represented by only a single living species. African cichlid fish, introduced in Chapter 26, and the 35,000 species of orchids are examples of species-rich lineages. In contrast, the red panda (**Figure 55.1a**) and the

Indian river dolphin (**Figure 55.1b**) have few close relatives. They are examples of species-poor lineages. Some biologists argue that it is particularly important to preserve populations from species-poor lineages, because they are the last living representatives of their lineages. While measures of species diversity focus on the number and relative abundance of species present, measures of taxonomic diversity incorporate how distinct each species is in its evolutionary history.

3. **Ecosystem diversity** is the variety of biotic communities in a region along with abiotic components, such as soil and water and nutrients. Ecosystem diversity is more difficult to define and measure than genetic diversity or species diversity, because ecosystems do not have sharp boundaries. Recall from Chapter 54 that ecosystems are complex and

(a) Red panda

(b) Indian river dolphin

FIGURE 55.1 Phylogenetically Distinct Species May Be High-Priority Targets for Conservation
(a) The red panda and **(b)** the Indian river dolphin have few close relatives and represent distinct branches on the tree of life. If conservation measures attempt to preserve taxonomic diversity, these species would be a high priority for preservation.

dynamic assemblages of organisms that interact with each another and their nonliving environment. Attempts to measure ecosystem diversity focus on capturing the array of biotic communities in a region, along with variation in the physical conditions present.

In addition to recognizing that biodiversity can be recognized and quantified on several distinct levels, it is important to note that biodiversity is dynamic. Mutations that create new alleles increase genetic diversity; natural selection, genetic drift, and gene flow may eliminate certain alleles or change their frequency, leading to an increase or decrease in overall genetic diversity. Speciation increases species diversity, while extinction decreases it (see Chapters 25 and 26). Changes in climate or other physical conditions can result in the formation of new ecosystems, as can the evolution of new species that interact in novel ways. Disturbances such as volcanic eruptions, human activities, and glaciation can destroy ecosystems.

Biodiversity is not static. It has been changing since life on Earth began.

Why Is Biodiversity Important?

If you live in a climate-controlled building, buy your food at a grocery store, and get your medicine from a pharmacy, it can be difficult to recogize the extent to which humans depend on biodiversity. But human health and well-being have always been closely tied to biodiversity.

When someone asks a biologist why biodiversity is important, the answers make up a long list that can be broken into two general categories: (1) ways that the presence of many species benefits humans directly and (2) ways that biodiversity aids humans indirectly by contributing to a healthy overall environment. Although most of these benefits are explained in detail in other chapters, it will be helpful to pull them together into a compact list.

Direct Benefits of Biodiversity: Providing Goods and Services The first large-scale, highly organized human societies began to form about 10,000 years ago, when people at several locations around the world began to domesticate wild plants and farm their food. Since that time, wild species have provided the raw material to fuel the continued development of human societies. Vast tracts of forest have been felled for building materials and fuel; wild plants have been selectively bred to yield food and fibers; animals have been domesticated to provide food, labor, and material goods; the oceans have been harvested for protein; and plants, animals, and fungi have been processed as sources of medicines. For thousands of years, humans have relied on a diversity of wild species to survive.

The direct use of biodiversity continues today:

- Research programs collectively known as *bioprospecting* focus on assessing bacteria, archaea, plants, and fungi as novel sources of drugs or ingredients in consumer products

(see Chapters 27, 29, and 30). Bioprospecting has benefited from the recent explosion of genetic and phylogenetic information, because biologists can now search genomes from a wide array of species to find alleles with desired functions. In addition, recent advances in biotechnology facilitated the development of a new painkiller from the paralyzing sting of tropical cone snails and a blood anticoagulant from the saliva of vampire bats.

- Agricultural scientists are preserving diverse strains of crop plants in seed banks and continue to use wild relatives of domesticated species in breeding programs aimed at improving crop traits (see Chapter 29). In addition, genetic engineering techniques are being used to transfer alleles from a diverse array of species into crop plants. In some cases these efforts have reduced dependence on pesticides, increased resistance to diseases and drought, and improved nutritional value and overall crop yields (see Chapter 19).

- The production of almonds, apples, cherries, chocolate, alfalfa, and an array of other crops depends on the presence of wild pollinators. In the United States alone, insect-pollinated crops produce $40 billion worth of products annually.

- Strategies for cleaning up oil spills, abandoned mines, and contaminated industrial sites are incorporating *bioremediation*—the use of bacteria, archaea, and plants to metabolize pollutants and render them harmless (see Chapters 27 and 29).

- Recreation based on visiting wild places, or *ecotourism*, is a major industry internationally; it is growing rapidly. In South Africa, for example, the number of tourists visiting wildlife preserves increased from 454,428 in 1986 to almost 6 million in 1997. In developing nations with well-designed nature preserves, it is not unusual for ecotourism to represent the first or second most important source of foreign currency earnings.

- High ecosystem diversity—particularly the presence of forests or grasslands on steep slopes and the presence of wetlands in low-lying areas—dramatically reduces flood damage.

Indirect Benefits of Biodiversity: Ecosystem Services The benefits of biodiversity extend beyond the direct use of diverse genes and species by humans to include **ecosystem services**—processes that increase the quality of the abiotic environment. Recall from Chapter 10 that green plants and other photosynthetic organisms produce the oxygen we breathe, and from Chapter 29 that the presence of land plants builds soil, reduces soil erosion, moderates local temperature and wind conditions, and increases the volume of water retained in lakes, streams, and soils. Species from throughout the tree of life are involved in cycling nitrogen, carbon, and other nutrients through ecosystems (see Chapters 27 through 30 and 54).

Current research on ecosystem services is focused on two issues: Economists are attempting to quantify the dollar value of ecosystem services as a way of justifying the cost of preserving natural areas, and biologists continue to document how the loss of biodiversity is affecting the quality of the abiotic environment.

Loss of biodiversity also affects biotic interactions in ecosystems. Recall from Chapter 53 that the loss of keystone predators or herbivores has had a dramatic effect on species diversity and the species present in several types of habitats. Similarly, the food chains, food webs, and mutualistic and parasitic interactions introduced in Chapter 53 change markedly when species are lost. With respect to the biotic aspects of ecosystem functioning, eliminating a species can be like breaking a thread in a piece of cloth and pulling on it. The species itself is lost, and its interactions with other organisms and the physical environment are lost. When the broken thread is an organism, the possibility that it will continue to evolve is lost also.

55.2   How Do Biologists Study Biodiversity?

Our current understanding of Earth's biodiversity is poor. Scientists have a more accurate estimate of how many stars there are in the known universe than of how many species coexist with us on this planet. It is not even clear how many species have been formally described to date. There is no central registry or database for the estimated 1.4 to 1.75 million described species. Work is now under way to use the World Wide Web to build such an encyclopedia of life.

The knowledge base of biodiversity is also heavily biased. The most intensively studied lineages on the tree of life are the vertebrates, flowering plants, conifers, and other groups that are large, attractive, and relatively easy to study. Bacteria, archaea, insects, mites, and other groups are acknowledged to be species rich but are much less well known. In addition, species that inhabit the temperate regions of the world have been studied much more extensively than tropical species.

Even among the larger plants and animals, however, new species are described each year. The number of recognized amphibian species increased by 36 percent in the last 20 years. Since 2000, 30 new species of reptiles have been documented. Biodiversity surveys in less well studied parts of the world, especially the tropics and the oceans, regularly find new species of flowering plants, crustaceans, and mollusks.

Understanding the extent and importance of biodiversity is one of the most massive and exciting challenges facing biological science today. To get a feel for how biologists address this challenge, let's explore four questions about biodiversity: (1) How can biologists estimate levels of genetic diversity? (2) How many species are alive today? (3) What does a thorough understanding of species diversity entail? and (4) How does species diversity affect the function of ecosystems?

Quantifying Genetic Diversity

Recall that, at the genetic level, biodiversity means allelic diversity. When the first techniques for analyzing genes and proteins

from many individuals became available in the 1960s, biologists immediately began using those techniques to quantify the level of genetic diversity within species. The basic approach was to collect tissue samples from many individuals within a population or species, isolate proteins or DNA from the samples, and document how many different alleles existed at a particular gene or suite of genes. With some notable exceptions, the overall message of this work is that most species contain a remarkably high level of allelic diversity. This was an important result, because it meant that changes in natural selection or other evolutionary processes could cause rapid changes in the frequencies of existing alleles and thus a rapid evolutionary response to environmental change. It also meant that losing a species to extinction signifies losing a large number of unique alleles.

Research on the genetic aspects of biodiversity continues and has recently taken on a new goal: quantifying levels of genetic diversity in entire communities or ecosystems rather than in individual species. To appreciate how this work is done and why it is important, consider recent research on genetic diversity among bacteria that live in the Sargasso Sea. The Sargasso Sea is located in the Caribbean region near the islands of the Bahamas and is exceptionally nutrient poor and species poor. It is considered one of the ocean's great deserts in terms of the abundance and diversity of life present. Because it represents such a simple system, the Sargasso Sea was an attractive choice for one of the first studies of genetic diversity in an ecosystem.

To inventory the complete array of bacterial genes present, a research team of investigators collected cells from different water depths and locations. The team isolated DNA from the samples and used the "shotgun sequencing" approach, introduced in Chapter 20, to sequence the complete genome of each species present. They did the gene sequencing in the absence of any information about the nature of the species present—the researchers did not try to culture or name the cells but instead simply sequenced all the bacterial DNA present in the Sargasso Sea samples. After analyzing over 1 billion base pairs of distinct sequences and grouping them by gene and by species, the team concluded that at least 1800 bacterial species were present, of which 148 were previously undiscovered. The researchers identified more than 1.2 million alleles that had never before been characterized. This result suggests that an enormous amount of genetic diversity exists even in Earth's simplest ecosystems.

The research team also claims that this genetic diversity is important. For example, they found more than 780 new bacterial alleles that code for proteins similar to the rhodopsin molecules that are found in your eyes. Do these proteins function as light receptors, like rhodopsin does? If so, does the presence of so many distinct rhodopsin-like genes mean that different bacterial species have different ways of responding to light? Answers to questions like these will come only with continued research. Stay tuned.

Estimating the Total Number of Species Living Today

Biologists acknowledge that only a fraction of the number of species living on Earth has been discovered and described. Consider recent work by Philippe Bouchet and colleagues. This group conducted a massive survey of marine mollusks in coral-reef habitats along the west coast of New Caledonia, a tropical island in the southwest Pacific Ocean. The team spent more than a year collecting mollusks, using an array of techniques, at 42 sites over a total area of almost 300 square kilometers. The survey represented the most thorough sampling effort ever made to determine the species diversity of mollusks.

The survey produced more than 127,000 mollusks representing 2738 species. These numbers far exceed the mollusk diversity recorded for any comparable-sized area. The species total was 2 to 3 times the total number of mollusk species that had been reported for similar habitats in the region.

In reporting their findings, the biologists emphasized that 20 percent of the species found were represented by a single specimen. This observation suggests that many species are exceedingly rare and thus likely to be missed by less-intensive sampling efforts. And when the investigators compared their data with a survey in progress at a second site, different in reef structure but only 200 kilometers away, they found that only 36 percent of species were shared between the two sites. The results from the New Caledonia survey support the hypothesis that the global biodiversity of mollusks—currently thought to be about 93,000 species—is an underestimate.

Given that only a fraction of the organisms alive have been discovered to date, how can biologists go about estimating the total number of species on Earth? Two general approaches have been used. One is based on intensive surveys of species-rich groups at very small sites, and a second is based on attempts to identify all taxa in a particular region.

Taxon-Specific Surveys To estimate the number of insect species that live in the canopy of tropical trees, Terry Erwin and J. C. Scott used an insecticidal fog to knock down insects from the top of a *Luehea seemannii* tree. The researchers identified over 900 species of beetles among the individuals that fell. Most of these species were new to science. To use these data as an indicator of total arthropod species diversity, the researchers used the following train of logic: Based on earlier work with insects on this tree species, they estimated that 160 of the 900 beetle species live only on *L. seemannii*. Worldwide, beetles represent about 40 percent of all known arthropods. Thus, it was reasonable to suggest that 400 species of arthropods live only in the canopy of *L. seemannii*. By adding an estimate of arthropods specializing on the trunk and roots of this tree, Erwin projected that it is host to 600 specialist arthropods. If each of the 50,000 species of tropical tree harbors the same number of arthropod specialists, then the world total of arthropod species exceeds 30 million species. Based on such studies,

(a) Great Smoky Mountains National Park

(b) All-taxon survey

FIGURE 55.2 The First All-Taxon Survey Is Now Under Way The first attempt to document every species living in a prescribed area is under way at Great Smoky Mountains National Park, along the border of Tennessee and North Carolina.

biologists estimate that at least 10 million, and possibly as many as 100 million, species of all types exist today.

All-Taxon Surveys To obtain a more direct estimate of total species numbers, the first effort to find and catalog all of the eukaryotic species present in a large area is now under way. The location is the Great Smoky Mountains National Park in the southeastern United States (**Figure 55.2**). A consortium of biologists, volunteers, and research organizations initiated this all-taxon survey in 1999. To date, the survey has discovered over 230 species that are new to science and over 1750 species that had never before been found in the park. When the inventory is complete, in 2015, biologists will have a much better database to use in estimating the extent of global biodiversity.

Steps in Understanding Biodiversity

A thorough understanding of biodiversity involves several levels of research. The first step is to name a species. When a new species is discovered, researchers write a paper describing its morphology or other distinguishing traits and assign it a genus and species name. Usually this written description is associated with a preserved physical specimen that is deposited in a collection at an herbarium or natural history museum. These repositories serve as the world's biodiversity libraries. Museum collections and published descriptions provide the baseline of biodiversity knowledge.

Once a species is described, the second step is to ask what other species are related to it. Using techniques introduced in Chapter 25, researchers use molecular or morphological data to locate the organism on the tree of life. Understanding a species' evolutionary relationships is the single most powerful tool available for predicting its characteristics. For example, physicians in Australia administer antivenom to snakebite victims based on the phylogenetic position of the snake in question. Each type of antivenom is effective against poisons from closely related snakes (**Figure 55.3**). Phylogenetic analyses also alerted researchers to a threat from an alga that recently began growing at several sites in coastal California. Because the species is closely related to an invasive algal species that has taken over large areas of the Mediterranean Sea, officials from California immediately put an eradication program in place.

A third step in biodiversity studies is to assess the species' geographic range. Biogeographic studies that compare the distributions of a large number of species have been extremely useful in the design of conservation programs. For example, the observation that species richness increases from the poles to the equator, introduced in Chapter 53, has helped focus attention on the importance of preserving habitats in the tropics. In addition, biologists have been able to map specific locations that have a high proportion of **endemic species**—taxa that are found nowhere else—or a high proportion of endangered species. Conservationists have used these biogeographic analyses to set priorities for programs designed to protect the highest diversity of species possible.

A fourth level of biodiversity studies involves understanding a species' ecology, or how it interacts with other organisms and the physical environment. Ecological studies establish what an organism eats and what eats it, whether it is a parasite or a mutualist, and which species it competes with. In addition to deepening our understanding of what a particular species is, ecological studies are required to probe how the diversity of species present at a particular location affects how the ecosystem functions. Let's take a closer look.

Biodiversity and Ecosystem Function

Intact ecosystems control flooding, provide oxygen and clean water, build soil, reduce soil erosion, and maintain nutrient levels. It is legitimate to ask, however, whether the services that ecosystems provide for humans depend on a diversity of species being present. Do ecosystems function "better" in some sense when biodiversity is high?

The answer is not necessarily yes. For example, the *redundancy hypothesis* of ecosystem function holds that many species perform similar roles in an ecosystem, such as primary producer or decomposer. If so, then a minimal level of diversity

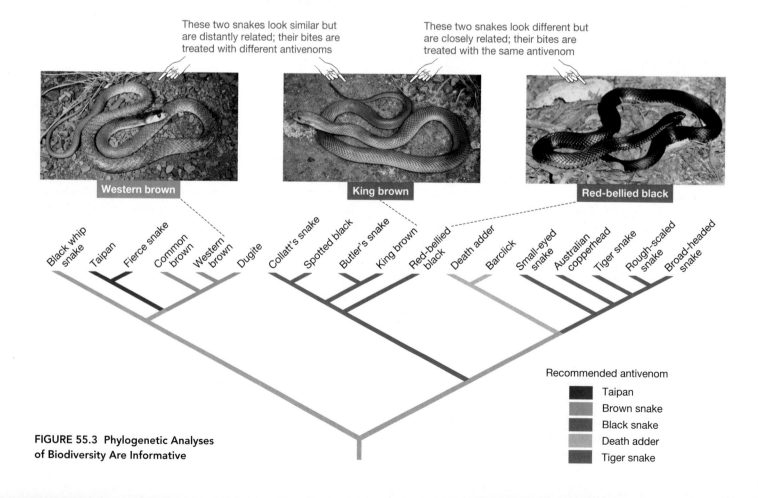

FIGURE 55.3 Phylogenetic Analyses of Biodiversity Are Informative

is required for an ecosystem to function properly. Additional species would likely be redundant and therefore not contribute to overall ecosystem products and services. In contrast, according to the *rivet hypothesis*, an ecosystem is like an airplane wing held together by rivets. If so, then an ecosystem's function may be compromised if certain species are lost, just as an airplane wing may function worse if particular rivets are missing.

To assess how biodiversity affects ecosystem function, David Tilman and colleagues classified 32 grassland plant species into the five types, or functional categories, described in **Figure 55.4**. Notice that some of the types differ by the timing of their growing season, while others are defined by whether they allocate most of their resources to manufacturing woody stems or seeds. To test the effect that these ecological differences have on productivity, the researchers planted plots measuring 13 m × 13 m with a mixture of between 0 and 32 randomly chosen species, representing from 0 to 5 functional categories. After 2 years of plant growth, the researchers harvested and weighed the aboveground tissues from the plots, as an index of net primary productivity. Recall

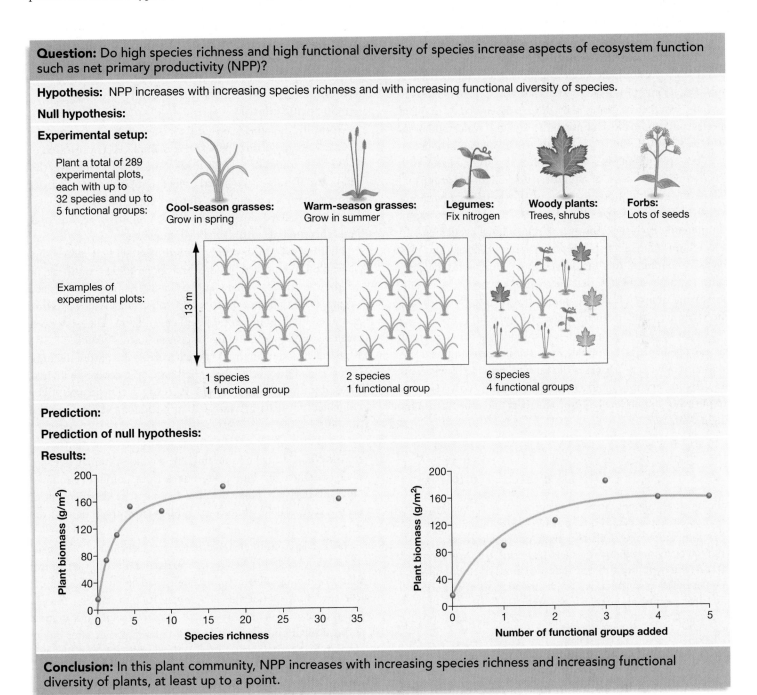

Question: Do high species richness and high functional diversity of species increase aspects of ecosystem function such as net primary productivity (NPP)?

Hypothesis: NPP increases with increasing species richness and with increasing functional diversity of species.

Null hypothesis:

Experimental setup:

Plant a total of 289 experimental plots, each with up to 32 species and up to 5 functional groups:

Cool-season grasses: Grow in spring

Warm-season grasses: Grow in summer

Legumes: Fix nitrogen

Woody plants: Trees, shrubs

Forbs: Lots of seeds

Examples of experimental plots:

13 m

1 species
1 functional group

2 species
1 functional group

6 species
4 functional groups

Prediction:

Prediction of null hypothesis:

Results:

Plant biomass (g/m²) vs Species richness

Plant biomass (g/m²) vs Number of functional groups added

Conclusion: In this plant community, NPP increases with increasing species richness and increasing functional diversity of plants, at least up to a point.

FIGURE 55.4 Evidence That the Productivity of Ecosystems Depends on the Number and Types of Species Present

EXERCISE Write in the null hypothesis and both sets of predictions.

from Chapter 53 that **net primary productivity (NPP)** is the total amount of photosynthesis per unit area per year that ends up in biomass, the total mass of living organisms.

Tilman's group compared the productivity of plots with different numbers of species and functional groups, and found that both the number and type of species present had important effects. Plots with more species and with a wider diversity of functional groups were more productive. Total biomass leveled off as species richness and functional diversity increased, however. This observation suggests that increasing species diversity improves ecosystem function only up to a point. The data support the hypothesis that at least some species in ecosystems are redundant.

Other experiments in grassland ecosystems are consistent with the conclusion that species richness causes an increase in productivity and thus in ecosystem function. One hypothesis to explain this pattern states that diverse assemblages of plant species make more efficient use of the sunlight, water, and other resources available and thus lead to greater overall productivity. A second hypothesis states that productivity increases with species richness because certain species or functional groups facilitate the growth of other species by providing them with nutrients, partial shade, or other benefits. These hypotheses are not mutually exclusive, however. Both could be occurring, and research on the relationship between species richness and productivity continues. And in addition to promoting productivity, some research indicates that increasing species richness improves the ability of plant communities to resist disturbance and to recover from disturbance (see Chapter 53).

The message of this research is that biodiversity does appear to provide benefits beyond human self-interest and that increased biodiversity increases the services provided by ecosystems. By implication, biologists can infer that if ecosystems are simplified by extinctions, then productivity and other attributes might decrease.

✔ CHECK YOUR UNDERSTANDING

Biologists document biodiversity at the genetic, species, and ecosystem levels. You should be able to (1) outline a study that would quantify levels of genetic, species, and ecosystem diversity on your campus and (2) design an experiment to test the hypothesis that species-rich areas are more effective at building soil, retaining soil nutrients, and minimizing soil erosion than are species-poor areas.

55.3 Threats to Biodiversity

No species lasts forever. Extinction, climate change, competition from newly arrived species, and habitat alteration are all natural processes. They have been happening since life began. Extinction, like death, is a fact of life.

But today, species are vanishing faster than at virtually any other time in Earth's history. Modern rates of extinction are 100 to 1000 times greater than the average, or "background," rate recorded in the fossil record over the past 550 million years. Either directly or indirectly, the extinctions are being caused by the demands of a rapidly growing human population, which is currently increasing by about 76 million people per year. If present trends in human population growth and species extinctions continue, a mass exinction on the scale of events described in Chapter 26 is almost certain to occur. Human impacts on the planet promise to equal or exceed those of the gigantic asteroid that smashed into Earth 65 million years ago. It's important to note, however, that human impacts on biodiversity are not a new phenomenon.

Humans Have Affected Biodiversity throughout History

Historically, humans have a poor record of conserving resources and protecting species. For example, recent research on fossil birds found on islands in the South Pacific suggests that about 2000 bird species were wiped out as people colonized this area between about A.D. 400 and 1600. Many of these extinctions occurred due to predation by humans or rats, pigs, and other animals introduced by humans.

Humans even have a poor record of preserving species they depend on for survival. To drive this point home, consider data on the easternmost island in the South Pacific—Easter Island (**Figure 55.5a**). When European explorers arrived on Easter Island in 1722, about 1000 people lived there. The island was treeless and dotted with gigantic stone statues (**Figure 55.5b**). Researchers who analyzed pollen cores taken from swamps on the island discovered that it had once been covered with lush forest dominated by palm trees. A steady decline in tree pollen coincided with the arrival of the first human settlers around A.D. 400.

Fossil digs confirmed that the fauna of the island underwent drastic changes at the same time. In the oldest human garbage piles excavated by biologists, bones from dolphins, seabirds, and land birds are abundant. But these species dropped out of the fossil record by about A.D. 1200—about the time that deforestation was complete.

Jared Diamond interpreted these data as evidence of an ecological disaster. In his view, people arrived to find a lush tropical island brimming with natural resources. The population flourished, possibly numbering as many as 7000 at its peak. People had the leisure time and resources to carve the gigantic statues and roll them into place on beds of palm logs. But after deforestation was complete and local extinctions had begun, the system collapsed. Without palm trunks to make canoes, Easter Island natives did less dolphin hunting and less fishing. Soil erosion may have cut into the productivity of banana and sweet potato plantations. The population crashed, and the great statues fell.

(a) Easter Island is the easternmost island in the South Pacific.

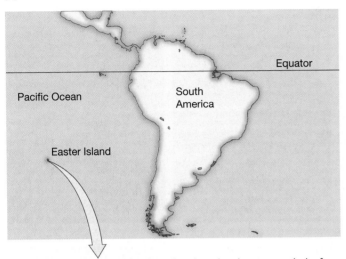

(b) Gigantic stone statues that dot the island were made before an ecological collapse occurred.

FIGURE 55.5 On Easter Island, Social Decline Coincided with the Exhaustion of Natural Resources
(a) As humans spread across the South Pacific, Easter Island was one of the last islands to be colonized. **(b)** The prosperity that supported the carving of these statues was based on resources that were overexploited.

Current Threats to Biodiversity

Over the past 1000 years, most recorded extinctions have occurred on islands and have been associated with overhunting or human-assisted introductions of competitors, diseases, or predators that were not native to the area. Starting in the twentieth century, however, this situation began to change. Human impacts now include entire continents as well as islands and they go far beyond overhunting and species introductions. As a growing human population consumes Earth's resources to fulfill its needs and desires, biodiversity is being degraded or destroyed at the genetic, species, and ecosystem levels.

The direct exploitation of plants and animals for food, medicines, trophies, or pets continues to be a problem for some species in some regions (**Figure 55.6a**, page 1274). Overexploitation is a particular concern for the world's major marine fisheries, two-thirds of which are currently fully exploited or depleted. A recent study of the global fishing industry over the last 50 years concluded that 90 percent of large ocean fish—such as tuna, swordfish, and marlin—and the same percentage of important bottom-dwelling fish—such as cod, halibut, and flounder—have been removed from the world's oceans.

In both aquatic and terrestrial ecosystems, human introductions of *exotic*, or nonnative, species are still causing severe problems (**Figure 55.6b**). **Invasive species**—exotic species that are introduced to a new area, spread rapidly, and eliminate native species (see Chapter 50)—are implicated in the demise of one-third to one-half of all species currently listed as endangered or threatened in the United States. These exotic species are largely free of their normal diseases, competitors, and parasites, and either directly consume native species or outcompete them.

Figure 55.6c highlights the problems caused by pollution such as sewage, toxic chemicals in fertilizers, pesticides and industrial wastes, oil spills, and acid rain. Pollution can kill organisms directly or cause reproductive abnormalities and failure. Some toxic compounds are highly persistent and can be transported long distances from their point of origin. Pollutants found in polar ice and in tissues from oceangoing whales reveal that chemical contaminants have reached every corner of the globe.

By burning forests and fossil fuels, humans have caused carbon dioxide levels in the atmosphere to rise enough to cause dramatic global warming (see Chapter 54). While there is no scientific consensus about the exact effects resulting from higher average temperatures worldwide, global climate models suggest that sea levels will increase and regional climate patterns will change (**Figure 55.6d**). Coral reefs, alpine tundras, and low-lying islands are among the most vulnerable ecosystems. The impact of climate change on biodiversity is unknown and represents a major research frontier in conservation biology.

With respect to the numbers of species and ecosystems affected, human alteration of natural habitat is now the dominant cause of biodiversity decline. Humans cause **habitat destruction** by logging and burning forests, damming rivers, dredging and trawling the oceans, plowing prairies, grazing livestock, filling in wetlands, excavating and extracting minerals, and building housing developments, golf courses, shopping centers, office complexes, airports, and roads (**Figure 55.6e**).

In addition to destroying natural areas outright, human activities fragment large, contiguous areas of natural habitats into small, isolated fragments. **Habitat fragmentation** concerns biologists for several reasons. First, it can reduce habitats to a size that is too small to support some species (**Figure 55.6f**). This is especially true for predators such as mountain lions, grizzly bears, and bluefin tuna that need vast natural spaces in which

W E B T U T O R I A L 5 5 . 1
Habitat Fragmentation

(a) Overexploitation

Overhunting led to the extinction of the passenger pigeon.

(b) Introduced species

When species that are not native are introduced to an area, a number of problems can occur.

Competition: In North American marshes, purple loosestrife is crowding out native organisms.

Disease: An introduced fungus has virtually wiped out the American chestnut.

Predation: The brown tree snake has extinguished dozens of bird species on Guam.

(c) Pollution

In the late 1940s, DDT was widely used to control mosquitoes in marshy areas. By the mid-1950s, fish were accumulating large quantities of DDT in their bodies. Fish-eaters such as bald eagles accumulated even larger quantities of DDT, which turned out to be toxic.

(d) Global warming

Due to increased levels of CO_2 in the atmosphere, average temperatures are expected to increase by 5.5°C over the next 100 years. Species that disperse slowly and take a long time to grow to maturity, such as beech trees, may not be able to shift their ranges quickly enough to survive.

(e) Habitat destruction

Logging in the Pacific Northwest of North America has removed so many large nesting trees that the northern spotted owl is now endangered.

(f) Habitat fragmentation

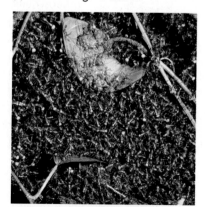

Army ants range widely in search of suitable food. Small forest plots are not large enough to support them.

(g) Domino effects

Antbirds feed on the insects stirred up by army ants. When army ants disappear from forest fragments, so do antbirds.

FIGURE 55.6 Seven Deadly Sins That Affect Biodiversity

[Photo (b) middle ©Gary Braasch; (g) ©Doug Wechsler/VIREO.]

to feed, find mates, and reproduce successfully. Second, habitat fragmentation reduces the ability of individuals to disperse from one habitat to another by creating islands of habitat in a sea of human-dominated landscapes. Small, isolated populations can suffer from inbreeding depression and random loss of alleles due to genetic drift (see Chapter 24). Third, fragmentation creates large amounts of "edge" habitat that is subject to invasion by weedy species and exposed to more intense sunlight and wind.

The decline in habitat quality caused by fragmentation was documented in a recent experiment in a tropical wet forest. In an area near Manaus, Brazil, that was slated for clear-cutting, a research group set up 66 square, 1-hectare experimental plots that remained uncut. (A hectare, abbreviated ha, is 100 meters by 100 meters, approximately the size of two football fields.) Thirty-nine of these study plots were located in fragments designed to contain 1, 10, or 100 hectares of intact forest. Twenty-seven of the plots were set up nearby, in continuous wet forest. As the "Experimental setup" section of **Figure 55.7** shows, the distribution of the study plots allowed the research team to monitor changes inside forest fragments of different sizes and to compare these changes with conditions in unfragmented forest.

When the research group surveyed the plots 10–17 years after the initial cut, they recorded two predominant effects: a rapid loss of species diversity, especially from the smaller fragments, and a startling drop in **biomass**, or the total amount of fixed carbon, in the study plots located near the edges of logged fragments. In tropical wet forests, most of the biomass is concentrated in large trees. The edges of the experimental plots contained many downed and dying trees. Based on this observation, the researchers inferred that the decrease in biomass occurred because large trees near the edges of fragments died from exposure to high winds and dry conditions. This edge effect is important because it eliminates treetop, or canopy, habitats. As a result, the quality as well as the quantity of habitat in the study area declined drastically.

Researchers are also beginning to document population reductions due to **domino effects**—that is, impacts on a given species as a result of the loss of a different species (**Figure 55.6g**). When one species goes extinct, any other species that had specialized in parasitizing or preying on that species will also go extinct.

Now, how many extinctions are the seven problems highlighted in Figure 55.6 projected to cause?

How Can Biologists Predict Future Extinction Rates?

One strategy for predicting the future of Earth's biodiversity focuses on the consequences of the most urgent problem: habitat destruction. Given reasonable projections of habitat loss, biologists can estimate rates of extinction based on well-documented **species–area relationships.**

Question: How does fragmentation affect the quality of tropical wet forest habitats?

Hypothesis: Fragmentation reduces the quality of wet forest habitats.

Null hypothesis: Fragmentation has no effect on the quality of wet forest habitats.

Experimental setup:

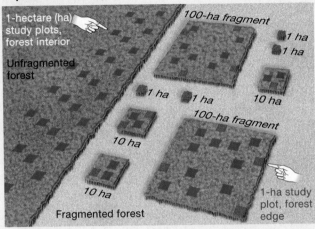

Prediction: Species diversity and biomass will decline in forest fragments compared with those of the forest interior, particularly along edges of fragments.

Prediction of null hypothesis: Species diversity and biomass will be the same inside forest fragments and along edges as in the forest interior.

Results:

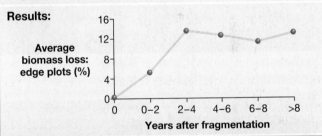

Conclusion: Biomass declines sharply along edges of forest fragments. To interpret this result, hypothesize that large trees near forest edges tend to die.

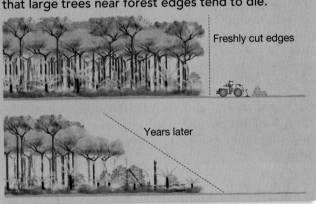

FIGURE 55.7 Experimental Evidence for Edge Effects in Fragmented Forests
Researchers tracked 66 study plots among four 1-hectare fragments, three 10-hectare fragments, and two 100-hectare fragments.

To understand how this work is done, consider recent estimates of habitat destruction in the world's largest continuous wet forest and most species-rich region: the Amazon River basin (**Figure 55.8a**). In the tropics, deforestation is the result of logging, the conversion of forest to pasture or cropland, and wildfires set by humans. To appreciate the extent of deforestation in some areas, consider the satellite images of Rondônia, Brazil, that were made in 1975 and in 2001 (**Figure 55.8b**). Analyses of satellite photos such as these have shown that an average of 5000 km², an area roughly the size of Connecticut, was deforested in the Amazon each year during this decade. In the 1990s losses of Amazonian wet forest increased to as much as 30,000 km² annually. On the basis of such studies, conservationists estimate that about 10 percent of the original area covered by tropical forests worldwide is being lost each decade. It is not unreasonable, then, to project that 90 percent of the world's tropical forest habitat will be lost over the next century.

To predict how many species will be driven to extinction as a result, biologists use species–area curves like the one shown in **Figure 55.9**. This graph was generated by a biologist who analyzed the number of bird species found on islands in the Bismarck Archipelago near New Guinea, in the South Pacific. The graph shows the number of species on islands of various sizes. Notice that both axes on the graph are logarithmic. The solid line drawn through the points is described by the function $S = (18.9)A^{0.15}$, where S is the number of species and A is the area. In this island group, the function indicates that each tenfold increase in island area increases the number of bird species present by about 40 percent; a tenfold decline in area reduces bird species by about 30 percent.

These data turn out to be typical for other habitats and taxonomic groups as well. When biologists have plotted species–area relationships for plants, butterflies, mammals, or birds from islands or continental habitats around the globe, the relationship is consistently described by a function of the form $S = cA^z$. The c term is a constant that scales the data. Its value is high in species-rich areas, such as coral reefs or tropical wet forests, and low in species-poor areas, such as arctic tundras.

(a) The devastation of deforestation

(b) Satellite view of deforestation in Rondônia, Brazil

FIGURE 55.8 Rates of Deforestation in Tropical Wet Forests Are Approaching 1 Percent per Year
(a) Deforestation is occurring in the tropics due to logging, the conversion of forest to pasture or cropland, and wildfires set by humans. (b) The satellite photos show the extent of deforestation in Rondônia, Brazil, in less than three decades.

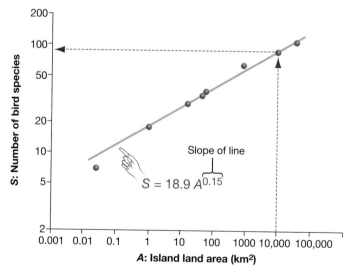

FIGURE 55.9 Species–Area Plots Quantify the Relationship between Species Richness and Habitat Area
Graph plotting the sizes of islands in the South Pacific (horizontal axis) versus the number of bird species present on each island (vertical axis). Notice the log-log scale. The dashed lines show the number of bird species that are expected to live on an island with an area of 10,000 km².

The exponent z represents the slope of the line on a log-log plot. Thus, z describes how rapidly species numbers change with area. In analyses of species–area relationships in island groups, z is typically 0.25 to 0.30. For continental habitats, z usually ranges from 0.10 to 0.20.

To understand how biologists use these analyses to project extinction rates, study Figure 55.9 again. Ask yourself, If 90 percent of the habitats were destroyed—for example, if A were reduced from 10,000 km^2 to 1000 km^2—what percentage of species would disappear? According to the graph, the number of bird species in the Bismarck Archipelago would drop from about 75 to about 53. Thus, the answer is roughly 30 percent. This is an important conclusion. If z is about 0.15 for taxa that inhabit tropical wet forests and other threatened habitats, the prediction is that 30 percent of all species will be wiped out in the next 100 years.

This prediction is consistent with projections based on analyses of species currently threatened with extinction. For example, each year the International Union for the Conservation of Nature (IUCN) publishes a book summarizing data on the status of virtually all species whose conservation status has been studied. If the total population in the wild is declining rapidly enough to threaten a species with extinction, then the species is "redlisted." By studying how fast these red lists are growing over time, biologists can project extinction rates. Consider that IUCN red-listed 5205 mammals in 1996 but 5435 in 2000. If this rate of growth in the number of threatened species continues, and if all these threatened species go extinct, then 60 percent of the mammal species living today will be wiped out in less than 500 years. Similar trends are occurring in listings of threatened birds, mollusks, butterflies, conifers, and flowering plants.

Recall from Chapter 26 that a mass extinction event is defined as the loss of 60 percent of all species in less than 1 million years. Based on analyses of habitat destruction and red lists, biologists have concluded that a mass extinction event is now well under way.

✓ CHECK YOUR UNDERSTANDING

Until recently, most human-induced extinctions have occurred on islands and were due to overhunting or the impact of introduced species. Although invasive species remain a major threat to biodiversity, the majority of current problems are on continents and are due to habitat destruction and fragmentation. You should be able to (1) explain why fragmentation of habitats reduces the habitats' ability to support biodiversity and provide ecosystem services, (2) respond to individuals who argue that the biodiversity crisis began with the emergence of technologically sophisticated industrial societies, and (3) cite data relevant to the hypothesis that the biodiversity crisis is overblown and may already have peaked in intensity.

55.4 Conserving Biodiversity: Biology, Sociology, Economics, and Politics

The conservation of biodiversity is among the most urgent but least-recognized issues facing us today. From subsistence fishermen in Pacific islands to Wall Street CEOs, humans depend on the intricate web of life that is biodiversity. Unlike ozone holes, acid rain, phosphorus pollution, and other environmental problems that humans have addressed recently, extinction is irreversible. Preventing the loss of alleles, species, and ecosystems is the only conceivable solution to the biodiversity crisis.

Solving any global problem requires a common goal to be defined. In the case of preserving biodiversity, the objective is to sustain a diversity of alleles, populations, species, habitats, and ecosystems in natural landscapes while supporting the extraction of resources and environmental services required to maintain the health and well-being of an increasingly large human population.

Fundamentally, the conservation of biodiversity boils down to maintaining viable populations of all species. Because this is a biological issue, biologists are the anchors of any sound conservation effort. One of the most important biological goals in conservation work is to establish a **baseline**, or reference point, that provides a standard for evaluating changes in the diversity of alleles, species, and ecosystems. In establishing baselines, it is crucial to recognize that ecosystems around the world have already deteriorated significantly. For example, if you are a fisheries biologist charged with maintaining salmon diversity in the Pacific Northwest of North America, you might think you are doing an excellent job if salmon numbers rise to double what they were in the 1930s. But by the 1930s salmon in this region had already been reduced by 90 percent of the populations present in the 1800s. By using the 1930s as a baseline, you have accepted a highly degraded state for measuring the success of conservation programs.

Ideally, the baseline for all conservation efforts would be the state of genetic, species, and ecosystem diversity prior to significant human impacts. But with each passing generation, this baseline becomes less and less realistic. Baselines are shifting. As a result, biologists often have to accept impaired ecosystems as reference points. In addition, biologists are realizing that there may be no such thing as a state prior to significant human impacts. Around the world, people have been changing ecosystems for thousands of years. These impacts range from the intentional burning of prairies and woodlands by Native Americans, to grazing of domestic livestock by aboriginal peoples in Africa, to the slash-and-burn agricultural systems of native peoples in the Amazonian forest.

Finally, biologists realize that they do not work in isolation when they are establishing baselines and working on management plans for threatened species and ecosystems. In almost every case, the underlying causes of the biodiversity crisis are socioeconomic factors that encourage the short-term overexploitation of land and other resources and discourage the long-term sustainability of resources. **Sustainability** is the planned use of resources at a rate only as fast as the rate at which they are replaced. For this reason, successful conservation practice requires a complex mixture of economic, social, and political action based on robust scientific knowledge. Let's take a closer look at some of the organizations involved in conservation action.

The Role of Governmental and Private Agencies

Because most species live in more than one country and because changes to the atmosphere, surface water, groundwater, and soils tend to affect large geographic areas, efforts to preserve biodiversity are often international in scope. At the same time, conservation programs tend to have a disproportionate impact on specific individuals, even though the general public may benefit. For example, it is possible for restrictions on fishing quotas or logging contracts to threaten families and businesses with bankruptcy if these programs are not managed carefully. Elected officials like to say that "All politics is local." The same might be said of conservation programs.

Implementing conservation programs effectively requires a combination of global support and local support. As a result, a blend of international, regional, and local agencies are often involved.

International Agreements   The biodiversity crisis is recognized as a global problem, and many international programs have emerged to help solve it. By far the most important is an international agreement called the *Convention on Biological Diversity* (CBD). The CBD was proposed in Brazil in 1992 at the United Nations Conference on Environment and Development, which still ranks as the largest meeting of world leaders ever held. The CBD was the first global agreement on the conservation and sustainable use of biological diversity and has now been formally ratified by over 175 countries.

The CBD has three main goals: (1) to conserve biodiversity, (2) to promote sustainable human use of biodiversity, and (3) to share the commercial benefits of biodiversity equitably, especially with respect to genetic diversity. The CBD was hailed as a landmark international law, because it recognized that the conservation of biological diversity is a "common concern of mankind" and because it laid the groundwork for a new philosophy of environmentally sound economic development.

Even though the CBD established goals to which all signatories agreed, the responsibility for achieving those goals rests with each nation. Under the CBD, each government agreed to undertake the conservation and sustainable use of biodiversity and to develop and implement national biodiversity action plans. Countries vary widely in the resources, expertise, and political will they devote to carrying out their obligations, however. Although much has been achieved, successful implementation of the CBD will require significant and sustained efforts by world governments.

National Agencies   Most countries have an array of offices or programs designed to preserve biodiversity. In the United States, for example, biodiversity is managed by a broad range of government agencies at the federal, state, and local levels. The federal government is by far the country's largest landholder, however. The U.S. Department of the Interior has several agencies that are charged with managing biological resources, including the Bureau of Land Management, the Fish & Wildlife Service, and the National Park Service. The U.S. Forest Service, under the Department of Agriculture, has primary responsibility for managing huge tracts of forested land, especially in the Southeast and Northwest.

In the United States, most federal land is managed for multiple uses, including extractive industries, such as timber logging and mining, gas and oil development, cattle grazing, and hunting and fishing. The major exception to this rule is wilderness areas, which are large tracts of land—usually greater then 5000 acres—left in their natural state. Currently, the United States has 662 wilderness areas that total over 100,000,000 acres; more than half of that total acreage is in Alaska. These wilderness areas are considered critical for research focused on establishing baselines for conservation programs.

In the ocean realm, the National Marine Fisheries Service, under the U.S. Department of Commerce, is charged with managing and conserving marine organisms. The ocean territory of the United States encompasses more than 4 million square miles of coastal waters and open seas, but less than 0.1 percent of this total is protected from extractive uses, such as fishing, oil drilling, dredging, or sea-floor mining. A large body of recent research points to negative trends in marine biodiversity, from declining biomass in many commercial fisheries, to massive coral-bleaching events, growing anoxic dead zones, and outbreaks of invasive species. Biologists know far less about all aspects of marine biodiversity than about its terrestrial counterpart, and governments are doing much less to conserve it.

Nongovernmental Organizations   *Nongovernmental organizations* (NGOs) play a vital role in drawing public attention to specific conservation issues and generating grassroots awareness and support. Many of these organizations have staffs made up of scientists, economists, and attorneys who work together to create comprehensive solutions to biodiversity problems. Some of the largest and most familiar NGOs involved in biodiversity conservation are Conservation International, The Nature Conservancy, and the World Wildlife Fund. In addition to sponsoring research and coordinating efforts to set up nature preserves or even purchasing land outright, NGOs may monitor legislation that affects conservation and lobby governmental authorities to fulfill their commitment to the CBD and other biodiversity agreements.

Another important role for NGOs is to organize collaborations among different levels of government, university-based scientists, local businesses and institutions, and individuals. The IUCN, for example, is a unique union of government agencies, individual scientists, and conservation organizations working together on many aspects of biodiversity conservation, including the production of the global red list of endangered species. The World Conservation Monitoring Center acts as a centralized information resource for policy and action to conserve biodiversity.

Conservation Strategies

Biologists are documenting the extent of biodiversity and identifying threats to species and ecosystems; an array of international and national public and private organizations are focused on solving the biodiversity crisis. What general strategies are being used to preserve biodiversity, and what is the outlook for the future?

In Situ Conservation: Protected Areas Preserving a significant portion of the world's biodiversity requires in situ ("in place") conservation programs that maintain organisms in their wild state and within their existing range. Conserving species where they already exist is more practical and less expensive than ex situ ("out of place") conservation approaches, in which organisms are maintained in zoos, aquaria, or botanical gardens. In situ conservation is usually achieved by government protection of designated areas from development. In theory, it allows wild populations to continue to evolve and to be used by indigenous people who depend on them for food, fuel, or medicine.

The IUCN monitors the success of in situ conservation programs by sponsoring the World Commission on Protected Areas, which holds the World Parks Congress every decade. In 1992 the Congress met in Caracas, Venezuela, and established a goal of setting aside 10 percent of Earth's land surface in protected areas. In 2003 the Congress met in Durban, South Africa, and announced that this goal had been surpassed: Protected areas covered 11.5 percent of Earth's terrestrial surface.

How efficiently is the existing network of protected areas protecting biodiversity? Researchers are attempting to answer this question via a geographic approach called the **Gap Analysis Program (GAP)**. A GAP analysis tries to identify gaps between geographic areas that are particularly rich in biodiversity and areas that are actually managed for conservation. One recent GAP analysis combined data sets on the distribution of mammals, birds, amphibians, and freshwater turtles with a map of world protected areas. The analysis revealed that many species' ranges occur completely outside any protected areas. It also pinpointed regions in Mexico, Madagascar, and elsewhere where the gap between species range and protected areas is particularly high. To date, most GAP analyses suggest that, because relatively few "biodiversity hotspots" are included in existing protected areas, the 11.5 percent of Earth's surface area that is now being managed for biodiversity will not be enough to conserve many species.

Ex Situ Conservation: Zoos, Aquaria, and Botanical Gardens In some cases, it is not possible to save species in the wild, and only intensive management efforts can save them from extinction. By propagating species in captivity, zoos, aquaria, and botanical gardens function like the emergency room of a hospital. To date, over 20 animal species that were exterminated in the wild, including the California condor and Pere David's deer, have been reintroduced into their native habitats after successful captive-breeding programs. Many aquaria are now implementing similar efforts. Because plants are easier to cultivate in captivity than animals are, the world's approximately 1500 botanic gardens can maintain a large percentage of existing plant species in captivity. One of the largest—the Royal Botanic Gardens, Kew, in England—grows representatives of one in eight of Earth's known plants.

In addition to pursuing research and captive propagation, zoos, aquaria, and botanic gardens play a key role in educating the general public about the importance of biodiversity and the need to conserve it. These institutions host tens of millions of visitors every year and in many cases offer visitors their only direct experience of biodiversity, its importance, and its threats. Experience has shown that people have to care about biodiversity before they will work to protect it.

Looking to the Future

For many years, the conservation community's common goal was **sustainable development,** or economic progress for local communities based on using certain species or resources carefully enough that they could regenerate and not decline over time. This concept has recently been expanded to a new community-based conservation paradigm called **adaptive management,** which recognizes that the design, management, and monitoring of conservation programs are inseparable. Under adaptive management, information from monitoring programs is used to review and modify the project's original goals and implementation continuously.

Under both sustainable development and adaptive management, biologists and conservation officials emphasize that, for biodiversity protection programs to be successful, local people must be involved and must benefit from them. If you are a biologist charged with reintroducing the black-footed ferret to the Montana prairies, you are much more likely to be successful if ranchers and other local residents are informed and committed to the project. If you were working in Myanmar to increase the amount of area protected for one of the last remaining Asian tiger populations, it would be essential that you get approval of and participation from local villagers who hunt and collect firewood in the forest, and that local people benefit from any increases in ecotourism based on viewing tigers.

Another cornerstone of sustainable development and adaptive management is active involvement by economists, politicians, and educators, as well as by biologists. Worldwide, there is an urgent need for economists who are trained in biology and who can (1) design profitable but sustainable ways to harvest resources and (2) estimate the dollar value of ecosystem services such as flood control, soil erosion control, and groundwater recharge. In the political arena, elected officials with a background in biology are needed to create new legislation aimed at promoting the sustainable use of biodiversity and to support the enforcement of existing regulations and international conventions. Environmental educators are needed at all levels, from kindergarten to adult education, to increase awareness of biodiversity and commitment to its preservation.

Biology will continue to be the scientific discipline that is central to our understanding of biological diversity and to our wise use of biodiversity. Millions of species are yet to be discovered, named, and studied, and the organized effort to survey genetic diversity within species and ecosystems has only just begun. Much more research is required for us to understand how species interact within ecosystems and how we can best preserve the ecosystem services and functions that support life. In addition to studying life, biologists are now charged with helping save it.

ESSAY Models for the Future: The Malpai Borderlands and Guanacaste

Emotionally, studying biodiversity and implementing conservation programs can be a tough business. The historical record is discouraging, and predictions for the future are not optimistic. It would be misleading, though, to conclude that the future of biodiversity is hopeless.

Recall from the essay at the end of Chapter 54 that governments, corporations, and private individuals banded together to take effective action after researchers identified serious problems with phosphorus pollution, acid rain, and depletion of ozone in the atmosphere. In at least some small pockets around the world, the same thing is happening in response to the biodiversity crisis.

In the United States, one of the greatest threats to biodiversity is suburban sprawl. In response to threats from development in southeast Arizona and southwest New Mexico, a group of ranchers banded together to form an association they called the

Malpai Borderlands group. To preserve biodiversity and their livelihood, the ranchers (1) protected 42,000 acres of their land with conservation easements, which are legal agreements that prevent land from being developed; (2) set up cooperative "grassbanks," or pastureland that is made available to ranchers whose own grazing areas are suffering from short-term drought and need relief from continued grazing; and (3) reintroduced fire to the area via prescribed (intentional) burns that remove encroaching woody shrubs and encourage the growth of native grasses (**Figure 55.10**). The ranchers have also been cooperating with each other and with biologists to monitor or

In addition to studying ecosystems, biologists have begun to restore them.

(a) Before prescribed burn

(b) After prescribed burn: 2 days after ... and 2 months later

FIGURE 55.10 Using Fire to Restore Grassland Ecosystems in the Malpai Borderlands
The same region, in the Malpai Borderlands of the southwestern United States **(a)** before and **(b)** after a prescribed burn. The burn occurred in a wet year and resulted in a dramatic increase in native grasses.

reintroduce native vegetation and animals such as the thick-billed parrot, bighorn sheep, Mexican jaguar, Yaqui chub, and Chiricahua leopard frog into the region.

Today, the Malpai Borderlands group is considered a model of innovative action by private individuals, in cooperation with governmental agencies and NGOs, that benefits local people as well as biodiversity. One aspect of the Malpai Borderlands program—restoration of native vegetation through active management—is turning out to be an important theme in efforts to preserve biodiversity. In many areas of the world, ecosystems are already heavily degraded or lost. The state of Illinois, for example, was estimated to contain 22 million acres of prairie when European settlers arrived; but only about 2000 acres of prairie remain there today. It is much easier to save intact ecosystems than to restore damaged ones. But when natural areas are badly degraded or gone, biologists turn to restoration.

Hundreds of large-scale and small-scale restoration and reforestation projects are now occurring around the globe. In the tropics, one of the most successful is the restoration of dry forest at the Area de Conservación Guanacaste in northwestern Costa Rica. The primary task facing conservation area staff there was to stop human-caused fires. In just 15 years, their efforts have succeeded in transforming a vast swath of marginal ranching land into an increasingly popular tourist destination and a water source for neighboring farms and ranches (**Figure 55.11**).

Conservationists contend that, to preserve biodiversity over the long term, it will not be enough to save patches of undisturbed habitat in parks and preserves; extensive work to restore damaged ecosystems is also crucial. In addition to studying ecosystems, biologists have begun to restore them. From small-scale restorations of woodland and prairie habitats in North America to efforts the size of the Malpai Borderlands and Guanacaste projects, the field of "restoration ecology" is burgeoning.

(a) Start of restoration project

(b) 17 years later

FIGURE 55.11 Using Fire Prevention to Restore Forest Ecosystems in the Guanacaste Reserve
(a) Before and **(b)** after views of the same region, 17 years apart, in the Area de Conservación Guanacaste, Costa Rica. To appreciate how quickly the trees grew, notice the person standing near the middle of the photo in (b).

CHAPTER REVIEW

Summary of Key Concepts

■ **Biodiversity can be analyzed at the genetic, species, and ecosystem levels.**

Genetic diversity is well characterized within some species, but biologists are only beginning to use genome sequencing techniques to explore the extent of genetic diversity in ecosystems. Research on the total number of species alive today is also at a preliminary stage, with estimates ranging from 10 to 100 million. At the ecosystem level, early studies support the hypothesis that high species diversity increases aspects of ecosystem function such as productivity and resistance to disturbance. To date, the message from efforts to characterize

biodiversity is that it is much more extensive than expected and that a great deal remains to be learned.

■ **Humans depend on biodiversity for the products that wild species provide and for ecosystem services that protect the quality of the abiotic environment.**

Humans gain direct economic benefits from fishing, forestry, agriculture, tourism, and other activities that depend on biodiversity. Species diversity is also important for maintaining the productivity of natural ecosystems and their ability to build and hold soil, moderate local climates, retain and cycle nutrients, retain surface water and recharge groundwater, prevent flooding, and produce oxygen.

■ **If present trends in human population growth and habitat destruction continue, a mass extinction event is under way.**

Historically, most human-caused extinctions have occurred on islands because of direct exploitation or the introduction of exotic herbivores and predators. Habitat destruction is currently the leading cause of extinctions, however. Experiments in the Brazilian Amazon have shown that habitat loss leads not only to a rapid decline in biodiversity but also to a decline in the quality of the remaining habitats due to fragmentation.

To estimate how many species will go extinct in the near future, biologists combine data on current rates of habitat loss—usually estimated from satellite images taken over time—with data on the average number of species found in habitats of a given size. These species–area analyses suggest that if 90 percent of habitats are destroyed as expected during the next century, then over 30 percent of all species will become extinct. According to data on the rate at which lists of threatened mammals and other species are growing, it is likely that 60 percent of all species will be wiped out within 500 years.

Web Tutorial 55.1   Habitat Fragmentation

■ **Effective conservation programs start with sound biological research and involve economists, politicians, and local residents.**

Preserving biodiversity may well be the most important—and difficult—enterprise our species has ever undertaken. Although 11.5 percent of Earth's total land area has been officially designated as protected from development, many areas with extraordinarily high biodiversity remain vulnerable and relatively few marine reserves exist.

The Convention on Biological Diversity has set broad goals for the preservation of biodiversity and has been endorsed by most countries. Various governmental agencies and nongovernmental organizations have been created to address the biodiversity crisis, through programs that protect species and ecosystems while encouraging the sustainable use of resources and promoting the welfare of local people.

Biologists play a key role in conservation programs by collecting the data required to make sound political and economic decisions. In addition to documenting the extent and benefits of biodiversity, biologists alert officials and the general public to biodiversity problems, establish baselines for biodiversity, evaluate priorities for conservation action, and monitor the effectiveness of conservation measures.

Questions

Content Review

1. What does a species–area plot show?
   a. The overall distribution, or area, occupied by a species.
   b. The relationship between the body size of a species and the amount of territory or home range it requires.
   c. The number of species found, on average, in tropical versus northern areas.
   d. The number of species found, on average, in a habitat of a given size.

2. What does a GAP analysis do?
   a. It compares the current distributions of species with the locations of preserved habitats.
   b. It quantifies *gross aboveground productivity.*
   c. It uses data on the rates at which lists of threatened species are growing to project the rate of future extinctions.
   d. It uses genome sequencing techniques to quantify genetic (allelic) diversity in an ecosystem.

3. Why do most existing data tend to support the redundancy hypothesis of ecosystem function?
   a. It is common to observe that many species in an ecosystem belong to the same functional group.
   b. Productivity and resistance to disturbance increase with increasing species richness but quickly level off.
   c. Species function like rivets in an airplane wing, because the loss of even one species alters the function of the entire ecosystem.
   d. Ecosystems with low species richness still provide high levels of ecosystem services.

4. What types of data suggest that Easter Island once had high biodiversity and productivity?
   a. comparisons with islands of similar size, latitude, and soil type
   b. historial records from the first Europeans to visit the island
   c. oral traditions of the indigenous people
   d. fossil pollen and animal bones

5. Why is the Convention on Biological Diversity considered a landmark achievement in efforts to preserve biodiversity?
   a. It halted international trade in wildlife pets and products from endangered species.
   b. It established the goal of designating 10 percent of Earth's land surface as protected area.
   c. It is a formal commitment by the signatory nations to preserve biodiversity.
   d. It set up the first international system of marine reserves devoted to preservation of biodiversity.

6. What are biologists doing when they establish a baseline for preservation of biodiversity?
   a. providing documentation for a conservation goal in the management or restoration of a species or an ecosystem
   b. estimating the current number of species alive today, as a basis for measuring change in the future
   c. creating a coalition of international, national, and regional governmental agencies and NGOs focused on a particular conservation goal
   d. documenting the direct and indirect benefits of biodiversity, as a justification for conservation programs

Conceptual Review

1. List two direct benefits to humans for preserving biodiversity and two indirect benefits from ecosystem services. Next to each entry in your list, state which members of society receive this benefit when biodiversity is preserved. Then state who, in your opinion, should be responsible for paying the costs associated with preserving biodiversity in return for this benefit.

2. Compare and contrast biodiversity at the genetic, species, and ecosystem levels. How do biologists analyze the extent of biodiversity at each of these levels?

3. Biologists claim that the all-taxa survey now under way at the Great Smoky Mountains National Park in the United States will improve their ability to estimate the total number of species living today. Discuss the benefits and limitations that this data set will provide in understanding the extent of global biodiversity.

4. How are species–area curves used to relate rates of habitat destruction to projected extinction rates?

5. Discuss evidence to support the hypothesis that species richness increases ecosystem functions such as productivity and resistance to disturbance, and thus the ecosystem services that benefit humans.

6. Explain why the fragmentation of habitats reduces their ability to support biodiversity.

Group Discussion Problems

1. Projections for an impending mass extinction are contingent on the continuation of present trends in human population growth and in habitat destruction. Do you believe that these trends will continue? In general, are you optimistic or pessimistic about the future of biodiversity? Explain your logic.

2. You are helping design a series of reserves in a tropical country. List the steps you would recommend for gathering data and creating a plan that would protect a large number of species in a small amount of land.

3. The maps that follow chronicle the loss of old-growth forest (>200 years old) that occurred in Warwickshire, England, and in the United States. In your opinion, under what conditions is it ethical for conservationists who live in these countries to lobby government officials in Brazil, Indonesia, and other tropical countries to slow the rate of loss of old-growth forest?

4. Make a list of characteristics that would render a species particularly vulnerable to extinction by humans. Make a list of characteristics that would render a species particularly resistant to pressure from humans. Try to think of an example of each type of species.

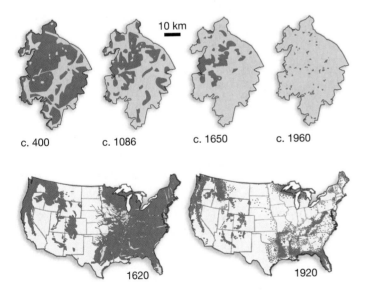

10 km

c. 400   c. 1086   c. 1650   c. 1960

1620   1920

Answers to Multiple-Choice Questions   **1.** d; **2.** a; **3.** b; **4.** d; **5.** c; **6.** a

Appendix

TABLE A.1 The Metric System

Measurement	Unit of Measurement and Abbreviation	Metric System Equivalent	Converting Metric Units to English Units
Length	kilometer (km)	$1 \text{ km} = 1000 \text{ m} = 10^3 \text{ m}$	$1 \text{ km} = 0.62 \text{ miles}$
	meter (m)	$1 \text{ m} = 100 \text{ cm}$	$1 \text{ m} = 1.09 \text{ yards} = 3.28 \text{ feet}$
			$= 39.37 \text{ inches}$
	centimeter (cm)	$1 \text{ cm} = 0.01 \text{ m} = 10^{-2} \text{ m}$	$1 \text{ cm} = 0.3937 \text{ inch}$
	millimeter (mm)	$1 \text{ mm} = 1000 \ \mu\text{m} = 10^{-3} \text{ m}$	$1 \text{ mm} = 0.039 \text{ inches}$
	micrometer (μm)	$1 \ \mu\text{m} = 1000 \text{ nm} = 10^{-6} \text{ m}$	
	nanometer (nm)	$1 \text{ nm} = 10^{-9} \text{ m}$	
	angstrom (Å)	$1 \text{ Å} = 0.1 \text{ nm} = 10^{-10} \text{ m}$	
Area	hectare (ha)	$1 \text{ ha} = 10{,}000 \text{ m}^2$	$1 \text{ ha} = 2.47 \text{ acres}$
	square meter (m^2)	$1 \text{ m}^2 = 10{,}000 \text{ cm}^2$	$1 \text{ m}^2 = 1.196 \text{ square yards}$
	square centimeter (cm^2)	$1 \text{ cm}^2 = 100 \text{ mm}^2 = 10^{-4} \text{ m}^2$	$1 \text{ cm}^2 = 0.155 \text{ square inches}$
Mass	kilogram (kg)	$1 \text{ kg} = 1000 \text{ g}$	$1 \text{ kg} = 2.20 \text{ pounds}$
	gram (g)	$1 \text{ g} = 1000 \text{ mg}$	$1 \text{ g} = 0.035 \text{ ounces}$
	milligram (mg)	$1 \text{ mg} = 1000 \ \mu\text{g} = 10^{-3} \text{ g}$	
	microgram (μg)	$1 \ \mu\text{g} = 10^{-6} \text{ g}$	
Volume	liter (L)	$1 \text{ L} = 1000 \text{ mL}$	$1 \text{ L} = 1.06 \text{ quarts}$
	milliliter (mL)	$1 \text{ mL} = 1000 \ \mu\text{L} = 10^{-3} \text{ L}$	$1 \text{ mL} = 0.034 \text{ fluid ounces}$
	microliter (μL)	$1 \ \mu\text{L} = 10^{-6} \text{ L}$	
Temperature	†Kelvin (K)		$K = {}^\circ C + 273.15$
	Degrees Celsius (°C)		${}^\circ F = \frac{9}{5} {}^\circ C + 32$
	Degrees Fahrenheit (°F)		${}^\circ C = \frac{5}{9}({}^\circ F - 32)$

†Absolute zero is $-273.15\,^\circ C = 0 \text{ K}$

TABLE A.2 Prefixes Used in the Metric System

Prefix	Abbreviation	Definition	
micro-	μ	0.000001	$= 10^{-6}$
milli-	m	0.001	$= 10^{-3}$
centi-	c	0.01	$= 10^{-2}$
deci-	d	0.1	$= 10^{-1}$
		1	$= 10^{0}$
kilo-	k	1000	$= 10^{3}$

Questions

(1) Some friends of yours just competed in a 5-kilometer run. How many miles did they run?

(2) An American football field is 100 yards long, while rugby fields are 140 meters long. In yards, how much longer is a rugby field than a football field?

(3) What is your normal body temperature in degrees Celsius? (Normal body temperature is 98.6 °F.)

(4) What is your current weight in kilograms?

(5) A friend asks you to buy a gallon of milk. How many liters would you buy to get approximately the same volume?

Glossary

–10 box The six-pair base sequence (usually TATAAT) found in most prokaryotic promoters 10 bases upstream from the start of mRNA transcription. Also called Pribnow box.

30-nanometer fiber A fiber of DNA and protein found in eukaryotic nuclei and consisting of many nucleosomes wound together and tightly packed.

–35 box The six-pair base sequence (usually TTGACA) in most prokaryotic promoters 35 bases upstream from the start of mRNA transcription.

ABC model A model of pattern formation of flowers. Proposes that three genes (A, B, and C) are necessary to build a normal flower containing sepals, petals, stamens, and carpels.

abdomen The region of the body that houses most of the digestive tract. In insects, the third and hindmost of the three major body regions.

abiotic Not alive (e.g., air, water, and soil).

aboveground biomass The total mass of living plants in an area, excluding roots.

abscisic acid (ABA) A plant hormone that inhibits cell elongation and stimulates leaf shedding and dormancy.

abscission The shedding of leaves from a plant.

abscission zone The region of a petiole that thins and breaks during dropping of leaves.

absorption The uptake of the ions and small molecules from food across the lining of the intestine and into the bloodstream.

absorption spectrum A graph depicting how well a pigment absorbs various wavelengths of light.

absorptive feeding The absorption of nutrients directly from the environment, across a cell's plasma membrane.

accessory fluids The fluid portion of semen, containing nutrients, enzymes, and other substances. Added to spermatozoa by male reproductive glands.

accessory pigments Pigments, such as carotenoids, that absorb light in plant cells and pass the energy on to chlorophyll to drive photosynthesis.

acclimatization Gradual physiological adjustment of an organism to new environmental conditions.

acetylation Addition of an acetyl group (CH_3COO^-) to a molecule.

acetylcholine (ACh) A neurotransmitter that is released by nerve cells onto muscle cells to trigger muscle contraction. Also used as a neurotransmitter in the nervous system.

acetyl CoA The final product of glycolysis, produced by oxidation of pyruvate and binding to CoA. Can enter the Krebs cycle.

acid Any compound that gives up protons or accepts electrons during a chemical reaction or that releases hydrogen ions when dissolved in water.

acid-base reaction A chemical reaction that involves a transfer of protons.

acid-growth hypothesis The hypothesis that auxin's effect on plant cell elongation occurs via installation of proton pumps that make the cell wall more acidic, causing the cell wall to expand.

acid rain Rain that is substantially more acidic than normal (usually with a pH below 5), enough to adversely affect plants and other species.

acoelomate Not having a coelom (internal body cavity). Diploblasts (sponges, cnidarians, and ctenophores), flatworms, and the Acoelomorpha are all acoelomate.

acquired immune deficiency syndrome (AIDS) A human disease characterized by death of immune system cells and subsequent vulnerability to other infections. Caused by the human immunodeficiency virus (HIV).

acquired immunity Immunity to a particular pathogen conferred by antibodies produced by previous exposure to the pathogen or to a vaccine or acquired passively through the placenta or breast milk.

acrosomal process A protrusion that forms on the head of a sperm cell during fertilization in some species, such as some echinoderms.

acrosomal reaction A set of events occurring in a sperm cell upon encountering an egg cell, including release of acrosomal enzymes and formation of microfilaments that help sperm reach the egg.

acrosome A packet of enzymes found on the head of a sperm cell that help dissolve the zona pellucida or jelly layer.

actin A globular protein that can be polymerized to form actin filaments, which are involved in cell movement.

actin filament A long, thin fiber composed of two intertwined strands of polymerized actin. Involved in cell movement. Also called a microfilament.

action potential A rapid, temporary change in electrical potential across a membrane, from negative to positive and back to negative.

action spectrum The range of wavelengths of light that can drive photosynthesis.

activation energy The amount of energy required to cause a chemical reaction; specifically, the energy required to reach the transition state.

active site The location on an enzyme where substrates (reactant molecules) bind and react.

active transport The movement of ions or molecules across a plasma membrane against a concentration gradient or an electrochemical gradient. Requires energy, such as ATP.

adaptation A heritable trait that increases the fitness of an individual with that trait compared with individuals without it, in a certain environment.

adaptive management A conservation strategy in which information from environmental monitoring is continuously used to review and modify ongoing conservation programs.

adaptive radiation Rapid evolutionary diversification within one lineage, producing numerous descendant species with a wide range of adaptive forms.

addiction Compulsive use of a substance despite recognition of its undesirable consequences, coupled with increasing resistance to the substance (tolerance) and withdrawal symptoms upon cessation of use.

adenosine diphosphate (ADP) A molecule consisting of adenine, a sugar, and two phosphate groups. Addition of a third phosphate group produces adenosine triphosphate (ATP).

adenosine triphosphate (ATP) A molecule consisting of adenine, a sugar, and three phosphate groups that can be hydrolyzed to release energy. Universally used by cells to store and transfer energy.

adenylyl cyclase An enzyme that can catalyze the formation of cyclic AMP (cAMP) from ATP. Involved in control of the gene transcription of various operons in prokaryotes.

adhesion The tendency of certain dissimilar molecules to cling together due to attractive forces.

adipocyte A fat cell.

adipose tissue A connective tissue whose cells store fats.

ADP See adenosine diphosphate (ADP).

adrenal glands Small glands that sit above each kidney. The outer portion (cortex) secretes several steroid hormones; the inner portion (medulla) secretes epinephrine and norepinephrine.

adrenaline A catecholamine hormone from the adrenal medulla. Triggers rapid responses of the fight-or-flight response. Also called epinephrine.

adrenocorticotropic hormone (ACTH) A peptide hormone from the anterior pituitary gland that stimulates release of cortisol, corticosterone, and aldosterone from the adrenal cortex.

adventitious root A root that develops from a plant's shoot system instead of from the plant's root system.

aerobe An organism that uses aerobic respiration (i.e., uses oxygen as a final electron acceptor).

aerobic respiration Cellular respiration using oxygen as the electron acceptor, typically using the Krebs cycle and the electron transport chain.

afferent division The part of the nervous system that transmits sensory information to the central nervous system. Consists mainly of sensory neurons.

agar A gelatinous mix of polysaccharides that is commonly used in labs to grow cell cultures.

age class A group of individuals of a specific age.

agent A nonliving infectious entity, such as a virus. Also called a particle.

age pyramid A graph of a population's age structure, with horizontal bars representing the numbers of males and females, stacked against age on the vertical axis.

age-specific fecundity The average number of female offspring produced by a female in a certain age class.

age structure The proportion of individuals in a population that are of each possible age.

agglutination Clumping together of cells, typically caused by antibodies.

agonist A compound that can bind to and activate a receptor such as a hormone receptor or neurotransmitter receptor.

AIDS See acquired immune deficiency syndrome (AIDS).

air sacs Thin-walled sacs of air that extend throughout most of a bird's body and connect to the lungs. Function as bellows that continuously feed air to the lungs.

albumen A solution of water and protein (particularly albumins), found in amniotic eggs, that nourishes the growing embryo. Also called egg white.

albumin A class of large proteins found in plants and animals, particularly in the albumen of eggs and in blood plasma.

alcohol fermentation Fermentation (production of ATP without oxygen) in which the end product of glycolysis is converted to ethanol.

aldosterone A hormone from the adrenal cortex that stimulates the kidney to conserve salt and water and promotes retention of sodium.

aleurone layer In a seed, a layer that releases the starch-digesting enzyme α-amylase during germination.

alkaptonuria A medical condition characterized by an accumulation of homogentisic acid in the body, caused by a defect in the enzyme that metabolizes homogentisic acid.

allantois In an amniotic egg, the membrane-bound sac that holds waste materials.

allele A particular version of a gene.

allergen Any molecule that can trigger an allergic response.

allergy An abnormal immune response to a non-pathogenic substance, usually involving production of the IgE class of antibodies.

allometry A disproportionate relationship of body size to another feature, such as limb size or heart rate. Also, the general study of the effects of body size on biological processes.

allopatric speciation The divergence of populations into different species by physical isolation of populations in different geographic areas.

allopatry Living in an area different from the area in which some other population lives.

allopolyploid Polyploid (having multiple copies of chromosomes) due to hybridization between different species.

allosteric regulation Regulation of enzymatic catalysis via a change in the enzyme's shape, usually caused by a regulatory molecule binding at a specific site.

α-amylase A plant enzyme that digests the starch in a seed so that the plant embryo can use the resulting sugars for growth.

alpha chain (α chain) One of the two polypeptide chains that make up a T-cell receptor protein, enabling the T cells of the immune system to bind to antigens on other cells.

alternation of generations A life cycle involving alternation of a multicellular haploid stage (gametophyte) with a multicellular diploid stage (sporophyte). Occurs in most plants and some protists.

alternative splicing In eukaryotes, the splicing of the same primary RNA transcript in different ways to produce different mature mRNAs and thus different proteins.

altruism Any behavior that has a cost to the individual (such as lowered survival or reproduction) and a benefit to the recipient.

alveolus (plural: alveoli) One of the tiny air-filled sacs of a mammalian lung.

ambisense virus A virus whose genome contains both positive-sense and negative-sense sequences.

amino acid A small organic molecule with a central carbon atom bonded to an amino group ($-NH_3$), a carboxyl group ($-COOH$), a hydrogen atom, and a side group. Proteins are polymers of 20 common amino acids.

aminoacyl tRNA A transfer RNA molecule that is covalently bound to an amino acid.

aminoacyl tRNA synthetases Enzymes responsible for catalyzing the addition of amino acids to tRNA molecules.

ammonia The molecule NH_3, a toxin produced by the breakdown of proteins and nucleic acids.

ammonium ion The ion NH_4^+.

amnion The membrane in an amniotic egg that surrounds the embryo and encloses it in a protective pool of fluid (amniotic fluid).

Amniota A major lineage of vertebrates that reproduce with amniotic eggs. Includes all reptiles, birds, and mammals.

amniotic egg An egg containing a membrane-bound supply of water (the amnion), a membrane-bound supply of food (yolk sac), and a waste sac (allantois), all encased in a leathery or hard shell.

amniotic fluid The fluid inside the amnion of a developing mammalian embryo.

amoeba Any unicellular protist that lacks a cell wall and is extremely flexible in shape.

amphipathic Containing hydrophilic and hydrophobic elements.

amylase Any enzyme that can break down starch by catalyzing hydrolysis of the glycosidic linkages between the glucose monomers.

amyloplast Dense, starch-storing organelles that settle to the bottom of plant cells and that may be used as gravity detectors.

anabolic pathway A set of chemical reactions that synthesizes larger molecules from smaller ones.

anadromous Having a life cycle in which adults live in the ocean (or large lakes) but migrate up freshwater streams to breed and lay eggs.

anaerobic respiration Cellular respiration using an electron acceptor other than oxygen, such as nitrate or sulfate.

analogous trait Similarity between different species due to adaptation to a similar way of life rather than inheritance from a common ancestor. Often the result of convergent evolution.

anal pore An organelle in some protists used to expel wastes from the cell.

anaphase A stage in nuclear division (mitosis or meiosis) during which chromosomes are moved to opposite ends of the cell.

anaphylactic shock A life-threatening condition in which an allergic response constricts respiratory passages and dilates blood vessels, reducing blood pressure to dangerously low levels.

anatomy The study of the physical structure of organisms.

androgens A class of steroid hormones that generally promote male-like traits (although females have some androgens, too). Secreted primarily by the gonads and adrenal glands.

aneuploidy The state of having an abnormal number of copies of a certain chromosome.

angiosperm A flowering plant; a member of the lineage of plants that produces seeds within mature ovaries (fruits).

animal hemisphere The upper half of an amphibian egg cell, containing little of the yolk. Gives rise to most of the animal's body.

animal model Any disease that occurs in a non-human animal and has many parallels to a similar disease of humans. Studied by medical researchers in hope that findings may apply to human disease.

animals A major lineage of eukaryotes. Typically have a complex, large, multicellular body, eat other organisms, and are mobile.

anion A negatively charged ion.

annual plant A plant whose life cycle normally lasts only one growing season—less than one year.

anoxic For an environment, lacking oxygen.

anoxygenic photosynthesis Photosynthesis that does not include photosystem II and hence does not split water or produce oxygen.

antagonistic muscle groups Sets of muscles whose actions oppose each other and whose control must be coordinated. Often a pair of muscles that move a body part back and forth.

antenna (plural: antennae) A long appendage that is used to touch or smell.

antenna complex An array of chlorophyll molecules that receives energy from light and directs the energy to a reaction center. Part of a photosystem of the light-dependent reactions of photosynthesis.

anterior Toward an animal's head and away from its tail. The opposite of posterior.

anterior pituitary The anterior part of the pituitary gland, containing endocrine cells that release a variety of peptide hormones in response to other hormones from the hypothalamus.

anther The pollen-producing structure at the end of a flower stamen of an angiosperm plant.

antheridium (plural: antheridia) The sperm-producing structure in most land plants except angiosperms.

anthropoids One of the two major lineages of primates, including apes, humans, and all monkeys, but not prosimians.

antibiotic Any substance, such as penicillin, that can kill or inhibit the growth of bacteria.

antibiotic resistance The ability of a microorganism to grow and reproduce effectively in the presence of a certain antibiotic.

antibody A Y-shaped immunoglobulin protein, secreted from B cells, that can bind to a specific part of an antigen, tagging it for attack by the immune system.

anticodon The three bases of a transfer RNA molecule that can bind to a messenger RNA codon with a complementary sequence.

antidiuretic hormone (ADH) A peptide hormone from the posterior pituitary gland that stimulates water retention by the kidney.

antigen Any foreign molecule, often a protein, that can stimulate a specific response by the immune system.

antigen presentation Display of an antigen on the surface of an immune system cell to recruit other immune system cells.

antihistamines Drugs that oppose histamine release from mast cells, reducing inflammatory and allergic symptoms such as swelling and itching.

antiporter A membrane protein that allows an ion to diffuse down a concentration gradient, using the energy of that process to transport some other substance against its concentration gradient and in the opposite direction.

anus In a multicellular animal, the end of the digestive tract where wastes are expelled.

aorta In terrestrial vertebrates, the major artery carrying oxygenated blood away from the heart.

aphotic zone Deep water receiving no sunlight.

apical Toward the top. In plants, at the tip of a branch. In animals, on the side of an epithelial layer that faces the environment and not other body tissues.

apical bud A bud at the tip of a stem, where growth occurs to lengthen the stem.

apical dominance Inhibition of lateral bud growth by the apical meristem at the tip of the plant branch.

apical meristem A group of undifferentiated cells at the tip of a stem or root of a vascular plant. Responsible for primary growth.

apoplast A transport pathway in plant roots through the porous cell walls of adjacent cells.

apoptosis Programmed cell death. Occurs frequently during embryonic development; may occur later in response to infections or cell damage.

aquaporin A membrane protein through which water can move by osmosis.

arbuscular mycorrhizal fungi (AMF) Fungi whose hyphae enter the root cells of their host plants.

Archaea A domain consisting of unicellular prokaryotes distinguished by cell walls made of certain polysaccharides, unique plasma membranes but ribosomes and RNA polymerase similar to those of eukaryotes.

archegonium (plural: archegonia) The egg-producing structure in most land plants except angiosperms.

arteriole Tiny arteries that deliver blood to capillaries. Has muscular walls.

artery Any blood vessel that carries blood from the heart to capillaries. Has thick muscular walls that withstand and control blood pressure.

articulation A movable joint of a skeleton.

artificial selection Deliberate manipulation by humans, as in animal and plant breeding, of the hereditary traits of a population by allowing only certain individuals to reproduce.

asci (singular: ascus) Specialized spore-producing cells found at the ends of hyphae in sac fungi.

ascocarp A large, cup-shaped reproductive structure produced by some ascomycete fungi. Contains many microscopic asci, which produce spores.

asexual reproduction Any form of reproduction that results in offspring that are genetically identical to the parent.

A site The site in a ribosome where an aminoacyl transfer RNA pairs with a messenger RNA codon, in preparation for adding an amino acid to the growing peptide chain.

assimilation hypothesis The hypothesis that modern humans (*Homo sapiens*) evolved in Africa but acquired modern traits after interbreeding with other *Homo* species in Europe and Asia.

asthma A chronic lung condition characterized by periodic episodes of difficulty in breathing, due to smooth-muscle contractions in the respiratory passages.

astrobiology The study of biological molecules that occur in space (e.g., on meteors) and of possible extraterrestrial life.

asymmetric competition Ecological competition between two species in which one species suffers a much greater fitness decline than the other.

atherosclerosis A disease in which lipids accumulate on artery walls, eventually resulting in blockage of the artery.

atom The smallest unit of a chemical element that retains the characteristics of that element.

atomic mass unit (amu) A unit of mass equal to 1/12 the mass of one carbon-12 atom; about the mass of 1 proton or 1 neutron. Also called a dalton.

atomic number The number of protons in the nucleus of an atom, giving the atom its identity as a certain chemical element.

ATP See adenosine triphosphate (ATP).

ATP synthase A membrane-bound enzyme in chloroplasts and mitochondria that uses the energy of protons flowing through it to synthesize ATP.

atrial natriuretic hormone A hormone from the heart that stimulates the kidney to excrete sodium, lowering blood pressure.

atrioventricular (AV) node A location in the heart between the atria and ventricles where electrical signals from the atria are slowed before spreading to the ventricles, allowing them to fill with blood before contracting.

atrium (plural: atria) A thin-walled chamber of the heart that receives blood from veins and pumps it to a neighboring chamber (the ventricle).

attenuated virus A virus that is functional but that has been rendered nonvirulent for a certain species, usually by culturing in cells of a different species. Used for vaccines.

attenuation Gradual reduction in strength. In bacterial genetics, fine-tuning of gene expression such that genes produce only as much of their proteins as required.

australopithecines Species of early hominins that appear in the fossil record shortly after the split from chimpanzees. Generally bipedal, but retained a chimpanzee-size brain.

autocrine signal A chemical signal that affects the same cell that released it.

autoimmunity A pathological condition in which the immune system attacks part of the body.

autonomic nervous system The part of the peripheral nervous system that controls internal organs and involuntary processes, such as stomach contraction, hormone release, and heart rate. Includes parasympathetic and sympathetic nerves.

autophagy The process by which damaged organelles are surrounded by a membrane and delivered to a lysosome to be destroyed.

autopolyploid Polyploid due to a mutation that doubled the chromosome number.

autosomal inheritance The inheritance patterns that occur when genes are located on autosomes rather than on sex chromosomes.

autosome One of any pair of chromosomes that do not carry the gene(s) that determine gender.

autotroph Any organism that can synthesize its own food—complex organic compounds—from simple inorganic sources such as CO_2 or CH_4, including most plants and some bacteria and archaea.

auxin Indoleacetic acid, a plant hormone that stimulates phototropism and some other responses.

avirulence (*avr*) loci Genes in pathogens that trigger a defense response in plants.

axon A long projection of a neuron that can propagate an action potential.

axoneme An arrangement of two central microtubules surrounded by nine doublet microtubules. Found in eukaryotic cilia and flagella and responsible for their motion.

axon hillock The location at which an axon joins the cell body of a neuron; the site at which action potentials are first triggered.

BAC (bacterial artificial chromosome) A lab-created loop of DNA that can be inserted into a living bacterial cell, which can then be grown in a cell culture to create many copies of the BAC.

background extinction The average rate of extinction that occurs normally, as opposed to the higher rate of extinction that occurs during mass extinction events.

BAC library A set of bacterial colonies containing BACs (bacterial artificial chromosomes) that each have a section of the DNA being studied. Commonly used in genome sequencing.

Bacteria A taxonomic domain that includes all bacteria—prokaryotes that have cell walls made of peptidoglycan, distinct forms of ribosomes and RNA polymerase, and certain other traits.

bacteriophage Any virus that infects bacteria.

baculum A bone inside the penis. Usually present in mammals that do not have erectile tissue.

bark The protective outer layer around woody plants, containing cork cells, cork cambium, and secondary phloem.

baroreceptors Specialized nerve cells in the walls of the heart and certain major arteries that detect changes in blood pressure and trigger appropriate responses by the brain.

basal body A structure of nine pairs of microtubules arranged in a circle at the base of eukaryotic cilia and flagella where they attach to the cell.

basal cell The lower cell produced by an angiosperm zygote. Gives rise to a structure that transfers nutrients from the parent plant to the developing embryo.

basal lamina A thick, collagen-rich extracellular matrix found in animal skin.

basal meristem A group of undifferentiated plant cells located below ground, from which the plant can regenerate the shoot system if necessary.

basal metabolic rate (BMR) The total energy consumption by an organism at rest in a comfortable environment. For aerobes, often measured as the amount of oxygen consumed per hour.

basal transcription complex A multi-protein structure that initiates transcription of eukaryotic genes.

basal transcription factors Proteins that bind to eukaryotic promoters and help initiate transcription.

base Any compound that acquires protons or gives up electrons during a chemical reaction or accepts hydrogen ions when dissolved in water.

baseline A measurement that represents normal values before a change or disturbance and that serves as a reference point for future comparisons.

basidia (singular: basidium) Specialized spore-producing cells at the ends of hyphae in club fungi.

basilar membrane The membrane in the vertebrate cochlea on which the hair cells sit.

basolateral Toward the bottom and sides. In animals, the side of an epithelial layer that faces other body tissues and not the environment.

basophil A circulating leukocyte involved in the inflammatory response and in allergies.

Batesian mimicry A type of mimicry in which a harmless species resembles a harmful species.

B cells A class of leukocytes that can produce antibodies. Produced by the bursa of birds and the bone marrow of mammals. Also called B lymphocytes.

B-cell receptor (BCR) A Y-shaped immunoglobulin protein found on the surfaces of B cells and to which antigens bind.

behavior Any action by an organism.

benign tumor A mass of abnormal tissue that grows slowly or not at all, does not disrupt surrounding tissues, and does not metastasize to other organs. Benign tumors are not cancers.

benthic Living at the bottom of an aquatic environment.

benthic zone The area along the bottom of an aquatic environment.

beta chain (β chain) One of the two polypeptide chains that make up a T-cell receptor protein, enabling the T cells of the immune system to bind to antigens on other cells.

bilateral symmetry An animal body pattern in which there is one plane of symmetry dividing the body into a left side and a right side. Typically, the body is long and narrow, with a distinct head end and tail end.

Bilateria A major lineage of animals that are bilaterally symmetrical at some point in their life cycle, have three embryonic germ layers, and have coeloms. Includes protostomes and deuterostomes.

bile A fluid produced by the liver, stored in the gall bladder, and secreted into the intestine, where it emulsifies fats during digestion.

bile salts Small lipids in bile that can emulsify fats.

binary fission Division of a bacterial cell to produce two daughter cells. Similar to mitosis of eukaryotic cells.

binomial nomenclature A system of naming species by using two-part Latinized names with a genus name and a species name. Always italicized, with genus name capitalized.

biodiversity The variety and relative abundance of species present in a certain area.

biofilm A tough, polysaccharide-rich substance secreted by certain prokaryotes. Encases and protects cells and attaches them to the organism's surface.

biogeochemical cycle The pattern of circulation of an element or molecule among living organisms and the environment.

biogeography The study of how species and populations are distributed geographically.

bioinformatics The field of study concerned with managing, analyzing, and interpreting biological information, particularly DNA sequences.

biological species concept The concept that species are best identified as groups that are reproductively isolated from each other; thus, different species cannot crossbreed in nature to produce viable and fertile hybrid offspring.

bioluminescence The emission of light by a living organism.

biomass The total mass of all organisms in a given population or geographical area; usually expressed as total dry weight.

biome A major category of ecosystem, characterized by a distinct type of vegetation and climate.

bioprospecting The search for naturally occurring compounds that can be useful to humans (e.g., as drugs, fragrances, or insecticides). Also called chemical prospecting.

bioremediation The use of living organisms, usually bacteria and archaea, to degrade environmental pollutants.

biotechnology The application of biological techniques and discoveries to medicine, industry, and agriculture.

biotic Living, or produced by a living organism.

bipedal Walking primarily on two legs.

bipolar cell Cells in the vertebrate retina that receive information from one or more photoreceptors and pass it to other bipolar cells and ganglion cells.

bivalves A lineage of mollusks that have two shells, such as clams and mussels.

bladder A mammalian organ that holds urine until it can be excreted.

blade The wide, flat part of a plant leaf.

blastomere The small cells created by cleavage divisions in early animal embryos.

blastopore A small pore in the surface of an early vertebrate embryo, through which cells move during gastrulation.

blastula An early stage of embryonic development in vertebrates, consisting of a ball of cells (a blastomere) enclosing a fluid-filled space. Immediately precedes gastrulation.

blood A type of liquid connective tissue consisting of red blood cells and leukocytes suspended in the fluid plasma.

blood doping The practice of giving an athlete a transfusion of his or her own red blood cells (removed and stored earlier) to increase aerobic capacity for a competition.

body mass index A mathematical relationship of weight and height used to assess obesity in humans. Calculated as weight (kg) divided by the square of height (m^2).

body plan The basic architecture of an animal's body, including the number and arrangement of limbs, body segments, and major tissue layers.

bog A wetland that has no or almost no water flow, resulting in very low oxygen levels and acidic conditions.

Bohr shift The rightward shift of the oxygen-hemoglobin dissociation curve that occurs with decreasing pH. Results in hemoglobin being more likely to release oxygen in the acidic environment of exercising muscle.

bone A vertebrate tissue consisting of living cells and blood vessels within a hard extracellular matrix of calcium phosphate ($CaPO_4$) with small amounts of calcium carbonate ($CaCO_3$) and protein fibers.

bone marrow The soft tissue filling the inside of long bones. Produces red blood cells and leukocytes and also contains fat.

Bowman's capsule The hollow, sphere-like end of a kidney nephron that surrounds a glomerulus.

braincase A bony, cartilaginous, or fibrous case that encloses and protects the brain of vertebrates. Forms part of the skull. Also called the cranium.

branch (1) A part of a phylogenetic tree that represents populations. (2) Any extension of a plant's shoot system.

bronchi (singular: bronchus) In mammals, the large tubes that lead from the trachea to each lung.

bronchioles The small tubes in mammalian lungs that carry air from the bronchi to the alveoli.

brown adipose tissue A specialized form of fat tissue whose cells have a high density of mitochondria as well as stored fats and that can produce extra body heat. Found in some mammals.

bryophytes Several phyla of green plants that lack vascular tissue. Includes liverworts, hornworts, and mosses. Also called nonvascular plants.

budding Asexual reproduction via growth of a small individual from part of the parent, eventually breaking free as an independent individual.

buffer A substance that, in solution, acts to minimize changes in the pH of that solution.

bulbourethral glands Small, paired glands at the base of the urethra in male mammals that secrete pre-ejaculatory fluid for lubrication during copulation. In humans, also called Cowper's glands.

bulk flow The directional movement of a substantial volume of fluid due to pressure differences, such as movement of water through plant phloem and movement of blood in animals.

bundle-sheath cell A type of cell that is located around the vascular tissue in the interior of plant leaves.

Burgess shale fauna A characteristic set of soft-bodied animal fossils found in several early Cambrian rock formations (525–515 million years old), particularly the Burgess Shale in Canada and the Chengjiang deposits in China.

bursa An immune system organ of birds that produces B cells.

C_3 photosynthesis A common form of photosynthesis in which atmospheric carbon is used to form 3-phosphoglycerate, a three-carbon compound.

C_4 photosynthesis A variant of photosynthesis in which carbon from CO_2 is first fixed into four-carbon compounds. Enhances photosynthetic efficiency in hot, dry environments by reducing loss of oxygen due to photorespiration.

cadherins Cell-surface proteins involved in cell adhesion and important for coordinating movements of cells during embryological development.

calcitonin A hormone from the thyroid gland that lowers blood calcium by preventing calcium and phosphorus withdrawal from bone.

callus A mass of undifferentiated plant cells that can generate roots and other tissues necessary to create a mature plant.

Calorie A unit of energy, often used to measure energy content of food. Also called a kilocalorie.

Calvin cycle In photosynthesis, the set of reactions that use the NADPH and ATP formed earlier (in the light-dependent reactions) to drive the reduction of atmospheric CO_2, ultimately producing sugars. Also called light-independent reactions.

CAM See **crassulacean acid metabolism (CAM)**.

cambium (plural: cambia) A layer of undifferentiated plant cells found in woody plants. Responsible for secondary growth. Also called secondary meristem or lateral meristem.

Cambrian explosion The rapid diversification of animal body types that began about 543 million years ago, during approximately 40 million years.

camera eye A type of eye in vertebrates and cephalopods, consisting of a hollow chamber with a hole at one end (through which light enters) and a sheet of light-sensitive cells against the opposite wall.

cAMP See **cyclic AMP (cAMP)**.

canopy The uppermost layers of branches in a forest (i.e., those fully exposed to the Sun).

5′ cap A structure added to the 5′ end of newly transcribed messenger RNA molecules. Consists of 7-methylguanylate and three phosphate groups.

CAP binding site A DNA sequence upstream of certain prokaryotic operons, to which catabolite activator protein can bind, increasing gene transcription.

capillarity The tendency of water to move up a narrow tube due to surface tension, adhesion, and cohesion.

capillary The smallest blood vessel, where gases and other molecules are exchanged between blood and tissues.

capillary bed A thick network of capillaries.

capsid A shell of protein enclosing the genome of a virus particle.

carapace In crustaceans, a large platelike section of the exoskeleton that covers and protects the cephalothorax (e.g., a crab's "shell").

carbohydrates A class of molecules, including sugars, starches, glycogen, and cellulose, that have a carbonyl group, several hydroxyl groups, and several to many carbon-hydrogen bonds.

carbonic anhydrase An enzyme that catalyzes the formation of carbonic acid (H_2CO_3) from carbon dioxide and water.

carbon sink A reservoir that holds carbon for very long time periods; e.g., fossil fuels, sedimentary rocks.

carcinogen Any cancer-causing agent.

cardiac cycle One complete heartbeat cycle, including systole and diastole.

cardiac muscle The muscle tissue of the vertebrate heart. Consists of long branched fibers that are electrically connected and that initiate their own contractions; not under voluntary control.

cardiac output The total volume of blood leaving the left ventricle per minute. Calculated as stroke volume times heart rate.

carnivore (1) Any animal whose diet consists predominantly of other animals. (2) Any member of the mammalian taxon Carnivora. (Most members of the Carnivora are meat eaters.)

carotenoids A class of plant pigments that absorb wavelengths not absorbed by chlorophyll. Typically appear yellow, orange, or red. Includes carotenes and xanthophylls.

carpel In a flower, the reproductive structure that produces female gametophytes; consists of the stigma, the style, and the ovary.

carrier (1) A heterozygous individual carrying a normal allele and a recessive allele for an inherited condition; does not display the phenotype of the condition but can pass the recessive gene to offspring. (2) A transmembrane protein that assists with facilitated diffusion by binding a specific ion or molecule and transporting it across the membrane.

carrying capacity The maximum population size of a certain species that a given habitat can support. Symbolized by K.

cartilage A type of connective tissue in the skeletons of vertebrates, consisting of relatively few cells scattered in a stiff matrix of polysaccharides and protein fibers.

Casparian strip A waxy layer that prevents movement of water through the walls in plant roots.

catabolic pathway Any set of chemical reactions that breaks down larger, complex molecules into smaller ones, releasing energy in the process.

catabolite The end product of a catabolic pathway; the final breakdown product of a substance.

catabolite activitor protein (CAP) A protein that can bind to the CAP binding site upstream of certain prokaryotic operons, facilitating binding of RNA polymerase and stimulating gene expression.

catabolite repression A type of inhibition of gene transcription in which a gene codes for an enzyme in a catabolic pathway and the end product of that pathway inhibits further transcription of that gene.

catalysis Acceleration of the rate of a chemical reaction, produced by lowering the potential energy of the transition state.

catalyst Any substance that increases the rate of a chemical reaction without itself undergoing any permanent chemical change. Functions by lowering the activation energy of the reaction.

catecholamines A class of small compounds, derived from the amino acid tyrosine, that are used as hormones or neurotransmitters. Includes epinephrine, norepinephrine, and dopamine.

cation A positively charged ion.

cation exchange The release of cations such as magnesium and calcium from soil particles, due to displacement by protons in acidic soil water.

CD4 A membrane protein on the surface of some T lymphocytes of the human immune system; these CD4+ T cells can give rise to helper T cells.

CD8 A membrane protein on the surface of some T lymphocytes of the human immune system; these CD8+ T cells can give rise to cytotoxic T cells.

Cdk See **cyclin-dependent kinase (Cdk)**.

cDNA (complementary DNA) DNA created in a lab from an RNA transcript, using the enzyme reverse transcriptase. Corresponds to a certain gene but lacks introns. Is also created naturally by some viruses (retroviruses).

cDNA library A set of cDNAs from a particular organism or cell type. Usually exists as a set of bacteria colonies, each with a plasmid containing a particular cDNA sequence.

cecum A blind sac between the small intestine and the colon. Used in some species as a fermentation vat for digestion of cellulose.

cell A highly organized living entity that is bounded by a thin, flexible structure called a plasma membrane and that contains concentrated chemicals in an aqueous (watery) solution.

cell body The part of a neuron that contains the nucleus and where incoming signals are integated. Also called the soma.

cell-cell signal A molecule that is secreted from one cell and affects the activity of a nearby cell.

cell crawling A form of cellular movement in which the cell produces bulges (pseudopodia) that stick to the substrate and pull the cell forward.

cell cycle The sequence of stages that a dividing eukaryotic cell goes through from the time it is created (by division of a parent cell) to the time it undergoes mitosis.

cell-cycle checkpoint A regulated point at which the cell cycle can be stopped.

cell division Creation of new cells by division of pre-existing cells (i.e., by mitosis or meiosis).

cell enlargement Growth in plants in which individual cells expand in size in a certain direction.

cell extract or cell homogenate A solution created by breaking cells apart. Cell extracts contain organelles, free macromolecules, and small membrane-bound vesicles.

cell-mediated response Defense mounted by cytotoxic T lymphocytes against infections.

cell plate A double layer of new plasma membrane that appears in the middle of a dividing plant cell; ultimately divides the cytoplasm into two separate cells.

cell sap An aqueous solution found in the vacuoles of plant cells.

cell theory The theory that all organisms are made of cells and that all cells come from pre-existing cells.

cellular respiration A common pathway for production of ATP, involving transfer of electrons from compounds with high potential energy (often NADH and $FADH_2$) to an electron transport chain and ultimately to an electron acceptor (often oxygen).

cellulases Enzymes that can digest cellulose.

cellulose A polysaccharide that is the major component of plant cell walls. Consists of β-glucose monomers joined with β-1,4-glycosidic linkages in long, straight chains.

cell wall A protective layer located outside the plasma membrane and usually composed of polysaccharides. Found in algae, plants, bacteria, fungi, and some other groups.

central dogma The long-accepted hypothesis that information in cells flows in one direction: DNA codes for RNA, which codes for proteins. Exceptions are now known (e.g., retroviruses).

central nervous system (CNS) The brain and spinal cord of vertebrate animals.

centrifugation A lab technique for separating substances by density and size by spinning them rapidly in a circle, so that larger or denser particles are flung to the outside of the sample container.

centriole One of two small cylindrical structures found together near the nucleus of a eukaryotic cell. Collectively called the centrosome, they serve as a microtubule hub for the cell's cytoskeleton.

centromere The structure that joins two sister chromatids during meiosis.

centrosome Structure in animal and fungal cells, consisting of two centrioles together near the nucleus. Serves as a microtubule organizing center for the cell's cytoskeleton and for the mitotic spindle during cell division.

cephalization The formation of a distinct head—an anterior region where sense organs and a mouth are clustered.

cephalochordates A lineage of small, mobile chordates also called lancelets or amphioxi.

cephalopods A lineage of mollusks including the squid, octopuses, and nautiluses. Distinguished by large brains, excellent vision, tentacles, and a reduced or absent shell.

cerebellum A posterior section of the vertebrate brain, involved in coordination of complex muscle movements, such as in locomotion and maintaining balance.

cerebrum The anteriormost section of the vertebrate brain. Divided into left and right hemispheres and involved in memory, interpretation of information, decision making, and (in humans) conscious thought.

cervix The narrow passageway between the vagina and the uterus of female mammals.

channel protein, or channel A membrane protein that forms a pore that selectively allows a specific ion or molecule to cross the membrane via diffusion. Channel proteins provide passive transport (facilitated diffusion) and do not require energy.

chelicerae Appendages found around the mouth of the chelicerates (spiders, mites, and allies).

chemical bond A strong attractive force binding two atoms together. Covalent bonds, ionic bonds, and hydrogen bonds are types of chemical bonds.

chemical carcinogen Any chemical that can cause cancer.

chemical energy The potential energy stored in covalent bonds between atoms.

chemical equilibrium A dynamic but stable state of a reversible chemical reaction in which the forward reaction and reverse reactions proceed at the same rate, so that the concentrations of reactants and products remain constant.

chemical evolution The hypothesis that simple chemical compounds in the ancient atmosphere and ocean combined by spontaneous chemical reactions to form larger, more complex substances, eventually leading to the origin of life and the start of biological evolution.

chemical reaction An event in which one compound or element is combined with others or is broken down.

chemiosmosis The production of ATP via proton movement, through ATP synthase, across a membrane, driven by a proton gradient.

chemiosmotic hypothesis The hypothesis that ATP synthesis in mitochondria and chloroplasts occurs indirectly via proton movement across a membrane.

chemokine A chemical signal that attracts leukocytes to a site of tissue injury or infection.

chemoreceptor A sensory cell or organ specialized for detection of specific molecules or classes of molecules.

chemotaxis Movement toward or away from a certain chemical.

chemotherapy Treatment with anticancer chemicals or drugs. Most chemotherapy drugs kill dividing cells or stop the cell cycle.

chiasma (plural: chiasmata) The X-shaped structure formed during meiosis by crossing over between adjacent chromatids of a pair of homologous chromosomes.

chitin A polysaccharide consisting of monomers of *N*-acetylglucosamine joined end to end in long, straight chains. Found in cell walls of fungi and many algae, and in external skeletons of insects and crustaceans.

chitons Marine mollusks that have a protective shell formed of eight calcium carbonate plates.

chloride cells In the gills of marine fish with bony skeletons, cells that excrete excess salt and maintain electrolyte balance.

chlorophyll A green pigment molecule that absorbs light energy to power photosynthesis. Found in plant cells and in photosynthetic protists.

chloroplast In plants, a chlorophyll-containing organelle in which photosynthesis occurs. Also, the location of amino-acid, fatty-acid, purine, and pyrimidine synthesis.

chloroplast DNA DNA found within a chloroplast of a eukaryotic plant cell. Consists of a single circular chromosome.

choanocytes The specialized feeding cells found in sponges.

choanoflagellates A phylum of protists thought to be the closest living relatives of animals.

cholecystokinin A peptide hormone from the small intestine that stimulates the secretion of digestive enzymes from the pancreas and bile from the liver and gallbladder.

cholera A human infectious disease characterized by watery diarrhea that can lead to dehydration and death. Caused by the bacterium *Vibrio cholerae*.

cholesterol A steroid that is a major component of plasma membranes. Required for synthesis of steroid hormones. In excess, can contribute to atherosclerosis.

chordate An animal of the phylum Chordata, distinguished by such traits as a dorsal hollow nerve cord, pharyngeal gill slits, and a post-anal tail. Includes vertebrates.

chorion In an amniotic egg, a highly vascularized membrane across which gas exchange occurs with the environment.

chromatid One of the daughter strands of a chromosome that has recently been copied (prior to mitosis or meiosis) and that is still connected to the other daughter strand.

chromatin The material that makes up eukaryotic chromosomes: a DNA molecule complexed with histone proteins. Can be highly compact (heterochromatin) or filamentous (euchromatin).

chromatin remodeling The process by which the DNA in chromatin is unwound from its associated proteins to allow transcription or replication.

chromatin-remodeling complex A protein involved in restructuring chromatin (DNA packed with proteins) in eukaryotic cells. The remodeling requires ATP.

chromosome A single long molecule of DNA and any associated proteins (e.g., histones).

chromosome inversion A mutation in which a segment of a chromosome breaks from the rest of the chromosome, flips, and rejoins with the opposite orientation as before.

chromosome painting A technique for producing high-resolution karyotypes by "painting" chromosomes with fluorescent tags that bind to particular regions of certain chromosomes. Also called spectral karyotyping (SKY).

chromosome theory of inheritance The theory that Mendel's rules of inheritance can be explained by the independent segregation of homologous chromosomes during meiosis.

chylomicron A ball of protein-coated lipids, used to transport the lipids through the bloodstream.

cilia (singular: cilium) Short, numerous, filamentous projections of some eukaryotic cells, used to move the cell and/or to move fluid or particles along a stationary cell.

circadian Lasting approximately one day.

circadian clock An internal mechanism found in most organisms that regulates many body processes (sleep-wake cycles, hormonal patterns, etc.) in a roughly 24-hour cycle.

circadian rhythm Any biological process that occurs on an approximately 24-hour cycle and that is controlled by an internal circadian clock.

circulation In physiology, mass movement of blood throughout the body.

cisterna (plural: cisternae) A flattened, membrane-bound compartments of the Golgi apparatus.

clade An evolutionary unit that includes an ancestral population, all of its descendants, and only its descendants. Also called a monophyletic group or a lineage.

cladistic approach A method for constructing a phylogenetic tree that is based on identifying the unique traits of each monophyletic group.

class A classic taxonomic rank above the order level and below the phylum level.

Class I MHC protein A type of MHC protein that is present on the plasma membrane of every body cell, marking the cells as "self" to the immune system.

Class II MHC protein A type of MHC protein, present only on the plasma membranes of macrophages, B cells, and T cells, that helps them join together during B-cell activation.

classical conditioning A type of learning in which an animal learns to associate two stimuli, so that a response originally given to just one stimulus can be evoked by the second stimulus as well.

cleavage Rapid cell division without production of new cytoplasm. Seen only in early embryonic development in animals. Cleavage transforms a zygote into a blastula.

cleavage furrow A pinching-in of the plasma membrane that occurs as the cytoplasm of an animal cell begins to divide; one of the final events of cell division in animals, fungi, and slime molds.

climate The prevailing long-term weather conditions in a particular region.

climax community The stable, final stage of an ecological community that develops from ecological succession.

clitoris A small rod of erectile tissue in the external genitalia of female mammals. Forms from the same embryonic tissue as the male penis and has a similar function in sexual arousal.

clonal expansion Rapid cell division by a particular T cell or B cell of the immune system in response to a particular antigen. Produces a large population of descendant cells that can attack the antigen.

clonal-selection theory The dominant theory of the development of acquired immunity in vertebrates, proposing that the immune system retains a vast pool of inactive lymphocytes, each with a unique receptor for a unique antigen. Lymphocytes that encounter their antigens are stimulated to divide (selected and cloned), producing daughter cells that combat infection and confer immunity.

clone An individual that is genetically identical to another individual.

closed circulatory system A circulatory system in which the circulating fluid (blood) is confined to blood vessels and flows in a continuous circuit.

clot A mass of red blood cells, platelets, and protein fibers that forms to plug a hole in the circulatory system and stop blood loss.

clutch size The number of eggs laid by a female in a single nest or a single breeding effort.

cnidocyte A specialized stinging cell found in cnidarians.

coactivators A class of regulatory proteins that help initiate transcription by bringing together the necessary transcription factor proteins.

cochlea A coiled, fluid-filled tube in the inner ear of mammals, birds, and crocodilians. Contains nerve cells that detect sounds of different pitches.

codominance An inheritance pattern in which heterozygotes exhibit both of the traits seen in either kind of homozygous individual.

codon A sequence of three nucleotides of DNA or RNA that codes for a certain amino acid or that initiates or terminates protein synthesis.

coefficient of relatedness A measurement of how closely two individuals are related. Calculated as the probability that an allele in two individuals is inherited from the same ancestor. Symbolized as *r*.

coelom An internal, usually fluid-filled, body cavity that forms within the mesoderm.

coelomate Possessing a coelom.

coenocytic Containing many nuclei and a continuous cytoplasm through a filamentous body, without the body being divided into distinct cells. Some fungi are coenocytic.

coenzyme Any non-protein molecule or ion that is a required cofactor for an enzyme-catalyzed reaction. Often transfers or receives electrons or functional groups.

coenzyme A (CoA) A coenzyme that is required for many cellular reactions involving transfer of acetyl groups ($-COCH_3$). In glycolysis, pyruvate reacts with CoA to produce acetyl CoA, which can then enter the Krebs cycle.

coenzyme Q A molecule that transfers electrons in the electron transport chain of cellular respiration. Also called ubiquinone or Q.

coevolutionary arms race A pattern of evolution in which one species evolves a defense against another species (e.g., a predator or parasite), which then evolves a counterdefense, and so on.

cofactor A non-protein molecule or ion that is required for an enzyme to function normally. May be a metal ion or a coenzyme.

cognition The mental processes involved in recognition and manipulation of facts about the world, particularly to form novel associations or insights.

cohesion A phenomenon seen in some liquids, such as water, in which attractive forces between molecules cause the liquid molecules to cling together and resist disruption.

cohesion-tension theory The theory that water movement upward through plant vascular tissues is due to transpiration (loss of water from leaves), which pulls a cohesive column of water upward.

cohort A group of individuals that are the same age and can be followed through time.

coleoptile A modified leaf that covers and protects the stems and leaves of young grasses.

collagen A fibrous, cable-like protein secreted by animal cells into the extracellular matrix. Forms a pliable, strong substance in which the cells sit.

collenchyma cell An elongated type of plant cell with cell walls thickened at the corners that provides support to growing plant parts; usually found in strands along leaf veins and stalks.

colonial growth Growth in which unicellular or multicellular individuals join to form a structure in which the multiple individuals are not specialized and each retains the ability to reproduce.

colony An assemblage of individuals. May refer to an assemblage of semi-independent cells or to a breeding population of multicellular organisms.

combination therapy Medical therapy that involves dosing an infected patient with several drugs simultaneously, to lessen the chances of the pathogen evolving resistance.

commensal A species that is dependent on another species but does not harm that species.

commensalism A species interaction in which one species benefits and the other is not harmed.

communication Any process in which a signal from one individual modifies the behavior of another individual.

community All of the species that interact with each other in a certain area.

companion cell Cell in phloem tissue that provides materials and nutrients to sieve-tube members.

compass orientation Motion in a specific compass direction.

competition The effect of two species or two individuals trying to use the same limited resource.

competitive exclusion principle The principle that two species cannot coexist in the same ecological niche in the same area because one species will out-compete the other.

competitive inhibition Inhibition of an enzyme's ability to catalyze a chemical reaction via a non-reactant molecule that competes with the substrate(s) for access to the active site.

complementary base pairing The specific pairing that occurs between nitrogenous bases of nucleic acids: Adenine pairs only with thymine (in DNA) or uracil (in RNA), and guanine pairs only with cytosine. Allows accurate replication of DNA and RNA sequences.

complementary DNA (cDNA) DNA created in the lab from an RNA transcript, using reverse transcriptase; corresponds to a particular gene but lacks introns. Created naturally by retroviruses.

complementary strand A new strand of RNA or DNA that has a base sequence complementary to that of the template strand.

complement system A class of proteins that circulate in the bloodstream and attack plasma membranes of bacteria.

compound eye An arthropod eye formed of many independent light-sensing columns (ommatidia).

concentration gradient Variation across space in the concentration of a dissolved substance, from a region of high concentration to a region of low concentration.

condensation reaction A type of chemical reaction involving the bonding together of two molecules by removal of an –OH from one and an –H from another to form water. Also called a dehydration reaction. The reverse of hydrolysis.

conditional strategy A behavioral response that varies with the current environmental and social conditions.

conduction Direct transfer of heat between two objects that are in physical contact.

cone (1) A photoreceptor cell with a cone-shaped outer portion that is particularly sensitive to bright light of a certain color. Found in eyes of vertebrates and some other animals. (2) The reproductive structure found in conifers.

confocal microscopy A technique for obtaining a focused image of a certain plane within a live cell.

conformational homeostasis Homeostasis (steady internal body conditions) that is achieved by the body's passively matching the conditions of a stable external environment.

conjugation The process by which DNA is exchanged between unicellular individuals. Occurs in bacteria, archaea, and some protists.

conjugation tube A connection between two prokaryotes that are in the process of transferring plasmids to each other.

connective tissue A class of animal tissue consisting of scattered cells in a liquid, jellylike, or solid extracellular matrix. Includes bone, cartilage, tendons, ligaments, and blood.

conservation biology The effort to study, preserve, and restore threated populations, communities, and ecosystems.

conservative replication A now-disproven hypothesis of DNA replication that proposed that one of the daughter molecules retains both original strands of DNA, while the other daughter molecule is built with two new strands.

constant (C) region A section of the light chains of antibodies that has the same amino acid sequence in every B cell of an individual.

constitutive Always occurring; always present (as in enzymes that are synthesized continuously).

constitutive defense A defensive trait that is always present, regardless of need. Also called standing defense.

constitutive mutant A mutant in which certain genetic loci are constantly transcribed due to flaws in gene regulation.

consumer An organism that consumes food created by other organisms; not a primary producer.

continental drift The motion of continents over large periods of time due to plate tectonics.

continental shelf The shallow, gently sloping portion of the ocean floor near continents.

continuous strand In DNA replication, the strand of new DNA synthesized in one continuous piece, with nucleotides added to the 3′ end of the growing molecule. Also called leading strand.

contraception Any method to prevent pregnancy.

control In an experiment, a group of organisms or samples that do not receive the experimental treatment but are otherwise identical to the group that does.

convection Transfer of heat by movement of large volumes of a gas or liquid.

convergent evolution Evolution of similar traits in distantly related organisms due to adaptation to similar environments and a similar way of life. Often produces analogous traits.

convergent trait Similarity between different species that is due to adaptation to a similar way of life rather than inheritance from a common ancestor. Often the result of convergent evolution. Also called an analogous trait.

cooperative binding The tendency of the protein subunits of hemoglobin to affect each other's oxygen binding such that each bound oxygen molecule increases the likelihood of further oxygen binding.

copulation The act of transferring sperm from a male directly into a female's reproductive tract.

coral reef A large assemblage of colonial marine corals that usually serves as shallow-water, sunlit habitat for many other species as well.

co-receptor Any membrane protein that acts with some other membrane protein in a cell interaction or cell response.

core enzyme The part of a holoenzyme that contains the active site for catalysis.

cork cambium A ring of undifferentiated plant cells found just under the cork layer of woody plants. Produces new cork cells on its outer side.

cork cells The waxy cells in the protective outermost layer of a woody plant.

corm A rounded, thick underground stem that can produce new plants via asexual reproduction.

cornea The transparent sheet of connective tissue at the very front of the eye in vertebrates and some other animals. Protects the eye and helps focus light.

corolla All of the petals of a flower.

corona The cluster of cilia at the anterior end of a rotifer.

coronary heart disease Progressive weakening of the heart muscle due to chronic oxygen deprivation caused by blocked coronary arteries.

corpus callosum A thick band of neurons that connects the two cerebral hemispheres of the mammalian brain.

corpus luteum A yellowish structure in an ovary, formed from a follicle that has recently ovulated. Secretes progesterone.

cortex (1) The outermost region of an organ, such as the kidney or adrenal gland. (2) In plants, a layer of ground tissue found outside the vascular bundles and pith of a plant stem.

cortical granules Small enzyme-filled vesicles in the cortex of an egg cell. Involved in formation of the fertilization envelope after fertilization.

corticosterone The major glucocorticoid hormone released by the cortex of the adrenal gland in most reptiles, birds, and many mammals. Increases blood glucose and prepares the body for stress.

corticotropin-releasing hormone (CRH) A peptide hormone from the hypothalamus that stimulates the anterior pituitary gland to release ACTH.

cortisol The major glucocorticoid hormone released by the cortex of the adrenal gland in some mammals. Increases blood glucose and prepares the body for stress. Also called hydrocortisone.

cotransporter A membrane protein that allows an ion to diffuse down a previously established concentration gradient and uses the energy of that process to transport some other substance in the same direction. Also called a symporter.

cotyledon The first leaf, or seed leaf, of a plant embryo. Used for storing and digesting nutrients and/or for early photosynthesis.

countercurrent exchange A particularly efficient mechanism for the exchange of heat or a soluble substance, based on transfer between parallel tubes carrying fluids in opposite directions.

countercurrent heat exchanger A specialized network of blood vessels that recirculates body heat within a certain part of the body. A type of countercurrent exchanger.

covalent bond A type of molecular bond in which two atoms share one pair of electrons.

cranium A bony, cartilaginous, or fibrous case that encloses and protects the brain of vertebrates. Forms part of the skull. Also called braincase.

crassulacean acid metabolism (CAM) A variant of photosynthesis in which CO_2 is stored in organic compounds at night when stomata are open and released to enter the Calvin cycle during the day when stomata are closed. Helps reduce loss of water and oxygen by photorespiration in hot, dry environments.

crista (plural: cristae) Membranous sacs that articulate with the inner membrane of a mitochondrion. Location of the electron transport chain and ATP synthase.

critical period A short time span in a young animal's life during which the animal can learn certain things, such as song, language, or imprinting. Also called the sensitive period.

Cro-Magnon A prehistoric European population of modern humans (*Homo sapiens*) known from fossils, paintings, sculptures, and other artifacts.

crossing over The exchange of segments of non-sister chromatids between a pair of homologous chromosomes that occurs during meiosis.

cross-pollination Pollination of a flower by pollen from another individual, rather than by self-fertilization.

cryptochromes A class of plant photoreceptors that detect blue light and affect stem growth and flowering in shady conditions.

cud A partially digested package of food and symbiotic bacteria from the rumen that ruminants regurgitate for further chewing.

culture A collection of cells growing under controlled conditions in a lab, usually in suspension or on the surface of a dish on solid growth medium.

cup fungi A monophyletic lineage of fungi that produce large, often cup-shaped reproductive structures that contain spore-producing asci. Also called sac fungi.

current A flow of electrical charge past a point. Also called electric current.

Cushing's disease A human endocrine disorder caused by loss of feedback inhibition of cortisol on ACTH secretion. Characterized by high ACTH and cortisol levels and wasting of body protein reserves.

cuticle A protective coating secreted by the outermost layer of cells of an animal or a plant.

cyanobacteria A lineage of photosynthetic bacteria formerly known as blue-green algae. Likely the first life-forms to evolve by oxygenic photosynthesis.

cyclic AMP (cAMP) Cyclic adenosine monophosphate; a small molecule, derived from ATP, that is widely used in cells for signaling and regulation (e.g., in gene transcription, enzyme control, and hormone signal transduction).

cyclic photophosphorylation An alternative pathway in the light-dependent reactions of photosynthesis, in which excited electrons from photosystem I are transferred back to the electron transport chain of photosystem II to increase ATP generation.

cyclin-dependent kinase (Cdk) A protein kinase that is active only when bound to a cyclin. Involved in control of the cell cycle.

cyclins A class of proteins whose concentrations fluctuate cyclically, following the cell cycle.

cyst In some species, a protective structure containing a diploid cell in a resting state.

cystic fibrosis A human disease caused by a defective chloride channel. Causes thickened mucus in the respiratory tract and deterioration of the gastrointestinal and reproductive tracts.

cytochrome C A protein that helps transfer electrons between the parts of the electron transport chain in mitochondria.

cytokine Generally, any substance that stimulates cell division. Many cytokines are secreted by macrophages and helper T cells during an immune response, stimulating leukocyte production, tissue repair, and fever.

cytokinesis Division of the cytoplasm to form two daughter cells. Typically occurs immediately after division of the nucleus by mitosis or meiosis.

cytokinins A group of plant hormones that stimulate cell division.

cytoplasm All of the contents of a cell, excluding the nucleus.

cytoplasmic determinant A regulatory molecule that is distributed unevenly in the cytoplasm of the egg cells of many animals and that directs the differentiation of embryonic cells.

cytoplasmic streaming The directed flow of cytosol and organelles around the interior of a plant or fungal cell. Occurs along actin filaments and is powered by myosin.

cytoskeleton A network of protein fibers embedded in the cytoplasm of eukaryotic cells. Involved in cell shape, support, locomotion, and transport of materials within the cell. Includes microtubules, intermediate filaments, and actin filaments.

cytosol The fluid portion of the cytoplasm in a cell (i.e., not including the organelles).

cytotoxic T lymphocyte (CTL) A T cell that destroys infected cells and cancer cells. Cytotoxic T lymphocytes are descendants of an activated $CD8^+$ T cell. Also called killer T cell.

Darwinian fitness The ability of an organism to produce surviving fertile offspring. Also called fitness.

day-neutral plant A plant whose flowering time is not affected by photoperiod (relative length of day and night).

dead space Portions of the air passages that are not involved in gas exchange with the blood, such as the trachea and bronchi.

deciduous Shedding leaves annually.

decomposer A species that feeds on the dead bodies of other organisms. Decomposers include various bacteria, fungi, and protists.

decomposer food web An ecological network of detritus, decomposers that eat detritus, and predators and parasites of the decomposers.

definitive host The host species in which a parasite reproduces sexually.

dehydration reaction A type of chemical reaction involving the bonding of two molecules by removal of an –OH from one subunit and an –H from another to form water. Also called a condensation reaction. The reverse reaction of hydrolysis.

deleterious allele Any allele that reduces an individual's fitness.

deleterious mutation Any mutation that reduces an individual's Darwinian fitness.

demography The study of factors that determine the size and structure of populations through time.

denatured For a protein's three-dimensional structure, unfolded; usually due to breakage of hydrogen bonds and disulfide bonds.

dendrite A short extension from a neuron's cell body that receives neurotransmitters from other neurons.

dendritic cell A type of leukocyte that ingests foreign antigens, moves to a lymph node, and presents the antigens on its membrane to other immune system cells.

density-dependent factors Factors that affect birth rates or death rates differently, depending on population size.

density-dependent growth Population growth that is limited by increasing population size.

density-independent factors Factors that change birthrates and death rates irrespective of population size.

density-independent growth Population growth that is not affected by the population size.

dental plaque A biofilm of bacteria that can form on mammalian teeth.

deoxynucleoside triphosphate (dNTP) A monomer that can be polymerized to form DNA. Consists of deoxyribose, a base (A, T, G, or C), and three phosphate groups; similar to a nucleotide, but with two more phosphate groups.

deoxyribonucleic acid (DNA) A polymer consisting of two strings of deoxyribonucleotides, wound together in a double helix. Contains the genetic information of a cell.

deoxyribonucleotide A nucleotide consisting of the sugar deoxyribose, a phosphate group, and one of four nitrogen-containing bases (adenine, cytosine, guanine, or thymine). A subunit of deoxyribonucleic acid (DNA).

dephosphorylation Removal of a phosphate group from a molecule. A common mechanism of controlling protein shape or function.

depolarization A change in membrane potential from its resting negative state to a less negative or a positive state.

deposit feeder An organism that eats its way through a substrate.

dermal tissue Tissue forming the outer layer of an organism. In plants, also called epidermis; in animals, forms two distinct layers: dermis and epidermis.

descent with modification The phrase used by Darwin to describe his hypothesis of evolution by natural selection.

desmosome A complex physical connection between two animal cells, consisting of proteins that bind the cells' cytoskeletons together. Found where cells are strongly attached to each other.

detergent An amphipathic molecule that forms micelles in water and that can cleanse by suspending hydrophobic molecules (such as oily dirt) in water.

determination The irreversible commitment during development of a cell to becoming a particular cell type (e.g., liver cell, brain cell).

detritivore An organism whose diet consists mainly of detritus.

detritus A layer of dead organic matter that accumulates at ground level or on seafloors and lake bottoms.

deuterostomes A major lineage of animals that share a pattern of embryological development, including radial cleavage, formation of the anus earlier than the mouth, and formation of the coelom by pinching off of layers of mesoderm from the gut. Includes echinoderms and chordates.

developmental homology A similarity in embryonic form, or in the fate of embryonic tissues, that is due to inheritance from a common ancestor.

diabetes insipidus A human disease caused by defects in the kidney's system for conserving water. Characterized by production of large amounts of dilute urine.

diabetes mellitus A human disease caused by defects in insulin production or response. Characterized by abnormally high blood glucose levels and large amounts of glucose-containing urine.

diaphragm An elastic, sheetlike structure. In mammalian ventilation, the muscular sheet that separates the chest from the abdominal cavities and that can expand the chest cavity by moving downward.

diastole The portion of the heartbeat cycle during which the atria or ventricles of the heart are relaxed.

diastolic blood pressure Blood pressure in arteries during relaxation of the heart's left ventricle.

dicot A dicotyledonous plant (i.e., any plant that has two cotyledons upon germination).

dideoxy sequencing A lab technique for determining the exact nucleotide sequence of DNA. Relies on the use of dideoxynucleotide triphosphates (ddNTPs), which terminate DNA replication.

differential centrifugation Separation of cellular components by spinning a cell homogenate in a series of centrifuge runs at progressively higher velocities.

differential gene expression Expression of different genes in different cell types.

differentiation The process by which a cell becomes a particular cell type (e.g., liver cell, brain cell) by differential gene expression.

diffusion Spontaneous movement of molecules and ions along a concentration gradient, from an area of high concentration to one of low concentration.

digestion The breakdown of food into pieces that are small enough to be absorbed.

digestive tract The long tube that begins at the mouth and ends at the anus. Also called alimentary canal, gastrointestinal tract, or the gut.

dihybrid cross A mating between two parents that are heterozygous for both of the two genes being studied.

dimer An association of two identical molecules.

dioecious Having either male flowers or female flowers but not both.

diploblast An animal whose body develops from two basic embryonic cell layers—ectoderm and endoderm; includes cnidarians and ctenophores.

diploid With two sets of chromosomes. Most animals and many plants are diploid.

directional selection A pattern of natural selection in which individuals with a particular extreme phenotype have higher fitness than individuals with average phenotypes or with the other extreme of the phenotype.

direct sequencing A lab technique for discovery and study of unknown microscopic organisms that will not grow easily in the lab. Relies on detecting and amplifying copies of their DNA.

disaccharide A carbohydrate consisting of two monosaccharides (sugar subunits) linked together.

discontinuous replication In DNA replication, the process by which the lagging strand is synthesized in separate pieces (Okazaki fragments) that are joined together later.

discontinuous strand In DNA replication, the strand of new DNA that is synthesized discontinuously in a series of short pieces that are later joined together. Also called lagging strand.

discrete trait A phenotypic trait that exhibits distinct forms rather than the continuous variation seen in quantitative traits such as body height.

dispersal The movement of individuals from their place of origin (birth, hatching) to a new location.

dispersive replication A now-disproven hypothesis of DNA replication, in which each strand in the daughter molecules was proposed to consist of a mixture of old and new segments of DNA.

disruptive selection A pattern of natural selection in which individuals with extreme phenotypes (at either end of the range of phenotypic variation) have higher fitness than do individuals with an average phenotype.

dissociation curve The graphical relationship that depicts the percentage of hemoglobin in the blood that will bind to oxygen at various partial pressures of oxygen.

distal Away from the center of the body; toward the furthest tip of an appendage. The opposite of proximal.

disturbance Any event that removes some individuals or biomass from a community.

disturbance regime The characteristic disturbances that affect a given ecological community.

disulfide bond A covalent bond between two sulfur atoms, typically in the side groups of some amino acids (e.g., cysteine). Often contributes to tertiary structure of proteins.

DNA See deoxyribonucleic acid (DNA).

DNA fingerprinting Any of several methods for identifying individuals by unique features of their genomes. Commonly involves using PCR to create many copies of certain simple sequence repeats and then analyzing their lengths.

DNA footprinting A technique used to find and sequence stretches of DNA that are bound by particular regulatory proteins.

DNA ligase An enzyme that can connect pieces of DNA by catalyzing formation of a phosphodiester bond between the different pieces.

DNA microarray A lab tool involving the use of cDNAs to investigate whether any of several thousand genes are expressed in a certain cell or tissue.

DNA polymerase Any enzyme that catalyzes synthesis of DNA from deoxyribonucleotides.

domain (1) A section of a protein that has a distinctive tertiary structure and function. (2) A fundamental taxonomic group of organisms sharing similarities in basic cellular biochemistry, such as Bacteria, Archaea, and Eukarya.

dominant An allele that determines the phenotype of a heterozygous individual (i.e., one that can hide the presence of a recessive allele).

domino effect Progressive loss of species diversity in a fragmented habitat, in which the loss of one species causes the loss of further species.

dopamine A catecholamine neurotransmitter that functions mainly in a part of the mammal brain involved with muscle control.

dormancy A temporary state of greatly reduced, or no, metabolic activity.

dorsal Toward an animal's back and away from its belly. The opposite of ventral.

dorsal hollow nerve cord A bundle of nerves extending from the brain along the dorsal (back) side of a chordate animal, with cerebrospinal fluid inside a hollow central channel. One of the defining features of chordates.

double fertilization An unusual form of reproduction seen in flowering plants, in which one sperm nucleus fuses with an egg to form a zygote and the other sperm nucleus fuses with two polar nuclei to form the triploid endosperm.

double helix The three-dimensional shape of a molecule of DNA, consisting of two antiparallel DNA strands wound around each other.

Doushantuo fossils A characteristic assemblage of microscopic fossils found in pre-Cambrian rocks in China from about 580 million years ago, including evidence of sponges, different cell types, and multicellular embryos.

downstream In genetics, the direction in which RNA polymerase moves along a DNA strand.

Down syndrome A human developmental disorder caused by trisomy of chromosome 21.

duct A thin tube through which an exocrine gland secretes some substance.

dyad symmetry A type of symmetry in which an object can be superimposed on itself if rotated 180°. Occurs in some regulatory sequences of DNA. Also called two-fold rotational symmetry.

dynein A motor protein that produces movement of cilia and flagella. Dynein bridges use the chemical energy of ATP to "walk" along the adjacent microtubule doublets.

early endosome A membrane-bound vesicle, formed by endocytosis, that is an early stage in the process of becoming a lysosome.

ecdysone An insect hormone that triggers either molting (to a larger larval form) or metamorphosis (to the adult form), depending on the level of juvenile hormone.

Ecdysozoa A lineage of protostomes that grow by shedding their external skeletons and expanding their bodies, rather than by increasing the length of their skeletons. Includes arthropods and nematodes.

echolocation The use of echoes from vocalizations to obtain information about locations of objects in the environment.

ecology The study of how organisms interact with each other and with their surrounding environment.

ecosystem   All organisms that live in a geographic area, together with abiotic components that affect or exchange materials with the organisms.

ecosystem diversity   The variety of biotic components in a region along with abiotic components, such as soil, water, and nutrients.

ecosystem services   Alteration of the physical components of an ecosystem by living organisms, especially beneficial changes in the quality of the atmosphere, soil, water, etc.

ectoderm   One of the three basic embryonic cell layers of a triploblast animal. Forms the outer covering and nervous system.

ectomycorrhizal fungi (EMF)   Fungi whose hyphae form a dense network that covers their host plant's roots but do not enter the root cells.

ectoparasite   A parasite that lives on the outer surface of the host's body.

ectotherm   An animal that does not use internally generated heat to regulate its body temperature.

Ediacaran fauna   A characteristic set of animal fossils found in various pre-Cambrian rocks around 565–544 million years old, containing sponges, jellyfish, comb jellies, and other filter-feeding, shallow-water marine animals.

effector   Any structure, cell, or organ with which an animal can respond to external or internal stimuli. Usually under control of the nervous system.

effector T cells   T lymphocytes—descendants of activated T cells—that are actively involved in combating an infection. Includes helper T cells and cytotoxic T lymphocytes.

efferent division   The part of the nervous system that carries commands from the central nervous system to the body. Consists primarily of motor neurons.

egg   A mature female gamete and any associated external layers (such as a shell). Larger and less mobile than the male gamete. In animals, also called an ovum.

ejaculation   The release of semen from the copulatory organ of a male animal.

ejaculatory duct   A short duct connecting the vas deferens to the urethra, through which sperm move during ejaculation.

elasticity   The ability to stretch and then spring back to the original shape.

electrical potential   Potential energy created by a separation of electric charges between two points. Also called voltage.

electric current   A flow of electrical charge past a point. Also called current.

electrocardiogram (EKG)   A recording of the electrical activity of the heart, as measured through electrodes on the skin.

electrochemical gradient   The combined effect of a concentration gradient and an electrical gradient. Affects the movement of ions across plasma membranes.

electrolyte   Any compound that dissociates into ions when dissolved in water. In nutrition, refers to the major ions necessary for normal cell function.

electromagnetic spectrum   The full range of wavelengths of electromagnetic radiation (light).

electron   An extremely tiny, negatively charged particle that usually occupies an orbital around an atomic nucleus. Exhibits wave as well as particle characteristics.

electron acceptor   A reactant that gains an electron and is reduced in a reduction-oxidation reaction.

electron carrier   A molecule that readily donates electrons to other molecules.

electron donor   A reactant that loses an electron and is oxidized in a reduction-oxidation reaction.

electronegativity   The tendency of an atom to attract electrons toward itself.

electron microscope   A microscope that uses beams of electrons instead of light to produce images of specimens. Can magnify hundreds of thousands of times.

electron shell   Atomic orbitals of similar energies. Arranged in roughly concentric layers around the nucleus of an atom, with electrons in outer shells having more energy than those in inner shells.

electron transport chain (ETC)   A set of molecules involved in a coordinated series of redox reactions in which the potential energy lost by electrons is used to pump protons from one side of a membrane to the other.

electroreceptor   A sensory cell or organ specialized to detect electric fields.

element   A fundamental chemical entity consisting of atoms with a specific number of protons. Elements preserve their identity in chemical changes.

elemental ion   An ion that is a single element, such as K^+ or Cl^-, and not a molecule.

ELISA (enzyme-linked immunosorbent assay)   A lab technique that can measure the concentration of a substances present in very small amounts.

elongation   (1) The process by which messenger RNA lengthens during transcription. (2) The process by which polypeptide chains lengthen during translation.

elongation factors   Proteins involved in the elongation phase of translation, assisting ribosomes in the synthesis of the growing peptide chain.

elongation phase   The phase of DNA transcription in which RNA polymerase moves along the DNA molecule and synthesizes messenger RNA.

embryo   A young developing organism; the stage after fertilization and zygote formation.

embryogenesis   The process by which a single-celled zygote becomes a multicellular embryo.

embryophyte   A plant, including a land plant, that nourishes its embryos inside its own body.

emergent vegetation   Any plants in an aquatic habitat that extend above the surface of the water.

emerging disease   Any infectious disease that suddenly afflicts significant numbers of individuals, often due to changes in host species or host population movements.

emerging virus   Any of several pathogenic viruses that that suddenly afflict significant numbers of individuals, often due to changes in host species or host population movements.

emigration   The movement of individuals from one population into another.

emphysema   A lung disease caused by breakdown of alveoli and loss of elasticity of the lungs.

emulsified   Broken up; the result of mixing of fat into an aqueous solution, usually with the aid of an emulsifying agent (an amphipathic substance such as a detergent) that can break large fat globules into microscopic fat droplets.

endemic species   A species that occurs only in one limited area.

endergonic   Said of a chemical reaction that will not occur spontaneously but requires an input of energy. For such a reaction, ΔG (Gibbs free-energy change) > 0.

endocrine   Relating to hormones.

endocrine gland   Any organ that secretes hormones into the blood.

endocrine system   All of the glands that produce and secrete hormones into the bloodstream.

endocytosis   Uptake of extracellular material by engulfing and pinching-off the plasma membrane to form a small membrane-bound vesicle in the cell.

endoderm   One of the three basic embryonic cell layers of a triploblast animal. Forms the digestive tract and organs that connect to it (liver, lungs, etc.).

endodermis   In plant roots, a cylindrical layer of cells that separates the cortex from the vascular tissue.

endomembrane system   A system of organelles in eukaryotic cells that performs most protein and lipid synthesis. Includes the endoplasmic reticulum (ER), Golgi apparatus, and lysosomes.

endoparasite   Any parasite that lives inside the host's body.

endophytic   Living inside of a plant.

endoplasmic reticulum (ER)   A network of interconnected membranous sacs and tubules found inside eukaryotic cells. Either rough ER (with ribosomes attached) or smooth ER.

endoskeleton   An internal skeleton, such as that found in vertebrates.

endosome   A membrane-bound vesicle created by endocytosis. Gradually transformed into a lysosome.

endosperm   A triploid tissue in the seed of a flowering plant. Serves as food for the plant embryo.

endosymbiont   An organism that lives in a symbiotic relationship inside the body of its host.

endosymbiosis   A type of symbiotic relationship in which individuals of one species live inside the bodies of individuals of another species.

endosymbiosis theory   The theory that mitochondria and chloroplasts evolved from prokaryotes that were engulfed by host cells and took up a symbiotic existence within those cells.

endotherm   An animal whose primary source of body heat is internally generated heat.

endothermic   (1) Said of a chemical reaction that absorbs heat. (2) Able to maintain a high body temperature using internally generated heat.

energy   The capacity to do work or to supply heat. May be stored (potential energy) or available in the form of motion (kinetic energy).

enhancer   In eukaryotes, a regulatory DNA sequence to which certain proteins can bind, enhancing the transcription of certain genes.

enrichment culture   A method of growing cells in the lab that involves providing cells from the environment with a very specific set of conditions and isolating those that grow rapidly in response.

enterokinase   An intestinal enzyme that converts trypsinogen (from the pancreas) to active trypsin, which then activates protein-digesting enzymes.

entomology   The study of insects.

entropy   The amount of disorder in any system, such as a group of molecules. Commonly symbolized by S.

envelope   A membrane-like covering that encloses some viruses and their capsid coats, shielding them from attack by the host's immune system.

enveloped   In a virus, having an envelope surrounding its capsid coat.

environmental variation   The proportion of phenotypic variation in a trait that is due to environmental influences rather than genetic influences, in a certain population in a certain environment.

enzyme   A protein catalyst used by living organisms to speed up and control biological reactions.

epicotyl   In some embryonic plants, a portion of the embryonic stem that extends above the cotyledons.

epidemic   The spread of an infectious disease throughout a large population in a short time.

epidermis   The outermost layer of cells of any multicellular organism.

epididymis   A coiled tube wrapped around the testis in reptiles, birds, and mammals. The site of the final stages of sperm maturation and storage.

epinephrine   A catecholamine hormone from the adrenal medulla. Triggers rapid responses relating to the fight-or-flight response. Also called adrenaline.

epiphyte   A plant that grows on trees or other solid objects and is not rooted in soil. Not parasitic—they do not harm the host.

epistasis   An interaction of independently inherited genes, such that alleles at one locus alter the phenotypic effect of alleles at another locus.

epithelial tissue, or epithelium   A class of animal tissues consisting of layers of tightly packed cells that line an organ, a duct, or a body surface. Also called epithelium (plural: epithelia).

epitope   The unique region of a particular antigen to which antibodies or lymphocytes bind.

equilibrium   A state of balance between forward and reverse processes, such as forward and reverse chemical reactions, or between diffusion rates from one side to the other of a selectively permeable membrane.

equilibrium potential   The membrane potential at which there is no net movement of a particular ion into or out of a cell.

ER signal sequence   A specific sequence of 20 amino acids, found at the beginning of certain newly synthesized proteins, that marks them for transport to the endoplasmic reticulum.

erythropoietin (EPO)   A peptide hormone, released by the kidney in response to low blood oxygen levels, that stimulates the bone marrow to produce more red blood cells.

E site   The site in a ribosome where an unattached transfer RNA is shunted after its amino acid is added to the growing polypeptide chain.

esophagus   The muscular tube that connects the mouth to the stomach.

essential amino acid   An amino acid that an animal cannot synthesize and must obtain from the diet. May refer specifically to one of the eight essential amino acids of adult humans: isoleucine, leucine, lysine, methionine, phenylalanine, threonine, tryptophan, and valine.

essential element   A chemical element that an organism must obtain from its environment.

essential nutrient   Any chemical element or compound required for normal growth, reproduction, and maintenance of a living organism.

ester linkage   The covalent bond that joins a fatty acid to glycerol to form a fat or phospholipid.

estradiol   The major estrogen produced by the ovaries of female mammals. Stimulates development of the female reproductive tract, growth of ovarian follicles, and growth of breast tissue.

estrogens   A class of steroid hormones, including estradiol, estrone, estriol, and others, that generally promote female-like traits. Secreted by the gonads, fat tissue, and some other organs.

estuary   An environment of brackish (partly salty) water where a river meets the ocean.

ethnobotanists   Plant biologists who study how humans use plants, often focusing on indigenous cultures' knowledge of medical uses for plants.

ethylene   A gaseous plant hormone that induces fruit to ripen, flowers to fade, and leaves to drop.

euchromatin   Chromatin (a eukaryotic chromosome and its histone proteins) that is unwound in a long, filamentous structure.

eudicot   A member of the lineage of angiosperms that includes complex flowering plants and trees. Eudicots include some of the plants classically called dicots.

Eukarya   The taxonomic domain that includes all eukaryotes (protists, fungi, plants, animals, etc.), which share a cell nucleus, a cytoskeleton, and other features.

eukaryote   A member of the domain Eukarya; an organism with complex cells with distinctive traits such as a nucleus, membrane-bound organelles, a cytoskeleton, and the presence of introns in genes. May be unicellular or multicellular.

eumetazoans   Animals whose cells are organized into distinct tissues. All animals except sponges.

eutherians   A lineage of mammals whose young develop in the uterus and are not housed in an abdominal pouch. Also called placental mammals.

eutrophication   Deoxygenation of an aquatic ecosytem due to a bloom of photosynthetic algae that produces a bloom of decomposers, which use up all the oxygen.

evaporation   The energy-absorbing phase change from a liquid state to a gaseous state. Many organisms evaporate water as a means of heat loss.

evo-devo   The study of the developmental and molecular causes of major evolutionary changes such as novel body parts.

evolution   (1) The theory that all organisms on Earth are related by common ancestry and that they have changed over time, predominantly via natural selection. (2) Any change in the genetic characteristics of a population over time; especially, a change in allele frequencies.

excision repair system   A set of enzymes that identify and repair damaged sections of DNA.

excitable membrane   A plasma membrane that is capable of generating an action potential.

excitatory postsynaptic potential (EPSP)   A change in membrane potential at a neuron dendrite that makes an action potential more likely. Usually a depolarization.

exergonic   Said of a chemical reaction that will occur spontaneously, releasing heat and/or increasing entropy. For such reactions, ΔG (Gibbs free-energy change) < 0.

exocrine gland   A gland that secretes some substance through a duct into a space other than the circulatory system, such as the digestive tract or the skin surface.

exocytosis   Secretion of cellular contents to the outside of the cell by fusion of vesicles to the plasma membrane.

exon   A region of a eukaryotic gene that is translated into a peptide or protein.

exonuclease   Any enzyme that can remove nucleotides from the end of a strand of DNA or RNA.

exoskeleton   A hard covering secreted on the outside of the body, used for body support, protection, and muscle attachment.

exothermic   Said of a chemical reaction that releases heat.

exotic   From a different area. May refer specifically to exotic species introduced into a new area.

expansins   A class of plant proteins that actively increase the length of the cell wall when the pH of the wall falls below 4.5.

exponential growth   A constantly accelerating increase in population size that occurs when growth rate is constant (not affected by population size).

exportins   A class of intracellular proteins whose function is to transport certain large molecules out of the nucleus.

extensor   A muscle that pulls two bones further apart from each other, as in the extension of a limb or the spine.

extinct   Said of a species that has died out.

extracellular digestion   Digestion that takes place outside of an organism, as occurs in many fungi.

extracellular matrix (ECM)   The substance that animal cells secrete and in which they are embedded; often has a fiber-composite structure with protein fibers (e.g., collagen, elastin).

extremophiles   Any of several groups of bacteria and archaea that thrive in "extreme" (e.g., high-salt, high-temperature, low-temperature, or low-pressure) environments.

F_1 generation   First filial generation. The first generation of offspring produced from a mating (i.e., the offspring of the parental generation).

facilitated diffusion   Diffusion of a substance across a plasma membrane down a concentration gradient with the assistance of carrier proteins.

facilitation   Ecological succession in which early-arriving species make conditions more favorable for later-arriving species.

facultative aerobe   Any organism that can perform aerobic respiration when oxygen is available to serve as an electron acceptor but can switch to fermentation when it is not.

fallopian tube   A narrow tube connecting the uterus to the ovary in humans, through which the egg travels after ovulation. Site of fertilization and cleavage. In nonhuman animals, called oviduct.

family   A classic taxonomic rank above genus and below order. In animals, usually ends in the suffix -idae.

fast-twitch fibers   A class of muscle fibers that contract rapidly and powerfully but fatigue quickly.

fate   The likely developmental path that a certain embryonic cell will follow (i.e., which tissue types it will give rise to).

fate map   A diagram of an embryo, showing the eventual fate of cells in that embryo—the ultimate location and the tissues each cell will give rise to.

fats   A class of lipids consisting of three fatty acids joined to a glycerol molecule. Also called triacylglycerols or triglycerides.

fatty acid   A type of lipid consisting of a hydrocarbon chain bonded to a carboxyl group (–COOH) at one end. Used by many organisms to store chemical energy; a major component of animal and plant fats.

fatty-acid binding protein   A membrane protein of intestinal cells that binds to lipids from food and takes them into the cell.

feather   A specialized skin outgrowth in all birds and only in birds. Composed of β-keratin. Used for flight, insulation, display, and other purposes.

feces   The waste products of digestion.

fecundity   The average number of female offspring produced by a single female per unit time.

feedback inhibition   Inhibition of a process by high concentrations of the product of that process.

female gametophyte   A multicellular haploid structure that can produce haploid eggs, in a species that exhibits alternation of generations. In flowering plants, consists of a sac (the embryo sac) containing haploid nuclei.

fermentation   Any of several metabolic pathways that allow continued production of ATP via glycolysis by transferring electrons from a reduced compound such as glucose to an electron acceptor other than oxygen.

ferredoxin   A molecule involved in the electron transport chain of photosystem I of photosynthesis. Passes electrons to the enzyme NADP$^+$ reductase, which catalyzes formation of NADPH.

fertilization   Fusion of the nuclei of two haploid gametes to form a zygote with a diploid nucleus.

fertilization envelope   A physical barrier that forms around a fertilized egg in amphibians and some other animals. Formed by an influx of water under the vitelline membrane.

fetal alcohol syndrome   A condition thought to be caused by exposure to high blood alcohol concentrations during embryonic development.

fetus   A later developmental stage of an embryo, usually developed sufficiently to be recognizable as a certain species.

fever   A sustained, regulated elevation of body temperature in an endotherm, typically as part of a response to infection.

fiber   In botany, a type of elongated sclerenchyma cell that provides support to vascular tissue.

fibronectins   A class of proteins in the extracellular matrix to which cells bind to stay in place.

Fick's law of diffusion   A law that gives the rate of diffusion of a gas into a liquid as a function of gas solubility, temperature, surface area, difference in partial pressures, and the thickness of the diffusion barrier.

fight-or-flight response   Rapid physiological changes that prepare the body for emergencies. Includes increased heart rate, increased blood pressure, and decreased digestion. Triggered by catecholamines.

filament   Any thin, threadlike structure, particularly (1) the threadlike extensions of a fish's gills or (2) the slender stalk that bears the anthers in a flower.

filter feeder   Any organism that obtains food by filtering small particles or small organisms out of water or air. Also called suspension feeder.

filtrate   Any fluid produced by filtration, such as the fluid in kidney nephrons.

filtration   A process of removing large components from a fluid by forcing it through a filter, such as that in a renal corpuscle of the kidney.

finite rate of increase   The rate of increase of a population over a given period of time. Calculated as the ending population size divided by the starting population size. Symbolized by lambda (λ).

first law of thermodynamics   A fundamental principle of physics stating that energy is conserved; it cannot be created or destroyed.

fission   The splitting of an organism into two daughter organisms.

fitness   The ability of an organism to produce surviving fertile offspring. Also called Darwinian fitness.

fixed action patterns (FAPs)   Highly stereotyped behavior patterns that occur in a certain invariant way in a certain species. A form of innate behavior.

flaccid   Limp, as a result of low internal pressure; e.g., a wilted plant leaf.

flagellum   (plural: flagella) A long, cellular projection that undulates (in eukaryotes) or rotates (in prokaryotes) to move the cell through an aqueous environment.

flavonoids   A class of accessory pigments, found in many plants, that can absorb ultraviolet (UV) radiation and hence protect cells from damage by UV light.

flexor   A muscle that pulls two bones closer together, as in the flexing of a limb or the spine.

floral meristem   A group of undifferentiated plant cells that can produce flowers.

florigen   A hypothesized hormone that may stimulate flowering but that has not yet been isolated.

flower   The reproductive structure of angiosperm plants, containing microsporangia and/or megasporangia. Typically includes a calyx, a corolla, and one or more stamens or carpels.

flowering plant   An angiosperm; a member of the lineage of plants that produces seeds within mature ovaries (fruits).

fluid-mosaic model   The widely accepted hypothesis that plasma membranes consist of proteins embedded in a phospholipid bilayer and that these components move fluidly around the membrane.

fluorescence   The spontaneous emission of light from an excited electron falling back to its normal (ground) state.

follicle   An egg cell and its surrounding ring of supportive cells in a mammalian ovary.

follicle-stimulating hormone (FSH)   A peptide hormone from the anterior pituitary that stimulates (in females) growth of eggs and follicles in the ovaries or (in males) sperm production in the testes.

follicular phase   The first major phase of a menstrual cycle, when follicles are growing and estrogen levels are increasing. Ends with ovulation.

food   A digestible material that contains nutrients.

food chain   A simple pathway of energy through a few species in an ecosystem; e.g., a primary producer, a primary consumer, a secondary consumer, and a decomposer.

food vacuole   A membrane-bound organelle containing food engulfed by the cell.

food web   Any complex pathway along which energy moves among many species at different trophic levels of an ecosystem.

foot   A muscular appendage of mollusks, used for movement and/or burrowing into sediment.

forb   Herbaceous flowering plants such as daisies and coneflowers.

forensic science   The use of scientific analyses to help solve crimes.

formed elements   Cells and cell fragments found in blood, including red blood cells, leukocytes and platelets.

fossil   Any trace of an organism that existed in the past. Includes tracks, burrows, fossilized bones, casts, etc.

fossil record   All of the fossils that have been found anywhere on Earth and that have been formally described in the scientific literature.

founder effect   A change in allele frequencies that often occurs when a new population is established from a small group of individuals, due to chance variations in gene frequency in the small group.

founder event   The establishment of a new population from a small group of individuals.

fovea   The small region of the vertebrate retina in which the photoreceptors are very tightly packed, producing the most acute vision.

F-plasmid   A particular type of bacterial plasmid that gives the bacterium the ability to initiate conjugation with another bacterium.

fragile-X syndrome   The most common form of inherited mental retardation in humans. Caused by an increase in the number of copies of a CGG codon at a certain location on the X chromosome.

free radical   An atom with an unpaired electron. Unstable and highly reactive.

freeze-fracture electron microscopy   A technique in which plasma membranes are frozen and split to obtain a highly magnified view of both the outer surface and the interior of the bilayer.

frequency   The number of wave crests per second traveling past a stationary point. Determines the pitch of sound and the color of light.

fronds   The large leaves of ferns.

frontal lobe   The anteriormost region of each cerebral hemisphere of the mammal brain. In humans, involved in complex decision making.

fruit   A mature, ripened plant ovary (or group of ovaries), along with the seeds it contains and any adjacent fused parts.

fruiting body   A structure formed in some fungi and prokaryotes for spore dispersal, usually consisting of a base, a stalk, and a mass of spores at the top.

functional genomics   The study of how a genome works (how the genes identified in genome sequencing interact to produce a functional organism).

functional group   A small group of atoms bonded together in a precise configuration. Each group has particular chemical properties that it imparts to any organic molecule in which it occurs.

fundamental asymmetry of sex   The fact that eggs are larger and more expensive to produce than sperm, and the resulting evolutionary consequences for males and females.

fundamental niche   The ecological space that a species occupies in its habitat in the absence of competitors.

fungi   A major lineage of eukaryotes that typically have a filamentous body (mycelium) and obtain nutrients by absorption.

fungicide   Any substance that can kill fungi or slow their growth.

fur   An insulative layer that covers most mammals' bodies, consisting of longer rain-shedding hairs covering a shorter layer of soft underfur.

G$_1$ phase   The phase of a cell cycle that constitutes the first part of interphase before DNA synthesis (S phase).

G$_2$ phase   The phase of a cell cycle between synthesis of DNA (S phase) and mitosis (M phase); the last part of interphase.

gall   A tumorlike growth that forms on plants that are infected with certain bacteria or parasites.

gallbladder   A small pouch that stores bile from the liver and releases it to the small intestine during digestion for emulsification of fats.

gametangium   (plural: gametangia) (1) The gamete-forming structure found in all land plants except angiosperms. Contains a sperm-producing antheridium and an egg-producing archegonium. (2) The gamete-forming structure of chytrid fungi.

gamete A haploid reproductive cell that can fuse with another haploid cell to form a zygote. Most multicellular eukaryotes have two distinct forms of gametes: egg cells (ova) and sperm cells.

gametogenesis The production of gametes (eggs or sperm).

gametophyte The multicellular haploid phase in a species that exhibits alternation of generations. Arises from a single spore and produces gametes.

ganglion cells Neurons in the vertebrate retina that collect visual information from one or several bipolar cells and send it to the brain via the optic nerve.

Gap Analysis Program (GAP) An analysis aimed at identifying the gaps between geographic areas that are rich in biodiversity and areas that are being protected or managed for conservation.

gap genes A class of fruit-fly segmentation genes that organize embryonic cells into major groups of segments. Active in broad regions of the embryo.

gap junction A direct physical connection between two animal cells, consisting of a gap in the ECM and plasma membranes lined by specialized proteins that allow passage of water, ions and small molecules between the cells.

gastrin A hormone produced by the stomach in response to the arrival of food or to a signal via nerves from the brain. Stimulates other stomach cells to release hydrochloric acid.

gastropods The slugs and snails; mollusks distinguished by a large muscular foot and a unique feeding structure, the radula.

gastrula A vertebrate embryo just after gastrulation, containing the three embryonic germ layers (ectoderm, mesoderm, and endoderm) but with no nerve cord yet.

gastrulation The process by which some cells on the outside of a young embryo move to the interior of the embryo, resulting in the three distinct cell layers of endoderm, mesoderm, and ectoderm.

gated channel A channel protein that opens and closes in response to a certain stimulus, such as the binding of a particular molecule or a change in the electrical charge on the outside of the membrane.

gel electrophoresis A technique for separating molecules on the basis of size and electric charge, which affect their differing rates of movement through a gelatinous substance in an electric field.

gemma (plural: gemmae) A small reproductive structure produced in some liverworts by a gametophyte and can grow into another gametophyte.

gene A section of DNA (or RNA, for some viruses) that encodes information for building a polypeptide or a functional molecule of RNA.

gene duplication The creation of an additional copy of a gene, typically by misalignment of chromosomes during crossing over.

gene expression The transcription and translation of a gene, producing a protein.

gene family A set of genetic loci whose DNA sequences are extremely similar. Thought to have arisen by duplication of a single ancestral gene.

gene flow The movement of alleles between populations.

gene-for-gene hypothesis The hypothesis that there is a one-to-one correspondence between the resistance (*R*) loci of plants and the avirulence (*avr*) loci of pathogenic fungi; particularly, that *R* genes produce receptors and *avr* genes produce molecules that bind to those receptors.

gene pool All of the alleles of all of the genes in a certain population.

generation The average time between a mother's first offspring and her daughter's first offspring.

generative cell A small haploid cell within the male gametophyte of a flowering plant. Gives rise to two haploid sperm cells.

gene therapy The treatment of an inherited disease by introducing normal alleles.

genetic bottleneck A reduction in allelic diversity via genetic drift due to a population bottleneck.

genetic cloning A technique for producing many identical copies of a certain gene, usually by inserting a cDNA version of the gene into a bacterial plasmid and growing the bacteria.

genetic code The set of all 64 codons of DNA and the particular amino acids that each specifies.

genetic correlation A type of evolutionary constraint in which selection on one trait causes a change in another trait as well, because the same gene(s) affect both traits.

genetic diversity The diversity of alleles in a population, species, or group of species.

genetic drift Any change in allele frequencies due to random events. Causes allele frequencies to drift up and down randomly over time, and eventually can lead to the fixation or loss of alleles.

genetic engineering The field of study of the manipulation of DNA sequences in living organisms.

genetic homology Similarities among certain organisms in DNA sequences or amino acid sequences of proteins that are due to inheritance from a common ancestor.

genetic map A map of the relative locations of specific genes on a certain chromosome. Also called a linkage map or meiotic map.

genetic marker A genetic locus that can be identified and traced in populations by lab techniques or by a distinctive phenotype. Includes reporter genes.

genetic model A set of hypotheses that explain how a certain trait is inherited.

genetic recombination A change in the combination of genes or alleles on a given chromosome or in a given individual. Also called recombination.

genetics The field of study concerned with the inheritance of traits.

genetic screen Any of several techniques for identifying individuals with a particular type of mutation. Also called a screen.

genetic variation (1) The number and relative frequency of alleles present in a particular population. (2) The proportion of phenotypic variation in a trait that is due to genetic rather than environmental influences in a certain population in a certain environment.

genitalia External copulatory organs.

genome All of the hereditary information in an organism, including not only genes but also other non-gene stretches of DNA.

genomic library A set of all the DNA sequences in a particular genome, split into small segments, each inserted into a vector for further study.

genomics The field of study concerned with sequencing, interpreting, and comparing whole genomes.

genotype All of the alleles of every gene present in a given individual. May refer specifically to the alleles of a particular set of genes under study.

genus (plural: genera) A taxonomic category of closely related species. Always italicized and capitalized to indicate that it is a recognized scientific genus.

geologic time scale The sequence of eons, epochs, and periods used to describe the geologic history of Earth.

geometric isomer A molecule that shares the same molecular formula as another molecule but differs in the arrangement of atoms or groups on either side of a double bond or ring structure.

germination The process by which a seed becomes a young plant.

germ layer In animals, one of the three basic types of tissues of early embryonic development that give rise to all other tissues: endoderm, mesoderm, or ectoderm.

germ theory of disease The theory that infectious diseases are caused by bacteria, viruses, and other microorganisms.

gestation The duration of embryonic development from fertilization to birth, in those species that have live birth.

gibberellic acid (GA) A plant hormone that stimulates elongation of shoots; a gibberellin.

gibberellins A class of plant hormones that stimulate growth.

Gibbs free-energy change A measure of the change in potential energy and entropy that occurs in a given chemical reaction. Calculated as $\Delta G = \Delta H - T\,\Delta S$, where ΔH is the change in potential energy, T is the temperature in kelvins, and ΔS is the change in entropy. Determines whether a reaction will be spontaneous.

gill Any organ in aquatic animals that exchanges gases and other dissolved substances between the blood and the surrounding water. Typically, a filamentous outgrowth of a body surface.

gill arch In aquatic vertebrates, curved regions of tissue between the gills. Gills are suspended from the gill arches.

gill filament The thin, pink strands of fish gills that extend from the gill arches into the water and across which gas exchange occurs.

gill lamellae (singular: lamella) Tiny, crescent-shaped flaps on the gill filaments of fish that serve to increase surface area for gas exchange.

gland An organ whose primary function is to secrete some substance, either into the blood (endocrine gland) or into some other space such as the gut or skin (exocrine gland).

glia Several types of cells in nervous tissue that are not neurons and do not conduct electrical signals but provide support, nourishment, and electrical insulation and perform other functions.

global carbon cycle The movement of carbon among terrestrial ecosystems, the oceans, and the atmosphere.

global nitrogen cycle The movement of nitrogen among terrestrial ecosystems, the oceans, and the atmosphere.

global warming A sustained increase in Earth's average surface temperature.

global water cycle The movement of water among terrestrial ecosystems, the oceans, and the atmosphere.

glomerulus (1) A ball-like cluster of capillaries at the beginning of a kidney nephron. Surrounded by Bowman's capsule. (2) The ball-shaped clusters of neurons in the olfactory bulb of the brain.

glucagon A peptide hormone produced by the pancreas in response to low blood glucose. Raises blood glucose by triggering breakdown of glycogen and stimulating gluconeogenesis.

glucocorticoids A class of steroid hormones released from the adrenal cortex that increase blood glucose and prepare the body for stress.

gluconeogenesis Synthesis of new glucose from non-carbohydrate sources, such as proteins and fatty acids. Occurs in the liver in response to low insulin levels and high glucagon levels.

glyceraldehyde-3-phosphate (G3P) A phosphorylated sugar produced during the Calvin cycle of photosynthesis.

glycerol A three-carbon molecule that forms the "backbone" of phospholipids and most fats.

glycogen A polysaccharide that is the major form of stored carbohydrate in animals. Consists of α-glucose monomers joined end to end in highly branched chains.

glycolipid Any lipid molecule that is covalently bonded to a carbohydrate group.

glycolysis A series of 10 chemical reactions that oxidize glucose to produce pyruvate and ATP. Used by all organisms as part of fermentation or cellular respiration.

glycoprotein Any protein with one or more covalently bonded carbohydrate groups.

glycosidic linkage The covalent bond between two sugar subunits of a polysaccharide.

glycosylation Addition of a carbohydrate group to a molecule.

glyoxisomes Specialized peroxisomes found in plant cells and that contain enzymes for processing the products of photosynthesis.

goblet cells The cells in the stomach lining that secrete mucus.

goiter A pronounced swelling of the thyroid gland in the neck, usually caused by a deficiency of iodine in the diet.

Golgi apparatus A stack of flattened membranous sacs (cisternae) in eukaryotic cells; processes proteins and lipids that will be secreted or directed to other organelles.

gonad An organ that produces reproductive cells; e.g., a testis or an ovary.

gonadotropin-releasing hormone (GnRH) A peptide hormone from the hypothalamus that stimulates release of FSH and LH from the anterior pituitary.

G proteins A class of peripheral membrane proteins that are important in signal transduction, typically by binding to GTP and activating a second messenger.

gracile australopithecines Several species of slender, lightly built hominins that appear in the fossil record shortly after the split from chimpanzees.

grade A group of species that share a position in an inferred evolutionary sequence of lineages but that are not a monophyletic group.

gram-negative bacteria Bacteria that look pink when treated with a Gram stain. Have a cell wall composed of a thin layer of peptidoglycan and an outer membrane.

gram-positive bacteria Bacteria that look purple when treated with a Gram stain. Have thick cell walls containing peptidoglycan.

Gram stain A dye that stains different types of bacteria different colors.

granum (plural: grana) A stack of the flattened, membrane-bound thylakoid disks inside plant chloroplasts.

gravitropism The growth or movement of a plant in a particular direction in response to gravity.

gray crescent A region of an amphibian zygote that becomes visible shortly after fertilization, opposite the point of sperm entry. Eventually gives rise to the blastopore and the organizer, which determines major body axes.

grazing food web The ecological network of herbivores and the predators and parasites that consume them.

great apes The hominids; members of the family Hominidae, including humans and extinct related forms, chimpanzees, gorillas, and orangutans. Distinguished by large body size, no tail, an exceptionally large brain, and a tendency toward bipedalism.

green fluorescent protein (GFP) A jellyfish protein that spontaneously emits green light after stimulation. Widely used in lab research to mark the location of certain molecules or cells.

greenhouse gas An atmospheric gas that absorbs and reflects infrared radiation, so that heat radiated from Earth is retained in the atmosphere instead of being lost to space.

green plants A lineage of eukaryotes that includes green algae and land plants.

gross photosynthetic efficiency The efficiency with which all the plants in a given area use the light energy available to them to produce sugars.

gross primary productivity The total amount of carbon fixed by photosynthesis, including that used for cellular respiration, in a given area over a given time period.

ground meristem The middle layer of a young plant embryo. Gives rise to the ground tissue.

ground tissue A plant tissue consisting of all cells beneath the outer protective layers of epidermis and cork, except for vascular tissue.

groundwater Any water below the land surface.

growth factor Any of several compounds that are secreted by certain cells and that stimulate other cells to divide or to differentiate.

growth hormone A peptide hormone produced by the mammalian pituitary gland. Involved in lengthening the long bones during childhood and in muscle growth, tissue repair, and lactation in adults.

growth-hormone-inhibiting-hormone (GHIH) A peptide hormone from the hypothalamus that inhibits release of growth hormone from the anterior pituitary.

guanosine triphosphate (GTP) A molecule consisting of guanine, a sugar, and three phosphate groups. Can be hydrolyzed to release free energy. Commonly used in many cellular reactions.

guard cell A specialized, crescent-shaped cell forming the border of a plant stoma. Changes shape to open or close the stoma.

gustation The perception of taste.

guttation Excretion of water droplets from plant leaves in the early morning, due to root pressure.

gymnosperms Four lineages of green plants that have vascular tissue and make seeds but that do not produce flowers. Includes cycads, ginkgoes, conifers, and gnetophytes.

H1 The histone protein associated with DNA in the "linker" stretches between the nucleosomes.

habitat destruction Human-caused destruction of a natural habitat with replacement by an urban, suburban, or agricultural landscape.

habitat fragmentation The breakup of a large region of a habitat into many smaller regions, separated from others by a different type of habitat.

Hadley cell An atmospheric cycle of large-scale air movement in which warm equatorial air rises, moves north or south, and then descends at approximately 30°N or 30°S latitude.

hair Mammalian fur that lacks an insulative layer of underfur; or a single strand of hair or fur consisting of keratin, dead cells, and pigments.

hair cell A pressure-detecting sensory cell that has tiny "hairs" (stereocilia) jutting from its surface.

hairpin A stable loop formed in an RNA molecule by hydrogen bonding between purine and pyrimidine bases on the same strand.

half-life The characteristic time taken for half of any amount of a particular radioactive isotope to decay.

halophile A bacterium or archaean that thrives in high-salt environments.

halophyte A plant that thrives in salty habitats.

Hamilton's rule The proposition that an allele for altruistic behavior will be favored by natural selection only if $Br > C$, where B = the fitness benefit to the recipient, C = the fitness cost to the actor, and r = the coefficient of relatedness between recipient and actor.

haploid Having one set of chromosomes. Bacteria, archaea, animal gametes, plant gametophytes, and many algae are haploid.

haploid number The number of different types of chromosomes in a cell. Symbolized as n.

Hardy-Weinberg principle A principle of population genetics stating that, if mutation, migration, genetic drift, random mating, and selection do not occur, then genotype frequencies will not occur in predictable ratios.

harmful algal bloom An extreme increase in the abundance of a toxin-producing protist in a particular aquatic environment.

HDL See **high-density lipoprotein (HDL)**.

head A distinct anterior region of an organism's body, usually bearing a cluster of sensory organs and/or a mouth.

heart A muscular pump that circulates blood throughout the body.

heart attack An episode of cramping or death of heart muscle due to oxygen deprivation, usually caused by a blood clot in one or more partially blocked coronary arteries. Also called a myocardial infarction (MI).

heart murmur A distinctive sound caused by backflow of blood through a defective heart valve.

heartwood The older xylem in the center of an older stem or root, containing protective compounds and no longer functioning in water transport.

heat Thermal energy that is transferred between two objects.

heat of vaporization The energy required to vaporize 1 gram of a liquid into a gas.

heavy chain One of two identical polypeptides that together form the base of immunoglobulin proteins. Differences in heavy chains determine the different classes of immunoglobulins (IgA, IgE, etc.).

helicase An enzyme that catalyzes the breaking of hydrogen bonds between nucleotides of DNA, "unzipping" a double-stranded DNA molecule.

helix-turn-helix motif A motif seen in many repressor proteins in prokaryotes, consisting of two α-helices connected by a short stretch of amino acids that form a turn.

helper T cell A T cell that assists with the activation of other lymphocytes. Helper T cells are descendants of an activated CD4+ T cell.

hemagglutinin A protein that juts out from the surface of influenza viruses, enabling them to bind to host cells.

heme group A small molecule containing an iron atom that can bind to oxygen. Myoglobin and hemoglobin contain heme groups held within large specialized heme-carrying proteins (globins).

hemimetabolous metamorphosis A type of metamorphosis in which the animal increases in size from one stage to the next, but does not dramatically change its body form.

hemocoel A body cavity of arthropods and some mollusks, containing a pool of circulatory fluid (hemolymph) bathing the internal organs.

hemoglobin An oxygen-binding protein consisting of four polypeptide subunits, each containing a heme group. The major oxygen carrier of mammalian blood.

hemolymph The circulatory fluid of animals with open circulatory systems (e.g., insects) in which the fluid is not confined to blood vessels.

hemophilia A human disease characterized by defects in the blood-clotting system. Caused by an X-linked recessive allele.

herb A seed plant that lacks wood and has a relatively short-lived stem.

herbaceous Said of a plant that is not woody.

herbivore An animal that eats primarily plants and rarely or never eats meat.

herbivory The practice of eating plant tissues.

hereditary nonpolyposis colorectal cancer (HNPCC) A type of colon cancer that is associated with a mutated version of a mismatch repair gene on chromosome 2.

heredity The transmission of traits from parents to offspring via genetic information.

heritable Refers to traits that are influenced by hereditary genetic material (DNA, or RNA for some viruses).

hermaphroditic Producing both eggs and sperm.

heterochromatin Chromatin (a eukaryotic chromosome and its histone proteins) that is highly compact and supercoiled.

heterokaryotic Containing nuclei that are genetically distinct. Occurs naturally in many fungi.

heterospory The production of two distinct types of spore-producing structures and thus two distinct types of spores. Occurs in seed plants, which produce both microspores (which become the male gametophyte) and megaspores (which become the female gametophyte).

heterotherm An animal whose body temperature varies markedly with environmental conditions.

heterotroph Any organism that cannot synthesize reduced organic compounds from inorganic sources and that must obtain them by eating other organisms. Some bacteria, some archaea, and virtually all fungi and animals are heterotrophs.

heterozygote advantage A pattern of natural selection in which heterozygotes have greater fitness than either parental homozygote. Also called heterozygote superiority.

heterozygous Having two different alleles of a certain gene.

hexose A monosaccharide (simple sugar) with six carbons.

Hfr strain Any strain of bacteria in which an F-plasmid has been incorporated into the main chromosome. These bacteria have a high frequency of recombination.

hibernation A period of torpor (including a decreased body temperature) that continues for weeks or months.

high-density lipoprotein (HDL) Balls of protein and fat that transport cholesterol from body tissues to the liver for breakdown. Associated with decreased risk of heart disease.

hindgut The posterior portion of the digestive tract of an animal. Often functions to reabsorb water from wastes.

hinge helix A motif found in many repressor proteins in bacteria. Involved in locking the helix-turn-helix motif onto DNA.

histone A globular protein that is tightly associated with DNA in eukaryotic cells.

histone acetyl transferases In eukaryotes, a class of enzymes that loosens chromatin structure by adding acetyl groups to histone proteins.

histone deacetylases In eukaryotes, a class of enzymes that recondense chromatin by removing acetyl groups from histone proteins.

HIV See **human immunodeficiency virus (HIV)**.

holoenzyme A multipart enzyme consisting of a core enzyme (containing the active site for catalysis) along with other required proteins.

holometabolous metamorphosis A type of metamorphosis in which the animal completely changes its form.

homeobox A 180-base-pair sequence present in the HOM/Hox genes of animals. Codes for a helix-turn-helix motif in the proteins produced by these genes.

homeosis The occurrence of extra, fully formed segments or appendages normally found elsewhere in the body, replacing the structures usually formed at that location.

homeostasis The maintenance of a relatively constant physical and chemical environment within an organism.

homeotherm An animal that has a constant or relatively constant body temperature.

homeotic complex (HOM-C) A set of eight fruit-fly genes, closely linked on the same chromosome, that are active in different regions of the fly embryo, specifying the identity of body segments.

homeotic genes A class of genes that specify a location within an embryo, leading to the development of structures appropriate for that location.

homeotic mutant An individual with a mutation in a homeotic gene, causing the development of extra body parts or body parts in the wrong places.

hominids The great apes; members of the family Hominidae, which includes humans and extinct related forms; chimpanzees, gorillas, and orangutans. Distinguished by large body size, no tail, and an exceptionally large brain.

hominins Humans and extinct related forms; species in the lineage that branched off from chimpanzees and eventually led to humans.

homologous chromosomes In a diploid organism, chromosomes that are similar in size, shape, and gene content. Also called homologs.

homologous trait Any trait showing marked similarity between different species, due to inheritance from a common ancestor.

homology Similarity between organisms or DNA sequences that is due to inheritance from a common ancestor.

homospory The production of just one type of spore. Occurs in the seedless vascular plants, in contrast to the heterospory of seed plants.

homozygous Having two identical alleles of a certain gene.

hormone A signaling molecule that circulates throughout the body in blood or other body fluids; can trigger pronounced responses in distant target cells at very low concentrations.

hormone-response elements Sites on DNA to which a steroid hormone-receptor complex can bind and affect gene transcription.

host An individual or a species in or on which a parasite lives.

host cell A cell that has been invaded by an organism such as a parasite or a virus.

Hox genes A class of homeotic genes found in several animal phyla, including vertebrates. Controls pattern formation in early embryos.

human Any member of the genus Homo. Includes modern humans (Homo sapiens) and several extinct species.

human chorionic gonadotropin (hCG) A glycoprotein hormone produced by the human placenta from about week 3 to week 14 of pregnancy. Maintains the corpus luteum, which produces hormones that preserve the uterine lining.

Human Genome Project The multinational research project that sequenced the human genome.

human immunodeficiency virus (HIV) A retrovirus that causes AIDS (acquired immune deficiency syndrome) in humans.

humoral response Defense against infections and cancers via antibodies produced by B cells.

humus The completely decayed organic matter in soils.

Huntington's disease A degenerative brain disease of humans, caused by an autosomal dominant allele.

hybrid The offspring of parents from two different strains, populations, or species.

hybridoma A mass of cells produced in a lab from a myeloma cell that fused with an antibody-producing B cell. Used to produce large amounts of a monoclonal antibody.

hybrid zone A geographic area where interbreeding occurs between two species, sometimes producing fertile hybrid offspring.

hydrocarbon A molecule that contains only hydrogen and carbon, usually with the carbon atoms bonded covalently to form chains or rings.

hydrogen bond A weak interaction between two molecules, due to attraction between a hydrogen atom with a partial positive charge on one molecule and another atom (usually O or N) with a partial negative charge on the other molecule.

hydrogen ion A proton (H^+); typically, one that has dissolved in solution or that is being transferred from one atom to another in a chemical reaction.

hydrolysis A type of chemical reaction in which a compound reacts with water to break down into smaller molecules. In biology, most hydrolysis reactions involve polymers breaking down into monomers. The reverse of a condensation, or dehydration, reaction.

hydrophilic Mixing readily with water. Hydrophilic compounds are typically polar compounds with charged or electronegative atoms.

hydrophobic Not mixing readily with water. Hydrophobic compounds are typically nonpolar compounds without charged or electronegative atoms and often contain many C–C and C–H bonds.

hydroponic growth Growth of plants in liquid cultures instead of soil.

hydrostatic skeleton A system of body support involving fluid-filled compartments that can change in shape but cannot easily be compressed.

hydroxide ion An oxygen atom bonded to a hydrogen atom and carrying a negative charge (OH^-).

hyperpolarization A change in membrane potential from its resting negative state to an even more negative state.

hypersensitive reaction An allergic immune response in which previously sensitized mast cells and basophils produce large quantities of histamine, cytokines, etc., in response to a small amount of allergen. Triggers symptoms of allergies.

hypersensitive response The rapid death of a plant cell that has been infected by a pathogen. Thought to be a strategic mechanism that protects the rest of the plant.

hypertension Abnormally high blood pressure.

hypertonic Having a greater solute concentration, and therefore a lower water concentration, relative to another solution.

hypha (plural: hyphae) One of the strands of a fungal mycelium (the meshlike body of a fungus). Also found in some protists.

hypocotyl The stem of a very young plant; the region between the cotyledon (embryonic leaf) and the radicle (embryonic root).

hypothalamic-pituitary axis The combination of the hypothalamus and the pituitary gland, which together regulate most of the other endocrine glands in the body.

hypothalamus A part of the brain that regulates the body's internal physiological state, such as the autonomic nervous system and endocrine system.

hypothesis A proposed explanation for a phenomenon or for a set of observations.

hypotonic Having a lower solute concentration, and therefore a higher water concentration, relative to another solution.

immigration The movement of individuals into a certain population from some other population.

immune system In vertebrates, the system of tissues and organs whose primary function is to defend the body against pathogens. Includes lymphocytes, lymph nodes, and other small organs.

immunity Protection against infection by disease-causing pathogens.

immunization The conferring of immunity to a particular disease.

immunoglobulins (Ig) A class of Y-shaped proteins capable of binding to specific antigens. Responsible for acquired immunity.

immunological memory The ability of the immune system to "remember" an antigen—i.e., to mount a rapid, effective response to a pathogen encountered years or decades earlier.

impact hypothesis The hypothesis that the mass extinction at the end of the Cretaceous period, in which dinosaurs died out, was due to an asteroid impact.

imperfect For a flower, containing male parts (stamens) or female parts (carpels) but not both.

implantation The process by which an embryo buries itself in the uterine wall and forms a placenta. Occurs in mammals and a few other vertebrates.

importins A class of intracellular proteins whose function is to transport certain large molecules into the nucleus.

imprinting A type of rapid, irreversible learning in which young animals learn the distinctive appearance of the individual caring for them.

inactivated virus A virus that has been deliberately damaged to render it incapable of causing infection, for use in vaccines.

inbreeding Mating between closely related individuals or within a genetically homozygous strain.

inbreeding depression A loss of fitness that occurs when homozygosity increases in a population.

incomplete dominance An inheritance pattern in which the heterozygote phenotype is a blend or combination of both homozygote phenotypes.

independent assortment The inheritance of the alleles of one gene independently of other genes. True only for genes on different chromosomes.

indeterminate growth A pattern of growth in which an organism continues to increase its overall body size throughout its life.

indicator plate A laboratory technique for detecting mutant cells by observing color changes due to cells cleaving (or not cleaving) the bonds in a pigmented substance.

induced defense A defensive structure or compound produced by plants only in direct response to attack by pathogens or herbivores.

induced fit The phenomenon whereby initial weak binding of a substrate to the active site of an enzyme causes the enzyme to change shape so as to bind the substrate more tightly.

inducer A molecule that triggers transcription of a specific gene.

inducible defense A defensive trait that is produced only when needed—i.e., in response to the presence of a predator or pathogen.

induction The process by which one embryonic cell, or group of cells, alters the differentiation of neighboring cells.

infection thread An invagination of a root hair membrane, through which beneficial nitrogen-fixing bacteria enter the roots of their host plants, legumes.

infectious disease Any disease that can be transmitted from infected individuals to uninfected individuals.

infertile Unable to produce offspring.

inflammatory response An aspect of the innate immune response, seen in most cases of infection or tissue injury, in which the affected tissue becomes swollen, red, warm, and painful.

infrared light Light with a wavelength longer than visible red light.

infrasound Sound frequencies that are too low for humans to hear—lower than about 20 hertz.

ingestion Taking nutrients or other substances into the body; feeding.

inheritance of acquired characters The now-disproven theory that traits acquired during the lifetime of an organism due to noninherited causes will be passed genetically to its offspring.

inhibition Ecological succession in which early-arriving species make conditions less favorable for the establishment of certain other species.

inhibitory hormone A hormone that inhibits the release of some other hormone.

inhibitory postsynaptic potential (IPSP) A change in membrane potential at a neuron dendrite that makes an action potential less likely. Usually a hyperpolarization.

initiation The first stage of a molecular reaction, in which the necessary molecules are brought together. (1) In an enzyme-catalyzed reaction, the stage during which enzymes orient reactants precisely as they bind at specific locations within the enzyme's active site. (2) In DNA transcription, the binding of RNA polymerase to the promotor sequence. (3) In RNA translation, the binding of the ribosome to the mRNA molecule.

initiation factors A class of proteins that assist ribosomes in binding to a messenger RNA molecule to initiate translation. Occur both in prokaryotes and in eukaryotes.

innate behavior Behavior that is inherited genetically and does not have to be learned.

innate immune response The body's nonspecific response to pathogens; the action of leukocytes.

innate immunity A set of nonspecific defenses against pathogens that occurs even without previous exposure to the pathogen and that does not involve antibodies. Includes responses by mast cells, neutrophils, and macrophages; typically results in an inflammatory response.

inner ear The innermost portion of the mammalian ear, consisting of a fluid-filled system of tubes that includes the cochlea (which receives sound vibrations from the middle ear) and the semicircular canals (which function in balance).

inoculation Introduction of some substance into the body to increase immunity to disease. Usually involves vaccination.

in situ hybridization A lab technique for revealing the locations of specific mRNAs in organisms (i.e., revealing which tissues transcribe certain genes).

insulin A peptide hormone produced by the pancreas in response to high levels of glucose (or amino acids) in blood. Enables cells to absorb glucose and coordinates synthesis of fats, proteins, and glycogen.

integral membrane protein Any membrane protein that spans the entire membrane. Also called transmembrane protein.

integration Processing of information from many sources.

integrator A component of an animal's nervous system that evaluates sensory information and triggers appropriate responses.

integrins A class of transmembrane proteins that binds to fibronectins in the extracellular matrix, thus holding cells in place.

intercalated discs The physical connections between adjacent heart muscle cells. Contain gap junctions to allow electrical signals to pass between the cells.

intermediate disturbance hypothesis The hypothesis that moderate ecological disturbance causes high species diversity.

intermediate filament Any of various long, thin cellular fibers composed of thin filaments of various protein polymers wound into thicker cables. Forms part of a cell's cytoskeleton.

intermediate host The host species in which a parasite reproduces asexually.

interneuron A neuron that passes information between two other neurons.

internode The section of a plant stem between two nodes (sites where leaves attach).

interphase The part of the cell cycle during which no cell division occurs. Includes the G_1 phase, the S phase, and the G_2 phase.

interspecific competition Competition between members of different species for the same limited resource.

interstitial fluid The plasma-like fluid found in the spaces between cells.

interstitial space Any area between two cells. Usually filled with fluid called interstitial fluid.

intertidal zone The region between the low-tide and high-tide marks on a seashore.

intracytoplasmic sperm injection (ICSI) A technique used to help infertile couples bear children, by injecting a single sperm directly into the cytoplasm of the egg.

intraspecific competition Competition between members of the same species for the same limited resource.

intrinsic rate of increase The rate at which a population will grow under optimal conditions (i.e., when birthrates are as high as possible and death rates are as low as possible). Symbolized by r_{max}.

intron A region of a eukaryotic gene that is transcribed into RNA but is later excised from the mRNA transcript before translation into protein.

invasive species An exotic species that, upon introduction to a new area, spreads rapidly and competes with native species.

invertebrates All multicellular animals that are not vertebrates. This is a paraphyletic group.

in vitro Outside the living body. Refers to a process or experiment done "in glass" (i.e., in a dish or test tube rather than in an intact living body).

in vitro fertilization (IVF) A technique for fertilizing mammalian eggs with sperm in a lab dish.

in vivo In a living body. Refers to a process or experiment done "in life."

ion An atom or a molecule that has gained or lost electrons and carries an electric charge.

ion channels A class of membrane proteins that allow certain ions to diffuse across a plasma membrane, via passive transport.

ionic bond A bond between atoms that is formed when an electron is completely transferred from one atom to another so that the atoms remain associated due to their opposite electric charges.

ionophore Any compound that increases membrane permeability for a certain ion. Many ionophores are membrane proteins that bind to the ion and carry it through the membrane.

iris A ring of pigmented muscle just inside the vertebrate eye that contracts or expands to control the amount of light entering the eye.

islets of Langerhans Clusters of cells in the pancreas that secrete insulin and glucagon directly into the blood.

isomer A molecule that has the same molecular formula as another molecule but differs from it in three-dimensional structure.

isometric (1) Characterized by scaling that is proportionate to some other feature, such as body size. (2) A type of muscle contraction in which the muscle exerts force without shortening.

isotonic Having the same solute concentration and water concentration relative to another solution.

isotope Any of several forms of an element that have the same number of protons but differ in the number of neutrons.

jelly layer A gelatinous layer that encloses the egg cells of some vertebrates.

jet propulsion Movement accomplished by forcibly ejecting water in the opposite direction. Used by various aquatic invertebrates.

joint A place where two pieces (bones, cartilages, etc.) of a skeleton meet. May be movable (an articulated joint) or immovable (e.g., skull sutures).

junctional diversity Genetic diversity created by variations in the joining of gene segments; occurs during gene recombination in lymphocytes of the immune system.

juvenile hormone An insect hormone that prevents larvae from metamorphosing into adults.

karyogamy Fusion of two haploid nuclei to form a diploid nucleus. Occurs in many fungi, and in animals and plants during fertilization of gametes.

karyotype The distinctive appearance of all of the chromosomes in an individual, including the number of chromosomes and their length and banding patterns (after staining with dyes).

keystone species A species that has an exceptionally great impact on the other species in its ecosystem relative to its abundance.

kidneys Paired organs situated at the back of the abdominal cavity that filter the blood, produce urine, and secrete several hormones.

kilocalorie (kcal) A unit of energy, often used to measure energy content of food. Also called a Calorie.

kinesin A motor protein that uses the chemical energy of ATP to transport vesicles, particles, or chromosomes along microtubules.

kinetic energy The energy of motion, as opposed to potential energy (stored in position or shape).

kinetochore The structure on chromatids where spindle fibers attach. Contains motor proteins that move the chromosome along the microtubule.

kingdom A classic taxonomic rank that is above the phylum level and below the domain level.

kinocilium A single cilium that juts from the surface of many hair cells and functions in detection of sound or pressure.

kin selection A form of natural selection that favors traits that increase survival or reproduction of an individual's kin at the expense of the individual.

Klinefelter syndrome A syndrome seen in humans who have an XXY karyotype. People with this syndrome have male sex organs, may have some female traits, and are sterile.

knock-out mutant A mutant allele that does not function at all, or an organism homozygous for such a mutation. Also called null mutant or loss-of-function mutant.

Koch's postulates Four criteria used to determine whether a putative infectious agent causes a particular disease.

Krebs cycle A series of chemical reactions, during which acetyl CoA (from glycolysis) is broken down to CO_2, producing ATP and reduced compounds for the electron transport chain. Occurs in the mitochondria.

labia majora (plural: labium majus) One of two outer folds of skin that protect the labia minora, clitoris, and vaginal opening of female mammals.

labia minora (plural: labium minus) One of two inner folds of skin that protect the opening of the urethra and vagina.

labor The strong muscular contractions of the uterus that expel the fetus during birth.

lac operon The operon in *E. coli* that includes genes responsible for metabolism of lactose.

lactation Production of milk from mammary glands of mammals.

lacteal Lymphatic vessels in the center of the villi of the small intestine. Receive chylomicrons containing fat absorbed from food and send them into the lymph system.

lactic acid fermentation Production of ATP without an electron transport chain, and the end product of glycolysis (pyruvate) is converted to lactic acid.

lagging strand In DNA replication, the strand of new DNA that is synthesized discontinuously in a series of short pieces that are later joined together. Also called discontinuous strand.

lamellae (singular: lamella) Any set of parallel platelike structures (e.g., the crescent-shaped flaps on the gill filaments of fish gills that serve to increase surface area for gas exchange).

large intestine The posterior part of the intestine, between the small intestine and the rectum.

larva (plural: larvae) An immature stage of a species in which the immature and adult stages have different body forms.

late endosome A membrane-bound vesicle, formed by endocytosis, that is a late stage in the process of becoming a lysosome.

lateral bud A bud on a plant stem that is capable of producing a new side branch.

lateral gene transfer Transfer of DNA between two different species, especially distantly related species. Commonly occurs among bacteria and archaea via plasmid exchange; also can occur in eukaryotes via viruses and some other mechanisms.

lateral meristem A layer of undifferentiated plant cells found in older stems and roots. Responsible for secondary growth. Also called cambium or secondary meristem.

lateral root A plant root extending from another, older root.

LDL See low-density lipoprotein (LDL).

leaching Loss of nutrients from soil via percolating water.

leading strand In DNA replication, the strand of new DNA that is synthesized in one continuous piece, with nucleotides added to the 3′ end of the growing molecule. Also called continuous strand.

leaf primordium (plural: leaf primordia) Small protuberances that form around an apical meristem of a plant shoot and that will develop into leaves.

leak channels The potassium channels in the membranes of nerve cells, which allow potassium ion to leak out of the cell in its resting state.

learning An enduring change in an individual's behavior that results from specific experience(s).

leghemoglobin An iron-containing protein similar to hemoglobin. Found in root nodules of legume plants where it binds oxygen, preventing it from poisoning a bacterial enzyme needed for nitrogen fixation.

legumes Members of the pea plant family. Form symbiotic associations with nitrogen-fixing bacteria in their roots.

lens A transparent, crystalline structure that focuses incoming light onto a retina or other light-sensing apparatus of an eye.

leptin A hormone produced by fat cells (adipocytes) that signals how much body fat is stored. Inhibits appetite.

leukocytes Immune system cells, including neutrophils, macrophages, B cells, and T cells, that circulate in blood or lymph and function in defense against disease. Also called white blood cells.

lichen A symbiotic association of a fungus and a photosynthetic alga.

life cycle The sequence of developmental events and phases that occurs during the life span of an organism, from fertilization to offspring production.

life history The sequence of events in an individual's life, from birth to reproduction to death. Also, the study of how the organism allocates resources and energy to these different activities.

life table A data set that summarizes the probability that an individual in a certain population will survive and reproduce in any given year over the course of its lifetime.

ligand Any molecule that binds to a specific site on a receptor molecule.

ligand-gated channel An ion channel that opens or closes in response to binding by a certain molecule.

light chain One of two identical short polypeptides forming part of an immunoglobulin protein (antibody, B-cell receptor, etc.). The light chains determine which antigen the molecule will bind to.

light-dependent reactions In photosynthesis, the set of reactions that use the energy of sunlight to split water, producing ATP, NADPH, and oxygen.

light-independent reactions In photosynthesis, the set of reactions that use the NADPH and ATP formed earlier (in the light-dependent reactions) to drive the reduction of atmospheric CO_2, ultimately producing sugars. Also called the Calvin cycle.

lignin An extremely strong polymer in the secondary cell walls of plant cells in woody plant parts. Composed of six-carbon rings joined.

lignin peroxidase A fungal enzyme that can digest lignin.

limiting nutrient An essential nutrient whose scarcity in the environment significantly affects growth and reproduction.

limnetic zone Open water (not near shore) that receives enough light to support photosynthesis.

lineage An evolutionary unit that includes an ancestral population, all of its descendants, and only its descendants. Also called a monophyletic group or a clade.

linkage A physical association between two genes because they are on the same chromosome; the inheritance patterns resulting from this association.

linkage map A map of the relative locations of specific genes on a certain chromosome. Also called a genetic map or meiotic map.

lipase Any enzyme that can digest fats.

lipid A carbon-containing subtance that is hydrophobic and thus does not dissolve in water, but dissolves well in nonpolar organic solvents. Lipids include fats, oils, phospholipids, and waxes.

lipid bilayer A double layer of phospholipid molecules, with hydrophobic tails oriented toward the inside and hydrophilic heads toward the outside. The fundamental component of membranes.

liposome An artificially formed tiny membrane-bound structure composed of a phospholipid bilayer membrane.

lithotroph An organism that produces ATP by oxidizing inorganic molecules with high potential energy, such as ammonia (NH_3) or methane (CH_4).

littoral zone Shallow waters near shore that receive enough sunlight to support photosynthesis. May be marine or freshwater.

liver An abdominal organ of vertebrates that performs many biochemical processes, including storage of glycogen, processing and conversion of food and wastes, and production of bile.

live virus vaccine A vaccine containing intact viruses that have been rendered nonvirulent for a

certain species. Used for vaccines. Also called an attenuated virus vaccine.

lobe-finned fishes A lineage of fishes with fins supported by an arrangement of bones and muscles similar to that seen in tetrapod limbs. Includes two living groups: coelacanths and lungfishes.

lock-and-key model The hypothesis that enzymes have a precise three-dimensional structure into which substrates (reactant molecules) fit. Modified to reflect knowledge that enzymes are not rigid and may change shape.

locomotion Movement of an organism under its own power.

locus (plural: loci) A gene's physical location on a chromosome.

logistic growth equation The mathematical equation that defines how fast a population will grow over time, given a certain carrying capacity.

logistic population growth Changes in growth rate that occur as a function of population size.

long-day plant A plant that blooms in midsummer, in response to short nights.

long interspersed nuclear element (LINE) A type of parasitic DNA sequence commonly found in eukaryotic genomes. Contains genes for reverse transcriptase, integrase, and a promoter and can create copies of itself, inserted elsewhere in the genome.

loop of Henle A long loop of the nephrons in mammalian kidneys. Functions to set up a concentration gradient that allows reabsorption of water from a subsequent section of the nephron.

loose connective tissue A type of connective tissue consisting of fibrous proteins in a soft matrix. Often functions as padding for organs.

lophophore A specialized feeding structure found in three phyla of the Lophotrochozoa and used in filter feeding.

Lophotrochozoa The lineage of protostomes that includes mollusks and annelids. Many phyla have lophophore feeding structures, have trochophore larvae, and grow by extending the size of their skeletons rather than by molting.

loss-of-function mutant A mutant allele that does not function at all, or an organism homozygous for such a mutation. Also called knock-out mutant or null mutant.

low-density lipoprotein (LDL) Balls of protein and fat that transport cholesterol from the liver to the rest of the body for storage and use. Associated with increased risk of heart disease.

lumen The interior space of any hollow structure (e.g., the rough ER) or organ (e.g., the stomach).

lung Any respiratory organ used for gas exchange between blood and air.

luteal phase The second major phase of a menstrual cycle, after ovulation, when the progesterone levels are high and the body is preparing for a possible pregnancy.

luteinizing hormone (LH) A peptide hormone from the anterior pituitary that stimulates estrogen production, ovulation, and formation of the corpus luteum in females and testosterone production in males.

lymph The mixture of fluid and lymphocytes that circulates through the ducts and lymph nodes of the lymphatic system in vertebrates.

lymphatic duct One of the thin-walled tubes that collect excess fluid from body tissues and return it to the circulatory system, passing through lymph nodes along the way.

lymphatic system In vertebrates, a network of thin-walled tubes that collects excess fluid from body tissues and returns it to the veins of the circulatory system. Includes lymph nodes.

lymph nodes Small oval structures through which lymph ducts run. Filter the lymph and screen it for infection.

lymphocyte A type of leukocyte that circulates through the bloodstream and lymphatic system and that is responsible for the development of acquired immunity. Includes B cells and T cells.

lysogeny A type of viral replication in which the viral DNA is inserted into the host's chromosome, remaining there indefinitely and passively replicating whenever the host cell divides.

lysosome A small organelle in an animal cell containing acids and enzymes that catalyze hydrolysis reactions and can digest large molecules. In plants, fungi and some other groups, may be called vacuoles.

lysozyme An enzyme that acts as an antibiotic by digesting bacterial cell walls. Occurs in saliva, tears, mucus, and egg white.

lytic replication cycle A type of viral replication in which new virus particles are made inside a host cell and eventually burst out of the cell, killing it.

macromolecule Any very large molecule, usually made up of smaller molecules joined together. Include proteins, nucleic acids, and polysaccharides.

macronutrient An essential nutrient required in large quantities. Usually a major component of many organic molecules.

macrophage A type of leukocyte, capable of amoeboid movement through body tissues, that engulfs and digests pathogens and other foreign particles. Also plays roles in secreting cytokines and presenting foreign antigens to other immune system cells.

MADS box A DNA sequence found in some genes of fungi, animals, and plants (genes that control pattern formation in flowers). Codes for a stretch of 58 amino acids that can bind to DNA.

major histocompatibility proteins (MHC proteins) Mammalian cell-surface glycoproteins involved in immunity and in marking cells as "self" to the immune system.

maladaptive For a trait, reducing the fitness of individuals.

malaria A human disease caused by four species of the protist *Plasmodium*, which is passed to humans by mosquitoes.

male gametophyte A multicellular haploid structure that can produce haploid sperm cells in a species that exhibits alternation of generations. In flowering plants, consists of two sperm cells within a larger tube cell.

malignant tumor A tumor that is actively growing and disrupting local tissues and/or is spreading to other organs. A cancer consists of one or more malignant tumors.

Malpighian tubules A major excretory organ of insects, consisting of blind-ended tubes that extend from the gut into the hemocoel. Filter hemolymph to form pre-urine and then send it to the hindgut for further processing.

mammary glands Specialized skin glands that can produce milk for nursing offspring. A diagnostic feature of mammals.

mandibles Any chewing mouthparts. In vertebrates, the lower jaw. In insects, crustaceans, and myriapods, the first pair of mouthparts.

mantle In mollusks, the thick outer tissue that protects the visceral mass and that may secrete a calcium carbonate shell.

Marfan syndrome A human syndrome involving increased height, long limbs and fingers, an abnormally shaped chest, and heart disorders.

mark-recapture study A method of estimating population size involving release of marked individuals into a population and an assessment of how many individuals captured later are marked vs. unmarked.

marsh A wetland that lacks trees and usually has a slow but steady rate of water flow.

marsupial A member of the taxon Marsupiala, a lineage of mammals that nourish their young in an abdominal pouch after a very short period of development in the uterus.

mass extinction Rapid extinction of an unusually large number of diverse evolutionary groups across a wide geographic area. May occur due to sudden and extraordinary environmental changes.

mass number The total number of protons and neutrons in an atom.

mast cell A type of leukocyte that is stationary (embedded in tissue) and that helps trigger the inflammatory response to infection or injury, including secretion of histamine. Particularly important in allergic responses and defense against parasites.

master plate In replica plating, the original plate of bacteria that is used as the source for producing replica plates for further study.

maternal chromosome A chromosome inherited from the mother.

maternal effect inheritance A pattern of inheritance in which an individual's phenotype is determined by its mother's genotype. Common in embryological development, during which egg components made by the mother can influence development of the offspring.

mating type A form of fungal hyphae that carries certain alleles for the genes involved in mating and mates only with hyphae of a different mating type.

matrix A general term for any liquid or semisolid that fills or surrounds some structure and that has some role in maintaining the shape or function of organelles, cells, or tissues.

medulla The innermost part of an organ (e.g., kidney or adrenal gland). In the brain, the posteriormost portion, responsible for rhythmic body functions such as heart rate, respiration, and digestion.

medullary respiratory center An area at the base of the brain (in the region known as the medulla oblongata) that stimulates breathing.

medusa (plural: medusae) The free-floating stage of a cnidarian life cycle.

megapascal (MPa) A unit of pressure (force per unit area), equivalent to 1 million pascals (Pa).

megaphyll A leaf of a fern, horsetail, or vascular plant.

megasporangium (plural: megasporangia) In seed plants, a spore-producing structure that produces megaspores, which can grow to become the female gametophytes.

megaspore A type of small spore produced by the megasporangia of seed plants and that can grow to become a female gametophyte.

megasporocyte A diploid cell contained in the megasporangium of a flower ovule. Undergoes meiois to produce four megaspores.

meiosis A type of cell division in which one diploid parent cell produces four haploid reproductive cells (gametes). In meiosis, chromosome pairs synapse and can exchange genes via crossing over.

meiosis I The first cell division of meiosis, in which synapsis and crossing over occur, and homologous chromosomes are separated from each other, producing daughter cells with half as many chromosomes as the parent cell.

meiosis II The second cell division of meiosis, in which sister chromatids are separated from each other. Similar to mitosis.

meiotic map A map of the relative locations of specific genes on a certain chromosome. Also called a genetic map or linkage map.

membrane potential A difference in electric charge across a cell membrane; a form of potential energy. Also called membrane voltage.

membrane protein Any protein found in a cell membrane. May span the entire membrane (transmembrane or integral membrane proteins) or may be found on only one side of the membrane (peripheral membrane proteins).

memory Retention of learned information.

memory cell A type of lymphocyte responsible for maintenance of immunity for years or decades after an infection. Descendants of B or T cells activated during a previous infection.

meniscus (plural: menisci) The concave boundary layer formed at most air-water interfaces, due to surface tension.

menstrual cycle A female reproductive cycle seen in Old World monkeys and apes (including humans), consisting of a follicular phase, ovulation, a luteal phase, and then (if no pregnancy occurs) menstruation.

menstrual synchrony The phenomenon in which human women living in close proximity experience synchronization of their menstrual cycles.

menstruation The periodic shedding of the uterine lining through the vagina that occurs in females of Old World monkeys and apes, including humans.

meristem A group of undifferentiated plant cells that can produce cells that differentiate into specific adult tissues.

mesoderm One of three basic embryonic cell layers of a triploblast animal. Forms the middle tissues between skin and gut: muscles, bones, blood, and some internal organs (kidney, spleen, etc.).

mesoglea A gelatinous material with scattered ectodermal cells, found in cnidarians between ectoderm and endoderm.

mesophyll cell A type of cell found near the surfaces of plant leaves and where most photosynthesis occurs.

messenger RNA (mRNA) An RNA molecule that carries encoded information, transcribed from DNA, and that can be used for the synthesis of one or more proteins.

meta-analysis An analysis that combines and compares the results of many smaller, previously published studies.

metabolic pathway A series of distinct chemical reactions that build up or break down a particular molecule. Often, each reaction is catalyzed by a different enzyme.

metabolic rate The total energy consumption of all the cells of an individual. For aerobic organisms, often measured as the amount of oxygen consumed per hour.

metabolic water The water that is produced as a by-product of cellular respiration.

metabolism All the chemical reactions occurring in a living cell or organism.

metallothioneins Small plant proteins that bind to and prevent excess metal ions from acting as toxins.

metamorphosis A dramatic change from the larval to the adult form of an animal.

metaphase A stage in cell division (mitosis or meiosis) during which chromosomes line up in the middle of the cell.

metaphase plate The plane along which chromosomes line up during metaphase of cell division (mitosis or meiosis).

metapopulation A population of a single species that is divided into many smaller populations.

metastasis The process by which cancerous cells leave the primary tumor and establish additional tumors elsewhere in the body.

methanogen A group of archaea that produce methane (CH_4; natural gas) as a by-product of cellular respiration.

methanotroph An organism that uses methane (CH_4; natural gas) as its primary electron donor and source of carbon.

methylation The addition of a methyl ($-CH_3$) group to a molecule.

MHC See **major histocompatibility complex proteins (MHC proteins)**.

micelle A small droplet of similar molecules clumped together in a solution.

microbe Any microscopic organism. Includes bacteria, archaea, and various tiny eukaryotes.

microfilament A long, thin fiber composed of two intertwined strands of polymerized actin. Involved in cell movement. Also called actin filament.

micrograph A photograph of an image produced by a microscope.

micronutrient An essential nutrient required in very small quantities—usually an enzyme cofactor.

microphyll A type of small leaf on lycopods.

micropyle The tiny pore in a plant ovule through which the pollen tube reaches the embryo sac.

microsatellite A noncoding stretch of eukaryotic DNA that contains a repeating sequence one to five base pairs long. A type of simple sequence repeat.

microsporangium (plural: microsporangia) In seed plants, a spore-producing structure that produces microspores, which can grow to become the male gametophytes.

microspore A small haploid spore produced by the microsporangia of seed plants and that can grow to become a male gametophyte.

microsporidians A lineage of single-celled, parasitic eukaryotes that are closely related to fungi.

microsporocytes Diploid cells contained within the microsporangium of a flower anther. Undergo meiosis to produce microspores.

microtubule A long, tubular polymer of protein subunits α-tubulin and β-tubulin. Involved in cell movement and transport of materials within the cell.

microtubule organizing center Any structure that organizes microtubules.

microvilli (singular: microvillus) Tiny protrusions from the surface of an epithelial cell that increase the surface area for absorption of substances.

middle ear The air-filled middle portion of the mammal ear, connecting to the throat via the Eustachian tube. Transmits and amplifies sound from the tympanic membrane to the inner ear.

middle lamella A central layer of gelatinous pectins between the primary cell walls of adjacent plant cells. Helps hold the cells together.

migration (1) A cyclical movement of large numbers of organisms from one geographic location or habitat to another. (2) In population genetics, movement of individuals from one population to another.

millivolt (mV) A unit of voltage equal to 1/1000 of a volt.

mimicry A phenomenon in which one species has evolved (or learns) to look or sound like another species.

minisatellite A noncoding stretch of eukaryotic DNA that contains a repeating sequence 6 to 500 base pairs long. A type of simple sequence repeat.

mismatch repair The process by which mismatched base pairs in DNA are fixed.

missense mutation A point mutation (change in a single base pair) that causes a change in the amino acid sequence of a protein. Also called replacement mutation.

mitochondrial DNA DNA found inside the mitochondria of eukaryotic cells.

mitochondrial matrix The solution inside the inner membrane of a mitochondrion. Contains the enzymes of the Krebs cycle.

mitochondrion (plural: mitochondria) A eukaryotic organelle that is the site of aerobic respiration.

mitosis Nuclear division of a eukaryotic cell producing two daughter nuclei that are genetically identical to the parent.

mitosis-promoting factor (MPF) A complex of two proteins (cyclin and cyclin-dependent kinase) that causes eukaryotic cells to initiate mitosis.

mitotic phase (M phase) The phase of the cell cycle during which cell division occurs. Includes mitosis and cytokinesis.

mitotic spindle An array of microtubules that moves chromosomes to opposite sides of the cell during cell division (mitosis or meiosis).

model organism An organism selected for intensive scientific study based on features that make it easy to work with (e.g., body size, life span), in hope that the findings will apply to other species.

molarity The number of moles of a dissolved solute in 1 liter of solution; a unit of concentration.

mole An amount of any substance containing 6.022×10^{23} atoms, ions, or molecules. This number of molecules of a compound will have a mass equal to the molecular weight of that compound expressed in grams.

molecular chaperones A class of proteins that facilitate the three-dimensional folding of newly synthesized proteins.

molecular clock The hypothesis that certain types of mutations tend to reach fixation in populations at a steady rate over large spans of time. As a result, comparisons of DNA sequences can be used to infer the timing of evolutionary divergences.

molecular formula A notation that indicates the numbers and types of atoms in a molecule, such as H_2O for the water molecule.

molecular ion An ion consisting of a group of several different atoms (rather than one element).

molecular weight The sum of the mass numbers of all of the atoms in a molecule; roughly, the total number of protons and neutrons in the molecule.

molecule Two or more atoms held together by covalent bonds.

molting A method of body growth, used by ecdysozoans, that involves the shedding of an external protective cuticle or skeleton, expansion of the soft body, and growth of a new external layer.

monoclonal antibody An antibody produced in the lab from a hybridoma derived from a single B cell. Such an antibody has a unique amino acid sequence and thus the ability to bind to a specific site of a particular antigen.

monocot A plant that has a single cotyledon (embryonic leaf) upon germination.

monoecious Having both male and female flowers.

monohybrid cross A mating between two parents that are both heterozygous for a given gene.

monomer A small molecular subunit that can bond to other subunits to form long macromolecules, or polymers.

monophyletic group An evolutionary unit that includes an ancestral population, all of its descendants, and only its descendants. Also called a clade or a lineage.

monosaccharide A single sugar monomer, such as glucose. Formally, a small carbohydrate of the chemical formula $(CH_2O)_n$ that cannot be hydrolyzed to form any smaller carbohydrates.

monosomy Having only one copy of a particular type of chromosome.

monotremes A member of the Monotremata, a lineage of mammals that lay eggs and then nourish the young with milk. Includes just three living species: the platypus and two species of echidna.

morphogenesis A process of embryologic development during which cells become organized into recognizable tissues, organs, and other structures.

morphology The shape and appearance of an organism's body and its component parts.

morphospecies concept The concept that species are best identified as groups that have measurably different anatomical features.

motif A repeating theme. In molecular biology, a domain (a section of a protein with a distinctive tertiary structure) in many different proteins.

motile Not sessile (not permanently attached to a substrate); capable of moving to another location.

motor neuron A nerve cell that carries signals from the central nervous system (brain and spinal cord) to an effector, such as a muscle or gland.

motor protein A protein whose major function is to convert the chemical energy of ATP into motion.

mRNA See **messenger RNA (RNA)**.

mucigel A slimy substance secreted by plant root caps to ease passage of the growing root through the soil.

mucins Glycoproteins, produced by salivary glands, that form mucus when mixed with water.

mucus A slimy mixture of glycoproteins and water, secreted by many organs for lubrication.

Müllerian inhibitory substance A peptide hormone secreted by the embryonic testis that causes regression (withering away) of the female reproductive ducts.

Müllerian mimicry A type of mimicry in which two (or more) harmful species resemble each other.

multicellular growth Growth in which individual cells join together to form a multicelled body and cells differentiate to perform specialized roles.

multicellularity The condition whereby an organism's body contains more than one cell and only certain cells pass genes to the next generation.

multiple allelism The occurrence of more than two alleles of a gene in a given population.

multiple sclerosis (MS) A human autoimmune disease caused by the immune system attacking the myelin sheaths that insulate nerve axons.

muscle fiber A single muscle cell.

muscle tissue A class of animal tissue consisting of bundles of long, thin contractile cells (muscle fibers).

mutagen Anything that can increase the rate of mutation.

mutant An individual that carries a mutation, particularly a new or rare mutation.

mutation Any change in the hereditary material of an organism (DNA in most organisms, RNA in some viruses).

mutualism A relationship between two species that benefits both species.

mutualist An organism that lives in a close relationship with a host and that benefits its host.

Myb proteins DNA-binding proteins that turn gene transcription on or off, acting as transcription activators or transcription repressors, respectively.

mycelium (plural: mycelia) A mass of underground filaments (hyphae) that form the body of a fungus. Also found in some protists and bacteria.

mycorrhiza (plural: mycorrhizae) A mutualistic association between certain fungi and most vascular plants, sometimes visible as nodules or nets in or around plant roots.

myelin sheath Multiple layers of myelin, a lipid, that are wrapped around the axons of neurons to provide electrical insulation.

myeloma A cancer or tumor of the cells of bone marrow. Myelomas of B cells are used in lab production of monoclonal antibodies.

MyoD A regulatory protein involved in differentiation of muscle cells during embryological development. Enhances transcription of muscle-specific genes.

myofibril A bundle of strands of contractile proteins organized into repeating units (sarcomeres) in vertebrate heart muscle and striated muscle.

myoglobin An oxygen-binding muscle protein consisting of a single globin and one heme group.

myosin A eukaryotic protein that can be polymerized to form thick filaments that are used in muscle contraction and intracellular movement.

natural history The branch of biology concerned with describing what exists in nature (i.e., primarily observational and not experimental).

natural selection The process by which individuals with certain heritable traits tend to produce more surviving offspring than do individuals without those traits, resulting in a change in the genetic makeup of the population. A major mechanism of evolution.

nauplius A distinct planktonic larval stage seen in many crustaceans.

Neanderthal A recently extinct European species of hominid, *Homo neanderthalensis*, closely related to but distinct from modern humans.

nectar The sugary fluid produced by flowers to attract and reward pollinating animals.

nectary The nectar-producing gland at the base of a flower.

negative control A type of gene regulation in which a transcription can occur only when a certain substance—typically a repressor protein that binds to a control sequence in the DNA and prevents transcription—is removed.

negative feedback A self-limiting, corrective response in which a deviation in some variable (e.g., body temperature, blood pH) triggers responses aimed at returning the variable to normal.

negative pressure ventilation Ventilation of the lungs that is accomplished by "pulling" air into the lungs by expansion of the rib cage.

negative result An experimental result that fails to show a predicted difference between two groups or conditions.

negative-sense virus A virus whose genome contains sequences complementary to those in the mRNA required to produce viral proteins.

nephron One of the tiny tubes within the vertebrate kidney that filter blood and concentrate salts to produce urine.

neritic zone Shallow marine waters beyond the intertidal zone, extending down to about 200 m, where the continental shelf ends.

Nernst equation A formula that converts the energy of a concentration gradient to the energy of an electrical potential, for a particular ion or ions.

nerve A long, tough strand of nervous tissue typically containing thousands of axons that carry information to or from the central nervous system.

nervous tissue A class of animal tissue consisting of nerve cells (neurons) and various supporting cells, and functioning in rapid transmission of complex information.

net primary productivity (NPP) In ecology, the amount of primary productivity that is stored in new biomass (not used for cellular respiration).

net reproductive rate The growth rate of a population per generation; equivalent to the average number of female offspring that each female produces over her lifetime. Symbolized by R_0.

neural Relating to nerve cells.

neural tube A folded tube of ectoderm that forms along the dorsal side of a young vertebrate embryo and that will give rise to the brain and spinal cord.

neuroendocrine Refers to nerve cells that release hormones into the blood.

neuron A nerve cell; a cell that is specialized for the transmission of nerve impulses. Typically has dendrites, a cell body, and a long axon that forms synapses with other neurons.

neurosecretory cell A neuron that secretes hormones into the blood; a neuroendocrine cell.

neurotoxin Any poison that specifically affects neuron function.

neurotransmitter A molecule that conveys information from one neuron to another or from a neuron to a muscle or gland. Released from the end of an axon and diffuse a very short distance to the next cell, where they can trigger an action potential.

neutral Any mutation that has no effect on an individual's fitness.

neutron An uncharged particle found in atomic nuclei. Variations in neutron number (with no change in proton number) produce different isotopes of the same element.

neutrophil A type of leukocyte, capable of amoeboid movement through body tissues, that engulfs and digests pathogens and other foreign particles and secretes various compounds that attack bacteria and fungi.

niche The particular set of habitat requirements of a certain species and the role that species plays in its ecosystem.

niche differentiation The tendency of competing species to use different ecological niches because of competition.

nitrogen fixation The incorporation of atmospheric nitrogen (N_2) into forms such as ammonia (NH_3) or nitrate (NO_3^-), which can be used to make many organic compounds. Occurs in only a few lineages of bacteria and archaea.

nociceptor A sensory cell or organ specialized to detect tissue damage, usually producing the sensation of pain.

node (1) Any small thickening (e.g., a lymph node); (2) The part of a stem where leaves or leaf buds are attached. (3) The point on a phylogenetic tree where two branches diverge, representing the point in time when an ancestral group split into two or more descendant groups.

node of Ranvier A point on a neuron's axon between sections of myelin sheath, where an action potential can be regenerated.

Nod factor Molecules produced by nitrogen-fixing bacteria that help them recognize and bind to legume roots.

nodules The lumplike structures in roots of the pea family, containing symbiotic nitrogen-fixing bacteria.

nondisjunction An error that can occur during meiosis or mitosis in which both homologous chromosomes of a pair move to the same side of the dividing cell. One daughter cell receives two copies of the chromosome, and the other daughter cell receives none.

nonenveloped For a virus, lacking an envelope surrounding its capsid coat.

nonpolar covalent bond A symmetrical covalent bond (i.e., one in which electrons are equally shared between the two atomic nuclei).

non-self Property of a molecule or cell whereby immune system cells will recognize it as "different" and attack it.

nonsense mutation A point mutation (change in a single base pair) that results in an early stop codon, resulting in a truncated polypeptide.

non-sister chromatids The chromosome copies of homologous chromosomes. Crossing over occurs between non-sister chromatids.

non-template strand The strand of DNA that is not transcribed to create RNA.

nonvascular plants Several phyla of green plants that lack vascular tissue. Includes liverworts, hornworts, and mosses. Also called bryophytes.

norepinephrine A catecholamine used as a neurotransmitter in the sympathetic nervous system and also released as a hormone from the adrenal medulla. Stimulates increased heart rate, increases blood pressure, decreases digestion, and produces other effects.

norm of reaction The range of phenotypes that are possible for an individual of a given genotype.

Northern blotting A lab technique for identifying the RNA produced by a particular gene. Involves separating RNAs by gel electrophoresis, transferring to a filter paper, and hybridizing to a labeled DNA probe.

notochord A long, gelatinous, supportive rod down the back of a chordate embryo, below the developing spinal cord. Replaced by vertebrae in adult vertebrates. A defining feature of chordates.

nuclear envelope The double-layered membrane enclosing the nucleus of a eukaryotic cell.

nuclear lamina A lattice-like sheet of fibrous nuclear lamin proteins that line the inner membrane of the nuclear envelope. Stiffens the envelope and helps organize the chromosomes.

nuclear lamins A class of fibers that form a dense mesh (the nuclear lamina) just inside the nuclear envelope. A form of intermediate filament.

nuclear localization signal (NLS) A certain sequence of amino acids that tags a protein for delivery to the nucleus by importins.

nuclear pore An opening in the nuclear envelope that connects the inside of the nucleus with the cytoplasm and through which molecules such as mRNA and some enzymes can pass.

nuclear pore complex A large complex of dozens of proteins lining a nuclear pore, defining its shape and transporting substances through the pore.

nuclease Any enzyme that can break down RNA and DNA molecules.

nucleic acids Polymers consisting of a chain of nucleotides. Generally used by cells to store or transmit hereditary information. Includes ribonucleic acid and deoxyribonucleic acid.

nucleoid The region of a prokaryotic cell that contains chromosomes.

nucleolus The structure in a eukaryotic nucleus where ribosomal RNA processing occurs and ribosomal subunits are assembled.

nucleoside A purine or pyrimidine base attached to a five-carbon sugar (ribose or deoxyribose).

nucleosome A repeating, bead-like structure of a eukaryotic chromosome, consisting of about 200 nucleotides of DNA wrapped twice around eight histone proteins.

nucleotide A monomer that can be polymerized to form the nucleic acid DNA or RNA. Consists of a five-carbon sugar (ribose or deoxyribose), a phosphate group, and one of several nitrogen-containing bases. Equivalent to a nucleoside plus one phosphate group.

nucleus (1) The center of an atom, containing protons and neutrons. (2) In eukaryotic cells, the membrane-bound organelle that contains DNA. (3) A discrete clump of neuron cell bodies in the brain, usually sharing a distinct function.

null hypothesis A hypothesis that specifies what the results of an experiment will be if the main hypothesis being tested is wrong. Often states that there will be no difference between experimental groups.

null mutant A mutant allele that does not function at all; or an organism homozygous for such a mutation. Also called knock-out mutant or loss-of-function mutant.

nutrient A substance that an organism requires for normal growth, maintenance, or reproduction.

nutritional balance A state in which an organism is taking in enough nutrients to maintain normal health and activity.

nymph An immature stage of an insect species in which the immature form looks like a miniature adult, such as that in dragonflies.

obesity The condition of having extremely and abnormally high reserves of body fat.

occipital lobe In the mammal brain, the posteriormost region of each cerebral hemisphere. Receives and interprets visual information.

oceanic zone The waters of the open ocean beyond the continental shelf.

oil A lipid that is liquid at room temperature.

Okazaki fragments Short fragments of DNA produced during DNA replication. Now known to be pieces of the lagging strand.

olfaction The perception of odors.

olfactory bulb A bulb-shaped projection of the brain just above the nose. Receives and interprets odor information from the nose.

oligopeptide A polypeptide with fewer than 50 amino acids. May also be referred to simply as "peptides."

ommatidium (plural: ommatidia) A light-sensing column of an arthropod's compound eye.

omnivore An animal whose diet regularly includes both meat and plants.

oncogene An allele that has mutated so as to stimulate cell growth at all times and thus promotes cancer development.

one-gene, one-enzyme hypothesis The hypothesis that each gene is responsible for making one (and only one) particular protein, in most cases an enzyme that catalyzes a specific reaction. Many exceptions to this hypothesis are now known.

oocyte A cell in the ovary that can undergo meiosis to produce an ovum.

oogenesis The production of egg cells (ova).

oogonia (singular: oogonium) The diploid cells in an ovary that can divide by mitosis to create more oogonia and primary oocytes, which can undergo meiosis.

open circulatory system A circulatory system in which the circulating fluid (hemolymph) is not confined to blood vessels.

open reading frame (ORF) Any DNA sequence that is suspected to be a functional gene because it has a start codon and a stop codon separated by a long stretch of DNA (and sometimes has other characteristic features, such as promoters).

operator The DNA binding site for a repressor protein in a prokaryotic operon.

operculum The stiff flap of tissue that covers the gills of teleost fishes.

operon A region of bacterial DNA that codes for a series of functionally related genes.

opsin One of several proteins involved in animal vision. An opsin joins with retinal to form the light-detecting pigment rhodopsin in rod cells.

optical isomer A molecule that shares the same molecular formula as another molecule but differs in the arrangement of atoms or groups around a carbon atom; left-handed or right-handed form of a molecule.

optic nerve A nerve that carries information from the eye to the brain. Vertebrates have two.

orbital The region around an atomic nucleus in which an electron orbits.

order A classic taxonomic rank above the family level and below the class level.

organ A group of tissues organized into a functional and structural unit.

organelle Any discrete, membrane-bound structure in the cytoplasm of a cell (e.g., mitochondrion).

organic For a compound, containing carbon and hydrogen and usually containing carbon-carbon bonds. Organic compounds are widely used by living organisms.

organism Any living entity that contains one or more cells.

organizer A region of an amphibian embryo (around the upper side of the blastopore) that can organize the development of the entire embryo.

organogenesis A stage of embryonic development just after gastrulation in vertebrate embryos, during which major organs develop from the three embryonic cell layers.

organotroph An organism that produces ATP, oxidizing organic molecules with high potential energy, such as sugars.

orientation A deliberate movement that results in a change in position relative to some external cue, such as toward the Sun or away from a sound.

origin of replication The place on a chromosome at which DNA replication begins.

osmoconformer An animal that does not actively regulate the osmolarity of its tissues but conforms to the osmolarity of the surrounding environment.

osmolarity The concentration of dissolved substances in a solution, measured in moles per liter.

osmoregulation The process by which a living organism controls the concentration of water and salts in its body.

osmoregulator An animal that actively regulates the osmolarity of its tissues.

osmosis Diffusion of water across a selectively permeable membrane from areas of high water concentration (low solute concentration) to areas of low water concentration (high solute concentration).

osmotic stress A condition in which there are abnormal concentrations of water and salts in an organism's cells or tissues.

ouabain A plant toxin that poisons the sodium-potassium pumps of animals.

outcrossing Reproduction by fusion of the gametes of different individuals, rather than self-fertilization. Typically refers to plants.

outer ear The outermost portion of the mammal ear, consisting of the pinna (ear flap) and the ear canal. Funnels sound to the tympanic membrane.

outgroup A taxon known to have diverged earliest from all the other taxa under study. Used to determine the root (most ancient node) of a phylogenetic tree.

out-of-Africa hypothesis The hypothesis that modern humans (*Homo sapiens*) evolved in Africa and spread to other continents, replacing other *Homo* species without interbreeding with them.

oval window A membrane separating the fluid-filled cochlea from the air-filled middle ear. The stapes, a middle ear bone, transmits sound vibrations to the cochlea by vibrating on the oval window.

ovary The egg-producing organ of a female animal, or the seed-producing structure of the female part of a flower.

oviduct A narrow tube that connects the uterus to the ovary, and through which the egg travels after ovulation. Fertilization and cleavage occur in the oviduct. In humans, called the fallopian tube.

oviparous Reproducing by laying eggs, rather than giving live birth.

ovoviviparous Reproducing by retaining eggs inside the body until they are ready to hatch. The embryos are nourished by egg yolk, not via a placenta.

ovulation The process by which an ovum is released from the ovary of a female vertebrate.

ovule A structure inside a flower ovary that produces the female gametophyte and eventually (if fertilized) becomes a seed.

ovum (plural: ova) An egg cell; a mature female gamete and any associated external layers. Larger and less mobile than the male gamete.

oxidation The loss of electrons from an atom during a redox reaction, either by donation of an electron to another atom or by the shared elec-trons in covalent bonds moving farther from the atomic nucleus.

oxidative phosphorylation Production of ATP molecules from the redox reactions of an electron transport chain, starting with reduced compounds (such as NADH or $FADH_2$) and ending with oxygen as the final electron acceptor.

oxygenic photosynthesis Photosynthesis that involves photosystem II, which catalyzes the splitting of water and produces oxygen. Occurs in cyanobacteria, algae, and plants.

oxytocin A peptide hormone from the posterior pituitary that triggers labor and milk production in females and that stimulates pair bonding, parental care, and affiliative behavior in both sexes.

p53 A tumor-suppressor protein that responds to DNA damage by stopping the cell cycle and/or triggering apoptosis. Codes for a protein with a molecular weight of 53 kilodaltons.

pacemaker cells Cells with an inherent rhythm that can set a rhythm for other cells.

pair-rule genes A class of fruit-fly segmentation genes that organize embryonic cells into particular segments. Active in alternating segments.

paleontology The study of organisms that lived in the distant past.

palisade mesophyll Elongated parenchyma cells found in the ground tissue of leaves. Contain many chloroplasts and perform most photosynthesis.

pancreas A gland attached to the small intestine that secretes digestive enzymes into the intestine and several digestion-related hormones (notably, insulin and glucagon) into the bloodstream.

pancreatic amylase A carbohydrate-digesting enzyme secreted by the pancreas into the small intestine.

pancreatic lipase A fat-digesting enzyme secreted by the pancreas into the small intestine.

paper chromatography A technique for separating molecules on the basis of their size and solubility, by wicking them through a piece of filter paper using a certain solvent.

parabiosis An experimental technique for determining whether a certain physiological phenomenon is regulated by a hormone, by surgically uniting two individuals so that hormones can pass between them.

parabronchi (singular: parabronchus) The tiny parallel air tubes that run through a bird's lung.

paracrine signal A chemical signal that is released by one cell and affects neighboring cells.

parafollicular cells The cells of the thyroid gland that release calcitonin.

paraphyletic group A group of organisms that is not monophyletic; i.e, a group that includes some but not all descendants of the last common ancestor of the group. Paraphyletic groups are not meaningful evolutionary groups.

parasite An organism that lives on or in a host species and that damages its host.

parasitism A long-term relationship between two organisms that is beneficial to one organism (the parasite) but detrimental to the other (the host).

parasitoid An organism that has a parasitic larval stage and a free-living adult stage. Most parasitoids are insects that lay eggs in the bodies of other insects.

parasympathetic nervous system The part of the autonomic nervous system that stimulates activities of relaxation, repair, and rebuilding, such as reduced heart rate and increased digestion.

parathyroid glands Four small glands that are near or embedded in the thyroid gland of vertebrates. Secrete parathyroid hormone, which increases blood calcium.

parathyroid hormone (PTH) A peptide hormone from the parathyroid glands that increases blood calcium.

parazoans Animals whose cells are not organized into distinct tissues. Contains one living group: the sponges.

parenchyma cell A thin-walled type of plant cell found in leaves, the centers of stems and roots, and fruits. Involved in photosynthesis, starch storage, and new growth.

parental care Any action by which an animal expends energy or assumes risks to benefit its offspring (e.g., nest-building, feeding of young, defense).

parental generation The adult organisms used in the first experimental cross in a formal breeding experiment.

parietal cells The cells in the stomach lining that secrete hydrochloric acid.

parietal lobe In the mammal brain, the region of each cerebral hemisphere that is behind and above the frontal lobe. In humans, involved in integrating sensory and motor control.

Parkinson's disease A human neurological disorder that causes progressive deterioration of motor function. Due to the inactivation or destruction of dopamine-secreting neurons at the base of the brain.

parsimony The principle that the phylogenetic tree most likely to be correct is the one that requires the fewest evolutionary changes.

parthenogenesis Development of offspring from unfertilized eggs. A form of asexual reproduction.

partial pressure The pressure of one particular gas in a mixture; the contribution of that gas to the overall pressure.

particle A nonliving infectious entity, such as a virus. Also called an agent.

pascal (Pa) A unit of pressure (force per unit area).

passive transport Diffusion of a substance across a cell membrane down a concentration gradient. When this occurs with the assistance of carrier proteins, it is also called facilitated diffusion.

patch clamping A lab technique for studying the electrical currents that flow through individual ion channels, by sucking a tiny patch of membrane to the hollow tip of a microelectrode.

paternal chromosome A chromosome inherited from the father.

pathogen Any entity capable of causing disease, such as a microbe, virus, or prion.

pathogenic Capable of causing disease.

pattern formation The series of events that determines the spatial organization of an embryo, including alignment of the major body axes and orientation of the limbs.

pattern-recognition receptor Leukocyte membrane proteins that bind to molecules in many bacteria. Part of the innate immune response.

PCR See **polymerase chain reaction (PCR)**.

peat Semidecayed organic matter that accumulates in moist, low-oxygen environments such as bogs.

pectins A class of gelatinous polysaccharides found in the primary cell walls of plant cells. Attract and hold water, forming a gel that helps keep the cell wall moist.

pedigree A family tree of parents and offspring, showing inheritance of particular traits of interest.

pellet Any solid material that collects at the bottom of a test tube below a layer of liquid (the supernatant) during centrifugation.

penis The copulatory organ of male mammals, used to insert sperm into a female.

pentaradial symmetry A form of radial symmetry, found in adult echinoderms, in which the body has exactly five planes of symmetry. Typically, five (or multiples of five) body parts radiate from a central hub.

pentose A monosaccharide (simple sugar) with five carbons.

PEP carboxylase An enzyme that catalyze addition of CO_2 to three-carbon compounds, forming four-carbon compounds. Occurs in mesophyll cells of plants that perform C_4 photosynthesis.

pepsin A protein-digesting enzyme produced in the stomach.

pepsinogen The precursor of the digestive enzyme pepsin. Converted to pepsin by the acidic environment of the stomach.

peptide bond The C–N bond between two amino acid residues in a peptide or protein.

peptide hormone A hormone that is a chain of two or more amino acids.

peptidoglycan A polysaccharide found in bacterial cell walls.

per-capita rate of increase The growth rate of a population, expressed per individual. Calculated as the per-capita birthrate minus the per-capita death rate and symbolized r. Also called per-capita growth rate.

perennial plant A plant that normally lives for more than one year.

perfect For a flower, containing both male parts (stamens) and female parts (carpels).

perforations In plants, small holes in the primary and secondary cell walls of vessel elements that allow passage of water.

pericarp The part of a fruit that surrounds the seeds, formed from the ovary wall. The flesh of most edible fruits; the hard shells of most nuts.

pericycle In plant roots, a layer of cells that give rise to lateral roots.

peripheral membrane protein Any membrane protein that is found only on one side of the membrane, rather than spanning the entire membrane.

peripheral nervous system (PNS) All the components of the nervous system that are outside the central nervous system (the brain and spinal cord). Includes the somatic nervous system and the autonomic nervous system.

peristalsis Rhythmic waves of muscular contraction that push food along the digestive tract.

permafrost A permanently frozen layer of icy soil found in most tundra and some taiga.

permeability The tendency of a structure, such as a membrane, to allow a given substance to diffuse across it.

peroxisome A eukaryotic organelle that contains oxidative enzymes, usually for degrading fatty acids and amino acids and the resulting hydrogen peroxide.

petal One of the modified leaves arranged around the reproductive structures of a flower. Often colored to attract pollinators.

petiole The stalk of a leaf.

phagocytosis The engulfment and uptake of a small particle or cell by an extension of another cell's plasma membrane.

pharyngeal gill slits A set of parallel openings from the throat through the neck to the outside. One of the diagnostic traits of chordates.

pharyngeal jaw A secondary jaw in the back of the mouth, found in some fishes. Derived from modified gill arches.

phenetic approach A method for constructing a phylogenetic tree by computing a statistic that summarizes the overall similarity among populations, based on the available data.

phenotype Any of the observable traits of an individual. Commonly includes physical, physiological, and behavioral traits.

phenotypic variation The total observable variation in a particular trait in a certain population in a certain environment.

phenylketonuria (PKU) A human genetic disease caused by lack of the enzyme that converts the amino acid phenylalanine to tyrosine.

pheophytin A molecule that acts as an electron acceptor in the light-dependent reactions of photosynthesis, accepting excited electrons from chlorophyll and passing them to an electron transport chain.

pheromone A chemical signal, released by one individual into the external environment, that can trigger responses in a different individual.

phloem A plant vascular tissue that conducts sugars. Contains sieve-tube members and companion cells.

phloem loading The movement of sugars into plant phloem.

phloem sap The sugary fluid found in phloem tissue of plants.

phloem unloading The movement of sugars out of plant phloem.

phonotaxis Orientation toward or away from sound.

phosphate The functional group $-OPO_3^{2-}$. Breaking the O–P bonds between adjacent phosphate groups releases large amounts of energy.

phosphodiester bond The type of bond that links the nucleotides in DNA or RNA. Joins the phosphate group of one nucleotide to the hydroxyl group on the sugar of another nucleotide.

phosphofructokinase The enzyme that catalyzes the synthesis of fructose-1,6-bisphosphate from fructose-6-phosphate, a key reaction (step 3) in glycolysis.

phospholipid A type of lipid having a hydrophilic head (a phosphate group) and a hydrophobic tail (one or more fatty acids), often linked by a glycerol molecule. Major components of plasma membranes.

phosphorylase An enzyme that breaks down glycogen, by catalyzing hydrolysis of the α-glycosidic linkages between the glucose monomers.

phosphorylation The addition of a phosphate group to a molecule. Commonly, refers to phosphorylation of proteins to control protein shape or function.

phosphorylation cascade A sequence of events in which one enzyme phosphorylates other enzymes, which in turn phosphorylate many more, leading to phosphorylation of thousands of proteins. Commonly used in signal transduction of hormone messages; also called signal transduction cascade.

photic zone In an aquatic habitat, water that is shallow enough to receive some sunlight (whether or not it is enough to support photosynthesis).

photon   A discrete packet of light energy; a particle of light.

photoperiodism   Any response by an organism to photoperiod—the relative lengths of day and night.

photophosphorylation   Production of ATP molecules by using the energy of light to excite electrons, which are then passed down an electron transport chain. Occurs during photosynthesis.

photoreceptor   A molecule, a cell, or an organ that is specialized to detect light.

photorespiration   A series of light-driven chemical reactions that consumes oxygen and releases carbon dioxide, basically reversing photosynthesis. Usually occurs when there are high O_2 and low CO_2 concentrations inside plant cells, often in bright, hot, dry environments when stomata must be kept closed.

photoreversibility   A change in conformation that occurs in certain plant pigments when they are exposed to their preferred wavelengths of light and that triggers responses by the plant.

photosynthesis   A series of chemical reactions and electron transfer events that converts the energy of light into chemical energy stored in glucose.

photosynthesis-transpiration compromise   The balance that plants must strike between maximizing photosynthesis and conserving water.

photosystem   A system of 200–300 chlorophyll molecules, accessory pigments, and proteins, found in plant chloroplasts and involved in the light-dependent reactions of photosynthesis.

photosystem I   A system of molecules and enzymes in chloroplasts that absorbs light energy, using it to produce NADPH.

photosystem II   A system of molecules and enzymes in plant chloroplasts that absorbs light energy, using it to produce ATP and to split water into protons and oxygen.

phototaxis   Orientation toward or away from light.

phototroph   An organism that produces ATP through photosynthesis.

phototropins   A class of plant photoreceptors that detect blue light and initiate phototropic responses.

phototropism   Growth or movement in a particular direction in response to light.

pH scale   A measure of the concentration of protons in a solution and thus of acidity or alkalinity. Defined as the negative of the base-10 logarithm of the proton concentration: $pH = -\log[H^+]$.

phylogenetic species concept   The concept that species are best identified as the smallest monophyletic group in a phylogenetic tree.

phylogenetic tree   A diagram that depicts the evolutionary history of a group of organisms.

phylogeny   The evolutionary history of a group of organisms.

phylum   (plural: phyla) A classic taxonomic rank above the class level and below the kingdom level. (In plants, sometimes called a division.)

physical map   A map of a chromosome that shows the number of base pairs between various genetic markers.

physiology   The study of how an organism's body functions.

phytoalexin   Any small plant compound produced to combat an infection (usually a fungal infection).

phytochromes   A class of light-sensitive plant proteins involved in detecting light and timing cer-

tain physiological processes, such as flowering and germination.

phytoplankton   Plankton (small drifting aquatic organisms) that are photosynthetic.

phytoremediation   The use of plants to clean contaminated soils.

pigment   Any molecule that absorbs only certain wavelengths of visible light and reflects or transmits other wavelengths.

piloting   Finding one's way by using familiar landmarks (i.e., without a specific compass direction).

pinocytosis   Uptake of extracellular fluid by endocytosis (i.e., by pinching off the plasma membrane to form small membrane-bound vesicles).

pioneering species   Species that are often the first to appear in a recently disturbed area.

pitch   The sensation produced by a particular frequency of sound. Low frequencies are perceived as low pitches, high frequencies as high pitches.

pith   The center of a plant stem.

pits   In plants, small holes in the secondary cell walls of tracheids that allow passage of water.

pituitary dwarfism   Dwarfism (abnormally small body size) in mammals that is caused by defects in production of growth hormone by the pituitary gland.

pituitary gland   A small gland directly under the brain, close to the hypothalamus. Releases hormones that affect many other glands and organs.

placenta   An organ formed by a union of maternal and fetal tissues. Exchanges nutrients and wastes between mother and fetus, anchors the fetus to the uterine wall, and produces some hormones. Occurs in most mammals and in a few other vertebrates.

placental mammals   Members of the Eutheria, a major lineage of mammals whose young develop in the uterus and are not housed in an abdominal pouch. Also called eutherians.

planar bilayer   A lipid bilayer (double-layered membrane) constructed across a hole in a glass or plastic wall separating two aqueous solutions.

plankton   Any small organism that drifts near the surface of oceans or lakes and swims little if at all.

plant   A red alga, green alga, glaucophyte alga, or green plant.

plant-defense hypothesis   The hypothesis that rates of herbivory are limited by plant defenses such as toxins and spines.

plantlet   A small plant, particularly one that forms on a parent plant via asexual reproduction and drops, becoming an independent individual.

plasma   The fluid portion of blood that remains when red blood cells, leukocytes, and platelets are removed. (Equivalent to serum plus clotting factors.)

plasma cell   A type of leukocyte that produces large quantities of antibodies to combat an ongoing infection. A descendant of an activated B cell.

plasma membrane   A membrane that surrounds a cell, separating it from the external environment and selectively regulating passage of molecules and ions into and out of the cell.

plasmid   A small, usually circular, supercoiled DNA molecule independent of the cell's main chromosome(s) in prokaryotes and some eukaryotes.

plasmodesmata   (singular: plasmodesma) Physical connections between two plant cells, consisting of gaps in the cell walls through which the two cells' plasma membranes, cytoplasm, and smooth ER can connect directly.

plasmogamy   Fusion of the cytoplasm of two individuals. Occurs in many fungi.

plastocyanin   A small protein that shuttles electrons from photosystem II to photosystem I during photosynthesis.

plastoquinone   The molecule involved in the light-dependent reactions of photosynthesis that receives excited electrons from pheophytin and passes them to more electronegative molecules in the chain. Also carries protons to the lumen side of the thylakoid membrane.

platelet   A small membrane-bound cell fragment in vertebrate blood, important in blood clotting.

platelet-derived growth factor (PDGF)   A protein secreted by platelets and some other cells at the site of an injury. Promotes wound healing.

plate tectonics   The theory that Earth's crust is made up of separate plates that have moved throughout geologic history.

pleiotropy   A pattern of genetic expression in which one gene affects more than one phenotypic trait.

ploidy   The number of each type of chromosome present. Haploid cells have a ploidy of 1; diploid cells have a ploidy of 2.

pneumonia   Inflammation of the lungs.

podium   (plural: podia) The part of the tube foot of an echinoderm that extends outside of the body and makes contact with the substrate.

point mutation   A mutation that results in a change in a single nucleotide pair of DNA.

polar   (1) Asymmetrical or unidirectional. (2) Carrying a partial positive charge on one side of a molecule and a partial negative charge on the other. Polar molecules are generally hydrophilic.

polar bodies   The tiny, nonfunctional cells produced during meiosis of a primary oocyte, due to most of the cytoplasm going to the ovum.

polar covalent bond   A covalent bond that is asymmetrical, such that the electrons spend more time near one atomic nucleus than the other. Often results in the molecule being polar.

polar nuclei   (singular: polar nucleus) The nuclei in the female gametophyte of a flowering plant that will fuse with one sperm nucleus to produce the endosperm. Most species have two.

pollen grain   In flowering plants, a male gametophyte enclosed within a protective coat.

pollen tube   In flowering plants, a tube that grows out of a pollen grain and toward the ovule and through which the two sperm nuclei move.

pollination   The process by which pollen reaches the carpel of a flower (in flowering plants) or reaches the ovule directly (in conifers and their relatives).

poly (A) tail   In eukaryotes, a long sequence of 100–250 adenine nucleotides added to the 3' end of newly transcribed messenger RNA molecules.

polycistronic   For an mRNA molecule, containing more than one protein-coding segment, each with its own start and stop codons and each coding for a different protein. Common in prokaryotes.

polyclonal antibody   A mix of several different antibodies that all bind to the same antigen, typically produced by a living animal after injection with the antigen.

polygenic inheritance   The inheritance patterns that result when many genes influence one trait.

polymer   Any long molecule composed of small repeating subunits (monomers) bonded together.

polymerase chain reaction (PCR) A lab technique for rapidly generating millions of identical copies of a specific stretch of DNA. Involves incubating the original DNA sequence with primers, free nucleotides, and DNA polymerase.

polymerization The process by which one monomer (a small subunit molecule) is bound to others to form a polymer (a long chain molecule).

polymorphism (1) The occurrence of more than one allele at a certain genetic locus in a population. (2) The occurrence of more than two distinct phenotypes of a trait in a population.

polyp The sessile stage of a cnidarian life cycle.

polypeptide A peptide of three or more amino acids linked together in a chain. Very large polypeptides may be called proteins.

polyploid Having more than two chromosome sets.

polyribosome A structure consisting of one messenger RNA molecule along with many attached ribosomes and their growing peptide strands. Occurs in prokaryotes and eukaryotes.

polysaccharide A large carbohydrate polymer consisting of many monosaccharides linked together in a chain.

polyspermy Fertilization of an egg by multiple sperm. This is usually an abnormal situation.

polytomy A multibranched node on a phylogenetic tree. Represents a time when an ancestral population split into descendant populations.

pons A small region of the brain that relays information to the cerebellum and is also involved in control of breathing.

poor-nutrition hypothesis The hypothesis that herbivore populations are limited by the poor nutritional content of plants, especially low nitrogen.

population A group of individuals of the same species living in the same area at the same time.

population bottleneck An extreme reduction in population size, followed by re-expansion to a larger population size. May cause genetic bottleneck.

population cycles Regular fluctuations in size exhibited by certain populations.

population density The number of individuals of a population per unit area.

population dynamics Changes in population size through time.

population ecology The study of how and why the number of individuals in a population changes over time.

population viability analysis (PVA) A method of estimating the likelihood that a population will avoid extinction for a given time period.

pore Any small opening, such as the small opening in the stoma of a plant leaf or in the septa of fungal filaments.

positive control A type of gene regulation in which transcription can occur only when a certain substance—typically an activator protein that binds to a control sequence in the DNA and promotes transcription—is present.

positive feedback Stimulation of a reaction or process by the end result of that process. Tends to accelerate processes rapidly.

positive pressure ventilation Ventilation of the lungs that is accomplished by "pushing" air into the lungs by positive pressure in the mouth.

positive-sense virus A virus whose genome contains the same sequences as the mRNA required to produce viral proteins.

posterior Toward an animal's tail and away from its head. The opposite of anterior.

posterior pituitary The posterior part of the pituitary gland, consisting of the ends of neurosecretory cells from the hypothalamus, which secrete oxytocin and antidiuretic hormone.

postsynaptic neuron A neuron that receives neurotransmitters from another neuron at a particular synapse.

post-translational control Regulation of gene expression by modification of proteins after translation.

post-translational modification Any chemical alteration of a protein after it has been synthesized; includes cleavage of side chains, phosphorylation.

postzygotic isolation Result of mechanisms that prevent gene flow between different species even if mating occurs between them, typically due to death of hybrid embryos or reduced fitness of hybrids.

potential energy Energy stored in matter through its position or its shape, as opposed to kinetic energy (the energy of motion).

power stroke During muscle contraction, the motion of a myosin "head" that pulls myosin and actin filaments further along each other.

prairie An extensive grassland. Typically found in the dry interiors of continents in the temperate latitudes. Also called a steppe.

prebiotic soup A hypothetical solution of sugars, amino acids, nitrogenous bases, and other building blocks of larger molecules that may have formed in shallow waters or deep-ocean vents of ancient Earth and given rise to larger biological molecules.

predation The killing and eating of one organism by another.

predator Any organism that kills other organisms for food.

presentation Display of an antigen on the surface of an immune system cell to recruit other immune system cells.

pressure-flow hypothesis The hypothesis that sugar movement through phloem tissue is due to differences in the turgor pressure of phloem sap.

pressure potential Potential energy of water caused by pressure differences. Equals the sum of all the types of pressure that affect water, such as atmospheric pressure, wall pressure, and tension.

presynaptic neuron A neuron that releases neurotransmitters to another neuron at a synapse.

pre-urine The fluid in kidney nephrons that is formed by filtration of blood. Upon further processing, becomes urine. Also called filtrate.

prey A species that is commonly attacked and eaten by a predator species.

prezygotic isolation Anything that prevents individuals of two different species from mating.

primary cell wall The outermost layer of a plant cell wall, made of cellulose fibers with gelatinous polysaccharides, that defines the shape of the plant cell and withstands the turgor pressure of the plasma membrane.

primary consumer An herbivore; an organism that eats plants, algae, or other primary producers.

primary decomposer A decomposer that consumes detritus from plants.

primary growth Plant growth that results in an increase in length of stems and roots. Produced by apical meristems.

primary immune response An immune response to a pathogen that the immune system has not encountered before.

primary oocyte The large diploid cell in an ovarian follicle that can initiate meiosis to produce a haploid ovum.

primary producer An autotroph; a species that creates its own food through photosynthesis or from reduced inorganic compounds and that is a source of food for other species in its ecosystem.

primary productivity The total amount of carbon fixed by photosynthesis per unit area per year, including that used for cellular respiration.

primary RNA transcript In eukaryotes, a newly transcribed messenger RNA molecule that has not yet been processed—i.e., that still contains introns and has not received a 5′ cap or a poly (A) tail.

primary sex determination The process by which an embryonic gonad becomes either a testis or an ovary.

primary spermatocyte A diploid cell in the testis that can initiate meiosis I to produce two secondary spermatocytes.

primary structure The sequence of amino acids in a peptide or protein; also the sequence of nucleotides in a nucleic acid.

primary succession The gradual colonization of a habitat of bare rock or gravel, usually after an environmental disturbance that removes all soil and previous organisms.

primase An enzyme that synthesizes a short stretch of RNA to use as a primer during DNA replication.

primates The lineage of mammals that includes prosimians (lemurs, lorises, etc.), monkeys, and apes (including humans).

primer A short, single-stranded sequence of RNA that enables the start of replication of a DNA sequence.

principle of independent assortment The concept that each pair of hereditary elements (alleles of the same gene) behaves independently of other genes during meiosis. One of Mendel's two principles of genetics.

principle of segregation The concept that each pair of hereditary elements (alleles of the same gene) separate from each other during the formation of offspring (i.e., during meiosis). One of Mendel's two principles of genetics.

prion An infectious protein that is thought to cause disease by inducing other proteins to assume an abnormal three-dimentional structure. Likely cause of spongiform encephalopathies, such as mad cow disease.

probe A single-stranded fragment of a labeled, known DNA sequence that will bind to a complementary sequence in the sample being analyzed.

proboscis A long, narrow feeding appendage through which food can be obtained.

procambium A group of cells in the center of a young plant embryo that will give rise to the vascular tissue.

product The final atoms or molecules after a chemical reaction has occurred.

productivity The amount of energy used by a certain component of an ecosystem (e.g., a community or a trophic level) in a given area over a given time period.

progesterone A steroid hormone produced, along with estrogens, in the ovaries. Secreted by the corpus luteum after ovulation; causes the uterine lining to thicken.

prokaryote A member of the domain Bacteria or Archaea; a unicellular organism lacking certain complex cell features, such as a membrane-bound nucleus and gene introns.

prolactin A peptide hormone from the pituitary gland that promotes milk production in female mammals and that has a variety of effects on parental behavior and seasonal reproduction in other vertebrates.

prolactin-inhibiting hormone (PIH) A peptide hormone from the hypothalamus that inhibits release of prolactin from the anterior pituitary.

prometaphase A stage of cell division (mitosis or meiosis), during which the nuclear envelope breaks down and spindle fibers attach to chromatids.

promoter A short sequence of DNA that facilitates binding of RNA polymerase to enable transcription of downstream genes.

promoter-proximal elements In eukaryotes, regulatory sequences that are close to a promoter and that can bind regulatory transcription factors.

prophase The first stage of cell division (mitosis or meiosis) during which chromosomes become visible and the mitotic spindle forms. Synapsis and crossing over occur during prophase of meiosis I.

proplastid A colorless organelle found in undifferentiated plant cells that matures to become a plastid.

prosimians One of the two major lineages of primates, including lemurs, tarsiers, pottos, and lorises but not monkeys or apes.

protease An enzyme that can break apart proteins, by cleaving the peptide bonds between amino acids.

protein A long chain of 50 or more amino acids linked together; a large polypeptide.

proteinase inhibitors Defensive compounds produced by plants that block the enzymes in animal digestive tracts responsible for digesting proteins.

protein kinase An enzyme that catalyzes the addition of a phosphate group to another protein, typically activating or inactivating the other protein.

proteoglycans A type of glycoprotein commonly found in the extracellular matrix. Proteoglycans have a greater carbohydrate content and are larger than most other glycoproteins.

proteomics The study of the three-dimensional structure and function of proteins.

protist Any microscopic eukaryote that is not a green plant, animal, or fungus.

protoderm The exterior layer of a young plant embryo. Gives rise to the epidermis.

proton A small, positively charged particle in atomic nuclei. The number of protons in an atom gives that atom its characteristics as an element.

proton-motive force The combined effect of a proton gradient and an electric potential gradient across a membrane, which can drive protons across the membrane. Used by mitochondria and chloroplasts to power ATP synthesis.

proton pump A membrane protein that uses the energy of ATP to transport protons across the membrane against an electrochemical gradient.

proto-oncogene Any gene that encourages cell growth, typically by triggering specific phases in the cell cycle.

protostomes A major lineage of animals that share a pattern of embryological development, including spiral cleavage, formation of the mouth earlier than the anus, and formation of the coelom by splitting of a block of mesoderm. Includes arthropods, mollusks, and annelids.

proximal Toward or from the center of the body; away from the furthest tip of an appendage. The opposite of distal.

proximal tubule The convoluted section of a kidney nephron into which filtrate moves from Bowman's capsule. Involved in active reabsorption of certain solutes and water.

proximate causation In biology, the immediate, mechanistic cause of a phenomenon (how it happens), as opposed to the ultimate cause (why it evolved). Also called proximate explanation.

pseudocoelom A body cavity that forms between the endoderm and mesoderm layers. Occurs in roundworms (nematodes) and rotifers.

pseudogene A DNA sequence that closely resembles a working gene but is not transcribed. Thought to have arisen by duplication of the working gene followed by accidental inactivation due to a mutation.

pseudopodium (plural: pseudopodia) A mobile, outward bulging of a cell's plasma membrane, used in cell crawling or ingestion of food.

P site The site in a ribosome where peptide bonds are formed between amino acids.

puberty The process by which an immature animal attains reproductive maturity.

pulmonary artery A short, thick-walled artery that carries oxygen-poor blood from the heart to the lungs.

pulmonary circulation The part of the circulatory system that sends oxygen-poor blood to the lungs. In many vertebrates, the pulmonary circulation is separate from the rest of the circulatory system (the systemic circulation).

pulmonary vein A short, thin-walled vein that carries oxygen-rich blood from the lungs to the heart.

pulse-chase experiment A lab technique for marking a population of cells or molecules at a particular moment in time by means of a labeled molelcule and then following their fate over time.

Punnett square A diagram that depicts the genotypes and phenotypes that should appear in offspring of a certain cross.

pupa A metamorphosing insect that is enclosed in a protective case.

pupation A developmental stage of many insects, in which the body metamorphoses from the larval form to the adult form while enclosed in a protective case.

pupil The hole in the center of the iris through which light enters a vertebrate or cephalopod eye.

pure line In animal or plant breeding, a strain of individuals that produce offspring identical to themselves when self-pollinated or crossed to another member of the same population. Pure lines are homozygous for most, if not all, genetic loci.

purines A class of small, nitrogen-containing, double-ringed bases (guanine, adenine) found in nucleotides.

pyramid of productivity The characteristic pattern of productivity in ecosystems, in which productivity declines with each higher trophic level.

pyrimidines A class of small, nitrogen-containing, single-ringed bases (cytosine, uracil, thymine) found in nucleotides.

pyruvate dehydrogenase A large enzyme complex, located in the inner mitochondrial membrane, that is responsible for conversion of pyruvate to acetyl CoA during glycolysis.

Q A molecule in mitochondria that helps transfer electrons between the complexes of the electron transport chain during cellular respiration. Also called coenzyme Q or ubiquinone.

quadrat A small rectangular plot set up in a habitat to mark an area under intensive study.

quantitative trait A phenotypic trait that exhibits variation along a smooth, continuous scale of measurement (for example, human height), rather than the distinct forms seen in discrete traits.

quantitative variation Variation that exhibits differences in degree across a smooth, continuous scale of measurement.

quaternary structure The three-dimensional shape of several polypeptide chains and sometimes other small functional groups, arranged together to form a large, multiunit protein.

race A population that has different characteristics from another population of the same species, whether or not there are significant genetic differences between them.

radial cleavage The pattern of embryonic cleavage seen in protostomes, in which cells divide at right angles to each other to form tiers.

radial symmetry An animal body pattern in which there are least two planes of symmetry. Typically, the body is in the form of a cylinder or disk, with body parts radiating from a central hub.

radiation Production of electromagnetic energy (light). One of the mechanisms of heat transfer between organisms and the environment.

radicle The root of a plant embryo.

radioactive decay A spontaneous change in the mass number of an atomic nucleus via emission of radiation or a particle.

radioactive isotope An isotope that spontaneously emits radiation (gamma rays) and/or subatomic particles. In the latter case, the isotope decays to form a different isotope or element.

radiometric dating A technique for determining the age of a rock by measuring the amount of radioactive decay that has occurred since the rock solidified, usually by measuring the amount of new daughter element that has formed.

radula A rasping feeding appendage in gastropods.

rain shadow The dry region on the side of a mountain range away from the prevailing wind.

rays Lines of parenchyma cells that extend laterally through the xylem of plant wood.

reactant The atoms or molecules in a chemical reaction, in their starting states.

reaction center A central area in a photosystem of the light-dependent reactions of photosynthesis. Surrounded by chlorophyll molecules of the antenna complex and receives energy from them.

reactive oxygen intermediates (ROIs) Oxygen-containing compounds that are highly reactive and that are used in plant and animal cells to kill infected cells and for other purposes.

reading frame The division of a sequence of DNA or RNA into a particular series of three-nucleotide codons. There are three possible reading frames for any sequence.

realized niche The ecological niche that a species occupies in the presence of competitors.

receptor (1) A molecule, cell, or group of cells specialized for detecting environmental signals. (2) A molecule that binds to a particular chemical (e.g., hormone, sperm protein) and triggers a cellular response. (3) A cell-surface molecule necessary for a virus to gain entry to a cell.

receptor-mediated endocytosis Endocytosis triggered by the binding of certain macromolecules outside the cell to membrane proteins.

receptor tyrosine kinases Transmembrane proteins involved in signal transduction. Typically bind to a signalling molecule, triggering a phosphorylation cascade of other proteins inside the cell.

recessive Property of an allele whereby its influence on phenotype can be entirely hidden by the presence of another, dominant allele.

reciprocal altruism Altruistic behavior that is exchanged between a pair of individuals at different points in time (i.e., sometimes individual A helps individual B, and sometimes B helps A).

reciprocal cross A breeding experiment in which the mother's and father's phenotypes are the reverse of that examined in a previous breeding experiment.

recognition sequence A sequence of amino acids that binds to a specific sequence of DNA and that is found within the helix-turn-helix motif of many repressor proteins.

recognition site The specific sequence of DNA bases cut by a certain restriction endonuclease.

recombinant Possessing a new combination of alleles. May refer to a single chromosome or an entire organism.

recombinant DNA Any DNA altered by exchange with, or inclusion of, foreign DNA. May be produced via meiosis, viruses, or lab manipulation.

recombinant DNA technology A variety of lab techniques for isolating specific DNA fragments and introducing them into different regions of DNA and/or a different host organism. Also called biotechnology or genetic engineering.

recombination A change in the combination of genes or alleles on a given chromosome or in a given individual. Also called genetic recombination.

rectal gland A salt-excreting gland in the digestive system of sharks, skates, and rays.

rectum The last part of the digestive tract; a short tube that holds feces until they are expelled.

red blood cells Hemoglobin-containing cells that circulate in the blood and deliver oxygen from the lungs to the tissues.

redox reactions A class of chemical reactions that involve the loss and gain of electrons. Also called reduction-oxidation reactions.

reduction An atom's gain of electrons during a redox reaction, either by acceptance of an electron from another atom or by the electrons in covalent bonds moving closer to the atomic nucleus.

reduction-oxidation reactions A class of chemical reactions that involve the loss and gain of electrons. Also called redox reactions.

redundant code A code in which different sequences can represent the same information. The genetic code is redundant because some amino acids are coded for by two or three different codons.

reflex An involuntary response to environmental stimulation. May involve the brain (e.g., conditioned reflex) or not (e.g., spinal reflex).

refractory No longer responding to stimuli that previously elicited a response. For example, the tendency of voltage-gated sodium channels to remain closed immediately after an action potential.

regeneration Growth of a new body part to replace a lost body part.

regulatory cascade In embryonic development, a progressive series of interactions among genes and/or cytoplasmic determinants that organizes the body plan of the embryo.

regulatory evolution The evolution of new body parts and patterns via changes in the regulatory genes that affect pattern formation in embryos.

regulatory homeostasis Steady internal body conditions achieved by active physiological processes.

regulatory protein Any protein that affects gene transcription by binding to specific enhancers, silencers, or other sites in DNA.

regulatory sequence Any section of DNA that is involved in controlling the activity of other genes.

regulatory site A site on an enzyme to which a regulatory molecule can bind and affect the enzyme's activity, separate from the active site where catalysis occurs.

regulatory transcription factors Proteins that bind to eukaryotic enhancers, silencers, and promoter-proximal elements, but not to the promoter itself.

reinforcement Natural selection for traits that prevent interbreeding between recently diverged species.

release factors Proteins that can trigger termination of RNA translation when a ribosome reaches a stop codon.

releaser A simple stimulus that elicits an invariant, stereotyped behavioral response (fixed action pattern) from an animal. Also called a sign stimulus.

releasing hormone A hormone that stimulates release of some other hormone.

renal corpuscle The ball-like structure at the beginning of a kidney nephron, consisting of Bowman's capsule surrounding a glomerulus.

replacement mutation A point mutation (change in a single base pair) that causes a change in the amino acid sequence of a protein. Also called missense mutation.

replacement rate The reproductive rate at which each female produces two surviving offspring over her entire life—enough to exactly replace herself and her mate, resulting in zero population growth.

replica plate In replica plating, a copy of the master plate produced by transferring bacteria from it with a velvet-covered block.

replica plating A method of identifying bacterial colonies that have certain mutations by observing their growth on a plate that is exposed to different conditions.

replicated chromosome A chromosome that has been copied; consists of two chromatids joined at the centromere.

replication The exact copying of something—e.g., DNA replication.

replication fork The Y-shaped site at which a double-stranded molecule of DNA is separated into two single strands for replication.

repolarization A return to a normal membrane potential after a depolarization.

repressor Any regulatory protein that inhibits transcription of certain genes, typically by binding to a silencer upstream of the promoter.

reservoir In biogeochemical cycles, a location in the environment where elements are stored for a time.

resilience A measure of how quickly a community recovers following a disturbance.

resistance (1) The ability of an organism to defend itself against drugs, pathogens, or parasites. (2) A measure of how much a community is affected by a disturbance.

resistance (R) loci Genes associated with disease resistance in plants.

respiration The biochemical pathways that produce ATP from compounds with high potential energy, via an electron transport chain and using an inorganic final electron acceptor. Also called cellular respiration.

respiratory distress syndrome A syndrome in which premature infants can suffocate due to insufficient surfactant in their lungs.

resting potential The membrane potential of a cell in its resting, or normal, state.

restriction endonucleases Bacterial endonucleases that cut DNA at a specific base-pair sequence (recognition site). Also called restriction enzymes.

restriction fragment length polymorphisms (RFLPs) Variations in the size of DNA fragments that are produced by restriction endonucleases, due to differences in the DNA sequences at recognition sites.

retina A thin layer of light-sensitive cells (rods or cones) and neurons at the back of a camera-type eye, such as that of cephalopods or vertebrates.

retinal A carotenoid pigment derived from vitamin A that, with opsin, forms rhodopsin, the light-detecting pigment in rods and cones of animal eyes.

retrovirus A virus with an RNA genome that reproduces by transcribing its RNA into a DNA sequence and then inserting that DNA into the host's genome for replication.

reverse transcriptase A enzyme of retroviruses (RNA viruses) that can synthesize double-stranded DNA from a single-stranded RNA template.

rhizobia Members of the bacterial genus *Rhizobia*, nitrogen-fixing bacteria that live in root nodules of members of the pea family.

rhizoid The hairlike structure that anchors a bryophyte (nonvascular plant) to the substrate.

rhizome A plant stem extending horizontally underground.

rhodopsin A combination of two molecules (retinal and one of various opsins) instrumental in detection of light by rods and cones of vertebrate eyes.

ribonucleic acid (RNA) A polymer of ribonucleotides, usually single stranded. RNAs function as structural components of ribosomes (rRNA), transporters of amino acids (tRNA), and translators of the message of the DNA code (mRNA).

ribonucleotide A nucleotide consisting of the five-carbon sugar ribose, a phosphate group, and one of several nitrogen-containing bases (adenine, guanine, thymine, or uracil). Can be polymerized to form ribonucleic acid (RNA).

ribosomal RNAs (rRNAs) A class of RNA molecules that form part of the structure of a ribosome.

ribosome A molecular machine that synthesizes proteins by using the genetic information encoded in messenger RNA strands. Consists of two subunits, each composed of ribosomal RNA and proteins.

ribosome binding site In bacteria, the sequence at the beginning of an mRNA molecule to which a ribosome binds to initiate translation. Also called the Shine-Dalgarno sequence.

ribozyme Any RNA molecule that can act as a catalyst for a chemical reaction.

ribulose bisphosphate (RuBP) A five-carbon compound that is the initial reactant in the Calvin cycle of photosynthesis.

rickets A human disorder characterized by malformed, soft bones. Usually caused by environmental factors (e.g., inadequate vitamin D intake), but also may be caused by an X-linked dominant allele.

RNA See **ribonucleic acid (RNA)**.

RNA polymerases A class of enzymes that catalyze synthesis of RNA from ribonucleotides, using a DNA template. Also called RNA pols.

RNA processing In eukaryotes, the changes that a primary RNA transcript undergoes to become a mature mRNA molecule. Includes splicing and the addition of 5′ caps and poly (A) tails.

RNA replicase A viral enzyme that can synthesize RNA from an RNA template.

RNA world hypothesis The hypothesis that life on Earth began as a polymer of ribonucleic acid.

robust australopithecines Several species of comparatively large, strong hominins that appear in the fossil record shortly after the split from chimpanzees.

rod A type of photoreceptor cell with a rod-shaped outer portion. Found in vertebrate retinas. Particularly sensitive to dim light, but not used to distinguish colors.

root (1) An underground appendage of a plant that anchors the plant and absorbs water and nutrients. (2) In a phylogenetic tree, the bottom, most ancient node.

root apical meristem A group of undifferentiated plant cells at the tip of a plant root.

root cap A small group of cells that covers and protects the tip of a plant root. Senses gravity and determines the direction of root growth.

rooted For a phylogenetic tree, oriented so that its bottom-most node is the most ancient node. Usually, the root is identified through the use of an outgroup.

root hair A long, thin outgrowth of the epidermal cells of plant roots, providing increased surface area for absorption of water and nutrients.

root pressure Positive pressure that is generated in plant roots during the night, due to accumulation of ions from the soil and subsequent movement of water into the root.

root system The belowground part of a plant.

rosette In plants, a compact growth form in which leaves pile up on each other in whorls.

rough ER (rough endoplasmic reticulum) A type of endoplasmic reticulum dotted with ribosomes. Involved in processing of membrane proteins and secretory proteins.

rRNAs See ribosomal RNAs (rRNAs).

rubisco The enzyme that catalyzes the first step of the Calvin cycle of photosynthesis: the addition of a molecule of CO_2 to ribulose bisphosphate. Also called ribulose 1,5-bisphosphate carboxylase/oxygenase.

rumen The largest chamber of a ruminant's stomach, containing a large vat of symbiotic cellulose-digesting bacteria. Creates cud, which is sent back to the mouth for further chewing.

ruminants A group of hoofed mammals that have a four-chambered stomach specialized for digestion of plant cellulose, with one chamber containing symbiotic cellulose-digesting bacteria.

sac fungi A monophyletic lineage of fungi that produce large, often cup-shaped reproductive structures that contain asci. Also called cup fungi.

sage-steppe An arid, shrub-dominated habitat with some characteristics of deserts and grasslands.

salicylic acid A compound thought to play a role in the systemic acquired resistance (SAR) defensive mechanism of plants; a component of aspirin.

salivary glands Mammalian glands that secrete saliva (a mixture of water, mucins, and digestive enzymes) into the mouth.

sampling error The accidental selection of a nonrepresentative sample from some larger population, due to chance.

saprophyte An organism that feeds primarily on dead plant material. Usually, a fungus or plant.

sapwood The younger xylem in the outer layer of wood of a stem or root, functioning primarily in water transport.

sarcomere A single contractile unit of a skeletal muscle cell.

sarcoplasmic reticulum Sheets of smooth endoplasmic reticulum in a muscle cell. Contains high concentrations of calcium, which can be released into the cytoplasm to trigger contraction.

SARS See severe acute respiratory syndrome (SARS).

saturated fat A fat that contains the maximum number of hydrogen atoms because all the carbon atoms in its fatty acid chains are joined by single bonds. Such fats have relatively high melting points.

scanning electron microscope (SEM) A microscope that produces images of the surfaces of objects by reflecting electrons from a specimen coated with a layer of metal atoms.

scarify Scraping, rasping, or other damage to the coat of a seed. Necessary in some species to trigger germination.

schizophrenia A human psychological disorder characterized by delusions, hallucinations, social withdrawal, and other symptoms.

Schwann cells Specialized cells that wrap around axons of neurons outside the brain and spinal cord, providing electrical insulation.

sclereid A type of plant sclerenchyma cell that usually functions in protection, such as in seed coats and nutshells.

sclerenchyma cells A thick-walled class of plant cells that provide support and typically contain the tough structural polymer lignin. Usually dead at maturity.

screen Any of several techniques for identifying individuals with a particular type of mutation. Also called genetic screen.

scrotum A sac of skin, containing the testes, suspended just outside the abdominal body cavity of many male mammals.

secondary antibody An antibody that will bind to another antibody. Used in ELISAs and in some other lab tests.

secondary cell wall An inner layer of the cell wall, formed by certain types of plant cells as they mature.

secondary consumer A carnivore; an organism that eats herbivores.

secondary endosymbiosis The presence of an organelle that originated with ingestion of a cell containing that organelle, which in turn was originally derived from ingestion and symbiosis.

secondary growth Plant growth that results in an increase in width of stems and roots.

secondary immune response The immune response, using memory cells, to an infection that the immune system has encountered before.

secondary meristem A layer of undifferentiated plant cells found in older stems and roots. Responsible for secondary growth (increase in width). Also called cambium or lateral meristem.

secondary metabolite Any poison produced by a plant that is synthesized by a variation of a biosynthetic pathway used for other purposes.

secondary phloem Phloem tissue produced by a lateral meristem rather than by an apical meristem.

secondary production The total amount of new body tissue produced by animals that eat plants. May involve growth and/or reproduction.

secondary spermatocyte A cell produced by meiosis I of a primary spermatocyte in the testis. Can undergo meiosis II to produce spermatids.

secondary structure A type of protein structure created by hydrogen bonding between C=O and N–H groups of the polypeptide backbone; most notably, the α-helix and β-pleated sheet structures.

secondary succession Gradual colonization of a habitat after an environmental disturbance (e.g., fire, windstorm, logging) that removes some or all previous organisms but leaves the soil intact.

secondary xylem Xylem tissue produced by a lateral meristem, rather than by an apical meristem.

second law of thermodynamics A fundamental principle of physics stating that, in an isolated system, entropy always increases during any chemical reaction.

second-male advantage The reproductive advantage of a male who mates with a female last, after other males have mated with her.

second messenger A nonprotein signaling molecule produced or activated inside a cell in response to stimulation at the cell surface. Commonly used to relay the message of a protein hormone.

secretin A peptide hormone produced by the small intestine in response to the arrival of food from the stomach. Stimulates secretion of bicarbonate (HCO_3^-) from the pancreas.

sedimentary rock A type of rock formed by gradual accumulation of sediment, as in riverbeds and on the ocean floor. Most fossils are found in sedimentary rocks.

seed A plant embryo with nutritive tissue (endosperm) to fuel its early growth, surrounded by an outer protective layer (seed coat). In angiosperms, forms from the fertilized ovule of a flower.

seed coat A protective layer around a seed that encases both the embryo and the endosperm.

seed dormancy A state of suspended development of a plant seed. Can be terminated by cues indicating that favorable environmental conditions have arrived or that unfavorable ones have passed.

seedless vascular plants Several phyla of green plants that have vascular tissue but do not make seeds. Include horsetails, ferns, lycophytes, and whisk ferns.

seedling A young plant that has emerged from a seed.

seed plants A group of several phyla of green plants that have vascular tissue and make seeds. Includes gymnosperms and angiosperms.

segment A well-defined region of the body along the anterior-posterior body axis, containing similar structures as other, nearby segments.

segmentation A body plan involving division of the body into many similar segments that bear similar or identical structures.

segmentation genes Genes that affect body segmentation in embryonic development. Includes gap genes, pair-rule genes, and segment polarity genes.

segment polarity genes A class of fruit-fly segmentation genes that establish the anterior-posterior orientation of each embryonic segment. Active in particular regions of each segment.

selective adhesion The tendency of cells of one tissue type to adhere to other cells of the same type.

selectively permeable membrane A membrane that some solutes can cross more readily than other solutes can.

selective permeability The property of a structure, such as a membrane, that allows some substances to diffuse across it much more readily than other substances.

self Property of a molecule or cell such that immune system cells do not attack it, due to certain molecular similarities to other body cells.

self-fertilization The fusion of two gametes from the same individual to form a diploid offspring. Also called selfing.

self-incompatible Incapable of self-fertilization.

selfing The fusion of two gametes from the same individual to form a diploid offspring. Also called self-fertilization.

selfish genes DNA sequences that survive and reproduce but reduce the fitness of the host genome.

self-pollination Pollination in which pollen from a certain individual lands on a flower stigma of that same individual.

semen The combination of sperm and accessory fluids that is released by male mammals and reptiles during ejaculation.

semiconservative replication The type of replication used by cells to copy DNA, in which each daughter DNA molecule is composed of one old strand and one new strand.

seminal vesicles Paired reproductive glands that, in mammals, secrete an alkaline fluid into semen to counteract the acidic environment of the vagina. In other vertebrates and invertebrates, often stores sperm.

senescence The process of aging.

sensitive period A short time span in a young animal's life during which the animal can learn certain things, such as song, language, or imprinting. Also called the critical period.

sensor A cell, organ, or structure that senses some aspect of the external or internal environment.

sensory neuron A nerve cell that carries sensory information to the central nervous system.

sepal One of the protective leaflike structures enclosing a flower bud and later supporting the blooming flower.

septa (singular: septum) Any wall-like structure. In fungi, cross-walls that divide fungal filaments into cell-like compartments.

serotonin A neurotransmitter involved in many brain functions, including sleep, pleasure, and mood.

serum The liquid that remains when clotted cells are removed from blood; plasma without clotting factors. Contains water, dissolved gases, hormones, food molecules, and other soluble substances.

sessile Permanently attached to a substrate; not capable of moving to another location.

set point A normal or target value for a regulated internal factor, such as body heat or blood pH.

severe acute respiratory syndrome (SARS) A human disease characterized by sudden and intense flu-like symptoms.

severe combined immune deficiency (SCID) A human disease characterized by an extremely high vulnerability to infectious disease, due to a genetic defect in the immune system.

sex chromosome One of the pair of chromosomes carrying the gene(s) that determine sex.

sex-linked inheritance Inheritance patterns observed in genes carried on sex chromosomes, so females and males have different numbers of alleles of a gene and may pass its trait only to one sex of offspring. Also called sex-linkage.

sexual dimorphism Any trait that differs between males and females.

sexual reproduction Reproduction in which genes from two parents are combined via fusion of gametes, producing offspring that are genetically distinct from both parents.

sexual selection The process by which individuals with certain heritable traits leave more offspring than other individuals specifically due to superiority in competing for mating opportunities.

Shannon-Weaver index A common measurement of species diversity, calculated as $H' = -\Sigma p_i \log(p_i)$, where p_i is the proportion of individuals in the community that belong to species i.

shell A hard protective outer structure. In protists, also called a test.

Shine-Dalgarno sequence In bacteria, the sequence at the beginning of an mRNA molecule to which a ribosome binds to initiate translation. Also called the ribosome binding sequence.

shoot The aboveground portion of a young plant, including stem and leaves.

shoot apical meristem A group of undifferentiated plant cells at the tip of a plant stem.

shoot system The aboveground part of a plant.

short-day plant A plant that blooms in response to long nights.

shotgun sequencing A method of sequencing genomes that is based on breaking the genome into small pieces, sequencing each piece separately, and then figuring out how the pieces are connected.

sieve plates The pore-containing structure at one end of a sieve-tube member in plant phloem tissue.

sieve-tube member An elongated sugar-conducting cell in phloem. Has sieve plates at both ends, allowing sap to flow to adjacent cells.

sigma A detachable protein subunit of RNA polymerase that binds to the −35 and −10 boxes to initiate transcription of DNA in prokaryotes.

signal Any information-containing behavior.

signal 1 The first step in activation of a T cell, in which the T cell engulfs an antigen presented by a dendritic cell.

signal 2 The second step in the activation of a T cell, in which additional receptors on the T cell bind to additional MHC-antigen complexes on a dendritic cell.

signal hypothesis The hypothesis that proteins destined for secretion are directed to the rough ER by a certain amino acid sequence at the beginning of the proteins.

signal receptor Any cellular protein that binds to a particular signaling molecule (such as a hormone or neurotransmitter) and triggers a response by the cell, usually by changing conformation or activity upon binding.

signal recognition particle (SRP) A complex of RNA and protein that transports certain newly synthesized proteins to the endoplasmic reticulum.

signal transducers and activators of transcription (STATs) In mammals, a group of regulatory transcription factors that, upon phosphorylation, can activate transcription of certain genes.

signal transduction The process by which a stimulus (e.g., a hormone, a neurotransmitter, or sensory information) outside a cell is translated into a response by the cell.

signal transduction cascade A sequence of events in which one enzyme phosphorylates other enzymes, which in turn phosphorylate many more, ultimately leading to phosphorylation of thousands of proteins. Commonly used in signal transduction of hormone messages. Also called phosphorylation cascade.

signal transduction pathway The exact sequence of molecular events by which signal transduction occurs.

sign stimulus A simple stimulus that elicits an invariant, stereotyped behavioral response (fixed action pattern) from an animal. Also called a releaser.

silencer A regulatory DNA sequence in eukaryotes to which repressor proteins can bind, inhibiting transcription of certain genes.

silent mutation A mutation that does not detectably affect the phenotype of the organism. Typically, a point mutation in the third position of certain codons that does not alter the amino acid coded for.

simian immunodeficiency viruses (SIVs) A family of lentiviruses that infect monkeys and apes and that are thought to be closely related to human immunodeficiency virus (HIV).

simple eye An eye with only one light-collecting apparatus (e.g., one lens), as in vertebrates.

simple sequence repeat A stretch of eukaryotic DNA consisting of repeats of a short, simple sequence that does not code for any protein or RNA.

single nucleotide polymorphism (SNP) A genetic locus where a single base pair often varies between individuals of a certain species. Used as a genetic marker to help track the inheritance of nearby genes.

single-strand DNA-binding proteins A class of proteins that attach to separated strands of DNA during replication or transcription, preventing them from re-forming a double helix.

sink A location where an element or a molecule is consumed or taken out of circulation.

sinoatrial node (SA node) A cluster of heart muscle cells that initiates the heartbeat and determines the heart rate. In the wall of the right atrium.

siphon A tubelike appendage of many mollusks. Often used for feeding or propulsion.

sister chromatids The paired strands of a recently replicated chromosome that has not yet divided.

sister groups or sister taxa Closely related taxa.

skeletal muscle The muscle tissue attached to the bones of the vertebrate skeleton. Consists of long, unbranched muscle fibers with a characteristic striped (striated) appearance; controlled voluntarily. Also called striated muscle.

sliding clamp A doughnut-shaped structure that holds the enzyme DNA polymerase in place during DNA replication.

sliding-filament model The hypothesis that the contraction of muscle cells is caused by filaments of actin and myosin sliding past each other.

slow-twitch fibers A class of muscle fibers that contract relatively slowly, but do not fatigue easily.

slug (1) A member of a certain lineage of terrestrial gastropods, closely related to snails but lacking a shell. (2) A mobile aggregation of cells of a cellular slime mold.

small intestine   The first section of the intestine, immediately after the stomach. The site of the final stages of digestion and of most nutrient absorption.

smooth ER (endoplasmic reticulum)   Endoplasmic reticulum that does not have ribosomes attached to it. Involved in synthesis and secretion of lipids.

smooth muscle   The unstriated muscle tissue that lines the intestine, blood vessels, and some other organs. Consists of tapered, unbranched cells that can sustain long contractions. Not voluntarily controlled.

snRNPs (small nuclear ribonucleoproteins)   A complex of proteins and small RNA molecules that catalyze splicing (removal of introns from mRNA). Components of spliceosomes.

sodium-potassium pump   A membrane protein that uses the energy of ATP to move sodium ions out of the cell and potassium ions in. Formally known as Na^+/K^+-ATPase.

soil erosion   The removal of soil from an area by wind or water.

soil organic matter   A mixture of partially and completely decomposed detritus.

solute   Any substance that is dissolved in a liquid.

solute potential   The potential energy of water caused by a difference in solute concentrations at two locations. Also called osmotic potential.

solution   A liquid containing one or more dissolved solids or gases in a homogeneous mixture.

solvent   Any liquid in which some substance will dissolve.

soma   The part of a neuron that contains the nucleus and where incoming signals are integrated. Also called the cell body.

somatic cell   Any type of cell that does not pass its genes on to the next generation. In a multicellular organism, all cells except eggs, sperm, and their parent cells (oogonia and spermatogonia) are somatic cells.

somatic hypermutation   Rapid DNA mutation that occurs in a somatic cell. Occurs in certain cells of the immune system, such as memory cells that are fine-tuning antibody performance.

somatic nervous system   The part of the peripheral nervous system (outside the brain and spinal cord) that controls skeletal muscles and is under voluntary control.

somatostatin   A hormone secreted by many organs including the pancreas and hypothalamus, with a wide variety of effects.

somite   A block of mesoderm on both sides of the developing spinal cord in a vertebrate embryo. Gives rise to muscle tissue, vertebrae, ribs, limbs, etc.

soredia   (singular: soredium) Small reproductive structures produced by lichen. Contain both symbionts of the lichen (the fungus and green alga).

sorus   (plural: sori) One of the small dots on the underside of fern fronds. Consists of many sporangia, each of which contains spores.

source   A location where a substance is produced or enters circulation (e.g., in plants, the tissue where sugar enters the phloem).

Southern blotting   A lab technique for the isolation and analysis of pieces of DNA. Involves cleavage with restriction enzymes, separation with gel electrophoresis, hybridization to a labeled probe, and visualization with autoradiography.

speciation   The evolution of two or more distinct species from a single ancestral species.

species   A distinct, identifiable group of populations that is thought to be evolutionarily independent of other populations. Generally distinct from other species in appearance, behavior, habitat, ecology, genetic characteristics, etc.

species-area relationship   The mathematical relationship between the area of a certain habitat and the number of species that it can support.

species diversity   The variety and relative abundance of the species present in a given ecological community.

species richness   The number of species present in a given ecological community.

specific heat   The amount of energy required to raise the temperature of 1 gram of a substance by 1°C; a measure of the capacity of a substance to absorb energy.

spectral karyotyping (SKY)   A technique for producing high-resolution karyotypes by "painting" chromosomes with fluorescent tags that bind to particular regions of certain chromosomes. Also called chromosome painting.

spectrophotometer   A lab instrument used to measure the wavelengths of light that are absorbed by a particular pigment.

sperm   A mature male gamete, smaller and more mobile than the female gamete.

spermatid   An immature sperm cell.

spermatogenesis   The production of sperm.

spermatogonia   (singular: spermatogonium) The diploid cells in a testis that can give rise to primary spermatocytes.

spermatophore   A gelatinous package of sperm cells that is produced by males of species that have internal fertilization without copulation.

sperm competition   Competition between sperm from different males to fertilize the eggs of the same female.

sperm nuclei   The two haploid nuclei within the pollen grain of a flowering plant. One sperm nucleus fuses with the egg cell to form a zygote; the other fuses with two other nuclei to form triploid endosperm.

sphincter   A muscular valve that can close off a tube, as in a blood vessel or a part of the digestive tract.

spicules   Stiff spikes of silica or calcium carbonate found in the bodies of sponges.

spindle fibers   Groups of microtubules that attach to chromosomes and pull them to opposite sides of the cell during cell division.

spine   A modified plant leaf that functions as a sharp protective structure.

spiracle   In insects, the small openings that connect air-filled tracheae to the external environment.

spiral cleavage   The pattern of embryonic cleavage seen in deuterostomes, in which cells divide at oblique angles to form a spiral coil of cells.

spleen   A dark red organ, found near the stomach of most vertebrates, that filters blood, stores extra red blood cells in case of emergency, and plays a role in immunity.

spliceosome   In eukaryotes, an organized complex of snRNPs (small nuclear ribonucleoproteins) that catalyzes removal of introns from primary RNA transcripts.

splicing   The process by which introns are removed from messenger RNA molecules and the remaining exons are connected together.

spongy mesophyll   Rounded parenchyma cells found in the ground tissue of leaves near stomata. The site of most gas exchange.

spontaneous   For a chemical reaction, occurring on its own, without any continuous external influences such as added energy.

spontaneous generation   The disproven hypothesis that living organisms can develop spontaneously and rapidly from nonliving, noncellular materials under certain conditions.

sporangium   (plural: sporangia) A spore-producing structure found in some plants, such as liverworts, and in some fungi, such as chytrids.

spore   A single cell produced by mitosis or meiosis (not by cell fusion) that is capable of developing into an adult organism.

sporophyte   The multicellular diploid phase of a species that exhibits alternation of generations; arises from two fused gametes; produces spores.

sporopollenin   A watertight material that encases spores and pollen of modern land plants.

stabilizing selection   A pattern of natural selection in which individuals with an average phenotype have higher fitness than those with extreme phenotypes.

stamen   The male reproductive structure of a flower. Consists of an anther, which produces pollen grains, and a filament, which supports the anther.

standing defense   Any defensive mechanism that is always present, regardless of need. Also called constitutive defense.

stapes   A stirrup-shaped bone in the middle ear of vertebrates. Receives vibrations from the tympanic membrane and passes them to the cochlea.

starch   A mixture of polysaccharides amylose and amylopectin; used primarily for food storage in plants. Both are helical polymers of α-glucose subunits, branched in amylopectin.

start codon   The mRNA sequence AUG, which induces the beginning of protein synthesis and codes for the amino acid methionine.

statocyst   A sensory organ of many arthropods that detects the animal's orientation in space (i.e., whether the animal is flipped upside down).

statolith   A tiny stone or dense particle that can be used to sense gravity.

statolith hypothesis   The hypothesis that amyloplasts (dense, starch-storing plant organelles) serve as statoliths (i.e., gravity detectors).

STATs   See **signal transducers and activators of transcription (STATs)**.

stem   A vertical aboveground part of a plant, usually bearing leaves, fruit, or flowers.

stem cell   Undifferentiated cells that have the potential to give rise to any tissue type.

steppe   An extensive grassland. Typically found in the dry interiors of continents in the temperate latitudes. Also called a prairie.

stereocilia   (singular: stereocilium) Stiff outgrowths from the surface of a hair cell that is involved in detection of sound by terrestrial vertebrates or of waterborne vibrations by fishes.

steroids   A class of lipids with a characteristic four-ring structure.

sticky ends   The short, single-stranded ends of a DNA molecule cut by a restriction endonuclease. Tend to form hydrogen bonds with other sticky ends that have complementary sequences.

stigma   The moist tip at the end of a flower carpel, to which pollen grains adhere.

stolon   Modified stems that run horizontally over the soil surface.

stomata (singular: stoma) Generally, pores or openings. In plants, microscopic pores on the surface of a leaf or stem, through which gas exchange occurs for photosynthesis.

stomach A tough, muscular pouch in the digestive tract that breaks up food and delivers it to the intestine.

stop codon One of three messenger RNA triplets (UAG, UGA, or UAA) that cause termination of protein synthesis. Also called a termination codon.

strain A population of genetically similar or identical individuals.

striated muscle The muscle tissue attached to the bones of the vertebrate skeleton. Consists of long, unbranched muscle fibers with a characteristic striped (striated) appearance; controlled voluntarily. Also called skeletal muscle.

stroke An episode of oxygen deprivation to part of the brain that results in death of some neurons; usually due to a blood clot that blocks blood vessels of the brain.

stroke volume The volume of blood ejected from the left ventricle of the heart in one contraction.

stroma The fluid between a chloroplast's outer membrane and its thylakoid disks.

structural formula A graphical representation of the atoms and bonds in a molecule, often with covalent bonds represented by straight lines.

structural gene A stretch of DNA that codes for a functional protein or functional RNA molecule—i.e., not a promoter, enhancer, etc.

structural homology Similarities in organismal structures (e.g., limbs, shells, flowers) that are due to inheritance from a common ancestor.

structural isomer A molecule that shares the same molecular formula as another molecule but differs in the order in which covalently bonded atoms are attached.

style The slender stalk of a flower carpel, connecting the stigma and the ovary.

suberin A water-repellent compound that is a major component of the waxy Casparian strip in plant roots.

subspecies A population that has distinctive traits and some genetic differences relative to other populations of the same species but that is not distinct enough to be called a separate species.

substrate (1) A reactant that interacts with an enzyme in a chemical reaction. (2) A surface on which a cell or organism sits.

substrate-level phosphorylation Production of ATP molecules via transfer of a phosphate group from an intermediate substrate directly to ADP, unmediated by an electron transport chain. Occurs in glycolysis and in the Krebs cycle.

subzonal injection (SUZI) A technique used to help infertile couples bear children by selecting a few healthy sperm and injecting them directly into the zona pellucida of the woman's egg.

succession Gradual colonization of a habitat after an environmental disturbance (e.g., fire, flood), usually by a series of species assemblages.

successional pathway The specific sequence of species that appears over a period of time in an environment undergoing primary succession or secondary succession.

sugars A class of small, water-soluble organic compounds containing a carbonyl (–C=O) group and several hydroxyl (–OH) groups. Include monosaccharides and disaccharides.

sulfate-reducers A group of archaea that produce hydrogen sulfide (H_2S) as a by-product of cellular respiration.

summation The additive effect of different post-synaptic potentials at a nerve or muscle cell, such that several subthreshold stimulations can cause an action potential.

supernatant The liquid above a layer of solid particles (the pellet) in a tube after centrifugation.

surface tension The attractive force between liquid molecules that causes the liquid to form a rounded surface at an air-liquid interface.

surfactant A mixture of phospholipids and proteins produced by lung cells that reduces surface tension, allowing the lungs to expand more.

survivorship The average proportion of offspring produced that survive to a particular age in a certain population.

survivorship curve A graph depicting the percentage of a population that survives to different ages.

suspension culture A population of cells grown in the lab in a liquid flask of nutrients that is rotated continuously to keep the cells suspended and the nutrients mixed.

suspension feeder Any organism that obtains food by filtering small particles or small organisms out of water or air. Also called filter feeder.

sustainability The planned use of environmental resources at a rate no faster than the rate at which they are naturally replaced.

sustainable agriculture Agricultural techniques that are designed to maintain long-term soil quality and productivity.

sustainable development Economic development based on sustainability (i.e., that uses natural resources no faster than they are naturally replaced).

swamp A wetland that has a steady rate of water flow and is dominated by trees and shrubs.

swim bladder A gas-filled organ of many ray-finned fishes that regulates buoyancy.

symbiosis Any close and prolonged physical relationship between individuals of two different species. May be mutualistic, parasitic, or commensal.

symmetric competition Ecological competition between two species in which both suffer similar declines in fitness.

sympathetic nervous system The part of the autonomic nervous system that stimulates fight-or-flight responses, such as increased heart rate, increased blood pressure, and decreased digestion.

sympatric speciation Speciation that occurs while the two diverging species are living in the same area.

sympatry Living in the same geographic area as some other population.

symplast A pathway for water transport in plants roots through the cytoplasm of adjacent cells that are connected by plasmodesmata.

synapomorphy A shared, derived (altered from the ancestral state) character that can be used to infer evolutionary relationships.

synapse The connection between two neurons, or between a neuron and a muscle cell: a tiny space into which neurotransmitters are released.

synapsis The physical pairing of two homologous chromosomes during prophase I of meiosis. Crossing over occurs during synapsis.

synaptic cleft The gap of a synapse; the space between two communicating nerve cells, across which neurotransmitters diffuse.

synaptic plasticity Long-term changes in the responsiveness or physical structure of a synapse that can occur after particular stimulation patterns. Thought to be the basis of learning and memory.

synaptic vesicles Tiny neurotransmitter-containing vesicles at the end of an axon. Can fuse with the axon membrane to release a neurotransmitter into a synapse, stimulating the next neuron or effector cell.

synaptonemal complex A network of proteins that hold non-sister chromatids together during synapsis in meiosis.

syndrome A group of medical symptoms that often occur together and that are suspected to have the same underlying cause.

syngamy Fusion of two gametes to create a diploid zygote. In multicellular species, also called fertilization.

synthesis phase (S phase) The phase of the cell cycle during which DNA is synthesized and chromosomes are duplicated.

system A set of interacting elements, such as (1) a set of reactants and products at chemical equilibrium and (2) a group of organs that work together to perform a function.

systemic acquired resistance (SAR) A set of events through which plant tissues prepare to combat a possible infection.

systemic circulation The part of the circulatory system that sends oxygen-rich blood from the lungs out to the rest of the body. In mammals and birds, separate from the pulmonary circulation.

systemin A peptide hormone, produced by plant cells damaged by herbivores, that initiates a protective response in undamaged cells.

systole The portion of the heartbeat cycle during which the heart muscles are contracting.

systolic blood pressure Blood pressure in arteries during ventricular systole (heart contraction).

taiga A vast forest biome throughout subarctic regions, consisting primarily of short conifer trees. Characterized by intensely cold winters, short summers, and high annual variation in temperature.

taproot A large vertical main root of a plant.

Taq **polymerase** A DNA polymerase commonly used in PCR due to its stability at high temperatures. Derived from *Thermus aquaticus*, a bacterium found in hot springs.

taste buds Sensory structures, found chiefly in the mammalian tongue, that are responsible for the sense of taste.

taste cells Spindle-shaped cells found in a taste bud of the mammalian tongue. Respond to certain chemical stimuli.

TATA box A DNA sequence in many eukaryotic promoters 30 base pairs upstream from the transcription start site. Recognized by RNA polymerase II.

taxis Movement toward or away from some external cue.

taxon (plural: taxa) Any named group of species at any level of a classification system.

taxonomy The branch of biology concerned with the classification and naming of organisms.

TBP (TATA-binding protein) A protein that binds to eukaryotic promoters and helps initiate transcription.

T-cell receptor (TCR) A T-cell membrane protein that can bind to antigens displayed on the surfaces of other cells.

T cells Lymphocytes that mature in the thymus and are involved in acquired immunity in vertebrates.

May be involved in activation of other lymphocytes (helper T cells) or destruction of infected cells (cytotoxic T cells). Also called T lymphocytes.

tectorial membrane A membrane in the vertebrate cochlea that takes part in the transduction of sound by bending the stereocilia of hair cells in response to sonic vibrations.

telomerase An enzyme that replicates telomeres by catalyzing DNA synthesis from an RNA template.

telomere The region at the end of a linear chromosome.

telophase The final stage in cell division (mitosis or meiosis), during which chromosomes finish moving and new nuclear envelopes begin to form around each set of daughter chromosomes.

temperate Having a climate with pronounced annual fluctuations in temperature (i.e., warm summers and cold winters) but typically neither as hot as the tropics nor as cold as the poles.

template strand (1) The strand of DNA that is transcribed by RNA polymerase to create RNA. (2) An original strand of RNA used to make a complementary strand of RNA.

temporal lobe In the mammalian brain, the region of each cerebral hemisphere near the ears. Functions in memory, interpretation of information from the ears, and, in humans, language.

tendon A band of tough, fibrous connective tissue that connects a muscle to a bone.

tension Any pulling force; the opposite of compression.

tentacle A long, thin, muscular appendage of gastropod mollusks.

terminal cell The upper cell produced by an angiosperm zygote. Gives rise to the plant embryo.

termination (1) In enzyme-catalyzed reactions, the final stage in which the enzyme returns to its original conformation and products are released. (2) In DNA transcription, the dissociation of RNA polymerase from DNA. (3) In RNA translation, the dissociation of ribosomes from mRNA.

territory An area that is actively defended by an animal from others of its species.

tertiary structure The overall three-dimensional shape of a single polypeptide chain, created by a variety of interactions among R-groups and the peptide backbone.

test A hard protective outer structure seen in some protists. Also called a shell.

testcross The breeding of an individual of unknown genotype with an individual having only recessive alleles for the traits of interest, in order to infer the unknown genotype from the phenotypic ratios seen in offspring.

testis (plural: testes) The sperm-producing organ of a male animal.

testosterone A steroid hormone from the testes that stimulates sperm production and various male traits and reproductive behaviors.

tetrad A pair of synapsed homologous chromosomes, each containing two chromatids.

tetraploid With four sets of chromosomes ($n = 4$).

tetrapod Any member of the taxon Tetrapoda, including all descendants of the first four-footed animals to move onto the land.

texture In soil biology, the proportions of different-sized particles present in soil.

theory A proposed explanation for a very general class of phenomena or observations.

thermal energy The kinetic energy of molecular motion.

thermocline A gradient (cline) in environmental temperature across a large geographic area.

thermophiles Bacteria or archaea that thrive in very hot environments.

thermoreceptor A sensory cell or an organ specialized for detection of changes in temperature.

thermoregulation Regulation of body temperature.

thick filaments Strands of myosin found in the middle of a muscle sarcomere. Bind to thin filaments (actin) and cause muscle contraction.

thin filaments Strands of actin found at the two ends of a muscle sarcomere. Bind to thick filaments (myosin) during muscle contraction.

thorax In mammals, the anterior region of the torso, containing the lungs. In insects, the middle of the three major body regions.

thorn A modified plant stem shaped as a sharp protective structure.

threshold The membrane potential at which a neuron's voltage-gated sodium channels will trigger an action potential.

thylakoid A flattened, membrane-bound disk inside a plant chloroplast. A stack of thylakoids is a granum.

thymus An organ, located in the anterior chest or neck of vertebrates, that, in young animals, processes T cells and sends them to lymph nodes.

thyroid gland A gland in the neck that releases thyroid hormone (which increases metabolic rate) and calcitonin (which lowers blood calcium).

thyroid-stimulating hormone (TSH) A peptide hormone from the pituitary gland that stimulates release of thyroid hormones from the thyroid gland.

thyroxine An iodine-containing peptide hormone from the thyroid gland that increases metabolic rate, both directly and via conversion to the more active hormone triiodothyronine.

tight junction A physical attachment between two adjacent animal cells, consisting of proteins that "stitch" the cells' plasma membranes together.

tip The end of a branch on a phylogenetic tree. Represents a specific population or species that has not (yet) produced descendants—either a group living today or a group that ended in extinction.

Ti plasmid A plasmid carried by *Agrobacterium* (a bacterium that infects plants) that can integrate into the plant cell's chromosomes and induce formation of a gall.

tissue A group of similar cells that function as a unit, such as muscle tissue or epithelial tissue.

tissue culture A collection of cells of a certain tissue type grown in a lab, typically in liquid suspension or in a petri dish on a solid food medium.

tolerance Ecological succession in which early-arriving species do not affect the probability that subsequent species will become established.

tonoplast The membrane surrounding a plant vacuole.

top-down control The limitation of herbivore population size by predation or disease rather than by limited or toxic nutritional resources.

topoisomerase An enzyme that cuts and rejoins DNA downstream of the replication fork, to ease the twisting that would otherwise occur as the DNA "unzips."

torpor A regulated, "deliberate" decrease in metabolic rate and body temperature in an endotherm.

totipotent Capable of dividing and developing to form a complete, mature organism.

tracheae (singular: trachea) A system of small air-filled tubes that extends throughout an insect's body and functions in gas exchange.

tracheid An elongated water-conducting plant cell that has gaps (pits) in its secondary cell wall to allow water movement from one cell to the next.

trade-off An inescapable compromise between two traits that cannot be optimized simultaneously.

trait Any characteristic of an individual.

transcription The process by which messenger RNA is made from a DNA template.

transcriptional control Regulation of gene expression via changes in the rate at which genes are transcribed to form messenger RNA.

transcription termination signal A DNA sequence that can terminate messenger RNA synthesis, usually because the RNA transcribed from this sequence forms a hairpin that prevents further transcription.

transduction Conversion of information from one mode to another. For example, the process by which a stimulus outside a cell is translated into a response by the cell.

transfer cells Cells that transfer nutrients from a parent plant to a developing plant seed. Occur in land plants.

transfer RNAs (tRNAs) A class of RNA molecules with an anticodon at one end and an amino acid binding site at the other. Match amino acids to messenger RNA codons during translation.

transformation Incorporation of DNA obtained directly from the environment into the genome. Occurs naturally in some bacteria; can be induced in the lab by certain processes.

transgenic Containing DNA that has been modified by genetic engineering.

transitional form A fossil species or population with traits that are intermediate between older and younger species.

transition state A high-energy intermediate state occurring during a chemical reaction that determines the activation energy necessary for the reaction to proceed.

translation The process by which proteins and peptides are synthesized from messenger RNA.

translational control The regulation of gene expression by altering the life span of messenger RNA or the efficiency of translation. A type of post-transcriptional regulation.

translocation (1) The movement of sugars through a plant by bulk flow. (2) A type of mutation in which a piece of a chromosome moves to a nonhomologous chromosome. (3) The process by which a ribosome moves down a messenger RNA molecule during translation.

transmembrane domain A structure that anchors a B-cell receptor molecule to the plasma membrane of the B cell.

transmembrane protein Any membrane protein that spans the entire membrane. Also called integral membrane protein.

transmission electron microscope (TEM) A microscope that forms an image from electrons that pass through a specimen.

transmission genetics The study of the genotypic and phenotypic patterns that occur as genes pass from one generation to the next.

transpiration Water loss from aboveground plant parts. Occurs primarily through the stomata.

transport protein Any membrane protein that enables specific molecules to cross plasma membranes, sometimes by causing a conformational change in the protein. Also called transporter.

transposable elements Any of several kinds of parasitic DNA sequences that are capable of moving themselves, or copies of themselves, to other locations in the genome.

tree of life A diagram depicting the genealogical relationships of all living organisms on Earth, with a single ancestral species at the base.

triacylglycerols Lipids consisting of three fatty acids joined to a glycerol molecule. Also called triglycerides or fats.

trichomes Protective hairlike appendages of the epidermal cells of some plants.

triglycerides Lipids consisting of three fatty acids joined to a glycerol molecule. Also called triacylglycerols or fats.

triiodothyronine An iodine-containing peptide hormone, secreted by the thyroid gland, that increases metabolic rate. Has a stronger effect than does the related hormone thyroxine.

trilobite A member of an extinct lineage of arthropods that were abundant from 550–440 million years ago.

trimester In humans, one of three major stages of gestation, each three months long.

triose A monosaccharide (simple sugar) that has three carbons.

triplet code A code in which a "word" of three letters encodes one piece of information. The genetic code is a triplet code because a codon is three nucleotides long and encodes one amino acid.

triploblast An animal whose body develops from three basic embryonic cell layers: ectoderm, mesoderm, and endoderm. All animals except sponges, cnidarians, and ctenophores are triploblasts.

triploid With three sets of chromosomes ($n = 3$).

trisomy The state of having three copies of one particular type of chromosome.

tRNAs See **transfer RNAs (tRNAs)**.

trochophore A type of larva that has a ring of cilia around its middle. Occurs in several phyla of the Lophotrochozoa lineage of protostomes.

trophic level A feeding level in an ecosystem.

tropomyosin A muscle protein that blocks the myosin-binding sites on actin filaments, preventing muscle contraction. Can be moved out of the way by troponin when intracellular calcium is high.

troponin A muscle protein that can trigger muscle contraction by moving tropomyosin off the myosin-binding sites on actin filaments. Activated by high intracellular calcium.

trp operon An *E. coli* operon that includes five cotranscribed genes involved in the synthesis of the amino acid tryptophan.

true navigation Navigation by which an animal can reach a specific point on Earth's surface.

trypsin A protein-digesting enzyme produced by the pancreas, secreted into the intestine, and activated by enterokinase. Trypsin in turn activates several other protein-digesting enzymes.

trypsinogen The precursor of protein-digesting enzyme trypsin. Secreted by the pancreas and activated by the intestinal enzyme enterokinase.

T tubules Membranous tubes that extend into the interior of muscle fibers. Propagate action potentials throughout the muscle cell and trigger release of calcium from the sarcoplasmic reticulum.

tube cell A large cell in a male gametophyte that will give rise to the pollen tube.

tube feet Small, mobile, fluid-filled extensions of the water vascular system of echinoderms. Used in locomotion.

tuber Plant rhizomes that are modified to function as carbohydrate-storage organs.

tuberculosis (TB) A highly contagious human disease caused by the bacterium *Mycobacterium tuberculosis*.

tube-within-a-tube Describing the basic body plan of all triploblast animals (i.e., an inner tube of endoderm within an outer tube of ectoderm, with mesoderm between the two tubes).

tumor A mass of cells formed by uncontrolled cell division. Can be benign or malignant.

tumor suppressor A gene that prevents cell division, particularly when the cell has DNA damage. Mutated forms are associated with cancer. Also, the protein produced by such a gene.

tundra The treeless biome in polar and alpine regions, characterized by short, slow-growing vegetation, permafrost, and a climate of long, intensely cold winters and very short summers.

turgid Refers to a plant cell that is firm (i.e., containing enough water for the cell cytoplasm to press against the cell wall).

turgor pressure The outward pressure exerted by the fluid contents of a plant cell against its cell wall.

Turner syndrome A human syndrome caused by the presence of only one X chromosome and no Y chromosome ("XO"). Individuals with this condition are female but sterile.

turnover In lake ecology, the complete mixing of upper and lower layers of water that occurs each spring and fall in temperate-zone lakes.

two-fold rotational symmetry A type of symmetry in which an object can be superimposed on itself if rotated 180°. Occurs in some regulatory sequences of DNA. Also called dyad symmetry.

tympanic membrane The eardrum; a membrane separating the middle ear from the outer ear in terrestrial vertebrates, or similar structures in insects.

type I diabetes mellitus Diabetes mellitus that is caused by insufficient secretion of insulin from the pancreas.

type II diabetes mellitus Diabetes mellitus that is caused primarily by insufficient responsiveness of tissues to insulin, despite normal secretion of insulin from the pancreas.

typhoid fever A human disease caused by the bacterium *Salmonella typhi*.

ubiquinone A molecule that transfers electrons in the electron transport chain of cellular respiration in mitochondria. Also called coenzyme Q or Q.

ulcer A hole or thin spot in the stomach or intestinal wall.

ultimate causation In biology, the reason that a trait or phenomenon is thought to have evolved; the adaptive advantage of that trait. Also called ultimate explanation.

ultrasound Sound frequencies that are too high for humans to hear—higher than about 20,000 Hz (20 kHz).

ultraviolet light Light with a wavelength shorter than visible blue light.

umami The taste of glutamate, responsible for the "meaty" taste of most proteins and of monosodium glutamate.

umbilical cord The cord that connects a developing mammal embryo or fetus to the placenta and through which the embryo or fetus receives oxygen and nutrients.

undershoot The brief phase after an action potential when a cell's membrane potential temporarily becomes more negative than the resting potential.

unequal crossover An error in crossing over during meiosis I in which the two chromatids match up at different sites. Results in gene duplication and gene loss in the two resulting chromatids.

uniformitarianism The concept that the laws of the physical universe are constant throughout time and space.

unsaturated fat A fat with one or more carbon-carbon double bonds in its fatty acid chains and thus fewer than the maximum number of hydrogen atoms. Double bonds produce kinks in the fatty acid chains and decrease the compound's melting point.

upstream Opposite to the direction in which RNA polymerase moves along a DNA strand.

urea A water-soluble excretory product of mammals and sharks, used to excrete excess nitrogen from amino acids.

ureter In mammals, a tube that transports urine from one kidney to the bladder.

urethra The tube that drains urine from the bladder to the outside environment. In male vertebrates, also used for passage of sperm during ejaculation.

uric acid A whitish excretory product of birds, reptiles, and terrestrial arthropods, used to get rid of excess nitrogen derived from amino acids.

urochordates The tunicates or sea squirts. A lineage of sessile, filter-feeding chordates.

uterus The organ in which developing embryos are housed in those vertebrates that give live birth. Common in most mammals and in some lizards, sharks, and other vertebrates.

vaccination Injection with weakened, killed, or altered pathogens to stimulate development of immunity against those pathogens.

vaccine Preparation containing pieces of a pathogen, or entire killed or weakened pathogens, administered to stimulate immunity without causing illness.

vacuole An organelle usually used for bulk storage of substances such as pigments, oils, carbohydrates, water, or toxins. In animal cells, these organelles are smaller and are called lysosomes.

vagina The birth canal of female mammals; a muscular tube that extends from the uterus through the pelvis to the exterior.

valence The number of electrons in the valence (outermost) electron shell of an atom; determines how many covalent bonds the atom can form.

valence electron An electron in the valence (outermost) electron shell of an atom. Valence electrons tend to be involved in chemical bonding.

valence shell The outermost electron shell of an atom.

valves In circulatory systems, flaps of tissue that prevent backward flow of blood, particularly in veins and between the chambers of the heart.

van der Waals interaction A weak electrical attraction between two hydrophobic side chains. Often contributes to tertiary structure in proteins.

variable (V) region A section of the light chains of antibodies that has a highly variable amino acid sequence, unique to each B cell.

vasa recta A network of blood vessels that runs alongside the loop of Henle of a kidney nephron, reabsorbing water and solutes.

vascular bundle A cluster of xylem and phloem strands in a plant stem.

vascular cambium A ring of undifferentiated plant cells inside the cork cambium of woody plants. Produces secondary xylem and secondary phloem.

vascular tissue In plants, a tissue that is involved in conducting water or solutes from one part of a plant to another, or that gives rise to such tissues.

vas deferens A pair of muscular tubes that store and transport semen from the epididymus to the ejaculatory duct. In nonhuman animals, called the ductus deferens.

vector (1) A plasmid or other vehicle used to transfer recombinant genes to a new host. (2) A biting insect or other organism that transfers pathogens between two other species.

vegetal hemisphere The lower half of an amphibian egg cell, containing most of the yolk. This portion eventually becomes part of the gut.

vein (1) Any blood vessel that carries blood from capillaries to the heart (oxygenated or not). Has thinner walls and lower blood pressure than an artery. (2) A strip of vascular tissue in a plant leaf. (3) A supporting filament of an insect wing.

veliger A unique form of larva in bivalves.

venae cavae (singular: vena cava) Large veins that return oxygen-poor blood to the heart.

ventilation Movement of air or water through the lungs or gills.

ventral Toward an animal's belly and away from its back. The opposite of dorsal.

ventricle (1) A thick-walled chamber of the heart that receives blood from an atrium and pumps it to the body or to the lungs. (2) One of several small fluid-filled chambers in the vertebrate brain.

vertebra (plural: vertebrae) One of the cartilaginous or bony elements that form the spine of vertebrate animals.

vertebrate An animal in the lineage Vertebrata, characterized by a skull, usually a spinal column, usually an endoskeleton of bone, and other traits.

vessel element An elongated water-conducting plant cell found in the xylem of certain advanced plants. Has gaps through both the primary and secondary cell walls, allowing unimpeded passage of water from one cell to the next.

vestigial trait Any rudimentary structure of unknown or minimal function that is homologous to functioning structures in other species. Vestigial traits are thought to reflect evolutionary history.

vicariance The physical splitting of a population into smaller, isolated populations by a geographic barrier.

villi (singular: villus) Small, fingerlike projections of the lining of the digestive tract. Function to increase surface area for absorption.

virion A single mature virus particle.

virulence The ability of a pathogen or parasite to cause disease and death.

virulent Tending to cause severe disease rather than mild disease.

virus A tiny infectious parasitic entity consisting of DNA or RNA enclosed in a protective covering (a capsid and sometimes an envelope). The DNA or RNA contain the necessary instructions to make more viruses but must use the machinery of a host cell to do so.

visceral mass The part of a mollusk containing most of the internal organs and external gill.

visible light The range of wavelengths of electromagnetic radiation that humans can see, from about 400 to 700 nanometers.

vitamin Any micronutrient that is a carbon-containing compound rather than a single chemical element. Usually functions as a coenzyme.

vitelline envelope A fibrous sheet of glycoproteins that surrounds a mature egg cell. Found in many vertebrates.

viviparous For animals, reproducing by live birth rather than by laying eggs.

volt A unit of voltage (electrical potential).

voltage Potential energy created by a separation of electric charges between two points. Also called electrical potential.

voltage clamping A lab technique for imposing a certain constant membrane potential on a cell. Widely used to investigate ion channels.

voltage-gated channel An ion channel that opens or closes in response to changes in membrane voltage.

wall pressure The inward pressure exerted by a cell wall against the fluid contents of a plant cell.

water potential The potential energy of water in a certain environment. In living organisms, the sum of solute potential and pressure potential. Determines movement of water into or out of cells.

water potential gradient A difference in water potential in one location compared with another. Determines the movement of water through plant tissues.

watershed The area drained by a single stream or river.

water table The upper limit of the underground layer of soil that is saturated with water.

water vascular system A system of fluid-filled tubes and chambers in echinoderms. Functions as a hydrostatic skeleton.

wavelength The distance between two successive wave crests in any regular wave, such as light waves, sound waves, or waves in water.

wax A type of lipid with long hydrocarbon tails, usually combinations of long-chain alcohols with fatty acids. Harder and less greasy than fats.

weather The specific short-term atmospheric conditions of temperature, moisture, sunlight, and wind in a certain area.

weathering The gradual wearing down of large rocks by rain, running water, and wind; one of the processes that transform rocks into soil.

weed A plant that is adapted for growth in disturbed soils.

Western blotting A lab technique for identifying proteins that will bind to a certain antibody. Involves separation of proteins with gel electrophoresis, transfer to a filter, and probing with a labeled antibody.

wetland A shallow-water habitat where the soil is saturated with water for at least part of the year.

white blood cells Immune system cells that circulate in the blood or lymph and function in the defense against disease. Also called leukocytes.

wild type The most common phenotype seen in a population; especially the most common phenotype in wild populations compared with inbred lab strains of the same species.

wilt To lose turgor, in a plant tissue.

wobble hypothesis The hypothesis that some tRNA molecules can pair with more than one codon of mRNA, tolerating some variation in the third base, as long as the first and second bases are correctly matched.

wood The secondary xylem of older woody stems and roots.

xanthophylls Carotenoid pigments found in many algae and some plants, typically appearing yellow.

xenotransplantation or xenografting The transplantation of tissues from one species to another.

xeroderma pigmentosum (XP) A human disease characterized by extreme sensitivity to ultraviolet light. Caused by an autosomal recessive allele that results in a defective DNA repair system.

X-linked inheritance Inheritance resulting from a gene being located on the mammalian X chromosome. Also called X-linkage.

X-ray crystallography A lab technique used to infer the three-dimensional structure of a molecule, on the basis of the diffraction patterns produced by X-rays beamed at the crystallized molecule.

xylem A plant vascular tissue that conducts water and ions; contains tracheids and/or vessel elements.

xylem sap The watery fluid found in xylem tissue of plants.

yeast Any fungus growing as a single-celled form. Also, a specific lineage of ascomycetes.

Y-linked inheritance Inheritance resulting from a gene being located on the mammalian Y chromosome. Also called Y-linkage.

yolk The nutrient-rich cytoplasm inside an egg cell, used as food for the growing embryo.

yolk sac In an amniotic egg, the membrane-bound sac that contains the yolk.

zeaxanthin A carotenoid pigment of plants that initiates the opening of stomata in response to blue light so that carbon dioxide can diffuse into photosynthesizing cells.

zero population growth (ZPG) A state of stable population size due to fertility staying at the replacement rate for at least a generation.

zona pellucida The gelatinous layer around the egg cell of a mammal.

zone of cellular division A group of plant cells just behind the root cap. Contains the apical meristem, where cells are actively dividing.

zone of cellular elongation A group of plant cells behind the apical meristem in plant roots. Consists of young cells that are increasing in length.

zone of cellular maturation A group of plant cells several centimeters behind the root cap that are differentiating into mature tissues.

Z scheme A widely accepted model for the passage of electrons from photosystem II to photosystem I during photosynthesis. Electrons on a graph of energy level versus location trace a Z shape.

zygosporangium (plural: zygosporangia) The spore-producing structure in fungi that are members of the Zygomycota.

zygote The diploid cell formed by the union of two haploid gametes. Capable of undergoing embryological development to form an adult.

Image Credits

Frontmatter

xv Jacob Halaska/Index Stock Imagery, Inc. **xvii** ©E.H. Newcomb & W.P. Wergin/Biological Photo Service **xviii** Dr. Gopal Murti/Science Photo Library/Photo Researchers, Inc. **xxi** Kent & Donna Dannen/Photo Researchers, Inc. **xxii** ©Beth Donidow/Visuals Unlimited **xxv** David Nunuk/Science Photo Library/Photo Researchers, Inc. **xxvi** ©Rob Nunnington/Foto Natura/Minden Pictures **xxix** ©Thomas Mangelsen/Minden Pictures **xxx** ©J.P. Ferrero/Peter Arnold **53.1** ©Christian Ziegler

Chapter 1

Opener ©Frans Lanting/Minden Pictures**1.1a** Burndy Library/Omikron/Photo Researchers, Inc. **1.1b** M.I. Walker/Photo Researchers, Inc. **1.3a** Photo by Kelly Buono/Courtesy of Dr. Richard Amasino, University of Wisconsin **1.3c** Bruce Forster/Getty Images Inc. - Stone Allstock **1.5a** Samuel F. Conti and Thomas D. Brock **1.5b** Kwangshin Kim/Photo Researchers, Inc. **1.6/1** ©Dr. David Phillips/Visuals Unlimited **1.6/2** Dennis Kunkel/Dennis Kunkel Microscopy, Inc. **1.6/3** Kolar, Richard/Animals Animals/Earth Scenes **1.6/4** Biophoto Associates/Photo Researchers, Inc. **1.6/5** Darwin Dale/Photo Researchers, Inc. **1.9b** Michael Hughes/Aurora & Quanta Productions Inc. **1.10a** Joshua J. Tewksbury **1.10b** William Weber/Visuals Unlimited **1.10c** Robert Dobbs **1.11** Data from Tewksbury and Nabhan. 2001. *Nature* 412: 403–404. Fig. 1a. **1.12** Data from Tewksbury and Nabhan. 2001. *Nature* 412: 403–404. Fig. 1b.

Unit 1

©Eric Meola/Image Bank/Getty Images

Chapter 2

Opener Science Photo Library/Photo Researchers, Inc. **2.1b** Mitchell Layton/Duomo Photography Incorporated **2.2** US Department of Energy/SPL/Photo Researchers, Inc. **2.4b** Minik Rosing, Geological Museum, Copenhagen **2.9c** Albert Copley/Visuals Unlimited **2.14a** P.W. Lipman/U.S. Geological Survey/U.S. Department of the Interior **2.14b** Richard Megna/Fundamental Photographs **2.17** Don Farrall/Getty Images Inc. - PhotoDisc **2.24** Robert and Beth Plowes Photography **2.27c** Geostock/Getty Images Inc. - PhotoDisc **2.29a** TSADO/NASA/Tom Stack & Associates, Inc. **2.29b** Jet Propulsion Laboratory/NASA Headquarters **T2.2** Source: Table 8.1, p. 312, in John McMurry and Robert C. Fay, *Chemistry*, 4th Edition, ©2004. Reprinted by permission of Pearson Education, Inc., Upper Saddle River, NJ

Chapter 3

Opener Jacob Halaska/Index Stock Imagery, Inc. **3.2** NOAA Vents Program **3.3** Photo courtesy R. Kempton, New England Meteoritical Services **3.6** Martin Bough/Fundamental Photographs **3.7b** Clare Sansom, Birkbeck College, University of London, London, England **3.12** Clare Sansom, Birkbeck College, University of London, London, England **3.13a** ©Microworks/Phototake **3.13b** ©Walter Reinhart/Phototake **3.15b** Clare Sansom, Birkbeck College, University of London, London, England **3.16** Clare Sansom, Birkbeck College, University of London, London, England **3.22** T.A. Steitz, Yale University **3.25** T.A. Steitz, Yale University **3.27a** Data from Nawani and Kapadnis. 2001. *Journal of Applied Microbiology* 90: 803–808. Fig. 3. Data also from Nawani et al. 2002. *Journal of Applied Microbiology* 93: 965–975. Fig. 7. **3.27b** Data from Hansen et al. 2002. *FEMS Microbiology Letters* 216: 249–253. Fig. 1. **3.29** ©Paul Fievez/Hulton Archive/Getty Images

Chapter 4

Opener Micheal Simpson/Getty Images, Inc. - Taxi **4.6** Reproduced by permission from J.P. Ferris et al., Synthesis of long prebiotic oligomers on mineral surfaces. Nature 381:59–61 (1996), Fig. 2. Copyright ©1996 Macmillan Magazines Limited. Image courtesy of James P. Ferris, Rensselaer Polytechnic Institute. **4.7** Omikron/Photo Researchers, Inc. **4.8** A. Barrington Brown/Science Source/Photo Researchers, Inc. **4.15** Reprinted with permission from *Science* 292: 1319–1325 Fig. 4B (2001) by Wendy K. Johnston, Peter J. Unrau, Michael S. Lawrence, Margaret E. Glasner, David P. Bartel "RNA-Catalyzed RNA Polymerization: Accurate and General RNA-Templated Primer Extension."

Chapter 5

Opener Dr. Jeremy Burgess/Photo Researchers, Inc. **5.8a** ©Biophoto Associates/ Photo Researchers, Inc. **5.8b** Dr. Jacob S. Ishay, and Dr. Eyal Rosenzweig. ©2000 E. Rosenzweig **5.8c** M. Jericho, Dept of Physics, Dalhousie University, Halifax, Nova Scotia and T.J. Beveridge, Dept of Microbiology, University of Guelph, Guelph, Ontario

Chapter 6

Opener Kit Pogliano and Marc Sharp, University of California at San Diego **6.1aL** JEOL USA Inc. **6.1aR** Photo courtesy of Peter M. O'Day, Juan Bacigalupo, Joan E. Haab, and Cecilia Vergara. The Journal of Neuroscience, October 1, 2000. 20(19): 7193–7198, Fig. 1C. ©2004 by the Society for Neuroscience. **6.1bL** SEM photo courtesy of JEOL-USA, Peabody, MA **6.1bR** ©Dennis Kunkel/Phototake **6.2a** Alec D. Bangham, M.D., F.R.S. **6.2b** Fred Hossler/Visuals Unlimited **6.8aL** James J. Cheetham, Carleton University **6.12a** ©Dorling Kindersley **6.12b** Clive Streeter ©Dorling Kindersley **6.12c** Phil Degginger/Color-Pic, Inc. **6.20** Don W. Fawcett/Photo Researchers, Inc. **6.25b** Kovacs, F., Quine, J. & Cross, T.A. (1999) "Validation of the Single-Stranded Channel Formation of Gramicidin A by Solid-State NMR" *Proc. Natl. Acad. Sci. U.S.A.* 96:7910–7915. **6.26L** Andrew Syred/Getty Images Inc. - Stone Allstock **6.26M** ©Dr. David Phillips/Visuals Unlimited **6.26R** Joseph F. Hoffman, Yale University School of Medicine **6.32** ©Science Photo Library/Photo Researchers, Inc.

Unit 2

©Yorgos Nikas/Stone/Getty Images

Chapter 7

Opener ©Albert Tousson/Phototake **7.1** ©Dr. T.J. Beveridge/Visuals Unlimited **7.2** ©Stanley C. Holt/Biological Photo Service **7.3** ©Dr. Gopal Murti/Visuals Unlimited **7.4** COVER (upper left image), from *Science* vol 285, no. 5436, September 1999. Image: J.H. Cate, M.M. Yusupov, G. Zh. Yusupova, T.N. Earnest, H.F. Noller. Copyright 1999 American Association for the Advancement of Science. **7.5** Wanner/Eye of Science/Photo Researchers, Inc. **7.7** ©Fawcett/Photo Researchers, Inc. **7.8** Photo Researchers, Inc. **7.9** ©Dr. Don Fawcett/Photo Researchers, Inc. **7.10** ©Biophoto Associates/Photo Researchers, Inc. **7.11** ©Don W. Fawcett/Photo Researchers, Inc. **7.12** ©Dr. Don Fawcett, Daniel Friend, and Richard Wood/Photo Researchers, Inc. **7.13** ©Dr. Gopal Murti/Visuals Unlimited **7.15** E.H. Newcomb & W.P. Wergin/Biological Photo Service/Getty Images Inc. - Stone Allstock **7.16** K.R. Porter/Photo Researchers, Inc. **7.17** W.P. Wergin/Biological Photo Service **7.18** ©E. H. Newcomb/Biological Photo Service **7.19a** ©Dr. Don Fawcett/S. Ito & A. Like/Photo Researchers, Inc. **7.19b** ©Dr. Don Fawcett/Photo Researchers, Inc. **7.19c** Courtesy of Dr. Julian R. Thorpe, Electron Microscope Division, The Sussex Centre for Advanced Microscopy, School of Life Sciences, University of Sussex. **7.19d** ©Jan Robert Factor/Photo Researchers, Inc. **7.21** ©Michael W. Davidson, Florida State University/Molecular Expressions (http://www.microscopy.fsu.edu) **7.22b** Don W. Fawcett/Photo Researchers, Inc. **7.27** Jamieson, J.D. and Palade, G.E. (1967). Intracellular transport of secretory proteins in the pancreatic exocrine cell. II. Transport to condensing vacuoles and zymogen granules. Reproduced from *The Journal of Cell Biology*, 1967, v. 34 , pp. 597–615. **7.33a** K.G. Murti/Visuals Unlimited **7.33b** Peter Dawson/Science Photo Library/Photo Researchers, Inc. **7.33c** ©Dr. Gopal Murti/Science Photo Library/Photo Researchers, Inc. **7.35** ©Dr. Conly L. Rieder/Biological Photo Service **7.36** Reproduced by permission of the American Society for Cell Biology from *Molecular Biology of the Cell* 9(12), December 1998, cover. Copyright ©1999 by the American Society for Cell Biology. Image courtesy of Bruce J. Schnapp, Oregon Health Sciences. **7.37a** John E. Heuser, M.D., Washington University School of Medicine, St. Louis, Missouri **7.38** Dennis Kunkel/Phototake NYC **7.39a** Dr. Gopal Murti/Science Photo Library/Photo Researchers, Inc. **7.40** Reproduced from C.J. Brokaw, Microtubule sliding in swimming sperm flagella: direct and indirect measurements on sea urchin and tunicate spermatozoa. *Journal of Cell Biology* 114:1201–1215 (1991), cover. Reproduced by copyright permission of The Rockefeller University Press.

Chapter 8

Opener ©E.H. Newcomb & W.P. Wergin/Biological Photo Service **8.2a** ©Spencer Grant, Photo Researcher, Inc. **8.2b** ©Biophoto Associates/Photo Researchers, Inc. **8.3a** Ken Eward/Science Source/Photo Researchers, Inc. **8.3b** Barry King, University of California, Davis/Biological Photo Service **8.5** C..T. Huang, Karen Xu, Gordon McFeters, and Philip S. Stewart **8.6** ©K.R. Porter/Photo Researchers, Inc. **8.8** Biophoto Associates/Science Source/Photo Researchers, Inc. **8.9a** ©Dr. Don Fawcett/Photo Researchers **8.10a** ©Dr. Don Fawcett/Gida Matoltsy/Photo Researchers, Inc. **8.13a** ©E. H. Newcomb & W.P. Wergin/Biological Photo Service **8.13b** ©Dr. Don Fawcett/Photo Researchers, Inc.

Chapter 9

Opener ©2004 Richard Megna/Fundamental Photographs **9.1/1** Peter Anderson ©Dorling Kindersley **9.1/2** ©Dorling Kindersley **9.1/3** ©Darren McCollester/CORBIS **9.2c** Clare Sansom, Birkbeck College, University of London, London, England **9.9** Clare Sansom, Birkbeck College, University of London, London, England. **9.19b** Modified from Wang and Oster. 1998. *Nature* 396: 279. Fig. 1.

Chapter 10

Opener J.P. Nacivet/Photo Researchers, Inc. **10.2a** John Durham/Science Photo Library/Photo Researchers, Inc. **10.2b** ©Dr. J. Burgess, Science Photo Library/Photo Researchers, Inc. **10.3L** Biophoto Associates/Photo Researchers, Inc. **10.3M** ©Dr. George Chapman/Visuals Unlimited **10.3R** Richard Green/Photo Researchers, Inc. **10.5b** Sinclair Stammers/Science Photo Library/Photo Researchers, Inc. **10.10** David Newman/Visuals Unlimited **10.19** James A. Bassham, Lawrence Berkeley Laboratory, UCB (retired) **10.21** Andersson I. Journal of Molecular Biology, May 1996, vol. 259, iss. 1, pp. 160–174(15). Courtesy of Rolf Bergmann, University of Hamburg **10.23a** Dr. Jeremy Burgess/Science Photo Library/Photo Researchers, Inc. **10.26a** Adam Hart-Davis, Science Photo Library/Photo Researchers, Inc. **10.26b** ©David Muench/CORBIS **10.28** ©David Woodfall/DRK Photo

Chapter 11

Opener Professor G. Schaten/Science Photo Library/Photo Researchers, Inc. **11.2a** Originally published in Walter Flemming, Zellsubstanz, Kern, und Zelltheilung. Leipzig: Verlag von F.C.W. Vogel, 1882. Image courtesy of Conly Rieder, Wadsworth Center, New York State Department of Health. **11.2b** ©Photo Researchers, Inc. **11.3** Mark E. Warchol, Central Institute for the Deaf, Washington University, and Jeffrey T. Corwin, University of Virginia School of Medicine **11.10a** ©Dr. Richard Kessel/Visuals Unlimited **11.10b** R. Calentine/Visuals Unlimited **11.11** Micrographs by Conly L. Rieder, Division of Molecular Medicine, Wadsworth Center, Albany, New York 12201–0509. **11.17** Source: American Cancer Society's *Cancer Facts and Figures – 2003*. Reprinted with permission. **11.18** Dr. E. Walker/Science Photo Library/Photo Researchers, Inc.

Unit 3

©Eye of Science/Photo Researchers, Inc.

Chapter 12

Opener David Phillips/The Population Council/Photo Researchers, Inc. **12.2a** Applied Image Corp., San Jose, CA **12.2b** ©Addenbrookes Hospital/Photo Researchers, Inc. **12.2c** ©Wessex Reg. Genetics Centre /Wellcome Photo Library **12.7** David A. Jones **12.10** Doug Sokell/Visuals Unlimited **12.16** ©Robert and Beth Plowes Photography

Chapter 13

Opener Margaret H. Peaslee, Vice President for Academic Affairs and Professor of Biology, University of Pittsburgh at Titusville. **13.7** Dr. Madan K. Bhattacharyya **13.10a** Robert Calentine/Visuals Unlimited **13.10b** Carolina Biological Supply Company/Phototake NYC **13.11a** ©Addenbrookes Hospital/Photo Researchers, Inc. **13.17a** Robert Calentine/Visuals Unlimited **13.18a** David Cavagnaro/Peter Arnold, Inc. **13.19a** Albert F. Blakeslee, *Journal of Heredity*, 1 Fig., 1914. Reproduced by permission of Oxford University Press. **13.23** ©Bettmann/CORBIS

Chapter 14

Opener Dr. Gopal Murti/Science Photo Library/Photo Researchers, Inc. **14.1** Reproduced from O.T. Avery, C.M. MacLeod, and M. McCarty, Studies on the chemical nature of the substance inducing transformation of pneumcoccal types, *The Journal of Experimental Medicine*, 1944, 79:137–157, plate 1, by copyright permission of The Rockefeller University Press. **14.4b** Oliver Meckes/Max Planck Institut-Tubingen/Photo Researchers, Inc. **14.11a** Dr. Gopal Murti/Science Photo Library/Photo Researchers, Inc. **14.20** Vesalius Studios

Chapter 15

Opener Halaska, Jacob/Index Stock Imagery, Inc. **15.11** Omikron/Photo Researchers, Inc.

Chapter 16

Opener Oscar Miller/Science Photo Library/Photo Researchers, Inc. **16.2** Reproduced with permission from Structural basis of transcription initiation: an RNA polymerase holoenzyme-DNA complex. Murakami KS, Masuda S, Campbell EA, Muzzin O, Darst SA. *Science*. 2002 May 17;296(5571):1285–90. Fig 2a pg 1287. Copyright ©2004 **16.4** Hans Reinhard/Bruce Coleman Inc. **16.5** Bert W. O'Malley, M.D., Baylor College of Medicine **16.13** Reprinted from *Cell Press*, "Ribosome Structure and the Mechanism of Translation" by V. Ramakrishnan from Vol. 108, 557–572, Fig 1b, February 22, 2002, with permission from Elsevier. **16.16T** ©E.V. Kiseleva and Donald Fawcett/Visuals Unlimited **16.19b** Bill Longcore/Photo Researchers, Inc. **16.21a** "Structural basis for the interaction of antibiotics with peptidyl transferase centre in eubacteria: by Schluenzen et al." Figure 5 in *Nature* (2001) 413: 814–821 by Schluenzen et al.

Chapter 17

Opener EM Unit, VLA/Science Photo Library/Photo Researchers, Inc. **17.11** Reproduced by permission from A. Schmitz and D.J. Galas, The interaction of RNA polymerase and lac repressor with the lac control region. *Nucleic Acids Research* 6:111–137 (1979), fig. 2b. Copyright ©1979 by Oxford University Press. Image courtesy of David J. Galas, Keck Graduate Institute. **17.15** ©Dr. Dennis Kunkel/Visuals Unlimited **p. 381** Michael Gabridge/Visuals Unlimited

Chapter 18

Opener Prof. Oscar L. Miller/Science Photo Library/Photo Researchers, Inc. **18.2a** Ada Olins/Don Fawcett/Photo Researchers, Inc. **18.4b** Copyright ©2002 from *Molecular Biology of the Cell* 4/e fig. 4.23 by Bruce Alberts et al. Reproduced by Permission of Garland Science Taylor & Francis Books, Inc., and Dr. Barbara Hamkalo. **18.4c** Copyright ©2002 from *Molecular Biology of the Cell* 4/e fig. 4.23 by Bruce Alberts et al. Reproduced by Permission of Dr. Victoria E. Foe, and Garland Science/Taylor & Francis Books, Inc. **18.16** Walter J. Gehring

Chapter 19

Opener Work of Atsushi Miyawaki, Qing Xiong, Varda Lev-Ram, Paul Steinbach, and Roger Y. Tsien at the University of California, San Diego. **19.2** ©Dennis Kunkel/ Phototake **19.12** Baylor College of Medicine/Peter Arnold, Inc. **19.14** ©Wayne P. Armstrong, Palomar College **19.16a** Brad Mogen/Visuals Unlimited **19.17b** Peter Beyer, University of Freiburg, Germany

Chapter 20

Opener ©Sanger Institute/Wellcome Photo Library **20.7** ©David Parker/Photo Researchers **20.8** AP Wide World Photos **20.10** Camilla M. Kao and Patrick O. Brown, Stanford University.

Unit 4

©Ted Levin/Animals Animals

Chapter 21

Opener Photo courtesy Dr. Andrew Ewald **21.2a** Microworks/Phototake NYC **21.2b** Cabisco/Visuals Unlimited **21.4a** Gregory Ochocki/Photo Researchers, Inc. **21.6a** Michael Whitaker/Science Photo Library/Photo Researchers, Inc. **21.6b** Victor D. Vacquier, Scripps Institution of Oceanography, University of California at San Diego **21.7b** Biodisc/Visuals Unlimited **21.8c** David M. Phillips/Visuals Unlimited **21.10a** David S. Addison/Visuals Unlimited **21.11a** Michael V. Danilchik, Oregon Health Sciences University **21.11b** Douglas A. Melton, Harvard University **21.12** Douglas A. Melton, Harvard University **21.15** Richard Hutchings/Photo Researchers, Inc. **21.18a** Holt Studios International/Photo Researchers, Inc. **21.24** ©Jochen Tack, Das Fotoarchiv/Peter Arnold

Chapter 22

Opener Gary C. Schoenwolf, University of Utah School of Medicine **22.1** F. Rudolf Turner, Indiana University **22.3a** Christiane Nusslein.Volhard, Max Planck Institute of Developmental Biology, Tubingen, Germany **22.3b** Wolfgang Driever, University of Freiburg, Germany **22.4a** Jim Langeland, Stephen Paddock, and Sean Carroll, University of Wisconsin at Madison **22.4b** Stephen J. Small, New York University **22.4c** Jim Langeland, Stephen Paddock, and Sean Carroll, University of Wisconsin at Madison **22.6** F. Rudolf Turner, Indiana University **22.7a/1** ©Oliver Meckes/Photo Researchers, Inc. **22.7a/2** Copyright ©Carolina Biological Supply Company/ Phototake **22.7b/1** ©Oliver Meckes/Photo Researchers, Inc. **22.7b/2** Edward B. Lewis, California Institute of Technology **22.11** John L. Bowman, University of California at Davis **22.12** Reprinted by permission from *Nature* (Vol 424, July 24, 2003, pg 439, by Kim et al) copyright (2004) Macmillan Publishers Ltd. Photograph supplied by Dr. Neelima Sinah. **22.13b** Roslin Institute/PA Photos Ltd **22.15** Reproduced by permission of Elsevier Science from K. Kuida et al., Reduced apoptosis and cytochrome *c*.mediated caspase activation in mice lacking caspase 9. *Cell* 94:325–337 (1998), figs. 2E and 2F. Copyright ©1998 by Elsevier Science Ltd. Image courtesy of Keisuke Kuida, Vertex Pharmaceuticals. **22.16a** Kathryn W. Tosney, University of Michigan

Unit 5

©Gary Meszaros/Photo Researchers

Chapter 23

Opener Kent & Donna Dannen/Photo Researchers, Inc. **23.1a** ©Bettmann/Corbis **23.1b** Alfred Russel Wallace by unknown artist, after a photograph by Thomas Sims, fl. 1860s. Reg. No.: 1765. National Portrait Gallery, London. **23.2a** ©Ken Lucas/Visuals Unlimited **23.2b** ©T. A. Wiewandt/DRK PHOTO **23.2c** Courtesy photographer, Knut Finstermeier, MPI.EVA Leipzig **23.3aL** Based on Gingerich et al. 2001. *Science* 293: 2239–2242 and Thewissen et al. 2001. *Nature* 413: 277–281. **23.3aR** Robert Lubeck/Animals Animals/Earth Scenes **23.4aL** Vincent Zuber/Custom Medical Stock Photo, Inc. **23.4aR** CMCD/Getty Images Inc./PhotoDisc **23.4bL** Custom Medical Stock Photo, Inc. **23.4bR** Mary Beth Angelo/Photo Researchers, Inc. **23.5aTL** Marie Read/Animals Animals/Earth Scenes **23.5aTR** Tui De Roy/Bruce Coleman Inc. **23.5aBL** George D. Lepp/Photo Researchers, Inc. **23.5aBR** Mickey Gibson/Animals Animals/Earth Scenes **23.7L** Photo by Michael K. Richardson, reproduced by permission from *Anatomy and Embryology* 305, Fig. 7. Copyright ©Springer-Verlag GmbH & Co KG, Heidelberg, Germany. **23.7M** Photo k by Professor R. O'Rahilly. Photos a–j by Dr. Michael K. Richardson. Reproduced by permission *from Anatomy and Embryology* 305, Fig. 8. Copyright ©Springer-Verlag GmbH & Co KG, Heidelberg, Germany. **23.7R** From: Richardson, M. K., et al. *Science*. Vol 280: Pg 983c, Issue # 5366, May 15, 1998. Embryo from Professor R. O'Rahilly. National Museum of Health and Medicine/Armed Forces Institute of Pathology **23.10** Arthur C. Aufderheide, M.D., University of Minnesota School of Medicine, Duluth **23.12a** Frans Lanting/Minden Pictures **23.11** Modified from Grant and Grant. 2001. *Science* 296: 707–711. Fig. 1.

Chapter 24

Opener ©Wayne Lynch/DRK Photo **24.5** Otorohanga Kiwi House, New Zealand **24.6** Reprinted with permission from Fig. 1B, C from B Faivre et al., *Science* 300:103 (2003). Copyright 2004 American Association for the Advancement of Science. Photo courtesy of Bruno Faivre. **24.7a** Cyril Laubscher ©Dorling Kindersley **24.7b** Data from Blount et al. 2003. *Science* 300: 125–127. Fig. 1a. **24.8** David H. Funk **24.9a** Marc Moritsch/National Geographic Image Collection **24.10aT&B** Robert & Linda Mitchell/Robert & Linda Mitchell Photography **24.10bT** ©B. Schorre/VIREO **24.10bB** ©R. & A. Simpson/VIREO **24.10cT** Jeremy Woodhouse/Getty Images, Inc./Photodisc. **24.10cB** Van Os, Joseph/Getty Images Inc. - Image Bank **24.11a** ©James Hughes/Visuals Unlimited **24.14** Data from Johnston, M. 1992. *Evolution* 46: 688–702.

Chapter 25

Opener ©Wayne Lankinen/DRK PHOTO **25.2** Credit is to Don McGranaghan in Tattersall, I. 1995. *The Fossil Trail.* (Oxford: Oxford University Press) **25.4** Diane Pierce/NGS Image Collection **25.7a** Joseph T. Collins/Photo Researchers, Inc. **25.7b** Alvin E. Staffan/National Audubon Society/Photo Researchers, Inc. **25.12** H. Douglas Pratt/NGS Image Collection **25.13** Jason Rick/Loren H. Rieseberg **25.15a** ©Anthony Bannister; Gallo Images/CORBIS **25.15b** ©Barbara Cushman Rowell/DRK PHOTO **25.15c** ©Bob Krist/CORBIS **25.15d** ©Nik Wheeler/CORBIS **25.15e** ©Matthew McKee; Eye Ubiquitous/CORBIS **25.15f** ©Richard Cummins/CORBIS

Chapter 26

Opener ©1985 David L. Brill **26.7a** Reproduced by permission from P.S. Herendeen, W.L. Crepet, and K.C. Nixon, Chloranthus like stamens from the Upper Cretaceous of New Jersey, *American Journal of Botany* 80(8):865–871. ©Botanical Society of America. Micrograph courtesy of William L. Crepet, Cornell University. **26.7b** Martin Land/Science Photo Library/Photo Researchers, Inc. **26.7c** Monte Hieb & Harrison Hieb/Geocraft/www.geocraft.com **26.7d** John Gerlach/DRK Photo **26.8B** Neg. no. 325097, shot from Sanford Bird Hall. Courtesy Department of Library Services, American Museum of Natural History. **26.12b** Reproduced by permission from S. Xiao, Y. Zhang, and A.H. Knoll, Three-dimensional preservation of algae and animal embryos in a Neoproterozoic phosphorite. *Nature* 391:553–558 (February 5, 1998), fig. 5. Copyright ©1998 Macmillan Magazines Limited. **26.12cT** Simon Conway Morris, University of Cambridge, Cambridge, United Kingdom **26.12cB** Ken Lucas/Visuals Unlimited **26.12dT** B. Miller/Biological Photo Service **26.12dB** ©Ed Reschke/Peter Arnold, Inc. **26.14** Denis Duboule, University of Geneva, Geneva, Switzerland **26.16a** Jonathan B. Losos, Washington University in St. Louis **26.17a** Colin Keates ©Dorling Kindersley, Courtesy of the Natural History Museum, London **26.17b** ©Tim Fitzharris/Minden Pictures **26.17d** ©Joe Tucciarone/SPL/Photo Researchers, Inc. **26.19b** Glen A. Izett/U.S. Geological Survey, Denver **26.19c** Peter H. Schultz, Brown University; Steven D'Hondt, University of Rhode Island Graduate School of Oceanography

Unit 6

©John Gerlach /DRK Photo

Chapter 27

Opener ©Beth Donidow/Visuals Unlimited **27.2** ©Science VU/Visuals Unlimited **27.3** J. Robert Waaland/Biological Photo Service **27.4** ©Richard L. Carlton/Visuals Unlimited **27.6a** ©T. Bannor/Custom Medical Stock Photo **27.6b** ©Carolina Biological/Visuals Unlimited **27.7** Reprinted with permission of the American Association for the Advancement of Science from S.V. Liu et al., Thermophilic Fe(III)-reducing bacteria from the deep subsurface. *Science* 277:1106–1109 (1997), fig. 2. Copyright©1997 American Association for the Advancement of Science. Image courtesy of Yul Roh, Oak Ridge National Laboratory. **27.11aL** CNRI/Science Photo Library/Photo Researchers, Inc. **27.11aR** Reprinted with permission from H.N. Schulz, et al., Dense populations of a giant sulfur bacterium in Namibian shelf sediments. *Science* 284:493–495, Fig. 1b, August 4, 1999. Copyright 1999 American Association for the Advancement of Science. Image courtesy of Dr. Heide Schulz, Max-Planck-Institute for Marine Microbiology, Bremen, Germany. **27.11bL** ©Gary Gaugler/Visuals Unlimited **27.11bR** David M. Phillips/Visuals Unlimited **27.11cL** Linda Stannard, University of Cape Town/Science Photo Library/Photo Researchers, Inc. **27.11cR** Richard W. Castenholz, University of Oregon **27.12c** ©Jack Bostrack/Visuals Unlimited **27.16** ©Eric Grave/Phototake **27.17** ©National Cancer Institute/SPL/Photo Researchers, Inc. **27.18** ©Michael Abbey/Visuals Unlimited **27.19** ©R. Calentine/Visuals Unlimited **27.20** ©J. C. Revy/Phototake **27.21a** Photo by Yves B. Brun **27.21b** Hans Reichenbach, Gesellschaft fur Biotechnologische Forschung mbH, Braunschweig, Germany **27.23** Photograph courtesy Dr. Kenneth M. Stedman/Portland State University **27.23 (inset)** ©Corale Brierley/Visuals Unlimited **27.24** Marli Miller **27.24 (inset)** Photograph courtesy Dr. Bonnie K. Baxter/Westminster College

Chapter 28

Opener Dr. Ann Smith/Photo Researchers, Inc. **28.1a** Norman T. Nicoll **28.1b** Gregory Ochocki/Photo Researchers, Inc. **28.1c** Tom and Therisa Stack/Tom Stack & Associates, Inc. **28.1R** John Anderson/Animals Animals/Earth Scenes **28.2** Dr. Gopal Murti/Photo Researchers, Inc. **28.3** Sanford Berry/Visuals Unlimited **28.3 (inset)** J. Robert Waaland/Biological Photo Service **28.6** Biophoto Associates/Photo Researchers, Inc. **28.12L** M.I. Walker/Photo Researchers, Inc. **28.13** David L. Kirk, Washington University **28.14a–b** Andrew Syred/Science Photo Library/Photo Researchers, Inc. **28.14c** David M. Phillips/Visuals Unlimited **28.15a** Biophoto Associates/Photo Researchers, Inc. **28.15b** Bruce Coleman Collection **28.16** Eric Grave/Photo Researchers,

Index